ROUTLEDGE HANDBOOK OF OCEAN RESOURCES AND MANAGEMENT

Edited by
Hance D. Smith, Juan Luis Suárez de Vivero
and Tundi S. Agardy

Routledge
Taylor & Francis Group

LONDON AND NEW YORK

earthscan
from Routledge

First published 2015
by Routledge
2 Park Square, Milton Park, Abingdon, Oxon OX14 4RN

and by Routledge
711 Third Avenue, New York, NY 10017

Routledge is an imprint of the Taylor & Francis Group, an informa business

© 2015 Hance D. Smith, Juan Luis Suárez de Vivero and
Tundi S. Agardy, selection and editorial material; individual chapters,
the contributors

British Library Cataloguing-in-Publication Data
A catalogue record for this book is available from the British Library

Library of Congress Cataloging in Publication Data
Routledge handbook of ocean resources and management/edited by
Hance D. Smith, Juan Luis Suárez de Vivero, and Tundi S. Agardy.
pages cm
Includes bibliographical references and index.
1. Marine resources – Handbooks, manuals, etc. 2. Marine resources –
Management – Handbooks, manuals, etc. I. Smith, Hance D., editor
of compilation. II. Suárez de Vivero, Juan Luis, editor of compilation.
III. Agardy, Tundi, editor of compilation.
GC1015.2.R68 2016
333.91′64 – dc23
2015017714

ISBN: 978-0-415-53175-7 (hbk)
ISBN: 978-0-203-11539-8 (ebk)

Typeset in Bembo
by Florence Production Limited, Stoodleigh, Devon, UK

Printed and bound in Great Britain by
TJ International Ltd, Padstow, Cornwall

ROUTLEDGE HANDBOOK OF OCEAN RESOURCES AND MANAGEMENT

This comprehensive handbook provides a global overview of ocean resources and management by focusing on critical issues relating to human development and the marine environment, their interrelationships as expressed through the uses of the sea as a resource, and the regional expression of these themes. The underlying approach is geographical, with prominence given to the biosphere, political arrangements and regional patterns – all considered to be especially crucial to the human understanding required for the use and management of the world's oceans.

Part 1 addresses key themes in our knowledge of relationships between people and the sea on a global scale, including economic and political issues, and understanding and managing marine environments. Part 2 provides a systematic review of the uses of the sea, grouped into food, ocean space, materials and energy, and the sea as an environmental resource. Part 3 on the geography of the sea considers management strategies especially related to the state system, and regional management developments in both core economic regions and the developing periphery. The primary themes within each chapter are governance (including institutional and legal bases); policy – sets of ideas governing management; and management, both technical and general.

Hance D. Smith was a Reader in the School of Earth and Ocean Sciences at Cardiff University, UK until 2011. He is Editor of *Marine Policy*, the leading academic journal of ocean affairs. His academic interests include marine geography, marine resources and environmental management, and marine policy.

Juan Luis Suárez de Vivero is a Professor in the Department of Human Geography, University of Seville, Spain. His special interests are social science aspects of maritime policy, ocean governance and integrated coastal zone management.

Tundi S. Agardy is Executive Director of Sound Seas, a US-based group focused on marine protected areas, ecosystem-based management, and coastal planning. She is author of *Ocean Zoning* (Earthscan, 2010).

CONTENTS

Author affiliations ix

Acknowledgements xiii

Introduction 1
Hance D. Smith, Juan Luis Suárez de Vivero and Tundi S. Agardy

1 The world ocean and the human past and present 5
Hance D. Smith, Juan Luis Suárez de Vivero and Tundi S. Agardy

PART 1
The world ocean 15

The globalisation of governance 15

2 Changing geopolitical scenarios 17
Juan Luis Suárez de Vivero, Juan Carlos Rodríguez Mateos, David Florido
del Corral and Fernando Fernández Fadón

3 State ocean strategies and policies for the open ocean 33
Patricio A. Bernal

4 International marine governance and protection of biodiversity 55
Jeff A. Ardron and Robin Warner

5 Regional ecosystem-based imperatives within global ocean
governance 73
Lee A. Kimball

Understanding marine environments   87

6   Blue planet: the role of the oceans in nutrient cycling, maintaining
    the atmospheric system, and modulating climate change   89
    Susan M. Libes

7   Ocean health   108
    Fiorenza Micheli, Giulio De Leo, Francesco Ferretti, A. Margaret Hines,
    Kristen Honey, Kristy Kroeker, Rebecca G. Martone, Douglas J. McCauley,
    Jennifer K. O'Leary, Daniele F. Rosim, Susanna Sokolow, Andy Stock
    and Chelsea L. Wood

8   Marine scientific research: overview of major issues, programmes
    and their objectives   127
    Montserrat Gorina-Ysern

Managing marine environments   143

9   Marine conservation   145
    Guiseppe Notarbartolo-di-Sciara

10 Science and policy   157
    Rebecca Koss and Geoffrey Wescott

11 Ecosystem services and their economic and social value   176
    Jason Scorse and Judith Kildow

12 Strategic environmental assessment   188
    Richard Kenchington and Toni Cannard

13 Greening the ocean economy: a progress report   199
    Linwood Pendleton, Megan Jungwiwattanaporn, Yannick Beaudoin,
    Christian Neumann, Anne Solgaard, Christina Cavaliere and Elaine Baker

PART 2
The uses of the sea   **213**

Living resources   213

14 Global fisheries: current situation and challenges   215
    Yimin Ye

15 The high seas and IUU fishing   232
    Henrik Österblom, Örjan Bodin, Anthony J. Press and U. Rashid Sumaila

16 Rethinking small-scale fisheries governance   241
    Ratana Chuenpagdee and Svein Jentoft

17 Mariculture: aquaculture in the marine environment   255
    Selina Stead

Energy and materials   267

18 Oil and gas   269
    Hance D. Smith and Tara Thrupp

19 Renewables: an ocean of energy   283
    Sean O'Neill, Carolyn Elefant and Tundi Agardy

20 Ocean minerals   296
    James R. Hein and Kira Mizell

21 Making progress with marine genetic resources   310
    Salvatore Aricò

Ocean space   329

22 Shipping and navigation   331
    Jeanette Reis and Kyriaki Mitroussi

23 Subsea telecommunications   349
    Lionel Carter and Douglas R. Burnett

24 Sea-power   366
    Steven Haines

The marine environment   379

25 Waste disposal and ocean pollution   381
    Michael O. Angelidis

26 Marine leisure and tourism   396
    Michael Lück

27 Maritime heritage conservation   408
    Juan-Luis Alegret and Eliseu Carbonell

PART 3
The geography of the sea   **423**

Spatial organisation   423

28 State maritime boundaries   425
    Chris M. Carleton

29 The deep seabed: legal and political challenges   449
   Tullio Scovazzi

30 Surveying the sea   462
   Robert Wilson

31 Marine protected areas and ocean planning   476
   Tundi S. Agardy

Regional developments: key core maritime regions   493

32 Maritime boundaries: the end of the Mediterranean exception   495
   Juan Luis Suárez de Vivero and Juan Carlos Rodríguez Mateos

33 Marine spatial planning in the United States: triangulating between
   state and federal roles and responsibilities   507
   Stephen B. Olsen, Jennifer McCann and Monique LaFrance Bartley

34 The East Asian seas: competing national spheres of influence   524
   Sam Bateman

Regional developments: the developing periphery   539

35 Africa: coastal policies, maritime strategies and development   541
   *Francois Odendaal, Zvikomborero Tangawamira, Bernice McLean
   and Joani November*

36 South Pacific and small island developing states: Oceania is vast,
   canoe is centre, village is anchor, continent is margin   560
   Peter Nuttall and Joeli Veitayaki

37 Polar oceans: sovereignty and the contestation of territorial and
   resource rights   576
   Klaus Dodds and Alan D. Hemmings

38 The world ocean and the human future   592
   Hance D. Smith, Juan Luis Suárez de Vivero and Tundi S. Agardy

Index   599

AUTHOR AFFILIATIONS

Tundi S. Agardy, Sound Seas, Bethesda, Maryland, USA.

Juan-Luis Alegret, Maritime Studies, University of Girona, Girona, Spain.

Michael O. Angelidis, International Atomic Energy Agency, Department of Nuclear Sciences and Applications, Marine Environmental Studies Laboratory, Monaco; formerly UNEP Mediterranean Action Plan.

Jeff A. Ardron, Commonwealth Secretariat, London, UK; formerly, Institute for Advanced Sustainability Studies (IASS), Potsdam, Germany.

Salvatore Aricò, Division of Science Policy, Natural Sciences Sector, UNESCO, Paris, France.

Elaine Baker, GRID-Arendal, Arendal, Norway.

Sam Bateman, Australian National Centre for Ocean Resources and Security, University of Wollongong, Wollongong, NSW, Australia and S. Rajaratnam School of International Studies (RSIS), Nanyang Technological University, Singapore.

Yannick Beaudoin, GRID-Arendal, Arendal, Norway.

Patricio A. Bernal, Faculty of Biological Sciences, Pontifical Catholic University of Chile, Santiago, Chile.

Örjan Bodin, Stewardship Research, Stockholm Resilience Centre, Stockholm, Sweden.

Douglas R. Burnett, Squire Sanders (US) LLP, New York, USA.

Toni Cannard, CSIRO Oceans & Atmosphere Flagship and Australian National Centre for Ocean Resources and Security, University of Wollongong, based in Brisbane, Queensland, Australia.

Eliseu Carbonell, Catalan Institute for Cultural Heritage Research, Girona, Spain.

Chris M. Carleton, Formerly of the Law of the Sea Division, UK Hydrographic Office.

Lionel Carter, Antarctic Research Centre, Victoria University of Wellington, New Zealand.

Christina Cavaliere, Hospitality and Tourism Management Studies and Sustainability, The Richard Stockton College of New Jersey, Galloway, New Jersey, USA.

Ratana Chuenpagdee, Department of Geography, Memorial University of Newfoundland, St John's, Canada.

Klaus Dodds, Department of Geography, Royal Holloway University of London, London, UK.

Carolyn Elefant, Law Offices of Claire Elefant (LOCE), Washington, DC, USA.

Fernando Fernández Fadón, Spanish Naval Academy, International Relations, Marin, Pontevedra, Spain.

Francesco Ferretti, Stanford University, Hopkins Marine Station, Pacific Grove, California, USA.

David Florido del Corral, Department of Social Anthropology, University of Seville, Seville, Spain.

Montserrat Gorina-Ysern, Healthy Children–Healthy Oceans Foundation, Chevy Chase, Maryland, USA.

Steven Haines, Royal United Services Institute for Defence and Security Studies, University of Greenwich, London, UK.

James R. Hein, U.S. Geological Survey, Santa Cruz, California, USA.

Alan D. Hemmings, Gateway Antarctica, University of Canterbury, Canterbury, New Zealand.

A. Margaret Hines, Stanford University, Hopkins Marine Station, Pacific Grove, California, USA.

Kristen Honey, American Association for the Advancement of Science (AAAS), Washington, DC, USA.

Svein Jentoft, Norwegian College of Fishery Science, The Arctic University of Norway, Tromsø, Norway.

Megan Jungwiwattanaporn, Coastal Resources Management Office, Saipan, Commonwealth of Northern Mariana Islands.

Richard Kenchington, Australian National Centre for Ocean Resources and Security, University of Wollongong, Wollongong, New South Wales, Australia.

Judith Kildow, National Ocean Economics Program, Center for the Blue Economy, Monterey Institute for International Studies, Monterey, California, USA.

Lee A. Kimball, Policy Analyst, Washington, DC, USA.

Rebecca Koss, School of Life and Environmental Sciences, Deakin University, Burwood, Victoria, Australia.

Kristy Kroeker, University of California – Santa Cruz, California, USA.

Monique LaFrance Bartley, Graduate School of Oceanography, University of Rhode Island, Coastal Resources Center, Narragansett, Rhode Island, USA.

Giulio De Leo, Stanford University, Hopkins Marine Station, Pacific Grove, California, USA.

Susan M. Libes, School of Coastal and Marine Systems Science, Coastal Carolina University, Conway, South Carolina, USA.

Michael Lück, School of Hospitality and Tourism, Auckland University of Technology, Auckland, New Zealand.

Jennifer McCann, Graduate School of Oceanography, University of Rhode Island, Coastal Resources Center and Rhode Island Sea Grant College Program, Narragansett, Rhode Island, USA.

Douglas J. McCauley, University of California – Santa Barbara, Department of Ecology, Evolution, and Marine Biology, Santa Barbara, California, USA

Bernice McLean, International Knowledge Management, Cape Town, South Africa.

Rebecca G. Martone, Center for Ocean Solutions, Monterey, California, USA.

Fiorenza Micheli, Stanford University, Hopkins Marine Station, Pacific Grove, California, USA.

Kyriaki Mitroussi, Cardiff Business School, Cardiff University, Cardiff, UK.

Kira Mizell, US Geological Survey, Santa Cruz, California, USA.

Christian Neumann, GRID-Arendal, Arendal Norway.

Guiseppe Notarbartolo-di-Sciara, Tethys Research Institute, Milan, Italy.

Joani November, International Knowledge Management, Cape Town, South Africa.

Peter Nuttall, School of Economics, University of the South Pacific, Suva, Fiji.

Francois Odendaal, Faculty of Engineering, Cape Peninsula University of Technology (CPUT), Bellville, Cape Town, South Africa and EcoAfrica Environmental Consultants.

Jennifer K. O'Leary, California Sea Grant, University of California – San Diego, and California Polytechnic University, San Luis Obispo, California, USA.

Stephen B. Olsen, Graduate School of Oceanography, University of Rhode Island, Coastal Resources Center, Narragansett, Rhode Island, USA.

Sean O'Neill, For Ocean Energy, Foundation for Ocean Renewables, Washington, DC, USA.

Henrik Österblom, Docent and Associate Senior Lecturer in Environmental Sciences and Deputy Science Director, Stockholm Resilience Centre, Stockholm University, Stockholm, Sweden.

Linwood Pendleton, International Chair of Excellence, UMR-AMURE, Brest, France.

Anthony J. Press, Antarctic Climate and Ecosystems Cooperative Research Centre, University of Tasmania, Hobart, Tasmania, Australia.

Jeanette Reis, School of Earth and Ocean Sciences, Cardiff University, Cardiff, UK.

Juan Carlos Rodriguez Mateos, Department Human Geography, University of Seville, Seville, Spain.

Daniele F. Rosim, Stanford University, Hopkins Marine Station, Pacific Grove, California, USA.

Jason Scorse, Center for the Blue Economy and International Environmental Policy Program, Monterey Institute of International Studies, Monterey, California, USA.

Tullio Scovazzi, University of Milano-Bicocca, Milan, Italy.

Susannae H. Sokolow, Stanford University, Hopkins Marine Station, Pacific Grove, California, USA

Anne Solgaard, ChangeLab, Oslo, Norway.

Hance D. Smith, School of Earth and Ocean Sciences, Cardiff University, Cardiff, UK.

Selina Stead, School of Marine Science and Technology, Newcastle University, Newcastle upon Tyne, UK.

Andy Stock, Emmett Interdisciplinary Program in Environment & Resources, Stanford University, Stanford, California, USA.

Juan Luis Suárez de Vivero, Department of Human Geography, University of Seville, Seville, Spain.

U. Rashid Sumaila, Fisheries Economics Research Unit, University of British Columbia Fisheries Centre, Vancouver, British Columbia, Canada.

Zvikomborero Tangawamira, EcoAfrica Environmental Consultants, Clydesdale, Pretoria, South Africa.

Tara Thrupp, The Institute for the Environment, Brunel University London, London, UK.

Joeli Veitayaki, School of Marine Studies, University of the South Pacific, Suva, Fiji.

Robin Warner, Australian National Centre for Ocean Resources and Security, University of Wollongong, Wollongong, New South Wales, Australia.

Geoffrey Wescott, School of Life and Environmental Science, Deakin University, Burwood, Victoria, Australia.

Robert Wilson, International Hydrographic Organization.

Chelsea L. Wood, Department of Ecology and Evolutionary Biology, University of Michigan and Michigan Society of Fellows, University of Michigan, Ann Arbor, Michigan, USA.

Yimin Ye, Marine and Inland Fisheries Branch (FIRF), Fisheries and Aquaculture Department, Food and Agriculture Organization of the United Nations, Rome, Italy.

ACKNOWLEDGEMENTS

The conception and preparation of a Handbook dealing with the world ocean is necessarily a long-term project involving much thought, research and writing by many people. We are first and foremost indebted to our willing and long-suffering authors for both undertaking and delivering their manuscripts over an extended period. Lead and co-authors have responded to continual editorial nagging with great patience, for which we are grateful.

The project has also benefited from the equally long-suffering editorial team at Earthscan, now part of Routledge. We are especially grateful to Tim Hardwick as Senior Commissioning Editor for encouraging us to embark on the considerable voyage of thought and investigation required; and to Ashley Wright, Senior Editorial Assistant for the editorial tasks involved in bringing the work to fruition.

Finally, we are indebted to Azmath Jaleel for the cover design.

Hance D. Smith
Juan Luis Suárez de Vivero
Tundi S. Agardy
Editors
1 May 2015

INTRODUCTION

Hance D. Smith, Juan Luis Suárez de Vivero and
Tundi S. Agardy

The concept of resources is primarily economic and, in the case of the world ocean, environmental, thus focusing on the interrelationships between human activities concerned with the uses of the sea on one hand, and the 71 per cent of the Earth's surface covered by the sea on the other. This is, of course, a vast subject, and this Handbook thus necessarily deals with selected themes, which have been chosen with particular reference to the current stage of economic development of the world ocean. There are three first-order themes corresponding to the three major sections of the book, with selected second-order themes within these. First is the human dimension of globalisation of economic activity, understanding and management of marine environments. Second are the uses of the sea, considering in turn living resources, energy and materials, ocean space and the marine environment considered as a resource. Finally attention turns to the complexity of regional patterns engendered by the first two themes – the geography of the sea in terms of spatial organisation and development of both core and peripheral maritime regions of the global economy.

A first priority throughout is to place the book firmly in the present geographical and historical context of the world ocean, in the belief that humankind's relationship with the sea and, for that matter, the Earth as a whole, is passing through a crucial juncture in its development on at least two timescales, with characteristic regional development patterns linking land and sea. This is in part the focus of the first chapter in the book. With this in mind, the final chapter looks forward with particular reference to trends in the major fields considered in various chapters. In between are the three themes, further elaborated below.

The world ocean

The world ocean encompasses many human worlds – environmental, technological, economic (the main focus of this Handbook), social and political – as well as the historical and geographical worlds that link all these together. At the present juncture in human history the sea above all represents the power of globalisation that permeates all these worlds, exemplified by the conclusion of the United Nations Conference on the Law of the Sea in 1982, and its entry into force in 1994. It is thus with globalisation that the Handbook begins. The primary focus is concerned primarily with ocean governance with particular reference to the system of states, and the relationships of governance on the one hand with the management and protection of biodiversity and ocean ecosystems. Thus Chapter 2 highlights the relatively dynamic world of changing geopolitical scenarios at the turn of the twenty-first century. Chapter 3 then focuses on the open ocean through discussion of state ocean strategies with especial reference to the international dimensions of treaties and organisations concerned with the sea. Chapter 4 then deals with the implications of this governance system for the protection of biodiversity; while Chapter 5 underlines the critical regional ecosystem-based dimension of governance that is needed to manage the increasingly stressed ecosystems of life in the world ocean.

The challenge that naturally follows that of governance is that revolving around knowledge and understanding. This challenge, which is based on scientific understanding, has many dimensions, including those concerned with the ocean–atmosphere system and its relationships with the Earth itself; life in the ocean, which is a major theme in the foregoing section on governance as well as the fisheries and aquaculture chapters in Part 2; the human impacts that result in substantial modification of the ocean environment; and the driving force of marine scientific research that is necessary not only to extend knowledge, but also to prioritise management issues. Chapter 6 begins the section by focusing on the ocean itself and its relationships with the atmosphere and life, as well as the implications for climate change, which has always been a central characteristic of what is an integrated ocean–atmosphere system – the Blue Planet. Chapter 7 highlights the relatively new concept of ocean health as a means of measuring the profound changes wrought by human influences on the ocean environment. Chapter 8 provides an overview of marine scientific research that underpins human understanding, including the nature of objectives; the issues with which research is concerned; and the programmes through which the research is accomplished.

The permutations and combinations of politics and governance allied to understanding lead to the practical challenges of management. Many would regard marine conservation, which is the topic of Chapter 9, as the most pressing issue in the pantheon of management activities. However, underlying all of these management concerns is the 'gyro' of science and policy which heavily influences and to a substantial extent steers management priorities, and which is the subject of Chapter 10. Many understand ocean science simply as being concerned with the natural sciences: Chapter 11, dealing with ecosystem services and their economic and social value, demonstrates the crucial importance of social science inputs. Chapter 12 on strategic environmental assessment then considers an all-important technical management dimension through which practical management must be attained. Finally, Chapter 13 on greening the ocean economy provides something of an audit of human activity and impacts on the ocean environment, primarily from the point of view of uses of the sea, leading naturally to the second main part of the book.

The uses of the sea

The uses of the sea section deals with four main themes. The first of these is living resources – the provision of food from the sea by exploitation of marine ecosystems through both the catching of wild fish and the enormous expansion of fish farming or aquaculture in the marine environment that has characterised the last several decades. The second theme is that of energy and materials: these are best considered together for a number of reasons, including the large capital investment, mainly transnational companies, and advanced technologies required; marine genetic resources, while living in the strict sense, are primarily the use of living resources as materials, rather than food. The use of ocean space is perhaps the best indicator of the geographical and economic globalisation of the world ocean; while the final section on the marine environment as a resource highlights the importance of the 'total', integrated and functioning ocean environment for material waste disposal, marine recreation and heritage.

The four principal large-scale geographical patterns in the use of marine resources are considered, beginning with Chapter 14, which deals with what is by far the greatest part of the world fishing industry – that concentrated over the continental shelves and, since the advent of the negotiations leading up to the 1982 Law of the Sea Convention, falling directly within coastal state jurisdiction. Chapter 15 then considers the fisheries of the High Seas beyond coastal state jurisdiction, where Illegal, Unreported and Unregulated (IUU) fishing remains rampant – perhaps the most graphic downside to current human impact on and mismanagement of the world ocean. The vast majority of the world's fishers are to be found in the small-scale fisheries of the developing world, the subject of Chapter 16. Finally, Chapter 17 highlights mariculture in the marine environment, which has been regarded wrongly by many for a long time as the ultimate antidote to overfishing.

The second theme is that of energy and materials. This section begins in Chapter 18 with offshore oil and gas, which is by far the most important, and remains at the leading edge of ocean technology; while Chapter 19 focuses on marine renewable energy which, by comparison, is hardly developed at all, although in some regions has great promise. The next two chapters deal respectively with minerals in Chapter 20, where there is a long history of exploitation of a range of minerals as well as considerable potential, most famously focusing during the Law of the Sea negotiations on ferro-alloy minerals from the deep sea bed; and marine genetic resources in Chapter 21, the exploitation of which, like marine renewables, would appear to lie mainly in the future.

The section on ocean space considers the three most critical components in the open ocean, beginning with Chapter 22 on shipping and navigation, now principally concerned with the movement of goods among the geographical nodes of the world economy, as well as being, together with living resources, the earliest use of the sea. By contrast, subsea telecommunications considered in Chapter 23 is one of the relatively recent major uses, with origins in the nineteenth century, and also primarily reflects the geographical organisation of the global economy. Chapter 24 on seapower merges the themes of geopolitics and shipping links, as seapower depends as always upon ships, both on and below the sea surface, albeit since the Second World War ever more closely integrated with the land and air dimensions of military activities in both peace and war.

The section on the marine environment considered as a resource emphasises the importance of the world ocean considered as an integrated functional natural environment. The first dimension of this is material – the use of the sea for waste disposal: the world's ultimate sink, and the implications for marine pollution, all considered in Chapter 25. Chapter 26 then deals with marine leisure and tourism, probably the world's biggest marine industry measured in terms of employment and participation, and certainly the biggest in many marine regions, albeit in coastal waters. Chapter 27 highlights the increasingly important world of maritime heritage – important not only as a component of the leisure and tourism business, but also the bedrock of a diversity of maritime cultures around the world.

The geography of the sea

The true complexity of human–ocean interactions is perhaps best demonstrated by the geography of the sea. This has been done in this book by considering three themes, beginning with spatial organisation of state governance and law; the practicalities of surveying the sea upon which the law and governance depend; and the challenges of place-based management that have emerged largely since the 1990s in the form of Marine Protected Areas and marine spatial planning. There follows the twin themes of regional development, which brings the reader back to the centrality of economic affairs, dealing respectively with developments in key regions, both in the urban industrial cores of the world economy and the far-flung peripheral regions.

The Law of the Sea Convention of 1982 has provided the contemporary framework for the agreement of state maritime boundaries discussed at length in Chapter 28, although a little over one-fifth of all the possible boundary settlements have been made. The most significant starting point is that coastal states' maritime rights in effect extend over most of the continental shelves, which is where most of the living, mineral and energy resources are located, not to mention the greatest overall intensity of sea uses generally. In the past decade much attention has accordingly shifted to the deep ocean beyond the limits of coastal state jurisdiction under the 1982 Convention, which is the focus of Chapter 29. This includes the extension of coastal states' jurisdiction beyond 200 miles under Article 76; as well as the problems of IUU fishing (Chapter 15); the maintenance of marine biodiversity (Chapter 4); management of ecosystems (Chapter 9); exploitation of minerals (Chapter 20); and marine genetic resources (Chapter 21). The technical management starting point is the surveying of the sea considered in Chapter 30. The technical management endpoint is the spatial management of areas with complex patterns of use, pioneered in the development of Marine Protected Areas and marine spatial planning, the latter now being actively developed by a significant number of coastal states, as discussed in Chapter 31.

The consideration of core maritime regions is based upon a limited selection of key case studies that exemplify the themes considered not only in individual chapters earlier in the book, but also the complex interrelationships of these themes. Thus Chapter 32 deals with maritime boundary delimitation in the Mediterranean, arguably the most complicated part of the world ocean for the settlement of these boundaries. Chapter 33 then shifts the focus to perhaps the most complex field of marine spatial planning, that of federal states with devolved internal responsibilities, using the United States as an example. Finally, in Chapter 34 on the East Asian seas, where the region is at an earlier stage of development, the emphasis is on the political dimensions of competing state influences.

The final subsection of the book focuses on key issues in the vast periphery of the global ocean, beginning in Chapter 35 with Africa, still the least developed global region overall, and underlining the importance of local economic development planning. Chapter 36 then deals with a large part of the world's greatest ocean, in this case the South Pacific, highlighting the issues affecting Small Island Developing States (SIDS). Even more remote – at least in the Southern Hemisphere – and least developed of all are the polar maritime regions of the Arctic and Southern Oceans respectively discussed in Chapter 37.

The final topic is that of the world ocean and the human future dealt with in Chapter 38, which returns to the regional and temporal timescales and patterns introduced in Chapter 1, and focuses upon some of the key considerations concerned.

1

THE WORLD OCEAN
AND THE HUMAN PAST
AND PRESENT

*Hance D. Smith, Juan Luis Suárez de Vivero and
Tundi S. Agardy*

Introduction

In the rapidly expanding pantheon of planets currently being discovered within our galaxy, Earth may be regarded as particularly unusual in that just over seven-tenths of the surface is ocean, and yet the dominant life form – human beings – can only realistically inhabit the remaining three-tenths of the planet that is land surface.

Although humankind does not naturally inhabit the marine environment, nearly all the uses of the land have their counterparts in the uses of the sea. Around these sea uses are woven an apparently bewildering range of ways of life – a sea of many worlds, each with economic, environmental and social components. Thus this Handbook begins with a closer look at the cultural worlds of the sea. Once this has been established, the discussion turns to the temporal dimension, exemplified by the ceaseless quest over the past several centuries for, above all, economic development, which has so often been equated with human progress (Paine 2013).

To many the drive for progress seems to be unstoppable, not to say uncontrolled. However, this is not really the case. Rather there is a dynamic interplay between the impetus for development on the one hand, and the influences of governance and management of development on the other, which is the third theme of this chapter. Underlying the processes of both development and governance is the nature of the human mind, which at the present juncture of human history continues to advance especially in scientific knowledge and understanding and apply it through innumerable technologies to further both development and governance.

Finally the discussion returns full circle to the virtual kaleidoscope of regional patterns of sea use, impacts, regional development, governance and management of the marine environment, to conclude this introductory chapter to the Handbook.

The worlds of the sea

The starting point for understanding ocean resources and management revolves around the twin ideas of sea-based ways of life or maritime cultures on the one hand, and the uses of the sea viewed primarily as an economic phenomenon on the other. Maritime cultures have three

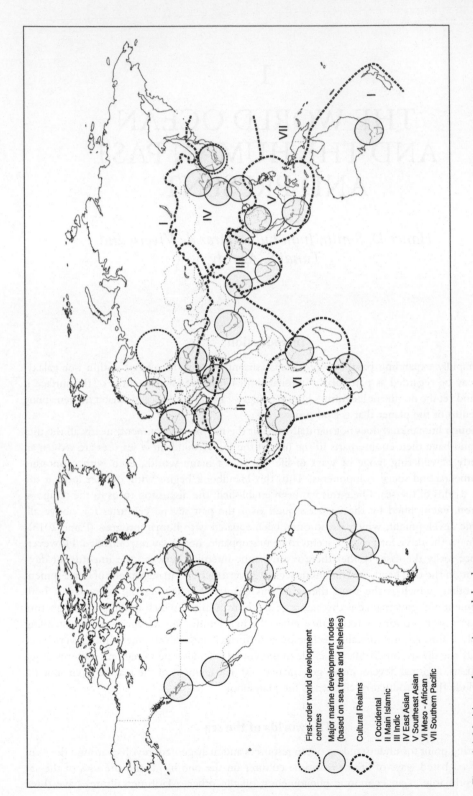

First-order world development centres

Major marine development nodes (based on sea trade and fisheries)

Cultural Realms

I Occidental
II Main Islamic
III Indic
IV East Asian
V Southeast Asian
VI Meso - African
VII Southern Pacific

Figure 1.1 Maritime culture regions and development

Source: Map created by Azmath Jaleel, based on Broek and Webb 1973; and Smith 1994.

essential components, namely, an economic dimension focused on sea uses; an environmental dimension focused on the sea itself; and a social dimension focused on human societies. As clearly seen by the World Commission on Environment and Development, these three components lie at the heart of the processes of development, sustainable or otherwise (WCED 1987). The interactions are dynamic, but there are discernible patterns on a range of timescales, which have resulted in a distinctive global pattern of culture regions encompassing both land and sea.

Critical to understanding is the timescale since the end of the last glaciation, and especially the past 6,000 years, at the beginning of which global eustatic sea level reached its present level. By around 4,000 years ago there had emerged a geographical pattern of culture regions – in history these may be termed civilisations (Figure 1.1). These are the traditional societies of Broek and Webb (1973), based either on pre-agricultural tribal forms of social organisation on the one hand, not infrequently associated with hunting and gathering economic organisation; or settled agricultural societies on the other. These agricultural societies were also partly based on urban settlements and long-distance trade networks over both land and sea.

Perhaps not surprisingly, traditional societies were and continue to be closely tied to the environmental geography of the planet. Tribal forms of social organisation and low population densities were and remain characteristic of land regions in particular, where living natural resources were and remain limited: the tropical rain forests, savannas, mangrove coasts, tropical and sub-tropical deserts, oceanic islands, sub-polar forest regions and coasts, and the truly polar settlements of the Arctic Ocean region. The later development of settled agricultural societies was concentrated in large river valleys and adjacent coasts, mainly in North Africa, south west and south Asia, and the monsoon lands to the east; together with the tropical highlands of the Americas.

Importantly, despite the enormous pace of development, especially since the 'industrial revolution', both tribal and traditional settled agricultural societies have often remained remarkably resilient in terms of economic and social organisation and relationships with the environment, despite being under enormous pressure from development. Only in the Americas and Australasia, where settlement by European peoples has been most intense, have these societies come closest to obliteration – but even here the resurgence of 'First Nations' is evident. Even more importantly, these traditional societies remain closely aligned to the evolution of language groups and belief systems – including the world's great religions (Smart 1998) – that are of comparable temporal provenance. This remains central to contemporary understanding of both economic and political developments on both land and sea.

The contemporary economic development of the world ocean – indeed the world – has its roots in the Western European region during the second half of the fifteenth century. Ocean exploration and domination which began then was particularly associated with the gradual ending of the medieval European world order, accompanied by the expansion of European influence around the world through population movement, exploration, trade and warfare, although comparable maritime expansion was also taking place in China. For the first 250 years or so of this period, economic expansion and associated empire building was based mainly on trade, and was primarily pre-industrial in its nature. However, beginning in the second half of the eighteenth century a process of industrialisation of economic activity accelerated through a series of stages – the first of these was the 'industrial revolution'.

From an economic standpoint, the crucial component was the application of science and technology to the industrialisation of production. From a social point of view, the industrialisation of the economy became associated with a rapidly increasing population enhanced by the same application of science and technology to the reduction of epidemic diseases, which lowered the death rate, a circumstance not immediately paralleled by a reduction in the birth rate; and the beginning of large-scale rural–urban migration. From an environmental point of view, industrialisation was based in part both initially and subsequently on use and progressive depletion of natural resources of land, sea and atmosphere. It is at this point that the evolution of sea use diverged from its pre-industrial cultural basis to intensified industrialised exploitation.

The worlds of the sea as illustrated in Figure 1.1 can thus be understood at two geographical scales. The first of these is the evolution of the maritime culture regions over the past four millennia that, although seemingly replaced by industrialisation later on, gave form to the major traditional societies that remain characteristic of most of what has become known as the 'developing world'. The second feature is the geography of industrialisation: although the processes of industrialisation are global, the regionalisation of industrialisation is characterised by a pattern of nodal regions in which both land and marine activities are generally tightly focused. These core regions are surrounded by vast peripheral regions – in effect most of the world ocean and land areas. Especially notable are the groups of core regions on both sides of the North Atlantic and North Pacific respectively. Until nearly the end of the twentieth century the North Atlantic system of core regions was in effect the core region of the global economy (Smith 1994). Now the world is transitioning to a system in which the Pacific system of core regions has begun to overtake that of the North Atlantic, most probably to eventually produce a much larger global core region.

The development of the world ocean

Industrialisation during and since the 'industrial revolution', can be viewed from a maritime perspective. As already noted, the pre-industrial traditional societies were symptomatic of the organisation of the world economy before the mid-eighteenth century, even in Europe, where economic development had been based to a remarkable degree on maritime trade and empire building since the 'age of exploration' in the late-fifteenth and sixteenth centuries (Padfield 1999). At sea, apart from the deep sea shipping trades linking the European empires with Europe itself, and the associated naval activities, by far the most important activity was traditional fisheries. A few fisheries were based on large ocean-going vessels, notably whaling and cod fishing; most were traditional open boat fisheries using lines and nets, operating close to shore.

The industrial revolution changed all that, although initially by a simple scaling up of traditional technologies and economies, rather than by application of new science and technology. The 'industrial revolution', properly speaking, encompassed the decades between 1780 and 1830. During this period both merchant ships and warships tended to increase in size and sophistication of design, and significant merchant shipping regulation was introduced in Britain, already by then the world's leading maritime trading nation, gaining full military mastery of the seas after the defeat of the French at the Battle of Trafalgar in 1805. But ships remained sailing ships. In commercial fisheries – still powered by sail and oar – there was enormous expansion of activity in Western Europe associated with economic expansion, population increase and urbanisation, but little real sign of overfishing on any significant scale, despite not infrequent politically inspired protestations to the contrary. The exception was in whaling and sealing, where adverse impacts on whale and seal populations were evident in regions as far apart as the Arctic and Southern Ocean (Tonnessen and Johnsen 1982).

The next stage of development, from around 1830 until around 1870 witnessed first a prolonged economic downturn in the 1830s and 1840s in the core economic region of the preceding industrial revolution, namely, the British Isles (between 1801 until 1922 the British Isles were a single state – the United Kingdom of Great Britain and Ireland) and adjacent coastlands of north west Europe. There were major changes, notably the liberalisation and subsequent expansion of international seaborne trade through the repeal of the UK Navigation and Corn Laws in the 1840s. There were also significant technological changes in shipping, pioneered as always by the military, and notably the introduction of the ironclad warships powered by both sail and steam, but also including the development of fast sailing ships called clippers for both deep and short sea trades, as well as continued scaling up of commercial vessels. However, although the first crossings of the Atlantic by steam-powered ships took place in the 1830s, steam power was not widespread. When steamships were widely adopted, this took the form of paddle steamers, rather than screw steamers. In fisheries also, sail remained supreme, and gear remained conventional lines and drift nets.

It was during the next stage of economic development, between the 1870s and 1930s, that the transformational influence of technology upon both sea uses and the marine environment became undeniable. As during the previous stage, there was a long-run economic downturn in the later 1870s and early 1880s. However, this time the processes of change were more profound. Large-scale rural–urban migration occurred in Europe and North America, associated with similarly large-scale emigration of Europeans to the temperate and subtropical lands of their empires in North and South America, southern Africa and Australasia, with millions of migrants carried on the new ocean liner trades. In the world of merchant shipping the balance between sail and steam shifted decisively to steam after the mid 1880s, associated with the introduction of steel shipbuilding, boilers and triple expansion steam engines (Graham 1956). The key trans-oceanic canals were opened: Suez in 1869 and Panama in 1914. Ocean-going sailing ships remained economical only on long ocean routes for bulk cargoes, up to the 1930s. By the outbreak of the First World War warships were being equipped with steam turbines.

In fisheries, the introduction of steam was decisive: steam engines for both drifters and trawlers, together with steam capstans and winches respectively for handling the gear led to clear evidence of overfishing from the 1890s onwards. This was undeniable, despite the fact that as late as 1883 the Chief Inspector of Fisheries for the United Kingdom averred that the fish resources of the ocean were unlimited (Huxley 1883). Overfishing triggered the establishment of the International Council for the Exploration of the Sea in 1902 (Rozwadowski 2002). There was a short-lived recovery of north west European fisheries in the 1920s, after the interregnum of the First World War, but it was not to last. In whaling, the combination of steam whalers and the Svend Foyn harpoon led to the collapse of the centuries-old Great Northern Whale Fishery in the North Atlantic in the 1880s. Whaling in the Southern Ocean began in 1904, and peaked with the introduction of factory ships in the mid 1920s. By the 1930s decline had already set in. Meanwhile, the beginnings of large-scale coastal and marine recreation began in Europe especially with the twin large-scale introduction of the railways associated with the development of seaside resorts; and the introduction of cruising by steamship along the Norwegian coast, in the Baltic, and throughout the Mediterranean.

From the later 1940s until the 1990s a new stage of economic development took place. As with the previous stages, economic depression characterised the transition from the previous stage throughout the 1930s, complicated by the Second World War. Although a true geographical globalisation had been evident from the second stage – around 1850 (Ashworth 1967), this new stage witnessed rapid economic expansion and associated technological change worldwide, in all the nodal regions illustrated in Figure 1.1. Full economic integration of the global economic system had not yet taken place (Korotayev and Tsirel 2010). In commercial shipping key developments were the replacement of steam by diesel; specialisation of ship types; the disappearance of the ocean liner trades, lost to air travel after 1960; and the introduction of unitised shipping or containerisation in the 1970s, allied to the efficient logistical systems of intermodal transport. Ship sizes increased enormously, especially cruise ships, oil tankers and dry bulk ships, while roll-on/roll-off ferries began to dominate short sea shipping. Naval power came to depend upon the nuclear powered and armed submarine and the aircraft carrier in a world in which integrated land, sea and air combined operations became the norm, having been pioneered by the invasion of Normandy in 1944, towards the end of the Second World War (Gardiner 1992–2004).

In fisheries, steam was also replaced by diesel, and ever improving fishing gears – especially trawls and seines. Global fish production increased between three and four times between the late 1940s and the 1990s, then levelled off; aquaculture further increased fish production, especially from the 1980s onwards. By this time overfishing had become a serious issue in many fisheries. Whaling from the United Kingdom and Norway ceased in the early 1960s, and a moratorium was finally agreed and implemented worldwide in 1986. Meanwhile, traditional seaside holidays were increasingly supplanted by marine activity holidays, including boating; cruising expanded enormously; and air travel made possible seaside holidays thousands of miles from the industrial core regions, especially to the Mediterranean, Caribbean and tropical oceanic islands. Other uses of the sea came into their own on a large scale, notably offshore hydrocarbon exploitation; waste disposal; large-scale marine science; and marine natural and human heritage conservation.

The world is now entering the early part of a new stage of development, considered further in Chapter 38. These changes may be viewed as cycles exhibiting common features. A key driving feature of these stages or cycles is the interplay between economic and technological factors (Smith 2000; Korotayev and Tsirel 2010). Thus the early sequence of events is associated with rapid economic expansion, accompanied by varying degrees of technological innovation. This process is very clearly seen with regard to the most important maritime uses: shipping, naval activity and fisheries – from the industrial revolution onwards. In the later twentieth century stage the process can be seen spreading to all the major uses of the sea. The successive peaks in the cycles or stages are determined by the limits of economic expansion, after which economic decline sets in, a process that ends in a 'gale of creative destruction' (Schumpeter 1942) affecting all sectors of the economy, including all uses of the sea. During this stage there is profound economic re-structuring including the enterprises that make up the private sector, and this sets the scene for the beginnings of new industries that characterise the next stage. Inevitably these changes also are associated with varying degrees of social and political change, including changes affecting the nature of governance by the state.

The importance of these stages for the present lies with their significance for relationships between sea uses on the one hand, and the marine environment on the other. Although there has been throughout the process of industrialisation of maritime activity much emphasis in some quarters on the negative impacts on the environment of economic expansion and the need for conservation – especially with regard to the fisheries – the extent of environmental modification was initially limited. Thus during the industrial revolution there were significant impacts on cetacean populations, and the beginnings of coastal sea pollution, especially originating in rivers. Major changes to marine ecosystems associated with overfishing became evident from the 1880s onwards, in the next stage, when fisheries were largely mechanised; the whale populations collapsed – first in the North Atlantic, then in the Southern Ocean; and serious overfishing of both pelagic and demersal commercial fish stocks became apparent after the First World War (Cushing 1988; Roberts 2007; Starkey *et al.* 2008; Thurston *et al.* 2010). The later twentieth-century stage witnessed the ultimate limits of exploitation of the global stock of wild fish, and the restructuring of entire marine ecosystems by human agency. At the same time, enormous expansion of the global economy and all uses of the sea has resulted in high levels of marine pollution of all kinds, countered by an increasingly high level of awareness of the ultimate value of the marine environment, evidenced by the expansion of the marine leisure industries; marine conservation and, to some extent, marine science; and the extent to which both overexploitation and environmental degradation is putting this value at risk.

Governance and human society

The sheer transformative power of the economic and technological dimensions of the historical processes outlined above, and the enormous pressure that has been put upon the marine environment – as well as the land surface of the Earth and the atmosphere – by these processes cannot be overestimated, especially in the period since the beginning of the industrial revolution (Steinberg 2001). Responsibility for controlling these impacts has fallen mainly upon coastal states. Thus the half-millennium since the end of the European Middle Ages has been a period when states have assumed an ever greater role in the governance of maritime affairs, including the uses of the sea. A key juncture in this development was the emergence of the modern state system at the end of the European Thirty Years War in 1648, marked by the Treaty of Westphalia (Kissinger 2014).

An important characteristic of the Westphalian state system was the continual divestment of state power in Europe away from monarchies towards evolving parliamentary systems of government. A crucial maritime expression of this was the promotion of a mercantilist philosophy in the management of shipping from the middle of the seventeenth century until the end of the industrial revolution stage, when the United Kingdom repealed the Navigation Laws and Corn Laws, thereby giving an enormous boost to international seaborne trade. The economic, military and political expansion of Europe after the mid-seventeenth century ensured that the Westphalian state system has come to be adopted in all the major culture regions outlined earlier in this chapter, based on the expansion of European empires. However, there remain profound cultural differences in the adoption of this state system both within and among the respective culture regions. The state continued to gain governance functions in the maritime world as elsewhere through the successive long stages of development. By the 1940s–1990s stage a minority of centralised states with direct state management of the economy had even emerged, including the Union of Soviet Socialist Republics (USSR) – which came to an end in 1989; the People's Republic of China, North Korea, Vietnam, Venezuela and Cuba. However, 'mixed economy' models remained characteristic of most states.

In a maritime context, particular interest attaches to these 'mixed economy' systems both at national and international levels. Extensive state roles in maritime affairs, although becoming particularly significant from the industrial revolution onwards, mainly emerge during the late twentieth century stage. Apart from the military, which may be regarded primarily as a state sea use at national level, states have extensive governance systems covering commercial shipping, marine mineral and energy development, fisheries, waste disposal at sea, marine science and marine conservation. These functions are replicated at supra-national level by the European Union and, in the military case, the North Atlantic Treaty Organisation (NATO); and at international level by both the United Nations system and a whole series of regional state-sponsored organisations mainly in the fields of fisheries, environmental management and, in the case of marine science, the International Council for the Exploration of the Sea (ICES) and the Intergovernmental Oceanographic Commission (IOC) of the United Nations Educational, Scientific and Cultural Organisation (UNESCO). The continuing role of the state at these levels will be returned to in the final chapter of this book.

Beyond the role of the state there are other dimensions of human social organisation of particular relevance in a maritime context. The first element to consider is the evolution of the human organisations directly responsible for the uses of the sea. These range from traditional community-based organisations characteristic of the traditional societies – both tribal and traditional settled agricultural – still prevalent in the fisheries of developing countries, and occasionally found also in the developed world, for example in France and Spain; to a whole range of company-type organisations, from small businesses to large transnational corporations. Each fundamental group of sea uses has characteristic sets of company organisations, which collectively constitute the private sector (Holtus 1999), although a limited number of companies are owned by the state, especially in countries where centralised state control has developed. Private sector organisations representing various uses, such as the World Ocean Council, have evolved in parallel in ways that are distinctive to each of the several development stages discussed; a significant portion of these have been internationalised, especially during the course of the later twentieth-century stage, for example in shipping and offshore hydrocarbon exploitation, which require high levels of capital investment and the application of advanced technologies.

The second element is the emergence of civil society groups that concern themselves with ocean use, marine access and protection, which had its origins in the stage dating from the 1830s and 1840s. This portion of civil society grew to prominence in the late twentieth-century stage, especially in the English-speaking developed world. The voluntary sector, broadly defined, was influential in the establishment of marine science in the nineteenth century and, in the United Kingdom's case, the establishment of the lifeboat service. Environmental NGOs started to push state governance of maritime uses, as well as investment in conservation. By the end of the twentieth century, volunteerism was also internationalised to a substantial degree, and was exerting substantial influence and political pressure – as well as much practical work – in marine conservation and international trade related to marine resources.

In essence civil society interest in maritime uses and affairs has expanded greatly in modern times. It comprises a vast range of special interests both directly as users of the sea, for example in marine leisure and tourism; and indirectly as, for example, consumers of goods from the sea ranging from fish to offshore oil and gas. Interest groups also arise from time to time in many fields for specific, often relatively local concerns.

It is thus useful to envisage the governance and management of the world ocean in terms of specific interest groups, often termed stakeholders, belonging respectively to the state, the private sector or civil society. These consist of a range of maritime communities of interest, defined primarily in terms of sea uses, of which shipping and commercial ports, naval, fisheries, offshore oil and gas, marine leisure, marine science and marine conservation are perhaps the most notable. It is especially interesting to note that conventional leading roles of the state in marine governance and management are increasingly being supplanted by partnership approaches in which the private sector and civil society groups are playing more influential governance roles.

The discussion of how this management actually takes place is largely confined to the individual chapters of this book. However, it is useful to distinguish between technical management and general management dimensions (Smith 1995). The former is concerned with the physical interactions between uses of the sea and the marine environment, and includes information management including monitoring, surveillance and information technology; assessment of environmental, technological, economic, social and political influences; and professional practice including especially science, technology, financial management, law and planning. The general management dimension includes coordination of technical management; the management of organisations; policy; and strategic planning.

Ocean resources and management

In conclusion, it is possible to conceive of human relationships with the world ocean in certain key ways. The first is an appreciation of the fundamental purposes of sea uses. Despite an apparently innumerable number of sea uses, there are in reality only a very limited number of prominent groups of maritime users, including commercial shipping and seaports, and the naval uses; the material groups of uses – mineral and energy resources; living resources; and waste disposal; and the non-material uses that are directly related to certain characteristics of the human mind and that use the marine environment as a total entity: leisure; research and education; and conservation of both the natural and human maritime heritage.

The second important factor in the relationship between humans and the world ocean is the governance dimension of maritime uses and values. Governance has evolved through time, reflecting expanding sea uses and expanding and contracting economies, but also reflecting shifts in political organisation and innovative sharing of responsibility among government entities, the private sector and civil society.

References

Ashworth, W. (1967) *A short history of the international economy since 1850.* London, Longman.

Broek, J. O. M., Webb, J. W. (1973) *A geography of mankind.* Second edn. New York, McGraw-Hill.

Cushing, D. H. (1988) *The provident sea.* Cambridge, Cambridge University Press.

Gardiner, R. (ed.) (1992–2004) *Conway's history of the ship.* 12 vols. London, Conway Maritime Press.

Graham, G. S. (1956) The ascendancy of the sailing ship, 1850–1885. *Economic History Review* 9(1) 74–88.

Holtus, P. (1999) Sustainable development of oceans and coasts: the role of the private sector. *UN Natural Resources Forum Journal* 23(2) 169–176.

Huxley, T. H. (1883) Inaugural Address, Fisheries Exhibition. London.

Kissinger, H. (2014) *World order: reflections on the character of nations and the course of history.* New York, Penguin.

Korotayev, A. V., Tsirel, S. V. (2010) A spectral analysis of world GDP dynamics: Kondratieff waves, Kuznets swings, Juglar and Kitchin cycles in global economic development, and the 2008–2009 economic crisis. *Structure and Dynamics* 4 1–57.

Padfield, P. (1999) *Maritime supremacy and the opening of the Western mind: naval campaigns that shaped the modern world 1588–1782.* London, John Murray.

Paine, L. (2013) *The sea & civilization: a maritime history of the world.* New York, Knopf.

Roberts, C. (2007) *The unnatural history of the sea: the past and future of humanity and fishing.* Washington DC, Island Press.

Rozwadowski, H. M. (2002) *The sea knows no boundaries: a century of marine science under ICES.* Seattle, WA, University of Washington Press.

Schumpeter, J. A. (1942) *Capitalism, socialism and democracy.* London, Routledge.

Smart, N. (1998) *The world's religions.* Cambridge, Cambridge University Press.

Smith, H. D. (1994) The development and management of the world ocean. *Ocean & Coastal Management* 24(1) 3–16.

Smith, H. D. (1995) The role of the state in the technical and general management of the oceans. *Ocean & Coastal Management* 27(1–2) 5–14.

Smith, H. D. (2000) The industrialisation of the world ocean. *Ocean & Coastal Management* 43(1) 11–28.

Starkey, D. J, Holm, P., Barnard, M. (2008) *Oceans past: management insights from the history of marine animal populations.* London, Earthscan.

Steinberg, P. E. (2001) *The social construction of the ocean.* Cambridge, Cambridge University Press.

Thurston, R. H., Brockington, S., Roberts, C. A. (2010) The effects of 118 years of industrial fishing on UK bottom trawl fisheries. *Nature Communications* 1(15) 1–6.

Tonnessen, J. N., Johnsen, A. O. (1982) *The history of modern whaling.* London, Hurst.

World Commission on Environment and Development (WCED) (1987) Report of the World Commission on Environment and Development: our common future. Annex to document A/42/427 – Development and International Co-operation: Environment. New York United Nations. www.un-documents. net/wced-ocf.htm (accessed 12 January 2015).

PART 1

The world ocean
The globalisation of governance

2

CHANGING GEOPOLITICAL SCENARIOS

Juan Luis Suárez de Vivero, Juan Carlos
Rodríguez Mateos, David Florido del Corral and
Fernando Fernández Fadón

Introduction

The shift from the twentieth to the twenty-first century is associated with deep structural changes in geopolitical balances. These, in turn, are connected with other aspects in the realms of the economy, the environment, science and technology. In geographical terms, these changes are resulting in a rearrangement of areas of power and the rise of new political actors; for their importance in the media, the so-called emerging countries, especially Brazil, Russia, India and China (BRIC) are a case-in-point of this type of change, where emerging actors displace those that 'historically' used to wield power. In political theory, the notion of territory has been closely linked to that of the State, the existence of which requires a geographical area over which to exercise power and a social body vitally rooted in emerged land that might result in a fortuitous maritime space of diffuse political substance. As regards geopolitical thought, from the time that it was founded modern geography instituted spatial schemes of political organisation for systems with relevance on the world scale that combined the large territorial areas of a recently explored world. Even recognising the importance of the oceans in shaping grand geostrategic visions, they were, for the most part, considered as no more than an *encasement* (as a result of which classical geographers, such as Carl Ritter and Richthofen, did not include them as part of the *oikouménē*) (Suárez de Vivero, 1979), lacking the territorial and political entity that would allow them to be regarded as the core subjects of geopolitical statements. It is since changes have begun to take place in the power balances on the global scale (including the decline in the maritime powers of the imperialist past) and the new law of the sea has consolidated (which occurred in parallel with the decolonisation process) that new spaces of geographical interest characterised by their consisting predominantly of marine areas have begun to be defined. In this way, emerging countries have not only changed the correlation of economic and political forces, but have also reconfigured the geopolitical chessboard (Brzezinski, 1997), to which they add their own maritime spaces as part of the general 'maritimisation' process, not only in the economic (Vigarié, 1990), but also in the geopolitical sense. States, as territorial units used as the basis for the construction of the political world, gradually take in the marine domain through jurisdictional expansion (i.e. exclusive economic zones) while geographical features, such as islands, archipelagos and the continental shelf, enable rights of sovereignty to be expanded. This chapter describes some of the geographical features that as a result of the formulation of the United Nations Convention on the Law of the Sea (UNCLOS) are transforming States' territorial bases and changing hierarchies hereto based exclusively on emerged land, and identifies new areas of geostrategic interest linked to the incorporation of new political actors and their areas of influence over the ocean. With this exercise in maritime geography it is also hoped to provide an interpretation of the scenarios where power is exercised and of some of the key concepts and historical background that refer back to the early approaches of classical geopolitics.

Conceptual framework

Relatively speaking, the world ocean order had consolidated by the time that UNCLOS came into force, but in the last decade the geopolitical and geo-economic scenario has seen a marked change due to two key processes: the emergence of the BRIC powers and the process for delimiting the continental shelf beyond 200 miles (Art. 76 UNCLOS).

The starting point is the so-called critical or post-structuralism geopolitical perspective (Agnew, 2003) that underscores the nation State's loss of specific weight in the globalisation scenario. The shaping of a new geo-economic–political order from the 1960s onwards is characterised by being a world-system moulded as a social reality comprised of interconnected nations, firms, households, classes, and identity groups of all kinds (Wallerstein, 2004). The traditional framework of modern States has been superseded in this new context. There has not only been a displacement of scales, but a change in the global power relationships between the developed and the developing countries.

If the so-called globalisation process is not the result of a consciously designed and established order, but of a number of interconnected processes (international trade, financial liberalisation, technological and communication revolution, alternative social movements, great population movements, etc.), it would be better for it to be referred to as *global governance* (Held and McGrew, 2002). It is characterised by new actors emerging onto the international political stage, such as large corporations, international organisations and non-governmental organisations, that create networks, alliances and strategies to position themselves alongside nation States in an increasingly competitive environment, and with no definitive hierarchy between them. The search for common rules and conflict resolution in a range of areas and on a range of scales is the major objective of this institutional process, which is global in nature and already underway.

It could be said that the political target of global governance is the planet as a whole in a wide range of aspects: energy resources, economic relationships, political disputes, environmental issues, and so on. This is a polycentric system in which nation States are gradually losing their ability to act, either due to the scale of the problems, or because they are part of transnational networks, or because globalised processes straddle borders that are more and more porous. However, the nation State continues to be a decisive actor. First, because the perspective is by necessity multisector and multilevel, and second, because it continues to be the agent that is responsible for agreements and for implementing the new global rule system.

Some have approached the transition from geopolitics to geo-economics (Lorot, 1997) in accordance with the State's limitations and with the emergence of business corporations as leading actors. However, the notion of geo-economics is understood to be applicable if it is understood as the search by the States themselves for economic objectives on the world scale and positions of hegemony in the new scenario of competitive economic relations led by corporations. States have become the international geo-economic agents (Glassman, 1999) for a privileged position to be gained with respect to emerging economic flows (Cowen and Smith, 2009). This is the line of interpretation that can be applied to the process of maritime nationalism over the last three decades.

New legal codification was instituted through the United Nations Conferences on the Law of the Sea that embraced the maritime environment as part of the nation State's territorial structure. However, this very same process served to create a new scale, the global scale, understood as an internationally contested *res communis*. The geographical reference points that were legally consolidated by this process – the territorial sea, the creation of the exclusive economic zone (EEZ) and the setting of the outer limit of the continental shelf beyond 200 miles – have meant that 36 per cent[1] of the oceans and between 85 and 90 per cent of commercially viable fisheries resources and known exploitable oil reserves in the sea have been nationalised (Lucchini and Voelckel, 1990). And this from a prior position in which the oceans had been treated as a *res nullius* or an area of free access. But, as a complement to this, a commitment was made to novel forms of management that pointed to a different, communitarian model of the definition of the seabed and marine subsoil and its mineral resources (Art. 133 UNCLOS) outside the boundaries of national jurisdiction and as the common heritage of mankind (Art. 136 UNCLOS). These apparently opposing dynamics dominate the global ocean stage: nationalisation/internationalisation, a regime of free access/a regime of global commons.

The submission of proposals for delimiting the outer edge of the continental shelf – which peaked in 2009 – evidences a new world geopolitical map in which the category of maritime State arises as an emerging entity.[2] This joint action gives rise to three interconnected processes:

i)   The transformation of States' territorial bases establishes that their territorial projection is reinforced in the ocean. As Blake recognises, not only do a large number of territorial disputes persist between States in the twenty-first century, but maritime boundaries have assumed an unprecedented primary role. This is the best proof that 'the territorial instincts of the modern state are still alive and well' (Blake, 2000, p. 7). Notwithstanding, does this formal legal process change the position of each State in the economic and power relationship system worldwide? Can the implementation of Art. 76 UNCLOS change the power relationships that have come before?

ii)   The second process is connected with the economic possibilities of exploiting emerging strategic resources (energy and mineral sources, biogenetic resources) that are becoming available to coastal and island States through the expansion of their State sovereignty. It is basically the micro States in the southern hemisphere and in 'non-western' areas that are being referred to. However, the current scenario in the oceans is embedded in a geo-economic context in a hotly disputed international arena. As such, the possibility of exploiting these new resources might exclusively reinforce the strong actors, mirroring what occurred with the EEZ episode in the 1970s and 1980s, when the developing countries saw their hopes and expectations for advancing their economic positions frustrated. The large-scale, free-trade and economic specialisation framework of peripheral territories – especially with regard to raw materials – might signify now, as then, that their position of dependence is reinforced.

iii) The dynamic of strong States vs weak States still persists to a great extent in the international framework and this is confirmed by the emergence of a new political arena with respect to mineral–metallurgical and biogenetic resources, the Area, the space situated outside the limits of national jurisdiction and therefore defined as global commons. Once more it is the hegemonic industrial States who have opposed placing the management of resources that are today strategic in the hands of an international institution (the International

Sea-Bed Authority) that could implement criteria of equity or support for the developing countries. The dispute over these resources is proof that some States and large companies have the technological and financial means needed for undertaking the activities required for their exploration and commercial exploitation. Their management is subject to a range of interpretations (see Scovazzi, in this volume). On the one hand, a communitarian model, through the legal precept of the Common Heritage of Mankind, which has been proposed with the aim of equity to the benefit of the developing countries, and on the other, a mercantilist model emerges that favours the great economic and political powers and that is closer to the concept of free access and based on the notion of *res nullius*. The positions that States take up in this respect depend on their possessing sufficient means to carry out bio prospecting activities.

In short, the new ocean scenarios do not entail the disappearance of States' economic, territorial, military or political goals, but it is not enough to maintain the State-centric perspective of conventional geopolitics. As Agnew so famously stated, the *territorial trap* has to be avoided. The new focus requires 'the need to keep state-centred ways of thinking open as a counter discourse to that of global networked society and borderless worlds' (Newman, 2003, p. 2) to be shown. In the global geo-strategic framework, pre-existing asymmetries will be repeated, but States' – both hegemonic and emerging powers – behaviours need to be interpreted anew – bearing in mind the capabilities, objectives and interests of large corporations, and also analysing the ability of disadvantaged States and NGOs with a worldwide reach to forestall the most powerful actions of State and business agencies.

Geopolitics, maritime space and naval doctrines

Decades ago, Carl Schmitt put forward the idea that universal history is the struggle between maritime and terrestrial powers (Schmitt, 1942) and the French Admiral, Castex, also entitled his book on strategy *La mer contre la terre* (Castex, 1935). In the following, an attempt is made to analyse how Geopolitics has approached this topic and to this end the theories of terrestrial power (more characteristic of continental powers, such as Germany) and the theories of sea power (characteristic of maritime powers, such as Great Britain and the US) will be set against each other. The main changes in direction that are currently being seen in the naval doctrine and strategies developed by the principal actors on the international stage will also be addressed.

Although there are isolated precedents in previous centuries, concern for addressing geopolitical problems and the balance of world power basically arose in the last third of the nineteenth century at the time when the great colonial empires were taking shape and a complex and conflictive world order was developing that was eventually to disintegrate at the end of the century due to a number of wartime crises. In this context, authors such as Mahan, Mackinder and Fairgrieve took up markedly geostrategic standpoints with the aim of assuming positions in the then crucial debate between the doctrines of naval power and terrestrial power. The maritime position gradually but forcefully entered into these geopolitical debates and experts from the field of geography, international relations, diplomacy and the realm of the military analysed and diagnosed the way in which controlling the seas, routes, straits and ocean resources was the key to the struggle for world or regional power.

At the same time, some German authors linked directly or indirectly with German expansionist politics, also paid special attention to the issues of political and military expansion over the oceans, creating a singular branch of science called *Political Oceanography*. This term (*Politische*

Ozeanographie in German) was introduced by Ratzel and frequently used by Haushofer (1924), an author who would focus above all on a geopolitical analysis of the Pacific Ocean (which he referred to as 'the Political Ocean'), which was to become the new centre of world power.

At the risk of appearing somewhat simplistic, it can be stated that there are a number of 'common places' in what could be termed the *Anglo Saxon theory of sea power*. First, the importance of establishing a range of control mechanisms (setting up naval bases, military occupation of key strategic areas, strengthening of naval resources, etc.) with the aim of controlling ocean routes and spaces. It was the approach of a major forerunner on ocean geopolitics, A. T. Mahan (1890), for example, whose ideas on the importance of maritime power influenced the navies of the entire world and led to the rapid development of navies before the First World War. This American Admiral noted the rivalry between continental (especially Russia) and maritime powers (Great Britain, the German Empire, Japan and the USA) and the importance of isthmuses and international straits. The Briton, H. Mackinder, whose approach is set out in *Britain and the British Seas* (1902) and, above all, in 'The geographical pivot of history' (1904), also focuses on these same strategic elements. He states the importance of naval power (as does Mahan) and of strategic bases to support fleets and to control straits, although he nonetheless believes that the age of maritime powers would reach its end at the end of the twentieth century, which would become the century of terrestrial power. His geopolitical model establishes the following areas: the *central area* (between the Oder River and the Urals, roughly speaking); the *'inner or marginal crescent'* (Germany, Austria, Turkey, India and China), and the *'outer or insular crescent'* (Great Britain, South Africa, Australia, the USA, Canada and Japan).

Linked to the above, it can be indicated that a second general aspect of Anglo Saxon theories is, precisely, establishing explanatory models of the world power balance in which three basic spatial components appear, albeit with certain nuances or differences: 1) a centre of terrestrial power (Eurasia, the Heartland), 2) a wide-ranging maritime space (Pacific–Atlantic–Indian Oceans) where the naval powers (principally the USA and Great Britain) deploy their forces to control the area, and 3) a world space that acts as a kind of gateway or transition area between the two.

The development of these geopolitical models in which the maritime aspect continues to play an especially major role is revitalised after the Second World War and during the Cold War. In the new positions taken up in the USA, aspects once again arise, such as the need to bolster naval military means and create networks of air and naval bases in various parts of the ocean, and to split the world geopolitical space into three parts between Eurasia/the Heartland (now dominated by the USSR), the oceans dominated by western maritime powers and the transition areas susceptible to possible disputes. A more ideological angle must be added to these 'traditional' aspects of Anglo Saxon geopolitical theory which, having already existed in the nineteenth and at the beginning of the twentieth centuries (justification of colonialism and imperialism), is now linked to the defence of the western-capitalist system and rivalry with the block headed by the soviets. Some theories, among others, will be of great interest as, even though they will not be strictly devoted to a geopolitical analysis of the ocean, they will nonetheless allude to the topic: N. Spykman's *Rimland* theory (a transition zone very similar to Mackinder's *inner or marginal crescent*), and S. Cohen's model, which returns to the traditional composition between Eurasian continental power and a 'world dependent on sea trade'.

It is possible to talk of the rise of a large number of strategic proposals in more recent years that, in one way or another, are returning once more to elements of the old naval doctrines (the tendency of the main traditional powers to maintain major fleets that enable them to possess a degree of sea power, for example). At the same time, the areas that had previously been considered as transition or buffer zones (*marginal crescent*, *rimland*, etc.) are being turned into the great geopolitical scenarios of today's world. This is how the new maritime powers (Brazil, India, China, etc.) that are playing an ever greater role are emerging alongside the traditional actors.

The western powers – which previously concentrated on the direct management of counter-insurgency in continental scenarios – are developing standpoints and actions (the last Libyan crisis) that point to a return to the indirect maritime approach (Nieto and Fernández, 2012). This can also be seen in aspects such as: i) the reformulation of strategic interest in safeguarding the free use of the seas and its routes – linking economic dominion and survival of the political–economic system with security and the control of sea routes (U.S. Department of Defense, 2012a, p. 9); ii) the appearance in numerous strategic documents –such as the recently revised *British Maritime Doctrine* (UK Ministry of Defence, 2011) – of the advantages of controlling areas near the shore that are disposed to being influenced and affected by operations to project naval power over the land (Prats, 2003, p. 579; Nieto and Fernández, 2012); iii) the development of strategic options developed with the goal of avoiding other powers' options for projection before they can materialise (Anti-Access, A2[3]) or limiting freedom of action in the nearest maritime *approaches* (Area Denial, AD[4]).

The big change, however, is that the gradual inclusion of the emerging powers in the 'club' of naval power projection signifies a break with the historical monopoly that the western navies held in this capability. The repercussions of this will depend on the pre-eminence of focuses closer to cooperation (effective multilateralism) or to schemes closer to geopolitical competition (multi-polarism) by some very significant emerging powers. Meanwhile, protecting the lines of maritime communication and the task of controlling the seas do not remain unaffected by the reshaping of the international order, as these activities will no longer be the exclusive domain of western navies due to the 'maritimisation' of the new centres of power's national security. A clear example of this is the fact that geopolitical rivals, such as China, the United States, Russia and India cooperate to repress piracy in the Indian Ocean. However, it cannot be taken as read that there will be similar focuses in scenarios where geopolitical rivalries are at stake, such as the Russian and Chinese positions on Iran and its threat to the Strait of Hormuz.

Further proof of the growing prominence of the emerging maritime powers can be garnered in Asia, where China's efforts to become a 'world class sea power' (Kondapalli, 2000, p. 2040) in the coming decades is combined with the efforts of smaller powers that are in conflict with China, with sovereignty/resources disputes being a major part of the backdrop. In other latitudes, a broad variety of reasons, ranging from the possibility of resorting to crises of limited trials of strength in adverse domestic situations, to the point of shutting out any chance of western powers intervening to avoid interference, involving the control of spaces and natural resources – can be found in the AD thrusts of countries such as Iran and Morocco.

As such, at the dawn of the twenty-first century, maritime space is becoming a privileged space where, and from which, the dynamics associated with a redistribution of power in international relations can be clearly perceived.

Change factors

Ocean geopolitics, as a projection of States' political and territorial power (as has been stated in the preceding sections) is experiencing changes with the result that new geographical maritime spaces or scenarios are being shaped that reflect a new territorial order, the most outstanding feature of which is the displacement of the strategic centre of gravity from the continental mass to the ocean basins. These would no longer be considered as off-centre with respect to the power bases (Mackinder's *pivot area*) and would progressively acquire centrality and political entity (emergence of maritime policies).

There are a number of different factors that are contributing to this transformation ranging from those that are legal in nature (new codification of the law of the sea), the emergence of new economic powers of major geographical importance (the BRIC countries) and the building of new supranational political bodies (the European Union), to technological innovation and climate change. A general effect that to a great extent is caused by these factors is the creation of what could be referred to as a new maritime paradigm. After defining this concept three of these factors will be focused on that are characterised by their marked spatial dimension: the law of the sea, emerging countries and the creation of the European Union.

A new maritime paradigm

During the transition to the new century, oceans have taken on progressively greater importance on the political agenda: two examples of this, because of their impact in the media, are the Obama Administration initiative (2009) and the European Union's Integrated Maritime Policy – launched in 2006. The wide range of international initiatives on this topic that have particularly marked the last two decades can together be interpreted as the markers that separate two concepts or visions of the oceans. The old paradigm, linked to discoveries, the creation of colonial empires and the alliance between trade and naval power, is giving way to a model firmly based on competition, innovation and knowledge. This new vision entails the displacement of strategic interest from traditional activities towards the new technologies, energy security and regional (international) leadership based on greater domination of the seas (spatial and economic). These initiatives have been expressed in different ways: formulations of new strategic visions, development of policies or new laws of the oceans of an inclusive nature. In general terms, the aim of creating the bases for a new way of understanding the role of the oceans, and the way in which States rely on the seas to contend with the challenges of a new world, are encouraging. The most developed countries are shaping a new vision of the oceans that leaves behind the 'navalist' mentality of the nineteenth century and in which, at least on a formal level, the environment is one of the fundamental lynchpins (sustainability and ecosystem management; *blue growth*) and a new order of priorities arises: food security has been displaced by energy security and leadership is based more on innovation, knowledge and the new technologies than the naval power-expansion of trade alliance.

The Law of the Sea

The Third United Nations Conference on the Law of the Sea (1973–1982) is well known as marking the beginning of a process whose great importance for geography has perhaps not been sufficiently stressed, or whose powerful spatial aspect has not been properly represented, at least. New jurisdictional concepts, such as the exclusive economic zone and the continental shelf have altered the political map and introduced a greater complexity into the way in which States exercise rights of sovereignty (and likewise assume responsibilities), although unlike in the case of the regime of the territorial sea, the implications are not heritage-related, but economic; these rights of sovereignty obtain in an area with geographical boundaries, the result of a projection from the territorial base of the coastal State. As economic interest grows, so jurisdictional spaces become more important for national security as a whole (food security, energy security, environmental security), which translates either into responses of a conventional defensive nature (naval power), or actions that are proactive and multidimensional (defence, safety, research) and are in modern times undertaken by the Coast Guard services (Paleri, 2009). In this way, geo-maritime scenarios are shaped by a dual process: i) land to sea projection: security is pivoting and moving towards the marine environment due to the growing importance of its resources and the growth in the maritime economy; ii) a reduction in the size of the High Seas and, as part of the effect of communicating vessels, the transfer of strategic priorities towards areas of influence on the national and regional scales, in keeping with the shift from the bi-polar to the regional (multi-polar) model.

The evolution of maritime scenarios is thus marked by the evolution of the law of the sea and its geographical formalisations, particularly those that have wide-ranging spatial repercussions, such as the EEZ and the continental shelf, although the latter is still in the claims submission process (implementation Art. 76 UNCLOS). Archipelagic seas have to be alluded to in the same way, however (Table 2.1) (Part IV UNCLOS), as does the potential of islands for extending their jurisdictional projections (Part VIII UNCLOS). The spatial impact is such that it has changed the iconography of the geopolitical picture of the world (Figure 2.1) and significantly altered the global jurisdictional balance (Figure 2.2). Largely speaking, the introduction of the concept of the EEZ involves some 30 per cent of the surface area of the planet, and the delimitation of the extended continental shelf (taking into account claims submitted up to 31 July 2012), a further 4.6 per cent (Table 2.2). From a geopolitical point of view, this means an increase in the part of the Earth subject (under a variety of legal regimes) to national jurisdictional control (around 60 per cent of its surface area) although the global commons still comprise some 40 per cent and, as such, (ocean) governance must comply with legal norms and institutions and international cooperation.

Table 2.1 Archipelagic States

Country	Km²	Date	Define coordinates
Antigua and Barbuda	–	17 August 1982	No
Kiribati	–	16 May 1983	No
Marshall Islands	–	1984	No
Saint Vincent and the Grenadines	–	19 May 1983	No
Seychelles	–	18 November 2002	No
Indonesia	3,100,109	25 March 2009	Yes
Philippines	604,461	28 July 2008	Yes
Papua New Guinea	562,333	28 February 1977	Yes
Bahamas	237,539	23 December 2008	Yes
Fiji	150,051	17 December 2007	Yes
Solomon Islands	130,427	1978	Yes
Tuvalu	84,591	1983	Yes
Maldives	71,688	1996	Yes
Vanuatu	70,981	29 July 2009	Yes
Dominican Republic	49,680	22 May 2007	Yes
Cape Verde	36,103	21 December 1992	Yes
Jamaica	22,032	16 October 1996	Yes
Comoros	15,895	13 August 2010	Yes
Trinidad and Tobago	7,186	27 May 2004	Yes
Sao Tome and Principe	6,012	31 March 1998	Yes
Mauritius (Chagos Archipelago)	5,011	27 August 2005	Yes
Grenada	554	22 December 2009	Yes

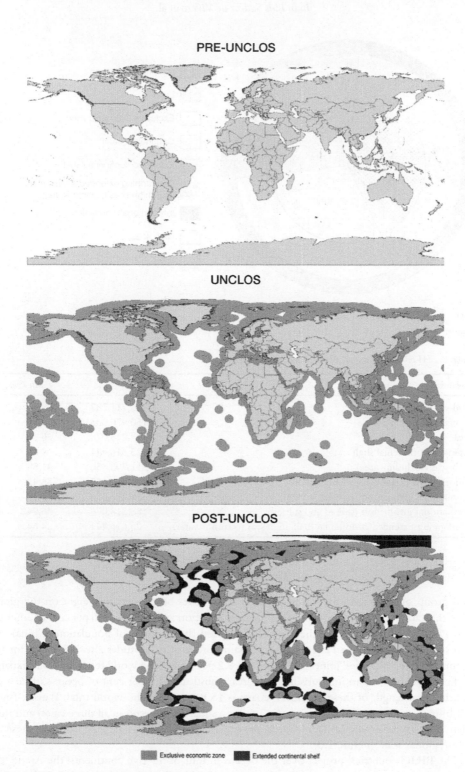

PRE-UNCLOS

UNCLOS

POST-UNCLOS

Exclusive economic zone   Extended continental shelf

Figure 2.1 Evolution of jurisdictions
Source: Author.

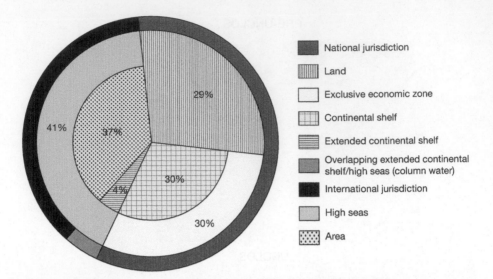

Figure 2.2 World jurisdictions
Source: Author.

Table 2.2 World national and international jurisdictions

Domains	Km²	Earth (%)
Land	146,711,282	28.79
Exclusive economic zone	150,970,333	29.63
Area	188,218,076	36.94
Extended continental shelf	23,616,441	4.64
High seas (including area)	211,703,856	41.58
National jurisdiction (land+EEZ+ECS)	297,681,615	58.42
International jurisdiction (column water)	211,703,856	41.58
International jurisdiction (seabed and subsoil)	188,218,076	36.94
Overlapping extended continental shelf/high seas (column water)	23,616,441	4.64

The BRIC countries

As a group of geographical entities, the BRIC countries bear a certain weight on the global scale: they are home to 42 per cent of the world population and occupy 25.6 per cent of emerged land (Table 2.3). Individually they respectively head the rankings of population (China) and land surface area (Russian Federation). This same order of magnitudes also holds true for the oceans: the four countries' joint EEZ represents 9.2 per cent of the world total;[5] the total expanse of the BRIC block maritime jurisdictional area[6] constitutes 4.5 per cent of ocean surface area, and the total length of these countries' coasts is 18 per cent of the world total. Russia, Brazil and India generate extensive EEZs (over 2 million square kilometres in all three cases) and their relative locations, as well as that of China, provide them with a wide-ranging maritime presence on a regional scale (Figure 2.3).

The BRIC countries' project over the oceans in four of the five continents: the Arctic (the Russian Federation), the Atlantic (Brazil), the Pacific (the Russian Federation and China) and the Indian Ocean (India). Their greatest presence in the oceans can be found around the Eurasian

Table 2.3 Basic geographical data

Countries	Population (000)[a]	Land area (km²)[b]	EEZ km²	EEZ World ranking[c]	Coast length (km)[b]	EEZ/ CL	Extended CS (km²)/ Dated[d]	ECS/ EEZ (%)
Brazil	195,493	8,514,877	3,191,827	14	7,491	426	911,847[e] 17.05.2004	28.5
Russian Fed.	140,367	17,075,200	7,566,673	7	37,653	200.9	1,279,800[f] 20.12.2001	16.9
India	1,214,464	3,287,263	2,305,143	21	7,000	329.3	628,327[g] 11.05.2009	27.2
China	1,354,146	9,326,410	879,666	36	14,500	60.6	–	–

Source: Author.

a UN Population Division. b CIA Factbook. c Author from VLIZ (www.vliz.be/); Sea Around Us Project (www.seaaroundus.org/). d CLCS website (www.un.org/Depts/los/clcs_new/clcs_home.htm). e Continental Shelf and UNCLOS Article 76. Brazilian Submission. Executive Summary. Brazilian Continental Shelf Survey Project, 2004. f Yenikeyeff, S. M., Krysiek T. F. *The battle for the next energy frontier: The Russian polar expedition and the future of Arctic hydrocarbons.* Oxford Institute for Energy Studies. Oxford Energy Comment, August 2007. g Author.

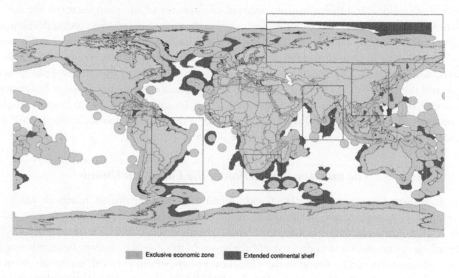

Exclusive economic zone ■ Extended continental shelf

Figure 2.3 BRIC countries
Source: Author.

continent. This is especially due to the Russian Federation's rights of sovereignty over the Arctic Ocean and India's dual seaboards (the Bay of Bengal and the Arabian Sea). This compacting of the Eurasian mass and its outlying waters forms a connection between Chinese and Indian strategic areas and the maritime traffic flows that link these countries with the global network and, especially, with the mega-port and harbour system of Eastern Asia.

However, Brazil and its maritime space are not located in the mainstream and have no links with the principal strategic maritime world scenarios. It does introduce the southern hemisphere into the block of emerging powers, however, due to its territorial size and its resource potential, turning the south-western Atlantic into an area of new global strategic interest.

China and India form the territorial nucleus that is to be the seat of greatest power on a global scale, with the former playing the primary role, possibly outstripping the United States by the year 2050.[7] In this context, both countries, with their intense economic growth rates, are increasingly dependent on the maritime transport flows that are fundamental to guaranteeing them a supply of vital energy and raw materials. India, meanwhile, also benefits from its geographical position with regard to the transport flows that pass through the Indian Ocean and connect the Persian Gulf and Middle East with Japan and Southeast Asia, turning the country into the 'natural sentry' standing guard over these traffic flows (Prakash, 2006).

And so, the new emerging phenomenon associated with the BRIC countries is that of the developing world (and countries where poverty is still a constituent feature) joining the club of naval powers (with the exception of the Russian Federation which, when it was part of the Soviet Union, was already a world naval leader during the Cold War[8]).

Although it easily evokes the first theorists of naval doctrines (Raja, 2009), this new sea power is, as might be anticipated, situated in a new, historically post-imperialist context and cloaked in modern globalisation. Naval power is now both the effect and the consequence of a maritime potential that is, for the first time, underpinned by territorial dominion over large continental and ocean masses (the latter of vast proportions) that, in turn, allow access to great quantities of resources thanks to scientific and technological development.

Territory and resources, the strategic value of which demands and justifies naval means, explains the lack of tradition and the previously unheard-of character of an ocean vector in the defence policies of some of the BRIC countries, and even, in the case of Brazil, the modest development of its armed forces as a whole. Each of the BRIC countries creates a geo-maritime scenario that in some cases, such as that of Brazil and its surrounding ocean, is much more individualised; others, such as China and India, could comprise a macro scenario due to the links between them and the fact that they share a large geographical domain, while the Russian Federation is the dominant territory in the ocean region where it is located. These are three great scenarios at the heart of which the BRIC countries dominate: i) the Arctic scenario; ii) the Indo-Pacific scenario and iii) the Southern scenario (Figure 2.3).

The maritime projection of the European Union

The European Union, with a planet-wide (economic and territorial) maritime reach, has developed its strategic vision of the ocean in the so-called Integrated Maritime Policy (2007). The EU's political characteristics do not involve member States ceding territorial sovereignty. Nevertheless, common norms do exist and apply – as in this case – to European maritime space. A precedent of this can be found in the Common Fisheries Policy. In other respects, from the angle of security, an EU Naval Force (EU NAVFORCE) already exists within the Common Security and Defence Policy and was given responsibility for Operation ATALANTA in Somalia. The European Union's maritime interests therefore project beyond the European seas and are, for reasons of ocean issues and the type of leadership that this institution exercises, on a global scale. The International Dimension of the Integrated Maritime Policy (2009)[9] therefore covers global issues, such as international governance, biodiversity and the high seas, climate change and strengthening the EU in international forums. Additionally, in strictly jurisdictional terms, European Union countries have a presence in almost all the oceans on the planet, and this generates a combined exclusive economic zone of over 24 million km^2, of which 18 million correspond to overseas territories (Figure 2.4). As the UNCLOS development process is one that is still ongoing, the implementation of Article 76 concerning the delimitation of the continental shelf beyond 200nm is still changing the areas under the jurisdiction of some States, quite

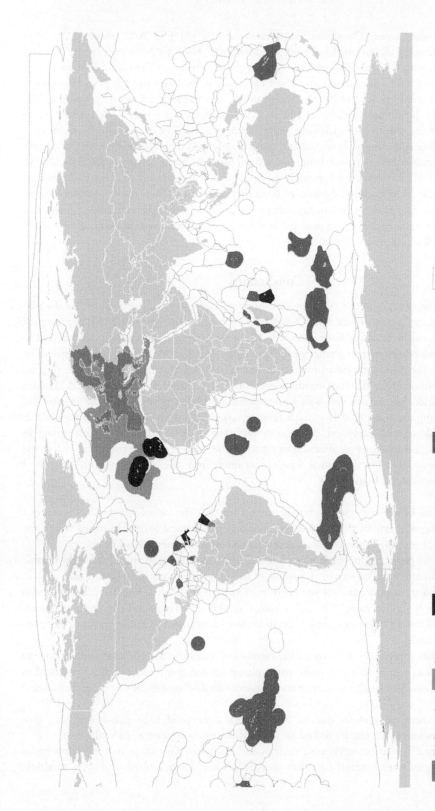

EU Countries ■ EU waters ■ EU Outermost regions waters ■ EU overseas territories waters ☐ Maritime jurisdictions (EEZ & ECS)

Figure 2.4 Overseas territories
Source: Author.

significantly in some cases. The delimitations made to date (31 July 2012) increase the EU's area of jurisdiction (in all oceans) by almost 8 million km². Almost 82 per cent of this is in the North Atlantic, where Portugal's jurisdiction alone extends to more than 2 million km². This gives the EU as a whole a dominant jurisdictional presence in the basin and has practically turned it into an internal European and North American sea. On the level of ocean geopolitics, the magnitude of the jurisdictional rights of EU member countries as a whole gives consistency to the view of it being the most widespread geographical organisation in the world; one space – the maritime space and its boundaries – that is more closely linked conceptually with the notions of deterritorialisation and network boundaries than those of fixed territory (Walters, 2004). In accordance with this aspect, what might be applicable is the notion of a maritime empire being a combination of traditional forms of territorial empire and new forms characterised by the diffuse expression of the territorial elements, the asymmetries between national States and new mechanisms for cooperation – a political structure between territorial empire and the current notion of the 'Wider Europe' (Scott, 2005).

Conclusions

In the above, the changes in the configuration of geopolitical scenarios seen in recent decades as a consequence of a new ocean order being constructed and the territorial effects prompted by the new law of the sea have been analysed. These are, therefore, circumstances that add to and complement other types of changes experienced in global geopolitics. Maritime space, by wit of it being the dominant environment on the planet, has an unquestionable bearing on spatial expressions of power that has become more evident as scientific and technological advances have enabled wider access and use. The jurisdictional expansion process over the oceans that has been ongoing since the mid-twentieth century, combined with the increased possibilities of it being explored and exploited, is not only changing its perception, but transforming its political essence and its entity as a substantive part of the large territorial pieces on the world geopolitical chessboard. The following aspects that result from the 'maritimisation' process of global geopolitics can be highlighted:

- displacement of geostrategic centres of interest towards the ocean space. This is a direct result of the new law of the sea that affects classic geopolitical formulations in which the oceans were no more than an *encasement* wrapped around emerged land;
- linked to the previous point, ongoing fragmentation of the ocean space, which is ceasing to be an undifferentiated space, as a result of the process of nationalising the seas;
- the incorporation into national jurisdiction of large marine areas that have to date been fringe areas (polar ice-caps). The combination of climate change and the implementation of Art. 76 UNCLOS is beginning to include these as new and valuable pieces of the geo-strategic puzzle;
- reshaping of national hierarchies with the creation of 'maritime States' (large and widespread areas of ocean – exclusive economic zones plus extended continental shelf – compared to the area of emerged land), on numerous occasions based around small States or micro island States;
- emerging maritime powers due to large maritime areas of high geo-economic value (natural resources, sea traffic) linked to States with strong economic growth;
- new political–territorial categories, such as the European Union, which exercises rights of sovereignty over extensive marine areas that could be described as a new maritime empire.

Notes

1   All calculations included in this chapter are based on data available at the United Nations (Division for Ocean Affairs and Law of the Sea, DOALOS) until 31 December 2013.
2   To 31 July 2012, 51 States had submitted 61 proposals to the Commission on the Limits of the Continental Shelf for defining the continental shelf, while another 40 States had presented 44 preliminary studies, especially African States, with technical and financial difficulties for undertaking the corresponding studies required for defining these limits.
3   Anti-Access (A2): 'Those actions and capabilities, usually long range, designed to prevent an opposing force from entering an operational area' (U.S. Department of Defense, 2012b, p. 6).
4   Area-Denial (AD): 'Those actions and capabilities, usually shorter range, designed not to keep an opposing force out, but to limit its freedom of actions within the operational area' (U.S. Department of Defense, 2012b, ibid.).
5   This calculation of the world total is made taking the surface area based on the 200nm line.
6   Exclusive economic zone plus extended continental shelf (continental margin beyond 200nm) claimed by this group of countries to date.
7   According to International Futures (Rogers, 2009), China's percentage share of world power in 2050 would be 23.07 per cent compared to 15.98 per cent for the United States; India's share would be 12.39 per cent. The United States, together with the European Union, Japan and Russia would develop negatively between 2009 and 2050. The 2025 Global Trends report (which cites the International Future Model) states that in 2025 the United States will maintain its share of power over China, despite negative development compared to China, India, Russia and Brazil, which all increase their share of world power (National Intelligence Council, 2008).
8   China, India and Brazil have already initiated programmes that tend towards the development of naval forces in keeping with their new status as global players. In April 2009 China celebrated the sixtieth anniversary of the creation of its naval fleet in an unprecedented manner (Chinese People's Liberation Army Navy (PLAN)) with the inclusion of two nuclear submarines. Its immediate goal is the construction of an aircraft carrier (Lai, 2009).
9   Communication from the Commission of the European Communities (COM(2009)536 final), developing the international dimension of the Integrated Maritime Policy of the European Union.

References

Agnew, J. A. (2003) *Geopolitics: Re-visioning World Politics*, Routledge, New York.
Blake, G. (2000) 'State limits in the early twenty-first century: Observations on form and function', *Geopolitics*, vol. 5, no. 1, pp. 1–18.
Brzezinski, Z. (1997) *The Grand Chessboard: American Primacy and Its Geostrategic Imperatives*, Basic Books, New York.
Castex, R. (1935) *La mer contre la terre*, Paris, Société d'Éditions Géographiques, Maritimes et Coloniales.
Cowen, D. and Smith, N. (2009) 'After geopolitics? From the geopolitical social to geo-economics', *Antipode*, vol. 41, no. 1, pp. 22–48.
Glassman, J. (1999) 'State power beyond the "territorial trap": The internationalization of the state', *Political Geography*, vol. 18, pp. 669–696.
Haushofer, K. (1924) *Geopolitik des Pazifischen Ozeans*, Kurt Vowinckel, Berlin.
Held, D. and McGrew, A. (eds) (2002) *Governing Globalization: Power, Authority and Global Governance*, Malden, MA: Polity, Cambridge.
Kondapalli, S. (2000) 'China's naval strategy', *Strategic Analysis*, vol. 23, no. 12, pp. 2037–2056.
Lai, D. (2009) China's maritime quest. Op-Ed. Strategic Studies Institute; June 2009.
Lorot, P. (1997) 'De la géopolitique à la géoéconomie', *Revue Française de Géoéconomie*, vol. 1, pp. 23–35.
Lucchini, L. and Voelckel, M. (1990) *Droit de la Mer*. Vol I. *La mer et son droit. Les espaces maritimes*, Ed. Pedone, Paris.
Mackinder, H. (1902) *Britain and the British Seas*, William Heinemann, London.
Mackinder, H. (1904) 'The geographical pivot of history', *The Geographical Journal*, vol. 23, pp. 421–437.
Mahan, A. T. (1890) *The Influence of Sea Power upon History, 1660–1783*, Little, Brown, Boston, MA.
National Intelligence Council (2008) *Global trends 2025: A Transformed World*, National Intelligence Council, Washington DC.

Newman, D. (2003) 'On borders and power: A theoretical framework', *Journal of Borderlands Studies*, vol. 18, no. 1, pp. 13–25.

Nieto, J. L. and Fernández, F. (2012) 'El poder naval a propósito del caso de Libia: ¿un punto de inflexión hacia una orientación marítima de la estrategia?', *Documento de Opinión* 02/12, Instituto Español de Estudios Estratégicos, Madrid, pp. 1–19.

Paleri, P. (2009) 'Coast guards of the world and emerging maritime threats', *Research Report*, Ocean Policy Research Foundation, Tokyo. Digital version: www.sof.or.jp/en/report/pdf/200903_ISSN1880–0017.pdf, accessed 29 October 2012.

Prakash, A. (2006) 'A vision of India's maritime power in the 21st century', *Air Power Journal*, vol. 3, no. 1, pp. 1–12.

Prats, J. M. (2003) 'Consideraciones sobre la guerra naval', *Revista General de Marina*, vol. 244, pp. 575–587.

Raja, C. (2009) 'Maritime power: India and China turn to Mahan', *ISAS Working Paper*, no. 71.

Rogers, J. (2009) 'From Suez to Shanghai: The European Union and Eurasian maritime security', *European Union Institute for Security Studies: Occasional Paper*, no. 77.

Schmitt, C. (1942): *Land und Meer. Eine weltgeschichtliche Betrachtung*, Reclam, Leipzig.

Scott, J. W. (2005) 'The EU and "Wider Europe": Toward an alternative geopolitics of regional cooperation?', *Geopolitics*, vol. 10, no. 3, pp. 429–454.

Suárez de Vivero, J. L. (1979) 'El espacio marítimo en la geografía humana', *Geocrítica*, no. 20.

U.S. Department of Defense (2012a) *Sustaining U.S. Global Leadership: Priorities for 21st Century*, Washington DC.

U.S. Department of Defense (2012b) *Joint Operational Access Concept*, Joint Chiefs of Staff, Washington DC.

UK Ministry of Defence (2011) *British Maritime Doctrine*. The Development, Concepts and Doctrine Centre, Shrivenham, UK.

Vigarié, A. (1990) *Géostratégie des océans*, Paradigme, Caen.

Wallerstein, I. (2004) *World-Systems Analysis: An Introduction*, Duke University Press, Durham, NC and London.

Walters, W. (2004) 'The frontiers of the European Union: A geostrategic perspective', *Geopolitics*, vol. 9, no. 3, pp. 674–698.

3

STATE OCEAN STRATEGIES AND POLICIES FOR THE OPEN OCEAN

Patricio A. Bernal

The ocean space and its resources in the twentieth century

This chapter addresses 'state ocean strategies and policies for the Open Ocean'. Open Ocean and Deep Seas are not legal categories included in the United Nations Convention on the Law of the Sea, UNCLOS, but are technical terms that encompass the large oceanic ecosystems extending thousands of kilometres offshore and the very inaccessible deep environments of the ocean, perhaps the most important part of the ocean in terms of the life support system of the planet.[1] It differs from the term 'High Seas', a clearly defined jurisdiction in Part VII of UNCLOS, since much of what is Open Ocean and Deep Seas lies within exclusive economic zones (EEZ) and Territorial Seas of coastal nations, many of them small islands and developing states. Since the international seabed beyond the continental shelf or EEZs of nations constitutes 'The Area', Part XI of UNCLOS puts it under the authority of the International Seabed Authority; much of the management needed in this oceanic domain can be done and is starting to be done under national law, existing legal international arrangements and regional coordination.

But not all oceans can be dealt with in this way. UNCLOS gives the responsibility for the governance, policy and management of the High Seas to the collective authority and action of nation states, with direct legal, political and economic control of nearly half of the world's ocean, excluding the corresponding seabed. In the final analysis individual nation states are the cornerstones of this system. Given the many shortcomings in implementing current regulations, it is a valid question to ask why this turns out to be so and here some possible answers are exposed.

Throughout the history of humanity, ocean space and its resources have been coveted by nations. As in the occupation and settlement of land, nations have tried to exert control and dominion over coastal seas and critical passages, and have regulated and taxed the free movement of vessels and merchandise along their coasts. Ocean-going capabilities and strong navies were and are the instruments of this dominion. In the first half of the twentieth century an increased interest in the use of the ocean space and its resources was accompanied by concomitant concerns on the security risks associated with an open and aggressive race for control and dominion of the open sea. The two world wars had demonstrated the importance of modern naval power and the advent of large aircraft carriers, the importance of the ocean to project military force to distant locations. The emergence of the Cold War as the dominant security scenario after the Second World War and the trauma over Hiroshima and Nagasaki put ocean affairs centre stage, at least from a global security stand.[2]

With the establishment of the United Nations in 1945, depositary of a strong and universal mandate to avert regional and global security risks, this state of affairs was considered problematic and eventually became untenable. Progress in marine technology and the impact of new scientific discoveries brought forward renewed interest on old and new civilian uses. The discovery of poly-metallic nodules at the bottom of the sea pre-figured a future with large deep-sea mining operations. This fact triggered a series of extensive deep-sea mining claims in the equatorial Pacific, registered by private companies under national law.[3]

The extension and intensification of whaling and fisheries in the High Seas, and the development of large fleets that were accompanied by factory vessels that could operate for many weeks in distant waters, highlighted the need to better define and protect fishery rights. Already facing a 'plethora of conflicting claims by coastal states' (Koh, 1983) and the expansion of world trade, made securing passage through international straits imperative, as well as the need to craft a common universal definition of maritime national jurisdictions. These concerns received varying levels of attention and reactions around the world. Due to the lack of capabilities to master sophisticated seagoing technologies, issues related to the open ocean were seen by many coastal states as issues for 'the rich and powerful countries'. Nations with large navies, commercial fleets and an increasing capability to conduct scientific research in the open ocean were clearly beneficiaries of the status quo based on the 'freedom of the sea' doctrine and might have had some level of reservations with regard to entering into a broad and complex negotiation under the UN. Nevertheless, global and regional security concerns were widely shared among all nations, regardless of their size and power.

This chapter describes first the scope of UNCLOS with a summary of its outcomes, discussing next the context in which UNCLOS was negotiated. Then a brief review of the series of international instruments and institutions available today to manage the open ocean is provided, including a description of the UN system and its key role in ocean governance. Next, the relationship of ocean governance and ocean law to civil society is critically analysed, followed by a brief analysis of the role surveillance and enforcement of regulations play. The challenges posed by the introduction of integrated management and the application of the ecosystem approach as standards of good practice are presented next. Finally, the chapter ends with a reasoned proposal laying down a '*feuille de route*' for organizing and improving action at the national level.

UNCLOS: the United Nations Convention on the Law of the Sea

Without any doubt, UNCLOS is the major legal instrument addressing the governance of the open ocean and High Seas. UNCLOS is often referred to as a constitution of the ocean and at least in one respect this is exact: UNCLOS is far from being a self-contained code. UNCLOS combines its norms and jurisdictions with 'a series of frameworks for developing specific rules in the context of other arrangements and organizations' (Oxman, 2002). For example, for all issues relating to shipping it is the International Maritime Organization (IMO) that acts as the 'international competent organization'; similarly, the United Nations Food and Agriculture Organization (FAO) does the same for fisheries and aquaculture, and the Intergovernmental Oceanographic Commission of UNESCO for ocean sciences and ocean observations.

The new international legal regime that emerged in 1982 was conceived to provide: first a framework to harmonize the rights and duties specified in the pre-existing Conventions into a single text, re-enforced by a common system for settling disputes; second, a common redefinition of ocean areas, the zonal or 'spatial' jurisdictions, i.e. the nature and extension of the new jurisdictions for the ocean space derived from the harmonized texts; and third, new functional principles and norms in domains such as protection of the marine environment, living marine resources and deep-sea mining.

In terms of zonal regimes, the common redefinition of ocean areas contained in UNCLOS provides precise definitions of *Baselines* at the coastal territorial boundary of states, *Internal Waters*, located behind the baselines, *Territorial Sea*, extending 12 miles offshore of the baselines, *Contiguous Zone* up to 20 miles, allowing control by coastal state of unlawful acts, *Exclusive Economic Zone*, a *sui generis* jurisdiction giving exclusive economic rights to coastal states up to 200 miles from baselines, the *Continental Shelf* regime giving exclusive access to the coastal state to its resources, the *High Seas* regime, preserving beyond EEZs the freedoms of the *'freedom of the sea'* doctrine for the water column and surface; and *The Area*, i.e. the international seabed area beyond EEZs or continental shelf boundaries, defined as *'the common heritage of humanity'*. It is interesting to note that due to the conflicting interests that needed to be accommodated, the definitions of these zonal regimes were far from being exclusively a matter of geographic boundaries alone. The jurisdictions attached to them are limited functionally or functional principles are affected by them.

The new functional principles that UNCLOS incorporated include environmental considerations on: Pollution Prevention (generally), Dumping at Sea; Fishing Rights, both in the EEZ and on the High Seas; Protection of Marine Mammals; Biodiversity; Land Based Pollution; Atmospheric Based Pollution; and Pollution from Ships.

UNCLOS also created four new Agencies:

- the Commission on the Limits of the Continental Shelf charged with the mandate of resolving claims for an 'extended continental shelf';
- the International Seabed Authority with its Assembly, Council, and Secretariat managing the Mining Code of part XIV and monitoring activities in The Area;
- the Enterprise, pertaining to the Mining Code established in part XIV; and
- the International Tribunal for the Law of the Sea, charged with the regime for the settlement of disputes.

UNCLOS also completed the preliminary work done in Geneva during the First UN Conference on the Law of the Sea *Concerning the Compulsory Settlement of Disputes* where an Optional Protocol (that entered into force in September of 1962) had been agreed, providing for the compulsory jurisdiction of the International Court of Justice, or for submission of the dispute to arbitration or conciliation. UNCLOS establishes a comprehensive regime for Settlement of Disputes, (Part XV) provides it with its own Tribunal and explicitly regulates three Enforcement regimes by: the Flag Nation, Port Nations, and more generally by Coastal Nations not acting as Port Nations.

It is a singularity in international affairs, that precisely during the bi-polar world of the Cold War, and during the discussion within the UN of a 'new international economic order', a sophisticated and complex negotiation on ocean affairs could succeed. In the 1960s, codifying 'a single body of rights and duties' for the use of the ocean space and its resources and agreeing on 'a precise allocation of jurisdictions applicable to all' (Oxman, 2002), was a pending issue in international law. Once in the first half of the twentieth century[4] and twice in the second half, in 1958 and 1960 under the UN, efforts to negotiate an integral agreement on the law of the sea had failed. The First UN Conference on the Law of the Sea in 1958 was unable to find a common base for the negotiation and concluded giving rise not to one but to four separated conventions: the Convention on the Territorial Sea and the Contiguous Zone; the Convention on the High Seas; the Convention on Fishing and Conservation of the Living Resources of the High Seas; and the Convention on the Continental Shelf. The Second UN Conference on the Law of the Sea in 1960, convened to address the breadth of the territorial seas and fishery limits, was unable to reach a consensus, adopting instead just two non-binding resolutions.

It was during the third Law of the Sea Conference (1973–1982) that the international community finally agreed on the current text of the United Nations Convention on the Law of the Sea, UNCLOS. The Success of the Third UN Conference on the Law of the Sea depended on the definition of a balanced negotiating package and on adopting a *sui generis* procedural rule. In the words of Bernard Oxman, 'many treaties, by permitting reservations, sacrifice uniformity of substance in order to promote universality of ratification' but UNCLOS aimed at 'uniformity of substance and universality of adherence' (Oxman, 2002). For achieving this, the Convention had to be negotiated and adopted as a whole, *in toto*, as a package deal, prohibiting reservations. To achieve this ambitious goal, the Third Conference was carefully prepared with a lot of informal preliminary work that outside or within the UN had started as far back as 1967.

The negotiating package did contain elements that would change the balance from the traditional 'freedom of the seas' doctrine that exclusively dominated ocean affairs for centuries to a more nuanced and modern system where the powers and rights of coastal states could be significantly extended, especially from an economic point of view, but at the same time preserving the basic rights of movement in the ocean, so essential from a sovereignty, commercial and security point of view.

Developing countries actively contributed to creating the base for a negotiating package between north and south. Triggered by the Truman declaration in September 1945 (i.e. the unilateral declaration by the USA of national jurisdiction over the resources of the continental shelf), Argentina with one of the widest continental shelves in the world did the same in 1946, extending its jurisdiction over the resources in overlying waters. The nations of western South America, neighbouring the subduction zone between the Antarctic and Nazca Plates and the American continent and having a series of deep-sea trenches off their coasts instead of an extensive continental shelf, established sovereign jurisdiction over 200 miles off their coasts regardless of the actual extension of their continental shelf: Chile in 1947, Peru in 1948 and Ecuador 1950. These early developments in South America were followed by Arab countries who asserted sovereign domain over the resources in the continental shelf, and later on by African countries. The archipelagic states of Indonesia and the Philippines also claimed jurisdiction over their 'interior waters'. This fast evolving legal landscape brought home to all coastal states the fact that regardless of their stage of development they also had a stake in the High Seas and some of these newly proposed jurisdictions became key negotiating points adopted by the new group of 77, that in the rhetoric of the time federated the 'non-aligned and developing countries of the third world'. Although South American countries were first in claiming sovereign jurisdiction over 200 miles, it was Kenya that in a working paper presented at the Thirteenth Session of the Asian–African Consultative Committee (Lagos, 18–25 January 1972),[5] proposed the term 'economic exclusive zone' for the 200 mile zone (Nandan, 1987). The evolution of the 200 mile zone concept from 1947 to 1982 reflects a compromise that legitimizes functional sovereign economic rights over the resources but not over the space, as a territory.

In contrast with the series of treaties and conventions negotiated a decade later in Rio de Janeiro, or the Antarctic Treaty designed primarily to solve territorial claims over the Antarctic territory and that protects what in 1959 was seen as a pristine continent, UNCLOS is far from being a Convention negotiated around the protection of the environment. It is a convention that deals with development and the potential benefits that nations could obtain from the ocean and its resources, and maintains the security equilibrium existing during the Cold War.

In the dynamic of the negotiations of a package with dissimilar contents, during the Third Conference *'quid pro quo'* transactions had to emerge. Parts XIII and XIV of the final text reflect very well this aspect of the negotiations. With the aim of protecting the economic rights vested on the coastal states by the new EEZ jurisdiction, Part XIII regulates the conduct of Marine Scientific Research inside the EEZ; in fact one of the freedoms preserved in Part VII for the High Seas. Since most developing nations do not have the capabilities to conduct scientific research in distant waters, the protections and guarantees in Part XIII affected predominantly research conducted by developed nations inside the EEZ of coastal states that could reveal or prospect for potential resources.

On the other hand, Part XIV tries to build a legal instrument to level the ground of capabilities for access to marine resources by exhorting all nations of the world to cooperate in the transfer of marine technology under *'fair and reasonable terms and conditions'* (Art. 266.1), and to *'foster favourable economic and legal conditions for the transfer'* (Art. 266.3).

At the time, for developing countries that had recently established the Group of 77, the context for Part XIV was provided by the larger discussion on equity in international economic affairs taking place under the UN Conference on Trade and Development (UNCTAD). The UN General Assembly had adopted on 1 May 1974 a strong Declaration containing a series of principles promoting a New International Economic Order (UN A/RES/S-6/3201, 1974). Part XIV was seen as a sort of promise of large country-to-country programmes of transfer of marine technology that would open access to marine resources, specially mining. The transfer of deep-sea mining technology is fully part of the mining code (Part XI Art. 144). This article turned out to be one of the final stumbling blocks to approving the whole treaty. Developed countries interpreted Part XIV as being part of the more general commitment to cooperate; something that they felt they already did under multilateral and bilateral agreements. Forty years on from the UNCLOS negotiations, it is fair to say that in most cases the key actor has been the private sector and not the states that negotiated the transfer of technology for the exploitation of marine resources, usually in compensation for being granted access to the resources of coastal states.

Are nations of the world assuming the challenges that UNCLOS put in their hands? For a great majority of coastal and developing nations the High Seas has been and remains a remote space. Most coastal nations do not have the specialized skills, expensive infrastructure and financial means to access the High Seas and its resources. Therefore, in practice, the effectiveness of the approach adopted relies on the commitment and engagement of those that possess the assets and capabilities to access the High Seas. However, for this very same reason the treatment that UNCLOS gives to the High Seas also raises basic questions of equity.

International instruments and institutions

The fact that UNCLOS combines its norms and jurisdictions with frameworks developed under other arrangements and organizations gives a special role to the UN in the implementation of UNCLOS, especially beyond territorial seas. This role is not exclusive, since in a court of law, all International Law will be called upon in the resolution of disputes, but gives the UN a priority central role that has been assumed and developed in time. Furthermore, a sort of tacit agreement exists among nations that governance of the sea is to be treated at the level of the General Assembly of the UN. Since 1982 this tacit agreement has been strictly adhered to, with the only exception the negotiation and adoption by UNESCO of the Underwater Cultural Heritage Convention in 2001 that entered into force on 2 January 2009, after its twentieth ratification.

On the other hand, there are 641 bilateral and multilateral agreements, including modifications, regulating the use of the ocean space and its resources (Mitchell, 2002–2014). Not all these agreements and treaties have the same relevance or importance, and many are bilateral arrangements written to solve ad hoc conflicts among neighbours; nevertheless their existence (or persistence) shapes international jurisprudence. This proliferation of international agreements reveals a low level of engagement or confidence on the global ocean governance system from the part of states. Many parts of UNCLOS have never been tested in the court of law, and many governments when given the opportunity by UNCLOS to choose, preferred the International Court of Justice in The Hague rather than the Law of the Sea Tribunal in Hamburg as their arbiter in the settlement of disputes. This means that when litigating a marine case, the legal advice on which tribunal or jurisdiction should take the case still remains a critical first step. The review below follows to a great degree the selection made by Haward and Vince (2008).

1 The 1995 UN Fish Stocks Agreement (UN)

In 1992, at the UN Conference on Environment and Development in Rio de Janeiro, nations agreed to call for a UN Conference to negotiate a complementary agreement for highly migratory and straddling fish stocks. Before 1992, several crises had developed in situations where a fish stock lying within national jurisdiction had a significant part of the same stock lying outside national jurisdiction, leaving it open to the exploitation by distant waters' fishing fleets. Real incidents in the 1980s resulted in the capturing of foreign fishing boats in the High Seas by coastal states trying to enforce fishing regulations and protect their stocks. These incidents could have been treated as acts of piracy or an unlawful act of aggression among nations, bordering on an act of war. Fortunately, diplomacy prevailed and parties concerned finally recognized that a real problem existed. After six sessions held between 1993 and 1995, the 'UN Agreement for the Implementation of the Provisions of UN Convention of the Law of the Sea of 10 December 1982 relating to the Conservation and Management of Straddling Fish Stocks and Highly Migratory Fish Stocks' was adopted. Better known as the UN Fish Stocks Agreement (UNFSA), it entered into force on 11 December 2001. As its long official name indicates, this agreement is understood to be an 'implementing agreement' for UNCLOS,[6] signalling a practical and low-risk way of complementing UNCLOS without opening its basic original text to lengthy negotiations, amendments and modifications.

Articles 19, 20, 21, 22 and 23 of UNFSA provides that states will ensure that vessels flying their flags shall comply with regional and sub-regional management measurements adopted by Regional Fisheries Management Organizations and Agreements (RFMO/A). A crucially important addition to international law contained in these articles was that nations were authorized to enforce those measures, 'regardless of where the violation occurs', including the High Seas. Some of these very same articles still remain today a major obstacle for several fishing nations to subscribe this agreement on the basis that they violate exclusive flag-state jurisdiction over fishing vessels in the High Seas (Balton and Koehler, 2006).

UNFSA contains a significant body of non-flag-state enforcement provisions that contribute significantly to strengthening port-states' authority (Rayfuse 2004, cited in Haward and Vince,

2008): i) member states' members of a RFMO can detain vessels that they suspect have acted to undermine the effectiveness of the conservation measures adopted by the RFMO, until the flag state concerned adopts 'appropriate' actions; ii) port-states have the right and duty to inspect gears, documents and cargo of fishing vessels calling voluntarily on their ports; iii) port-states can prohibit landings and trans-shipments when it can be established that the catch has been obtained undermining the effectiveness of management and conservation measures adopted by a RFMO/A.

Only states that belong to a RFMO or have agreed to apply the conservation measures adopted by a RFMO can fish the resources (species) regulated by those measures. Most RFMO/A were established to regulate single species stocks, such as the 17 tuna commissions. This feature of most RFMOs also makes the application of the ecosystem approach to fisheries difficult, where consideration of the catch of non-targeted species and the impact on other species should also be considered. Since RFMO/As allocate quotas or levels of fishing effort among their members, this in effect should limit fishing in the High Seas only to those countries that are members of a RFMO. This clause has become very contentious. 'Free rider' vessels that fish regardless of complying with regulations remain a real threat in many parts of the world. Some RFMOs manage different types of stocks, such as NAFO in the North West Atlantic. This is also the case of the new South Pacific RFMO that in its western region will manage essentially Deep Sea Fisheries, and in its Central and Eastern region will manage a large pelagic fishery. The bottom line is that each *RFMO is an autonomous international agreement (or treaty) binding a group of fishing nations, the parties to the agreement, that self-regulate their own behaviour.* Many RFMOs have a good record and adhere to acceptable international standards. However, many have resisted or grudgingly accepted efforts to establish a system of external review of their procedures (Balton and Koehler, 2006).

2 The Compliance Agreement (FAO)

Reflecting the level of difficulty with the enforcement of fisheries in the High Seas, 'the Agreement to promote Compliance with International Conservation and Management Measures by Fishing Vessels on the High Seas' had to be established to deter the practice of reflagging vessels to avoid compliance with the conservation and management measures adopted for the High Seas. In the words of Balton and Koehler (2006, p. 9) this is a problem because 'many developing States (and some developed States) allow High Seas fishing vessels to fly their flags without any meaningful ability or intention to control the operations of those vessels'. The agreement was intended to be applied to all vessels fishing or intending to fish in the High Seas, although parties can exempt vessels below 24m length. Entering into force in 2003, today it has 39 'acceptances', which in the context of FAO agreements have the same function as ratification or accession. The last three acceptances were given by Senegal, Mozambique and Brazil in 2009.

3 Code of Conduct (FAO)

A voluntary 'soft law' instrument, it contains a comprehensive prescription to guide fisheries practices towards a trajectory of sustainability. It deals with: i) fisheries management practices; ii) fishing operations; iii) aquaculture development; iv) integrating of fisheries into coastal area management; v) post-harvest practices and trade; and vi) fishery research. The compliance agreement (above) is an integral part of the Code of Conduct as well as several International Plans of Action (IPOAs) drafted within the FAO system answering a call contained in the Code to provide guidance for 'the formulation of international agreements, and other legal agreements'.

There are several IPOAs: i) 'for the management of fishing capacity'; ii) 'for the conservation and management of sharks'; iii) 'for reducing incidental catch of seabirds in longline fisheries'; and iv) 'to prevent, deter and eliminate illegal, unreported and unregulated fishing' (IPOA-IUU). As part of the 'Code' all these instruments are voluntary.

4 International Whaling Commission (IWC)

Established in 1946 under the International Convention for the Regulation of Whaling, the IWC was intended to serve as the primary international mechanism for the conservation of whales as a fishing (hunting) stock. Since 1970, *a contrario sensu* however, the Commission has become the primary international mechanism for the protection and conservation of all species of whales (Caron, 1995). In 1986 the IWC adopted a five-year moratorium on commercial whaling that has been extended to the present and has designated an Antarctic sanctuary for whales. 'The IWC power to "legislate" a moratorium or quotas is very restricted because any member state may opt out of a quota or moratorium simply by objecting to it' (Caron, 1995). Furthermore, the IWC has no authority or practical means to enforce them and any member might shield itself from the obligation of compliance by leaving the Commission, as Iceland did in 1992 until rejoining in 2002.

5 Marine Pollution and Safety of Life at Sea (IMO)

Over time the IMO has developed a total of 58 treaties and arrangements, dealing with safety issues related to cargo (shipping of dangerous substances), safety of life at sea, search and rescue procedures, dumping of substances into the ocean, ship construction standards (double hull tankers, safe containers, safe fishing vessels), and pollution from vessels, and more recently with rules dealing with the abatement of terrorism and terrorists acts.

According to the IMO itself, the three most important instruments are:

1 the International Convention for the Safety of Life at Sea (SOLAS), 1974, as amended;
2 the International Convention for the Prevention of Pollution from Ships, 1973, as modified by the Protocol of 1978 relating thereto and by the Protocol of 1997 (MARPOL); and
3 the International Convention on Standards of Training, Certification and Watchkeeping for Seafarers (STCW) as amended, including the 1995 and 2010 Manila Amendments.

Everything that happens at sea on board a ship falls under the responsibility of a state through the Flag-State regime of UNCLOS. The state that registers or licenses the vessel, granting it the right to carry its flag to navigate in international waters and to enter into the territorial waters of other states, has first and exclusive responsibility of what the vessels and the people on board do vis-à-vis compliance with international and national laws and regulations.

According to the UN Review of Maritime Transport (UNCTAD, 2011), 90 per cent of foreign controlled tonnage of the world fleet is registered in ten countries that maintain 'open and international registries'. These ten countries are: Antigua and Barbuda, Bahamas, Bermuda, Cyprus, Isle of Man, Liberia, Malta, Marshall Islands, Panama and Saint Vincent and the Grenadines. In fact four countries: Panama, Liberia, the Marshall Islands and Hong Kong (China) register 47.5 per cent of the world fleet (UNCTAD, 2011). A less elegant name for this universal practice is 'flags of convenience'.

In general terms, it can be said that a vessel flies a flag of convenience when it has no real economic connection (or no *'genuine economic link'*) with the country whose flag it flies. From a viewpoint of the countries of registration, an *'open registry'* country is one which accept vessels on its shipping register with which it has no genuine economic link . . . real owners live outside the jurisdiction of the Flag State.

(Benham, 2003, pp. 126, 128)

UNCTAD identifies the following elements as relevant for determining whether a genuine link exists (Benham, 2003):

1 the merchant fleet contributes to the national economy of the country;
2 revenues and expenditures of shipping, as well as purchases and sales of vessels, are treated in the national balance of payments accounts;
3 the employment of nationals on vessels; and
4 the beneficial ownership of the vessel.

The genuine economic link that appears broken by the practice of open registries is what UNCTAD was supposed to guard 'in order to help developing countries make headway in the development of shipping capabilities . . . in a participatory international economic system' (Behnam, 2003, p. 124). Today, developing countries have a dominant economic participation in the provision of seafarers, ship scrapping and in registration, although there is also progress in 'maritime sectors of higher business sophistication and technical complexity' (UNCTAD, 2011).

6 Convention on Biological Diversity (CBD)

The Convention on Biological Diversity was opened for signature on 5 June 1992 at the United Nations Conference on Environment and Development (the Rio 'Earth Summit') and entered into force on 29 December 1993. The objectives of the Convention as stated in Article 1 are:

[T]he conservation of biological diversity, the sustainable use of its components and the fair and equitable sharing of the benefits arising out of the utilization of genetic resources, including by appropriate access to genetic resources and by appropriate transfer of relevant technologies, taking into account all rights over those resources and to technologies, and by appropriate funding.

Early on, CBD adopted in 1995 the Jakarta Mandate on the Conservation and Sustainable Use of Marine and Coastal Biological Diversity, committing to:

a series of specific goals including the development of a global system of marine and coastal protected areas, the establishment of and implementation of a global program of making fisheries and mariculture sustainable, blocking the pathways of invasions of alien species, increasing ecosystem resilience to climate change, and developing, encouraging, and enhancing implementation of wide-ranging integrated marine and coastal area management that includes a broad suite of measures at all levels of society.

(Secretariat CBD, 2012)

The 2008 Conference of the Parties of CBD adopted criteria for the identification of Ecological and Biological Significant Areas, EBSAs (decision IX/20 annex 1) as well as guidance concerning the development of representative networks of marine protected areas (decision IX/20 annex 2). An inter-sessional CBD expert workshop reviewed the experience with the application of the CBD EBSA and the FAO's Vulnerable Marine Ecosystems (VME) criteria, concluding that the two sets of criteria were compatible. The inter-sessional workshop results fed into the 2010 COP decision X/29 that, *inter alia*, outlined regional processes to apply the criteria for the identification of EBSAs. These processes have already taken place in several regions of the world (NE Atlantic, SW Pacific, Tropical West Atlantic and Caribbean, South Indian Ocean, Eastern and Tropical Pacific, South East Atlantic and North Pacific), generating a list of 187 EBSAs (Dunn, *et al.* 2014).

Part of the Rio Convention's[7] CBD does encourage and invite a strong presence and participation of civil society in their proceedings. This is in contrast to the strict procedural rules that can govern the UN General Assembly proceedings, closing all discussions to all except state representatives as well as in state parties' meetings of UNCLOS. To correct this lack of participation of civil society in UNCLOS and based on a recommendation of the UN Commission on Sustainable Development, the UN General Assembly (UN A/RES/ 54/33, 1999) established the Informal Consultative Process on Oceans and the Law of the Sea (Simcock, 2010).

7 The Convention on International Trade in Endangered Species of Wild Fauna and Flora (CITES)

A multilateral treaty to protect endangered plants and animals, the convention was proposed in 1963 at a meeting of the International Union for Conservation of Nature (IUCN). Its text, adopted by 80 countries in Washington, was opened for signature in 1973, entering into force on 1 July 1975. CITES aims to ensure that international trade in specimens of wild animals and plants does not threaten the survival of the species in the wild. Roughly 5,000 species of animals and 29,000 species of plants are protected by CITES. Each protected species or population is included in one of three lists, called Appendices, that are afforded different levels or types of protection from over-exploitation. Appendix I lists species that are threatened with extinction and CITES prohibits international trade in specimens of these species, except when the purpose of the import is not commercial, for instance for scientific research. Appendix II lists species that are not necessarily now threatened with extinction but that may become so unless trade is closely controlled. Appendix III is a list of species included at the request of a Party that already regulates trade in the species and that needs the cooperation of other countries to prevent unsustainable or illegal exploitation. Parties may enter reservations with respect to any species listed in the Appendices in accordance with the provisions in the Convention.

The role of the United Nations system in the Law of the Sea

The United Nations is a complex system composed of a central system and a wider 'family' of UN specialized agencies and related organizations. *Sensu stricto*, the United Nations is the body constituted in 1945, around the UN Charter. This central system is built around the Governance provided by its General Assembly that meets annually in New York and incorporates programmes and other bodies created by the General Assembly.

Parallel to this central system there are 15 specialized agencies, including the World Bank and the International Monetary Fund, each of them constituted under their own independent international treaty or convention, that answer to the governance of their own supreme bodies. These two groups, given the common and shared principles and aspirations, cooperate and try to integrate and harmonize as much as practical their policies and actions. This combined group of institutions is what is generally called the 'UN family'.[8]

The closest to an executive board for this family is the UN System Chief Executives Board for Coordination (CEB), presided over by the UN Secretary General, that includes the CEOs of 30 entities: the United Nations central system; 15 specialized agencies established by inter-governmental agreement; the World Trade Organization and the International Atomic Energy Agency; and 12 funds and programmes.[9] What is important to retain is that the whole UN family is not a single organization, but many organizations.

Similarly to IMO for shipping (see above), for fisheries and aquaculture FAO is the inter-national competent authority setting the international technical standards. FAO advises and serves as an umbrella organization for the Regional Fisheries Management Organizations and Arrangements (RFMO/A). There are 20 RFMO/As, each of them constituted around an autonomous treaty or agreement whose parties or members are the fishing nations participating and benefiting from a given fishery or group of fisheries.

With 165 state members, the Intergovernmental Oceanographic Commission has responsibility for coordinating international research programmes (WCRP, GEOHAB) and the collection of oceanographic observations and data through its Global Ocean Observing System (GOOS). IOC has several regional sub-commissions: IOCARIBE in the Caribbean Region based in Cartagena de Indias and Kingston, WESTPAC based in Bangkok for the Western Pacific including Australia, and the IOC Sub-commission for Africa based in Nairobi serving both coasts of Africa. IOC also runs the Global Tsunami Warning System, with main operational centres in Hawaii, Alaska, San Juan, Perth, Tokyo, Jakarta and Hyderabad.

UNESCO having eliminated its Marine Sciences Division in 1990 in favour of concentrating ocean sciences under the IOC maintains other science programmes focusing on small island development states (SIDS), the secretariat for the World Heritage Convention with its 46 Marine World Heritage sites, the Underwater Cultural Heritage Convention and the Division on Education on Sustainable Development.

Other UN organizations are also involved with the ocean; the World Meteorological Organization (WMO) dealing with ocean–atmosphere interaction, marine meteorology and climate and its implications; the International Atomic Energy Agency (IAEA), monitoring marine pollution of radioactive substances; the United Nations Industrial Development Organization (UNIDO) with industrial marine technology, and active in the management of Large Marine Ecosystems; the International Labour Organization (ILO) in the protection of maritime workers in the shipping and fisheries industries; the World Health Organization (WHO) for ocean–related health problems and food safety; the United Nations Development program (UNDP) and the World Bank, financing the sustainable development of ocean and coasts.

Several Divisions of the central UN Secretariat also play a role: the Division of Economic and Social Affairs (UN-DESA) acted for 20 years as the secretariat for the Commission on Sustainable Development, coordinating programmes for coastal management, small island development states and oceans. In 2012, at the United Nations Conference on Sustainable Development (Rio+20), member states agreed to establish a High-level Political Forum on Sustainable Development,[10] to replace the Commission on Sustainable Development. The Division of Ocean Affairs and the Law of the Sea (UN-DOALOS), acts as the secretariat for UNCLOS, the Commission on the limits of the Continental Shelf and by default for any other meeting on oceans that is organized under the central UN system in New York, as is the case today for the Informal Consultative Process on Oceans and the Regular Process for the Assessment of

Table 3.1 Roles of agencies, programs and secretariats of the United Nations 'family' in relation to the ocean

	Direct roles: management & governance	Indirect roles	Standard setting roles
ILO		Capacity development on labour protection at sea	Shipping and fisheries labour
FAO	Fisheries and aquaculture management	Land nutrient inputs into the ocean	Fisheries and aquaculture. 'Codex alimentarium'
UNESCO			
UNESCO/IOC	World heritage – ocean science, services and observations	Education sustainable development	Open and free exchange of ocean data. Heritage preservation
ICAO		Overflight and innocent passage EEZ	
WHO			Food safety. Public Health 'Codex alimentarium'
WB	Coastal and ocean development		Financial management
IMF			Financial standards
UPU		Public awareness (Ocean Stamps editions)	
ITU		Frequency and bandwidth allocation for ocean communications and instruments	
WMO	Ocean observations for weather and climate forecasting	Operational warning of ocean extreme events to the commercial fleet. Meteorological and climate services	Meteorological data exchange
IMO	Shipping regulation. Marine pollution	Search and rescue operations at sea. Emergency communication system	Pollution of ocean or from ocean-going sources
WIPO		Property of data and databases	
UNIDO	Coastal industrial development	Management of LME projects	
UN-WTO			Coastal tourism
UN-DESA	Sustainable development of oceans, coasts and SIDS		
UN-DOALOS	Law of the sea. Extension of the continental shelf	Capacity development on the law of the sea	
UNDP	Coastal and ocean management and governance	Capacity development in ocean management and governance	
UNEP	Regional seas/coastal and ocean environment		Environmental protection standards
IAEA	Radioactive pollution monitoring and control		Radioactive pollution standards
UNISDR	Disaster prevention, disaster mitigation		Disaster preparedness
UNU	Coastal communities research/ awareness	Capacity development on ocean governance	
UNOPS		Management of LME ocean projects	
CTBTO		Data for tsunami warning	
UNCLOS	Framework convention for ocean. Limits of the continental shelf		
UNFCCC	Climate change adaptation – mitigation		
CBD	Biodiversity of oceans and coasts	Biodiversity-inclusive Environmental Impact Assessment (EIA). Strategic Environmental Assessments (SEA)	MPAs. Ecological and biologically significant areas (EBSAs). Access and benefit sharing

the Marine Environment. Table 3.1 contains a listing of the different UN competent organizations and their roles in relation to the Ocean.

Oceans, UNCLOS and civil society, a broken link

Ocean dwellers are a distinct, tiny minority of the human population.[11] There are very few human activities that are truly oceanic in nature: national navies and commercial shipping crews, mariners and long-distance fishermen are probably the human beings that spend most of their lives 'out at sea', roaming in the High Seas.[12] They constitute highly specialized, cohesive and isolated 'guilds' that follow old 'corporative' traditions. Sociologically and politically this fact has huge consequences both for our collective perception of the ocean and for the effectiveness of the institutions and jurisdictions created to provide governance and stewardship to the different ocean spaces.

The rights and responsibilities that modern states give to citizens close the loop of accountability for elected and designated officials, for the High Seas are certainly not embodied in these minority groups. There are no true citizens of the ocean empowered to exert that function. Faced with this reality, and lacking the political will to create a body empowered with the authority to exert at least some of the functions of modern states for the High Seas, the law of the sea entrusted these obligations collectively to nation states. This fundamental decision is in stark contrast to the treatment that UNCLOS gave to the bottom of the sea in Part XI under the principle of 'the common heritage of mankind'.

Although the exhortation contained in UNCLOS is for the collective, cooperative, concerted action of all nations, in practice this responsibility is delegated in different circumstances to coastal states, flag states and port states. This means that it is through individual national-state strategies, policies and actions in marine affairs that the system is supposed to operate. Few nations have evolved the institutions to deal with this challenge properly. Ocean Ministries or Departments with sufficient power to oversee marine affairs across the board do not exist and relatively competent substitutes exist only in a tiny minority of nations. This leaves the weakest link to establish the minimum standard. Nations are to provide the financial muscle and scientific know-how for the management of the marine environment as a whole and of the open oceans and deep seas in particular.

However, the entities that extract benefits from the High Sea, with the exception of defence activities, generally are not public entities but rather private individuals or private corporations. Depending on the effectiveness of national policies, laws, and institutions and of the associated capabilities, this arrangement allows for a wide range of behaviours, many of them at variance with international legal standards.

There is little doubt that nation states assume full responsibility and exert the monopoly in the use of force in terms of security, and that there is a one-to-one correspondence between these concerns and what is said and not said in UNCLOS about security. Negotiated during the Cold War, what is said about security in the text is as important as those security issues on which the Convention is 'silent'. There is no doubt that at least informal high-level consultations had to be conducted between the USA and the USSR, to define the envelope of the negotiations before engaging in the UNCLOS process. Even after both nations concurred to the consensus that adopted the text in December 1982 in Montego Bay, additional negotiations had to be conducted between the USA and the USSR to clarify the interpretation of some security-related language in the Convention, for example the USA–USSR Joint Statement with attached Uniform Interpretation of Rules of International Law Governing Innocent Passage protocol that was negotiated between 1986 and 1989 was adopted in Wyoming in September 1989, only six weeks before the fall of the Berlin wall. The new multi-polar world emerging in the twenty-first century will certainly witness new challenges to existing arrangements and interpretations. Innocent passage is a keystone of the security arrangements in UNCLOS. What is a military vessel supposed to do with all its non-destructive remote surveying equipment on board when undertaking innocent passage over an EEZ or territorial waters of another nation? Are military surveys subject to the previous consent regime applicable to Marine Scientific Research under Part XIII of UNCLOS, as some nations sustain they are?

Surveillance and enforcement

There is a legitimate argument around whether it is a lack of governance that generates sub-standard types of behaviours in the High Seas, or is it simply that we have the converse situation – there is plenty, or maybe even an excess of governance but we lack effective surveillance and compliance mechanisms.

A wide range of good and bad practices can be catalogued for each of the major conventions and agreements. For example, trans-shipment in the High Seas goes on regularly in distant-fleet fishing operations, a practice that makes enforcement of conservation measures extremely difficult. On the other hand, finding solutions is not easy, as the piracy crisis in the Horn of Africa and the North-Western Indian Ocean has shockingly showed the world. Expanding the role of the defence community for 'constabulary' and 'benign' roles on the coast always faces cultural and practical obstacles. Even in the case of prosecuting criminal acts, restrictions exist for most naval organizations undertaking 'police functions' in terms of law enforcement outside national territorial waters (Bernal, 2010). These can assume bizarre expressions: confronting a flagrant violation in the Indian Ocean, where intervention was clearly legitimated by UNCLOS, officers on board a naval vessel deployed to deter illegal operations in the High Seas found themselves inhibited from taking action because there was no equivalent rule under national law authorizing military personnel to take the type of action required by the circumstances. The bottom line is that naval personnel perform their duties under national jurisdiction, the jurisdiction of their flag. After signing and ratifying UNCLOS each country should conduct a process of harmonizing to the extent possible its precepts with national law to render it fully effective. Unfortunately, it seems that this has been done for all possible cases in very few countries.

When such action is possible, the tangle of legal arrangements necessary to render it effective soon becomes overwhelming. In the Caribbean Region, for example, with intensive illegal trafficking of drugs and people, the US Coast Guard works under more than 22 bilateral agreements, allowing for law enforcement within the territorial waters of other countries (Bernal, 2010).

But there are positive signs. The Malacca Strait is a critical and strategic waterway in the global trading system. It carries more than one fourth of the world's commerce and half the world's oil. In 2006, having rejected a previous offer of the USA to patrol the strait, Singapore, Malaysia and Indonesia signed the Straits of Malacca Patrol Joint Coordination Committee Terms of Reference and the Standard Operation Procedures to act jointly in order to tighten security in the Strait. In 2008 Thailand too became part of the joint committee for joint air and surface patrols. This combined effort had a decisive effect in limiting the piracy activity in the strait. In 2004 there were 38 cases of piracy but only two in 2008 (Bernal, 2010).

The challenge of integrated management

Since the early 1970s a more integrated approach to ocean management has been advocated, focusing first on the coastal zone. Integrated Coastal Zone Management (ICZM)[13] came first, probably because impacts on marine ecosystems from the development of the coastal zone are visible and easier to grasp. However, after the UNCED conference in 1992 in Rio de Janeiro, the concept of integrated management of ocean systems was extended beyond the coast to the formulation of ocean policy and governance in national and international jurisdictions. This trend was accompanied by other changes in emphasis through the emergence of the 'ecosystem approach to management' concept.

From a natural sciences point of view these changes recognize the biological, ecological and biogeochemical interconnectedness of natural ocean systems. From a sociopolitical point of view, this mutation recognizes that management is essentially the management of human behaviour associated with the extraction of human benefits and that in using the resources of the ocean there always will be conflicting interests that need to be resolved, hopefully through rational, consensual and peaceful means.

During the Second London Ocean Workshop, convened in 1998 by the Minister of Environment of Brazil and the Minister of Environment of the UK, Dr Meryl Williams, then Director of the International Centre for Living Aquatic Resources Management, stated the challenge in stark terms: 'Section A, Chapter 17 of Agenda 21 lays down a comprehensive prescription for integrated development of the ocean environment. However, six years after the Rio de Janeiro conference, most nations can demonstrate only limited progress towards filling this prescription'(Williams, 1998). In the same Workshop, Mr Atle Fretheim, then Minister of Environment for Norway, stated the issues around integration with clear precision:

> [T]here is a special need for better integrated international action to deal with 'offshore and deep sea environments', related in particular to action on 'marine activities' such as shipping, offshore oil and gas activities and fisheries. However, most marine ecosystems are open ecosystems with complex interactions. Consequently, impacts on the ecosystem in one part of the marine environment will influence on other parts. Clearly activities in the coastal zone may have considerable impact on the offshore environment. So will also land-based activities causing pollution of the marine environment, directly or indirectly. A truly integrated approach to international action to protect the offshore/deep sea environment should therefore look at all activities having negative effects on the marine environment, and not only 'maritime activities'.
>
> *(Fretheim, 1998)*

Coordination of policies across national ministries is not standard, nor a universal practice around the world. On the contrary, as Meryl Williams also pointed out in her paper there is a clear 'institutional inequality among sectors, especially between the sectors dealing with large economic activities such as ports and tourism and those dealing with natural resources and the environment' (Williams, 1998).

During the UN International Year of the Ocean in 1998, high-level national legislation with long-term management implications was initiated or promulgated, most notably in Australia, Canada and the USA (IOC, 2007). The majority of these texts introduce the concept of integrated management, defining standards that should guide policy development in a process leading to integration across sectors and jurisdictions. The experience shows that efforts to design horizontal integrated policies in ocean affairs had faced the challenges of their intrinsic complexity as well as the resistance from the strongly vertical structure of the political system of management currently in place. Nevertheless, and despite the difficulties, the nature of ocean process that calls for a horizontal treatment across sectors has made significant inroads in the institutional arrangements of nations.

A fascinating example comes from Australia's Ocean Policy, initiated in December 1995 and after broad national consultations, promulgated on 23 December 1998 during the International Year of the Ocean. This was an ambitious piece of legislation containing 390 initiatives that laid down a set of standards, created an independent agency, the National Oceans Office, a ministerial-level National Oceans Ministerial Board, and called for the development of integrated Regional Marine Plans (RMPs).

Implementation of the policy started with the development of the RMPs. Despite significant progress, this was a slow process and after five years a performance assessment review concluded that the initial implementation of the regional marine planning was 'very ambitious', adding that there 'was uncertainty about what will be delivered, how it will work and weather it will add value'. The review also criticized the policy for the lack of a legislative base and noted that among the major impediments was the fact that the Oceans Policy having been originated and promulgated at the federal level by the Commonwealth 'did not represent an agreed position with the States and Territories and it has not been subsequently endorsed by them' (TFG International, 2002). In 2005 Australian Oceans institutions were restructured. The National Oceans Office lost its executive agency status and was relocated inside the Marine Division of the Department of Environment and Heritage (Haward and Vince 2008, p. 114), and the Minister of Environment announced that RMPs would be established under section 176 of the Environment Protection and Biodiversity Conservation Act 1999, a pre-existing law providing a legislative basis for their implementation (Haward and Vince 2008, p. 115). These were hard lessons to be learned. Under this new institutional arrangement Australia has progressed significantly its Ocean Policy through the implementation of marine bioregional plans that also provide a platform for the National Representative System of Marine protected areas. Much of what has been used can be traced back to the initial work of the National Ocean Office and the first RMPs.

Canada, New Zealand, the USA and China have also laid down integrated ocean policies and can show interesting trajectories of their implementation, unfortunately beyond the scope of this review. The excellent book by Haward and Vince (2008) has a detailed account of the similarities and differences of the first three cases mentioned and an abundant literature in the social sciences and policy has documented the interesting shifts taking place in marine policy during the last 20 years.

The ecosystem approach to management

In terms of 'ecosystem approach to management' the way most fisheries are managed today offers a paradigmatic counterexample. Traditional management of fisheries deals with individual fish populations strictly in demographic terms, i.e. accounting for the input of individuals as population growth or immigration and the output in terms of natural and fishing mortality. In the ocean this is a highly complex task dealing with huge numbers and where measurements and the precision of population estimates are such that almost always they are close to the limits of empirical sufficiency. Nevertheless, the problem is that fish populations are also affected by changes in other external factors, such as predators and prey abundances and other changes in their bio-physical environments and at the same time what happens with their numbers will affect all the surrounding ecosystem of which fish are part. We can say that in traditional fish population dynamics, all this additional ecological complexity is subsumed in the error term of the demographic numerical estimates. The fact that a given population of fish could be maintained under these circumstances, despite a significant fraction being taken out of the system by fishing, doesn't mean that the situation created provides a stable ecological trajectory for the system as a whole, i.e. for all the series of accompanying fish in the community and other organisms participating in the same food chain.

A well-documented example is the anchovy (*Engraulis ringens*) and sardine (*Sardinops sagax*) fisheries off Central Peru and Northern Chile, where close to 30 per cent of the total throughput of solar energy in the pelagic ecosystem is taken out in the form of fish catch that is transformed into fishmeal and oil. This has been going on for at least the last 60 years. There is no question that all animal populations in the Humboldt ecosystem have already shifted their demographic equilibrium points from where they were before this fishery was developed in the 1950s. Decreases in the abundance of toothed whales, sea lions, birds, squid and other large predator populations are quite apparent and have been abundantly documented (Pauly and Tsukayama, 1987; Pauly *et al.* 1989). The fossilized excrement of the 'guanay' (*Phalacrocorax bougainvillii*) formed in the past the world-famous guano deposits in central Peru. A decrease of 30 per cent in its population made the 'guanay' a *near threatened* species and guano is no longer accumulating at the same rate. Abundance of other fish and crustacean components of the same food chain, although less visible, have also changed. This still seems to be an acceptable state of affairs in terms of human benefits and economic development in Chile and Peru, where revenues from the fish-meal industry are a significant fraction of their exports, but show clearly that the ecological changes in the system extend well beyond the demographics of the single species of anchovies or sardines. Wherever intense fishing operations have been sustained over stocks during decades, similar impacts must have happened, unfortunately less well documented than in Peru. Anchovies and sardines are closer to the bottom of the food chain than larger predators such as whales, sharks and tuna. The amount of biomass consumed by these large predators is huge and diminution in their numbers must have shifted dramatically the flux of organic matter in the ocean, favouring the demographic explosion or collapse of other organisms. In the real world there is no such thing as an 'ecological vacuum', as the BIOMASS project's robust negative result showed in the 1980s, assessing the expected huge size of 'Antarctic krill' populations due to the disappearance of its main predator, the 'blue whale' (Fraser *et al.* 1992; Kock and Shimadzu, 1994).

Une feuille de route

Ocean governance has been appropriately described as a 'two level game', insofar as the process of formulating its principles and building their institutions, the domestic political apparatus, accustomed to act without external limits, enters into contact with the international relations of nations. In other words, in a world that is increasingly more interdependent, ocean governance concerns do not emerge exclusively out of the domestic political and social dynamics of a nation; rather a nation maintains and projects its presence internationally, thereby emerging from its international engagements or at least being in contact with these international engagements. 'Two level game' is a valid description of these relationships creating a special domestic and international political scenario 'so long as . . . countries remain interdependent, yet sovereign' (Putnam, 1988, p. 434). Keeping in mind this two-level game dynamics, in what follows and as a form of conclusion, the different challenges for a nation to address their interests and responsibilities, challenges or opportunities vis-à-vis ocean governance are presented.

1 Full application of domestic norms and standards

Every nation has norms and standards, usually defined sector by sector (industrial, agricultural, environmental, health, etc.), whereby regulating activities on land has an impact on coastal and ocean waters and their living resources. Their full application is a prerequisite and their systematic harmonization with the International Treaties, Conventions and Agreements subscribed by each nation is a must. With regards to the ocean, the UNEP Global Programme of Action for the Protection of the Marine Environment from Land-based Activities (GPA) was adopted by the international community in Washington DC in 1995. The GPA is a tool that provides guidance and capacity development opportunities for coastal states to address this first-order challenge. It 'aims at preventing the degradation of the marine environment from land-based activities by facilitating the realization of the duty of States to preserve and protect the marine environment'. The GPA targets major threats to the health, productivity and bio-diversity of the marine and coastal environment resulting from human activities on land and proposes an integrated, multisectoral approach based on commitment to action at local, national, regional and global levels.

2   Coordination and harmonization of sectorial regulations and policies with international standards and obligations signed by the state

In fact the task of harmonizing national law and lower level regulations containing standards (e.g. levels of pollutants from industrial solid or liquid residues, treatment and disposal of radioactive medical material, levels of organic, inorganic pollutants and pesticide loads on rivers, sewage disposal, etc.) with international treaties is a major challenge that only a few nations have addressed in a comprehensive way. Much progress could be accomplished if such harmonization were to take place. Most nations have a designated authority charged with the administration of its maritime territory, usually in charge of maritime and related activities (shipping, ports, among others). Although all sovereign jurisdictions of the nation extend to its maritime territory, most functional regulations remain with the administration of origin and are not delegated or transferred to the designated maritime authority (e.g. health, environment, agriculture, fisheries, etc.). Although in principle the designated authority could address this harmonization, by its institutional identity, usually closely linked to the security or sovereign aspects of the maritime territory, it is frequently ill-equipped to the task and conflicts between authorities with different but overlapping functional mandates do emerge. With increasing pressures for the use of maritime space these inter-sectorial conflicts tend to become more acute and paralyse action.

Furthermore, it is frequent to find the situation where there is no predefined authority charged with monitoring new duties emanating from international treaties endorsed and ratified by the state and of leading the process of harmonization with national laws and regulations. Furthermore, regulations designed to be applied in the ocean should take into account the dynamic and fluid nature of the marine environment, an aspect absent from all land-based legislation.

3   Coordination across sectors

The new trends calling for integrated ocean policies on the coast and in the open ocean implicitly call for the definition of policies across productive sectors. This is a huge political challenge since it affects long-established practices. It is an additional challenge in strongly federal systems of government where many functions and jurisdictions cascade down to different levels of the administration: state or provincial, regional, county or departmental and cities. As reviewed above, for the case of Australia, the establishment of a high-level Ocean Policy act or instrument, appears to be a plausible strategy to introduce incremental but clearly directional change.

The establishment of National Institutions with the mandate of coordinating across all productive sectors benefiting from ocean resources or oceanic ecosystem services is another. For example in a spatial sequence:

- Coastal Commissions with the mandate of verifying the integrated character of policies impacting the coastal domain;
- National Administration for the EEZ, as in China; can improve the performance of the state in terms of their ability to provide stewardship to EEZs, beyond shipping and fishing, improving the ability of coastal states to benefit from their EEZ. This is becoming more important in cases where mining (coastal or deep-sea), oil and gas exploitation, extensive aquaculture and the establishment of large wind parks for harvesting energy exist. An administration or authority is needed for the EEZ: the successful application of marine spatial planning techniques, requires a minimum institutional framework to support it;
- a dedicated Agency answering to the highest levels of the Executive Branch with the mandate of verifying the integrated character of policies emanating from ministries and departments as an alternative. The National Accounting Offices reports assessing the progress on the integrated ocean policies in Canada and Australia are sobering examples of the challenges and limits of reports of this nature if other political issues are not well aligned with the mission of this agency.

4 Beyond the EEZ

In terms of compliance with international norms what is lacking is an effective mechanism to enforce national and international norms and standards over all individuals and entities operating under the jurisdiction of the 'flag state'. This is perhaps the weakest link of the current system. A nation could empower an administration to monitor the activity of everything that takes place under its 'flag-state' jurisdiction. However, it would be naive to expect flag states to exercise proper control unless some minimum requirements were laid down on an international instrument. Voluntary mechanisms at the hand of corporate interest might not be sufficient (fishing nations and fishing operators within RFMOs, or shipping nations and shipping operators through voluntary quality assurance protocols), and eventually some international agreement would be necessary to improve performance in the High Seas through the 'flag-state' regime.

5 National coordination for the formulation of national positions in international fora

In the words of Edgar Gold (1999), referring to the shipping industry: 'As international organizations had no enforcement power (which had traditionally been left to flag states), acceptance or adherence to international codes and conventions did not entail that the accepting State was willing or able to enforce such codes.' This is a strong description of the current state of affairs and not only in shipping.

'The international community' is a difficult concept. On one side the international community is the best instrument for nations to act in a coordinated and harmonious way under at least the spirit of the UN Charter. However, nations do not speak with one voice within the international community. They speak with the voice of their national interest in the more limited context of the forum in which they are participating. For example: The Ministry of Foreign Affairs with close advice from defence and intelligence represents nations in the Security Council of the UN, the Fishing and Agriculture Minister in the FAO, the Transport

Ministry in the IMO, the Education, Science and Culture minister in UNESCO, the Minister of Environment in UNEP and the Minister of Finance in the IMF, the WB and the GEF.

Small nations see in the International Community an opportunity to defend their interests collectively, since they can hardly confront bilaterally, through one-on-one negotiations, all issues of their interest. The UN family uses a sort of regional caucuses to organize its debates, and within those caucuses, usually in front of a smaller and more empathic audience, small nations can more easily find the bases for concerted action. Nevertheless, in smaller groups leadership can play a key decisive role. Some nations have developed sophisticated coordination mechanisms, usually entrusted to their Foreign Affairs ministries, to build a presence and leadership image when operating in the International Community. Surprisingly, these are not the most powerful nations in the world, but mid-size emerging powers.

So the least that a nation must do in order to engage in the Ocean Governance processes at the global level is to have an effective national coordination across sectors and stakeholders in order to represent appropriately their interests in international negotiations. Some of these mechanisms are permanent task groups with different stakeholders from the public and private sector. For example, for climate change, for international fisheries, or for Antarctic affairs, or crafting national positions for regional and sub-regional bodies. Others might respond to special high-level UN Conferences, such as the one for the Conservation and Sustainable Use of Marine Biodiversity in the High Seas. This mechanism requires an honest broker, usually the Minister of Foreign Affairs, and a clear political leadership from the top. Given the increasing importance of ocean affairs, this is and will continue to be the minimum requirement of the future.

Notes

1 The 'life support system' is the network of complex natural dynamic processes that maintain the conditions that make life possible on the planet. Key properties of the system are the heat capacity of the ocean, the oxygen production, the carbon capturing by water and the carbon sequestration by sediments, and the nutrient cycling and conversion of organic matter to inorganic nutrients on a global scale. The point usually missed is that these properties are not fixed and stable forever; they can and do change by being part of a complex set of dynamic equilibrium. The fact that every second breath of oxygen we take comes from the oxygen produced in the ocean by phytoplankton and that the accumulation of CO_2 and other gases in the lower atmosphere, changing the 'permeability' of the upper atmosphere to electromagnetic radiation and causing global warming, are just two examples of those interlinked dynamic processes.

2 During the Cold War the freedom to move naval assets across the ocean became a cornerstone of the strategic equilibrium between the USA and the USSR. Both nations developed a nuclear retaliatory capability and deployed it on board their submarine fleets. Because the ocean is essentially opaque to electromagnetic radiation, satellites can only see a few millimetres below the surface; the submarines carrying nuclear weapons were able to hide and avoid detection. A massive nuclear attack with intercontinental ballistic missiles could annihilate the response capabilities of the adversary, but would leave intact their submarine retaliatory power.

3 For example, poly-metallic nodules claims by Anaconda Co. under the State of Arizona (USA) law. Anaconda, at the time one of the largest producers of copper in the world, was probably using a strategy of co-opting the access of a new source of concentrated copper laced together with other minerals such as nickel and cobalt. Since the entry into force of UNCLOS, the status of these claims has never been tested in a court of law, a situation that is complicated by the fact that the USA, although signing UNCLOS and benefiting from all its zonal jurisdictions, has not ratified UNCLOS, precisely due to the strong opposition to part XI in some segments of American society. The chemical composition of nodules of economic interest varies but on average has the following constituents: Manganese 29 per cent; Iron 6 per cent; Silicon 5 per cent; Aluminum 3 per cent; Nickel 1.4 per cent; Copper 1.3 per cent; Cobalt 0.25 per cent.

4 The League of Nations Codification Conference held in The Hague from 13 March to 12 April 1930, addressed the issue of 'the extent of territorial waters', failing to reach agreement.

5   In this paper the rationale for the emergence of the exclusive economic zone concept was clearly
     stated from the point of view of developing nations: the 'present regime of the high seas benefits
     only the developed countries . . .'. The developed countries, because of their advanced
     technologies, were able to engage in distant-water fishing activities wherever and whenever they
     chose to do so. At the same time, developing countries were often incapable of exploiting the
     resources in waters closely adjacent to their own coasts much less in waters great distances away.
     (Cited from Nandan, 1987, p. 9)

6 Strictly speaking the first 'implementing agreement' of UNCLOS should be considered to be the one
   negotiated for Part XI, which significantly modified the original text of its 'mining code', opening
   the road for the signing of the Convention by the USA.
7 Around UNCED 1992 in Rio de Janeiro three environmental agreements were negotiated: The United
   Nations Framework Convention of Climate Change, the Montreal Protocol on Cloro-fluorocarbon
   Emissions and the Convention on Biodiversity.
8 Articles 57 and 63 of the UN Charter refer directly to and define the status of 'specialized agencies'
   and many aspects of the operation of the whole system are common to all. The employees of the
   central system and the wider family have a single system of remuneration and retirement, are protected
   while in mission by a single worldwide security system and utilize a single daily allowance scale, adjusted
   regularly to place and time.
9 There are 19 other UN entities and bodies that are not members of CEB, including all the UN regional
   economic and social commissions.
10 The High-level Political Forum on Sustainable Development provides political leadership and guidance
   in implementing sustainable development commitments and addresses new and emerging sustainable
   development challenges. It meets every four years at the level of Heads of State and Government
   under the auspices of the General Assembly and every year under the auspices of the UN Economic
   and Social Council.
11 Coastal dwellers and in particular many islanders could enlarge this number. These populations have
   an enhanced appreciation of ocean processes and usually know how to extract benefits from the
   sustainable use of ocean resources.
12 Maybe we could add today the crews operating day and night on the most distant offshore oil and
   gas platforms.
13 Perhaps one of the earliest initiatives was the California Coastal Initiative established by voter initiative
   in 1972 (Proposition 20) creating the California Coastal Commission that later was made permanent
   by the Legislature through adoption of the California Coastal Act of 1976.

References

Balton, D. A. and H. R. Koehler (2006) 'Reviewing the United Nations Fish Stocks Treaty'. *Sustainable
   Development Law and Policy*, 75: 5–9.
Benham, A. (2003) 'Ending Flag State Control?'. In Kirchner, A. (ed.), *International Marine Environmental
   Law*. The Hague: Kluwer Law International, pp. 123–135.
Bernal, P. A. (2010) 'For the Ocean'. In G. Holland and D. Pugh (eds) *Troubled Waters: Ocean science and
   governance*. Cambridge: Cambridge University Press, pp. 13–27.
Caron, D. D. (1995) 'The International Whaling Commission and the North Atlantic Marine Mammal
   Commission: the institutional risks of coercion in consensual structures'. *The American Journal of
   International Law*, 89: 154–174.
Dunn, D. C., J. Ardron, N. Bax, P. Bernal, J. Cleary, I. Cresswell, B. Donnelly, P. Dunstan, K. Gjerde,
   D. Johnson, K. Kaschner, B. Lascelles, J. Rice, H. von Nordheim, L. Wood and P. N. Halpin (2014)
   'The Convention on Biological Diversity's ecologically or biologically significant areas: origins,
   development, and current status'. *Marine Policy*, 49: 137–145.
Fraser, W. R., W. Z.Trivelpiece, D. G. Ainley and S. G. Trivelpiece (1992) 'Increases in Antarctic penguin
   populations: reduced competition with whales or a loss of sea ice due to environmental warming?'
   Polar Biology, 11: 525–531.
Fretheim, A. (1998) 'How can we best develop an integrated approach to international action to deal with
   pressure points arising from maritime activities, including shipping, oil and gas activities and fisheries?'.
   Paper 3, Panel 2: Focusing International Action on Pressure Points in the Wider Seas. The second
   London Ocean Workshop, 10–12 December 1998, Eaton House, London.

Gold, E. (1999) 'Learning from disaster: lessons in regulatory enforcement in the maritime sector'. *Review of European Community and International Environmental Law*, 8(1): 16–20.

Haward, M. and J. Vince (2008) *Ocean Governance in the Twenty-first Century: Managing the blue planet*. Cheltenham, Northampton, MA: Edward Elgar.

IOC (2007) 'National Ocean Policy. The basic texts from: Australia, Brazil, Canada, China, Colombia, Japan, Norway, Portugal, Russian Federation, United States of America'. Intergovernmental Oceanographic Commission of UNESCO; Paris, IOC Technical Series 75: 277 pp.

Kock, K.-H. and Y. Shimadzu (1994) 'Trophic relationships and trends in population size and reproductive parameters in Antarctic high-level predators'. In El-Sayed, S. Z.(ed.) *Southern Ocean Ecology: The BIOMASS perspective*. Cambridge: Cambridge University Press, pp. 287–312.

Koh, T. B. (1983) 'A constitution for the Oceans', Remarks by Tommy T. B. Koh of Singapore, President of the Third UN Conference on the Law of the Sea, in *The Law of the Sea*, official text of the United Convention on the Law of the Sea with Annexes and Index. New York, United Nations.

Mitchell, R. B. (2002–2014) The IEA Database Project, Oregon State University. http://iea.uoregon.edu/page.php?query=list_subject.php, accessed 29 December 2014.

Nandan, S. N. (1987) 'The Exclusive Economic Zone: a historical perspective'. In *FAO (1987) Essays in memory of Jean Carroz: The Law and the Sea*. Rome, FID, Book, pp. 1–23.

Oxman, B. H. (2002) 'The tools for change: the amendment procedure'. In *Commemoration of the 20th Anniversary of the Opening for Signature of the 1982, UN Convention on the Law of the Sea*. 57 UNGA, 9 December 2002.

Pauly, D. and I. Tsukayama (1987) 'The Peruvian Anchoveta and its Upwelling Ecosystem: Three decades of change'. *ICLARM Studies and Reviews*, 15: 351. Instituto del Mar del Perú, IMARPE, Callao, Perú; Deutsche Gesellschaft für Technische Zusammenarbeit (GTZ), GmbH, Eschborn, FRG; International Center for Living Aquatic Resources Management (ICLARM), Manila, Philippines.

Pauly, D., P. Muck, J. Mendo and I. Tsukayama (1989) 'The Peruvian upwelling ecosystem: dynamics and interactions'. *ICLARM Conference Proceedings*, 18: 438. Instituto del Mar del Perú, IMARPE, Callao Perú; Deutsche Gesellschaft für Technische Zusammenarbeit (GTZ), GmbH, Eschborn, FRG; International Center for Living Aquatic Resources Management (ICLARM), Manila, Philippines.

Putnam, D. (1988) 'Diplomacy and domestics politics: the logic of two-level games'. *International Organization*, 43(3): 427–460.

Secretariat CBD (2012) *Marine Biodiversity: One ocean, many worlds of life*. Montreal: Secretariat of the Convention on Biological Diversity.

Simcock, A. (2010) 'The United Nations, oceans governance and science'. In G. Holland and D. Pugh (eds), *Troubled Waters: Ocean science and governance*. Cambridge: Cambridge University Press, pp. 28–40.

TFG International (2002) 'Review of the implementation of Oceans Policy'. Hobart: TFG International. (cited by Haward and Vince, 2008, p. 113)

UN A/RES/S-6/3201 (1974). United Nations General Assembly Resolution 3201 (S-VI). Declaration on the Establishment of a New International Economic Order.

UN A/RES/54/33 (1999) United Nations General Assembly Resolution 54/33. Results of the review by the Commission on Sustainable Development of the sectoral theme of 'Oceans and seas': international coordination and cooperation.

UNCTAD (2011) 'Review of Maritime Transport 2011'. United Nations Conference on Trade and Development. United Nations, Geneva, 213 pp.

Williams, M. (1998) *Policy Integration in Ocean Management: Are we advancing?* Keynote paper. The second London Ocean Workshop, 10–12 December 1998, Eaton House, London.

4

INTERNATIONAL MARINE GOVERNANCE AND PROTECTION OF BIODIVERSITY

Jeff A. Ardron and Robin Warner

Introduction

Oceans governance is a relatively recent phenomenon, with the protection of marine biodiversity generally emerging later in the twentieth century, after the agreements that focused on oceans resources and management were established. The first such global institutions were developed post Second World War to regulate sectoral issues associated with increased international shipping and expanded global fisheries. Key institutions established in this period were the International Maritime Organization (IMO) for shipping and the Food and Agricultural Organization (FAO) for fisheries, both UN specialized agencies and treaty based bodies (FAO, 1945; IMO, 1948). These institutions have played pivotal roles in regulating their respective sectors, adopting a mix of hard and soft law instruments to achieve their objectives. Later, after these sectoral bodies were established, came the negotiations that led to the establishment of the 1982 United Nations Convention on the Law of the Sea (UNCLOS).

Since 1959, the IMO has been the primary regulatory body for international shipping, providing a proactive forum for member States and other stakeholders in the shipping industry to cooperate on a wide range of technical matters relating to maritime safety, navigation and vessel source marine pollution. As well as regulating the myriad routine issues associated with international shipping, the IMO has frequently taken the initiative in regulating emerging issues such as the protection of particularly sensitive sea areas from adverse shipping impacts, the elimination of toxic anti-fouling paints from ships and combatting the introduction of alien species from ships ballast water (IMO, 2001, 2004, 2005).

In contrast to the IMO, which has initiated multiple binding treaties, the FAO has exercised its regulatory influence primarily through developing soft law instruments such as Codes of Conduct, International Plans for Action and Guidelines setting best practice standards for responsible fishing and providing technical guidance for regional fisheries management organizations (RFMOs) and member States. The FAO Code of Conduct for Responsible Fisheries is an extensive primer on best practice in fisheries management including concepts and measures designed to lead to responsible and sustainable fisheries (Edeson, 2003). The FAO International Plans for Action have provided global level statements of aspiration and guidance on prominent problems in global fisheries.[1] In the fisheries sector, however, regional and national level institutions have been the focal point for formulating and implementing fisheries conservation and management measures with quite disparate levels of achievement in terms of responsible and sustainable fishing (Cullis-Suzuki and Pauly, 2010).

The United Nations began to examine ocean governance issues from a more holistic perspective in the 1950s through the work of the International Law Commission (ILC). The ILC's 1956 draft articles on the Law of the Sea provided the basis for negotiations at the First United Nations Conference on the Law of the Sea (UNCLOS I) in 1958. However, it was not until the Third United Nations Conference on the Law of the Sea (UNCLOS III) from 1973 until 1982 that a wider measure of consensus among States was achieved on a more comprehensive code of ocean governance. The 1982 United Nations Convention on the Law of the Sea was the first real attempt at holistic ocean governance and has been widely acclaimed as laying the foundation for oceans governance (UNCLOS, 1982).

Overview of the international law framework and institutions

Table 4.1 lists the key international treaties and their implementing institutions (henceforth, simply 'agreements') relevant to the protection of marine biodiversity, and Figure 4.1 illustrates these schematically. These various marine- and maritime-related governance bodies evolved largely independently of one another, resulting in what has been criticized as a fragmented regulatory framework (Tladi, 2011), which does not readily allow for the implementation of integrated conservation measures, such as marine protected areas (MPAs) (Gjerde and Rulska-Domino, 2012). At first appearance an 'alphabet soup', there are some patterns that emerge, most notably between the sectoral and conservation agreements. The agreements on the right of Figure 4.1 are those that regulate sectoral activities – fishing, mining and shipping – that are explicitly covered in UNCLOS and its two implementing agreements. However, the conservation, scientific and cultural agreements, on the left of Figure 4.1, that are not linked to specific sectoral activities are also not explicitly linked to UNCLOS or its implementing agreements, though they still fall under its general legal umbrella (as well as customary international law). Instead, many of them are administered under the UN. The governance bodies in the middle of Figure 4.1 share attributes of both sectoral and conservation agreements.

In general, the sectoral agreements focus on usage and exploitation, and rely heavily upon binding measures, such as fisheries closures and shipping discharge restrictions, though may also utilize voluntary measures such as recommended ship routeing and reporting requirements. The conservation agreements, on the other hand, rely mostly upon voluntary measures, and often lack the mandate to manage those activities that adversely affect the species or habitats that they seek to protect. The Convention on Migratory Species (CMS, below), for example, is a framework agreement under which both binding agreements and voluntary memoranda of understanding (MoUs) have been adopted relating to the protection of migratory species and their habitats in the marine environment. However, even in the case of the binding agreements, the CMS is limited in its regulatory options and must rely on its Parties to unilaterally adopt measures (which many are reluctant to do, since it could put their nationals at a competitive disadvantage), or to submit measures to the relevant sectoral agreements, which may or may not agree to adopt them.

The Convention on International Trade in Endangered Species of Wild Fauna and Flora (CITES, below), a conservation agreement, is an exception to the above generalization in that

Table 4.1 Summary of key international agreements relevant to marine resources management and the conservation of biodiversity

Short name	Full name	Year/in force	Parties	Governance/admin. bodies
Global framework agreement[a]				
UNCLOS (LOSC)	United Nations Convention on the Law of the Sea (also known as: Law of the Sea Convention)	1982/94	166 (including the European Union (EU))	UN Division for Ocean Affairs and Law of the Sea
Global sectoral agreements (to manage marine natural resource exploitation and maritime activities)				
Part XI Agreement	Agreement relating to the implementation of Part XI of the United Nations Convention on the Law of the Sea of 10 December 1982	1994/96	145	International Seabed Authority (ISA)
Fish Stocks Agreement	The United Nations Agreement for the Implementation of the Provisions of the United Nations Convention on the Law of the Sea of 10 December 1982 relating to the Conservation and Management of Straddling Fish Stocks and Highly Migratory Fish Stocks	1995/01	81	Regional fisheries management organizations/agreements and Flag States are expected to implement the agreement. No secretariat per se[b]
MARPOL and other agreements[c]	International Convention for the Prevention of Pollution from Ships, and other shipping agreements	1972 and 78/83 (Annex VI protocol 1997/05)	74[d]	International Maritime Organization (IMO)
LC/LP	Convention on the Prevention of Marine Pollution by Dumping of Wastes and Other Matter 1972 and 1996 Protocol Thereto	1972/75 (Protocol 1996/06)	87/42	Secretariat of the LC/LP is hosted by the IMO (see above)
IWC	International Convention for the Regulation of Whaling	1946/48	88	International Whaling Commission

continued

Table 4.1 continued

Short name	Full name	Year/in force	Parties	Governance/admin. bodies
Global conservation agreements (primarily to protect species, habitats, and/or biodiversity)				
CITES	Convention on International Trade in Endangered Species of Wild Fauna and Flora	1973/75	178	Autonomous secretariat
CMS	The Convention on the Conservation of Migratory Species of Wild Animals[e]	1979/83	119	Secretariat under the UN Environment Programme (UNEP)
CBD	Convention on Biological Diversity	1992/93	193 (including the EU)	Secretariat under UNEP
WHC[f]	Convention for the Protection of the World Cultural and Natural Heritage	1972/75	190	Secretariat under the UN Educational, Scientific and Cultural Organization (UNESCO)
Regional agreement bodies (summarized)				
RFMO/As	Regional fisheries management organizations/agreements[g]	various	various	In ABNJ: 5 tuna RFMOs (+1 dolphin agreement). 9 geographic RFMOs + 2 advisory. Some have FAO oversight
RSP/As	Regional Seas Agreements (including the UNEP Regional Seas Programme[h])	various	various	Four currently extend into ABNJ: Mediterranean (Barcelona Convention); Northeast Atlantic (OSPAR),South Pacific (Noumea Convention for high seas pockets), and Antarctic (CCAMLR – see below).
CCAMLR/ATS	Convention for the Conservation of Antarctic Marine Living Resources/Antarctic Treaty System	1982/82; 1959/61	36	The Convention, administered by a Commission of the same name, is part of the Antarctic Treaty System.

Source: Adapted from Ardron et al., 2014c.

a For resources management, UNCLOS is supplemented by two implementing agreements, below. b There have been two UN review conferences (2006 and 2010). Parties need not have ratified UNCLOS (e.g. the USA). c There are over 50 shipping-related IMO conventions/agreements. d The combined merchant fleets of these parties constitute approximately 94.73 per cent of the gross tonnage of the world's merchant fleet. www.imo.org/About/Conventions/StatusOfConventions/Documents/Status%20-%202013.pdf (accessed November 2013). e CMS is an umbrella agreement under which reside seven binding agreements, five of which are marine-related; and 19 voluntary memorandums of understanding (MoUs), of which six are marine. The binding agreements include: Cetaceans of the Mediterranean Sea, Black Sea and Contiguous Atlantic Area; Small Cetaceans of the Baltic, North-East Atlantic, Irish and North Seas; Seals in the Wadden Sea; African–Eurasian Migratory Waterbirds; Albatrosses; and Petrels. The voluntary MoUs include: Marine Turtles of the Atlantic Coast of Africa; Marine Turtles and their Habitats of the Indian Ocean and South-East Asia; Cetaceans and their Habitats of the Pacific Island Region; Dugongs and their Habitats; Eastern Atlantic Populations of the Mediterranean Monk Seal; Small Cetaceans and Manatees of West Africa; Sharks. f The World Heritage Convention is not currently applied to areas beyond national jurisdiction (ABNJ, 'high seas'). g In ABNJ there are five tuna RFMOs (+1 dolphin agreement) and nine geographic RFMOs (+ 2 advisory bodies). h www.unep.org/regionalseas/ (accessed September 2015).

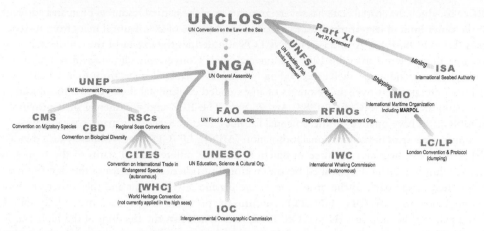

Figure 4.1 Constellation of marine agreements in ABNJ. On the left are those primarily concerned with conservation, and on the right are those primarily concerned with human extraction of marine species and resources, as well as the use of the marine environment
Source: Author.

it has the authority to adopt binding regulations and compliance mechanisms. However, these are trade measures only, focused on individual species, and CITES does not consider holistic conservation measures such as protected areas. Furthermore, sectoral agreements can, and often do, pass conservation measures, thus blurring the distinction made above. The International Whaling Commission (IWC), for example, was established after the Second World War originally as a sectoral body to manage whaling, much as fisheries are managed today. However, over time its character has changed until it more resembles a conservation agreement, reflecting the values of its membership that includes mostly non-whaling States. In 1986 it passed measures that amount to a moratorium on commercial whaling.[2]

Unlike the IWC, the membership of regional fisheries management organizations/agreements (RFMO/As) is comprised only of States engaged in fishing in the region. RFMO/As have often been criticized for not taking conservation under sufficient consideration (Cullis-Suzuki and Pauly, 2010). The one exception is the Commission for the Conservation of Antarctic Marine Living Resources (CCAMLR), which does manage fisheries, but under a strong conservation mandate, not unlike a regional seas organization/agreement. CCAMLR's membership reflects these two themes, comprised of nations engaged in fisheries as well as those engaged solely in scientific research, and disagreements within CCAMLR can often be traced to these differing national priorities (Brooks, 2013). Were other RFMO/As to allow non-fishing States to join them, it follows that their decisions would likely reflect interests of non-fishing States as well, which could be more conservation-oriented.

United Nations Convention on the Law of the Sea

UNCLOS outlines a vision for an integrated approach to ocean governance in its preamble which acknowledges that 'the problems of ocean space are closely interrelated and need to be considered as a whole' (UNCLOS, 1982, Preamble). The 'equitable and efficient utilization' and the 'conservation' of ocean resources are prominent elements in that vision and these objectives underpin many UNCLOS provisions (UNCLOS, 1982, Preamble). UNCLOS further elaborated the system of offshore maritime zones to include an exclusive economic zone

(EEZ) in which the coastal State has sovereign rights over the natural resources of an area adjacent to the outer limit of its territorial sea to a maximum distance of 200 nautical miles from its baselines (UNCLOS, 1982, Arts. 55–57). UNCLOS also defines the extent of the continental shelf in more precise terms than the 1958 Continental Shelf Convention, allowing for coastal States with physical continental shelves extending beyond 200 nautical miles from their baselines to claim sovereign rights over the resources of an extended continental shelf out to a maximum of 350 nautical miles (UNCLOS, 1982, Art. 76(8)).[3] The EEZ and continental shelf provisions of UNCLOS have opened up the potential for huge resource gains for many coastal States.

For marine areas beyond national jurisdiction (ABNJ), UNCLOS provisions create a divided rather than an integrated resource jurisdiction, and the different legal status of the high seas water column and the deep seabed beyond national jurisdiction complicates the development of a coherent approach to the protection of the marine environment and the conservation of marine biodiversity in ABNJ. UNCLOS confirms the principle that no State may validly appropriate parts of the high seas (UNCLOS, 1982, Art. 89). It lists the freedom of the high seas in Article 87(1) including the freedom of fishing, which is not an unfettered freedom, but rather is subject to provisions on the conservation and management of the living resources of the high seas in section 2 of Part VII.[4] These provisions require States to take both unilateral and cooperative measures to conserve the living resources of the high seas with a view to maintaining their populations above levels at which their reproduction may be seriously threatened (UNCLOS, 1982, Arts. 117–119). Conservation and management measures must be based on the best scientific evidence available that considers factors such as the interdependence of fish stocks and the impacts of fishing on associated and dependent species (UNCLOS, 1982, Art. 119(1)). These obligations were further elaborated in the UN Fish Stocks Agreement (1995) negotiated to address the over exploitation of highly migratory and straddling fish stocks transiting the high seas and zones within national jurisdiction and associated tensions arising between coastal and distant water fishing States. It provided the first comprehensive template for sustainable fisheries management in ABNJ and model provisions for cooperation between coastal States and flag States with high seas fishing fleets. In addition to codifying relevant international environmental law principles such as the precautionary and ecosystem based approaches for fisheries, it provided practical guidance for RFMOs on establishing cooperative compliance and enforcement measures on the high seas rather than relying solely on the individual efforts of flag States of fishing vessels to enforce compliance with conservation and management measures (Tahindro, 1997). Since the UN Fish Stocks Agreement was adopted in 1995, a complex pattern of high seas fishing regulation has emerged as existing RFMOs continue to adapt their agreements, and institutions to incorporate the Agreement's provisions and new RFMOs are established. Recent reviews of RFMO practice at the global level reveal several factors that have limited their effectiveness in implementing fisheries conservation and management measures in an ecologically sustainable manner (High Seas Task Force, 2006; Lodge *et al.*, 2007).

Juxtaposed with the high seas regime for marine living resources, is Part XI of UNCLOS which declares that the non-living resources of the deep seabed beyond national jurisdiction, known as the *Area*, are the common heritage of mankind and places them under the administration of a global institution, the International Seabed Authority (ISA) (UNCLOS, 1982, Arts. 136, 153(1)). Part XI of UNCLOS establishes an elaborate system to regulate exploration for and exploitation of deep seabed minerals and the distribution of derived profits among States Parties to UNCLOS on the basis of equity and need.

Extensive informal consultations sponsored by the UN Secretary General in the early 1990s resulted in the Part XI Implementation Agreement (1994). This agreement is an integral part of UNCLOS with its provisions prevailing over Part XI in the event of inconsistency for those

States that are parties to both instruments (ibid., Art. 2(1)). While reaffirming the common heritage of mankind principle, the Part XI Implementation Agreement removed many of the contentious provisions of Part XI, reduced some of the potential costs involved in deep seabed mining for industrialized States Parties and reflected a more market-oriented approach to the development of the deep seabed mining industry (Nelson, 1995; Oxman, 1999).

In its first two decades of operation, the ISA has presided over only exploration activities but commercial interest in deep seabed mining is gathering pace with a greater number of applications for exploration licences in recent years (Nelson, 1995; Oxman, 1999). It has developed detailed rules, regulations and procedures for exploration activities that include requirements for exploration contractors to conduct prior environmental impact assessment, baseline studies and ongoing monitoring of the impacts of their activities on the marine environment of the Area (ISA, 2010, 2012, 2013a, 2013b).

An integral component of the UNCLOS 'legal order for the seas' is the compulsory dispute settlement system created under Part XV of UNCLOS. The International Tribunal of the Law of the Sea (ITLOS) is a permanent institution established by Annex VI of UNCLOS to resolve disputes related to the law of the sea (UNCLOS, 1982, Annex VI, Art.1). ITLOS has a bench of 21 judges who have 'recognized competence in the field of the law of the sea' (UNCLOS, 1982, Art. 2(1)). Although the ITLOS case load has not been extensive it has developed a body of jurisprudence and principles particularly around prompt release and bonds in fisheries cases and delivered an important Advisory Opinion clarifying the environmental protection responsibilities of States sponsoring deep sea mining exploration contractors in the Area (ITLOS, 2011).

As well as the institutions created by UNCLOS itself, other multilateral institutions relevant to ocean governance are recognized in UNCLOS through the use of the term 'competent international organization'. In most contexts this has been interpreted as meaning the International Maritime Organization (IMO) but there are some articles where the term could be interpreted as referring to other organizations such as regional seas organizations/agreements (RSO/As) or regional fisheries management organizations/agreements (RFMO/As).[5]

Key marine conservation agreements

A considerable body of hard and soft law instruments have developed that complement and extend the framework for protection and preservation of the marine environment in Part XII of UNCLOS. Developments in international environmental law and policy over the past 40 years, since the 1972 Stockholm Conference on the Human Environment, have promoted a more integrated approach to the protection of the marine environment that has aligned environmental protection objectives with social and economic goals and focused on marine ecosystems rather than concentrating principally on a single sector.

Convention on Biological Diversity (CBD)

Although negotiated in a separate process, the objectives of the CBD are closely linked to the principles set out in the 1992 Rio Declaration (discussed below) and the action programme contained in Agenda 21 for integrated and ecosystem-based management of the environment including its marine components (Grubb *et al.*, 1993, pp. 75–76). The CBD was negotiated to assist States in arresting the alarming rate of extinction of species and the destruction of their habitats (Grubb *et al.*, 1993, p. 75; Joyner, 1995, p. 644). In the context of the marine environment, the concept of biodiversity was allied to the notion of large marine ecosystems forming an interconnecting web of marine living resources and their habitats (Joyner, 1995, p. 637). This multidimensional approach transformed law and policy for the protection of the marine environment which had previously focused on pollution control and the protection of single species (Joyner, 1995, p. 637). The conservation of marine biodiversity entails protection of a range of components of biodiversity including species, habitats, ecosystems and genetic material (Joyner, 1995, p. 646). The three broad objectives of the CBD, set out in Article 1, are the conservation of biodiversity, the sustainable use of its components and the fair and equitable sharing of the benefits arising out of the utilization of genetic resources. The jurisdictional scope provision in Article 4 of the CBD limited its application to components of biodiversity in areas within the limits of national jurisdiction, whereas beyond national jurisdiction only processes and activities carried out under the jurisdiction or control of the Contracting Parties are applicable. Article 5 of the CBD limits the obligations of Contracting Parties in relation to conservation and sustainable use of components of biodiversity in ABNJ to a duty to cooperate directly or through competent international organizations.

In contrast to UNCLOS, the CBD has an active Conference of the Parties (COP), which meets biennially and regularly takes decisions and sponsors initiatives concerning conservation of marine biodiversity. The CBD has laid some of the groundwork for area-based management in ABNJ at the regional level through the provision of expert advice on describing ecologically or biologically significant areas (EBSAs) and in addressing biodiversity concerns in sustainable fisheries. In 2008, the Ninth Meeting of the Conference of Parties (COP 9) of the CBD adopted the following scientific criteria for identifying EBSAs 'in need of protection in open ocean waters and deep sea habitats' (CBD, 2008, Annex 1):

*   uniqueness/rarity;
*   special importance for life history stages of species;
*   importance for threatened, endangered or declining species and/or habitats;
*   vulnerability, fragility, sensitivity or slow recovery;
*   biological productivity;
*   biological diversity; and
*   naturalness.

This decision also provided scientific guidance for selecting areas to establish a representative network of marine protected areas including in open ocean waters and deep sea habitats (CBD, 2008, Annex II). The Tenth CBD COP in 2010 agreed on a process of regional workshops for the *description* of EBSAs (CBD, 2010a, §36), to inform relevant regional and global organizations. The CBD also recognized that the *identification* of EBSAs and the selection of conservation and management measures (such as MPAs) is a matter for States and competent intergovernmental organizations (CBD, 2010a, §26). Regional workshops on describing EBSAs have been organized in most regions of the world (CBD, 2012a).

CBD COP has also investigated the scientific and technical aspects of environmental impact assessments (EIA) for activities in ABNJ. It convened an Expert Workshop on Scientific and Technical Elements of the CBD EIA Guidelines that focused on ABNJ in November 2009 (CBD, 2009). The Tenth Conference of Parties of the CBD in 2010 endorsed the development of voluntary guidelines for the consideration of biodiversity in EIAs for marine and coastal areas, drawing on the guidance from the Workshop (CBD, 2010a, §50). The Guidelines were developed for all marine and coastal areas rather than simply for ABNJ, emphasizing the interconnections between ocean ecosystems across jurisdictional boundaries, and endorsed by the Eleventh COP CBD in 2012 (CBD, 2012b, p. 7).

Convention on Migratory Species (CMS)

The objective of the CMS is to conserve migratory species of wild animals and their habitats, including marine species such as marine mammals and seabirds that migrate through marine areas within and beyond national jurisdiction (CMS, 1979). The CMS establishes a framework within which States can cooperate in conducting scientific research, restoring habitats and removing impediments to the migration of endangered species listed in its Appendix I. It also provides for the conclusion of formal conservation agreements between range States of particular migratory species listed in its Appendix II as having unfavourable conservation status. The blue, humpback, right and bow head whales are listed in Appendix I, and Appendix II includes white whales and certain populations of common grey and monk seals, various species of dolphin, seabirds and the dugong. Agreements have been concluded under Appendix II among range States dealing with seabirds (ACAP, 2001), seals and small cetaceans (ASCOBANS, 1992; ACCOBAMS, 1996). The effectiveness of the CMS for marine and other species has been limited by the non-party status of a number of important range States and regions (such as North America) although this is slowly changing (Boyle and Redgwell, 2009, pp. 684–685).

Convention on Trade in Endangered Species (CITES)

CITES (1973) has an indirect role in the conservation and management of ocean resources. Its objective is to control or prevent international commercial trade in endangered species or their products. It regulates international trade in species listed in its three appendices by means of a permit system. Trade is prohibited for species listed in Appendix 1 that are threatened with extinction. Trade is permitted subject to control for species listed in Appendix 2, that is those species not yet threatened with extinction but which could be so if trade is not monitored and controlled. Marine species including cetaceans, fish and seabirds appear in all the CITES Appendices. In recent years, CITES has become involved in the protection of endangered marine species of commercial value, including some species of sea horses, corals, eels and sharks. Nevertheless, as the opposition to the proposed listings of Bluefin Tuna and various shark species demonstrates, States engaged in those fisheries consider such listings to be inappropriate and have opposed CITES listing on the basis that the regulation of fisheries should remain the exclusive domain of existing fisheries agreements. On the other hand, it has been argued that adding a global trade component would support the regional decisions of the RFMO/As, and hence the invocation of CITES could be mutually beneficial for fisheries management (Vincent *et al.*, 2013). CITES has a large number of ratifying parties and is considered to be quite effective as it provides sanctions for non-compliance (Birnie, Boyle and Redgwell, 2009, p. 68).

World Heritage Convention

The World Heritage Convention is a well-established vehicle for protecting places of outstanding universal value, taking into consideration both cultural and natural heritage (WHC, 1972). Increasingly applied in national waters, the designated sites can cover vast areas (e.g. Great Barrier Reef (Australia), Papahānaumokuākea (USA), and Phoenix Islands (Kiribati)), as well as more typically smaller nearshore sites. To maintain World Heritage status, States must demonstrate that sites have operational management plans that are protecting the identified cultural and natural values. There is growing interest in considering how the WHC's marine coverage could be expanded (UNESCO, 2011, Recommendation 5). At a glance, the definitions of 'natural' and 'cultural' heritage in Articles 1 and 2 of the Convention show that its application is not restricted to the protection of heritage in areas under national jurisdiction. Certainly, before being applied in ABNJ, a number of issues would need to be resolved, such as identification and possible establishment of a responsible body for developing management plans and monitoring compliance, but these hurdles are not necessarily insurmountable.

Soft law developments regarding international marine conservation

Increasing international attention and acknowledgement of the global nature of many environmental problems have precipitated international conferences where States have made soft law commitments to better protect the environment. The 1972 UN Stockholm Conference, which focused on environmental degradation and pollution, set a precedent that was followed by the UN 'Earth Summits', beginning with the Conference on Environment and Development (UNCED) in Rio de Janeiro in 1992. Since then there have been Earth Summits every ten years, in Johannesburg in 2002 (World Summit on Sustainable Development (WSSD)), and again in Rio in 2012 (Rio+20). The range of global environmental problems considered has broadened considerably to include, *inter alia*, climate change, depletion of the ozone layer, deforestation, desertification and land degradation, hazardous waste, poverty alleviation, and loss of biological diversity on land, in freshwater, and in the marine environment. The 1992 Rio Earth Summit, arguably the most successful, served as a platform for several initiatives and aspirational statements, particularly Agenda 21 (UN, 1992), and the Rio Declaration on Environment and Development, which included 27 principles such as public participation (UNGA, 1992, Principle 10) and most famously, the Precautionary Principle (UNGA, 1992, Principle 15). The health of the marine environment is specifically covered in the 136 paragraphs of section 17 of Agenda 21 (UN, 1992). This first Earth Summit also launched three international conservation treaties: the Convention on Biological Diversity (CBD) (1992), the Framework Convention on Climate Change (1992), and the United Nations Convention to Combat Desertification (1994). With regard to the marine environment, the CBD has been the most relevant.

At WSSD in 2002, some of the aspirations developed at Rio were further consolidated in the form of the Johannesburg Plan of Implementation, which included many general commitments to action, as well as some explicit targets, such as maintaining or restoring fisheries 'stocks to levels that can produce the maximum sustainable yield with the aim of achieving these goals for depleted stocks on an urgent basis and where possible not later than 2015' (UN, 2002a, §31a); and 'the establishment of marine protected areas consistent with international law and based on scientific information, including representative networks by 2012' (UN, 2002a, §32c).

Unlike the first Rio Earth Summit, which impressively launched three environmental agreements, the third Earth Summit, Rio+20, committed only to making a decision by August 2015 on whether a new international instrument concerning the conservation and sustainable use of marine biological diversity in ABNJ should be developed or not (UNGA, 2012, §182). There was only one new marine target (set for 2025, to 'achieve significant reductions in marine debris' (UNGA, 2012, §163)). For the most part, Rio+20 consolidated and re-affirmed targets and commitments from the previous two Earth Summits and related international efforts.

In operationalizing the Earth Summit commitments, the CBD has been a key actor. It has, for example, elaborated guidelines for biodiversity-inclusive EIAs and strategic environmental assessments (SEAs) specifically for marine and coastal areas, including ABNJ (CBD, 2012b). In 2010, its Conference of Parties (COP) adopted 20 *Aichi Biodiversity Targets*, two of which are particularly relevant to marine conservation: Target 6 seeks to establish sustainable fisheries by 2020, and Target 11 seeks to protect through 'protected areas and other effective area-based conservation measures' at least 10 per cent of coast and marine areas (including ABNJ) by 2020 (CBD, 2010b). This latter target replaced the previous (unfulfilled) CBD MPA target of 10 per cent by 2012 (CBD, 2006), which had been inspired by the 2002 Johannesburg WSSD commitment. Nevertheless, as noted above, the CBD lacks the regulatory authority to directly implement its commitments, either within or beyond national jurisdiction. Rather, it relies on compliance through the actions of its Parties nationally and, in ABNJ, through their actions as flag States and their participation in the sectoral agreements.

The UN Millennium Development Goals (MDGs) and their planned successor, the Sustainable Development Goals (SDGs) arose out of the 2000 UN Millennium Declaration (UNGA, 2000). There are eight MDGs, one of which concerns ensuring environmental sustainability by 2015. Endorsed by 189 countries, a roadmap for the MDGs was completed in 2002, setting out goals to be reached by 2015 (UNGA, 2002). Beyond 2015, the Rio+20 Earth Summit agreed that a new set of goals, the SDGs, should be developed. With a stand-alone marine goal, marine sustainability figures more prominently in the SDGs. Goal 14 calls on states to 'conserve and sustainably use the oceans, seas and marine resources for sustainable development.' However, interpretation of this double-barrelled goal promoting both conservation and sustainable development could be challenging.

Global initiatives on cooperation in ocean conservation, management and science

United Nations Informal Consultative Process (ICP)

The disjunction between the global policy meetings and institutions considering marine environmental protection and the meeting of the States Parties of UNCLOS was acknowledged by the Commission for Sustainable Development (CSD) at its seventh session in 1999 (CSD, 1999, §38) where it recommended that UNGA set up a mechanism to provide more detailed and expert preparation for the UNGA oceans debates (CSD, 1999, §39). At its 54th session in 1999 the UNGA passed resolution 54/33 establishing the United Nations Informal Consultative Process (ICP) to facilitate annual review of developments in ocean affairs. ICP's annual meetings have raised the profile of issues associated with protection of marine environment beyond national jurisdiction and identified a variety of oceans management issues that could benefit from enhanced coordination between UN organizations and national governments. The fifth meeting of ICP in 2004 discussed the risks that would continue to arise from new and emerging uses of the high seas to the conservation and sustainable use of biodiversity beyond national jurisdiction in the absence of environmental safeguards (UNICPOLOS, 2004). Recommendations from that meeting to the UNGA resulted in the establishment of the BBNJ Working Group (below) (UNGA, 2004).

UN-Oceans

UN-Oceans was established in 2003 to provide a regular inter-agency coordination mechanism on ocean and coastal issues within the United Nations system. It includes representatives from all the relevant UN agencies with interests in ocean affairs and also involves intergovernmental organizations and non-governmental organizations in its work.[6] It operates as a flexible mechanism to review joint and overlapping ongoing activities and to support related deliberations of the ICP, coordinating as far as possible its meetings with ICP sessions. It has four current task forces, each coordinated by a lead UN institution: on Marine Biodiversity beyond National

Jurisdiction, Establishing a Regular Process for the Assessment of the Global Marine Environment, a Global Partnership for Climate Fisheries and Aquaculture, and Marine Protected Areas and Other Area Based Management Tools.[7]

UN BBNJ

Recognizing the need to better protect biodiversity in ABNJ, the UN General Assembly decided in 2004 to establish the Ad Hoc Open-ended Informal Working Group to Study Issues Relating to the Conservation and Sustainable use of Marine Biological Diversity beyond Areas of National Jurisdiction (BBNJ) (UNGA, 2004). BBNJ first met in 2006, where there and at subsequent BBNJ meetings it has been emphasized that more could be achieved through better cooperation, coordination and implementation of existing global and regional arrangements.

Additionally, in 2011, States at BBNJ recommended a process that

> would address the conservation and sustainable use of marine biodiversity in areas beyond national jurisdiction, in particular, together and as a whole, marine genetic resources, including questions on the sharing of benefits, measures such as area-based management tools, including marine protected areas, and environmental impact assessments, capacity-building and the transfer of marine technology.
>
> *(UNGA, 2011, §1b)*

This has been interpreted as a 'package deal'; i.e. until progress is made on all issues, including difficult questions related to benefit sharing of marine genetic resources in ABNJ, there will likely be reluctance at BBNJ to proceed on issues related to marine conservation (Druel *et al.*, 2013). Consideration of these issues led to the landmark decision in June 2015 by the UNGA to 'develop an international legally-binding instrument under the Convention [on the Law of the Sea] on the conservation and sustainable use of marine biological diversity of areas beyond national jurisdiction' (UNGA, 2015, §1). At least 40 days of preparatory committee meetings are to be held in 2016 and 2017, with negotiations expected to commence after that.

Intergovernmental Oceanographic Commission (IOC)

The IOC resides within the UN system under the United Nations Educational, Scientific, and Cultural Organization (UNESCO), and is responsible for ocean science, observatories, data storage and exchange, and specialized services such as the Pacific Tsunami Warning System. IOC coordinates ocean observation and monitoring through the Global Ocean Observing System (GOOS) which aims to develop a unified network providing information and data exchange on the physical, chemical, and biological aspects of the ocean. IOC sponsors the World Climate Research Programme and the IOC's GOOS serves as the ocean component of the Global Climate Observing System, which supports the Intergovernmental Panel on Climate Change (IPCC). UNESCO–IOC is co-convener with the World Meteorological Organization of the World Climate Change Conference which aims to systematically make the existing knowledge on climate science available to a wide variety of potential users.[8] IOC also co-led with the UN Development Programme (UNDP) the Assessment of Assessments (below).

UN World Ocean Assessment

At the 2002 Earth Summit in Johannesburg, it was decided that the health of the global ocean required regular monitoring (UN, 2002b). In 2005, the UN General Assembly endorsed a 'regular process for the global reporting and assessment of the state of the marine environment, including socio-economic aspects' (UNGA, 2005). To provide background information for this regular process, a global assessment of existing assessments of marine health was completed in 2009, the so-called 'Assessment of Assessments' (UNEP and UNESCO–IOC, 2009). A total of 1,023 assessments currently reside in the dedicated database.[9]

Under the direct authority of the UN General Assembly through an ad hoc Working Group of the Whole, the first World Ocean Assessment (WOA-I) began in 2010, ending in 2014. At the time of writing, the 57 chapters of the ambitious assessment are in the drafting stage, in the hands of teams of scientists nominated by UN Member States. WOA-I will assess: i) major marine ecosystem services; ii) food security and food safety; iii) other human activities and the marine environment; and iv) marine biological diversity and habitats; as well as providing an overall assessment. The draft WOA-I will be submitted to UN Member States for their comments, as well as to independent peer-reviewers who have not been involved in the assessment.[10] It is hoped that WOA-I, global in its breadth and ambition, will inform international maritime/marine decision making. However, it remains unclear how willingly the international governance institutions will take on board this scientific information, and other information such as the CBD's EBSAs, originating from outside their own dedicated processes. Currently, there is no obligation to use such knowledge, or to report back to the information providers on its usage.

Intergovernmental Platform on Biodiversity and Ecosystem Services (IPBES)

IPBES was established in 2012 as an independent intergovernmental body to synthesize, review, assess and critically evaluate relevant information concerning ecosystems globally (terrestrial and marine) from knowledge generated worldwide, encompassing a wide variety of scientific, traditional and indigenous knowledge. Its first session was held in October 2011, and the second in December 2013 where an initial work programme was presented. There are possible overlaps with WOI-I (above); however, it was noted at the December 2013 meeting that in the draft IPBES work programme there was a lack of substantive reference to marine and coastal ecosystems (IPBES, 2013, §26). Although still in its early stages, States have agreed to a fast-track assessment of scenarios and modelling of nature's benefits to people, in order to provide insights into the impacts of plausible future socio-economic development pathways and policy options and to help evaluate actions that can be taken to protect them in terrestrial, inland water and marine ecosystems (IPBES, 2013, Annex 6, p. 71). Once these early results are available, synergies with WOA-I may be better identified.

Regional governance in ABNJ

Regional governance has proven to be critical in the implementation of many global agreements. For example, fisheries agreements and commitments are to be implemented in ABNJ mainly by RFMO/As, which including those just coming into force will cover most of the globe (Ban *et al.*, 2014). However, practices among these RFMO/As vary considerably, and none appear to yet be fully meeting the sustainability objectives of the Straddling Fish Stocks Agreement, or similar conservation objectives for resident stocks (Cullis-Suzuki and Pauly, 2010). Furthermore, enforcement and compliance mechanisms within RFMOs remain incomplete, despite many RFMO/As having been in existence for several years (Gilman and Kingma, 2013; Koehler, 2013; Englender *et al.*, 2014). There have been some promising developments in the protection of *vulnerable marine ecosystems* (VMEs), a commitment that arose from UN General Assembly debates and the subsequent Resolutions 61/105 and 64/72 (UNGA, 2006, 2009). Though more work needs to be done in protecting VMEs, the progress to date is a reminder that properly worded global decisions, such as UNGA resolutions, though technically 'soft law' can nonetheless positively affect the protection of biodiversity in ABNJ (Ardron *et al.*, 2014a, 2014c).

Regional Seas organizations and agreements (RSO/As) were typically created with a shared interest among States in reducing marine pollution in a given region, but have since expanded their conservation mandates. Currently, they are largely confined to national waters, with just three extending into ABNJ – in the North-East Atlantic, South Pacific 'donut holes' (surrounded mostly by EEZs of neighbouring States), and the Mediterranean (where some States have not yet declared EEZs, leaving about 29 per cent as high seas – see Chapter 34). In all three cases, some steps have been taken to protect parts of ABNJ, with efforts in the North-East Atlantic arguably the furthest advanced, including the establishment there of seven MPAs (O'Leary *et al.*, 2012). However, full management plans for these MPAs have not yet been completed (Freestone *et al.*, 2014). The expansion of other RSO/As into ABNJ has been argued as desirable, representing a practical mechanism by which to implement integrated management and conservation (Rochette *et al.*, 2014).

Civil society involvement

Interest and participation by civil society in ABNJ governance has been increasing. In part, this is a reflection of the increasing level of activities in ABNJ, and the growing recognition of the expertise and assistance that non-governmental actors can provide. For international bodies such as the IMO, industry groups have long played a role (e.g. World Shipping Council, International Bunker Industry Association, etc.), as well as educational institutions (e.g. International Association of Maritime Universities); however, there is an increasing presence of sustainability- and conservation-oriented organizations as well (e.g. Clean Shipping Coalition, Friends of the Earth International, World Wide Fund For Nature, etc.). In general, the rules for public access to meetings and meeting documents have been slowly improving; nevertheless, transparency remains a topic that within the international governance bodies is more often discussed than practised (Ardron *et al.*, 2014b), and more could be done to encourage public participation.

The scientific community has long cooperated among themselves in oceanographic research and cruises, and is becoming increasingly better organized; for example, the International Network for Scientific Investigation of Deep-sea Ecosystems (INDEEP[11]) that arose, in part, at the conclusion of the ten-year Census of Marine Life[12] in 2010, which was itself a significant undertaking in global scientific cooperation. Science-based initiatives have also formed around specific policy processes. The Global Ocean Biodiversity Initiative[13] has played an active role in providing scientific and technical support to the CBD Secretariat in the operation of its regional EBSA workshops (discussed above). The Deep Ocean Stewardship Initiative (DOSI[14]) is focusing on providing advice concerning the impacts of deep seabed mining.

Similarly, non-governmental environmental organizations have also been organizing themselves globally around specific policy issues in ABNJ. The Deep Sea Conservation Coalition[15] is focused on reducing deep sea trawling and other damaging bottom fisheries, and has recently turned to deep-sea mining. The High Seas Alliance[16] is mainly focused on encouraging the development of an international agreement to better protect biodiversity in ABNJ. In both cases, these NGO coalitions have contributed substantially to the international discussions on these topics.

Conclusions

Under the umbrella of UNCLOS, governance and conservation of biodiversity in ABNJ is accomplished through a mix of hard- and soft-law instruments that continue to evolve. Increasingly, high seas governance is about more than just governments, and includes industry, science and civil society organizations. Nevertheless, States play the central role, and any new initiatives need 'champion' States to promote them (Freestone *et al.*, 2014). Cooperation among ABNJ governance institutions is still weak to non-existent, and will need to be increased significantly if integrated management reflecting an ecosystem approach is to be achieved (Ardron *et al.*, 2008, 2014a). Considering the deteriorating condition of high seas fish stocks, some have called for a hiatus to high seas fishing altogether, arguing that under UNCLOS the so-called 'right to fish' is contingent upon properly managing the stocks (White and Costello, 2014; Brooks *et al.*, 2014). Many argue for a new agreement under UNCLOS to better protect biodiversity (Gjerde and Rulska-Domino, 2012; Druel and Gjerde, 2014). However, whatever steps are taken in the future, most States and experts agree that concurrently making better use of existing mechanisms, including reforming those in need of reform, represents a necessary and sensible approach to protection of biodiversity in ABNJ, without in any way foreclosing the possibilities of new and improved legal instruments.

Notes

1 FAO, International Plans of Action, www.fao.org/fishery/code/ipoa/en, accessed September 2015.
2 Iceland and Norway have opted out of these measures and still continue to commercially hunt whales, and Japan exercises a hunt under the banner of scientific research. The Commission continues to set catch limits for aboriginal subsistence whaling.
3 UN, Commission on the Limits of the Continental Shelf, www.un.org/depts/los/clcs_new/clcs_home.htm, accessed September 2015.
4 UNCLOS, Art. 87(1) defines the freedom of the high seas as 'comprising, *inter alia*, both for coastal and land locked States:

(a) freedom of navigation;
(b) freedom of overflight;
(c) freedom to lay submarine cables and pipelines, subject to Part VI;
(d) freedom to construct artificial islands and other installations permitted under international law, subject to Part VI;
(e) freedom of fishing, subject to the conditions laid down in section 2;
(f) freedom of scientific research, subject to Parts VI and XIII.'

5 For example Article 205 of UNCLOS provides that 'States shall publish reports of the results obtained pursuant to Article 204 or provide such reports at appropriate intervals to the competent international organizations, which should make them available to all States.' Article 204 relates to monitoring of the risks or effects of pollution.
6 UN-Oceans, About UN-Oceans, www.unoceans.org/en/, accessed September 2015.
7 UN-Oceans, UN-Oceans Task Forces, www.unoceans.org/task-forces/en, accessed September 2015.
8 www.unesco.org/new/en/natural-sciences/ioc-oceans/about-us/, accessed March 2014.
9 www.unep-wcmc-apps.org/GRAMED/index.cfm, accessed March 2014.
10 www.worldoceanassessment.org/?page_id=6, accessed March 2014.
11 www.indeep-project.org, accessed March 2014.
12 www.coml.org, accessed March 2014.
13 www.GOBI.org, accessed March 2014.
14 www.indeep-project.org/deep-ocean-stewardship-initiative, accessed March 2014.
15 www.savethehighseas.org, accessed March 2014.
16 highseasalliance.org, accessed March 2014.

References

ACAP. (2001) *Agreement on the Conservation of Albatrosses and Petrels*, 19 June 2001, entered into force 1 February 2004, ATS 5.

ACCOBAMS. (1996) *Agreement on the Conservation of Cetaceans of the Black Sea, Mediterranean Sea and Contiguous Atlantic Area*, 24 November 1996, entered into force 1 June 2001, 36 ILM 777.

ASCOBANS. (1992) *Agreement on the Conservation of Small Cetaceans of the Baltic, North East Atlantic, Irish and North Seas, 17 March 1992*, entered into force 29 March 1994, 1772 UNTS 217.

Ardron, J. A., Gjerde, K., Pullen, S. and Tilot, V. (2008) 'Marine spatial planning in the high seas'. *Marine Policy*, 32(5), 832–839.

Ardron, J. A., Clark, M. R., Penny, A. J., Hourigan, T. F., Rowden, A. A., Dunstan, P. K., Watling, L. E., Shank, T. M., Tracey, D. M., Dunn, M. R. and Parker, S. J. (2014a) 'A systematic approach towards the identification and protection of vulnerable marine ecosystems'. *Marine Policy*, 49, 146–154.

Ardron, J. A., Clark, N., Seto, K., Brooks, C., Currie, D. and Gilman, E. (2014b) 'Tracking 24 years of discussions about transparency in international marine governance: where do we stand?' *Stanford Environmental Law Journal*, 33(2), 167–190.

Ardron, J. A., Rayfuse, R., Gjerde, K. and Warner, R. (2014c) 'The sustainable use and conservation of biodiversity in ABNJ: what can be achieved using existing international agreements?' *Marine Policy*, 49, 98–1.

Ban, N. C., Bax, N. J., Gjerde, K. M., Devillers, R., Dunn, D. C., Dunstan, P. K., Hobday, A. J., Maxwell, S. M., Kaplan, D. M., Pressey, R. L., Ardron, J. A., Game, E. T. and Halpin, P. T. (2014) 'Systematic conservation planning: a better recipe for managing the high seas for biodiversity conservation and sustainable use'. *Conservation Letters*, 7(1), 41–54.

Boyle, B. A. and Redgwell, C. (2009) *International Law and the Environment*, 3rd edn. Oxford, Oxford University Press, pp. 684–685.

Brooks, C. M. (2013) 'Competing values on the Antarctic high seas: CCAMLR and the challenge of marine protected areas'. *The Polar Journal*, 3(2), 277–300.

Brooks, C. M., Weller, J. B., Gjerde, K., Sumaila, R., Ardron, J., Ban, N. C., Freestone, D., Seto, K., Unger, S., Costa, D. P., Fisher, K., Crowder, L., Halpin, P. and Boustany, A. (2014) 'Challenging the "right to fish" in a fast-changing ocean'. *Stanford Environmental Law Journal*, 33(3), 289–324.

CBD. (1992) *Convention on Biological Diversity*, Nairobi 22 May 1992, entered into force 29 December 1993, 31 ILM 1455.

CBD. (2006) CBD. 'Framework for monitoring implementation of the achievement of the 2010 target and integration of targets into the thematic programmes of work'. Decision VIII/15; Annex II.

CBD. (2008) 'Report of the Ninth Meeting of the Conference of the Parties to the Convention on Biological Diversity'. UNEP/CBD/COP/9/29, Decision IX/20.

CBD. (2009) 'Report of the Expert Workshop on Scientific and Technical Aspects relevant to environmental impact assessment in marine areas beyond national jurisdiction'. UNEP/CBD/EW-EIAMA/2.

CBD. (2010a) 'Report of the Tenth Meeting of the Conference of the Parties to the Convention on Biological Diversity'. UNEP/CBD/COP/10/27; Annex, Decision X/29.

CBD. (2010b) 'Strategic Goals and the Aichi Biodiversity Targets'. Decision X/2, §IV.

CBD. (2012a) 'Briefing on organizing a series of regional workshops on describing ecologically or biologically significant marine areas (EBSAs)'. CBD Secretariat. www.cbd.int/doc/meetings/mar/ebsa-briefing/other/ebsa-briefing-oth-01-en.pdf, accessed March 2014.

CBD. (2012b) 'Report of the Eleventh Meeting of the Conference of the Parties to the Convention on Biological Diversity'. UNEP/CBD/COP/11/27; Annex, Decision XI/18.

CITES. (1973) *Convention on International Trade in Endangered Species of Wild Fauna and Flora*, Washington 1973, entered into force 1 July 1975, 992 UNTS 243.

CMS. (1979) *Convention on the Conservation of Migratory Species of Wild Animals*, Bonn 1979, entered into force 1 November 1983, 19 ILM 15.

CSD. (1999) 'Report of the Seventh Session of the Commission on Sustainable Development (19–30 April 1999)'. E/CN-17/1999/20.

Cullis-Suzuki, S. and Pauly, D. (2010) 'Failing the high seas: a global evaluation of regional fisheries management organizations'. *Marine Policy*, 34, 1036–1042.

Druel, E. and Gjerde, K. M. (2014) 'Sustaining marine life beyond boundaries: options for an implementing agreement for marine biodiversity beyond national jurisdiction under the United Nations Convention on the Law of the Sea'. *Marine Policy*, 49, 90–97.

Druel, E., Rochette, J., Billé, R. and Chiarolla, C. (2013) 'A long and winding road. International discussions on the governance of marine biodiversity in areas beyond national jurisdiction'. IDDRI, studies no. 07/2013.

Edeson, W. R. (2003) 'Soft and hard law aspects of fisheries issues: some recent global and regional approaches'. In M. H. Nordquist, J. N. Moore and S. Mahmoudi (eds), *The Stockholm Declaration and the Law of the Marine Environment*. The Hague, Kluwer Law International, pp. 165–182.

Englender, D., Kirschey, J., Stöfen, A. and Zink, A. (2014) 'Cooperation and compliance control in areas beyond national jurisdiction'. *Marine Policy*, 49, 186–194.

FAO. (1945) *Constitution of the United Nations Food and Agriculture Organization*, Quebec 16 October 1945, entered into force 16 October 1945, as amended in 1947, 12 UST 980, TIAS 4803.

Freestone, D., Johnson, D., Ardron, J. A., Morrison, K. K. and Unger, S. (2014) 'Can existing institutions protect biodiversity in areas beyond national jurisdiction? Experiences from two on-going processes'. *Marine Policy*, 49, 167–175.

Gilman, E. and Kingma, E. (2013) 'Standard for assessing transparency in information on compliance with obligations of regional fisheries management organizations: validation through assessment of the Western and Central Pacific Fisheries Commission'. *Ocean and Coastal Management*, 84, 31–39.

Gjerde, K. and Rulska-Domino, A. (2012) 'Marine protected areas beyond national jurisdiction: some practical perspectives for moving ahead'. *The International Journal of Marine and Coastal Law*, 27, 351–373.

Grubb, M., Koch, M., Thomson, K., Munson, A. and Sullivan, F. (1993) *The 'Earth Summit' Agreements: A guide and assessment*. London, Earthscan Publications.

High Seas Task Force. (2006) *Closing the Net: Stopping illegal fishing on the high seas*. Governments of Australia, Canada, Chile, Namibia, New Zealand and the United Kingdom, WWF, IUCN and the Earth Institute at Columbia University.

IMO. (1948) *Convention on the Intergovernmental Maritime Consultative Organization*, Geneva 6 March 1948, entered into force 17 March 1958, 289 UNTS 3.

IMO. (2001) *International Convention on the Control of Harmful Anti-fouling Systems on Ships*, London 5 October 2001, entered into force 17 September 2008.

IMO. (2004) *International Convention on the Control and Management of Ships Ballast Water Sediments*, London 13 February 2004 (not yet in force) IMO Doc BWMCONF/36.

IMO. (2005) 'Revised Guidelines for the Identification and Designation of Particularly Sensitive Sea Areas'. IMO Assembly Resolution A982(24).

IPBES. (2013). 'Report of the second session of the Plenary of the Intergovernmental Science-Policy Platform on Biodiversity and Ecosystem Services'. IPBES/2/17.

ISA. (2010) 'Decision of the Assembly of the International Seabed Authority relating to the regulations on prospecting and exploration for polymetallic sulphides in the Area'. ISBA/16/A/12/Rev.1.

ISA. (2012) 'Decision of the Assembly of the International Seabed Authority relating to the Regulations on Prospecting and Exploration for Cobalt-rich Ferromanganese Crusts in the Area'. ISBA/18/A/11.

ISA. (2013a) 'Decision of the Assembly of the International Seabed Authority regarding the amendments to the Regulations on Prospecting and Exploration for Polymetallic Nodules in the Area'. ISBA/19/A/9.

ISA. (2013b) 'Decision of the Council of the International Seabed Authority relating to amendments to the Regulations on Prospecting and Exploration for Polymetallic Nodules in the Area and related matters'. ISBA/19/C/17.

ITLOS. (2011) 'Advisory opinion on responsibilities and obligations of states sponsoring persons and entities with respect to activities in the area'. *International Tribunal of the Law of the Sea*. www.itlos.org/fileadmin/itlos/documents/cases/case_no_17/adv_op_010211.pdf, accessed March 2014.

Joyner, C. C. (1995) 'Biodiversity in the marine environment: resource implications for the Law of the Sea'. *Vanderbilt Journal of Transnational Law*, 28, 644.

Koehler, H. (2013) 'Promoting Compliance in Tuna RFMOS: A comprehensive baseline survey of the current mechanics of reviewing, assessing and addressing compliance with RFMO obligations and measures'. *ISSF Technical Report* 2013–02. International Seafood Sustainability Foundation, McLean, Virginia, USA.

Lodge, M. W., Anderson, D., Lobach, T., Munro, G., Sainsbury, K. and Willcock, A. (2007) *Recommended Best Practices for Regional Fisheries Management Organization*. London, Chatham House.

Nelson, L. (1995) 'The new deep sea-bed mining regime'. *International Journal for Marine and Coastal Law*, 10(2), 189–203.

O'Leary, B. C., Brown, R. L., Johnson, D. L., von Nordheim, H., Ardron, J., Packeiser, T. and Roberts, C. M. (2012). 'The first network of marine protected areas (MPAs) in the high seas: the process, the challenges and where next'. *Marine Policy*, 36, 598–605.

Oxman, B. (1999) 'The 1994 Agreement Relating to the Implementation of Part XI of the UN Convention on the Law of the Sea'. In D. Vidas and W. Ostreng (eds), *Order for the Oceans at the Turn of the Century*. The Hague, Kluwer Law International, pp. 15–36.

Part XI Implementation Agreement. (1994) *1994 Agreement relating to the Implementation of Part XI of the United Nations Convention on the Law of the Sea of 10 December 1982*, New York 28 July 1994, entered into force 28 July 1996, 33 ILM 1309.

Rochette, J., Unger, S., Herr, D., Johnson, D., Nakamuar, T., Packeiser, T., Proelss, A., Visbeck, M. and Wright, A. (2014) 'The regional approach and the conservation and sustainable use of ABNJ'. *Marine Policy*, 49, 109–117.

Tahindro A. (1997) 'Conservation and management of transboundary fish stocks: comments in light of the adoption of the 1995 Agreement for the Conservation and Management of Straddling Fish Stocks and Highly Migratory Fish Stocks'. *Ocean Development and International Law*, 28(1), 1–58.

Tladi, D. (2011) 'Ocean governance: a fragmented regulatory framework'. In P. Jacquet, R. Pachauri and L. Tubiana (eds), *Oceans: The new frontier – a planet for life*. Delhi, TeriPress, pp. 99–111.

UN. (1992) 'Agenda 21'. *United Nations Conference on Environment & Development*, Rio de Janerio, Brazil, 3 to 14 June 1992. http://sustainabledevelopment.un.org/content/documents/Agenda21.pdf, accessed March 2014.

UN. (2002a) 'Plan of Implementation of the World Summit on Sustainable Development'. www.un.org/esa/sustdev/documents/WSSD_POI_PD/English/WSSD_PlanImpl.pdf, accessed March 2014.

UN. (2002b) 'Report of the World Summit on Sustainable Development'. Johannesburg, South Africa, 26 August to 4 September 2002. A/CONF.199/20, §36b. ISBN 92–1–104521–5.

UN Fish Stocks Agreement. (1995) *1995 Agreement for the Implementation of the Provisions of the 1982 United Nations Law of the Sea Convention Relating to the Conservation and Management of Straddling Fish Stocks and Highly Migratory Fish Stocks*, New York 4 August 1995, entered into force 11 December 2001, 2167 UNTS 3.

UNCLOS. (1982) *1982 United Nations Convention on the Law of the Sea*, Montego Bay, 10 December 1982, entered into force 16 November 1994, 1833 UNTS 3.

UNEP and IOC-UNESCO. (2009) 'An Assessment of Assessments, Findings of the Group of Experts. Start-up Phase of a Regular Process for Global Reporting and Assessment of the State of the Marine Environment including Socio-economic Aspects'. ISBN 978–92–807–2976–4.

UNESCO. (2011) 'Final report of the Audit of the Global Strategy and the PACT initiative'. WHC-11/35.COM/INF.9A; p. 24.

UNGA. (1992) 'Rio Declaration on Environment and Development'. *Report of the United Nations Conference on Environment and Development*. A/CONF.151/26 (Vol. I) Annex 1.

UNGA. (2000) UN General Assembly Resolution 55/2.

UNGA. (2002) 'Road map toward the implementation of the United Nations Millennium Declaration'. United Nations General Assembly Document A56/326.

UNGA. (2004) 'Oceans and the Law of the Sea'. (Published 2005); A/RES/59/24 §73.

UNGA. (2005) UN General Assembly Resolution 64/71, §177.

UNGA (2006) UN General Assembly Resolution 61/105, §§83–87.

UNGA (2009) UN General Assembly Resolution 64/72, §§113,114, 119–123, 129.

UNGA (2011) 'Letter dated 30 June 2011 from the Co-Chairs of the Ad Hoc Open-ended Informal Working Group to the President of the General Assembly'. A/66/119, §1b.

UNGA. (2012) 'The Future We Want'. A/RES/66/288.

UNGA. (2015) UN General Assembly Resolution 69/292.

UNICPOLOS. (2004) 'Report on the work of the United Nations Open-ended Informal Consultative Process on Oceans and the Law of the Sea at its Fifth Meeting'. A/59/122.

Vincent, A. C. J., Sadovy de Mitcheson, Y. J., Fowler, S. L. and Lieberman, S. (2013) 'The role of CITES in the conservation of marine fishes subject to international trade'. *Fish and Fisheries*, 15(4), 563–592.

WHC. (1972) *Convention for the Protection of the World Cultural and Natural Heritage*, entered into force 1975. I-15511; UNTS1037.

White, C. and Costello, C. (2014) Close the high seas to fishing? *PLoS Biology*, 12(3), e1001826. http://journals.plos.org/plosbiology/article?id=10.1371/journal.pbio.1001826, accessed 10 July 2015.

5

REGIONAL* ECOSYSTEM-BASED IMPERATIVES WITHIN GLOBAL OCEAN GOVERNANCE

Lee A. Kimball

The purpose of this chapter is to revisit the rationales for and pathways toward effective regional ocean governance. It argues that the fundamental imperative for regional approaches is to strengthen ecosystem approaches to ocean assessment and management, including social and economic dimensions. Part 3 of this handbook covers in more detail developments in several ocean regions. This chapter provides background and context for evolving ocean regionalization more generally and ruminations over future directions and opportunities. It takes stock of the merits of regional approaches and then considers the implications of the following developments:

- changing ecological knowledge, including
  - a shift in international *political* support for ecosystem approaches to ocean management, and
  - improved methods and tools for ecosystem approaches;
- greater international attention on marine biodiversity in areas beyond national jurisdiction (ABNJ);
- new challenges and opportunities, including the effects of climate change; and
- growing appreciation of the need for focused analysis of international institutions, including
  - how to "promote the science/policy interface through inclusive, evidence-based and transparent scientific assessments", as called for in the Rio+20 outcome document (UN 2012), and
  - how to "manage" the interplay between regional and global bodies as they continue to multiply and intersect.

Background and context

The rationales for regional ocean approaches have evolved to some extent during the last century, but fundamentally they remain the same. While the list below may seem obvious, it is useful to reiterate and deconstruct them when considering new developments and challenges.

* "Regional" refers to *international* arrangements, not to large-scale areas within a single nation.

The most ancient and obvious is *shared resources*—whether fish stocks, marine mammals or sedentary marine species. Over the years, small initiatives between coastal communities to allocate and maintain these resources and avoid conflicts have been scaled up. Today regional fisheries agreements, most of which were concluded after World War II, aim to encompass the full range of target stocks.

The *protected areas* rationale related initially to shared resources. From the early sealing agreements (Conventions: 1911/1957 Seals, 1972 CCAS) to contemporary regional fisheries agreements, they allow areas to be set aside to prohibit harvesting for conservation purposes and/or encourage scientific study. Today, international agreements have substantially expanded the goals of protected area designations—in order to protect vital habitat for endangered or threatened species or sites of special ecological, scientific, cultural, aesthetic or other values. At the regional level these agreements are normally subsidiary instruments, or protocols, to the regional seas agreements (see below). When the parties to the Convention on Biological Diversity (Conventions: 1992 CBD) adopted a major program on marine and coastal protected areas in 1995 (Decision II/10), this set in motion a substantial push to match terrestrial protections in the marine area and to develop a more systematic, conservation-oriented approach. The concrete target set by the 2002 World Summit on Sustainable Development (WSSD), to establish representative networks of marine protected areas (MPAs) by 2012, shifted and strengthened momentum toward a more systematic approach, within and beyond regions and drawing on mandates under both regional and global agreements.

The transboundary effects of marine pollution led to a third focus of regional agreements, first in Europe in the early 1970s (Conventions: 1972 Oslo, 1974 Paris, 1974 Helsinki) and subsequently under the auspices of the UNEP regional seas programme, launched in 1975. The early agreements covered pollution from land-based sources that could flow easily from one nation to another, and the deliberate at-sea disposal of wastes (dumping) by one nation that could adversely affect another. The main dumping concern at the time was nuclear wastes disposal in the northeast Atlantic beyond national jurisdiction that might adversely affect other states in the region.

A fourth rationale, originally subsumed in the early resources agreements (e.g., CCAS), is to identify *protected species* that warrant special measures due to their conservation status. Since the mid 1980s, a number of the regional seas agreements have adopted subsidiary instruments to identify and conserve such species throughout their range. Additional agreements concluded under the 1979 global Convention on Migratory Species (Conventions: 1979 CMS) also cover populations of species throughout their range, whether restricted to a given region (e.g., seals, sea turtles), or those that migrate through two or more regions (e.g., small cetaceans, seabirds, sea turtles).

These "ecological" rationales for regional agreements are complemented by several functional rationales. Specifically, *cooperation in data collection, research and analysis* form a fifth rationale for regional agreements. In fact, the earliest regional marine agreement focused on just these issues—the 1902 International Council for Exploration of the Sea (ICES), concerned primarily with the North Atlantic Ocean. The exchange of scientific information on shared species is vital for understanding their biology and thus how to manage and conserve them, and for determining further research needs. It requires common standards and methods, so that data and analyses in different countries can be easily compared and integrated. Moreover, such cooperation helps establish trust between experts from different countries, which fosters a collective stake in research findings and management decisions.

A further rationale for functional regional collaboration is to share information and *work together on appropriate technical and policy response measures* for common problems. This emerged in early sealing and marine pollution control agreements (Kimball 1996), and today a prominent feature of regional agreements is to promote and expedite "learning," among both government representatives and at the expert level.

A final step in functional cooperation, of course, is for parties *to cooperate in harmonizing and enforcing response measures* so that all nations in the region share the burden and none is discriminated against.

Such functional cooperation can increase knowledge and skills in each country, foster personal relationships and trust among individuals from different countries, and pool technical and financial resources for common ends and priorities, reducing costs to individual governments.

Beyond the region

All these rationales for regional approaches have implications beyond the region. The ecological bases reflect the fact that many shared marine species do not range widely beyond a given region, that transboundary marine pollution is largely contained within regions, and that vital habitats such as estuaries, mangroves and coral reefs rarely span more than two countries. At the same time, fish species such as salmon or tuna may spawn or breed in one region yet spend a substantial part of their life cycle in another. Marine pollution may originate from shipping or airborne sources outside the region, and even river-borne and coastal pollution from one region can be carried long distances by ocean currents. The migratory range of certain species of sea turtles or seabirds also traverses more than one region, while some whale and seabird species roam vast ocean areas around the globe. Conservation of these species may need to include critical breeding, nursing and feeding grounds in different regions and in areas beyond national jurisdiction.

The functional bases for regional cooperation also extend beyond the region. Scientific and technical information exchanged within a region can advance ocean knowledge generally. The spread of innovative *methods* for scientific research and assessment helps increase skills and capabilities worldwide. Similarly, the exchange of information on appropriate technical and policy response measures for particular problems and conditions increases human knowledge and skills everywhere. Today regional agreements offer an important platform for exchanging knowledge, experience and best practices for dealing with ocean problems, not only within the region but drawing on and contributing to appropriate models around the world.

Global institutions can more easily design cost-effective support programs when neighboring countries have similar problems and capacities. They can mobilize worldwide skills and resources but target programs to address the unique combination of circumstances (environmental, social and economic) and priorities in each region. It is up to stakeholders in the region to take the initiative in determining priorities and to assert a major role in targeting external resources.

Shifting boundaries

Two major factors bear on the role and evolution of regional agreements: changing jurisdictional boundaries and growing ecological knowledge. The 1982 UN Convention on the Law of the Sea (Conventions: 1982 UNCLOS) vastly expanded coastal state rights and obligations in offshore zones. It extends coastal state sovereign rights over natural resources throughout a 200 nautical mile exclusive economic zone (EEZ); it further extends exclusive coastal state sovereign rights over mineral resources and sedentary species to the outer edge of the continental margin if the margin goes beyond 200 nm (articles 76 and 77); and it extends coastal state jurisdiction over the protection and preservation of the marine environment and marine scientific research throughout the EEZ as specifically provided. These changes do not alter the rationales for regional agreements but they shift the boundaries seaward for cooperation on shared species, habitat and transboundary pollution, including pollution originating in areas beyond national jurisdiction.

Following the conclusion of UNCLOS, coastal states asserted greater control in offshore zones; just as human activities began to intensify, with greater potential to affect neighboring countries, access for international research was curtailed. The result was that the extent of species decline, marine pollution and habitat degradation were not widely known within let alone beyond the regions. Numerous piecemeal projects supported by national and international bodies did not produce a coherent picture of the health of the marine environment and resources at any level. Although the Convention contemplated enhanced regional cooperation, it took some time for mounting problems to produce the necessary momentum.

The second major factor bearing on the role and evolution of regional agreements was growing understanding of ocean problems and marine ecology. The 1992 UN Conference on Environment and Development (UNCED) in Rio de Janeiro served as the impetus for an important oceans stock-taking—especially on marine pollution and fisheries management. Several studies, notably the 1990 global assessment of the state of the marine environment, were important in drawing attention to the *inter-connections* of oceans problems; specifically, that marine pollution stems not only from a variety of contaminants but also from physical degradation and sediments; that watershed development can significantly impact coastal environments and habitat; and that fisheries and marine species are heavily impacted by habitat degradation in addition to direct take. Also as a consequence of the 1990 assessment, the focus for action on marine pollution shifted from shipping and offshore oil and gas development to damage caused by land-based activities (GESAMP 1990).

Marine living resources were the subject of other studies, with growing understanding within the scientific community of relationships between harvested, dependent and associated species and the effects of environmental conditions on marine species. The first express international institutional manifestation of this "ecosystem" approach to marine species came with the 1980 adoption of the Convention on the Conservation of Antarctic Marine Living Resources (Conventions: 1980 CCAMLR), whose geographic scope was intentionally designed to encompass fully the Antarctic marine ecosystem. But it would be over 20 years before an ecosystem approach to fisheries was more widely promoted. Unfortunately, information on the status of stocks within national jurisdiction was seriously deficient or closely held in many regions. Beyond national jurisdiction, it was difficult to piece together a complete picture due to national practices on data collection and reporting and the limitations of statistics collected and assessed by FAO on a global basis. Systematic attempts to collect and assess inter-species and environmental data in relation to target fish stocks were even more limited. Finally, in 2001, the international community's Reykjavik Declaration on Responsible Fisheries in the Marine Ecosystem opened the door for FAO to support technical work on an ecosystem approach to fisheries (FAO 2003).

During the last two decades a crescendo of studies has drawn attention to the deteriorating health of the oceans and marine resources, especially in coastal and nearshore areas. They have documented ecological linkages and the important role of vital systems such as coral reefs, seagrasses and mangroves—highlighting how deteriorating *ecosystems* deprive humans of goods and services. Further studies on the conservation status of marine species have helped identify critical breeding, nursing and recruitment habitat, both within and beyond national jurisdiction. The expanding scale of oceans problems, within and across national boundaries, and growing perception that cumulative and interactive threats originate in different countries have spurred greater regional cooperation and agreement. For global bodies, regional agreements and programs have increasingly become the vehicle of choice to assist states in addressing oceans problems.

These studies and trends have, on the one hand, led to more systematic attempts to delineate and classify larger regional marine systems and the ecological roles of smaller, vital systems nested within (UNESCO 2009); and, on the other, underscored the urgency of more integrated approaches to ocean planning and management, scaled up small marine protected area and coastal zone initiatives. The Global Environment Facility (GEF) undertook the first practical steps to apply marine ecosystem approaches across sectors on a large scale. Its 1995 operational strategy uses large marine ecosystems (LMEs) as units for assessment and management in international waters. Building on the concept utilized in CCAMLR, these projects address not only marine living resources but also marine pollution and habitat degradation. Today the GEF, together with partners, funds some 16 LME projects involving more than 100 countries (Duda 2009).

International political endorsement of ecosystem approaches in the ocean arena came more slowly. Although the CBD adopted ecosystems as its primary implementation framework in 1995 (Decision II/8), it was not until 2001 that the UN General Assembly (UNGA), the overarching international body for ocean discussion, expressly acknowledged the importance of ecosystem approaches to ocean management (Resolution 56/12). (The delay was due to two opposing reasons: some distant-water fishing nations were worried that coastal states might use the concept to seek greater control over "straddling" stocks that move within and beyond the EEZ; conversely, some coastal states were concerned that distant-water fishing states might seek to expand the scope of international fisheries agreements, reaching international controls into areas within national jurisdiction.) Since 2001, there has been further political endorsement— the WSSD encouraged the application of ocean ecosystem approaches by 2010, and the UNGA regularly supports them in its annual discussions and in 2006 drew attention to agreed consensual elements of ecosystem approaches to ocean management (Resolution 61/222). This has finally opened the door wide to strengthening the application of ecosystem approaches to ocean assessment and management through international instruments.

Two remaining challenges for ecosystem approaches

Two major challenges remain. The first is technical—developing tools and methods to support marine assessment and management on an ecosystems basis (recognizing that significant data gaps remain in various regions of the world) (UNEP and IOC-UNESCO 2009). The second is institutional—to address the sectoral silos within which international oceans decision making normally takes place and design a more coherent process, with implications both for regional and global agreements. (The latter is addressed further below.)

The two are linked. Technical tools are needed to provide decision makers with a comprehensible, integrated picture of all the human activities affecting particular ocean ecosystems (species, habitat and environmental components) and associated social and economic aspects. Analytical findings need to address the policy questions posed by decision makers and present clearly any response options and the trade-offs involved. In addition, they must generate trust; that is, the tools and methods used to assess problems and evaluate response options, and the assessment process itself, must be considered credible and legitimate. Thus, for complex marine assessments, integrated on an ecosystems basis, an agreed assessment process is an integral part of the institutional landscape, both at regional and global scales.

During the last decade there has been major progress on the technical means for aggregating information; for example, a variety of mapping tools, marine spatial planning (MSP, see Agardy, chapter 31 this volume) and strategic environmental assessment (SEA, see Kenchington, chapter 12 this volume). Many additional tools and "apps" provide support for decision makers and other stakeholders to undertake ecosystem approaches and sort out conservation priorities.[1] On the social and economic front, there has been progress in valuing ocean products and accounting for the environmental costs of activities impacting the oceans; capturing the economic and social value of marine ecosystem services and their loss is a more difficult prospect (see Kildow and Scorse, chapter 11 this volume). As the world considers shifting to a new "green" economy, having tools fit for purpose is an essential prerequisite.

Greater attention on areas beyond national jurisdiction (ABNJ)

Scientific studies during the last decade, complemented by vivid underwater photography and computer graphics, have revealed major damage to important deepsea habitats caused by bottom trawling and produced greater understanding of the extent and role of deepsea habitats such as seamounts, hydrothermal vents and deepsea corals, new knowledge of the migratory routes of scores of marine species, and a more systematic accounting of marine biodiversity in different regions and habitats (CoML 2010). This has spawned two important and ongoing conservation initiatives in ABNJ:

1 to review and reform the regional fishery management organizations (RFMOs) so that they reflect ecosystem approaches and other provisions of the 1995 UN Fish Stocks Agreement (Conventions: 1995 UNFSA) and protect vulnerable marine ecosystems (VMEs) as agreed by the UN General Assembly in 2006 (Resolution 61/105), subsequently elaborated through FAO (FAO 2008); and
2 to develop MPA networks at both regional and global levels, including in ABNJ. In order to advance this goal, scientific criteria to identify ecologically or biologically significant areas (EBSAs) were adopted in 2008 under the auspices of the CBD, together with scientific guidance for designing representative networks (Decision IX/20, Annexes I and II). There are ongoing regional workshops to identify EBSAs, and a major report on biogeographic classification of the open oceans and deep seabed provides the basis for ensuring that networks are representative (Rice et al. 2011, UNESCO 2009).

The state of international fisheries in ABNJ makes (1) the most pressing concern. Nevertheless, international shipping, enormous agglomerations of marine debris, greater reliance on subsea cables for electronic communications, growing interest in deepsea minerals development and renewable ocean energy and other potential uses threaten to increase impacts in ABNJ.

As marine research continues to find linkages between regions and with ABNJ in the distribution and movements of marine species and habitat (CoML 2010, Toropova et al. 2010), this poses new challenges for regional agreements. *And it underscores the protected areas rationale as a keystone of regional, ecosystem-based approaches.* Marine species may follow different paths within and beyond national jurisdiction, concentrating to feed, spawn or breed in different locations, but those locations themselves are vital for the health of ocean species and ecosystems. Thus, achieving (2) above, *before* human activities intensify in ABNJ, becomes equally compelling.

The critical (dis)juncture in advancing ecosystem approaches—linking actions at regional levels with actions in ABNJ—lies at political and institutional levels. Geographically, the RFMOs have long encompassed ABNJ in order to encompass the full range of target stocks. The same is true of the regional and inter-regional CMS agreements. On the other hand, most of the regional seas agreements cover only areas within national jurisdiction, although for historical and other reasons, the 1992 OSPAR, 1986 South Pacific Convention and 1959 Antarctic Treaty include ABNJ (Conventions: 1992 OSPAR, 1986 SPC, 1959 AT). However, regional agreements are generally limited in the scope of activities covered; that is, RFMOs are restricted to fishing and related research, while the regional seas and CMS agreements do not regulate international shipping or fishing covered by other international agreements. Moreover, all of these agreements function within the Law of the Sea Convention framework, which strictly specifies the rights and obligations of coastal states in areas within national jurisdiction and the rights and obligations of all states in ABNJ.

New challenges and opportunities, including the effects of climate change

With growing understanding of marine ecosystems and linkages, and the critical role of essential habitat (EBSAs, as shorthand), new challenges arise for regional agreements:

- how to protect essential habitat for non-fish species (or fish species not covered by an RFMO) that migrate through two or more regions, whether through collaborative initiatives between regional seas agreements, between regional seas and CMS agreements, and/or between regional seas agreements and RFMOs;
- how to protect essential habitat such as cold water corals and other sedentary species of the extended continental shelf (beyond the EEZ) from high seas fishing activities, through collaboration between RFMOs and the relevant coastal state(s);
- how to protect essential habitat and ecosystems that "straddle" areas within and beyond national jurisdiction, such as the Sargasso Sea;[2] and
- how to protect essential habitat in ABNJ for non-fish species that migrate within and beyond national jurisdiction, including endangered and threatened species that may be subject to regional or inter-regional agreements.

Further challenges to existing regional arrangements are posed by climate change. Sea level rise, warming waters, extreme weather events, and ocean acidification due to increased CO_2 uptake will increasingly affect coastal and ocean ecosystems by changing the freshwater/saltwater mix in coastal areas, altering the distribution, range and productivity of marine species, physically damaging important coastal habitat, spreading disease or the introduction of new species, or causing die-offs of "foundation" species in the marine food web such as corals, shellfish and phytoplankton. Such changes may alter species composition and reduce ecosystem resilience, with consequent effects on a wider range of species and the coastal and island communities dependent upon them. They may also shift ecosystem boundaries.

The ocean ramifications of climate change have bolstered support for ecosystem approaches through "blue carbon" adaptation initiatives. The aim is to protect and restore marine ecosystems to enhance resilience and improve protective barriers against coastal storms, slow or reverse ongoing loss of carbon capture and storage in marine "sinks," help maintain fisheries and other marine biodiversity, and maintain services such as nutrient cycling and improved water quality. The potential benefits for both natural and human communities resonate in many coastal and island nations and such initiatives will most likely occur within the context of existing regional seas agreements (Herr and Galland 2008; Nellemann *et al.* 2009). At the same time, changes in species migration, habitat health and distribution, or even large-scale ecosystem boundaries have implications for the geographic scope of RFMOs and the regional seas and CMS agreements and pose challenges for EBSAs, noted above.

Various proposals have emerged during the last decade supporting regional ocean governance arrangements that can address more comprehensively multiple threats and are more aligned with large-scale regional ocean systems. The question is how to accomplish this within the UNCLOS framework:

- proposals to modify the geographic scope of RFMOs or the CMS agreements to encompass the full range of species are generally not problematic, although any resulting overlaps between RFMOs will need to be worked out;
- proposals to expand the geographic scope of the regional seas agreements are problematic if they seek to extend coastal state jurisdiction or alter the UNCLOS balance between the rights and obligations of coastal states and the high seas rights and obligations of all states; and
- proposals to expand and or merge the substantive scope of any of these agreements are problematic if they seek to alter that UNCLOS balance.

Yet the UNCLOS regime is by no means static. It expressly recognizes that its "constitutional" framework will continue to be developed and harmonized through new international instruments at regional and global levels—notably, to protect and preserve the marine environment and to conserve and manage high seas living resources. Numerous provisions offer scope for further measures, for example, to protect rare or fragile ecosystems and habitat (article 194(5)); to prevent and control pollution from the use of technologies (article 196); and to assess the potentially harmful environmental effects of planned activities (article 206).

In recognizing special situations, UNCLOS may open the door to other special circumstances. The Convention offers coastal states certain rights to regulate vessel-source pollution in vulnerable ice-covered areas within the EEZ (article 234). By analogy, in ABNJ the IMO has already adopted special measures to control marine pollution from ships in polar areas (Antarctic Special Area under 1973/1978 MARPOL) and is currently considering a Polar Code applicable to both Arctic and Antarctic areas. In addition, UNCLOS expressly defines the special situation of enclosed or semi-enclosed seas, urging bordering states to cooperate and coordinate on marine living resources, marine environmental protection and scientific research (articles 122–123). The definition is somewhat limiting, but the concept that in relatively closed seas a higher level of cooperation might be required for bordering states to avoid adverse impacts and enjoy benefits might be applied to other situations.

For example, is there a similar "special" concern of bordering states to encompass fully the high seas segment of "straddling" ecosystems in a regional seas agreement, or to include patches of ABNJ in order to cover fully a larger scale system encompassed by the regional agreement (e.g., 1986 SPC Convention)? When changes occurring beyond national jurisdiction might impact states within a region, whether due to marine debris, the effects of climate change, or other causes, what is their recourse? Couldn't the right of adjacent states to encompass ABNJ in their regional initiatives be recognized, without altering the UNCLOS balance of rights and obligations or the limits of national jurisdiction, giving them greater clout to pursue specific actions in other regional or global bodies when additional cooperation and coordination is required? Such a proposition would eliminate any presumption of coastal states extending the limits of their jurisdiction yet recognize a collective role for potentially affected states to take action and engage broader international cooperation as appropriate.

As noted above, three of the regional agreements already encompass ABNJ, but the rationales were not ecosystem based. Do widespread international endorsement of, and growing technical ability to apply, ecosystem approaches warrant additional steps to facilitate this? For example, the Wider Caribbean regional agreement (Conventions: 1983 WCR) is one that might benefit from a larger geographic scope, not least because, like the Mediterranean, many nations have yet to extend and resolve EEZ boundaries. More broadly, an ecosystems basis might lead to more logical ocean management units and help cohere the geographic scope of existing regional agreements. For example, although it does not appear to have been followed up, there was in 2009 a proposal in the Antarctic Treaty forum to extend existing vessel pollution control measures beyond the northern boundary of the Antarctic Treaty to the CCAMLR boundary so that they would cover the entire Antarctic marine ecosystem (i.e., to extend the boundary of the Antarctic Special Area under MARPOL 73/78 through the IMO). (Final Report of the 32nd ATCM 2009).

The results of the RIO+20 Conference in June 2012 suggest another opportunity to leverage ecosystem approaches. In calling for further work on broader measures of progress to complement gross domestic product (UN 2012), the conference encouraged ongoing efforts by the UN Statistical Commission, UNEP, UNDP, the UN University, OECD, IMF and the World Bank, in collaboration with academic and other experts, to develop a more integrated set of measures and indicators that incorporate ecological limits as well as measures of social and economic well-being. A recent report sums up progress toward more comprehensive accounting of the true wealth of nations, including ecosystem services. Among the remaining challenges it draws attention to are: incorporating into national accounts natural capital that cannot be privately held, using the example of the open oceans (within national jurisdiction); and "off-site" impacts (environmental externalities) that are international—that is, ecosystem service flows like pollution or the transfer of alien invasive species that spread regionally or globally. The report further notes that from a global perspective, wealth accounts should "cover all assets on which human wellbeing depends, including those beyond national jurisdiction" (UNU–IHDP and UNEP 2012).

Could the "keystone" role played by EBSAs in ecosystem approaches serve as a proxy for spurring international conservation decisions within and beyond national jurisdiction? The report noted above, through an example using mangrove loss in Thailand, illustrates progress in accounting for the depreciation or appreciation of these natural assets in national wealth. In addition to products produced, their "service" roles as nursery and breeding habitats for offshore fisheries and in carbon sequestration are among the values tallied. Today there should be some basis for habitat-focused analyses that estimate the value of certain vital open ocean areas vis-à-vis known fisheries, if not for carbon sequestration, or of depleted fisheries due to degraded deepsea habitat such as seamounts. In ABNJ, however rudimentary, the gains and losses would accrue to the world community, illustrating what is at stake and helping to build the case for conserving keystone habitat so as not to further undermine human welfare and the wealth of nations. It would make sense to tackle first the value of EBSAs, where recent scientific studies may offer substantial guidance. This will inform priority-setting by oceans decision makers and is likely to add to current compelling evidence that stronger institutional arrangements are needed to maintain and restore ocean ecosystems for present and future generations.

International institutions

With deeper understanding of ocean ecology and intensifying pressures from multiple human activities, integrated ocean management has grown increasingly compelling. It has grown in scale as relatively small MPAs or integrated coastal management initiatives are inevitably affected by development in surrounding areas. And it has grown in complexity with the many new global and regional agreements concluded following adoption of UNCLOS, each with its own decision-making body and, for the most part, sectoral focus (Kimball 2001). The twenty-first-century challenge is to address the scope of oceans problems at larger regional scales, taking into account ecological linkages, limits and effects on human welfare. Effective institutional arrangements remain elusive.

In 1997, this author suggested ways to realign institutional arrangements for ocean governance to respond to the changing problems and challenges of ocean space in order to:

- strengthen well-coordinated collective action within each region to advance ecosystem approaches, recognizing ecological linkages extending beyond the region;
- help decision makers maintain an overview of interconnected ocean problems and at the same time find practical solutions for individual sources of threat;
- give full rein to the comparative advantages of regional and global bodies; and
- devolve appropriate responsibilities from global to regional levels in the interest of truly vesting knowledge and capabilities in each region.

Within each region, it was suggested that governments either identify an existing regional body(ies) or periodically convene an informal overview forum in order to:

- ensure integrated regional marine assessments (environmental, social, economic) that take linkages among problems and potential response actions into account and enhance information resources and expert capabilities within the region;
- produce a roadmap of international institutional roles in the region, noting ecological and management linkages;
- articulate regional goals and strategies and define priorities clearly;
- determine the need to elaborate or reconcile policies and programs within the region to ensure that they are mutually reinforcing;
- coordinate regional views in preparing for global conventions and other intergovernmental processes so that regional concerns receive due attention; and
- review progress periodically.

The following comparative advantages of global institutions were identified:

- synthesize integrated regional assessments to provide an overview, identify emerging issues, and highlight linkages across regions and problems;
- extract the findings of regional assessments that have wider implications for international decision making, both norm-setting and programmatic, and leverage wider global concerns into regional action plans and capacity-building as appropriate;
- update worldwide knowledge with respect to particular issues, problems and response measures, and ensure recourse to accepted methods and tools for data quality control, data integration and problem diagnosis;
- address global-scale problems where uniform global measures should provide at least a bottom line, for example in international shipping; and
- elaborate norms and principles that can advance more effective and innovative approaches at all levels (Kimball 1997, 2001).

The essential underpinning for an ecosystem approach to ocean management is an ecosystem approach to ocean assessment. Marine regions are logical ocean management units for the reasons stated above. Integrated assessments at the regional level provide the basis for integrated management by offering decision makers *specific* information on issues and problems in the region, the relative importance of different sources and impacts, and the implications and trade-offs of different response options for ecosystem health and human welfare. *In order to advance ecosystem approaches*, regional assessments should also identify external ecological linkages so that decision makers can consider where collaboration and coordination with other regional or global bodies is needed to achieve regional and global objectives.

The launch by the UN General Assembly in 2010 of the Regular Process for global reporting and assessment, including socio-economic aspects (Resolution 65/37) offers a means for strengthening integrated, ecosystems-based regional assessments combined with a regular global overview. The first "baseline" global assessment will be considered by a specialized working group of the General Assembly in September 2015, including lessons learned and what next. That group's recommendations will be taken up later in the year by the full Assembly. Early workshops aim to strengthen regional capabilities and expert networks within a global framework that can advance quality, comparability and inter-regional connections. The process of sorting out institutional roles at the regional level to improve and cohere data collection and produce integrated assessments is only just beginning. It is vital that the Regular Process preserve a reputation for credibility, legitimacy and policy-relevant, non-prescriptive assessments (UNEP and IOC-UNESCO 2009).

Recent attempts to understand and analyze the fragmented oceans institutional complex, at national, regional and global levels, take two approaches. One simply uses a quantitative assessment of intersecting institutional responsibilities to provide decision makers with baseline information as they struggle with ecosystem-based management (Ekstrom *et al.* 2009, Fidelman and Ekstrom 2012). A more analytical, interdisciplinary ten-year research initiative looks broadly at governance of human impacts on the Earth's biophysical systems. The research framework of this new program of the International Human Dimensions Programme on Global Environmental Change is designed to address interlinked problems and cross-cutting themes and explore effective and equitable solutions. Its findings may have substantial implications for ocean institutions, inter alia, as "scale" and "knowledge" are two of its cross-cutting themes (Biermann *et al.* 2010).

Pending further analysis, the weak link in oceans institutions remains an "overview" mechanism at the regional level that has taken on board the imperative of ecosystem approaches: to consider the findings of integrated assessments, agree on priorities, and provide agreed guidance on responses to be taken at national levels and through collective action at regional and global levels. Nevertheless, growing emphasis on healthy coastal and marine ecosystems and the goods and services they provide is set to become a major driver of additional regional agreements and cooperation.

There are some signs of regional, ecosystem-based priority-setting and of ecosystem approaches leveraging more coherent, systematic international institutional support—through the GEF program, some of the regional seas arrangements, and large-scale regional projects such as the Coral Triangle and Caribbean Challenge Initiatives. More regional agreements are acting on the need to establish specific, problem-oriented linkages with other regional and global agreements; for example, RFMOs with CMS agreements to avoid adverse impacts of fishing on migratory seabirds; and, in order to coordinate protections for designated critical and sensitive sea areas in ABNJ, the OSPAR Convention with the IMO and ISA and the CCAMLR with the IMO. Efforts to strengthen protected areas networks at the regional level continue also in the Mediterranean and South Pacific (Toropova *et al.* 2010, Rice *et al.* 2011). The Arctic Council, an informal forum, continues to evolve toward a more robust "governance" mechanism in the Arctic marine realm.

Conclusions

The fundamental rationales and benefits of regional initiatives in ocean governance have not changed. Nor have the comparative advantages of regional and global bodies. What has changed with the growing imperative of ecosystem-based approaches is the need to highlight and address the connections between regions and with ABNJ. An overview oceans "forum" within each region could play a major role in cohering initiatives to support ecosystem approaches not only within the region but also in relation to linkages between regions and with ABNJ. While respecting the UNCLOS balance, such bodies could still address extra-regional sources and changes of direct concern to the region.

At the global level, the Regular Process is to provide an overview of integrated, ecosystem-based regional assessments and wider inter-regional and global linkages in 2015. This should provide a basis for considering where institutional coordination is needed between regions and in ABNJ to advance ecosystem-based management. Further institutional analyses may turn up new ideas, but it is important to bear in mind that tools for integrated assessment and planning do not necessarily require integrated implementation; they generally contemplate coordinated implementation across sectoral agencies within the context of integrated goals and objectives. A more effective coordination process at the global level between regional and global institutions is certainly needed.

Progress in identifying and valuing EBSAs, including in open ocean and deep sea areas, may provide focal points to catalyze international support (political, institutional) for ecosystem approaches both within and beyond national jurisdiction. They may function as a basis for pilot conservation initiatives in ABNJ and "straddling" systems, providing a testing ground for focused institutional coordination to protect and conserve valued ecosystem goods and services, and for focused international collaboration in scientific research and the exchange of scientific data and samples of potential value to the international community as a whole. The growing concern to address ecological limits and equity in the context of three-dimensional assessments (environmental, social, economic) only underscores the need for assessment processes considered credible, legitimate and policy relevant.

Notes

1 See Marine Ecosystem and Management bimonthly information service at www.meam.net and Ecosystem-Based Management Tools Network at www.natureserve.org (accessed July 16, 2015).
2 See Sargasso Sea Alliance at www.sargassoalliance.org (accessed July 16, 2015).

References

Biermann, F., Betsill, M. M., Gupta, J., Kanie, N., Lebel, L., Liverman, D., Schroeder, H., Siebenhuner, B., Zondervan, R. (2010) "Earth system governance: a research framework," *International Environmental Agreements*, vol. 10, pp. 277–298.

CoML (2010) *First Census of Marine Life 2010: Highlights of a Decade of Discovery*. Eds. J. H. Ausubel, D. T. Crist, and P. E. Waggoner. 2010 Census of Marine Life, USA.

Duda, A. M. (2009) "GEF support for the global movement toward the improved assessment and management of large marine ecosystems," in Sherman, K., Aquarone, M. C. and Adams, S. (Eds.), *Sustaining the World's Large Marine Ecosystems*, Gland, Switzerland, IUCN, pp. 1–12.

Ekstrom, J. A., Young, O. R., Gaines, S. D., Gordon, M., McCay, B. J. (2009) "A tool to navigate overlaps in fragmented ocean governance," *Marine Policy*, vol. 33, pp. 532–535.

FAO (2003) *The Ecosystem Approach to Fisheries*. Suppl. 2: 112 pp. in FAO Technical Guidelines for Responsible Fisheries, vol. 4, Rome, FAO.

FAO (2008) *International Guidelines for the Management of Deep-sea Fisheries in the High Seas*, Rome, FAO.

Fidelman, P. and Ekstrom, J. A. (2012) "Mapping seascapes of international environmental arrangements in the Coral Triangle," *Marine Policy*, vol. 36, pp. 993–1004.

Final Report of the 32nd Antarctic Treaty Consultative Meeting (ATCM) (2009) available at www.ats.aq (last accessed July 10, 2012).

GESAMP (1990) *The State of the Marine Environment*. UNEP Regional Seas Reports and Studies No. 115. IMO/FAO/Unesco/WMO/WHO/IAEA/UN/UNEP, Joint Group of Experts on the Scientific Aspects of Marine Pollution (GESAMP).

Herr, D. and Galland, G. R. (2008) *The Ocean and Climate Change: Tools and guidelines for action*. Gland, Switzerland, IUCN.

Kimball, L. A. (1996) *Treaty Implementation: Scientific and Technical Advice Enters a New Stage*. Studies in Transnational Legal Policy No. 28, American Society of International Law, Washington, D.C.

———. (1997) "Whither international institutional arrangements to support ocean law?" *Columbia Journal of Transnational Law*, vol. 36, pp. 307–339.

———. (2001) *International Ocean Governance: Using international law and organizations to manage marine resources sustainably*. Gland, Switzerland and Cambridge, UK, IUCN.

Nellemann, C., Corcoran, E., Duarte, C. M., Valdes, L., De Young, C., Fonseca, L., Grimsditch, G. (Eds.) (2009) *Blue Carbon*. A Rapid Response Assessment. United Nations Environment Programme, GRID-Arendal, www.grida.no.

Rice, J., Gjerde, K. M., Ardron, J., Arico, S., Cresswell, I., Escobar, E., Grant, S. Vierros, M. (2011) Policy relevance of biogeographic classification for conservation and management of marine biodiversity beyond national jurisdiction, and the GOODS biogeographic classification. *Ocean & Coastal Management*, vol. 54, pp. 110–122.

Toropova, C., Meliane, I., Laffoley, D., Matthews, E. and Spalding, M. (Eds.) (2010) *Global Ocean Protection: Present status and future possibilities*. Brest, France: Agence des aires marines protégées, Gland, Switzerland, Washington, DC and New York, USA: IUCN, WCPA, Cambridge, UK: UNEP-WCMC, Arlington, USA: TNC, Tokyo, Japan: UNU, New York, USA: WCS.

UN (2012) "The future we want," UN Document A/CONF.216/L.1, June 19, 2012.

UNEP and IOC-UNESCO (2009) An Assessment of Assessments, Findings of the Group of Experts. Start-up Phase of a Regular Process for Global Reporting and Assessment of the State of the Marine Environment, including Socio-economic Aspects. ISBN 978–92–807–2976–4.

UNESCO (2009) Vierros, M., Cresswell, I., Escobar Briones, E., Rice, J., Ardron, J. (Eds.), *Global Open Oceans and Deep Seabed (GOODS) Bioregional Classification*, IOC Technical Series, vol. 84, UNESCO-IOC.

UNU-IHDP and UNEP (2012) *Inclusive Wealth Report 2012. Measuring progress toward sustainability*. Cambridge, Cambridge University Press at www.ihdp.unu.edu/article.iwr (accessed July 10, 2012).

Conventions

1911/1957 Seals—Convention for the Preservation and Protection of Fur Seals (1911), superseded by the Interim Convention on Conservation of North Pacific Fur Seals (1957).

1959 AT—Antarctic Treaty.

1972 CCAS—Convention for the Conservation of Antarctic Seals.

1972 Oslo—Convention for the Prevention of Marine Pollution by Dumping from Ships and Aircraft (superseded by the 1992 OSPAR Convention).

1973/1978 MARPOL—International Convention for the Prevention of Pollution from Ships (1973) and 1978 Protocol.

1974 Paris—Convention for the Prevention of Marine Pollution from Land-Based Sources (superseded by the 1992 OSPAR Convention).

1974 Helsinki—Convention on the Protection of the Marine Environment of the Baltic Sea Area (superseded by 1992 Baltic Sea Convention).

1979 CMS—Convention on Conservation of Migratory Species of Wild Animals.

1980 CCAMLR—Convention on the Conservation of Antarctic Marine Living Resources.

1982 UNCLOS—United Nations Convention on the Law of the Sea.

1983 WCR—Convention for the Protection and Development of the Marine Environment of the Wider Caribbean Region.

1986 SPC—Convention for the Protection and Development of Natural Resources and Environment of the South Pacific Region.

1992 OSPAR—Convention for the Protection of the Marine Environment of the Northeast Atlantic.

1992 CBD—Convention on Biological Diversity.

1995 UNFSA (Fish Stocks Agreement)—Agreement for the implementation of the Provisions of the UNCLOS of 10 December 1982 relating to the Conservation and Management of Straddling Fish Stocks and Highly Migratory Fish Stocks.

Understanding marine environments

6

BLUE PLANET

The role of the oceans in nutrient cycling, maintaining the atmospheric system, and modulating climate change

Susan M. Libes

Introduction

This chapter explores the role of the oceans in nutrient cycling, maintaining the atmospheric system, and modulating climate change. The essential connections between these topics are: (1) nutrients are one of the most important controls on biological productivity in the oceans, (2) this biological activity generates and consumes gases that are exchanged across the air–sea interface and (3) these gases play an important role in the natural greenhouse effect that moderates global climate. Rising levels of these greenhouse gases (GHGs) have the potential to cause the temperature of the atmosphere to rise.

Readers are expected to understand basic concepts of biogeochemical cycling including fluxes, residence and turnover times, positive and negative feedbacks, sources, sinks, and reservoirs. See Chapter 1 in Libes (2009) for an explanation of these terms and processes. Sarmiento and Gruber (2006) provide a more mathematical treatment of these concepts.

The crustal–ocean–atmosphere factory

How humans use and affect ocean resources is a consequence of the interconnected biological, chemical, physical and geological processes that move materials and energy between the crust, ocean and atmosphere. These movements are quantitatively referred to as fluxes. In some cases, fluxes that supply materials are balanced by fluxes that remove materials from a given reservoir. The amount of material in that reservoir remains constant over time as long as the supply and removal fluxes are balanced. The reservoir is said to be in a "steady state." This condition provides stability to what humans experience and is a desirable feature as it enables reliance on that reservoir as a renewable resource. Stability also arises from negative feedbacks among interconnected processes in which a change in the flux from one process is countered by a compensating change in the flux of another process.

Figure 6.1 is a mechanistic representation of these material movements between the crust, ocean, and atmosphere. From an elemental perspective, such as that of carbon, material

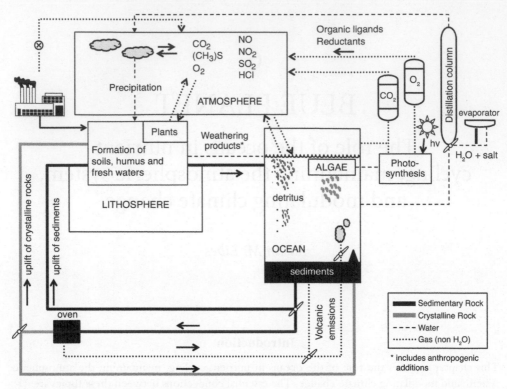

Figure 6.1 The crustal–ocean–atmosphere factory
Source: After Stumm and Morgan (1996), p. 874.

movement among the reservoirs forms a closed circuit when viewed over sufficiently long periods of time. These periods of time are typically much longer than that of a human lifetime and so are harder to appreciate as a renewable feature. Many of the negative feedbacks inherent in the crustal–ocean–atmosphere factory occur on much longer timescales, i.e., in the order of thousands to millions of years.

The Anthropocene

Human impacts to these material movements have reached such a large scale that we have greatly altered many of the material fluxes in the crustal–ocean–atmosphere factory, including ones that control biological productivity and climate stabilization. These changes have been so significant that geologists have named this period of time in which we live the "Anthropocene." Evidence of the changes we have caused in global material fluxes is now being recorded in marine sediments. These will eventually become sedimentary rocks and as part of the geological record, provide geochemical evidence of our impacts. Human impacts to the world ocean are being experienced from poles to equator and surface to bottom. As shown in the global map of human impacts to marine ecosystems of Halpern *et al.* (2008), these impacts are most intense in coastal waters, reflecting the concentration of human populations along coastlines and the role of these populations in mobilizing materials. These materials are, with some important exceptions, transported into the ocean via discharge into coastal waters.

Positive feedbacks and tipping points

In the crustal–ocean–atmosphere factory, not all changes in fluxes are counterbalanced by negative feedbacks. Some changes are amplified as a consequence of interactions between various interconnected material transport processes. These are termed "positive feedbacks" and are undesirable to humans as they destabilize biogeochemical cycling, leading to changes in the sizes of reservoirs that we rely on as natural resources. The amount of change in a flux that must occur before a positive feedback leads to significant destabilization has been termed a "tipping point." Tipping points are viewed as thresholds that once exceeded lead to a rapid destabilization and hence resource impairment. One of the tipping points that we appear to be reaching, or have reached, is associated with nutrient loading into the oceans, particularly for nitrogen (Rockström *et al.* 2009). The negative impacts of this loading are stimulation of algal growth (aka "cultural eutrophication"), followed by eventual decay of the resulting organic matter by aerobic bacteria. The latter process removes O_2 gas leading to O_2-deficient conditions, called hypoxia, which can kill benthic and pelagic animals.

Ocean–atmosphere linkages

The ocean and atmosphere are linked by processes that exchange heat and materials across the air–sea interface. Gases that are exchanged include ones produced and consumed by marine organisms, such as CO_2 and O_2. This exchange is important enough to impact the atmospheric levels of these gases. Other material exchanges include salts that are ejected from the ocean surface as sea spray and the deposition of atmospheric dust, mostly windborne clay minerals, as fallout onto the sea surface. The exchange of gaseous and liquid water via the processes of evaporation and precipitation is an important part of the global hydrological cycle.

Human inputs of gases to the atmosphere have indirect impacts to the oceans. For example, humans have injected gases, called chlorofluorocarbons, into the stratosphere that chemically remove ozone in the stratosphere. This "ozone hole" is most intense over the poles and has an important consequence for ocean biology as it permits more UV radiation to reach the sea surface. The additional UV radiation has the potential to negatively impact phytoplankton in some of the most productive waters of the world ocean. Since phytoplankton form the base of the marine food web, a decline in these plankton would significantly impact global ocean productivity.

Ocean–climate linkages

The fluxes of gases across the air–sea interface are of particular importance to Earth's climate as many of these gases help the atmosphere retain heat. Earth's climate is a consequence of how incoming solar energy is retained and distributed in our planet's atmosphere. This energy is retained as heat that we experience as temperature. Heat is transported within the atmosphere by physical processes including advection by vertical and horizontal winds, radiation, and conduction. Some of the solar energy is first absorbed by the crust and oceans and then radiated back as longwave infrared radiation into the atmosphere. This type of radiation is absorbed by some gases, leading to an enhanced thermal content and higher temperature. This phenomenon is termed the "Greenhouse Effect" and has led to Earth's present moderate climate, which is— at most latitudes—supportive of life. These natural GHGs include water, carbon dioxide (CO_2), ozone (O_3), methane (CH_4), and nitrous oxide (N_2O). An increase in the atmospheric level of these gases can lead to more retention of heat and hence a higher temperature. Humans are causing such increases, with the resulting temperature rise termed the "Greenhouse Gas Problem".

The degree to which the ocean removes or supplies GHGs to the atmosphere varies over space and time. Characterizing these fluxes is a major enterprise of chemical oceanographers as this is key to understanding important climate controls.

Likewise, the role of the oceans in moving and supplying heat to the atmosphere varies over space and time. This physical phenomenon is a consequence of the ability of water to absorb and release heat, which is due to its relatively high heat capacity. For example, surface waters at low latitudes tend to absorb heat. These waters can be transported poleward by surface currents where the heat is eventually released into the atmosphere. Global ocean circulation involving the movement of water from the sea surface to the sea floor also plays an important role in heat transport. Other heat transport processes involve sea ice, which influences global albedo (reflectance), the role of typhoons in episodic heat release, and even the salt content of the oceans as this influences the heat capacity of seawater.

A complete circuit of global ocean circulation takes about a thousand years. Shorter term changes in circulation, such as the El Niño Southern Oscillation, the North Atlantic Oscillation, and the Pacific Decadal Oscillation, can cause changes in regional climates at locations quite far from these ocean phenomena. These effects are termed "teleconnections." They include positive and negative climate feedbacks that appear to operate over timescales of decades.

Not all of the ocean–atmosphere–climate linkages are related to the Greenhouse Effect. For example, evidence suggests that a negative feedback stabilizing atmospheric temperature involves a biogenic gas, dimethylsulfoniopropionate (DMSP), which is created by some marine phytoplankton. These phytoplankton release DMSP into seawater where it breaks down into dimethylsulfide gas (DMS). DMS degasses from the surface waters of the ocean into the atmosphere where it is oxidized into $SO_2(g)$. This gas leads to the production of sulfate aerosols that serve as cloud condensation nuclei. The net effect is increased cloud cover that increases the reflectance of incoming solar radiation, thereby decreasing the flux of solar energy available to be retained by the atmosphere and lowering its temperature. This also reduces the growth rate of the phytoplankton, so their gas production tapers off until cloudy conditions dissipate under cooler atmospheric temperatures. The negative feedback is then reactivated following dissipation of the cloud cover as temperatures climb and the solar radiation necessary for photosynthesis increases. Since certain marine phytoplankton are essential to this feedback, other controls on their growth, namely nutrient availability, are influential. These controls are discussed below, including ones that involve more ocean–atmosphere–climate linkages.

The major biogeochemical cycles

Biological activity in the ocean is a major control on the distribution and transport of carbon and of the macronutrients, nitrogen and phosphorus. This biological control also causes the global elemental cycling of carbon, nitrogen, and phosphorous to be highly interdependent, featuring many inter-elemental feedbacks. Marine organisms influence the global cycling of other elements, but to a lesser degree. The global cycling of carbon is particularly important because of the role of CO_2 as a GHG and hence feedbacks with the nitrogen and phosphorus cycles have the potential to influence climate.

The part of the carbon cycle that humans interact with operates over timescales of millennia. This has been termed the "Fast Carbon Cycle" and is highly linked to the cycling of the macronutrients and to other essential elements, required in smaller quantities, and hence termed "micronutrients." The component of the carbon cycle that acts over timescales of millions of years is termed the "Geological Carbon Cycle." This involves processes that are part of the rock cycle, such as CO_2 uptake during weathering and release from deposition of limestone

(CaCO$_3$). Changes in this uptake flux arise from processes controlled by plate tectonics, such as mountain building and hydrothermal emissions of calcium ion into seawater.

Ocean macro- and micronutrient cycling

Autotrophy

The base of the marine food web is comprised primarily of single-celled photosynthetic algae called phytoplankton. Other primary producers that are of local importance in some marine ecosystems, such as hydrothermal vents, are chemoautotrophic bacteria that obtain energy from reduced inorganic compounds such as hydrogen sulfide, methane, and hydrogen (H$_2$). Although these collectively comprise less than 1 percent of the total marine autotrophic production, their metabolic processes help complete some of the elemental biogeochemical cycles.

Both types of primary producers have elemental requirements for the constituents that comprise their soft tissues and hard parts. The latter include structural entities comprised of either silica (SiO$_2$) or calcium carbonate (CaCO$_3$). The elements that are in scarce supply have the potential to limit productivity of the entire food web by controlling the rates of primary production. These elements are grouped into two categories: macronutrients and micronutrients. The former are required in higher amounts than the latter. The three macronutrients are nitrogen, phosphorus, and silicon. The latter is a macronutrient only for phytoplankton that deposit siliceous hard parts, the most abundant of which are the diatoms. Most silicon is found in molecules that include oxygen that are generically referred to as silica. Nitrogen and phosphorus are macronutrients because they are components of some of the most abundant biomolecules comprising the soft tissues, i.e., the ones that constitute cellular membranes, pigments, energy transport agents (ADP/ATP), and genetic material (DNA/RNA).

Most of the micronutrients are trace metals, such as iron, molybdenum, vanadium, cobalt, manganese, copper, zinc, and selenium. These trace metals are essential components of common enzyme systems. For example, iron plays a role in electron transport processes associated with photosynthesis, respiration, and nitrogen fixation, as well as the reduction of nitrate and nitrite to ammonium. Under some conditions, the growth of autotrophs is limited by more than one nutrient, but typically marine phytoplankton are considered to be limited by the availability of nitrogen. Exceptions to this are some oceanic regions where iron appears to be limiting. These include the subarctic Pacific Ocean, the Southern Ocean, and the equatorial Pacific Ocean, which collectively represent 20 percent of the sea surface area.

Molecular forms of the nutrients

The molecular form of the nutrients is important in determining the degree to which algae and microbes can utilize the elements. Since these organisms do not have mouths, all of their elemental needs must be met by active transport of small molecules through their cell membranes, a process called "nutrient assimilation." In the case of phosphorus, most of this element is present in seawater as the small inorganic molecule phosphate, which is readily assimilated by the algae and microbes. The situation with nitrogen is far more complex, as a variety of small inorganic forms can be present in seawater, i.e., nitrate, nitrite, and ammonium. Ammonium is taken up most rapidly, being energetically rich, since its nitrogen atom is in a redox state that matches the target biomolecules, i.e., proteins, chlorophyll, ADP, ATP, RNA, and DNA. Some dissolved organic nitrogen compounds, such as urea, are sufficiently small to be assimilated by autotrophs. For the micronutrients, several dissolved forms can also be present, with some being more bioavailable than others. The bioavailability of many trace metals is enhanced through complexation with dissolved organic matter as this serves to solubilize these elements.

Heterotrophy

Organic matter created by primary producers is consumed by heterotrophic microbes and animals. The microbes include bacteria, archaea, and protozoans. Some are obligate aerobes or anaerobes. Others are facultative. There are even examples of autotrophs that can engage in heterotrophy—these are called "mixotrophs." Some of the nutrients are incorporated into the biomass of the heterotrophs and the remainder are returned to extracellular soluble forms as excreta, secretions, or exudations.

Following death, the organic matter that constituted biomass is subject to sinking and microbial degradation. The dead biomass is termed detrital particulate organic matter (POM). Viruses play an important role in the degradation process as they cause cell lysis, which is the leading cause of "death" for phytoplankton and bacteria. Cell lysis releases the dissolved organic compounds present in the cellular matrix into seawater. Other sources of dissolved organic matter (DOM) in seawater include the secretions and exudations from marine organisms as well as the heterotrophic breakdown of POM by bacteria and archaea. Microbes continue the degradation process on DOM, such that some portion of the carbon, nitrogen, and phosphorus are transformed back into the small inorganic molecules required by the autotrophs, i.e., CO_2, ammonium, and phosphate. This occurs through a series of steps collectively called "remineralization." The resulting additional microbial biomass is consumed by protozoans, which serves to move some of the carbon and nutrients back into the marine food web.

Microbially-mediated heterotrophic processes occur anywhere that detrital organic matter is present. This detrital organic matter is present everywhere including the euphotic (aka "sunlit") zone where photoautotrophy occurs, the aphotic zone (generally depths greater than 200m in the open ocean), and in marine sediments. Microbial heterotrophy is particularly prolific at interfaces where detrital organic matter collects, including the air–sea interface, the sediment–water interface, on the surface of suspended and sinking POM, and at boundaries between ocean currents.

The nitrogen cycle includes several additional important microbial processes. Nitrogen fixation is conducted by free living and endosymbiotic microbes, some of which are photosynthetic. This process, in which the highly abundant $N_2(g)$, is converted into ammonium, plays an important role in regulating the degree to which nitrogen is the limiting nutrient. Conversely, nitrogen can be converted back into $N_2(g)$ via the anaerobic process of denitrification in which nitrate is reduced using organic matter as the electron donor. The GHG, N_2O, can be generated as a by-product. Ammonium is another by-product and is subject to anaerobic oxidation to N_2 via an autotrophic process called anammox. Ammonium is also a product of oxic heterotrophy. Under oxic conditions, ammonium is oxidized by specialized bacteria to nitrite and then to nitrate. This bewildering variety of processes is summarized in Figure 6.2 and includes another mechanism for nitrate loss called dissimilatory nitrate reduction in which nitrate is reduced to nitrite or ammonium. In some settings, this microbially mediated pathway is more important than denitrification. The take-home message here is that the processes involved in the nitrogen cycle are far more varied than those of the other nutrients, or even carbon, owing to the variety of oxidation states of nitrogen.

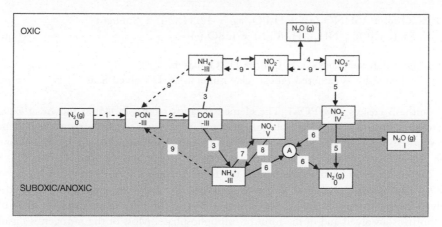

Figure 6.2 A simplified depiction of the marine nitrogen cycle illustrating redox and phase transformations mediated by microbes. The boxes contain the nitrogen species and its oxidation number. The arrows represent transformation reactions as follows: (1) nitrogen fixation, (2) solubilization, (3) ammonification, (4) nitrification, (5) denitrification, (6) anammox, (7) anaerobic nitrification mediated by manganese reduction, (8) dissimilatory nitrate reduction to ammonia (DNRA), (9) assimilatory nitrogen reduction. A = anammox microbes. The dashed lines represent processes in which nitrogen is incorporated into biomass.

Source: Libes (2009), p. 668.

The role of nutrients in the crustal–ocean–atmosphere factory

The "fast" carbon cycle

Humans have greatly increased nutrient fluxes into the ocean and GHG fluxes into the atmosphere. The impacts of these increased fluxes can be predicted by considering the global biogeochemical cycles of carbon, nitrogen, and phosphorus. As noted above, the carbon cycle can be characterized by two sets of biogeochemical processes that serve to cycle this element on timescales of thousands versus millions of years. In this chapter, we consider only the "fast" carbon cycle since it is closely linked with the nitrogen and phosphorus cycles and is more closely matched to human timescales.

Because of the interlinked nature of these cycles, some have proposed a geoengineering approach to solving the GHG problem, i.e., by fertilizing the oceans, algae would proliferate and in doing so, remove anthropogenic CO_2 from the atmosphere. If this could be realized, two pollution problems—nutrient pollution of the coastal oceans and GHG loading into the atmosphere—would seemingly work together to cancel each other out! Unfortunately this is not happening. To understand why, as well as other more promising means to address the two pollution problems, we need to consider in more detail the interconnected nature of the global biogeochemical cycles of the three elements.

The fast carbon cycle moves carbon between the atmosphere, biosphere, ocean, its surface sediments, and soils on land. Although humans have significantly increased the CO_2 content of the atmosphere, it is still the smallest carbon reservoir in the crustal–ocean–atmosphere factory. Most of the carbon resides in the crustal rocks, about 100,000 times more than is in the atmosphere, but does not participate in the fast carbon cycle. Most of the atmospheric carbon is in the form of CO_2 gas with a minor fraction present as gaseous CH_4 and CO. CO_2 is removed from the atmosphere by photosynthesis on land and in the ocean and via the chemical weathering of crustal rocks. The chemical reactions for these two processes are shown below:

$$106CO_2(g) + 16HNO_3(aq) + H_3PO_4(aq) + 122H_2O(l) + \text{Sunlight}$$
$$\rightarrow (CH_2O)_{106}(NH_4)_{16}H_3PO_4(s) + 138O_2(g) \qquad \text{(Eqn 6.1)}$$

$$CO_2(g) + \text{Crustal Rocks} + H_2O(l) \rightarrow HCO_3^-(aq) +$$
$$\text{Chemically Weathered Rocks (clay minerals)} + \text{Dissolved Salts} \qquad \text{(Eqn 6.2)}$$

Photosynthesis creates new POM. The elemental ratios of this POM, as shown in Equation 6.1, are representative of marine plankton collected in net tows. An important consequence of photosynthesis has been the provision of O_2 gas that degasses into the atmosphere. About half of the global O_2 production is currently supported by marine photosynthesis.

In the ocean, most of the carbon is in the form of inorganic molecules, i.e., bicarbonate (HCO_3^-), carbonate (CO_3^{-2}), and carbon dioxide (CO_2). These are collectively referred to as dissolved inorganic carbon (DIC). The vast majority of the DIC is bicarbonate. Most of the phosphorus is dissolved and in the form of orthophosphate. Most of the nitrogen is dissolved, with the organic form being nearly as abundant as the inorganic forms. The most abundant form of dissolved inorganic nitrogen is nitrate.

The biological pumps

The evolution of life on this planet had a profound impact on segregating carbon into the major modern-day reservoirs. Prior to human influences, the atmospheric and oceanic reservoirs are thought to have been in steady state over timescales of millennia. This condition was largely controlled by biotic processes in which uptake via photosynthesis was approximately balanced by respiration with a turnover time of carbon in the atmosphere of about 4.5 years.

The rates of photosynthesis on land and in the ocean are approximately equal but the effect of these processes on the global carbon cycle is quite different because the fate of the resulting organic carbon is quite different. On land, detrital POM tends to undergo oxidation via aerobic microbial respiration that occurs in the soils. This rapidly returns the CO_2 to the atmosphere. In the ocean, the detrital POM has a far better chance of escaping degradation by sinking into the deep sea with a minor fraction settling onto the seafloor and getting buried. Even if this POM undergoes remineralization, the carbon is trapped in the bottom waters until ocean circulation returns it to the sea surface. Hence the sinking process acts to store carbon, keeping CO_2 out of the atmosphere for a much longer period of time than is achieved on land. This sinking process is referred to as the "soft-tissue biological pump."

The soft-tissue pump operates as follows: primary production in the euphotic zone leads to the production of new biomass. This new POM has three fates: (1) it can settle into the deep sea, (2) it can be recycled in the surface waters by microbial heterotrophy, and (3) it can be consumed by protists or animals. In the case of the latter, death of the consumers and their excretory products both generate detrital POM that will either be recycled in the euphotic zone or sink into the deep sea. In the deep sea, the detrital POM is remineralized by microbes, thereby regenerating dissolved carbon and nutrients. A small fraction survives the trip to become permanently buried in the sediments.

The soft-tissue biological pump is augmented by a "hard-part" pump fueled by marine algae and zooplankton that create structural components comprised of calcium carbonate. The hard-part pump works the same way as the soft-tissue pump, except that the sinking hard parts are subject to dissolution as they settle into the deep sea. This dissolution is enhanced by the higher CO_2 content of the deep waters through the following chemical reaction.

$$CO_2(g) + CaCO_3(s) + H_2O(l) \rightarrow 2HCO_3^-(aq) + Ca^{2+}(aq) \qquad \text{(Eqn 6.3)}$$

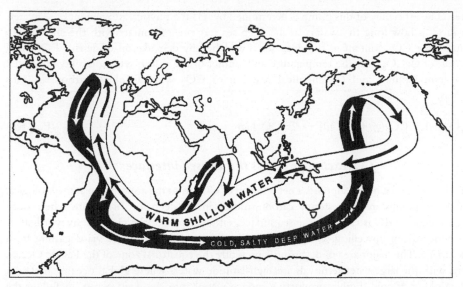

Figure 6.3 Simplified model of global ocean circulation emphasizing circuit of deep and surface water flows. Deep waters are created from sinking of cooled salty water at high latitudes. Water is returned to the surface by wind-driven upwelling and turbulent mixing.

Source: Shimokawa and Ozawa (2011). After Broecker (1987).

The recycling of organic matter in the surface and bottom waters is very efficient such that somewhat less than 3 percent of the primary production (as carbon) survives to be buried on the seafloor. Most of the burial is occurring on the continental shelves as their water depths are shallow and hence provide short sinking times for detrital POM to reach the seafloor. Because of this low percentage of burial, the residence time of carbon in the ocean, just prior to the Anthropocene, was on the order of 30,000 to 40,000 years.

The net effect of the biological pumps is to relocate carbon and the nutrients into the deep ocean. Global ocean circulation moves water from the surface to the deep sea and back to the surface on a circuit that passes through the entire ocean over a timescale of about a thousand years. This circuit is shown in Figure 6.3. It is commonly referred to as the "Global Conveyor Belt."

Thus carbon and nutrient enriched deep waters are eventually returned to the sea surface where both are re-utilized by algae. Any excess carbon will degas into the atmosphere. So on timescales greater than 1,000 years, a more permanent removal of carbon is needed to counter the GHG problem. This is achieved if the sinking soft and hard parts can survive to become permanently buried in the sediments. This happens in shallow waters, i.e., on the continental shelves. Benthic animals that deposit calcareous hard parts, such as coral reefs, also contribute to a more permanent removal of carbon, but overall do not represent a significant carbon sink. The efficiency of the biological pumps is influenced by many factors, such as the sinking rate of the particles, the length of the water column, and how corrosive the seawater is towards the calcareous hard parts. The latter effect is described below.

The solubility or gas-exchange pump

The ocean has one more pump that enhances its ability to store carbon. This is called the "solubility or gas exchange pump." It acts to remove atmospheric CO_2 by dissolution into seawater. The efficiency of this pump is determined by: (1) the physical movement of seawater, which dictates how long the waters in the deep sea stay out of contact with the atmosphere (remember the CO_2 content of the deep sea is enriched by degradation of detrital POM and dissolution of $CaCO_3$); (2) the temperature and salinity of the water, with solubility enhanced by low temperatures; and (3) chemical reaction of CO_2 with dissolved carbonate. This reaction is:

$$CO_3^-(aq) + CO_2(g) + H_2O(l) \rightarrow 2HCO_3^-(aq) \qquad \text{(Eqn 6.4)}$$

CO_2 exchange across the air–sea interface

In many locations, the ocean supports a net removal of CO_2 from the atmosphere as the biological and solubility pumps are sequestering carbon, mostly in the deep sea. In regions where deep water enriched in DIC is rising to the sea surface, either as part of the global conveyor belt or due to wind-driven upwelling, CO_2 is released back into the atmosphere via degassing from the sea surface. The major area of degassing is presently the equatorial zone of the Pacific Ocean. Regions with the largest net removals are high-latitude waters where global ocean circulation leads to sinking of cold, highly productive surface waters into the deep ocean, including the North Atlantic, North Pacific, and Southern Oceans. These fluxes can vary seasonally and over long timescales. Important examples of the latter are associated with the coming and going of Ice Ages with atmospheric CO_2 levels being lower during the cold periods.

About half of the CO_2 released by humans has remained in the atmosphere. (This includes carbon emissions arising from land-use changes such as deforestation.) About 30 percent has been taken up by the ocean and 20 percent has presumably been removed by the terrestrial biosphere. The ability of the ocean to continue to take up excess CO_2 will likely change over time as its buffer capacity decreases due to the titration of carbonate by the reaction in Equation 6.4. When carbonate ion concentrations are low, seawater becomes acidic as $H^+(aq)$ accumulates from the reaction in Equation 6.5.

$$CO_2(g) + H_2O(l) \rightarrow H_2CO_3(aq) \rightarrow HCO_3^-(aq) + H^+(aq) \qquad \text{(Eqn 6.5)}$$

This impact has already been detected as the pH of seawater has declined by 0.1 pH unit since preindustrial times. This has made seawater increasingly corrosive to calcareous hard parts, which has negative impacts on the survival of an important part of the phytoplankton and the function of the biological pumps.

The interconnected marine carbon, nitrogen, and phosphorus cycles

In the same way that carbon is pumped into the deep sea, leading to higher DIC concentrations in the bottom waters, so are the nutrients reflecting their concurrent remineralization during microbial respiration of organic matter. As the deep waters move around their global circuit, remineralized carbon and nutrients accumulate from the continued decay of POM raining out of the surface waters. So locations where these waters rise to the surface are enriched in both CO_2 and nutrients. Elsewhere, surface waters have very low concentrations of nutrients due to their rapid assimilation by phytoplankton.

The interconnected nature of the C, N, and P cycles is illustrated in Figure 6.4 which also highlights interactions with the global cycling of O_2. In the case of O_2, the euphotic zone is a

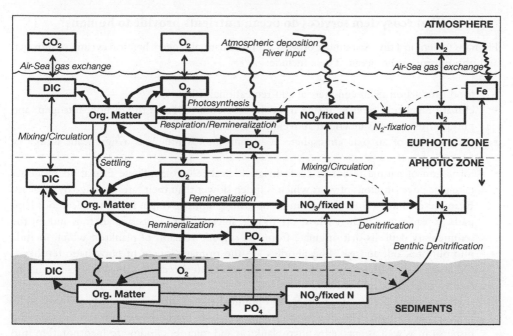

Figure 6.4 Schematic representation of the coupled marine carbon, nitrogen, phosphorus, and oxygen cycles. The solid lines represent biogeochemical processes and fluxes. The dashed lines point from a controlling property to a particular process where that control is highly important.
Source: Gruber (2004).

net source to the atmosphere. Remineralization of sinking detrital POM via microbially mediated aerobic respiration creates an oxygen minimum zone (OMZ) at mid depths in the open ocean. The existence of various feedback loops between C, N, and P has been postulated as well as the potential for nutrient availability to be a master variable in global climate control.

Feedbacks that stabilize the ocean nitrogen cycle appear to rely on linkages between N_2 fixation and denitrification. Since N_2 fixation relies on iron availability, processes that control iron fluxes into the ocean and its molecular form are likely to be important. N_2 fixation is notable because it represents an adjustable input of reactive nitrogen to the ocean that is under some degree of biotic control. On the other hand, iron fluxes appear to be controlled by climate. For example, increased fluxes from airborne dust occur during periods of drought.

Prior to the Anthropocene, the nitrogen cycle appears to have been in a steady state with a residence time for reactive nitrogen in the order of a few thousand years. In contrast, the marine cycle of phosphorus is largely controlled by riverine inputs and burial in the sediments with a residence time on the order of 10,000 to 20,000 years.

The marine cycling of carbon, nitrogen, and phosphorus is influenced by physical processes such as global ocean circulation and river runoff whose rates are influenced by feedbacks associated with climate change. Another influence on these cycles is exerted by species diversity in the phytoplankton. Not all phytoplankton deposit calcareous hard parts, some species sink more rapidly than others, and some are nitrogen fixers or harbor nitrogen-fixing endosymbiotic bacteria. Thus changes in species composition of phytoplankton communities have the potential to affect carbon and nutrient fluxes into the deep sea and burial of these elements in the sediments.

What ecosystem services do ocean nutrients provide to humans?

Humans benefit in a diverse number of ways from the natural biogeochemical cycling of nitrogen and phosphorus in the ocean. These include:

- food production from commercial and recreational fisheries;
- climate regulation services arising from the interconnection of the carbon, nitrogen, and phosphorus cycles as described above;
- maintenance of an oxic atmosphere as marine photosynthesis is a significant source of O_2 gas;
- formation of natural resources based in geological reservoirs. These formation processes take place over millions of years which is much slower than their rates of usage by humans. Examples include: (1) chalk and limestone whose ultimate origin is in the calcareous hard parts of plankton; (2) chert whose origin is in siliceous hard parts of plankton; and (3) the sedimentary deposits that originate from the organic remains of plankton which include phosphorites, oil, and gas. Phosphorites are the primary source of phosphatic fertilizers. Oil and gas are components of fossil fuel resources and also chemical feedstocks for manufacture of plastics, rubbers, etc.;
- waste treatment for sewage, septage, and polluted runoff that would otherwise have to be disposed of on land. The treatment processes include physical ones, such as dilution and dispersion, as well as uptake by phytoplankton and microbes leading to incorporation into marine food webs;
- support of marine organisms that synthesize unique organic molecules currently used as, or have the potential for use as, pharmaceuticals, food additives, nutritional supplements, cosmetics, and novel engineering materials;
- support of tourism and aesthetic enjoyment through the provision of healthy coastal systems and recreational fisheries;
- support of coastal marshes that provide: (1) storm protection; (2) pollutant removal; (3) seafood production; and (4) a carbon sink for GHG emissions.

Anthropogenic alteration to nutrient fluxes

Despite the many values that ocean nutrient cycling provides to humans, the Anthropocene has been characterized by significant alterations in several of the key fluxes of nitrogen, phosphorus, and iron.

Reactive nitrogen

Reactive nitrogen includes all forms of nitrogen except $N_2(g)$. The latter is the most abundant form of nitrogen in the crustal–ocean–atmosphere factory, but is biologically available only to N_2-fixing organisms. Anthropogenic sources of reactive nitrogen eventually make their way from the land into the ocean via stormwater runoff, submarine discharge, and atmospheric transport. Most of the anthropogenic mobilization of reactive nitrogen has resulted from the widespread adoption of the Haber-Bosch process, in which a high-temperature combustion reaction converts N_2 and H_2 into NH_3. This industrial version of N_2 fixation was discovered in 1913 and now supplies the nitrogenous fertilizers used on crops worldwide. It has been key to increasing food production and thus has supported human population growth. Humans have also increased biotic N_2 fixation through the planting of N_2-fixing crops and burning of fossil fuels. By 2005, the anthropogenic rate of N_2 fixation had risen (and the natural rate declined due to land-use change), such that human sources were close to 80 percent of the natural global rate (Galloway *et al.* 2008). By 2050, the anthropogenic rate of N_2 fixation is expected to rise to a level that exceeds the natural rate. Other anthropogenic processes that have mobilized reactive nitrogen are land-use change including deforestation and the burning of fossil fuel, wood, and grasslands. Combustion processes, particularly those associated with fossil fuels generate nitrogen oxide gases (NO_x).

Phosphorus

Increased fluxes of phosphate that eventually make their way from land into the ocean occur similarly to nitrogen except for the absence of gas phase transport. As with nitrogen, most of the increased fluxes are associated with the use of fertilizers. Phosphate is obtained by the mining of phosphate rocks, called phosphorite, or from bird guano deposits. The latter were found on remote oceanic islands that hosted seabird rookeries and had been largely depleted by 1900 AD. Phosphorites are sedimentary rocks formed under highly productive waters supported by wind-driven coastal upwelling that brought nutrient-enriched waters into the euphotic zone. The ensuing flux of detrital POM was large enough to enhance its own burial due to fast sedimentation rates. Over millions of years of overpressure, the sediments were physically and chemically consolidated into rock. Humans have mined phosphorite deposits that, due to geological uplift, are now accessible from the land surface. These deposits are limited in geographic scope and have been heavily mined. In countries that have not had, or no longer have, these mineral deposits and in those with concerns about nutrient loading of the oceans, efforts have turned to recycling of waste phosphate.

Iron

One of the most dramatic impacts of land-use change has been the mobilization of soils from the land. Humans became the prime agent of erosion sometime during the latter part of the first millennium AD. Humans are now an order of magnitude more important at moving sediment than the natural processes operating on the surface of the planet (Wilkinson 2005). The resulting fluxes of sediment, via river transport into the ocean, have the potential to increase the movement of iron into the ocean, which is a key micronutrient. As noted above, periods of drought can also lead to enhanced dust fluxes, especially from areas that have experienced land-use change.

Transport pathways of nutrients into the ocean

Nutrients mobilized by humans make their way into the ocean by several routes. River runoff is a major conduit. It receives nutrient inputs via stormwater runoff of excess nutrients applied to agricultural lands and from waste streams of livestock and humans. In many countries, these wastes are discharged into rivers with little or no treatment and hence include both dissolved nitrogen and phosphorus.

Various types of land-use change, including deforestation and draining of wetlands, has led to the mobilization of nutrients that had been naturally stored in these environments. In some cases, the rates of mobilization have been enhanced by the addition of acid supplied by acid rain. This acid has its origin in the excess NO_x, SO_x, and CO_2 gases that are injected into the atmosphere by anthropogenic burning of fossil and wood fuels.

Terrestrial waters that percolate into soils have the potential to carry mobilized nitrogen and phosphorus into the groundwaters. These subsurface waters can discharge into the coastal ocean as seepage called "submarine ground discharge," particularly along coastlines where limestone rock is common.

Atmospheric deposition is most important for nitrogen and perhaps iron. In the case of nitrogen, NO_x from burning activities and NH_3 volatilized from manure are transported in the gas phase. Both can dissolve across the air–sea interface. By the mid 1990s, atmospheric deposition constituted 44 percent of the terrestrial input to the ocean and is projected to supply almost 52 percent by 2050, with the impacts concentrated near the source, i.e., in coastal waters. Thus the major impacts from the anthropogenic mobilization of nitrogen and phosphorus are occurring in the coastal oceans.

Ship emissions are also recognized as a significant source of nutrients arising from discharge of wastes directly into the water and from atmospheric release of combustion products generated from burning of fossil fuel or wastes (aka "incineration at sea").

Technically, fishing represents a net removal of nutrients from the ocean. In regions where commercial stocks have been overfished, this can represent a significant local loss of nutrients.

Impacts to the crustal–ocean–atmosphere factory from altered nutrient fluxes

The following impacts have been observed or are hypothesized as a consequence of anthropogenic alteration to nutrient fluxes.

Impacts to the ocean

Rising CO_2 levels in the atmosphere have increased the uptake of CO_2 by the oceans. Although the carbonate–bicarbonate equilibrium reactions in seawater provide buffering against changes in pH, the rate and amount of acid addition has been large enough to cause a 0.1 unit decrease in global average pH since preindustrial times. As shown in Equation 6.3, this addition of acid can lead to the dissolution of calcareous hard parts even in living plankton. This consumption of buffering capacity has the potential to reduce the ability of the ocean to take up anthropogenic CO_2 emissions due to reductions in the functions of the biological and solubility pumps.

The impacts of nutrient pollution into the coastal ocean are recognized nationally and internationally as one of the most serious threats to ocean and human health. The primary impact is through cultural eutrophication in which organic matter production is increased and the species composition of algae is changed. As illustrated in Figure 6.5, oxygen-deficient conditions develop in the sub-surface waters as sinking detrital POM is respired by aerobic microbes. The frequency and duration of these low-oxygen events tends to increase over time as the impacts of nutrient loading can be cumulative if POM created by cultural eutrophication accumulates in the underlying sediments.

As nutrient loading into the coastal ocean has increased, so have the number of locations where oxygen deficient sub-surface waters occur for at least part of the year. The current coastal locations where these effects have been observed are distributed worldwide as mapped by Diaz and Selman (2010). The number of coastal sites where oxygen deficiency has been so low as to be classified as hypoxic (< 2 mg/L of O_2) has approximately doubled each decade since the 1950s and currently numbers around 500 (Diaz and Rosenberg 2008).

Nutrient loading also appears to be altering the species composition of marine phytoplankton, favoring organisms that exert harmful effects (Heisler *et al.* 2008). These phytoplankton

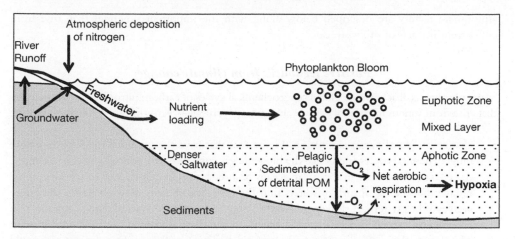

Figure 6.5 Conceptual diagram showing the linkage between external nutrient loading, nutrient enhanced algal bloom formation, and hypoxia in a vertically stratified water column
Source: Author.

proliferate in great numbers in response to the nutrient abundance. This condition is termed a "Harmful Algal Bloom" or HAB, in which the harmful effects can include chemical poisoning from toxins biosynthesized by these plankton or mechanical abrasion from hard parts that can cause fatal haemorrhaging in the gills of fish. The net effects are impairments to fishing and human health threats from ingestion of contaminated fish as well as dermal and respiratory contact to the toxins. The high concentrations of POM can decrease light availability, leading to a decline in submerged aquatic vegetation, such as sea grasses, thereby resulting in loss of habitat. Some species secrete copious amounts of mucilage that fouls fishing nets and beaches with a malodorous foam. All of these effects negatively impact commercial and recreational fishing, aesthetic values, and tourism.

Impacts on the atmosphere

Atmospheric impacts resulting from alterations to the marine biogeochemical cycles of carbon, nitrogen, and phosphorus involve changes to the fluxes of the biogenic gases that have already been discussed, namely CO_2, O_2, NH_3, and NO_x. Other gases affected include two potent greenhouse gases, CH_4 and N_2O. Both are produced in the ocean by microbes respiring organic matter under anaerobic conditions such as those present in sediments, coastal hypoxic zones, and OMZs. Thus production rates are likely to increase if oxygen levels decline. The latter appears to be occurring not just in coastal areas impacted by cultural eutrophication, but in the OMZ in the central North Pacific Ocean and tropical oceans worldwide, particularly on continental margins. The causes of this deoxygenation are not well understood nor are the time trends since long-term datasets are limited. Some of the likely causes are described in the next section.

Synergies

Many of the impacts of nutrient pollution have the potential to exert synergistic effects in response to global warming and sea level rise resulting from climate change. Some that represent positive feedbacks and hence have the potential to enhance negative impacts on the crustal–ocean–

atmosphere factory are described below along with the important synergies that arise from concurrent impacts of other stressors.

Positive feedbacks from climate change

Some positive feedbacks to the global biogeochemical cycles of carbon, nitrogen, and phosphorus that arise from various impacts of climate change to the crustal–ocean–atmosphere factory include:

- changes in global ocean circulation (Figure 6.3) that have the potential to affect the major biogeochemical cycles and hence the biogenic gas content of the atmosphere. For example, the effects of warming in the polar oceans or a reduction in salinity due to ice melt could reduce the sinking of surface waters into the deep sea, thereby slowing down the entire conveyor belt system. A slowdown has the potential to increase vertical density stratification enabling an increased build-up of nutrients in the deep water via the biological pumps. This leaves the surface waters with lowered levels of nutrients. Deoxygenation would intensify due to the relative stagnancy of the water and decreased solubility of O_2 at higher temperatures. Other circulation changes include impacts from increased storminess as this enhances the mixing in the surface waters, thereby decreasing its vertical density stratification. This would in turn negatively impact the phytoplankton by increasing the amount of time spent below the euphotic zone;
- more rapid rises in atmospheric temperatures due to accelerated emissions of GHGs from the ocean. This includes production of N_2O and CH_4 from newly deoxygenated waters and the release of CH_4 from the melting of methane clathrates in marine sediments and permafrost;
- increased weathering, predicted as a consequence of: (1) higher atmospheric CO_2 levels (as per Equation 6.2), (2) higher temperatures, and (3) and acceleration in the hydrological cycle. The combined effects would increase riverine fluxes of salts, alkalinity (carbonate and bicarbonate), and nutrients into the ocean;
- changes in phytoplankton species composition and overall production arising from short- to medium-term fluctuations in climate events, such as the El Niño Southern Oscillation, the North Atlantic Oscillation, and the Pacific Decadal Oscillation;
- changes in phytoplankton species over longer timescales associated with rising temperatures. Such a decline in global marine primary production is reportedly occurring at a rate of about 1 percent per year (Boyce *et al.* 2010);
- synergistic impacts of higher temperature on survival of tropical coral reefs already stressed by ocean acidification, increased particle loads, and spread of disease. Similar concerns arising from ocean acidification and warming exist for plankton that deposit calcareous hard parts and mollusks;
- loss of coastal marshes, one of the most productive habitats on Earth, due to the combined effects of rapid sea level rise, nutrient pollution, and changes in extreme storm surges;
- changes in the ranges of species that are functionally invasive, including pathogens.

Other stressors and the "rise of the slime"

The concurrent impacts of ocean acidification, deoxygenation, and warming are expected to have profound effects on the global marine cycles of carbon, nitrogen, and phosphorus by affecting the biogeochemical function of key organisms in the marine food web (Bijma *et al.* 2013; Doney 2010; Gruber 2011; Noone *et al.* 2012; Norris *et al.* 2013). This has the potential to reduce biological productivity and diversity and to alter animal behavior, leading to declines in fisheries. Other ocean-scale stressors that exert synergistic effects on marine life include the widespread dispersal of plastics and in polar waters, increasing levels of semi-volatile toxicants, such as mercury and DDT, which are migrating polewards via a fractional distillation process termed the "Grasshopper Effect."

Other ocean stressors are exerting their impacts on a local scale, particularly in coastal waters. These include coastal hypoxia, nutrient pollution, coastal habitat loss, overharvesting, especially of keystone species such as filter-feeding oysters, and the introduction of invasive species. The latter is partly a consequence of warming waters and changing circulation patterns. Still other local stressors include the spread of disease, genetic pollution from aquaculture operations, and hydrologic manipulations that have led to decreasing riverine fluxes of silica and increasing inputs of iron to coastal waters. The latter two have collectively diminished the populations of diatoms and favored an increasing dominance of cyanobacteria that are HAB species. Other species changes that have emerged include: (1) bioinvasions of macroalgae and microbial mats that have supplanted seagrass and kelp beds; (2) coral reef die offs; (3) infestations of jellyfish; and (4) blooms of mucilaginous phytoplankton. Overall coastal zones have seen a decline in plankton grazing and commercial fish production that has been described as a type of de-evolution of formerly complex marine food webs. This process has been termed the "Rise of the Slime."

Great unknowns and uncertainties

Developing management strategies for avoiding and reducing the negative impacts of nutrient pollution, including its synergies with climate change, requires a more detailed quantification of anthropogenic fluxes into the ocean and a better understanding of the following marine biogeochemical processes:

- for the nitrogen cycle: (1) determining what microbes are responsible for biotic N_2 fixation in the ocean, where they are living, and better quantifying their rates; (2) understanding the feedbacks between N_2 fixation and denitrification; (3) understanding the biotic controls on production of N_2O; and (4) quantifying the rates of anammox and DNRA;
- for the phosphorus cycle: determining the dynamics of the recent sedimentary reservoirs;
- for the micronutrients: (1) better defining the degree of limitation these metals exert on marine production and (2) the molecular forms that control their bioavailability;
- for the carbon cycle: (1) better identifying and quantifying sedimentary sinks, especially in coastal zones and esturies, (2) better quantifying the roles of various species of marine plankton in the biological pumps, and (3) determining the impact of ocean acidification on the function of the biological pumps.

To enable projections into the future of what ecosystem and interconnected climate impacts we can expect, more accurate biogeochemical models are needed. Specifically, existing models need a better mathematical description for the marine food webs including interactions between viruses, phytoplankton, bacteria, archaea, and their primary consumers, protists and microzooplankton. These interactions need to reflect responses to stressors that can cause species-specific shifts in growth rates, habitat ranges, and loss of photosymbionts. For bacteria and archaea, we need a better modelling of their rates of organic matter degradation of nutrient remineralization. Another input pathway that is not well quantified in these models is submarine ground discharge of nutrients. Finally, to test the hypothesis that tipping points exist, the models need to reflect nonlinear interactions that can give rise to unstable conditions.

Management implications

A particular controversy in managing the contributions of humans to global climate change and nutrient pollution is deciding at what stage intervention becomes cost effective. This seems critical given the interconnected nature of the global biogeochemical cycles of carbon and nutrient elements, their nonlinear behaviors and potential for numerous positive feedbacks as all of these suggest the existence of tipping points beyond which we risk inducing a mass extinction event. Scientific understanding is so limited that a precautionary approach is likely to be the only practical means of ensuring that the most negative environmental consequences are avoided. Part of the limited scientific understanding is due to insufficient observational measurements over space and time. These are costly and the recent global economic slowdown that began towards the end of the 2000s has further reduced resources dedicated to collecting observational measurements.

The other fundamental consideration is that multiple stressors are acting synergistically, but over different time and space scales. Some of these stressors are locally manageable; others must include management across national boundaries. Commercial fisheries are an example of the latter.

An alternative view is to consider that catastrophic mass extinction events have occurred numerous times in Earth's past as a result of geologic phenomena, such as volcanic eruptions that gave rise to the flood basalts that comprise large igneous provinces (LIPs), and from extra-terrestrial events, such as meteorite impacts. Regardless of anthropogenic causes, mass extinction events are inevitably going to happen in Earth's future. While we might be able to reduce our impacts or geoengineer the planet to maintain it in a form that is most conducive for human sustainability, beyond some point this will not be possible.

References

Bijma, J., Pörtner, H. O., Yesson, C., and Rogers, A. D. 2013. Climate change and the oceans: what does the future hold? *Mar Pollution Bulletin*, 74(2), pp. 495–505.

Boyce, D., Lewis, M., and Worm, B. 2010. Global phytoplankton decline over the past century. *Nature*, 466, pp.591–596.

Broecker, W. S. 1987. The Largest Chill. *Natural History*, 96 (10), pp. 74–82.

Diaz, R. J. and Rosenberg, R., 2008. Spreading dead zones and consequences for marine ecosystems. *Science*, 321, pp. 926–928.

Diaz, R. J. and Selman, M. 2010. *World hypoxic and eutrophic coastal areas.* World Resources Institute. www.wri.org/sites/default/files/world_21dec2010.jpg (accessed July 1, 2015).

Doney, S. C. 2010. The growing human footprint on coastal and open-ocean biogeochemistry. *Science*, 328, pp.1512–1516.

Galloway, J. N., Townsend, A. R., Erisman, J. W., Bekunda, M., Cai, Z., Freney, J. R., Martinelli, L. A., Seitzinger, S. P., and Sutton, M. A. 2008. Transformation of the nitrogen cycle: recent trends, questions and potential solutions. *Science*, 320, pp. 889–892.

Gruber, N. 2004. The dynamics of the marine nitrogen cycle and its influence on atmospheric CO_2 variations. In: *The Ocean Carbon Cycle and Climate*, M. Follows and T. Oguz (Eds.), pp. 97–148, Dordrecht, Kluwer Academic Publishers.

Gruber, N. 2011. Warming up, turning sour, losing breath: ocean biogeochemistry under global change, *Philosophical Transactions of the Royal Society A: Mathematical, Physical and Engineering Sciences*, 369, pp. 1980–1996.

Halpern, B. S., Walbridge, S., Selkoe, K. A., Kappel, C. V., Micheli, F., D'Agrosa, C., Bruno, J. F., Casey, K. S., Ebert, C., Fox, H. E., Fujita, R., Heinemann, D., Lenihan, H. S., Madin, E.M.P., Perry, M. T., Selig, E. R., Spalding, M., Steneck, R., and Watson, R. 2008. A global map of human impact on marine ecosystems. *Science*, 319, 948–952. www.nceas.ucsb.edu/globalmarine (accessed July 1, 2015).

Heisler, J., Glibert, P., Burkholder, J., Anderson, D., Cochlan, W., Dennison, W., Dortch, Q., Gobler, C., Heil, C., Humphries, E., Lewitus, A., Magnien, R., Marshall, H., Sellner, K., Stockwell, D., Stoecker, D., and Suddleson,M. 2008. Eutrophication and harmful algal blooms: A scientific consensus. *Harmful Algae*, 8, pp. 3–13. www.whoi.edu/fileserver.do?id=47045&pt=2&p=28251(accessed July 1, 2015).

Libes, S. 2009. *An Introduction to Marine Biogeochemistry*, 2nd ed. Burlington, MA: Academic Press.

Noone, K., Sumaila, R., and Diaz, R. J. (eds) 2012. *Valuing the Oceans*. Stockholm Environmental Institute, www.sei-international.org/publications?pid=2064 (accessed July 1, 2015).

Norris, R., Kirtland, D., Turner, S., Hull, P. M., and Ridgwell, A. 2013. Marine ecosystem responses to Cenozoic global change. *Science*, 341, pp. 492–498.

Rockström, J., Steffen, W., Noone, K., Persson, Å., Chapin, III, F. S., Lambin, E. F., Lenton, T. M., Scheffer, M., Folke, C., Schellnhuber, H. J., Nykvist, B., de Wit, C. A., Hughes, T., van der Leeuw, S., Rodhe, H., Sörlin, S., Snyder, P. K., Costanza, R., Svedin, U., Falkenmark, M., Karlberg, L., Corell, R. W., Fabry, V. J., Hansen, J., Walker, B., Liverman, D., Richardson, K., Crutzen, P. and Foley, J. A. 2009. A safe operating space for humanity. *Nature*, 461, pp. 472–475.

Sarmiento, J. L. and Gruber, N. 2006. *Ocean Biogeochemical Dynamics*. Princeton, NJ, Woodstock: Princeton University Press.

Shimokawa, S. and Ozawa, H. 2011. Thermodynamics of the oceanic general circulation: is the abyssal circulation a heat engine or a mechanical pump?, *Thermodynamics: Interaction Studies—Solids, Liquids and Gases*, Dr Juan Carlos Moreno Pirajan (Ed.), www.intechopen.com/books/thermodynamics-interaction-studies-solids-liquids-and-gases/thermodynamics-of-the-oceanic-general-circulation-is-the-abyssal-circulation-a-heat-engine-or-a-mech (accessed July 1, 2015).

Stumm, W. and Morgan, J. J. 1996. *Aquatic Chemistry*, 3rd ed. Wiley-Interscience.

Wilkinson, B. H. 2005. Humans as geologic agents: a deep-time perspective. *Geology*, 33(3), pp. 161–164.

7

OCEAN HEALTH

Fiorenza Micheli, Giulio De Leo, Francesco Ferretti,
A. Margaret Hines, Kristen Honey, Kristy Kroeker,
Rebecca G. Martone, Douglas J. McCauley, Jennifer K.
O'Leary, Daniele F. Rosim, Susanne H. Sokolow,
Andy Stock and Chelsea L. Wood

Introduction

Over the past decade, the benefits that healthy oceans provide have increasingly become the focus of science, management, and policy making (e.g., Halpern *et al.* 2012, Samhouri *et al.* 2013). Productive oceans enhance food security (Garcia and Rosenberg 2010), and marine habitats protect millions of people from floods, hurricanes, and typhoons (Barbier *et al.* 2014). At the same time, there is increasing recognition that human activities and uses of the oceans, including fishing, coastal development, and pollution, have profoundly altered many marine ecosystems and are eroding their function and ability to provide benefits (Jackson *et al.* 2001, Worm *et al.* 2006). These impacts are compounded by current and projected impacts of climate change (e.g., Doney 2010).

Given dependence on and substantial alteration of ocean ecosystems, rigorous assessments of ocean health (OH) are critically needed. Assessments of OH are needed to measure the state of ocean ecosystems, or of marine social–ecological systems, using some limited suite of indicators, in the same way that a doctor might assess a patient's health status by measuring body temperature or blood pressure. The indicators selected to reflect OH should convey how the ocean as a system is operating, considering multiple interacting forces and pressures. Because OH assessments may provide a holistic view of status, the concept can be used in conjunction with ecosystem-based management to manage a range of human activities and ultimately promote productive and resilient oceans that provide benefits to society. The OH concept is also important as a driver of attitude and policy change: focusing attention on the current state—or health—of the oceans will highlight areas that need improvement and convey the consequences of inaction.

Despite the clear need for comprehensive assessments of OH, developing scientific frameworks for conducting such assessments has been challenging. A first challenge lies in the definition of OH. Environmental health is a normative concept that implies judgment on the desirable state for an ecosystem. Such judgment is influenced by human values and needs, and thus definitions of OH have varied from human-centric views that focus primarily on the benefits that oceans provide to people (e.g., Halpern *et al.* 2012), to nature-centric views that would rate ecosystems with the fewest human pressures as the healthiest (e.g., McCauley *et al.* 2013).

Setting baselines for healthy oceans has also been problematic. The status of ocean ecosystems has undoubtedly changed substantially over time, making it difficult to assess what a healthy ocean is (Pauly 1995, Jackson *et al.* 2001). A further challenge is how to capture the complexity of relationships that drive the dynamics and functioning of marine social–ecological systems, where the complexity of biophysical interactions is further compounded by equally complex feedbacks with human systems. For example, it has proven difficult to relate the functioning and provision of services of an ecosystem to some of its attributes, such as its diversity (Micheli *et al.* 2014).

Finally, an additional key challenge is how to ensure that assessments are conducted with approaches and over spatiotemporal scales that effectively inform management actions. For ecological assessments to foster sustainable outcomes, near-real-time data collection and system monitoring may be needed at various scales of decision making (e.g., local, state, regional, national, and international). Technological, analytical, and communication advances are needed in order to provide this critical information to policy makers, environmental managers, and the public in a useful and timely fashion.

In the past decade, there have been important advances that aid in defining and measuring OH and implementing OH concepts as part of ecosystem-based management (EBM). Major advances in the quantification of cumulative impacts to marine ecosystems (e.g., Halpern *et al.* 2008), historical reconstructions of past changes of ocean ecosystems (e.g., Jackson *et al.* 2001), and projections of future scenarios (Cheung *et al.* 2009) can help in determining appropriate baselines and predicting future rates of change. Moreover, models linking ecosystem condition to the flow of ecosystem services have contributed critical information on parts of this complex set of feedbacks and interactions (e.g., marine Invest), www.naturalcapital project.org. Comprehensive quantitative frameworks that integrate drivers, ecosystem states, management actions, and services have also been developed (e.g., Halpern *et al.* 2012, Samhouri *et al.* 2013). These frameworks and analyses provide powerful tools and represent important first steps in ongoing efforts to include crucial social–natural feedbacks in assessments and management of OH.

This chapter reviews and synthesizes selected concepts and data that are relevant to assessments of OH. In particular, the following questions are addressed: (1) What is OH? (2) How is the condition of marine ecosystems affected by multiple drivers, and what are the major current and future threats to our oceans? (3) Are healthy oceans disease-free? (4) What indicators and targets can inform management aimed at maintaining healthy oceans? The chapter concludes by considering what recent advances and new frontiers can enable timely assessments of OH and effective use of this information to protect ocean and human health simultaneously.

What is OH?

The idea that natural systems can be healthy or unhealthy dates at least to Aldo Leopold's land ethic, published 1949 in his Sand County Almanac. Over the years, the health metaphor has been applied to various natural systems including the oceans. People intuitively understand the concept "health," even though health can be defined in different ways.

Based on review of the existing literature, Tett *et al.* (2013, p. 1) defined OH as:

> the condition of a system that is self-maintaining, vigorous, resilient to externally imposed pressures, and able to sustain services to humans. It contains healthy organisms and populations, and adequate functional diversity and functional response diversity. All expected trophic levels are present and well interconnected, and there is good spatial connectivity amongst subsystems.

However, this definition requires some clarification to make OH a practical tool for ecosystem managers. For example, what is a vigorous system? What are healthy organisms and populations? What is adequate functional diversity and what represents good spatial connectivity?

OH can be about ecosystem services, the structure and functioning of ecosystems, and human impacts—perhaps all at once. Two major attempts at evaluating OH at large spatial scales include the Ocean Health Index (OHI; Halpern *et al.* 2012), and the descriptors of good environmental status (GES) provided in the European Union's Marine Strategy Framework Directive (MSFD). Comparing these efforts illustrates the various definitions and attributes of OH (Table 7.1). Nine out of ten OHI attributes directly describe ecosystem services, or benefits to humans. Samhouri *et al.* (2012) summarize this view as "a coupled systems perspective, defining a healthy ocean as one that delivers benefits to people now and in the future." The MSFD GES, in contrast, describes a healthy ocean primarily as one without (human-caused) symptoms of altered structure and functioning, stating the symptoms primarily in biological and physical terms (Table 7.1).

Most people would agree on some general attributes or goals of a healthy ocean. These include clean waters, sustainable seafood production, and persistence of some key components of biodiversity, such as large charismatic species and species that provide critical habitat. However, finding a generally agreed-upon set of criteria for health of coastal or marine systems is at best difficult, if not impossible. Different stakeholders have different understandings, different needs, and different expectations of a healthy ocean, based on their own values, preferences, or level

Table 7.1 Characteristics of a healthy ocean according to the OH Index and the EU Marine Strategy Framework Directive descriptors of Good Environmental Status

Ecosystem structure or function	Ecosystem service
OHI goals	**OHI Goals**
• *Clean waters*	• Food provision
• Biodiversity	• Artisanal fishing
	• Natural products
GES descriptors	• Carbon storage
• Biodiversity (habitats, species)	• Coastal protection
• Normal abundance and diversity at all trophic levels	• Coastal livelihoods/Economies
	• Tourism/Recreation
• No major ecological changes due to non-indigenous species	• Sense of place (iconic species/places)
	• *Clean waters*
• Fish and shellfish populations within safe biological limits	
	GES descriptors
• No pollution effects due to contaminants	• Contaminants in seafood below legal thresholds
• Minimized effects of eutrophication	
• No environmental harm due to marine litter	
• High sea-floor integrity	
• No adverse effects of permanent hydrographical alterations	
• No adverse effects of energy introduction	

Source: OHI; Halpern *et al.* 2012; GES; ec.europa.eu/environment/marine/good-environmental-status/index_en.htm (accessed August 24, 2015).

Note: To facilitate comparison, we have grouped each indicator (i.e., the OHI goals and GES descriptors) under two categories: ecosystem structure or function, and ecosystem service. Italicized items fit under more than one category.

of well-being. For example, in a place where the local population suffers from malnutrition, food provision might be the most important function of marine ecosystems. Indeed, one of the major criticisms of the application of the health metaphor to ecosystems is that it inserts personal values under the guise of scientific impartiality (Lackey 2001). However, in reality, management of ecosystems is a human enterprise. If OH concepts and indices are meant to be useful to humans, then the humans using them should have the ability to insert their values into management goals, as long as the values are explicitly described in OH assessments.

Ultimately, it may be less important to develop a single, universally agreed-upon definition of OH than it is to foster global recognition of the range of its important attributes and stressors. The marine management and stakeholder community can constructively support numerous practical definitions of OH, employed by different management agencies or communities. Collective investment in the importance of the OH concept can help promote active and adaptive management (including policy change) to achieve certain health standards, even if the definition of health varies, if standards are transparently defined and consistently measured. When management is held accountable to standards, it creates a system with: 1) knowledge of ecosystem status, 2) active management to achieve status if performance is low, and 3) evaluation of the effectiveness of management actions. Perhaps the most important role of the OH concept is to move society toward positive change, much as we do for human health.

Human impacts on the oceans

Humans impact ocean ecosystems in multiple ways and cause change at a rapid pace. Human activities—ranging from sea-based activities such as fishing, aquaculture, and shipping to land-based activities such as development, agriculture, and mining—produce a suite of pressures on marine species and ecosystems. Frequently, the total cumulative impact of these activities on ocean ecosystems is greater than each activity's impact in isolation, and the combination of activities has the potential to cause severe environmental degradation (Crain *et al.* 2008). However, in evaluating human impacts to oceans, researchers have typically focused on the effects of a single activity or stressor on single ecosystem components (but see Halpern *et al.* 2008). This is, in part, because it is difficult to track the sources of multiple stressors, measure them at appropriate temporal and spatial scales, and determine baselines for measuring change in ecosystems.

Approaches for assessing the cumulative impacts of multiple drivers of ecosystem change are discussed next, with an examination of how historical data can provide baselines for modern management. We then look ahead at how we might expect OH to change in the near future with emergent issues of warming, hypoxia, and ocean acidification.

Cumulative impacts of multiple drivers

A broad range of human activities that impact OH can occur in coastal and marine regions. Each human activity produces a number of stressors or pressures that may occur at multiple scales and impact the surrounding environment, including sedimentation, nutrient input, contaminants, noise, acidification, and many others. These pressures may have impacts of varying degrees on a suite of species and habitats in ecosystems, ranging from mortality to behavioral and physiological change. Further, these pressures may manifest at different levels, from species to whole ecosystems. The environmental effects caused by human and natural activities and pressures do not occur independently of one another. Instead, each activity interacts with past and contemporary human activities and the broad range of stressors they produce, resulting in cumulative impacts to a suite of ecological components. Natural variability in ecosystem processes and interactions among components in the system add further complexity and affect the manifestation of resulting impacts.

Overfishing is arguably one of the most significant and pervasive human impacts on marine ecosystems (Jackson *et al.* 2001). Poorly managed marine harvest can lead to the direct depletion of targeted marine species as well as unintended, but equally severe effects on non-target marine species (e.g., bycatch) and the degradation of marine habitats. These impacts can unleash propagating indirect change on whole marine food webs (e.g., trophic cascades) and cause decreased resilience of marine ecosystems to climate variability (Micheli *et al.* 2012).

There are, however, other pressures beyond fishing that undermine OH. Halpern *et al.*'s (2008) quantification and mapping of cumulative impacts to the oceans revealed for the first time the global extent of human alteration of marine ecosystems. They combined spatial data on the distribution of 17 anthropogenic drivers of ecosystem change, including fishing, pollution, invasive species, and climate change, with the distribution of 20 marine ecosystem types and expert-based weights estimating the vulnerability of each ecosystem type to each driver. This systematic integration of the estimated cumulative human impact on marine ecosystems showed that large portions of the global oceans are heavily impacted (40 percent), most areas of the oceans are impacted by multiple drivers (typically ten or more in coastal areas), and less than 4 percent of the oceans remain relatively unimpacted, mostly in polar regions. The highest cumulative impacts occur in heavily populated areas, including the North and Norwegian Seas, South and East China Seas, Eastern Caribbean, North American eastern seaboard, Mediterranean, Persian Gulf, Bering Sea, and the waters around Sri Lanka (Halpern *et al.* 2008). Ecosystems with the highest predicted cumulative impact scores include hard and soft continental shelves. Coral reefs, seagrass beds, mangroves, rocky reefs and shelves, and seamounts have few to no areas remaining anywhere in the world with low cumulative impact scores. In contrast, shallow soft-bottom and pelagic deep-water ecosystems had the lowest scores, partly because of their lower vulnerability to a suite of activities and their associated pressures (Halpern *et al.* 2008).

The worrisome picture that this analysis unveils may yet be conservative, because some important activities and pressures, including coastal hypoxia, marine debris, and illegal or unreported fishing, could not be included in the analysis due to unavailability of global spatial data. Moreover, estimated impacts of individual pressures were combined using an additive model. However, pressures can interact to produce a variety of effects, including synergistic, antagonistic or compensatory effects (Crain *et al.* 2008). For example, stressors are considered synergistic when their combined effect is greater than predicted from the sizes of the responses to each stressor alone and antagonistic when the cumulative impact is less than expected (Crain *et al.* 2008). Additional and better quality data, and an enhanced understanding of how multiple pressures interact in their ecological impacts could be included in future analyses to present a more accurate assessment of the cumulative impact of anthropogenic drivers to the ocean.

Looking back: historical change

Over the past 10–20 years human perception of the status of the ocean, and consequently of a healthy ocean, has changed radically. In many cases, what researchers considered natural was indeed the results of centuries or millennia of unappreciated human impact on marine ecosystems. The "shifting baseline syndrome," an inter-generational loss of information on the states of ecosystem and animal populations, has affected scientists, managers, and ocean resource users due to humans' limited observational experience and capacity to access and retain historic information (Pauly 1995).

Historical analyses have revealed that human impacts on ocean ecosystem functioning can date back centuries, in most cases preceding the advent of monitoring programs and scientific investigations (e.g., Pauly 1995, Jackson *et al.* 2001). The cumulative effects of fishing, coastal development, pollution, and climate change have removed suitable habitats, depleted populations, caused species range contractions and regional extirpations, and facilitated species invasions (Lotze and Worm 2009). Large animals have been the most susceptible to declines because they are preferentially targeted by harvesters and easier to catch. Thus, even ancient fisheries with limited technological and spatial scope impacted populations of ecologically important large consumers (Jackson *et al.* 2001). Historical analyses have also shown that loss of diversity coincided with loss of services (Worm *et al.* 2006).

These patterns of ocean change have followed similar but asynchronous trajectories across world regions. Human societies proceed through stages of resource exploitation, and consequently ecosystems have responded in a similar fashion independent of context (Lotze *et al.* 2006, Worm *et al.* 2006). However, despite evidence of historical impact of humans on marine ecosystems, there are examples in which past societies have used resources from the ocean in a sustainable fashion. Historical reconstruction of coral reef ecosystems in the main and northwestern Hawaiian Islands indicates that, while human impacts to ecosystems can be cumulative and lead to environmental decline, there are records of recovery periods where degraded systems begin to show improved ecosystem health (Kittinger *et al.* 2011). These recovery periods are attributed to a complex set of changes in underlying social systems that served to release reefs from direct anthropogenic stressor regimes. These types of social–ecological interactions may be common. In British Columbia, archaeological records of midden sites illustrate that humans sustained continuous harvesting of herring populations for millennia, before modern industrial exploitation (McKechnie *et al.* 2014).

Marine managers need baselines to set recovery targets. In some instances, these reference points can be drawn from well-enforced marine protected areas, but such places are rare (Guidetti and Micheli 2011). Historical data sources can also provide some degree of understanding of how OH has changed over time and give insight into benchmarks for OH management. Selection of such baselines must be cautiously approached, as historical ecologists have demonstrated that human impact on the oceans reaches back, in some instances, thousands of years (Lotze *et al.* 2006). This process of critically examining the changing health of the oceans requires considering both what species and habitats have been impacted by human activity, as well as determining the functional significance of these losses. This process of critically looking back at OH is a constructive first step toward moving forward.

Looking forward: current threats and future scenarios

Climate change is a global-scale issue that jeopardizes many aspects of ocean ecosystem health and the sustainability and well-being of human communities. By absorbing heat and carbon dioxide, the oceans have played a critical buffering role against climate change. However, this service does not come without repercussions. Several pressures to the ocean result from climate change, including increasing sea surface temperature, altered circulation and stratification patterns, sea level rise, hypoxia, and ocean acidification. Below we outline how these stressors can fundamentally alter the structure and functioning and service provision of ocean and coastal ecosystems.

Global climate change alters environmental forcing mechanisms—or factors that impact ocean circulation—such as wind, precipitation, temperature, and salinity patterns. Thus, climate change is projected to cause the widespread warming of ocean surface and alter large-scale oceanic circulation patterns, which can alter the distribution and interactions among species. Marine fishes and invertebrates are expected to track changing ocean temperatures when movement or dispersal into new environments is possible. Species unable to move will either acclimatize, evolve, or suffer local extirpations (Pinsky *et al.* 2013). Poleward movement of fish into cooler waters has been documented for fish assemblages between 1970 and 2006 in 52 of 64 Large Marine Ecosystems (LMEs) (Cheung *et al.* 2013). Some species are already at thermal tolerance limits, particularly in tropical waters, so local extirpations and future declines in biodiversity are expected. Changes in tropical waters are probably irreversible, with negative implications for the future of marine ecosystems, as well as fisheries yields. Local extirpations of foundation species, such as the mortality of corals through bleaching due to climate warming, could also cause the loss of biodiversity dependent on these foundation species (Hoegh-Guldberg *et al.* 2007). However, several species are adapting to global temperature change. For example, by switching to heat-tolerant symbionts, coral species may be able to tolerate higher temperatures in some regions of the oceans (Barshis *et al.* 2013). Further studies of adaptation will provide greater understanding of how some species may evolve and maintain some aspects of ecosystem functioning in the face of global temperature rise.

In addition to mean ocean temperature increase, climate change is associated with increased temperature variability. Increased variability in ocean temperature can have profound effects on marine organisms, where impact of temperature fluctuations on fitness or other performance-related traits can amplify the impact of warming on species (Tewksbury *et al.* 2008). In fact, it has been recently suggested that increased variability has an even greater effect than increased mean temperatures (Vasseur *et al.* 2014).

Changes in forcing mechanisms associated with climate change may lead to changes in ocean circulation and stratification of surface waters. Intensified stratification of ocean waters is expected to diminish productivity of the upper layer, thereby reducing primary production and altering food webs dependent on these primary producers (Carlisle 2014). As a result of climate change and warming oceans, the productivity of many fish stocks and some ecosystems is expected to decrease, while others may increase in productivity (Cheung *et al.* 2013). Surface waters and coastal areas will also experience more intense storms and catastrophic events, such as extreme hurricanes that destroy mangroves, reefs, and other near-shore nursery habitats (Knutson *et al.* 2010).

The rise in anthropogenic carbon dioxide (CO_2) that is responsible for projected trends of ocean warming and stratification is also causing profound shifts in seawater chemistry. Ocean acidification is a series of predictable changes in seawater carbonate chemistry caused by the absorption of atmospheric CO_2 into the ocean (Feely *et al.* 2009). Since the industrial revolution, seawater acidity has already increased over 30 percent and is expected to increase 150 percent relative to the pre-industrial levels by the end of the century based on business-as-usual scenarios of CO_2 emissions (Feely *et al.* 2009).

Ocean acidification is predicted to increase the energy needed to maintain a number of important physiological processes, including calcification, across a wide range of marine species (Kroeker *et al.* 2013). While there is variability in species responses, syntheses suggest that ocean acidification may reduce the growth rates of many species, especially species that build their shells or skeletons out of calcium carbonate, from corals to marine snails to oysters (Kroeker *et al.* 2013). In contrast, the shifts in carbonate chemistry associated with ocean acidification may decrease the energy necessary for primary producers to grow, resulting in enhanced growth rates of marine algae and seagrasses (Koch *et al.* 2013). Understanding how the wide range of species' responses to ocean acidification will scale up to affect OH remains a challenge.

Studies in naturally acidified ecosystems suggest that ocean acidification could cause local reductions in the abundance of calcareous species, shifts towards dominance by algae, and overall reductions in biodiversity in the future (Kroeker *et al.* 2011, Fabricius *et al.* 2011). In ecosystems with calcareous foundation species, such as coral or oyster reefs, reductions in the abundance of these ecologically important calcareous species could cause subsequent losses of those species dependent upon them (Fabricius *et al.* 2013) with repercussions for local and global economies. Shellfish fisheries are also expected to suffer significant economic losses with continued ocean acidification (Cooley and Doney 2009), and modeling efforts suggest that OA-related reductions in the abundance of calcareous prey (such as sea urchins or bivalves) could reduce the abundance of groundfish species by up to 20–80 percent, depending on the magnitude of the reductions in the abundance of the prey (Kaplan *et al.* 2010).

There is some limited recent evidence that adaptation to ocean acidification is possible. In temperate systems, one study has demonstrated genome-wide selection on sea urchin larvae (*Strongylocentrotus purpuratus*) when larvae were exposed to acidified seawater. Selection led to few changes in larval development (Pespeni *et al.* 2013). Thus, some species may have the ability to tolerate broad pH fluctuations or have sufficient genetic diversity to allow rapid evolution in response to environmental change.

The combined effects of multiple pressures associated with climate changes, and the cascading impacts of warming, acidification, and changes to ocean circulation, are likely to result in reduced ecosystem service provisioning. For example, increased water temperatures are expected to affect fish species and fish stocks that people depend on for food, by reducing fish sizes due to physiological changes in fish metabolism with temperature, oxygen content, and other biogeochemical properties of water (Cheung *et al.* 2012). Fish sizes tend to be smaller in warmer temperatures, partly because warmer water contains less oxygen, which restricts gill size and ultimately metabolism and growth rates (Pauly 1984). Future modeling scenarios with high emissions estimate that the assemblage-averaged maximum body weight of 600 marine fish species will shrink by 14–24 percent between 2000 and 2050, and will be possibly more pronounced in tropical and intermediate latitudes (Cheung *et al.* 2012). These projections do not account for ocean acidification, hypoxia, or disease outbreaks, so projections likely underestimate the integrated impacts on fish stocks and marine ecosystems from climate change. Size shifts induced by changes in ocean chemistry may further exacerbate reductions in fish size caused by size-selective fishing.

Understanding how the ocean warming, stratification, acidification, and hypoxia will combine and interact with human activities to impact OH remains a major challenge for science and management. Mechanistic understanding of how individual drivers influence the physiological response of marine species to other drivers is very limited (Griffith and Fulton 2014). Ecological theory and data suggest, however, that declines in diversity due to any single driver are likely to reduce resilience to other environmental changes or human activities and cause subsequent reductions in ecosystem services (Micheli *et al.* 2012, Worm *et al.* 2006).

We have thus far considered factors that affect OH as being confined to the marine environment. This ocean-centric viewpoint of OH may in fact be short-sighted: the health of the terrestrial environment influences the health of the oceans and vice versa. Such links have been relatively well recognized in situations where terrestrial pollutants or other deleterious materials transit from land to sea. Runoff from intensively modified terrestrial areas generates high sediment loads in coastal water that can smother coral reefs. Fertilizers used in large-scale terrestrial agricultural projects that are directed by rivers into the marine environment can generate eutrophic conditions, such as the dead zone located in Gulf of Mexico near the outflow of the Mississippi River. Connections between the health of terrestrial and ocean ecosystems are also interlinked in more subtle, but equally or more important ways. Poor yields from marine fisheries, for example, have been demonstrated to increase hunting of terrestrial wildlife in nature reserves (Brashares *et al.* 2004). Such cryptic but important linkages demonstrate the intimacy of the connections that exist between ocean and terrestrial health and suggest that marine managers must engage colleagues in terrestrial ecosystem management.

Ocean health and marine diseases

A healthy person is, by definition, disease-free. Does the same hold for a healthy ocean? Marine ecosystems contain a broad diversity of parasites, defined here as organisms that live in an intimate (spatially close) and durable (temporally long) relationship with a host, where the host experiences disease (i.e., negative fitness effects; Combes 2001). For parasites, the OH metaphor breaks down: while outbreaks of pathogenic diseases sometimes decimate marine populations (Lessios 1988), disrupt fisheries (Wilberg *et al.* 2011), cause cascading effects in marine communities (Kennedy *et al.* 2013), and contribute to species extinctions (Powles *et al.* 2000), parasites can also be normal, natural, and even vital components of healthy, functioning ecosystems (Hudson *et al.* 2006; Gomez *et al.* 2012). Proposed definitions of OH have, to date, rarely included consideration of marine disease. Here, we outline the challenges faced when attempting to incorporate marine disease into the definition of OH, and suggest some promising metrics to integrate parasite- and disease-monitoring into ocean health assessments.

Disease ecologists have long recognized that "healthy ecosystems are rich in parasites" (Hudson *et al.* 2006), and recent studies in marine ecosystems have borne this out. More parasite species occur in restored salt marshes than in degraded ones (Hechinger and Lafferty 2005), in fishes inhabiting pristine coral reefs than in those inhabiting heavily fished reefs (Lafferty *et al.* 2008a; Wood *et al.* 2014), and in marine protected areas relative to open-access areas (Wood and Lafferty 2014). These rich parasite communities probably both arise from and contribute to ecosystem health. That is, parasites need hosts, and are therefore more likely to be found in ecosystems where hosts are diverse and abundant (Lafferty 2012). Parasites can also serve important ecological roles in ecosystems, by checking host populations (Hudson *et al.* 1998), affecting the species composition and increasing the diversity of marine communities (e.g., Poulin and Mouritsen 2005), and changing the flow of energy through the food web (Lafferty *et al.* 2008b), often by increasing the amount of energy flowing to apex predators (Lefevre *et al.* 2009). As just one example, marine viruses are major players in the global carbon cycle through their significant effects on marine microbial mortality (Suttle 2005).

But while parasite species richness appears to be positively associated with host species richness in many healthy ecosystems, a more difficult question is whether parasite abundance or the degree of pathology (i.e., loss of host fitness) caused by parasites should be lower in healthy than in unhealthy ecosystems. Recent evidence suggests that the response of parasite abundance to human disturbance varies, depending on parasite and host traits. For example, fish parasites with complex life cycles appear to be especially likely to decline in abundance in response to fishing pressure, while directly transmitted fish parasites (i.e., parasites transmitted directly between conspecifics, without other host species in the life cycle) can sometimes capitalize on fishing-driven community changes and experience dramatic increases in abundance if their hosts increase in abundance (Wood *et al.* 2014).

While most marine parasites are of little public health concern, some can spill over from their natural hosts to infect humans. One group of nematodes—*Anisakis* spp.—can infect humans who consume uncooked marine fish, usually as sushi (Overstreet 2013). The worm's natural life cycle includes a marine mammal definitive host: the human host is a "dead-end." Fortunately for humans and anisakids, anisakiasis is not common and usually causes only mild symptoms in the human host (Overstreet 2013). Several other marine helminths can infect humans, including cestodes in the genus *Diphyllobothrium*, and the trematodes *Heterophyes heterophyes* and *Metagonimus yokogawai* (Adams *et al.* 1997).

Although human infections from marine diseases are rare, bacteria and viruses encountered by consumption of raw or undercooked shellfish (e.g., *Vibrio* spp., noroviruses), during swimming in coastal waters, or from drinking water can cause substantial pathology and are a major public health concern (Shuval 2003). These parasites can occur naturally in marine and estuarine ecosystems, as in the case of *Vibrio* spp. bacteria, or may be introduced into coastal waters through sewage pollution (Bosch 1998). Every year an estimated 120 million human cases of gastroenteritis and 50 million cases of respiratory disease around the world are linked with exposure to waste-polluted coastal waters (Shuval 2003). Cholera, caused by the *Vibrio cholerae* bacterium, can be transported between coastal cities by attaching to the exoskeletons of marine zooplankton that are swept along by ocean currents or trapped in ballast water (Colwell 1996).

Just as ocean parasites can affect land animals, land parasites can sometimes affect ocean animals. For example, the protozoan *Toxoplasma gondii* is globally distributed and can use nearly any mammal or bird as an intermediate host (Lehmann *et al.* 2006). Because the parasite is dependent on cats as obligate, definitive hosts, *T. gondii* was long considered to be an exclusively terrestrial parasite (Jackson and Hutchison 1989). Recent work has revealed that a substantial proportion of sea otters (*Enhydra lutris*) in California are infected by *T. gondii*, their exposure probably due to ingestion of oocysts washed into the ocean by runoff contaminated with cat faeces (Conrad *et al.* 2005). Toxoplasmosis is a major cause of mortality and a contributor to the slow rate of population recovery for sea otters in California (Conrad *et al.* 2005).

Although parasites are clearly integral, natural components of healthy ecosystems, there is increasing evidence that stress to marine organisms, caused by natural or human disturbance such as climate change or pollution, can lead to disease outbreaks that are more severe than expected for healthy, unstressed populations (Sokolow 2009; Morley 2010). There is compelling evidence that coral diseases, including both stress-induced bleaching and several infectious diseases, are on the rise (Sokolow 2009) and some coral diseases have been correlated with periods of seasonal or inter-annual ocean warming (Harvell *et al.* 2009), increased nitrogen or sewage pollution (Bruno *et al.* 2003), and runoff (Haapkyla *et al.* 2011). However, whether environmental degradation, global warming, and human influences are promoting a general increase in ocean diseases is still hotly debated (Harvell *et al.* 2004; Wood *et al.* 2010).

Demonstrating the causal link between environmental degradation and disease is not straightforward since drivers of disease are complex and sometimes opposing (Sokolow 2009). Some marine diseases are caused by bacterial consortia or ubiquitous opportunists, rather than one pathogen, which can make identifying causative agent(s) difficult (Bourne *et al.* 2009; Burge *et al.* 2013). Moreover, even the most "pristine" locations often harbor diseases similar to those seen at more degraded locations, obscuring the connection with anthropogenic drivers (Williams *et al.* 2008).

Just as environmental degradation can influence disease risk, disease in marine populations can increase ecosystem susceptibility to other stressors. For example, during the 1980s, white band disease outbreaks rapidly killed large stands of Caribbean acroporid corals. The skeletons of these corals were then weakened by bioerosion, making the ecosystem more susceptible to hurricane damage (Aronson and Precht 2001). The disappearance of acroporids, along with that of the urchin *Diadema antillarum* (also caused in part by a disease outbreak; Lessios 1988) precipitated a change in the Caribbean-wide reef ecosystem from coral- to algal-dominance, which may be difficult or impossible to reverse (Mumby *et al.* 2007). This has had cascading effects for many reef species and even some of the human communities that rely on healthy reefs for their livelihoods (Bellwood *et al.* 2004).

Increasing global connectivity can result in the rapid spread of parasites across regions, with sometimes dire consequences for OH. Non-native, invasive parasites can be particularly harmful (Torchin *et al.* 2002). For example, *Anguillicola crassus*, a parasitic nematode of the Japanese eel, *Anguilla japonica*, was introduced into European waters in the 1980s. It became endemic in the European eel (*Anguilla anguilla*) and is believed to have contributed, along with other anthropogenic pressures, to the collapse of the species (Palstra *et al.* 2007). Shipping traffic and ballast water can move live bacteria, protists, algae, zooplankton, benthic invertebrates, and even fish across ocean basins (Carlton and Geller 1993). The rapid spread of a human cholera pandemic across continents in the 1990s has been linked to ballast water movements (Colwell 1996).

In this context, it is evident that disease outbreaks should be considered among the suite of potential stressors that can impact OH. The Caribbean coral reefs are one example where monitoring and prevention of disease in a few key populations could have impacted the health of the whole ecosystem. Other examples where disease monitoring might be a priority for OH assessments include ecosystems under threat of disease spillover from aquaculture, such as those near abalone (Lafferty and Ben-Horin 2013), salmon (Torrissen *et al.* 2013), and shrimp farms (Walker and Winton 2010).

Interest in using parasites as bioindicators (i.e., taxa or functional groups that are able to reflect the state of the environment; *sensu* McGeogh 1998) has grown in recent years for several reasons. The US Environmental Protection Agency (EPA) highlighted the value of parasites as bioindicators by demonstrating that the richness of trematodes in snails provides a rigorous indicator of the condition of estuarine tidal wetlands in the US (Weilhoefer 2011).

While some parasites can capitalize on human impacts to cause conspicuous and sometimes devastating epidemics, most are natural and even vital components of marine food webs. This dichotomy may make some parasites useful indicators of ocean health—with the presence of some species indicating health, and others indicating stress and even ecosystem collapse.

Measuring ocean health: indicators and targets

The above sections illustrate the complexity and diversity of issues and challenges that confront marine managers when trying to understand, measure, and address OH. To address these challenges and fulfill these mandates, marine and coastal managers need to determine the level of impact that is acceptable and minimizes the risk of serious environmental degradation and lost economic or cultural value. Legal mandates to assess the condition of marine ecosystems, and to take actions to improve such condition through EBM (e.g., US National Ocean Policy, EU MSFD and United Nations LME projects) confront managers with a need to clearly define indicators and targets for OH, and select specific management strategies for achieving the goals set by these policies. As previously discussed, specific OH goals may vary depending on local needs and values, but it is crucial that goals are clearly defined and measurable indicators and targets are specified.

Management goals are defined as broad statements about desired ocean conditions or health that are specific enough to allow concrete management decisions. A target is a point of reference on the specific status or amount of benefit that equals goal achievement. Targets should be informed by scientific evidence, and should follow SMART guidelines (Specific to management goal, Measurable, Ambitious, Realistic, and Time-Bound) (Niemeijer and de Groot 2008). Targets can be specified in three ways based on: 1) an ideal state, 2) historical status, or 3) maximum possible value (Figure 7.1; Samhouri *et al.* 2012). The ideal state can be derived from functional relationships between the indicator of OH for an ecosystem goal and natural or human pressure. Where a functional relationship is unavailable, either a historical status (time

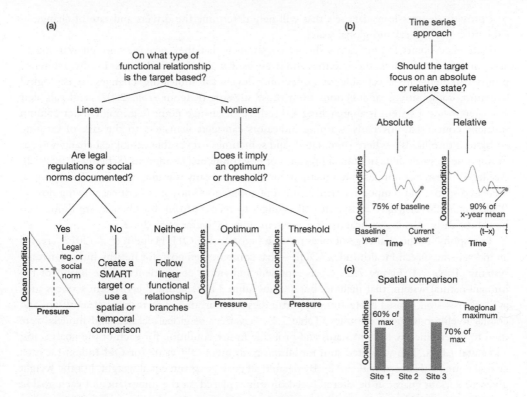

Figure 7.1 Decision trees for choosing between three types of health targets based on (a) functional relationships, (b) time series approaches, and (c) spatial comparisons

Source: Reproduced with permission from Samhouri *et al.* 2012.

series) or spatial comparison may be used to set a target. Time series are useful in providing an internal measure against which current conditions can be compared with historic conditions, and may involve either a baseline comparison or a moving window approach. Spatial comparisons gauge conditions in one area to those in a reference area elsewhere where maximum benefits are assumed to be achieved. This could be used, for example, to assess the status of a fished area where a fisheries closure (within a marine protected area) is used as a baseline.

The physical and ecological components and processes that provide effective warning signals for changed conditions in relation to management goals are commonly called indicators. Indicators of ecosystem health ideally reflect the broader ecosystem, thus eliminating the need to measure every variable or species of interest or concern. It can be difficult to select indicators that provide this accurate view of broader ecosystem structure and function, are easy to monitor, and provide early warning for managers. However, a number of EBM processes are developing suites of indicators to monitor ecosystem change and OH. For example, the Puget Sound Partnership developed indicators to assist managers monitoring ecosystem health in Puget Sound, Washington (Orians *et al.* 2012).

Long Term Ecological Research (LTER) and Ocean Observatory programs have emerged to simultaneously monitor biological or ecological processes. Coupling of long-term physical and biological observations has already detected climate-related change in sites such as the Palmer LTER on the western Antarctic Peninsula (Hofmann *et al.* 2013). In the long term, these programs will provide decades-long data sets that will help determine the drivers and rate of change in OH attributes at local and global scales.

Reference points for indicators should be set such that they provide enough warning to allow managers to address the activities and stressors that are causing changes. Ideally, reference points can delineate acceptable or concerning levels for indicating changes to ecological structure, composition, or functions from those stressors (Kilgour *et al.* 2007), and can alert managers when a system is approaching a threshold or change point (e.g., high water column nutrient concentration). Early warning indicators can alert managers to the risk of crossing ecosystem thresholds—where large, rapid, and sometimes irreversible ecological changes occur in response to small shifts in human pressures or environmental conditions (Scheffer *et al.* 2012). Academic research continues to test methods for identifying early warning indicators. For example, increased spatial and temporal variance, and a phenomenon known as "critical slowing down" (Dakos *et al.* 2012) where ecosystems take longer to recover from disturbance, are thought to be robust early warning indicators of ecosystem shifts (Scheffer *et al.* 2012).

In response to the challenge of measuring and monitoring OH, Halpern *et al.* (2012) created an index—the Ocean Health Index (OHI)—that combines ten goals for healthy human–natural systems (Table 7.1, Figure 7.2). Goals represent individual components of the health of the human–natural system, that include, but are not limited to established ecosystem services such as food provision, coastal protection, opportunities for tourism and recreation, clean water, and coastal livelihoods and economies (Table 7.1). A variety of global datasets and models were used to determine the current status and probable future condition for each of the goals across all coastal EEZs. The weighted sum for all ten goals gives OH value (or OH Index) for each coastal country (Halpern *et al.* 2012). By default, all goals are given equal weight, but the weight given to various goals can be altered based on values placed on the importance of each goal in contributing to OH.

Halpern *et al.*'s (2012) study represents the most comprehensive effort to date to operationally define and quantify OH. At a global scale, index score varied among countries from 36 to 86 out of 100 (mean score = 60). Many countries in Western Africa, the Middle East, and Central America scored poorly, while parts of Northern Europe, Canada, Australia, Japan, and some tropical islands and uninhabited regions scored well. Index scores are positively correlated with the Human Development Index, a measure of development status, indicating that countries with stronger economies, regulations, and capacity to manage pressures are able to maintain healthier oceans and associated social systems. Results depart from expectations based on a purely protectionist perspective, because health is assessed for human–natural systems, not for ecological systems alone. Thus, intense uses of marine resources, if sustainable, can result in high index scores. The long-term objective of the OHI is to provide a scorecard against which individual countries can measure themselves and monitor progress. If, for example, a nation scored low on the "clean waters" goal, and subsequently took steps to remove debris or stop marine pollution, their OHI score should rise, reflecting that progress.

To apply the OHI concept to specific regions, Halpern *et al.* (2013) worked with managers and experts in the California Current to assign regionally specific weights to the ten OHI goals. In this example, clean waters and sense of place were ranked higher than any other goals, including those that are most commonly discussed, such as livelihoods and food provision. This application of the OHI reflects how, even within a single set of OH criteria, different regions or user groups might place emphasis on different suites of criteria based on existing value systems.

Samhouri *et al.* (2012) outline the process by which OH criteria (or goals) can become part of an OH scorecard that can serve policy, management, and communication purposes. This work provides a roadmap for setting targets and evaluating current ecosystem conditions relative

Ten public goals: sub-goals

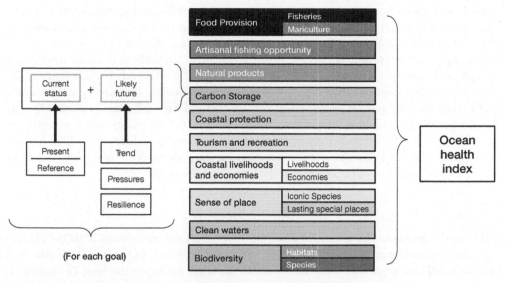

Figure 7.2 Conceptual framework for calculating the Ocean Health Index. Each dimension (status, trend, pressure, and resilience) is derived from a wide range of data. Dimensions combine to indicate the current status and likely future condition for each of ten public goals.

Source: Reproduced with permission from Halpern *et al.* 2012.

to them. As a general procedure, the authors recommend three steps: 1) precisely articulating management goals, 2) setting targets, and 3) scaling the current status of management goals relative to those targets. As described above, selection of indicators and targets is not a trivial task. Selecting the right type of target is critical because targets set the bar for achievement of management goals. Using functional relationships to define targets is preferred, but limited availability of data may prohibit this approach in some cases. Selecting a range of indicators is also critical. In a simulation done by Samhouri *et al.* (2012), the status of marine livelihoods across the US depended on which of three indicators was used: employment opportunities, job quality, or employment satisfaction. The preferred option for indicators (as for targets) will ultimately depend on the exact statement of the management goal.

Finally, in order to evaluate OH using indicators and associated targets, there should be an evaluation of the functional relationship between them. Understanding functional relationships and, in particular, non-linear functional relationships that identify ecological thresholds can help managers of marine ecosystems predict how new activities are likely to alter ecosystem health and improve the regulation of future development and use of ocean resources. However, ecological thresholds and functional relationships can be difficult to determine and assess due to the complexity of natural systems and the presence of multiple drivers and stressors. There are very few examples of established reference points that delineate an acceptable level of cumulative stressors or changes to ecological structure, composition, or functions from those stressors (Kilgour *et al.* 2007). Alternatively, when relationships between indicators and OH are not known, a range of values might be used to indicate when the goal has been achieved. Using reference points based on levels of socially acceptable or ecologically tolerable change may serve as placeholders until a better understanding of functional relationships and ecosystem thresholds can be developed.

Conclusions

Ocean health is intimately linked with human health and well-being. Legal mandates by several nations to achieve OH goals reflect a shift in thinking about the oceans from commodities to be exploited, to highly valuable systems upon which humans depend and that are in urgent need of protection. Healthy oceans are productive, resilient, and able to maintain services that people want and need. However, while the notion of health is intuitively obvious to most, and emotionally easy to relate to, definitions that can be operationalized for assessing and managing OH are challenging and are largely based on the values and needs of different people. Despite challenges in the definition, monitoring, and assessment of OH, major recent advances are providing tools and conceptual frameworks for effectively assessing and supporting OH. Moving this agenda forward will require continued efforts to integrate physical, biological, and social sciences, and greater political and public involvement and support.

Acknowledgments

This work was funded by grants from the US NSF-CNH program (award #DEB-1212124) and the Lenfest Ocean Program. D. F. Rosim is a Lemann fellow and received a postdoctoral research fellowship abroad from the National Council for Scientific and Technological Development, Brazil.

References

Adams, A. M., K. D. Murrell, and J. H. Cross. 1997. Parasites of fish and risks to public health. *Revue Scientifique et Technique* 16: 652–660.

Aronson, R. B. and W. F. Precht. 2001. White-band disease and the changing face of Caribbean coral reefs. *Hydrobiologia* 460: 25–38.

Barbier, E. B., H. M. Leslie, and F. Micheli. 2014. Services of Marine Ecosystems: A Quantitative Perspective. Invited chapter in M. Bertness, J. Bruno, B. Silliman and J. Stachowicz, Eds. *Marine Community Ecology and Conservation*, Sinauer Associates, Sunderland, MA, pp. 403–425.

Barshis, D. J., J. T. Ladner, T. A. Oliver, F. O. Seneca, N. Traylor-Knowles, and S. R. Palumbi. 2013. Genomic basis for coral resilience to climate change. *Proceedings of the National Academy of Sciences* 110(4): 1387–1392.

Bellwood, D. R., T. P. Hughes, C. Folke and M. Nyström. 2004. Confronting the coral reef crisis. *Nature* 429: 827–833.

Bosch, A. 1998. Human enteric viruses in the water environment: a minireview. *International Microbiology* 1: 191–196.

Bourne, D. G., M. Garren, T. M. Work, E. Rosenberg, G. W. Smith, and C. D. Harvell. 2009. Microbial disease and the coral holobiont. *Trends in Microbiology* 17(12): 554–562.

Brashares, J. S., P. Arcese, M. K. Sam, P. B. Coppolillo, A. R. Sinclair, and A. Balmford. 2004. Bushmeat hunting, wildlife declines, and fish supply in West Africa. *Science* 306(5699): 1180–1183.

Bruno, J. F., L. E. Petes, C. D. Harvell, and A. Hettinger. 2003. Nutrient enrichment can increase the severity of coral diseases. *Ecology Letters* 6: 1056–1061.

Burge, C. A., C. J. Kim,, J. M. Lyles, and C. D. Harvell. 2013. Special issue Oceans and Humans Health: the ecology of marine opportunists. *Microbial Ecology* 65(4): 869–879.

Carlisle, K. M. 2014. The large marine ecosystem approach: application of an integrated, modular strategy in projects supported by the Global Environment Facility. *Environmental Development* 11: 19–42.

Carlton, J. T. and J. B. Geller. 1993. Ecological roulette: the global transport of nonindigenous marine organisms. *Science* 261: 78–82.

Cheung, W. W. L., R. Watson, and D. Pauly. 2013. Signature of ocean warming in global fisheries catches. *Nature* 497: 365–368.

Cheung, W. W. L., C. Close, V. W. Y. Lam, J. Sarmiento, K. Kearney, R. Watson, and D. Pauly. 2009. Projecting global marine biodiversity impacts under climate change scenarios. *Fish and Fisheries* 10: 235–251.

Cheung, W. W. L., J. L. Sarmiento, J. Dunne, T. L. Frolicher, V. W. Y. Lam, M. L. D. Palomares, R. Watson, and D. Pauly. 2012. Shrinking of fishes exacerbates impacts of global ocean changes on marine systems. *Nature Climate Change* 3: 254–258.

Colwell, R. R. 1996. Global climate and infectious disease: the cholera paradigm. *Science* 274: 2025–2031.

Combes, C. 2001. *Parasitism: The ecology and evolution of intimate interactions.* Chicago, IL, University of Chicago Press.

Conrad, P. A., M. A. Miller, C. Kreuder, E. R. James, J. Mazet, H. Dabritz, D. A. Jessup, F. Gulland, and M. E. Grigg. 2005. Transmission of *Toxoplasma*: clues from the study of sea otters as sentinels of *Toxoplasma gondii* flow into the marine environment. *International Journal for Parasitology* 35: 1155–1168.

Cooley, S. R. and S. C. Doney. 2009. Anticipating ocean acidification's economic consequences for commercial fisheries. *Environmental Research Letters* 4: 024007. http://iopscience.iop.org/1748–9326/4/2/024007/ (accessed July 2, 2015).

Crain, C. M., K. Kroeker, and B. S. Halpern. 2008. Interactive and cumulative effects of multiple human stressors in marine systems. *Ecology Letters* 11: 1304–1315.

Dakos, V., E. H. Van Nes, P. D'Odorico, and M. Scheffer. 2012. Robustness of variance and autocorrelation as indicators of critical slowing down. *Ecology* 93: 264–271.

Doney, S. C. 2010. The growing human footprint on coastal and open-ocean biogeochemistry. *Science* 328(5985): 1512–1516.

Fabricius, K. E., G. De'ath, S. Noonan, and S. Uthicke. 2013. Ecological effects of ocean acidification and habitat complexity on reef-associated macroinvertebrate communities. *Proceedings of the Royal Society of London B* 281: 24–79.

Fabricius, K. E., C. Langdon, S. Uthicke, C. Humphrey, S. Noonan, G. De'ath, R. Okazaki, N. Muehllehner, M. S. Glas, and J. M. Lough. 2011. Losers and winners in coral reefs acclimatized to elevated carbon dioxide concentrations. *Nature Climate Change* 1: 165–169.

Feely, R. A., S. C. Doney, and S. R. Cooley. 2009. Ocean acidification: present conditions and future changes. *Oceanography* 22(4): 36–47.

Garcia, S. M. and A. A. Rosenberg. 2010. Food security and marine capture fisheries: characteristics, trends, drivers and future perspectives. *Royal Society Bulletin* 365(1554): 2869–2880.

Gomez, A., E. Nichols, and S. L. Perkins. 2012. Parasite conservation, conservation medicine, and ecosystem health. In A. A. Aguirre, R. S. Ostfeld, and P. Daszak, Eds. *New Directions in Conservation Medicine: Applied Cases of Ecological Health*, New York, Oxford University Press, pp. 67–81.

Griffith, G. P. and E. A. Fulton. 2014. New approaches to simulating the complex interaction effects of multiple human impacts on the marine environment. *ICES Journal of Marine Science* 71(4): 764–774.

Guidetti, P. and F. Micheli. 2011. Ancient art serving marine conservation. *Frontiers in Ecology and the Environment, Ecological Society of America* 9: 374–375.

Haapkyla, J., R. K. F. Unsworth, M. Flavell, D. G. Bourne, B. Schaffelke, and B. L. Willis. 2011. Seasonal rainfall and runoff promote coral disease on an inshore reef. *PLoS One* 6(2): e16893.

Halpern, B. S., C. Longo, D. Hardy, K. L. McLeod, J. F. Samhouri, S. K. Katona, and D. Zeller. 2012. An index to assess the health and benefits of the global ocean. *Nature* 488(7413): 615–620.

Halpern, B. S., S. Waldbridge, K. A. Selkoe, C. V. Kappel, F. Micheli, C. D'Agrosa, J. F. Bruno, K. S. Casey, C. Ebert, H. E. Fox, R. Fujita, D. Heinemann, H. S. Lenihan, E. M. P. Madin, M. T. Perry, E. R. Selig, M. Spalding, R. Steneck, and R. Watson. 2008. A global map of human impact on marine ecosystems. *Science* 319: 948–952. www.nceas.ucsb.edu/globalmarine (accessed July 2, 2015).

Harvell, D., S. Altizer, I. M. Cattadori, L. Harrington, and E. Weil. 2009. Climate change and wildlife diseases: When does the host matter the most? *Ecology* 90(4): 912–920.

Harvell, D., R. Aronson, N. Baron, J. Connell, A. Dobson, S. Ellner, L. Gerber, K. Kim, A. Kuris, H. McCallum, K. Lafferty, B. McKay, J. Porter, M. Pascual, G. Smith, K. Sutherland, and J. Ward. 2004. The rising tide of ocean diseases: unsolved problems and research priorities. *Frontiers in Ecology and the Environment* 2(7): 375–382.

Hechinger, R. and K. Lafferty. 2005. Host diversity begets parasite diversity: bird final hosts and trematodes in snail intermediate hosts. *Proceedings of the Royal Society B-Biological Sciences* 272: 1059–1066.

Hoegh-Guldberg, O., P. J. Mumby, A. J. Hooten, R. S. Steneck, P. Greenfield, E. Gomez, C. D. Harvell, P. F. Sale, A. J. Edwards, K. Caldeira, N. Knowlton, C. M. Eakin, R. Iglesias-Prieto, N. Muthiga, R. H. Bradbury, A. Dubi, M. E. Hatziolos. 2007. Coral reefs under rapid climate change and ocean acidification. *Science* 318(5857): 1737–1742.

Hofmann, G. E., C. A. Blanchette, E. B. Rivest, and L. Kapsenberg. 2013. Taking the pulse of marine ecosystems: the importance of coupling long-term physical and biological observations in the context of global change biology. *Oceanography* 26(3): 140–148.

Hudson, P. J., A. P. Dobson, and D. Newborn. 1998. Prevention of population cycles by parasite removal. *Science* 282: 2256–2258.

Hudson, P. J., A. P. Dobson, and K. D. Lafferty. 2006. Is a healthy ecosystem one that is rich in parasites? *Trends in Ecology and Evolution* 21: 381–385.

Jackson, J. B. C., M. X. Kirby, W. H. Berger, K. A. Bjorndal, L. W. Botsford, B. J. Bourque, R. H. Bradbury, R. Cooke, J. Erlandson, J. A. Estes, T. P. Hughes, S. Kidwell, C. B. Lange, H. S. Lenihan, J. M. Pandolfi, C. H. Peterson, R. S. Steneck, M. J. Tegner, and R. R. Warner. 2001. Historical overfishing and the recent collapse of coastal ecosystems. *Science* 293: 629–638.

Jackson, M. H. and W. M. Hutchison. 1989. The prevalence and source of *Toxoplasma* infection in the environment. *Advances in Parasitology* 28: 55–106.

Kaplan, D. M., S. Planes, C. Fauvelot, T. Brochier, C. Lett, N. Bodin, F. Le Loc'h, Y. Tremblay, and J. Georges. 2010. New tools for the spatial management of living marine resources. *Current Opinion in Environmental Sustainability* 2: 88–93.

Kennedy, E. V., C. T. Perry, P. R. Halloran, R. Iglesias-Prieto, C. H. L. Schonberg, M. Wisshak, A. U. Form, J. P. Carricart-Ganivet, M. Fine, C. M. Eakin, and P. J. Mumby. 2013. Avoiding coral reef functional collapse requires local and global action. *Current Biology* 23: 912–918.

Kilgour, B. W., M. G. Dube, K. Hedley, C. B. Portt, and K. R. Munkittrick. 2007. Aquatic Environmental Effects Monitoring Guidance for Environmental Assessment Practitioners. *Environmental Monitoring and Assessment* 130(1–3): 423–436.

Kittinger, J. N., J. M. Pandolfi, J. H. Blodgett, T. L. Hunt, H. Jiang, K. Maly, L. E. McClenachan, J. K. Schultz, and B. A. Wilcox. 2011. Historical Reconstruction Reveals Recovery in Hawaiian Coral Reefs. *PLoS ONE* 6(10): e25460.

Knutson, T. R., J. L. McBride, J. Chan, K. Emanuel, G. Holland, C. Landsea, I. Held, J. P. Kossin, A. K. Srivastava, and M. Sugi. 2010. Tropical cyclones and climate change. *Nature Geoscience* 3(3): 157–163.

Koch, M., G. Bowes, C. Ross, and X. H. Zhang. 2013. Climate change and ocean acidification effects on seagrasses and marine macroalgae. *Global Change Biology* 19: 103–132.

Kroeker, K. J., F. Micheli, M. C. Gambi, and T. R. Martz. 2011. Divergent ecosystem responses within a benthic marine community to ocean acidification. *Proceedings of the National Academy of Sciences of the United States of America* 108: 14515–14520.

Kroeker, K. J., R. L. Kordas, R. Crim, I. E. Hendriks, L. Ramajo, G. S. Singh, C. M. Duarte, and J. P. Gattuso. 2013. Impacts of ocean acidification on marine organisms: quantifying sensitivities and interaction with warming. *Global Change Biology* 19: 1884–1896.

Lackey, R. T. 2001. Values, policy, and ecosystem health. *BioScience* 51(6): 437–443.

Lafferty, K. D. 2012. Biodiversity loss decreases parasite diversity: theory and patterns. *Philosophical Transactions of the Royal Society B-Biological Sciences* 367: 2814–2827.

Lafferty, K. D. and T. Ben-Horin. 2013. Abalone farm discharges the withering syndrome pathogen into the wild. *Frontiers in Microbiology* 4: 373.

Lafferty, K. D., J. C. Shaw, A. M. Kuris. 2008a. Reef fishes have higher parasite richness at unfished Palmyra Atoll compared to fished Kiritimati Island. *EcoHealth* 5: 338–345.

Lafferty, K. D., S. Allesina, M. Arim, C. J. Briggs, G. De Leo, A. P. Dobson, J. A. Dunne, P. T.J. Johnson, A. M. Kuris, D. J. Marcogliese, N. D. Martinez, J. Memmott, P. A. Marquet, J. P. McLaughlin, E. A. Mordecai, M. Pascual, R. Poulin, and D. W. Thieltges. 2008b. Parasites in food webs: the ultimate missing links. *Ecology Letters* 11: 533–546.

Lefevre, T., C. Lebarbenchon, M. Gauthier-Clerc, D. Misse, R. Poulin, and F. Thomas. 2009. The ecological significance of manipulative parasites. *Trends in Ecology and Evolution* 24: 41–48.

Lehmann, T., P. L. Marcet, D. H. Graham, E. R. Dahl, and J. P. Dubey. 2006. Globalization and the population structure of *Toxoplasma gondii*. *Proceedings of the National Academy of Science* 103 (30): 11423–11428.

Leopold, A. 1949. The land ethic. In: *A Sand County Almanac*. New York, Oxford University Press, pp. 201–226.

Lessios, H. 1988. Mass mortality of *Diadema antillarum* in the Caribbean: what have we learned? *Annual Review of Ecology and Systematics* 19: 371–393.

Lotze, H. K. and B. Worm. 2009. Historical baselines for large marine animals. *Trends in Ecological Evolution* 24: 254–262.

Lotze, H. K., H. S. Lenihan, B. J. Bourque, R. H. Bradbury, R. G. Cooke, M. C. Kay, S. M. Kidwell, M. X. Kirby, C. H. Peterson, and J. B. C. Jackson. 2006. Depletion, degradation, and recovery potential of estuaries and coastal seas. *Science* 312: 1806–1809.

McCauley, D. J., E. Power, D. W. Bird, A. McInturff, R. B. Dunbar, W. H. Durham, F. Micheli, and H. S. Young. 2013. Conservation at the edges of the world. *Biological Conservation* 165: 139–145.

McGeogh, M. A. 1998. The selection, testing and application of terrestrial insects as bioindicators. *Biological Reviews* 73:181–201.

McKechnie, I., D. Lepofsky, M. L. Moss, V. L. Butler, T. J. Orchard, G. Coupland, F. Foster, M. Caldwell, and K. Lertzman. 2014. Archaeological data provide alternative hypotheses on Pacific herring (*Clupea pallasii*) distribution, abundance, and variability. *PNAS* 111(9): E807-E816. www.pnas.org/content/111/9/E807.abstract www.pnas.org/cgi/doi/10.1073/pnas.1316072111 (accessed July 2, 2015).

Micheli, F., A. Saenz, A. Greenley, L. Vazquez, A. Espinoza, M. Rossetto, and G. De Leo. 2012. Evidence that marine reserves enhance resilience to climatic impacts. *PLoS ONE* 7(7): e40832.

Micheli, F., P. Mumby, D. Brumbaugh, K. Broad, C. Dahlgren, A. Harborne, K. Holmes, C. Kappel, S. Litvin, and J. Sanchirico. 2014. High vulnerability of diversity and function in Caribbean coral reefs. *Biological Conservation* 171: 186–194.

Morley, N. J. 2010. Interactive effects of infectious diseases and pollution in aquatic molluscs. *Aquatic Toxicology* 96: 27–36.

Mumby, P. J., A. Hastings, and H. J. Edwards. 2007. Thresholds and the resilience of Caribbean coral reefs. *Nature* 450: 98–101.

Niemeijer, D. and R. S. de Groot. 2008. A conceptual framework for selecting environmental indicator sets. *Ecological Indicators* 8(1): 14–25.

Orians, G., M. Dethier, C. Hirschman, A. Kohn, D. Patten, and T. Young. 2012. *Sound Indicators: A review for the Puget Sound Partnership*. Washington, DC, Washington State Academy of Sciences.

Overstreet, R. M. 2013. Waterborne parasitic diseases in the ocean. In: P. Kanki and D. J. Grimes, eds. *Infectious Diseases: Selected entries from the Encyclopedia of Sustainability Science and Technology*, New York, Springer-Verlag, pp. 431–496.

Palstra, A. P., D. F. M. Heppener, V. J. T. Van-Ginneken, C. Szekely, and G. E. E. J. M. Van-Den-Thillart. 2007. Swimming performance of silver eels is severely impaired by the swim-bladder parasite *Anguillicola crassus*. *Journal of Experimental Marine Biology and Ecology* 352: 244–256.

Pauly, D. 1984. A mechanism for the juvenile-to-adult transition in fishes. *ICES Journal of Marine Science* 41: 280–284.

Pauly, D. 1995. Anecdotes and the shifting baseline syndrome of fisheries. *Trends in Ecological Evolution* 10: 430–430.

Pespeni M. H., B. T. Barney, and S. R. Palumbi. 2013. Differences in the regulation of growth and biomineralization genes revealed through long-term common-garden acclimation and experimental genomics in the purple sea urchin. *Evolution* 67(7): 1901–1914.

Pinsky, M. L., B. Worm, M. J. Fogarty, J. L. Sarmiento and S. A. Levin. 2013. Marine taxa track local climate velocities. *Science* 341: 1239–1242.

Poulin, R. and K. Mouritsen. 2005. Parasites boost biodiversity and change animal community structure by trait-mediated indirect effects. *Oikos* 108: 344–350.

Powles, H. and M. J. Bradford, R. G. Bradford, W. G. Doubleday, S. Innes, and C. D. Levings. 2000. Assessing and protecting endangered marine species. *ICES Journal of Marine Science* 57: 669–676.

Samhouri, J. F., A. J. Haupt, P. S. Levin, J. S. Link, and R. Shuford. 2013. Lessons learned from developing integrated ecosystem assessments to inform marine ecosystem-based management in the USA. *ICES Journal of Marine Science* 71(5): 1205–1215.

Samhouri, J. F., S. E. Lester, E. R. Selig, B. S. Halpern, M. J. Fogarty, C. Longo, and K. L. McLeod. 2012. Sea sick? Setting targets to assess OH and ecosystem services. *Ecosphere* 3(5): art41.

Scheffer, M., S. R. Carpenter, T. M. Lenton, J. Bascompte, W. Brock, V. Dakos, J. van de Koppel, I. A. van de Leemput, S. A. Levin, and E. H. van Nes. 2012. Anticipating critical transitions. *Science* 338: 344–348.

Shuval, H. 2003. Estimating the global burden of thalassogenic diseases: human infectious diseases caused by wastewater pollution of the marine environment. *Journal of Water Health* 1: 53–64.

Sokolow, S. 2009. Effects of a changing climate on the dynamics of coral infectious disease: a review of the evidence. *Diseases of Aquatic Organisms* 87(1–2): 5–18.

Suttle, C. A. 2005. Viruses in the sea. *Nature* 437: 356–361.

Tett, P., R. J. Gowen, S. J. Painting, M. Elliott, R. Forster, D. K. Mills, E. Bresnan, E. Capuzzo, T. F. Fernandes, J. Foden, R. J. Geider, L. C. Gilpin, M. Huxham, A. L. McQuatters-Gollop, S. J. Malcolm, S. Saux-Picart, T. Platt, M. F. Racault, S. Sathyendranath, J. van der Molen, and M. Wilkinson. 2013. Framework for understanding marine ecosystem health. *Marine Ecology Progress Series* 494: 1–27.

Tewksbury, J. J., R. B. Huey, and C. A. Deutsch. 2008. Putting the Heat on Tropical Animals. *Science* 320(5881): 1296–1297.

Torchin, M., K. D. Lafferty, and A. M. Kuris. 2002. Parasites and marine invasions. *Parasitology* 124: S137–S151.

Torrissen, O., S. Jones, F. Asche, A. Guttormsen, O. T. Skilbrei, F. Nilsen, T. E. Horsberg, and D. Jackson. 2013. Salmon lice: Impact on wild salmonids and salmon aquaculture. *Journal of Fish Disease* 36: 171–194.

Vasseur, D. A., J. P. DeLong, B. Gilber, H. S. Greig, C. D. G. Harley, K. S. McCann, V. Savage, T. D. Tunney, and M. I. O'Connor. 2014. Increased temperature variation poses a greater risk to species than climate warming. *Proceedings of the Royal Society Bulletin* 281: 2013–2612.

Walker, P. J. and J. R. Winton. 2010. Emerging viral diseases of fish and shrimp. *Veterinary Research* 41: 51–65.

Weilhoefer, C. L. 2011. A review of indicators of estuarine tidal wetland condition. *Ecological Indicators* 11: 514–525.

Wilberg, M., M. E. Livings, Barkman, J. S., Morris, B. T. 2011. Overfishing, disease, habitat loss, and potential extirpation of oysters in upper Chesapeake Bay. *Marine Ecology Progress Series* 436: 131–144.

Williams, G. J., G. S. Aeby, and S. K. Davey. 2008. Coral disease at Palmyra Atoll, a remote reef system in the Central Pacific. *Coral Reefs* 28(1): 207.

Wood, C. L. and K. D. Lafferty. 2014. How have fisheries affected parasite communities? *Parasitology* 142(1): 134–144.

Wood, C. L., K. D. Lafferty, and F. Michaeli. 2010. Fishing out marine parasites? Impacts of fishing on rates of parasitism in the ocean. *Ecology Letters* 13(6): 761–775.

Wood, C. L., S. Sandin, B. Zgliczynski, A. S. Guerra, and F. Micheli. 2014. Fishing drives declines in fish parasite diversity and has variable effects on parasite abundance. *Ecology* 95: 1929–1946.

Worm, B., E. B. Barbier, N. Beaumont, J. E. Duffy, C. Folke, B. S. Halpern, J. B. C. Jackson, H. K. Lotze, F. Micheli, S. Palumbi, E. Sala, K. A. Selkoe, J. J. Stachowicz, and R. Watson. 2006. Impacts of biodiversity loss on ocean ecosystem services. *Science* 314: 787–790.

8

MARINE SCIENTIFIC RESEARCH

Overview of major issues, programmes and their objectives

Montserrat Gorina-Ysern

Introduction

Today, the relevance of Part XIII, 1982 United Nations Convention on the Law of the Sea (hereinafter 1982 UNCLOS), governing Marine Scientific Research (hereinafter MSR), within public international law scholarship is understood and recognized. However, MSR remains a marginal area of legal scholarship generally because MSR activity triggers very few controversies on the ground worldwide and therefore it does not generate a case law and a corresponding jurisprudence. The satisfactory implementation on the international and domestic planes of the 1982 UNCLOS regime for MSR should be a cause for optimism.

MSR activity, however, is increasingly characterized by multifaceted layers of technological, technical, scientific, political, and industrial complexity. This complexity is compounded by the lack of a legal definition for MSR and an endemic lack of capacity at the worldwide base, both of which lend an esoteric character to MSR for those outside the field and, increasingly, for those within it as well. These layers of complexity thicken over a wide network of private and public institutions and stakeholder interests, because MSR does not operate in isolation from other fields of activity governed by other areas of international, regional, and domestic law. Each stakeholder pushes or pulls in a different direction, motivated by wider or narrower agendas, and therefore MSR activity is increasingly governed by other existing and evolving legal regimes outside the 1982 UNCLOS. This is relevant because the small circle of MSR legal experts no longer share a common working terminology with the massive number of climate, environmental, conservation, microbiology, intellectual property, industry, and other scientists, policy makers, and legal experts whose fields intersect with MSR activity. Within the United Nations system, funding for comprehensive studies and follow-ups on MSR regime implementation in areas within national jurisdiction have been limited; however, over the last decade funding for the study of "marine genetic resources" in areas beyond national jurisdiction has thrived, in spite of the statistically unrepresentative incidence of activity in those remotest ocean depths. The terminology challenges that the term "marine genetic resources," coined by policy makers in the marine conservation field, poses for microbiology (Heidelberg *et al.* 2008), intellectual property and law of the sea governance purposes, need to be addressed thoroughly.

Complexity of subject-matter and field-compartmentalization affect the agendas, dynamics, and content of institutional negotiations that are necessary to achieve an effective implementation of the 1982 UNCLOS MSR regime worldwide. Compartmentalization is a barrier to effective integration, for governance and law enforcement purposes, of 1982 UNCLOS provisions with climate, environmental, conservation, microbiology, intellectual property regimes—and their institutional dynamics. Other fields that also need to be integrated with the 1982 UNCLOS regime include relevant industries, scientific developments, governmental and inter-governmental policies, and legal scholarship across all the fields being integrated. Often, complexity and compartmentalization also affect the political will to reach a clear, holistic, and integrated understanding of the physical phenomena in order to regulate and to govern oceanic research in all its facets under transparent and predictable domestic and international laws. In light of the above, there have been calls for the adoption of new treaties, in spite of the practical limitations and delays that new treaty making faces pursuant to the norms and methodology of international law. Together, complexity, compartmentalization, the absence of a common terminology, conflicting motivations, high costs of organizing conferences and symposia to figure a way forward, and the lack of political will, undermine the chances that younger generations can be and are being adequately trained to become the stewards of the oceans through MSR activity and other oceanographic research. Because field capacity at the most basic level of technical and scientific knowledge is lacking in most regions of the worldwide base (i.e., elementary school grades, high schools, colleges and universities, laboratories and centers of study), and also within the MSR institutional inter-agency and inter-governmental maze (UN Secretary-General Report 2003, 2005; 2007; 2013), lofty efforts to protect coastal communities and global ocean health and wealth are not likely to succeed in the short term or near future. The task is daunting and the capacity to accomplish it is very limited.

To illustrate and to support the preceding observations four original figures are provided. Figure 8.1 depicts maritime zones where sovereignty, sovereign rights, and jurisdiction are exercised for MSR purposes, and divides ocean space between areas within and areas beyond national jurisdiction under the 1982 UNCLOS. Figure 8.2 (Oceanic Research Family) outlines five major technical and scientific areas of activity inter-twined with but different from MSR under Part XIII, 1982 UNCLOS. Compromise on a legal definition for MSR as distinct from these interrelated activities has not materialized (Bork *et al.* 2008), in spite of scholarly and practical efforts by Roach (2007). Table 8.1 is not set in alphabetical or hierarchical order, but reflects what oceanic and inter-governmental agencies have considered relevant and pressing to report for inclusion in the annual Reports on Oceans and Law of the Sea by the Secretary-General of the United Nations (2002–2013). Table 8.2 provides an original and comprehensive legal analysis toolbox on major legal frameworks governing MSR activities, taking into account core scientific interest, jurisdictional zone, rights and duties at stake, and applicable public as well as private international and municipal (or domestic) laws.

Background

To fully understand the Legal Analysis Toolbox provided in Table 8.2, some preliminary observations are necessary. It is not an exaggeration to say that from the perspective of international and comparative law, MSR legal scholarship has been a rare jewel. As Professor Alfred H. A. Soons, of the NILOS remarked (Soons 2007), since the publication of his 1982 study (Soons 1982), *only two books* on MSR in the English language have been published:

One by Dr. Florian H. Th. Wegelein (2005), and the other by this author (Gorina-Ysern 2003, 2004). For those outside this *Moses inner circle*, it is important to know that in 1982 Soons published a masterpiece on the international codification of the MSR regime under the 1982 UNCLOS. To understand the degree of achievement that this represented, it is fair to observe that a leading authority on the Law of the Sea in the early 1980s was D.P. O'Connell, whose two volumes had been edited by the great Australian law of the sea scholar, Professor Ivan Shearer, and in one volume there were exactly *eight pages* devoted to marine scientific research (O'Connell 1984). There appeared to be no other legal treatment of MSR in the English language in any other Law of the Sea manual at the time. There were about a dozen articles in the English language on the legal aspects of marine science or oceanographic research in the 1970s decade. This was the state of the art on which Soons could build his PhD thesis. After Soons' pioneering study, the literature on MSR in English increased modestly and, throughout the 1980s and 1990s, other experts published reports and analysis relating to MSR activity from different perspectives (Treves 1980, 2008; Yankov 1983; de Marffy 1985; Franckz 1986 and 1990; Ross and Landry 1987; Fenwick 1992; Glowka 1996; Roach 1996; Brandon 1997), though virtually none approached MSR from the perspective of the sources of public international law. In 1999 MSR was given sixteen pages in a leading manual on the Law of the Sea (Churchill and Lowe 1999). Lee Stevens produced a practical Handbook for International Operations of U.S. Scientific Research Vessels (UNOLS Handbook 1986). This was the legal state of the art in the English language for MSR. Since 2000, German scholarship on the MSR regime legal implementation has been considerable, thanks to the interest raised by the writings of Judge Rüdiger Wolfrum of the International Tribunal on the Law of the Sea (ITLOS), and Dr. Nele Matz, of the Max Planck Institute (Wolfrum and Matz, 2000); and the work by Bork, Karstensen, Visbeck, and Zimmermann (2008) on the legal regulation of floats and gliders, as well as by Hubert (2011) on potential environmental impacts of MSR. In the Spanish language, legal scholarship from Argentina and Italy has enriched the field. Tome Young has outlined, however, the magnitude of the scholarship gap on MSR activity and its implications for the Convention on Biological Diversity and the United Nations Environmental Program camps (Young 2009).

The usefulness of MSR activity is clearly stated by the Secretary-General of the United Nations in his 2010 Annual Report on Oceans and Law of the Sea. The generic concept of marine science is described *by its utility* as follows:

> a tool for exploring, understanding and using the marine environment in a sustainable manner. Enhancing humankind's knowledge of the natural processes of the oceans, marine science and its supporting technologies can support decision-making, contribute to improving integrated coastal management and the sustainable utilization of marine resources and provide effective means for the protection and conservation of the marine environment and its resources [making] a major contribution to the elimination of poverty, ensuring food security, supporting human economic activity, conserving the world's marine environment, helping predict and mitigate the effects of, and respond to, natural events and disasters, and generally promoting the use of the oceans and their resources.
>
> (*UN Secretary General Report 2010*)

Historically, MSR is anchored in scientific missions and voyages of exploration financed by Royal Houses and Kingdoms, carried out pre-eminently on board naval vessels under Admiralty rule. This characteristic lasted well into the twentieth century because civilian and philanthropic

Table 8.1 Major MSR/oceanic research issues/concerns expressed to the UN Secretary General (2002–2013)

Issue/concern	Description	UNSG Report
Scientific equipment and instrumentation vandalism creates gaps in knowledge and services; Damage to submarine cables/pipelines.	Since 2002, JCOMM warns of vandalism to ocean data buoys (drifting and moored). These are essential sources of meteorological and oceanographic data for GOOS, distributed freely and globally to shore in real time via satellite on the GTS of WMO's World Weather Watch. GOOS supports weather forecasts, warning services, global climate and global change monitoring, prediction, and research. UK model laws against damage to cables/pipelines.	2003 2010(3)
Massive oceanic research/MSR data/information gaps.	Scarce funding decreases observations by Voluntary Observing Ships (VOS) for maritime services by National Meteorological Services (i.e., safety of navigation and life and property at sea). GOOS supports UNFCCC global climate implementation by 2015, but GEOSS lacks enough ocean sampling.	2003
Lack of capacity, sufficient and effective marine scientific/technical training for individuals and institutions in many coastal States.	WMO, IOC, JCOMM (DBCP), IMO and IHO issue notices to mariners and VOS. Lack of capacity is a serious obstacle to the conduct of MSR and all modes of oceanic research. Many countries lack capacity to collect essential oceanographic data/operational observations. IOC undertakes massive efforts to increase capacity in ocean mapping in Africa and Asia (i.e., bathymetric charts for tsunami inundation modeling and training more hydrographers and marine scientists).	2003 2010(1) 2010(3)
Danger of radioactive material in food chain.	IAEA and research partners (under the London Dumping Convention) collect extensive information and research data on radioactive waste and materials dumped at sea, as well as harmful algal blooms potentially contaminating fish – so more research is needed in this critical field.	2003
Serious and extensive hydrographic survey gaps.	IMO and IHO under SOLAS Ch. V urge Governments to collect hydrographic data/nautical information for the safety of navigation (nautical charts, sailing directions, list of lights, tide tables, etc.). Under MARPOL (Annexes I, II and V), IMO has implemented PSSAs to combat marine pollution from vessels sources (ballast water, anti-fouling paint harmful effects, ship recycling, green house gas emissions, etc.). IMO and IHO call for greater chart data accuracy and electronic data display. Hydrographic survey gaps occur due to lack of capacity building, methodology training and better data management applied to marine pollution, coastal zone management, sensitive systems identification/ monitoring, and for legal disputes to extended continental shelf areas.	2003 2007
Importance of early warning systems.	Vandalism of tsunameters in Indian Ocean undermines array of drifting buoys, Argo profiling floats, sea level stations and other instruments deployed in support of tsunami warning systems under GOOS, GCOS, the foundation of global operational oceanography. UNESCO, IOC and IMSO initiate near real time seismic data in Pacific and Indian Oceans. Guidelines on hazard awareness adopted.	2007 2008 2010(3)

Table 8.1 continued

Marine pollution research incomplete.	HELCOM conducts Baltic oxygen depletion and eutrophication, agricultural, oil spill pollution research and vulnerable ecosystem (BSPA), fisheries research (IBSFC).	2003–13
MSR: a villain or a savior? Confusing reports.	In the early 1990s, Agenda 21 (Chapters 34–35) recognized availability of scientific information as an essential requirement for sustainable development. However 2003 to 2005 UNSGRs portray MSR activities as the villain of the oceans.	2002–3 2004–5
Limited access to remote sensing information worldwide; inadequate ocean sampling.	IODE (International Oceanographic Data Exchange) within the IOC. Limited access by developing States to oceanic remote sensing information under EOS-GEO, IGOS, COSMAR/NEPAD and other GOOS, JCOM, GODAE and ITSU collaborative and interdisciplinary programs applying an ecosystem approach to ocean management through ICAM and WOCE.; inadequate ocean sampling to achieve GEOSS goals.	2003–4 2010(3)
More research on sanctuaries needed.	IWC has undertaken extensive decadal research on marine mammal sanctuaries in collaboration with major programs and organizations (CMS, ASCOBANS, ACOBAMS, ICES, CCAMLR, GLOBEC, NANMCO, FAO, PICES, CITES, IUCN, ECCO).	2003
Better integrated efforts for marine science supporting sustainable development through ecosystem approach.	Since 1974 UNEP focuses on its Regional Sea Programs for sustainable development and integrated management of coastal areas, river basins, living and aquatic resources, in legal collaboration with OECD and IMO (adopting Regional Seas Conventions and Action Plans), to protect vulnerable ecosystems, assess economic importance of marine environment, and effective inter-agency cooperation to those ends. Carbon sequestration research needs to proceed under these principles. UNEP collaborates with OBIS and COLM on marine biodiversity. ISOM and OSPAR codes adopted.	2006 2008
Terminological chaos for biological and microbiological research (by the misnomer marine genetic resources).	European Commission funds extensive private and public biological, bio-chemical and microbiological research (OMARC cluster). Arctic research can be a model for bio-prospecting. Problematic definition of "marine genetic resources" and large cost of research require clarification and private/public funding. Scientific databases are indispensable and IFREMER offers Algae Base and BIOCEAN. IOC Assembly promotes chemical and biological ocean data IODE equivalent system for sharing collections and inventories (SPODIM, IODIE).	2003 2006 2007 2008
International Seabed Authority MSR data repository.	Vast amounts of MSR data on polymetallic nodules integrated with environmental data and deep sea ecosystems environmental research are stored at and used by the ISA in its Central Data Repository.	2003

Source: © 2013 M.Gorina-Ysern.

Table 8.2 Legal analysis toolbox for MSR activities

Analysis Toolbox for MSR activities	By core scientific interest	By UNCLOS areas within and beyond the limits of national jurisdiction	By rights and duties of coastal, researching states and third parties	By applicable private, municipal and public international law
Field of oceanic research activity	a) physical b) chemical c) biological d) geological e) geophysical f) satellite altimetry g) other	As a general rule, the low water mark along the coastline at low tide serves as baseline for the measurement of coastal State limits of national jurisdiction see below)	(There are binding and non-binding rights and duties; see below)	MSR and its Public International Law Sources: 1945 Statute of the International Court of Justice (Art. 38) a) Conventions b) Custom c) General Principles of Law d) Judicial decisions e) Doctrine f) Equity
Nature of oceanic research activity	Intended research use/output: • fundamental MSR • hydrographic/other surveys • applied research (bio-discovery, bio-surveying and bio-prospecting) • living resource exploration • mineral, oil and gas prospecting • industrial research and other	UNCLOS does not define key terms such as Jurisdiction; nor MSR, or provides limited definitions reflective of political compromise among States	(There are binding and non-binding rights and duties; see below)	MSR provisions interpreted consistently with Rules of International Law Interpretation: 1969 Vienna Convention on the Law of Treaties i) hierarchies among treaties ii) binding force iii) parties/non-parties iv) breach and non-performance v) remedies
Jurisdictional area (by distance to low-water mark) where oceanic or other marine research/survey activity is intended	(as above)	• internal waters/archipelagic waters • international straits • enclosed or semi-enclosed seas • territorial sea and contiguous zone • exclusive economic zone/fishery zone • continental shelf • extended continental shelf • high seas [including PSSA, LME, MPA enclosures consistent with UNCLOS] • The Area • airspace–atmosphere–sea interaction • special Arctic/Antarctic treaty areas	Legally binding; Coastal State exercises in different zones: • sovereignty • sovereign rights • jurisdiction • ancillary rights Researching State is bound by: • enforceable obligations • duties to consult and to cooperate • duty to enforce laws over physical/juridical persons, equipment, • instrumentation, vessels and platforms	• Municipal laws of Coastal State (local, regional, state, federal) • Bilateral, Regional, Private/Public Research/Industrial Agreements (may cover IPRs) • 1958 Geneva Convention on the High Seas • 1958 Geneva Convention on the Continental Shelf • 1982 UN Convention on the Law of the Sea • UNCLOS Implementing Agreements bearing on MSR activity • 1973 Convention on the International Protection of Endangered Species of Wild Fauna and Flora • 1992 Framework Convention on Climate Change

Table 8.2 continued

				• FAO, IMO, UNEP conservation/environmental agreements • 1992 Convention on Biological Diversity • Other Agreements impinging on MSR activity
Maritime zone (by depth) where oceanic or other marine research/survey activity is intended	(as above)	• coastal zone/Neritic zone (littoral/sublittoral zones with splash, inter-tidal and subtidal) oceanic zone (in meters) with: • Epipelagic/Photic zone (0–200m) • Mesopelagic zone (200–1000m) • Bathypelagic zone (1000–4000m) • Aphotic zone (2000–4000m) • Abyssopelagic zone (4000–6000m) • Hadal zone (beyond 6000m depth)	Non-binding Codes and Guidelines: • 1991 and 2010 DOALOS Guidelines • IOC/ABE-LOS Guidelines • ISOM/OSPAR/Inter-Ridge Codes of Conduct • Census of Marine Life Codes of Conduct • Venter Institutes Codes of Conduct • AIMS Codes of Conduct • Bonn Guidelines, Nagoya Protocol • other	(By UNCLOS areas; by existing and emerging Rights and Duties; by other areas of Public International and Domestic Laws)
Distribution of intellectual property rights (International Agreements)	Linked to intended research output: • fundamental MSR/other research • hydrographic/other surveys • applied research (i.e., bio-discovery, bio-surveying and bio-prospecting) • living resource exploration • mineral, oil and gas prospecting • industrial research and other	• internal waters/archipelagic waters • international straits • enclosed or semi-enclosed seas • territorial sea and contiguous zone • exclusive economic zone/fishery zone • continental shelf • extended continental shelf • high seas/PSSA, LME, MPA • The Area • airspace–atmosphere–sea interaction • special Arctic/Antarctic treaty areas	a) Open access mechanisms b) Shared and/or Proprietary: • samples, specimens • compilations of data • written output (reports, articles) • bio-active molecules and active compounds for isolation, characterization, extraction, purification and screening • other output subject to IPRs	Paris Convention (Protection of industrial property) Patent Cooperation Treaty Strasbourg Agreement (International patent classification) Budapest Treaty (Deposit of microorganisms for patent Procedure) International Convention (Protection of new plant varieties) European Patent Convention Berne Convention (Protection of literary and artistic works) Universal Copyright Convention Brussels Convention (Distribution of program-carrying signals transmitted by satellite) Trade related aspects of Intellectual Property (TRIPS) Agmt. Other relevant treaties and conventions

Source: © 2013 Montserrat Gorina-Ysern.

societies, in general, could not afford the cost of craft, instrumentation, and equipment. As a result of the public nature of traditional oceanographic research, much of the information gathered remained classified and was used in naval warfare up to the end of WWII, when a move by the US and other Western nations to de-classify such information resulted in the explosion of oceanographic data and research availability worldwide (Gorina-Ysern 1988–2010). New technologies have facilitated new fields of inquiry underpinning ocean uses – oil and gas extraction, long distance fisheries. Satellite altimetry applied to MSR and other oceanographic research activity has opened up since 2000 the vast field of operational oceanography applied to climate change. The expansive march of marine science has brought attention to the need for legally binding regional and global environmental laws, as well as hortatory codes of conduct calling for all scientific activities at sea to protect the environment and preserve its habitats (Defying Oceans' End 2004). Over a century after Gregor Mendel experimented with pea pods, the complex science of genomics has opened the DNA secrets of marine biota for the world to study, survey, prospect, collect and exploit (see Gorina-Ysern 1993–2009; Nicholl 1994/2002; Glowka 1996; National Research Council 2003; Leary 2007; Heidelberg *et al.* 2008; Young 2009; Fedder 2013; UN General Assembly 2013). The intersection of such complex fields calls for terminological clarity for educational and governance purposes under the 1982 UNCLOS and related regimes.

Major international programs and current issues of concern

In the last decade (2002/3 to 2013) inter-governmental agencies within the UN system with mandates bearing on the conduct of MSR and other oceanographic research, and other public institutions, have reported progress with major international programs as well as issues of concern to the Secretary-General of the United Nations for inclusion in his annual Report on Oceans and the Law of the Sea.

Table 8.1 outlines MSR/oceanic research issues and concerns under international programs. The list is not alphabetical, comprehensive, or by hierarchical order of issues. Each item outlined below is discussed in the box provided in Table 8.1, and the year of the report is provided on the right-hand side corner under *UNSG Report*:

- vandalism of scientific equipment and instrumentation used in early warning systems that aimed to minimize hazards for local communities;
- massive gaps in the collection of oceanographic research data and information;
- lack of capacity in terms of marine scientific technical training and for basic hydrographic surveys, undermining the ability of individuals and institutions to participate actively in the programs of major international, regional and national scientific organizations;
- need for research into radioactive materials entering the food chain though dumping at sea;
- need for research into protection of vulnerable ecosystems and fishing grounds against marine pollution;
- insufficient ocean sampling, exchange of real and near real time, and high quality information exchanged through international mechanisms;
- need for research into marine sanctuaries for mammals;
- implementation of newly adopted Codes of Conduct to achieve the goals of sustainable development through ecosystem approaches;
- better terminology and integration of biological and micro-biological research (under the misnomer "marine genetic resources");
- depository role of the International Seabed Authority for MSR data.

In the 2013 Report on Oceans and the Law of the Sea to the General Assembly, the Secretary-General of the UN announced the adoption of a "global strategy on capacity building" (paragraph 49) on MSR and ocean observations, and a revised strategic plan for IODE (2013–2016; paragraph 53), to respond to these concerns (UN Secretary-General Report 2013).

Governance and policy

Figures 8.1 and 8.2 are provided to illustrate the jurisdictional scope of the MSR regime under the 1982 UNCLOS and the major activities in the oceanic research family. Table 8.2 contributes an original legal analysis toolbox for a comprehensive study of MSR activity governance from private/public agreements to domestic and international law perspectives.

Figure 8.1, describes "legal regime within national jurisdiction" as the ocean space within 200 nautical miles (hereinafter nm) situated above the line, with the water column underneath the water surface line, and enclosing the continental shelf. On the bottom left-hand corner and middle of the figure, each jurisdictional zone is listed with relevant articles governing MSR for that zone under Part XIII, 1982 UNCLOS (Arts. 238 to 265). These articles govern the conduct of MSR activities in the world's oceans, building on principles established for the conduct of fundamental and other oceanographic research activities in the 1958 Geneva Conventions on the Law of the Sea. However, a more comprehensive analysis of the wider legal framework for MSR activities is provided in Table 8.2.

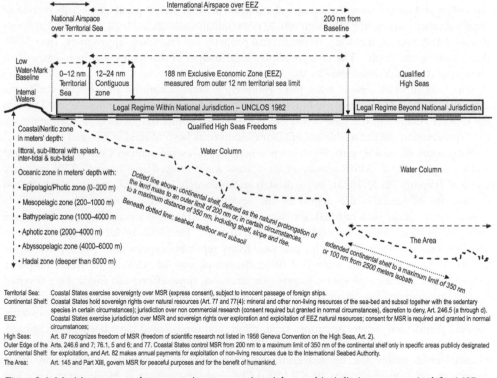

Figure 8.1 Maritime zones where sovereignty, sovereign rights, and jurisdiction are exercised for MSR purposes. Ocean space divided between areas within and areas beyond national jurisdiction under the 1982 UNCLOS.

Source: © Montserrat Gorina-Ysern and Alicia B. Gorina, 2013.

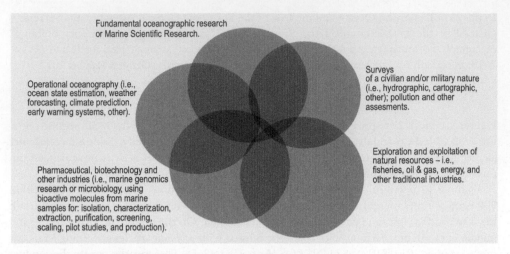

Figure 8.2 Oceanic research family
Source: © Montserrat Gorina-Ysern, 2013.

It is generally understood that for MSR activities to be conducted in areas within national jurisdiction, coastal State consent is required, subject to various provisions. Article 250, 1982 UNCLOS, functions as the executive MSR article whereby the coastal and researching State authorities and competent international organizations or other private parties, especially individual chief scientists, can negotiate through official government channels the legal and technical details of each MSR project, including its intellectual property aspects, where appropriate, over MSR data, samples, and results. The comprehensive legal analysis of this regime has been carried out elsewhere (Gorina-Ysern 1988–2010).

The bulk of MSR activity takes place routinely in areas of national jurisdiction within the 200 nm boundary. However, since November 2004 and throughout the past decade, the Division for Oceans Affairs and the Law of the Sea (DOALOS) agenda has focused pre-eminently on the "conservation and sustainable use of marine biodiversity beyond areas of national jurisdiction" (UN Secretary General Report, 2004; www.un.org/depts/los/index.htm). Commencing in 2003 and throughout 2005, MSR activity in those areas was portrayed as a "threat" (UN Secretary General Reports 2003, 2005), even though such research is minimal, due to the cost and complexity of reaching the depths of the seafloor, and turns out not to be statistically representative of the bulk of MSR activity worldwide. In Figure 8.1, areas beyond 200 nm are described as "legal regime beyond national jurisdiction." Hundreds of UN delegates have sat through lengthy discussions year in, year out. Many reputable experts have participated in the debates and the intellectual output of these discussions may be relevant for MSR purposes. Nevertheless, a strong perception lingers that attention, time, effort, and funds were diverted by sectoral interests (i.e., pharmaceutical and industrial companies engaged in biotechnology development, as well as other commercial uses of biogenetic material), away from areas within national jurisdiction, where attention, time, effort, and funds could have been better utilized, particularly for capacity building purposes. This costly distraction, coupled with political and financial crises within the Intergovernmental Oceanographic Commission's Advisory Body of Experts on Law of the Sea (IOC/ABE-LOS)—whose mandate is MSR governance and policy—has prevented DOALOS from pursuing two useful mandates: consolidating the legal analysis of developments in MSR State practice post 2010 Guidelines; and consideration of public/private MSR agreements reached through official channels (i.e., Art. 250 UNCLOS), dealing with intellectual property rights over MSR data, samples, and results in areas within national jurisdiction as depicted inside the 200 nm line in Figure 8.1.

On the left hand side of Figure 8.1 are listed coastal and oceanic zones based on marine habitats by depth under scientific criteria of study. A more realistic, comprehensive and integrated approach to MSR activity worldwide will require policy makers and legal scholars to incorporate these geographical and scientific criteria into their governance and policy studies, especially in the context of marine spatial planning tools. Major foundations have sustained the work of the Intergovernmental Oceanographic Commission in this area since 2005 (www.unesco-ioc-marinesp.be).

Table 8.2 provides a legal analysis toolbox for MSR activities. Though at first sight the table may look daunting, it is a useful reminder that MSR governance does not end with Part XIII 1982 UNCLOS. Generally, MSR governance and policy issues tend to be reduced to discussing Part XIII, 1982 UNCLOS. However, starting with the right-hand side boxes in the last column of the table, under the rubric "applicable private, municipal and public international law," emphasis is placed on the need to link MSR to public international law and its sources. This approach infuses any study with a sound legal method. Similarly, MSR governance and policy need to be interpreted consistently with other rules of international law, as required by the Vienna Convention on the Law of Treaties (1969) and by the 1982 UNCLOS itself, not least of which because important oceanographic stakeholders such as the US are not part of the 1982 UNCLOS and therefore the 1958 Geneva Conventions on the Continental Shelf and the High Seas may still apply in part to the relationships between the US and other States over MSR activity, on a case by case basis and depending on the circumstances. The third box down lists the major legal sources of MSR governance and policy when considered from a more comprehensive perspective. These include not only other major treaties but also more private and sectoral agreements and arrangements, in addition to municipal laws that affect oceanographic vessels Officers, scientific and ship crews, equipment, instrumentation, vessels, craft, and other aspects of MSR activity in domestic and foreign waters, on the high seas, and in the course of satellite communications between the vessel and the shore. In the bottom right-hand side box of the last column are listed major intellectual property treaties governing MSR data, samples and results in the context of Part XIII, 1982 UNCLOS. These aspects have been covered in great detail elsewhere (Gorina-Ysern 2003, 2004, 2006).

The legal sources of MSR governance vary depending on the field of oceanic research pursued, the nature of the research intended, the physical space where the activity is intended, the maritime zone applicable to that physical space, and the distribution of intellectual property arising from the activity for the scientists involved, their institutions, the coastal State and its own. The first, second, and third columns in Table 8.2 intend to depict the relevance of these elements for governance purposes.

Finally, the fourth column in Table 8.2 addresses the rights and duties of coastal and researching States as well as those of third parties in the governance of MSR activity in areas within national jurisdiction. A distinction is made between legally and non-legally binding rights and duties. An important development for MSR governance and policy is the adoption of codes of conduct and guidelines concerning a) vessel access issues, b) access to MSR data, samples and results, c) protection of habitats against bad sample collection practices, d) and access and benefit sharing best practices.

The 1982 UNCLOS does not define Marine Scientific Research; though it does not define other key concepts such as "sovereignty," "sovereign rights," "jurisdiction," or "resources" either. These concepts are to be interpreted using criteria under the Vienna Convention on the Law of Treaties and other principles of international law, including relevant jurisprudence (though there is none for MSR). The definition of MSR has been a vexing issue ever since the term itself was used from the 1970s onward to refer to what had previously been described as fundamental oceanographic research, distinct from exploration for resources, and experiments, and tests with hydrogen bombs in the Pacific (Gorina-Ysern 2003, 2004).

The vexing lack of a definition for MSR can be succinctly attributed to three major causes:

1   *Lack of capacity.* As depicted in Figure 8.2 (Oceanic Research Family), the wider concept of Oceanographic research includes purely scientific or fundamental marine research, maritime and fisheries industry-useful surveys (cartographic, hydrographic, etc.), operational surveys using de-classified military technologies and applied to climate change studies, and a plethora of industry-sensitive research activities (i.e., fisheries, oil and gas, energy generation, environmental and pollution studies, and relatively new industries such as genomics and microbiology engaging in bio-prospecting in areas within and beyond national jurisdiction). Very sophisticated human resources expertise, scientific, and technological infrastructures are needed to discern what oceanographic activities are purely scientific and which are not. Such capacity is not readily available outside a specific scientific discipline, neither among the vast majority of UN delegates adopting governance regimes for the activities described, nor in most parts of the world.

2   *Ownership of federally funded research.* Since the 1980s, in great part due to the US Supreme Court granting a patent in *Diamond v. Chakrabarty*, (447 U.S. 303, 1980; Sears 1980), and across the OECD region, talk of intellectual property rights has blurred the lines between fundamental, applied, and industrial research. This is particularly noticeable in the marine molecular microbiology field, where the phases of research and discovery, product and process development, and commercialization are intertwined. Commercial links between scientists, governments, companies needing to find whole organisms, hospitals and research institutes, and the commercial and entrepreneurial strategies involved, have blurred the traditional lines between non-commercial and commercially oriented MSR activity. Pride in commercial competition has replaced the independence and transparency that university laboratories and oceanography schools once offered through fundamental oceanographic research studies (Committee on Exploration of the Seas 2003). In the US, the Bayh-Dole Act of 1980 (37 C.F.R 401) required a new approach to intellectual property and ownership of federally funded research. These pressures have turned universities into competitive centers tied to and at the service of commercial and industrial interests. This is not in any way intended as a criticism of honorable and valuable goals pursued in the quest for new antimicrobial agents from ocean mud and samples that offer tangible benefit to humanity as a whole to combat drug resistant infectious diseases caused by an army of nasty microorganisms (see for example Professor William Fenical's lecture in marine microorganisms (Fenical 2008)). The preceding is intended to support what was indicated in Table 8.2, that data, samples, and results from MSR activity may be subject to "open access mechanisms" but also to "shared or proprietary" regimes. There is an urgent need to analyze how these are implemented in private and public agreements of a domestic as well as international nature.

3   *Politics.* In April 2009, DOALOS, Office of Legal Affairs, gathered at the United Nations Headquarters in New York, a Group of twenty-one Experts representing Europe (4), Africa (3), Middle East (1), South Pacific (5), North America (2) and South and Central America (4). Among them, eight had legal credentials in international law and represented Argentina, Canada, India, Indonesia, Jamaica, Peru, Russian Federation, and South Africa. The Group of Experts was tasked to finalize "A revised guide to the implementation of the relevant

provisions of the United Nations Convention on the Law of the Sea." The revised guide was published in 2010 as an update of the 1991 guide, and is a valuable addition to the scarce literature on MSR (*The Law of the Sea. Marine Scientific Research. A revised guide to the implementation of the relevant provisions of the United Nations Convention on the Law of the Sea*, 2010—hereinafter DOALOS revised guide 2010).

The DOALOS revised guide 2010 used a core sample of seventy-two State responses to three IOC questionnaires, representing slightly over one third of the UN membership, to outline "the experience of States" concerning the conduct of MSR and to identify trends. It mentions three major trends: emerging large-scale international collaboration programs in marine data acquisition and dissemination; increased MSR data acquisition from autonomous platforms using technology that increases the cost of ship borne research (as deployment-recovery of instrumentation increases) and the demand for continuous high resolution for long-term ocean observations to meet research as well as social needs; and finally, greater need for collaboration regarding access to and interpretation of large data sets, which requires data dissemination standards and protocols for the increased flow of data exchange at national, regional, and global oceanographic data centers. Underlying these trends, capacity building and the transfer of marine technology remain recurrent challenges in the field of MSR carried out within and beyond the limits of national jurisdiction.

The DOALOS revised guide 2010 advocates for a loose definition of MSR according to which "*the validation by the coastal State of a given [MSR] project is what, in practice, defines its nature*" (italics and brackets added; *para*. 99 at 29). Although the DOALOS revised guide 2010 distinguishes "survey activities," "prospecting," and "exploration" as separate categories of "marine research" (*para*. 14 at p. 6), distinct from MSR in Part XIII, 1982 UNCLOS, it is hardly rigorous for DOALOS experts to conclude that the coastal State has the final word on the scientific "nature" of a project. Such a reluctance to analyze in depth and with nuance the "experience of States" in the provided sample (72), limits the effectiveness of the revised 2010 guide, because it lacks the element of "*opinio juris*" which in international law is the indispensable twin companion of "State experience" when the pursued goal is to update State practice over time.

In essence, the DOALOS revised guide 2010 opted for omitting legal scholarship in the field. The omission prevented DOALOS from using the methodologies of international law appropriate for a more in-depth analysis of *opinio juris*. Technically, the DOALOS revised guide 2010 is also likely to disappoint marine scientists seeking a systematic road map from the IOC Questionnaires on how researching and coastal States approach MSR definitions in their experience with authorizations and clearance negotiations.

Conclusion

The lack of capacity is precisely that: the inability to understand, to engage, and to transform the physical world that surrounds us. In the UN system and in technologically developed countries there are intellectual *avant gardes*, environmentally aware, scientifically and legally trained elites, who live outside problem areas, but who understand the ever mounting and multilayered complexities surrounding MSR and oceanic research activity, and can tackle the list of concerns reported by institutional and specialized agency stakeholders to the UN Secretary General for inclusion in his Annual Reports to the General Assembly (2002–2013). Many of these concerns are tied to poverty and lack of capacity at all levels, especially for younger generations, and compound the challenges posed by terminological chaos. The gap is ever wider between the lofty goals and the inability to achieve them on the ground. Echoing the words of a great sage describing the previous decade: *"the growth of increasing disparities between the worlds of the 'haves' and the 'have-nots'"* in the oceanographic research field remains (Bernal 2001).

References

Bernal, P. (2001), *Intergovernmental Oceanographic Commission. Annual Report*, pp. 8–10. Available at: http://unesdoc.unesco.org/images/0012/001279/127943e.pdf (accessed September 2015).

Bork, K. Karstensen, J., Visbeck, M. and Zimmerman, A. (2008) "The Legal Regulation of Floats and Gliders: In Quest of a New Regime?" in *Ocean Development and International Law*, vol. 39, no. 3, p. 298.

Brandon, M. (1997) "The 1982 United Nations Convention on the Law of the Sea Consent Regime for Marine Scientific Research: Caribbean Coastal State Responses to United States Clearance Requests." Unpublished Master of Marine Affairs Paper, University of Rhode Island.

Churchill, R. R. and Lowe, A. V. (1999) *The Law of the Sea*, Manchester, Manchester University Press.

Committee on Exploration of the Seas (2003) "Exploration of the Sea: Voyage into the Unknown" Report. Ocean Studies Board, National Research Council, Division on Earth and Life Studies, The National Academies Press, Washington, DC. Available at: www.nap.edu/openbook.php?record_id=10844 (accessed July 2, 2015).

DOALOS revised guide (2010) "A revised guide to the implementation of the relevant provisions of the United Nations Convention on the Law of the Sea". Marine Scientific Research. UN Publication No. E.10.V.12 (2010), 71 pp. Available at: http://www.un.org/Depts/los/doalos_publications/publicationstexts/msr_guide%202010_final.pdf (accessed September 2015).

Fedder, B. (2013) *Marine Genetic Resources, Access and Benefit Sharing: Legal and Biological Perspectives*, Oxon, New York, Canada, Routledge.

Fenical, W. (2008) "Marine Microorganisms: The Antibiotic Era Revisited – Perspectives on Ocean Science," available at www.youtube.com/watch?v=bQZPA0Kn_A0 (accessed June 30, 2015).

Fenwick, J. (1992) *International Profiles on Marine Scientific Research Jurisdiction, and Research Histories for the World's Coastal States*, Woods Hole, MA, Woods Hole Oceanographic Institution Sea Grant Program.

Franckz, E. (1986) "Marine Scientific Research and the New USSR Legislation on the Economic Zone," *International Journal of Estuarine and Coastal Law*, vol. 1, pp. 367–90.

Franckz, E. (1990) "The Soviet Union Adapts its Legislation on the Conduct of Marine Scientific Research in the USSR Economic Zone," *International Journal of Estuarine and Coastal Law*, vol. 5, p. 406.

Glowka, L. (1996) "The Deepest of Ironies: Genetic Resources, Marine Scientific Research, and the Area," *Ocean Yearbook*, vol. 12, p. 96.

Gorina-Ysern, M. (1988) "Scientific Missions at Sea: From Immunity to Suspicion," *Maritime Studies*, vol. 43, November–December, pp.10–16.

Gorina-Ysern, M. (1991) "The USA/USSR Agreement on Ocean Studies," July–August, *Maritime Studies*, vol. 59, pp. 5–8.

Gorina-Ysern, M. (1993) "Marine Scientific Research and Intellectual Property Rights Over Research Results," in *Turning the Tide. Papers Presented at Conference on Indigenous Peoples and Sea Rights*, pp. 96–106, Darwin, Australia, Faculty of Law Northern Territory University,.

Gorina-Ysern, M. (1995) "Principles of International Law of the Sea Governing Coastal State Access to Marine Scientific Research Results," PhD thesis, University of New South Wales, Sydney, Australia.

Gorina-Ysern, M. (1997) (with Tsamenyi, M.) "Defense Aspects of Marine Scientific Research", 96 *Maritime Studies*, vol. 96, September–October, pp. 13–23.

Gorina-Ysern, M. (1998) "Marine Scientific Research Activities as the Legal Basis for Intellectual Property Claims?," *Marine Policy*, vol. 22, no. 4–5, pp. 337–357.

Gorina-Ysern, M. (2002) (*Rapporteur*) *Perspectives on International Oceanographic Research*. Bureau of Intelligence and Gorina-Ysern, M. Research *et al.*, Washington DC, US Department of State (Unclassified).

Gorina-Ysern, M. (2003) "Legal Issues Raised by Profitable Biotechnology Development Through Marine Scientific Research," *ASIL Insights*, September, available at www.asil.org/insights/volume/7/issue/22/legal-issues-raised-profitable-biotechnology-development-through-marine (accessed June 30, 2015).

Gorina-Ysern, M. (2004) *An International Regime for Marine Scientific Research*, Netherlands, Boston, MA and Singapore, Martinus Nijhoff (first imprint Transnational Publishers, New York).

Gorina-Ysern, M. (2005) (*Rapporteur*) *Capacity Building for the Protection and Sustainable Use of Oceans and Coasts*. Bureau of Intelligence and Research *et al.*, Washington, DC, US Department of State.

Gorina-Ysern, M. (2005) "Ocean Governance: A New Ethos through a World Ocean Public Trust" (with Gjerde, K. and Orbach, M.) in L. K. Glover and S. A. Earle (Eds.), *Defying Ocean's End: An Agenda for Action*, Washington, Covelo and London, Island Press, pp. 197–212.

Gorina-Ysern, M. (2006) "International Law of the Sea, Access and Benefit Sharing Agreements and the Use of Biotechnology in the Development, Patenting and Commercialization of Marine Natural Products as Therapeutic Agents" (with Captain J. Jones), *Ocean Yearbook*, vol. 20, pp. 221–281.

Gorina-Ysern, M. (2007) "Marine Technology, Oceanic Research Activities and Their Integration into the General Framework of International Law," *Marine Technology Society Journal*, vol. 41, no. 3, pp. 58–67.

Gorina-Ysern, M. (2009) "Book Review of D.K. Leary: *International Law and the Genetic Resources of the Deep Sea* (D. K. Leary, Martinus Nijhoff)," in *Ocean Yearbook*, vol. 23.

Gorina-Ysern, M. (2010) "Climate Change and Guidelines for Argo Profiling Float Deployment on the High Seas" (with A. Mateos), *ASIL Insights*, vol. 14, no. 8, available at www.asil.org/insights/volume/14/issue/8/climate-change-and-guidelines-argo-profiling-float-deployment-high-seas (accessed June 30, 2015).

Heidelberg, K., Allen, A., Stepanauskas, R., Yildiz, F., Murray, A., Sullivan, M., Yakimov and M. (2008) "Marine Molecular Microbiology. The Great Questions," *Marine Genomics Workshop*, Monaco.

Hubert, A. M. (2011) "The New Paradox in Marine Scientific Research: Regulating the Potential Environmental Impacts of Conducting Ocean Science," in *Ocean Development and International Law*, vol. 42, pp. 329–355.

Leary D. K. (2007) *International Law and the Genetic Resources of the Deep Sea*, Leiden, Martinus Nijhoff.

Marffy, A. de (1985) "La Recherche Scientifique Marine," in J-R. Dupuy and D. Vignes (Eds.), *Traite de Nouveau Droit de la Mer*, Paris, Economica, pp. 957–974.

Nicholl, D. S. T. (1994/2002) *An Introduction to Genetic Engineering*, Cambridge, Cambridge University Press.

O'Connell, D. P. (1982 and 1984) in I. A. Shearer (Ed.), *The International Law of the Sea*, Oxford, Clarendon Press.

Roach, J. A. (1996) "Marine Scientific Research and the Law of the Sea," *Ocean Development and International Law*, vol. 27, pp. 59–72.

Roach, J. A. (2007) "Defining Scientific Research: Marine Data Collection," in M. H. Nordquist, R. Long, T. H. Heidar and J. Norton Moore (Eds.), *Law, Science and Ocean Management*, Leiden, Martinus Nijhoff Publishers, pp. 541–573

Ross, D. and Landry, T., (1987) *Marine Scientific Research Boundaries and the Law of the Sea. Discussion and Inventory of National Claims*, Woods Hole, MA, Woods Hole Oceanographic Institution, available at http://nsgl.gso.uri.edu/whoi/whoit87001.pdf (accessed June 30, 2015).

Sears, M. H. (1980) Amicus brief, *Diamond v. Chakrabarty* 447 U.S. 303.

Soons, A. H. A. (1982) *Marine Scientific Research and the Law of the Sea*, The Hague, Kluwer.

Soons, A. H. A. (2007) "The Legal Framework for Marine Scientific Research: Current Issues," in N. Nordquist, R. Long, H. Heidar and J. Norton Moore (Eds.), *Law, Science and Ocean Management*, Leiden, University of Virginia Center for Oceans Law and Policy, pp. 139–166.

Stevens, L. (1986) *Handbook for International Operations of U.S. Scientific Research Vessels*, Seattle, WA, University of Washington.

Treves, T. (1980) "Principe du Consentement et Recherche Scientifique dans le Nouveau Droit de la Mer," *Revue General de Droit International Public*, vol. 84, pp. 253–268.

Treves, T. (2008), *Marine Scientific Research, Max Planck Encyclopedia of Public International Law*, Heidelberg, Max Planck Foundation for International Peace and the Rule of Law under the direction of Rüdiger Wolfrum.

Wegelein, F. H. T. (2005) *Marine Scientific Research: The Operation and Status of Research Vessels and Other Platforms in International Law*, Leiden and Boston, MA, Martinus Nijhoff Publishers.

Wolfrum, R. and Matz, N. (2000) "The Interplay of the United Nations Convention on the Law of the Sea and the Convention on Biological Diversity," in J. A. Frowein, R. Wolfrum, and C. E. Philipp (Eds.), *Max Planck United Nations Year Book*, Volume 4, The Netherlands, Kluwer Law International, pp. 445–480, available at www.mpil.de/files/pdf2/mpunyb_wolfrum_matz_4.pdf (accessed June 30, 2015).

Yankov, A. (1983) "A General Review of the New Convention on the Law of the Sea: Marine Science and its Applications," *Ocean Yearbook*, vol. 4, pp. 150–175.

Young, T. (Ed.) (2009) "Administrative and Judicial Remedies Available in Countries with Users under their Jurisdiction and in International Agreements," in *Covering ABS: Addressing the Need for Sectoral, Geographical, Legal and International Integration in the ABS Regime*. Papers and Studies of The ABS Project. IUCN, Gland, Switzerland. xxii + 201pp. Available at http://cmsdata.iucn.org/downloads/abs_5_eplp_no_67_5.pdf (accessed June 30, 2015).

UN Documents and Secretary-General Reports (2002–2013)

United Nations General Assembly (2013) A/AC.276/6★, Intersessional workshops aimed at improving understanding of the issues and clarifying key questions as an input to the work of the Working Group in accordance with the terms of reference annexed to General Assembly resolution 67/78, available at www.un.org/depts/los/general_assembly/general_assembly_reports.htm (accessed June 30, 2015).

Oceans and the Law of the Sea. Report of the Secretary-General, UN Doc. A/68/71, April 8, 2013; and Add,1, September 9, 2013.

Oceans and the Law of the Sea. Report of the Secretary-General, UN Doc. A/67/79, April 4, 2012; Add. 1 and Add. 2; Doc. A/66/70, March 22, 2011; Add. 1 and Add. 2; Doc. A/65/69, March 29, 2010; Add. 1 and Add. 2; Doc. A/64/66, March 13, 2009: and Add. 1 and Add. 2; Doc. A/63/63, March 10, 2008; and Add. 1; Doc. A/62/66, March 12, 2007; Add. 1 & Add. 2; Doc. A/61/63, March 9, 2006; and Add. 1; Doc. A/60/63, March 4, 2005; and Add.1 and Add. 2; Doc. A/59/62, March 4, 2004; and Add. 1; Doc. A/58/65, March 3, 2003; and Add. 1; Doc. A/57/57, March 7, 2002; and Add.1.

Managing marine environments

9

MARINE CONSERVATION

Guiseppe Notarbartolo-di-Sciara

Homme libre, toujours tu chériras la mer!
(Charles Baudelaire, 1858)

Suffering oceans

In human history, the oceans have come closer to the representation of infinity than any other component of Planet Earth. Many marine biologists are probably more familiar with Thomas H. Huxley's famously faulty prediction in 1883, 'Any tendency to over-fishing will meet with its natural check' (Roberts and Hawkins 1999), than with all of his substantive contributions to the theory of evolution and with his staunch defence of Darwin's revolutionary ideas. We now know well that the oceans and their resources are far from being infinite. The ecological integrity of the seas, as well as their physical, chemical and biological balance, are in jeopardy as a result of human actions that are causing fundamental changes worldwide, from coastal waters to the deep seas, and from the Tropics to the Poles. Humans have become the dominant environmental force on Earth, to the extent that the current geological epoch is now being referred to as the Anthropocene (Caro *et al.* 2011), and it is commonly recognised that the Sixth Great Extinction is in full swing (Chapin *et al.* 2000). Entire marine ecosystems, such as some tropical coral reefs, may be already condemned by overfishing, pollution, global warming and ocean acidification (Pandolfi *et al.* 2003).

While the proximate causes for the pervasive degradation of the world's environment, terrestrial and marine alike, obviously reside in the impacts on a finite planet from a growing and increasingly consumerist human population, a more fundamental consideration concerns the oceans. Unlike land, the oceans lie outside of *Homo sapiens*' habitat, and as such are a free-for-all global commons sanctioned by the still invoked principle of the freedom of the seas (Grotius 1609), largely alien to the notion of property rights. This condition is conducive to a mistaken and noxious dearth of sense of responsibility and stewardship for marine conservation, which totally contradicts the oceans' fundamental importance in maintaining the global balance. Thus the imperative of conserving the health and speaking up for this neglected, albeit major portion of our natural world, which is deprived of constituency, voice and rights, bestows special relevance on marine conservation.

Human impact on marine biodiversity deriving from various pressures (e.g., overexploitation of marine protein, habitat destruction, pollution, climate change) is most often measured in terms of the reduction in species abundance or diversity it can cause. Roberts (2007) suggested that once abundant aquatic wildlife has declined due to overfishing to the extent that less than 5 per cent of the total fish biomass that once swam in Europe's seas has remained. Jackson *et al.* (2001) argued that historical abundances of marine consumers were 'fantastically' large compared to today, with ecological extinctions caused by overfishing having preceded all other human disturbances to coastal ecosystems. Although marine population collapses and extinctions are improperly perceived today due to the rarity of accurate quantitative information from the past – a phenomenon known as the 'syndrome of shifting baselines' (Pauly 1995) – projections based on global fisheries data and long-term regional time series support the view that all taxa currently fished could be collapsed around the middle of our century (Worm *et al.* 2006).

Just as important as the reduction in species' abundance and diversity, albeit often more difficult to detect, is the unwitting elimination of ecological interactions caused by anthropogenic interference, which can occur even in the most remote of locations. McCauley *et al.* (2012) demonstrated that substituting native forests with coconut monoculture in the coastal area of a near-pristine remote Pacific atoll dramatically simplified the chain of trophic interactions, with cascading consequences ultimately affecting the distribution of manta rays feeding on zooplankton in the atoll lagoon's waters.

In a scenario of collapsing ecosystems and biodiversity (Butchart *et al.* 2010), with worldwide fisheries in decline, in some cases irreversibly, and marine habitats becoming extensively and increasingly degraded and dysfunctional, marine conservation is no longer optional or a luxury; it has become a pressing necessity (Roff and Zacharias 2011).

The many souls of marine conservation

Historically, marine conservation has been the remit of ecologists. However, considering that the need for conservation is caused by pressures on the marine environment solely derived from human activities, the core of conservation resides in affecting human behaviour, which is ultimately achieved through political, cultural, education and awareness actions. Marine conservation planning therefore is a multidisciplinary endeavour, involving a collaborative effort among ecologists and policy, economics, social and legal experts.

Marine conservation can be undertaken from many different angles. Roff and Zacharias (2011) broadly classify conservation approaches into 'species-based' and 'space-based', e.g., fisheries management (a species approach), coastal zone management (a species and space approach), ecosystem-based management (a species and space approach) and marine protected areas (MPAs: a space approach). The need for managing fished stocks started to be perceived only about a century ago, with the increased human ability of exploiting marine resources, derived from technological advances combined with exponential human growth. During the second half of the twentieth century increasing concern for the deteriorating quality of the marine environment and habitats, mostly due to pollution and coastal development, resulted in the adoption of a number of international conventions, such as the 1973 International Convention for the Prevention of Pollution from Ships (MARPOL) and the 1975 London Dumping Convention. Species-based conservation, emphasising the conservation of rare and endangered species, also became popular starting from the mid-twentieth century, when a number of international conventions and agreements were adopted for such purposes (e.g., the 1946 International Convention for the Regulation of Whaling, the 1973 Convention on International Trade in Endangered Species of Wild Fauna and Flora, and the 1979 Convention on Migratory Species).

During the past few decades marine conservation has increasingly shifted from an emphasis on species and on a 'traditional' approach to conserve the marine environment, perceived by many to be scarcely effective, to a more holistic approach whereby species (i.e., biodiversity) are protected within, and together with, their spaces (i.e., the ecosystems)(Agardy *et al.* 2011a). This often involves the establishment of MPAs. Just like with marine conservation in general, effective MPA practice requires harmonisation among ecological and social, political and economic considerations (Agardy 1997). MPAs are a testimony of our inability to properly manage human activities at sea, as there would be little need for them in a scenario of diffused, effective sustainability. In this sense, MPAs can be viewed as necessary but insufficient contributions to marine conservation, and ideally would become redundant once humans have learned to live in harmony with their environment and to treat it with respect (Roff and Zacharias 2011).

The success of MPAs largely depends on two fundamental components: proper designation and effective management. Once it is ascertained that space-based protection is the most adequate tool to address the specific threats to biodiversity that operate in the considered area, MPAs need to be designated systematically on the basis of a) our understanding of the structure and function of the ecosystems to be protected, and our often incomplete knowledge of the biological and physical relationships between organisms and their environment, and b) our knowledge of the existing and potential anthropogenic threats to the natural elements that we wish to protect. MPA designation has been increasingly benefiting from the development of methods to identify features to be protected, e.g., the process of selecting Ecologically or Biologically Significant Areas (EBSAs) (Convention of Biological Diversity 2008), and the use of decision-support software tools such as MARXAN (Leslie *et al.* 2003). Managing MPAs to reach the goals and objectives for which they were designated (Salm *et al.* 2000), including restoration of degraded environments, is just as important, because designation alone will not ensure the fulfilment of any MPA goal. This is a common problem because often MPAs get designated by politicians, but then are left without management (Abdulla *et al.* 2008, Notarbartolo-di-Sciara 2009), sometimes indefinitely; as such, they remain 'paper parks'.

Space-based protection through MPA creation has evolved during the past decades from a concept of isolated MPAs established in a piecemeal fashion to protect specific sites, to networks of MPAs systematically planned to protect representative portions of whole regions or seascapes, or of different, interconnected critical habitats of migratory species (IUCN-WCPA 2008). A further refinement of the design of MPA networks is systematic conservation planning, which applies quantifiable targets to spatially explicit surrogates of species and ecosystems to design networks to reach the goals of representation and persistence (Margules and Pressey 2000). Lately, MPAs are increasingly becoming integrated within greater schemes seeking to harmonise a wide range of human uses of the sea with the protection of biodiversity, such as Marine Spatial Planning (MSP) and Ocean Zoning (Agardy 2010, Agardy *et al.* 2011b).

The natural sciences component of marine conservation suffers from the complexity of ecological interactions in marine ecosystems, which we are able to understand still only in part. Scientific knowledge is affected by a number of biases, e.g. taxonomic (with an emphasis on metazoans, particularly vertebrates), and geographic, as scientific capacity is not equally distributed on the planet (Rands *et al.* 2010). However, knowledge limitation is not the main challenge to MPA practice. Policy, legal, social and economic issues, only apparently more readily addressed than scientific uncertainty, are by far more demanding. This is not only because space-based conservation involves opportunity costs that can be difficult to sustain in the short term by the rural poor, but also because short-term economical advantages often get precedence over long-term environmental benefits, and we have so far been unable to render the required changes in economies and behaviours acceptable and embraced by all the concerned stakeholders.

The imperative to conserve

Nature should be preserved because it makes the world a better place (Child 2009). Biodiversity and natural ecosystems are fundamental components of our planet, and there is no doubt in anyone's mind that their degradation and destruction is to be avoided, because not only the survival, but also the well-being and the quality of life of all species – including humans which are an integral part of the biosphere – depend on the maintenance of evolutionary processes and functioning ecosystems (Lavigne 2006). But numerous reasons exist underpinning why natural ecosystems and their biological diversity should be protected, and these reasons can be traced back to a range of different values. Roff and Zacharias (2011) list three main categories of values providing the rationale for protecting biodiversity: anthropocentric, intrinsic and ethical values. While classifying the different types of values into broad categories can be a slippery task, fraught in places with philosophical difficulties, identifying the diversity of reasons for protecting the marine environment is a useful exercise because it allows us to clarify who should do what, and why.

Anthropocentric values, often also referred to as utilitarian values, provide the rationale for protecting the marine environment and its biodiversity so that benefits are maintained to humans in terms of goods and services, known to environmental economists as 'natural capital' (Dasgupta 2010). Goods include renewable resources such as food and pharmaceutical products, the extraction of which can be sustainable if correctly performed, and non-renewable resources such as hydrocarbons and minerals, normally extracted at significant environmental cost in terms of habitat disruption, degradation and destruction. Services provided by coastal and marine ecosystems, such as oxygen, water, depuration, nutrient cycling, shoreline erosion protection, planetary homeostasis through climate regulation and carbon sequestration, but also including attributes leading to aesthetic and recreational benefits, are tightly connected with the health of the ecosystems themselves; unfortunately, many of the world's ecosystems and the services they provide are under threat, as more than 60 per cent of them are currently being used in an unsustainable fashion (Millennium Ecosystem Assessment 2005). Also, since ecosystems become more susceptible to perturbations when simplified by loss of biodiversity (e.g. Chapin *et al.* 2000, Duarte 2000), as biodiversity continues to decline the maintenance of ecosystem processes is becoming an essential imperative for human survival (Rands *et al.* 2010). An approach to create economic incentives for the sustainable use and restoration of ecosystems involves the encouragement of beneficiaries to contribute to maintaining the services flow by forcing non-market values into the marketplace through a scheme called Payment for Ecosystem Services or PES (International Institute for Environment and Development 2012); thereby a source of income can be generated that can be used to finance management, conservation and restoration activities. Another way of viewing utilitarian use is by subdividing it into consumptive and non-consumptive. Consumptive use of an endangered taxon – e.g. a population of whales – may bring it to extinction, and although non-consumptive use such as whale watching may serve to offset such process to some extent, in many cases even the combination of the two values (consumptive and non-consumptive) is insufficient to provide the capital needed to stem the tide of extinction (Alexander 2000). Adding existence value to the equation – which may be considerable for some charismatic species – could significantly improve the situation and turn the tables in favour of conservation. However, existence value is problematic because: a) it is difficult to assess; and b) mechanisms to ensure that those who would benefit from consumptive use are appropriated a sufficient proportion of existence value have not been sufficiently experimented yet (Alexander 2000).

Intrinsic values are based on the notion that natural systems and biodiversity have their own worth regardless of whether humans need them or not. This is a philosophical concept fraught with controversy, with many arguing that: a) values cannot exist without an evaluator, b) evaluators can only be humans, and c) therefore values can only be human-centred. According to Justus *et al.* (2009), the concept of intrinsic values is inherently confusing because non-human natural entities that are targets of conservation do not possess properties considered intrinsically valuable by traditional ethical theories, and the lack of clarity of what intrinsic value exactly means impairs decision making. At the opposite end of the spectrum lies the notion, known as biocentrism or ecocentrism (Naess 1986), that all species have intrinsic value and humans are no more important than the others. Given that all species are intrinsically valuable, they deserve protection regardless of their use to humans, and the onus to provide a justification for their destruction should be on who wants to exploit them. In spite of claimed philosophical difficulties, it is worth noting that the intrinsic value of biodiversity was recognised by the United Nation's 1982 'World Charter for Nature', which stated that 'every form of life is unique, warranting respect regardless of its worth to man'. This general concept, however, is even more ancient. Francis of Assisi (1181–1226) considered all elements of creation, regardless of whether alive or inanimate, as his 'brothers' and 'sisters'. He probably was the first Western thinker to emphasize the beauty and goodness of the natural world, and to introduce into Western thought the idea that humans have an obligation to care not only for each other but for all living beings and natural processes.

Ethical values are also based on the view that humans are an integral part of nature. Considering that we are the only species on the planet capable of simultaneously driving countless other species to extinction and of striving to protect the environment, we have the moral obligation to do the latter (Schweitzer and Notarbartolo-di-Sciara 2009). Aldo Leopold, the foremost champion of the extension of ethical criteria to the relationship between people and land (assumed generically as territory, thereby allowing for the inclusion of the sea), predicted that ethics will be eventually extended to land because economic criteria alone have proved to be insufficient to adjust men to society, and society to its environment (Leopold 1933).

Challenges

In spite of widespread agreement on the need to conserve the marine environment, progress is too slow to be able to match the rate of degradation. The persistence of humanity on a steep downward trajectory of habitat destruction, extinction of species and populations, and reduction in the quality and quantity of ecosystem services provided by the oceans contributes to fears that at least in part the damage will eventually be irreversible. Impediments to marine conservation progress therefore need to be urgently identified and addressed. These include: insufficient public awareness and concern, lack of assertiveness by the conservation community, lack of implementation of commitments by decision makers, insufficient attention given to practical planning, and lack of consideration for the value of biodiversity by the mainstream economic and political mechanisms.

Public awareness and concern must grow to match the gravity of the situation. The urgent need for marine conservation is still insufficiently understood by a large proportion of the general public, and conservation must become more widely embraced than in the past. Acceptance that humans are an integral part of organic evolution and that well-being, quality of life and survival of all species, people included, depends on the maintenance of evolutionary processes and functioning ecosystems, must become part of mainstream thinking and not the turf of a small portion of society, as it still is today. When using nature, humanity as a whole, and not just a small fraction of it, must resolve to reduce the risk of causing irreversible damage to the biosphere, and the concept of sustainability must be transformed from the realm of dreams to that of reality (Lavigne

2006). Current lack of environmental concern by the greater portion of the public opinion hampers the radical changes that are required to recognise biodiversity as a global public good that integrates biodiversity conservation into policies and decision frameworks for resource production and consumption (Rands *et al.* 2010).

Assertiveness by the conservation community is insufficient. Auster *et al.* (2009) advocated the development by the ocean conservation community of a wider constituency to reverse the degrading trends, through actions taken to shift public attitudes in ways that enhance marine conservation efforts resulting in local conservation action and increased political will. New pathways for effective communication with a much broader audience need to be opened to expand the public's understanding and motivation, so that the deterioration of the world's oceans is addressed and reversed. However, considering that at least part of the conservation community has been engaged for decades in the effort of communicating the need for protecting the oceans, one wonders whether the lack of substantial results so far is simply a matter of giving the process more time, or whether there are inherent obstacles that cause such view to be totally unrealistic. In fact, part of the problem resides in the nature of the conservation community itself, which is not monolithic in its advocacy action. In striving to avoid recommending to decision makers actions that they consider too hard to take, too often exponents of the conservation community are prone to water down their own recommendations, forgetting that the politicians will water them down further. Scientists should strive instead to issue recommendations that are solely based on the objective results of scientific analysis, and if such recommendations are challenging, so be it. Noss *et al.* (2012) also noted that those best equipped to say why biodiversity is continuing on its downward ride are not being sufficiently assertive. A case in point is 'Aichi Target 11', formulated during the 2010 meeting of the parties to the Convention of Biological Diversity in Nagoya, whereby 10 per cent of the world's marine and coastal areas should be fully protected by 2020. Noss *et al.* (2012) argue that a science-driven target of 50 per cent, derived from empirical data and rigorous analyses, poorly compares with the policy-driven target of 10 per cent, considered to be inadequate to maintain the ecosystem services and restore connectivity across large seascapes (Svancara *et al.* 2005, McCauley *et al.* 2012). Although arguing about targets in terms of percentages of protected marine surfaces to be attained by a certain date could be viewed as a conservationist foot in the door in the hard fight against political and industrial resistance, the argument risks missing the points that, sooner than later, it is not over part of the ocean, but over the whole, that human activities must be managed for sustainability, and effectively protecting biodiversity must remain one of the main goals of marine conservation (Dulvy 2013). Accordingly, such management must include ensuring effective protection to all the areas that are needed to conserve marine biodiversity, on the basis of rigorous ecologic reasoning, likely to be place-dependent, rather than on that of an arbitrary, one-size-fits-all number.

Commitments are not implemented. The commitment by the world's governments of reaching the meagre 10 per cent target of marine protection was already taken in 2002 at the World Summit on Sustainable Development, with the deadline set in 2012 (Wood *et al.* 2008). Having pushed forward in Nagoya this same commitment by eight years is emblematic of the governments' inability to fulfil their obligations, and one wonders whether simply postponing the 10 per cent target's deadline to 2020 will enable the world's institutions to do a better job. Failure to implement commitments is, however, not limited to decision makers, and should not be entirely attributed to political inertia. Knight *et al.* (2006) suggest that the responsibility of what is known as the 'implementation crisis' should at a minimum be shared with the conservation practitioners themselves, noting that systematic assessments to identify defensible priority conservation areas (e.g. gap analyses, MPA and network selection and design) are too rarely followed by systematic planning and management, whereby such assessments are linked to processes towards the development of implementation strategies and conservation actions. To avoid this 'knowing–doing' gap between assessment and planning, which hampers our ability to effectively apply our ecological savvy to pragmatic conservation problems and empower stakeholders to implement conservation action, Knight *et al.* (2006, p. 410) advocate that 'academic conservation planners' should 'climb down from their ivory towers to get their shoes muddy in the messy political trenches, where conservation actually takes place'.

Rands *et al.* (2010) also point to the need for creating enabling conditions for policy implementation to address the global loss of biodiversity. Based on their analysis, responses to biodiversity loss typically fall into three tiers: 1) foundational (knowledge about the social and biological dimensions of biological loss); 2) enabling (institutions/governance, social/behavioural patterns); and 3) instrumental (legislation, markets/incentives, technology). Existing efforts tend to jump from tier 1 to tier 3, with insufficient efforts to ensure that enabling conditions are in place, and radical changes are required to focus on wider institutional and societal changes to enable more effective implementation of policy. Unfortunately, conditions for the coexistence of an adequate knowledge base with effective institutional and governance action rarely occur across the world's nations, and failure to address enabling factors may be facilitated by entrenched practices of patronage or corruption in matters related to the use of natural resources (Rands *et al.* 2010).

Economic and political systems still ignore the value of biodiversity. Mainstream economic systems are still far from being amenable to integrating biodiversity – part of 'natural capital' – into macroeconomic forecasts (Bayon and Jenkins 2010); the depreciation of this natural capital caused by productive activities is never properly accounted for, because development policies ignore human reliance on such capital (Dasgupta 2010). Public and private sectors, as well as civil society, are insufficiently taking into account the benefits of conserving biodiversity and the costs of losing it, and the value of biodiversity is still not an integral element of social, economic and political decision making. Therefore, biodiversity loss will continue unless it is managed as a public good through conscious and collective choices (Rands *et al.* 2010). Things are even more problematic when moving from land to sea; the world is still far from achieving a regime of responsible ocean governance, capable of ensuring that 'the oceans are used for the benefit of all and in the interest of future generations' (Soares 2008, p. 2). There are fundamental reasons underlying the difficulties in accepting the simple paradigm of biodiversity managed as a public good, and one of the most important concerns a time-scale mismatch. Environmental and bio-diversity conservation are long-term policies, with a far longer horizon than what political institutions are ready to contemplate. Until the full set of drivers that move individuals toward or away from sustainable activities, rather than simply economic motivations, are pursued by political action (Steinberg 2009), and until biodiversity and the environment are included among our conventional measures of well-being, which now solely focus on wealth creation and internationally recognized estimates of GDP, drivers to protect biodiversity are too weak to make a difference (Rands *et al.* 2010). Against such background, arguments that humans in the present owe it to posterity to conserve the natural environment, although ethically and philo-sophically proper (Partridge 1992), are destined to remain in the realm of the good but unful-filled intentions. Efforts to conserve the natural environment and its biodiversity are still too often perceived by many as a selfish rich-world attitude which is oblivious of the needs of the poor, when it should rather be considered, quite uncontroversially, as the intergenerational ecological justice-based imperative of striking a balance between the interests of the current generation and those of its progeny (Weston 2012).

Understanding where we want to go

Many conservation ecologists argue that a most effective way of saving nature today involves its 'commodification' through the attribution of economic values to biodiversity and to ecosystem services. Such approach is thought to provide incentives to the beneficiaries for contributing to the maintenance of such services and the sustainable use of goods, also ultimately concurring to a generalised appreciation and protection of nature's non-market resources. If ecosystem services have quantifiable economic value, this can be used to stimulate investment in restoration and maintenance and to provide a source of income for such management and restoration activities (Daily *et al.* 2000, Lau *et al.* 2010).

Market-oriented mechanisms for conservation work under the assumption that by identifying ecosystem services and quantifying their economic value, it will be possible to induce decision makers to fully perceive the costs of environmental destruction and biodiversity loss, and to work to preserve nature (McCauley 2006). However, although PES schemes are seductive and increasingly popular within part of the conservation community, warnings are voiced that by involving the commodification of natural values these schemes entail the risk of significant backfire (e.g. Redford and Adams 2009). For instance, when ecosystem services are provided not by whole ecosystems, but by subsets of species that fulfil certain basic functions, the conservation of other less 'profitable' species may be discouraged or abandoned (Ridder 2008), thereby demolishing the economic justification of preserving biodiversity (Child 2009, Redford and Adams 2009). Other ecosystem payment shortcomings include the likelihood that rights to ecosystem services could become privately assigned, thereby entailing undesirable welfare implications (Redford and Adams 2009), or that when the perceived value of an ecosystem service is raised under a governance regime that is not just, it can become a value that is only benefiting an elite (Monbiot 2012). McCauley (2006) argued that the conservation relevance of ecosystem services has been grossly overstated, citing cases demonstrating their limitations as conservation tools, e.g.: a) ecosystem services are often offset by 'ecosystem disservices'; b) the fluctuations of market forces can hardly ensure the needed constancy of conservation commitments; c) technological advances may endanger conservation processes based on ecosystem services by providing economically more viable alternatives; and d) several other instances of unsolvable conflicts between profit and conservation, which can be mutually exclusive. By contrast, advocates of the benefits of ecosystem services warn against polarizing environment against economy, because 'if nature contributes significantly to human well-being, then it is a major contributor to the real economy' (Costanza 2006). Furthermore, conservation mostly relying on considerations of the intrinsic values of nature has been obviously ineffective, and 'it is time to add to the mix other approaches based on a fuller consideration of ecosystem services and options for distributing costs and benefits' (Reid 2006). Perhaps, rather than considering the environment as a sector of the economy that needs to be properly integrated into it, so that growth opportunities will not be missed, it seems that inverting the relationship would be more appropriate: 'The economy is part of the environment and needs to be steered so that opportunities to protect our world of wonders will not be missed' (Monbiot 2014).

Clearly, more progress is needed in the adoption of nature value-arguments based on economic rationalism, to extract the full potential of such approaches without unduly overshadowing the aesthetic and ethical arguments that originally inspired the conservation movement (Jepson and Canney 2003). Reverting to attributing deserved emphasis to ethical and aesthetic arguments in conservation policy could contribute to recreating connections between the conservation movement and the wider public at least in Western-style societies, ultimately leading to greater conservation effectiveness (Jepson and Canney 2003).

Although humans are the cause of environmental degradation on land and at sea, they also have the potential of being the solution, and this is why conservation action must focus on people, on their needs and on their values. One of the ultimate goals of conservation should be of enabling present and future generations to live healthy enriched lives, and be engaged with nature which they are part of (Knight *et al*. 2006). The weakness of the relationship between GDP and alternative indicators designed to measure economic welfare in broader terms such as the Genuine Progress Index (Kubiszewski *et al*. 2013), and ultimately between material wealth and happiness (Happy Planet Index 2012), should warn against attributing excessive importance to nature's material benefits to people. People's attitudes toward conservation, far from being solely determined by economic and utilitarian considerations, often involve valuing nature for its own sake (Noss *et al*. 2012). Facilitating reconnection between people and nature is a key factor in the progress of conservation, and this is particularly true for the marine environment considering the powerful attraction that the sea exerts on human beings.

As humans, we have not only the ability of identifying ourselves with other beings, but also the capacity of experiencing strong connections to other organisms and the nonorganic world (Naess 1986). Unfortunately, current governance mechanisms, strongly conditioned by economic, social and political forces, and lifestyles influenced by consumerist drives, make the expression of these attitudes impossible (Naess 1986). Nevertheless, societies in the developed world are experiencing an increasing trend of reconnection with nature, as exemplified by the growth of ecotourism (UNEP 2011) and of marine tourism in particular (Orams 1999); however, this still involves a very small and privileged portion of humanity. Global mass urbanisation is producing loss of contact and subsequent alienation from nature – the 'Extinction of Experience' – an inexorable cycle of disconnection, apathy, and progressive depletion, facilitated by a prevailing climate of corporate growth and by a condition of ecological illiteracy, both of which are inimical to sustainability (Pyle 2003).

The future of our environment, terrestrial and marine, certainly depends in large part on the rational decision by human societies of altering activities that contribute to the demise of Earth's ecosystems, and will require a clear vision of the goals and the roles of societies in achieving them, long-term planning, rigorous science, and an interdisciplinary, comprehensive approach to conservation (Reynolds *et al*. 2009). However, while pursuing our rational agendas and the imperative of being systematic and strategic in setting our priorities, we should be careful to keep alive the magic and beauty of nature, which is the cement of our allegiance to the natural world, and ultimately an important component of our recipe for happiness.

Obviously, humanity's condition is still too heterogeneous for much of this reasoning to have an extended practical validity at the moment, and this adds significantly to the difficulty of the task. While the developed world has the luxury – and at the same time the obligation – of searching its soul on ethical matters and on the intrinsic rights and values of nature (Doak *et al*. 2014), the majority of people on Earth are still forced to worry about feeding their children, and keeping them alive, on a day-to-day basis (Marvier *et al*. 2006). Things are even worse now, with the current global economic downturn, and it is an ironic tragedy that the environment supporting human life is pushed aside as a problem rather than being protected as a solution. The little attention humanity is ready to devote today to environmental problems is almost entirely dedicated to climate change, which in spite of its undisputable conservation relevance obscures other pressing problems such as the degradation of biodiversity – terrestrial and marine alike – and diverts funding from addressing them (Noss *et al*. 2012).

Humanity has certainly the capacity of achieving marine conservation, providing that the relevant elements of society will manage to work together towards the common goal. The path is difficult but practicable, and its direction known. The challenge resides in managing to influence political change rather than letting ourselves be overwhelmed by political destructiveness.

References

Abdulla, A., Gomei, M., Maison, E., Piante, C. 2008. *Status of marine protected areas in the Mediterranean Sea*. IUCN, Malaga and WWF, France. 152 pp.

Agardy, T. 1997. *Marine protected areas and ocean conservation*. Academic Press and R.G. Landes Co., Austin, TX. 244 pp.

Agardy, T. 2010. *Ocean zoning: making marine management more effective*. Earthscan, London. 220 pp.

Agardy, T., Davis, J., Sherwood, K., Vestergaard, O. 2011a. *Taking steps toward marine and coastal ecosystem-based management – an introductory guide*. UNEP Regional Seas Reports and Studies No. 189. UNEP, Nairobi. 68 pp.

Agardy, T., Notarbartolo-di-Sciara, G., Christie, P. 2011b. Mind the gap: addressing the shortcomings of marine protected areas through large-scale marine spatial planning. *Marine Policy* 35:226–232.

Alexander, R. R. 2000. Modelling species extinction: the case for non-consumptive values. *Ecological Economics* 35:259–269.

Auster, P. J., Fujita, R., Kellert, S. R., Avise, J., Campagna, C., Cuker, B., Dayton, P., Heneman, B., Kenchington, R., Stone, G., Notarbartolo-di-Sciara, G., Glynn, P. 2009. Developing an ocean ethic: science, utility, aesthetics, self-interest and different ways of knowing. *Conservation Biology* 23(1):233–235.

Bayon, R., Jenkins, M. 2010. The business of biodiversity. *Nature* 466:184–185.

Butchart, S. H. M., Walpole, M., Collen, B., van Strien, A., Scharlemann, J. P. W., Almond, R. E. A., Baillie, J. E. M., Bomhard, B., Brown, C., Bruno, J., Carpenter, K. E., Carr, G. M., Chanson, J., Chenery, A. M., Csirke, J., Davidson, N. C., Dentener, F., Foster, M., Galli, A., Galloway, J. N., Genovesi, P., Gregory, R. D., Hockings, M., Kapos, V., Lamarque, J.-F., Leverington, F., Loh, J., McGeoch, M. A., McRae, L., Minasyan, A., Hernández Morcillo, M., Oldfield, T. E. E., Pauly, D., Quader, S., Revenga, C., Sauer, J. R., Skolnik, B., Spear, D., Stanwell-Smith, D., Stuart, S. N., Symes, A., Tierney, M., Tyrrell, T. D., Vié J.-C., Watson, R. 2010. Global biodiversity: indicators of recent decline. *Science* 328:1164–1168.

Caro, T., Darwin, J., Forrester, T., Ledoux-Bloom, C., Wells, C. 2011. Conservation in the Anthropocene. *Conservation Biology* 26(1): 185–188.

Chapin, F. S., Zavaleta, E. S., Eviner, V. T., Naylor, R. L., Vitousek, P. M., Reynolds, H. R., Hooper, D. U., Lavorel, S., Sala, O. E., Hobbie, S. E., Mack, M. C., Diaz, S. 2000. Consequences of changing biodiversity. *Nature* 405:232–242.

Child, M. F. 2009. The Thoreau ideal as a unifying thread in the conservation movement. *Conservation Biology* 23(2):241–243.

Convention on Biological Diversity. 2008. Report on the expert workshop on ecological criteria and biogeographic classification systems for marine areas in need of protection. SBSTTA 13th Meeting, Rome, 18–22 February 2008. 25 pp.

Costanza, R. 2006. Nature: ecosystems without commodifying them. *Nature* 443:749.

Daily, G. C., Söderqvist, T., Aniyar, S., Arrow, K., Dasgupta, P., Ehrlich, P. R., Folke, C., Jansson, A. M., Jansson, B. O., Kautsky, N., Levin, S., Lubchenco, J., Mäler, K. G., Simpson, D., Starrett, D., Tilman, D., Walker, B. 2000. The value of nature and the nature of value. *Science* 289(5478):395–396.

Dasgupta, P. 2010. Nature's role in sustaining economic development. *Philosophical Transactions of the Royal Society B* 365:5–11.

Doak, D. F., Bakker, V. J., Goldstein, B. E., Hale, B. 2014. What is the future of conservation? *Trends in Ecology and Evolution* 29(2):77–81.

Duarte, C. M. 2000. Marine biodiversity and ecosystem services: an elusive link. *Journal of Experimental Marine Biology and Ecology* 250:117–131.

Dulvy, N. K. 2013. Super-sized MPAs and the marginalization of species conservation. *Aquatic Conservation: Marine and Freshwater Ecosystems* 23:357–362.

Grotius, H. 1609. *The freedom of the seas, or the right which belongs to the Dutch to take part in the East Indian trade*. Translated with a revision of the Latin text of 1633 by R. Van Deman Magoffin, Oxford University Press, Oxford.

Happy Planet Index. 2012. The Happy Planet Index, London. Available from www.happyplanetindex.org (accessed September 2012).

International Institute for Environment and Development. 2012. Payments for coastal and marine ecosystem services: prospects and principles. IIED Briefing. 4 pp. Available from http://pubs.iied.org/17132IIED (accessed 7 September 2012).

IUCN World Commission on Protected Areas (IUCN-WCPA). 2008. *Establishing marine protected area networks—making it happen.* IUCN-WCPA, National Oceanic and Atmospheric Administration and The Nature Conservancy, Washington, DC. 118 pp.

Jackson, J. B. C., Kirby, M. X., Berger, W. H., Bjorndal, K. A., Botsford, L. W., Bourque, B. J., Bradbury, R. H., Cooke, R., Erlandson, J., Estes, J. A., Hughes, T. P., Kidwell, S., Lange, C. B., Lenhian, H. S., Pandolfi, J. M., Peterson, C. H., Steneck, R. S., Tegner, M. J., Warner, R. R. 2001. Historical overfishing and the recent collapse of coastal ecosystems. *Science* 293:629–638.

Jepson, P., Canney, S. 2003. Values-led conservation. *Global Ecology and Biogeography* 12:271–274.

Justus, J., Colyvan, M., Regan, H., Maguire, L. 2009. Buying into conservation: intrinsic versus instrumental value. *Trends in Ecology and Evolution* 24(4):187–191.

Knight, A. T., Cowling, R. M., Campbell, B. M. 2006. An operational model for implementing conservation action. *Conservation Biology* 20(2):408–419.

Kubiszewski, I., Costanza, R., Franco, C., Lawn, P., Talberth, J., Jackson, T., Aylmer, C. 2013. Beyond GDP: Measuring and achieving global genuine progress. *Ecological Economics* 93:57–68.

Lau, W., Agardy, T., Hume, A. 2010. *Payments for ecosystem services. Getting started in marine and coastal ecosystems: a primer.* Forest Trends and the Katoomba Group, Washington DC. 80 pp. Available from http://www.forest-trends.org/publication_details.php?publicationID=2374 (accessed 16 September 2015).

Lavigne, D. M. 2006. Wildlife conservation and the pursuit of ecological sustainability: a brief introduction. Pp. 1–18 in: D.M. Lavigne (editor), Gaining ground: in pursuit of ecological sustainability. IFAW, London, ON, Canada. 425 pp.

Leopold, A., 1933. The conservation ethic. *Journal of Forestry* 31(6):634–643.

Leslie, H., Ruckelshaus, M., Ball, I. R., Andelman, S., Possingham, H. P. 2003. Using siting algorithms in the design of marine reserve networks. *Ecological Applications* 13(1):S185–S198.

McCauley, D. J. 2006. Selling out on nature. *Nature* (London) 443:27–28.

McCauley, D. J., DeSalles, P. A., Young, H. S., Dunbar, R. B., Dirzo, R., Mills, M. M., Micheli, F. 2012. From wing to wing: the persistence of long ecological interaction chains in less-disturbed ecosystems. *Scientific Reports* 2:409.

Margules, C. R., Pressey, R. L. 2000. Systematic conservation planning. *Nature* 405:243–253.

Marvier, M., Grant, J., Kareiva, P. 2006. Nature: poorest may see it as their economic rival. *Nature* 443:749–750.

Millennium Ecosystem Assessment. 2005. *Ecosystem and human well-being: synthesis.* Island Press, Washington, DC. 137 pp.

Monbiot, G. 2012. The great impostors. *The Guardian*, 7 August 2012. Available from www.monbiot.com/2012/08/06/the-great-impostors/ (accessed 8 August 2012).

Monbiot, G. 2014. Reframing the planet. *The Guardian*, 22 April 2014. Available from www.monbiot.com/2014/04/22/reframing-the-planet/(accessed 23 April 2014).

Naess, A. 1986. Intrinsic value: will the defenders of nature please rise? Pp. 504–515 in: M.E. Soulé (editor), *Conservation Biology: the science of scarcity and diversity.* Sinauer Associates, Sunderland, MA. 584 pp.

Noss, R. F., Dobson, A. P., Baldwin, R., Beier, P., Davis, C. R., Dellasala, D. A., Francis, J., Locke, H., Nowak, K., Lopez, R., Reining, C., Trombulak, S. C., Tabor, G. 2012. Bolder thinking for conservation. *Conservation Biology* 26(1):1–4.

Notarbartolo-di-Sciara, G. 2009. The Pelagos Sanctuary for the conservation of Mediterranean marine mammals: an iconic High Seas MPA in dire straits. 2nd International Conference on Progress in Marine Conservation in Europe 2009, 2–6 November 2009, OZEANEUM/DMM, Stralsund, Germany. 4 pp.

Orams, M. 1999. *Marine tourism: development, impacts and management.* Routledge Chapman & Hall, London and New York. 136 pp.

Pandolfi, J. M., Bradbury, R. H., Sala, E., Hughes, T. P., Bjorndal, K. A., Cooke, R. G., McArdle, D., McClenachan, L., Newman, M. J. H., Paredes, G., Warner, R. R., Jackson, J. B. C. 2003. Global trajectories of the long-term decline of coral reef ecosystems. *Science* 301:955–995.

Partridge, E. 1992. The moral uses of future generations. Pp. 33–38 in: Reeves, G. H., Bottom, D. L., Brookes, M. H. (Editors). 1992. *Ethical questions for resource managers.* Gen. Tech. Rep. PNW-GTR-288. U.S. Department of Agriculture, Forest Service, Pacific Northwest Research Station, Portland, OR. 39 pp.

Pauly, D. 1995. Anecdotes and the shifting baseline syndrome of fisheries. *Trends in Ecology and Evolution* 10(10):430.

Pyle, R.M. 2003. Nature matrix: reconnecting people and nature. *Oryx* 37(2):206–214.

Rands, M. R. W., Adams, W. M., Bennun, L., Butchart, S. H. M., Clements, A., Coomes, D., Entwistle, A., Hodge, I., Kapos, V., Scharlemann, J. P. W., Sutherland, W. J., Vira, B. 2010. Biodiversity conservation: challenges beyond 2010. *Science* 329:1298–1303.

Redford, K. H., Adams W. M. 2009. Payment for ecosystem services and the challenge of saving nature. *Conservation Biology* 23(4):785–787.

Reid, W. V. 2006. Nature: the many benefits of ecosystem services. *Nature* 443:749.

Reynolds, J. E., Marsh, H., Ragen, T. J. 2009. Marine mammal conservation. *Endangered Species Research* 7:23–28.

Ridder, B. 2008. Questioning the ecosystem services argument for biodiversity conservation. *Biodiversity and Conservation* 17:781–790.

Roberts, C. 2007. *The unnatural history of the sea.* Island Press, Washington DC. 456 p.

Roberts, C., Hawkins, J. P. 1999. Extinction risks in the sea. *Trends in Ecology and Evolution* 14(6):241–246.

Roff, J., Zacharias, M. 2011. *Marine conservation ecology.* Earthscan, London, Washington DC. 439 p.

Salm, R. V., Clark, J. R., Siirila, E. 2000. *Marine and coastal protected areas: a guide for planners and managers.* IUCN. Washington DC. xxi + 371pp.

Schweitzer, J., Notarbartolo-di-Sciara G. 2009. *Beyond cosmic dice: moral life in a random world.* Jacquie Jordan, Los Angeles. 246 pp.

Soares, M. 2008. Ocean Governance XXI. Speech delivered at the Commemorative Session of the Lisbon Declaration of 1998. Available from www.drgeorgepc.com/IOInforma0109SOARESOcean GovernanceXXI.pdf (accessed 19 October 2012).

Steinberg, P. 2009. Institutional resilience amid political change: the case of biodiversity conservation. *Global Environmental Politics* 9(3):61–81.

Svancara, L. K., Brannon, R., Scott, J. M., Groves, C. R., Noss, R. F., Pressey, R. L. 2005. Policy-driven vs. evidence-based conservation: a review of political targets and biological needs. *BioScience* 55(11): 989–995.

UNEP. 2011. Towards a Green Economy: pathways to a sustainable development and poverty eradication. Available from www.unep.org/greeneconomy (accessed 10 September 2012).

Weston, B. H. 2012. The theoretical foundations of intergenerational ecological justice: an overview. *Human Rights Quarterly* 34(1):251–266.

Wood, L. J., Fish L., Laughren J., Pauly D. 2008. Assessing progress towards global marine protection targets: shortfalls in information and action. *Oryx* 42(3):1–12.

Worm, B., Barbier E. B., Beaumont N., Duffy J. E., Folke C., Halpern B. S., Jackson J. B. C., Lotze H. K., Micheli F., Palumbi S. R., Sala E., Selkoe K. A., Stachowicz J. J., Watson R. 2006. Impacts of biodiversity loss on ocean ecosystem services. *Science* 314:787–790.

10

SCIENCE AND POLICY

Rebecca Koss and Geoffrey Wescott

Introduction

Global governments have come to recognise and acknowledge that an integrated approach to coastal and marine policy and management in order to govern human activity is necessary to maintain biodiverse marine ecosystems. This approach to policy and management requires integration rather than simply coordinating policy, planning and management across the coastal and marine interface (Wescott 2012). With just over a third of the global population living along the coastal zone (Millennium Ecosystem Assessment [MA] 2005), it is necessary to consider how natural and political boundaries, economic, social and environmental seascapes integrate concurrently for coastal and marine policy planning and implementation. This is because human activities take place on a continuous gradient across interconnecting catchment, coastal and marine ecosystems, where resource use as diverse as agriculture in coastal marshland and offshore non-renewable energy extraction have to be considered (MA 2003).

As highlighted by Cicin-Sain *et al.* (2011), the global marine environment is a quintessential sustainable development issue with ocean management needing to consider economic development, social development and environmental protection in parallel. Emphasis in marine science literature is often placed on using the best available physical and biological science to underpin marine policy; however, science alone cannot create effective policy planning and implementation. Science has progressed considerably over the past decades in identifying the anthropogenic drivers and impacts on coastal and marine ecosystems. However, it is how science integrates with social, political, legal and economic landscapes at local, national and regional geographies that will determine the success of policy implementation (Hinrichsen 2011). Connecting science to policy is a challenge widely acknowledged by scientists and policy makers. The aim of this chapter is to describe how science amalgamates with these landscapes to create an integrated framework for marine policy.

In order to address this aim, this chapter describes how each landscape is considered in the integrated approach and the pertinent challenges for successful marine policy implementation. Thus, this chapter will provide:

* a brief overview of the various landscapes and their context for integrated marine planning and decision-making;

- a chronological history of global environmental directives that were instrumental in shaping current marine policy to adopt the integrated approach; and
- frameworks and management tools currently in place to implement the integrated approach.

The various landscapes in marine planning and decision making

Science

Science is an important foundation in developing our understanding of impacts on coastal and marine ecosystems from anthropogenic drivers. Overfishing, coastal development, natural resource extraction, industrial waste, shipping, untreated sewerage and other pollution sources have altered marine and coastal ecosystems, with profound effects specifically on coral reefs, seagrass beds, mangrove swamps, continental shelves and seamounts (Hinrichsen 2011; Halpern *et al.* 2012). Human activities in the medium- to high-impact range have already caused damage to approximately 41 per cent of the ocean, equating to an area of nearly 170 million km^2 (Hinrichsen 2011). Only 4 per cent of our oceans remains relatively unaffected (Hinrichsen 2011). Marine scientists employ the best available scientific information and expertise to understand how these anthropogenic pressures alter our marine systems. Scientific marine research provides innovative, forward-thinking and comprehensive approaches to addressing complex challenges. Yet, these approaches take time to yield tangible impacts due to the non-linear integration of science into policy (Pietri *et al.* 2011).

We have progressed considerably over the past 30 years, with research yielding a greater understanding of coastal and ocean dynamics through monitoring, modelling and analysis. This positive progression in scientific understanding of coastal and marine systems has allowed, for example, Working Group 1 of the Intergovernmental Panel on Climate Change to predict impacts of climate change on coastal populations. Continuous emerging scientific research and information, for example, the Ocean Health Index (Halpern *et al.* 2012), employs comprehensive and quantitative methods to measure and monitor the health of coupled human–ocean systems, which can be used as a tool to facilitate policy and management approaches.

However, integrating scientific outputs into marine policy is difficult and challenging due to a number of reasons including:

1   the combination of a deficiency and limitation in our understanding of the complexity, functions and dynamics of marine ecosystems (Sorenson 1997; Kidd *et al.* 2011);
2   poor communication due to language and terminology differences, institutional and cultural differences between scientists and policy makers (Sorenson 1997; Pietri *et al.* 2011);
3   potential costs of future research to address questions raised by current research (Sorenson 1997);
4   how to keep research findings adaptive to develop effective policy (Sorenson 1997);
5   difficulties scientists face in using their research to inform policy planning while keeping their scientific integrity (Pietri *et al.* 2011); and
6   reluctance of scientists to become involved and advocate in policy processes (Gray and Campbell 2008).

Although the combination of these factors creates large barriers between science and its uptake in marine and coastal policy, there is an emerging community of marine scientists acting as advocates to address these science–policy interface barriers (Hopkins *et al.* 2012). With increased global accessibility to online social media, marine scientists and research organisations across the globe are promoting increased public awareness in an attempt to change behaviours and perceptions towards the challenges facing oceans.

Economics

The marine economic landscape was traditionally constructed from maritime trade, commerce and tourism. Sectors within marine trade and commerce fulfil a dual role as stakeholders and creators of global economic development. The success or failure of marine sectors and their operations has far reaching implications on businesses, human employment, livelihoods and well-being. The ability to calculate the economic value of coastal and marine resources has increased over the past decades with information provided by organizations such as the Fisheries and Agriculture Organization of the United Nations (FAO), The Economics of Eco-systems and Biodiversity (TEEB), International Maritime Organization (IMO) and World Trade Organization (WTO) giving us insight and understanding as to how, where, and what quantity of natural resource use is in demand by the global population. Over the past decade the intro-duction of ecosystem services as a means to assess the benefits humans derive from our natural systems, has become the new political buzzword. The uptake of this paradigm as form of economic assessment of our coastal and marine ecosystems, is slowly being established as the economic arm of the integrated approach.

Ecosystem services together with the traditional maritime trade have been recently tagged as the Blue Economy. The European Commission uses the term Blue Growth with contributions from marine sectors providing work for 5.4 million people and accounting for gross value added of just under €500 billion annually (European Commission (COM) 2012). This is indica-tive of the important contribution our oceans make towards economic development. A number of the main maritime sectors including: shipping, non-renewable energy, offshore renewable-energy, wild fisheries, tourism and mariculture, annually contribute to international trade, employ-ment and national gross domestic product (GDP). Raw materials extraction, telecoms, medical and genetic bio-prospecting and ornamental resources also contribute to stimulating global eco-nomic activity. With a growing global population there will be an increased reliance on natural resources from our oceans and this needs to be considered in how we shape marine policy.

Social

For generations, humans have depended upon oceans for their well-being and in more recent centuries, livelihoods. Of the current global population, 2.9 billion people depend on the sea for 15 per cent of their daily protein intake and 40 per cent of the world's population (equivalent to 3 billion people) occupy a coastal strip 100km wide, representing 5 per cent of the earth's land surface (Hinrichsen 2011). Although large strides have been made in acknow-ledging and recognising the importance of marine ecosystems in traditional customs, folklore of indigenous cultures, religious experiences and coastal communities, the mechanisms for their incorporation have been only successful at the local scale. This close connectivity humans have to oceans and seas is globally recognised; however, the social landscape to understand these beliefs, values and attitudes is the least researched sphere of the integrated approach.

Where science indicators such as the biodiversity index or spawning stock biomass for commercial fisheries species can inform us of the oceans' health and ecosystem elasticity, and economic indicators such GDP and market values can determine the monetary value of our marine resources, there are no overarching social indicators to provide a measure for our physical, spiritual and mental well-being that is derived from coastal and marine systems. Much work still needs to be done in understanding the values and beliefs of the public and their relationship to coastal and marine ecosystems to strengthen the social arm of the integrated approach for marine policy. Greater research on defining these links is desperately needed as more people move to coastal areas to improve their standard of living (National Research Council 2008). Research is needed to identify human connectivity to marine systems in creating environmental stewardship, community capacity, social and ecological connectedness, aesthetic value, inspiration for creativity, art and design, the provision for cultural heritage, and influencing our mental, spiritual and physical well-being. An exploration and definition of these linkages can assist in developing a suite of measurable and relevant social indicators to be employed in marine policy.

Law, governance and constituency

If the integrated approach is analogous to a three-legged table, with each leg representative of the science, economics and social landscape, then the tabletop is the overarching domain of law, governance and constituency (Figure 10.1). This tabletop will effectively determine the success or failure of policy implementation. Aside from national exclusive economic zones, oceans and seas do not have the same ecological delineations visual to the naked eye as terrestrial systems. This creates legal complications, and governing areas, such as the high seas, pose many challenges. Even transboundary areas are fraught with logistical and political issues (National Research Council 2008). By also including different governance models and their arrangements, further complicated by vertical and horizontal multi-agency co-ordination and power sharing, it is a wonder that any policies can be developed and implemented. Is this a classic case of reductionism? Policy success is also dependent on a supportive constituency through engaging and including stakeholders in decision making to overcome political opposition (Sorenson 1997; Wescott and Fitzsimmons 2011). These confusing, politically sensitive and difficult but important platforms will determine the success of the integrated approach in marine policy.

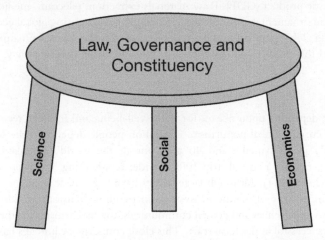

Figure 10.1 The three-legged table as an analogy to the integrated approach in marine policy

A history of the integrated approach in global environmental directives

The evolution of the integrated approach needs to be briefly described in order to understand how it has become mainstream in marine policy. It was during the 1960s and 1970s that technology played an important role in exploiting marine natural resources, perpetuating the 'tragedy of the commons' (Wescott, 2015). The declaration of exclusive rights to resources evolved with legislating areas under national jurisdiction. These former free areas available for resource extraction became under jurisdiction of the nation's government requiring sectors to obtain permission for any extractive activities. Although no longer a free for all, fisheries, oil and gas extraction and pharmaceutical discoveries spurred on the increasing unsustainable utilisation of these resources. These, in combination with land-based activities such as agricultural runoff, untreated sewerage, chemical pollutants from industrial processes and increasing coastal infrastructure to support growing populations, created a large number of pressures that caused detrimental impacts to coastal and marine ecosystems. It was apparent and paramount that marine policy consisting of a framework that accounted for resource extraction allocation and regulation, input controls, while recognising population growth and human development, was needed.

The World Commission on Environment and Development, established in 1984 as an independent body of the United Nations General Assembly, led the way in examining these critical issues of environment protection and human development to formulate innovative and realistic actions by strengthening international cooperation and raising the level of understanding and commitment to action (World Commission on Environment and Development 1987b). It was this Commission that introduced the integrated approach in global directives, by setting the foundation of the integrated framework. The Commission hold national governments to account in considering economic, political, social and environmental landscapes concurrently in environmental policy decision making and implementation.

Law of the Sea

Public awareness of global marine ecosystems for human livelihood and development grew in the 1950s and 1960s owing to the underwater cinematography of Jacques Cousteau, the iconic imagery of the earth from the Apollo spacecraft with the planet wrapped in blue water, an increasing number of fishery collapses and significant pollution events. This led to the UN General Assembly passing a resolution in 1970 to revive discussions on the Law of the Sea (Wescott 2012). These discussions, held from 1974 to 1982, were formally titled the Third United Nations Conference on the Law of the Sea (UNCLOSIII) and resulted in the Law of the Sea Convention (LOSC) (Wescott 2012).

Over the ensuing decades, there have been considerable efforts to develop governance frameworks in an attempt to manage and protect vast areas of the ocean. Marine expanses within a certain nautical mile limit of numerous national coastlines are now considered exclusive economic zones (EEZs) by nations who have ratified the LOSC. However, areas outside of these EEZs, termed the high seas, may or may not have jurisdiction rights; this varies from nation to nation based on their resource activities and conservation objectives.

As of January 2013, 165 nations have ratified the Law of the Sea Convention with 78 nations agreeing to implement the provisions relating to the conservation and management of fish stocks, specifically for migratory species that are more difficult to govern and manage (Wescott 2012). A number of land-locked countries have signed on to LOSC in order to guarantee access routes through neighbouring countries that host seaports, providing that nation is also a signatory to LOSC (Wescott 2012).

National directives

Canada was one of the first nations to create an Ocean Act in 1996, which was closely followed by Australia with their Ocean Policy formalised in 1998, the USA with their Oceans Act 2000 and more recently, the European Union's Integrated Maritime Policy adopted in 2007. Formal bodies governing and managing regional seas also adopt the principles of LOSC. For example, the Asia Pacific Economic Corporation (APEC) forum has a Marine Resources Conservation Working Group and the Partnerships in Environmental Management for the Seas of East Asia (PEMSEA) has operated for over a decade (Wescott 2012); HELCOM, the Helsinki Commission, works to protect the Baltic Sea marine environment from all sources of pollution through intergovernmental co-operation of nations surrounding this sea, and the OSPAR commission works to protect and conserve the North-East Atlantic and its resources.

The type of framework used to implement LOSC will vary from nation to nation, yet most governments use an integrated approach to incorporate social, cultural, political, economic and science landscapes concurrently in addressing marine ecosystem conservation and resource use. The acceptance and ratification of LOSC initiated an accountability system for nations. In conjunction with a number of other global directives, LOSC was a major step forward in conserving marine ecosystems while trying to manage resource exploitation.

Tokyo declaration and sustainable development

At the eighth and final meeting of the World Commission on Environment and Development (WCED), the sustainable development paradigm was presented as part of the Tokyo Declaration (WCED 1987a). Sustainable development is defined as (WCED 1987b, p. 41):

> Development that meets the needs of the present without compromising the ability of future generations to meet their own needs.

The Sustainable Development paradigm contains two key concepts (WCED 1987b, p. 41):

- the concept of 'needs', specifically pertaining to the world's poor and developing countries which should receive priority to provide opportunities to those for a better life; and
- the idea of limitations imposed by state of technology and social organization on the environment's ability to meet present and future needs, where perceived needs are socially and culturally determined and consumption values need to be within the bounds of what is ecologically possible.

Together, the above concepts set the foundation for a combined social, cultural, economic, political, legal and ecological approach to tackling overexploitation of our natural resources and assisting populations marginalized by over-development. It is clear from this directive, that the WCED were aware that a global paradigm was needed to address the rapid economic growth and development causing physical change in the ecosystem. This paradigm instigated a process of change where sustaining the overall integrity of ecosystems while progressing human development was paramount. This holistic and integrated approach recognised that focusing on one type of landscape would only contribute to a fragmented solution.

Agenda 21

The Programme of Action of Agenda 21, adopted by more than 178 national governments at the 1992 Rio Earth Summit, changed the future of environmental policy development by holding all nations to account by setting a blueprint for sustainable development in the twenty-first century (United Nations (UN) 1993). Agenda 21, in conjunction with the Rio Declaration on Environment and Development, provides a set of principles and strategies to halt and reverse the negative impact of human behaviour on the physical environment while simultaneously promoting environmentally sustainable economic development for all countries by recognising that human well-being is inextricable tied to nature (UN 1993). This led to declaring the Sustainable Development paradigm as the underpinning principle for managing the environment.

Agenda 21 consists of 40 chapters divided into four sections that set up an integrated framework to assist individual nations in developing their own Local Agenda 21 and include (UN 1993):

1   Social and Economic Dimensions
2   Conservation and Management of Resources
3   Strengthening the Role of Major Groups
4   Means of Implementation.

Chapter 17 exclusively focuses on the protection, rational use and development of living resources in oceans including all kinds of seas, whether enclosed or semi-enclosed and coastal areas. The chapter highlights the need for nations to improve fragmented processes to integrated approaches in marine management (UN 1993). Activities and implementation of the integrated approach are to improve human resource development and capacity building where specifically (UN 1993):

> Coastal states should promote and facilitate the organisation of education and training in integrated coastal and marine management and sustainable development for scientists, technologist, managers (including community-based managers) and users, leaders, indigenous peoples, fisherfolk, women and youth, among others.

This new integrated approach requires signatory nations to focus on the following programme areas:

*   integrated management and sustainable development of coastal areas, including exclusive economic zones;
*   marine environment protection;
*   sustainable use and conservation of living resources under national jurisdiction;
*   addressing critical uncertainties for the management of the marine environment and climate change;
*   strengthening international, including regional, cooperation and coordination;
*   sustainable development of small islands.

What ensued was a dramatic change in the marine policy landscape with nation states developing integrated marine policy containing the principles of sustainable development at its core. This influenced a shift from single-sector management, where top-down governance structures only consulted with the sector of interest, to a holistic approach with the inclusion of the community voice (indigenous, non-indigenous, general public, non-government organisations, local author-ities, the scientific community) in decision-making processes.[1] The success of Agenda 21 as a blueprint for an integrated approach, underpinned by the sustainable development paradigm, is reflected by its presence in national policies relevant to the environment and human development.

Convention on Biological Diversity

Alongside Agenda 21 and the Rio Declaration, the Convention on Biological Diversity (CBD), signed by 168 nations and was entered into force in 1993 (UN 1992). The CBD addresses the importance of global biodiversity to human and social development while simultaneously recognising the alarming rate at which species are disappearing. Its inception was inspired by global commitment to sustainable development, where biological diversity is considered a global asset for all. The objective of this Convention is (UN 1992, p. 3):

> the conservation of biological diversity, the sustainable use of its components and the fair and equitable sharing of the benefits arising out of the utilization of genetic resources, including by appropriate access to genetic resources and by appropriate transfer of relevant technologies, taking into account all rights over those resources and to technologies, and by appropriate funding.

Here, it is the wording of sustainable use in the CBD that adopts the sustainable development paradigm in its core objective where (UN 1992, p. 4):

> Sustainable use means the use of components of biological diversity in a way and at a rate that does not lead to the long term decline of biological diversity, thereby maintaining its potential to meet the needs and aspirations of future generations.

Similar to Agenda 21, the CBD promotes nature as being important to human well-being, with nature's products as the building blocks for the economic and physical health of communities and society. More recently, the focus of the CBD has turned to assessing the various ecosystem services derived from our natural systems as a framework to integrate social and economic dimensions in costing the loss of global biodiversity.

Millennium Ecosystem Assessment and ecosystem services

The release of the Millennium Ecosystem Assessment (MA) in 2005, under the auspices of the UN, was to identify how ecosystems and their services have changed due to natural and anthropogenic pressures. The objective of the MA is to assess the consequences of ecosystem change for human well-being and to establish the scientific basis for actions needed to enhance the conservation and sustainable use of ecosystem and their contributions to human well-being (MA 2005). This framework also adopts the sustainable development paradigm recognising that humans are an integral part of global ecosystems where social, economic, cultural and political landscapes can not only alter the human condition but can also influence ecosystem state (MA 2005).

The MA focuses on the relationship between ecosystem change, how this affects the supply of ecosystem services and what this means for human well-being. Here, an ecosystem service is defined as the benefits people obtain from ecosystems which are grouped into four broad categories:

* provisioning services – for example, food, raw materials, ornamental resources;
* regulating services – for example, climate regulation, air purification, coastal erosion protection;
* habitat services – for example, life-cycle maintenance and gene pool protection;
* cultural services – for example, cultural heritage, aesthetic information and recreation and tourism.

Ecosystems services can be incorporated into marine policies as a framework or be used within a management tool to economically assess service values provided by marine and coastal ecosystems. To do so requires an explicit description and adequate assessment of the links between the structures and functions of marine ecosystems, how they produce benefits for humans and what is their economic value, whether as a use or non-use value that contains direct, indirect, existence and bequest subcomponents (Barbier 2012).

Although the uptake of the ecosystem services in environmental economic assessments has been rapid, with both scientists and policy makers seeing the benefit of placing a monetary figure on marine and coastal natural capital, others are concerned that this approach will allow the global community to view nature as a commodity. According to Barbier (2012), the global community should consider marine and coastal ecosystems as assets, producing beneficial goods and services for human survival and well-being over time. This means marine and coastal ecosystems are no different from any other asset in the economy, and in principle, should be valued in a similar manner. There are two advantages in viewing marine and coastal ecosystems as capital assets in producing goods and services (Barbier 2012):

1   It allows application of the standard tools and analysis developed by natural resource economics for modelling of these complex systems.
2   It facilitates a focus on competing uses, such as conservation versus development of coastal and marine seascapes, and the need to account for the value of ecosystem services in order to make efficient choices between these uses.

Overall, assessing ecosystems services will highlight the connectivity of coastal and marine ecosystems across land and sea gradients and the need to manage these systems in a spatially and temporal integrated way to preserve their synergistic effects (Barbier 2012). Ecosystem service assessments address the integrated approach by using all landscapes from understanding the science of marine and coastal systems, how these functions and processes link to supplying an ecosystem service and how these services supply societal benefits that are economically valued.

In summary, the binding narrative for all of the above directives is sustainable development. For this paradigm to be realised in our oceans, it requires the application of the integrated approach in marine policy. To use the table analogy, the integrated approach requires all three legs to be equal. If the table lacks one leg, or a leg is shorter than the others, or breaks, or there is no table-top to attach the legs, then the table does not provide its function. Like a table, designing and planning marine policy needs to address all landscapes concurrently in an integrated manner for implementation to be successful. This clearly highlights that policy requires more than just good science. Policy makers working with scientists, economists, legal experts and social/cultural groups need to develop new frameworks to address all landscapes simultaneously.

Frameworks for integrated marine policy and management

The progressive development of global environmental policies supporting the integrated approach in achieving sustainable development has led to the creation of various frameworks over the past 25 years. Of relevance to ocean management, two such frameworks include: Integrated Coastal Management (ICM) and the Ecosystem-Based Approach (EA).

Integrated Coastal Management

Immediately following Agenda 21 and the Rio Declaration, Integrated Coastal Management (ICM) spearheaded the process of bringing together the different landscapes into a comprehensive framework for integrated coastal and marine policy. With the use of the word *integrated* in its title, the ICM framework was a popular new approach to managing the coastal and marine continuum. ICM is also termed Integrated Coastal Zone Management (ICZM), where the two can be used interchangeably. ICM is defined as (Sorenson 1997, p. 9): 'the integrated planning and management of coastal resources and environments in a manner that is based on physical, socioeconomic and political interconnections both within and among the dynamic coastal systems, which when aggregated together define a coastal zone.' Here, an integrated approach requires both horizontal (cross sectoral) and vertical (the levels of government and non-government organizations) coordination of those stakeholders whose actions significantly influence the quantity or quality of coastal resources and environment (Sorenson 1997; Wescott and Fitzsimmons 2011; Wescott 2012).

The management of coastal and ocean resources as a continuum is difficult as both areas contain different stakeholders, jurisdictions and ecosystem functions. This fragmentation within the large system, creates a large challenge for effective planning and coordination (Hinrichsen 2011). To address this, ICM requires five key operating principles (Cicin-Sain and Belfiore 2005):

1 intergovernmental (vertical) integration – between the tiers of government;
2 inter-sectoral (horizontal) integration – between the various sectors operating at the same level of government;
3 spatial integration – between catchment, coasts and marine environments;
4 science-management integration – management responses and directions should be based on the best available scientific information at the time;
5 international integration – co-operation and integration between and across national boundaries and systems.

ICM requires the above five principles to be implemented in practice.

As of 2002, 145 of 187 nations, territories and semi-sovereign states with a coastline have launched ICM programs and policies (Hinrichsen 2011). Due to the availability of ICM literature spanning over five decades, there has been much analysis on the challenges and successes of the framework, with most citing limited success of ICM. The most frequently cited failure across most nations is the gap from planning to implementation. This is caused by (Hinrichsen 2011; Sorenson 2011; Hopkins *et al.* 2012):

• complicated coordination of multi-agencies and the lack of creating appropriate institutions in managing coastal and ocean areas;
• allocating a budget to move plans into practice;
• retaining competent staff to implement the plans; and
• all landscapes are considered as separate dimensions from each other and have been rarely successfully integrated.

Additionally, the lack of stakeholder and public engagement in planning processes (Wescott and Fitzsimmons 2011), and more recently cuts across all public policy areas have created management implementation gaps (Sorenson 2011). Undercutting funding and undermining the support needed to move from planning into on-ground implementation has caused the international community to no longer consider ICM to be an effective and efficient means to achieve sustainable development goals (Sorenson 2011).

However, not all hope has been lost, with multidisciplinary teams addressing the shortfalls of ICM in practice. Recently, the Science and Policy Integration for Coastal Systems Assessment (SPICOSA) project in Europe developed the Systems Approach Framework (SAF) as a specific tool for ICM and to evaluate sustainable development (Hopkins *et al.* 2011; Hopkins *et al.* 2012). By incorporating ecological, social and economic landscapes, SAF addresses the how to gap between sustainable management goals and the information and actions needed to implement or modify them using holistic perspectives and issue-oriented investigation (Hopkins *et al.* 2012). The one drawback in the SAF methodological approach is that it does not include the dynamics of policy making, where it is stated that 'it is important to delineate the boundary between science and policy as it separates the objective role of science and the democratic role of governance'. It goes on to state that, 'SAF attempts to strengthen the science-policy interface by improving both the content of the information and the manner in which it is presented' (Hopkins *et al.* 2012, p. 5).

Is it possible to have successful implementation of the SAF if it does not consider the policy dynamics? As stated above, one of the major contributory failures of ICM is the lack of vertical and horizontal institutional arrangements, which one could interpret as policy dynamics. This leads to the question that although the SAF approach takes a positivist view in order to successfully apply the integrated approach in practice, will its shortcomings by not including policy dynamics be another failure in trying to achieve ICM in European seas? Although it is recognised that there is a co-dependency, this example highlights the continuing difficulties in bringing science and policy landscapes together. This continuing challenge has motivated scientists and policy practitioners to rethink the integrated approach.

Ecosystem-based approach

The ecosystem-based approach (EA) is the new environmental paradigm, incorporating into its framework environmental processes, environmental challenges, institutional frameworks and practices to better manage the current exploitation of marine resources and to tackle current fragmented approaches. The ratification of the CBD in 1992 initiated momentum for creating a unifying formal definition of the EA to be (CBD COP 2000 V/6):

> A strategy for the integrated management of land, water and living resources which promotes conservation and sustainable use in an equitable way.

In this context, the EA acts as a framework allowing integrated management practices to focus on application of appropriate scientific methodologies and recognising that humans are an important component of all ecosystems. There are 12 EA principles (Table 10.1), with the first principle being *management of all living resources is a matter of societal choice*. This highlights the hegemonic approach to societal engagement in environmental management. Here, the EA covertly persuades the public through institutional state apparatus (where the opposite is using repressive state apparatus such as law and enforcement) to do something willingly. The EA is trying to persuade society that it is their responsibility to look after our natural front yards through environmental stewardship. In doing so the EA, via society, tackles the objectives of the CBD through addressing the need to conserve biological diversity, the sustainable use of its components, and the fair and equitable sharing of benefits from the use of genetic resources.

The EA framework extends to marine ecosystems in recognition that human impact on marine resources is growing at an alarming rate. The EA has become the prevailing framework in ocean policies, guiding new developments in marine management and planning (Kidd *et al.* 2011). A 2005 scientific consensus signed by 221 USA scientists and institutions took the EA definition one step further for marine ecosystems by defining marine ecosystem-based management (EBM) to be (McLeod *et al.* 2005, p. 1):

Table 10.1 The ecosystem approach principles as defined by the IUCN Commission on Ecosystem Management

Group 1: Key stakeholders and area		
Stakeholders	Principle 1	The objectives of management of land, water and living resources are a matter of societal choice.
	Principle 12	The EA should involve all relevant sectors of society and scientific disciplines.
Area analysis	Principle 7	The EA should be undertaken at the appropriate spatial scale.
	Principle 11	The EA should consider all forms of relevant information.
	Principle 12	The EA should involve all relevant sectors of society and scientific disciplines.
Group 2: Ecosystem structure, function and management		
Ecosystem structure and function	Principle 5	Conservation of ecosystem structures and function, to maintain ecosystem services, should be a priority.
	Principle 6	Ecosystems must be managed within the limits of their functioning.
	Principle 10	The EA should seek the appropriate balance between, and integration of, conservation and use of biological diversity.
Ecosystem management	Principle 2	Management should be decentralized to the lowest appropriate level.
Group 3: Economic issues		
	Principle 4	There is usually a need to understand and manage the ecosystem in an economic context and to: 1) reduce market distortions that adversely affect biological diversity, 2) align incentives to promote biodiversity conservation and sustainable use, and 3) internalize costs and benefits in the given ecosystem.
Group 4: Adaptive management over space		
	Principle 3	Ecosystem managers should consider the effects of their activities on adjacent and other ecosystems.
	Principle 7	The EA should be undertaken at the appropriate spatial scale.
Group 5: Adaptive management over time		
	Principle 7	The EA should be undertaken at the appropriate temporal scale.
	Principle 8	Recognizing the varying temporal scales and lag-effects that characterize ecosystem processes, objectives for ecosystem management should be set for the long term.
	Principle 9	Management must recognise that change is inevitable.

Source: As cited in Kidd *et al.* 2011.

Ecosystem-based management is an integrated approach to management that considers the entire ecosystem, including humans. The goal of ecosystem-based management is to maintain an ecosystem in a healthy, productive and resilient condition so that it can provide the services humans want and need. Ecosystem-based management differs from current approaches that usually focus on a single species sector, activity or concern: it considers the cumulative impact of different sectors.

The emphasis of EBM is the application of the holistic approach to managing ocean systems, where societal values and choices are tightly interlinked with how we manage, govern, and use our natural marine resources. Marine EBM is about interactions and includes four key principles:

1 addressing multiple spatial and temporal scales in the design and implementation of EBM;
2 recognising the linkages between marine ecosystems and human communities that depend on these systems;
3 connecting environmental policy and management efforts across air, land and sea boundaries; and
4 meaningful engagement with stakeholders to create management initiatives that are credible, enforceable and realistic (Leslie and McLeod 2007).

Addressing each principle in practice and understanding the interlinkages between each principle will determine the success of EBM (Box 10.1).

Kidd *et al.* (2011) articulate significant challenges in the application of EBM in marine planning and management (Box 10.2). There is a synchronicity between each challenge and the different landscapes required for the integrated approach. For example, connecting to wider agendas relates to the landscape of law, governance and constituency while addressing key information challenges correlates to the uncertainties and unknowns in the field of science. In summation,

---

Box 10.1 The barriers and successful characteristics for implementation of the Marine Ecosystem-Based Approach (Kidd *et al.* 2011)

Barriers to EBM implementation implementation	Successful characteristics of EBM
• Ineffective stakeholder participation in planning and management.	• Good stakeholder engagement.
• Limited understanding of what the approach seeks to achieve.	• Good public awareness.
• The lack of capacity for decentralised and integrated management.	• Development of a management plan.
• Insufficient institutional cooperation and capacity.	• Good communication among stakeholders.
• The lack of dedicated organisations able to support delivery of EA.	• Adequate funding.
• The overriding influence of perverse incentives.	• Good communication among stakeholders and agencies.
• Conflicting political priorities, including those that arise when a more holistic approach to planning is adopted.	• Good information sharing.
	• The availability of scientific information.
	• Adequate personnel resources.
	• Subsequent changes in the management of activities.

Box 10.2 Significant challenges in applying the Ecosystem Approach

The significant challenges in applying the Ecosystem Approach in coastal and marine policy grouped under three themes, developing the human dimension, addressing key information challenges and connecting to wider agendas (Kidd *et al.* 2011)

Theme 1: Developing the human dimension	Theme 2: Addressing key information challenges	Theme 3: Connecting to wider agendas
• Stressing the holistic ambitions of the EA. • Human activity as the focus of planning and management action. • Developing objectives that reflect societal choice. • Stakeholder engagement.	• Informing societal choice. • Spatial dynamics and different planning and management responses. • Temporal dynamics: the importance of a long-term view and adaptive management. • Understanding structural and functional biodiversity. • Dealing with complexity and uncertainty.	• Connecting marine and terrestrial planning. • Challenging the ecological modernisation paradigm.

Kidd *et al.* (2011) stipulate that although EA brings many benefits for integration and trans-disciplinary discourse, its downfall is that it means all things to all people, that is, each individual has a very different idea about what the EA entails, what it aims to achieve and how it should be operationalized through EBM. Thus, EA application is not straightforward nor easy, requiring an understanding of human beliefs and attitudes and how these translate into decision making and acceptance of policy and management objectives.

Much of managing marine systems is understanding the factors that drive human behaviour and the choices that are made regarding the use of marine resources (Ruckelshaus *et al.* 2008), of which there is little information as compared to economics or even the sciences. This is partly due to the lack of social indicators in marine management and the EA being a relatively new paradigm. There is a paucity of evidence to demonstrate improvements in ecosystem outcomes as a result of long-term EBM applications (Tallis *et al.* 2010). Further research into human behaviours would value-add to the EA and identify social parameters that can be used as objectives or measurable indicators for tools to implement EA on-the-ground.

Tools for implementing an integrated approach

International Management Plans

As a precursor to many marine management tools, International Management Plans (IMPs) (Hinrichsen 2011) recognised the interdependent relationships between marine and near shore coastal ecosystems and their role as drivers for economical development and trade. As of 2002, 168 countries were participating in one or more international management plans at a regional level where there was sharing of marine and/or coastal ecosystems (Hinrichsen 2011). Of the

168 countries, there were generally two categories that captured the approach. The first category was regional integrated coastal management programmes, involving 84 countries across 40 regions. The second was categorised under the banner of United Nations Environment Programme (UNEP) Regional Sea Programs that included 13 large-scale regional efforts involving 140 countries (Hinrichsen 2011). However, of the 13 plans only a few are successful, namely: Baltic Sea Joint Comprehensive Environmental Action Program 1974, Mediterranean Action Plan 1975, Wider Caribbean Action Plan 1981, Black Sea Convention 1992, North-West Pacific Action Plan 1994 and the South Pacific Action Plan 1982.

Similar to ICM, the reduced success of the Regional Sea Programs is due to the gap between planning and on-the-ground implementation. Although International Management Plans initiated the integrated approach, there was much to learn for future tool development, namely (Hinrichsen 2011):

1   creating a framework for policies that govern oceans and coasts at the national level that is backed up by expertise and capacity (human and resource);
2   locally or regionally oriented site management programmes governing specific areas or ecosystems;
3   ability to integrate planning and implementation across economic sectors and other disciplines.

Large Marine Ecosystems (LMEs)

The Large Marine Ecosystems approach considers large spatial extents of coastal and marine ecosystems at the regional level and considered the next step from IMPs in the conservation and sustainable development of coastal and marine resources. LMEs provide a flexible approach to EBM by addressing and identifying the main drivers of ecosystem change and applying management and assessment strategies. The United States National Oceans and Atmospheric Administration (NOAA) led the way in the creation of LMEs in 1984, facilitated by a number of organisations including: the Intergovernmental Oceanographic Commission (IOC), UNESCO, IUCN, UNDP, UNEP and the World Bank's Global Environment Facility (GEF).

LMEs are defined by: bottom depth contours (bathymetry), currents and water mass (hydrography), marine productivity and food webs (trophic relationships). Each LME should be greater than 200,000 km^2 and encompass marine aspects such as continental shelves, oceanic currents and coastal features including river basins and estuaries (Sherman 1991). As of 2009, 121 nations have developed management strategies to meet ecosystem-related targets in an attempt to reduce anthropogenic pressures. Of these, 111 nations have been involved in transboundary diagnostics analysis to identify primary loss of marine biomass, coastal pollution, damaged habitats and depleted status (Sherman 1991; Hinrichsen 2011).

Unfortunately, only five out of 64 LMEs have an implementation mechanism in place that considers the ecosystem as a continuum, with the rest only implementing fisheries measures specific for the fisheries sector. Implementing ecosystem measures in LMEs can be tricky due to legal and political boundaries not necessarily delineating onto ecosystem boundaries. This can be frustrating for trans-boundary conservation efforts where for example, a marine habitat spacing across multiple EEZs may have different conservation values. Even the most carefully developed plans need to consider how and whether trans-boundary legal frameworks, national policies and various sector resource dependencies within EEZs will work with either: current institutional arrangements, the creation of new institutions or redefining institutional arrangements among involved nations. This complex process has created a gap between planning and implementation, and similar to IMPs, has caused limited successes for LMEs.

Marine Spatial Planning

Marine Spatial Planning (MSP) is a popular management tool for implementing the integrated approach. The use of MSP is dependent on the scope, objective and goals of the EA framework. MSP is a planning framework providing a means to improve decision making as it relates to the use of marine resources and space. The MSP framework focuses on the unique and dynamic spatial planning requirements in marine ecosystems to sustain the goods and services society needs or desires from these environments over time (GEF 2012). MSP is a spatial place-based management process and is not a substitute for ICM, rather it builds on important approaches and the policies that support them. The European Commission recognises MSP and ICM as complementary tools, where both improve sea–land interface planning and management when applied jointly (COM 2013).

The MSP framework supports temporal dimensions and can be employed at different spatial scales within different economic, social, legal and scientific landscapes. Although still quite a relatively new approach, there has been some success and it has the potential to greatly improve marine management. Within European Union waters, MSP is gaining support to address the increasing frequency of conflicts in sea space use where European Union Member States are required to establish a process that identifies the problems, collects information and uses these fields in planning and decision making while simultaneously monitoring implementation and management using stakeholder participation throughout all processes (COM 2013).

At the core, MSP addresses multiple management objectives specifically for the allocation of space and resources. It does so by incorporating the level of human use and activities that can operate in line with management objectives specific to a locale. Similar to the above tools, MSP needs to consider various governance structures and institutional arrangements and how stakeholders should be engaged in the decision-making processes for implementation to be successful.

In addition to addressing resource uses, MSP can include conservation tools, for example, marine protected areas (MPAs). A MPA is an umbrella term to define sections of the oceans set aside to protect and restore marine ecosystems from multiple stressors in order to support the sustainable use of marine resources via legal and institutional frameworks. The core objective of MPAs is biodiversity conservation. The IUCN defines a MPA to be (Laffoley 2008, p. 7; Worboys 2015, p. 15): 'A clearly defined geographical space, recognised, dedicated and managed, through legal or other effective means, to achieve long-term conservation of nature with associated ecosystem services and cultural values.'

The level of protection will be defined by the management objectives and the IUCN has compiled a list of protected area management categories to provide consistency and standardisation across national and international jurisdictions (Laffoley 2008). The beginning of this millennium saw MPA targets proposed and set, with 10 per cent of the world's ecological regions to be protected in 2010 and an extensive representative global system by 2012. Although these targets have not been met, coastal nations have been progressing forward in a bid to increase MPAs within their jurisdiction. Incorporating MPAs into a MSP framework may be a step forward in meeting future targets. A successful example of MSP implementation is the Great Barrier Reef Marine Park, Australia. Although titled as a park, the planning and decision-making framework incorporates marine spatial planning to identify areas important for sector use, recreation and tourism and conservation. The Great Barrier Reef Marine Park Authority (GBRMPA) provides a governance structure to accommodate institutional arrangements and regulatory bodies specific to the park. It is the creation of GBRMPA that makes this park successful, allowing autonomy from state and federal institutions and restrictions.

As with other tools, successful implementation of MSP requires momentum and support to jump the divide between creating a paper tool and seeing it through to actual on-the-ground implementation. Stakeholder engagement is required for implementation to be meaningful and successful, and sustainable financing is necessary to provide adaptive management and planning. Finally, MSP must take into account the entire ecosystem, not just certain habitats of interest for resource sectors.

Conclusion: the key challenge for successful implementation of the integrated approach

Connecting science to policy, and integrating it with other landscapes, is a challenge widely acknowledged by all actors involved in marine policy planning and implementation. Although science is important for underpinning marine policy in order to understand how anthropogenic pressures are impacting marine ecosystems and their response to management, science is only one of six landscapes to be considered in developing an integrated framework. The uncertainties that science raises are less significant as compared to the salient reoccurring theme of bridging the gap between policy development and on-ground implementation; realising the integrated approach is challenging, complex and requires investment in time, resources and capacity to move from policy development to implementation.

If we continue with the three-legged table analogy, the success of the integrated approach in marine policy requires:

- an agreed vision incorporating objectives;
- a flexible but defined framework;
- incorporation of a range of expertise across the various landscapes;
- constituency support by engaging stakeholders including cross-national governments, sectorial, cultural and community, in vertical and horizontal decision-making processes and planning;
- the creation of relationships and communication across the different landscapes;
- monetary and human resource support throughout the planning and implementation process; and
- implementing using adaptive processes and management measures.

The three-legged approach requires 100 per cent commitment from all actors from planning to implementation for the integrated approach to be successful. Marine policy success will only occur if all of these elements are rooted in an equitable cooperative within this global public sphere.

Note

1   Authors' note: Community inclusion in marine decision-making processes simultaneously addresses the objective of Chapter 23 in Agenda 21, the need for broad public participation in decision making to achieve sustainable development and here is interpreted as a social parameter under the social landscape in the integrated approach.

References

Barbier, E. B. 2012. Progress and challenges in valuing coastal and marine ecosystem services. *Review of Environmental Economics and Policy* 6(1):1–19.

Cicin-Sain, B. and Belfiore, S. 2005. Linking marine protected areas to integrated coastal and ocean management: a review of theory and practice. *Ocean and Coastal Management* 48(11–12): 847–868.

Cicin-Sain, B., Balgos, M., Appiott, J., Wowk, K., and Hamon, G. 2011. Oceans at Rio +20. How well are we doing in meeting the Commitments from the 1992 Earth Summit and the 2002 World Summit on Sustainable development? Summary for Decision Makers. Global Ocean Forum. Delware, USA.

European Commission (COM). 2012. Blue Growth: sustainable growth from the oceans, seas and coasts. Summary report of the online public consultation results. Directorate-General for Maritime Affairs and Fisheries. http://ec.europa.eu/dgs/maritimeaffairs_fisheries/consultations/blue_growth/blue-growth-consultation-report_en.pdf (accessed 22 August 2013).

European Commission (COM). 2013. Proposal for a Directive of the European Parliament and of the Council establishing a framework for maritime spatial planning and integrated coastal management. (SWD(2013) 64 final) (SWD(2013) 65 final). 2013/007 (COD). 133 Final.

GEF. 2012. (Secretariat of the Convention on Biological Diversity and the Scientific and Technical Advisory Panel). Marine Spatial Planning in the context of the Convention on Biological Diversity: A study carried out in response to CBD COP10 decision X/29, Montreal, Technical Series No. 68, 44 pages.

Gray, N. J. and Campbell, L. M. 2008. Science, policy advocacy, and marine protected areas. *Conservation Biology* 23(2):460–468.

Halpern, B. S., Longo, C., Hardy, D., McLeod, K. L., Samhouri, J. F., Katona, S. K., Kleisner, K., Lester, S. E., O'Leary, J., Ranelletti, M., Rosenberg, A. A., Scarborough, C., Selig, E. R., Best, B. D., Brumbaugh, D. R., Chapin, F. S., Crowder, L. B., Daly, K. L., Doney, S. C., Elfes, C., Fogarty, M. J., Gaines, S. D., Jacobsen, K. I., Karrer, L. B., Leslie, H. M., Neeley, E., Pauly, D., Polasky, S., Ri, B., St Martin, K., Stone, G. S., Sumaila, U. R., and Zeller, D. 2012. An index to assess the health and benefits of the global ocean. *Nature* 488: 615–622.

Hinrichsen, D. 2011. *The Atlas of Coasts & Oceans: Mapping ecosystems, threatened resources and marine conservation.* Earthscan, London.

Hopkins, T. S., Bailly, D. and Støttrup, J. S. 2011. A systems approach framework for coastal zones. *Ecology and Society* 16(4): 25.

Hopkins, T. S., Bailly, D., Elmgren, R., Glegg, G., Sandberg, A., and Støttrup, J. S. 2012. A systems approach framework for the transition to sustainable development: potential value based on coastal experiments. *Ecology and Society*, 17(3): 39.

Kidd, S., Maltby, E., Robinson, L., Barker, A., and Lumb, C. 2011. The ecosystem approach and planning and management of the marine environment. Chapter 1, pp.1–33, in: Kidd, S., Plater, A. and Frid, C. (eds) *The Ecosystem Approach to Marine Planning and Management.* Earthscan, London.

Laffoley, D. d'A. 2008. Towards networks of Marine Protected Areas. The MPA Plan of Action for IUCN World Commission on Protected Areas. IUCN, WCPA, Gland, Switzerland.

Leslie, M. and McLeod, K. L. 2007. Confronting the challenges of implementing marine ecosystem-based management. *Frontiers in Ecology and the Environment* 5(10): 540–548.

McLeod, K. L., Lubchenco, S. R., Palumbi, S. R. and Rosenberg, A. A. 2005. Scientific consensus statement on marine ecosystem-based management. The Communication Partnership for Science and the Sea (COMPASS) Signed by 221 academic scientists and policy experts with relevant expertise. www.compassonline.org/science/EBM_CMSP/EBMconsensus (accessed 22 August 2013).

Millennium Ecosystem Assessment (MA) 2003. *Ecosystems and human well-being: A framework for assessment.* Island Press, Washington, DC.

Millennium Ecosystem Assessment (MA) 2005. *Ecosystems and human well-being: Synthesis.* Island Press, Washington DC.

National Research Council (2008). *Increasing Capacity for Stewardship of Oceans and Coasts. A priority for the 21st Century.* National Research Council of the National Academies. The National Academic Press, Washington DC.

Pietri, D., Mcafee, S., Mace, A., Knight, E., Rogers, L., and Chorensky, E. 2011. Using science to inform controversial issues: a case study from the California Ocean Science Trust. *Coastal Management* 39: 296–316.

Ruckelshaus, M., Klinger, T., Knowlton, N., DeMaster, D. P. 2008. Marine ecosystem-based management in practice: scientific and governance challenges. *BioScience* 58(1): 53–63.

Sherman K. 1991. The large marine ecosystem concept: research and management strategy for living marine resources. *Ecological Applications* 1(4): 349–360.

Sorenson, J. 1997. National and international efforts at integrated coastal management: definitions, achievements and lessons. *Coastal Management* 25: 3–41.

Tallis, H., Levin, P. S., Ruckelshaus, M., Lester, S. E., McLeod, K. L., Fluharty, D. L., and Halpern, B. S. 2010. The many faces of ecosystem-based management: making the process work today in real places. *Marine Policy* 34: 340–348.

United Nations (UN). 1992. Convention on Biological Diversity. United Nations Environment Programme. Montreal, Canada.

United Nations (UN). 1993. Earth Summit Agenda 21. The United Nations programme of action from Rio. United Nations Department of Public Information, Geneva, Switzerland.

Wescott, G. 2012. Disintegration or disinterest? Coastal and marine policy in Australia. Chapter 7, pp. 88–101, in: Crowley, K. and Walker, K. (eds) *Environmental Policy Failure: The Australian story*. Tilde University Press, Prahran, Australia.

Wescott, G. 2015. Ocean Governance and Risk Management. Chapter 25, in: Paleo, F. P. (ed.) *Risk Governance: The articulation of hazard, politics and ecology*. Springer, Dordrecht.

Wescott, G. and Fitzsimons, J. 2011. Stakeholder involvement and interplay in coastal zone management and marine protected area planning. Chapter 17, pp. 225–238, in: Gullett, W., Schofield, C. and Vince, J. (eds) *Marine Resources Management*. LexisNexis Butterworths, Sydney, Australia.

Worboys, G. L. 2015. Concept, purpose and challenges. Pp. 9–42 in: Worboys, G. L., Lockwood, M., Kothari, A., Feary S. and Pulsford, I. (eds) *Protected Area Governance and Management*. ANU Press, Canberra, Australia.

World Commission on Environment and Development. 1987a. Tokyo Declaration. 0020t/W0006e/ 27.2.87–8/final/Tokyo (accessed 22 August 2013).

World Commission on Environment and Development. 1987b. *Report of the World Commission on Environment and Development: Our Common Future*. Geneva, Switzerland.

11

ECOSYSTEM SERVICES AND THEIR ECONOMIC AND SOCIAL VALUE

Jason Scorse and Judith Kildow

Introduction

The concept of "ecosystem services valuation" has become an important theoretical construct for linking ecosystem functions to human well-being, using basic principles of natural science combined with welfare economics (as well as contributions from psychology, sociology, and even more recently, neuroscience). It is important to recognize that ecosystem services are valued using inherently anthropocentric methods— nature is afforded no intrinsic value. The socioeconomic value of an ecological resource depends solely on the value humans derive from that resource. This value may be direct—in the case of fish harvested from the sea—or indirect—in the case of water filtration provided by wetlands. It is the indirect ecosystem services that are frequently overlooked in planning and decision-making since they often require sophisticated scientific understanding, and provide value to human society through complex mechanisms and interactions.

Understanding these links is becoming increasingly critical in a wide range of policy and management contexts, because when ecosystem services are degraded or destroyed the costs to society can be great, oftentimes eclipsing the benefits that are conveyed through the degrading activity. For example, in the wake of Hurricane Katrina in New Orleans in 2005 it has become apparent that the economic benefits that accrued over the decades from destroying the wetlands surrounding the city (for ports, canals, and other forms of industrial development), were less than the costs of the lost storm mitigation services that the wetlands naturally provided (Costanza *et al.* 2014). In other words, when accounting for the full value of the benefits derived from New Orleans' wetlands' ecosystem services, the decision to remove them in favor of industrial uses was uneconomic, and would not have passed a cost-benefit test.

It is vital to establish a consistent definition for ecosystem services, even if the concept can never be perfectly defined. The most recent attempt to standardize both a definition and to create categories of ecosystems services came out of the United Nations' Millennium Assessment (MEA) in 2005: "ecosystem services are the benefits people obtain from ecosystems" (MEA 2005). While a very simple definition, this has become the standard for many studies today. The Millennium Assessment goes on to describe specific categories of services that include provisioning services, such as food and water; regulating services such as flood and disease control; cultural services such as spiritual, recreational, and cultural benefits; supporting services, such as nutrient cycling that maintains the conditions for life on Earth, and biodiversity.

Although the Millennium Assessment helped to clarify current thinking about ecosystem services, this definition has not been universally accepted. After surveying many competing definitions of ecosystem services across many studies, Boyd and Banzhaf (2007) concluded that "ecology and economics have failed to standardize the definition and measurement of ecosystem service." They proposed the creation of a consistent definition that can be easily integrated into market accounting systems. To this end, they focus on defining "'final' ecosystem services, which are components of nature, directly enjoyed, consumed, or used to yield human well-being" (Boyd and Banzhaf 2007, p. 619).

Their definition makes a distinction between final ecosystem services and final economic goods. Ecosystem services can be thought of as inputs in a manufacturing process that contribute to the production of final goods enjoyed by people. Boyd and Banzhaf point out that much of the value of ecosystem services is already captured in economic models through normal measures of GDP, to the extent that ecosystem services contribute to the final value of many market goods (i.e., crop production or fisheries).

This is an important distinction because it is the "non-market" goods provided by ecosystem services that are almost completely ignored in traditional measures of GDP. Non-market values refer mostly to the indirect benefits of ecosystem services that are not currently priced in market systems, and include, everything from climate regulation to the quality of life benefits from clean air, water, open space, and the existence value of wildlife. In order to make more rational decisions about how to use and manage resources—and attempt to maximize their social and economic value—these non-market values must be captured in new models and incorporated into all levels of decision making.

The reality is that markets will never account for the full range of ecosystem services unless they are forced to. This point cannot be emphasized strongly enough. Even though measuring ecosystem services and the values they provide is a challenge—both methodologically and practically—the much greater challenge is using this information to change behavior and produce better societal outcomes.

It is likely that under current regulatory systems, the lack of proper accounting for ecosystem services is leading to many uneconomic decisions from the local level all the way to the international level, especially when longer time frames are considered and proper discount rates are used (Stern 2006). Climate change will likely prove the largest "meta" example of collective market failure—the damages that will be suffered due to the warming of the planet and the acidification of the oceans will eventually eclipse the benefits accrued by burning excess quantities of fossil fuel (Stern 2006).

But there are many more examples at a smaller scale as well, whether they are areas such as New Orleans where wetlands destruction ultimately proved disastrous, offshore mining that ends up fouling beaches and depressing tourism values, or the decimation of pelagic fisheries and excessive marine mammal by-catch that ultimately impoverishes nations rather than enrich them.

Though a detailed discussion of why markets fail to account for ecosystem services is beyond the scope of this paper (but see, for example, Scorse 2010), the primary reasons are relatively straightforward:

1   Externalities—pollution costs are not borne by those who cause the pollution, which leads to the underpricing of polluting goods. This distorts the entire economy in favor of resource-intensive production.

2   Lack of property rights—where there are no property rights, resources are exploited in a "free-for-all" manner. This is especially true in the open oceans, but also has implications near the shore.

3   Imperfect information—the complexity of ecosystem services and the fact that many are poorly understood leads to a bias in favor of what is known and easily quantifiable (i.e., direct market benefits over non-market ecosystem services). This also plays out when those who benefit from an ecosystem service may not be aware of it, and those who degrade the service may not be aware of the values they are destroying (but it's important to note that even with perfect information about these costs and benefits the parties may still have a very hard time using that information to create mutually beneficial transactions that protect the ecosystem services over the long term).

4   Discounting—people value the future a lot less than the present and this creates a bias in favor of short-term planning that is very difficult to overcome; this is partially driven by two-to-six-year election cycles, but also corporate quarterly reports that cause publicly held companies to invest for short-term gains to the detriment of long-term sustainability.

The field of environmental and natural resource economics has an extensive literature on how to "correct" these market failures that has been developed over many decades. But much of the world continues to ignore the bulk of this work largely because of very powerful socio-political forces that favor the status quo, rather than a collective lack of knowledge of how to make markets function more effectively for society. In addition, the high transaction costs inherent in dealing with large numbers of actors over large distances make incorporating ecosystem services into planning and management decisions very difficult, so absent institutions are needed to broker these arrangements.

The economic value of coastal and marine ecosystem services

Much of the market value of ecosystem services is derived through biological productivity that produces goods directly exploited for human use and consumption (i.e., fish, kelp, coral, or even shark fins), but there is also tremendous market value derived indirectly through all sorts of non-consumptive activities, with tourism and real estate values being the most prominent. In many coastal areas and island nations, the natural beauty, coral reefs, beaches, surf breaks, and wildlife provide an important source of economic value that is directly measureable through market prices (i.e., hotel, travel, restaurant, and diving revenue).

Non-market values exist on a continuum, with some that are completely divorced from market prices and others that are actually embedded in them. For example, the ecosystem value of carbon cycling through the oceans and atmosphere is not something that is recognizable in market prices (although it is critical for life), whereas homes adjacent to beaches are much more expensive than equivalent homes inland due to their proximity to the natural beauty of the coasts. In essence, what at first appears to be a non-market value—an ocean view or a nearby beach entrance—is actually incorporated into real estate prices in the market.

There is also a category of non-market value that is derived from the benefits that individuals receive freely (or for very low cost) from ecosystem services, but for which they would be willing to pay. Economists refer to this type of value as "consumer surplus," and contend that it should be considered when assessing a resource's overall value. The logic is that if taking a walk on the beach on a sunny day provides an individual with the equivalent of $10 in value (perhaps akin to the value of going to a movie on a rainy day), then just because the beach has free access doesn't mean that its value should be counted as zero; the $10 in value that individual receives "internally" (since no actual dollars are exchanged) should be considered as real as if they had paid an admissions ticket to enter the beach.

This example illustrates that many ecosystem services have both market and non-market values. Coastal zones are the prime example; they contain market values from the fish that is harvested, the tourist revenue, and also a portion of nearby home values, but they also provide consumer surplus to many millions around the world just by their very existence, which is not captured by market systems or traditional means.

In the U.S., reasonably good data exist to measure the market values of ocean and coastal resources. Since 1933, the U.S. has kept national income and product accounts (NIPA), which measure total income and output of the nation, and provide a comprehensive picture of the nation's economy (NRC 1999). These accounts measure production and income that arise primarily from the market economy, where goods and services are traded openly through the interaction of supply and demand. With the NIPA datasets it is possible to ascertain the value the oceans and coasts generate for national, regional, state, and local economies. The national accounts are divided into sectors, such as shipbuilding and repair, coastal construction, marine transportation, fishing, offshore minerals, and coastal tourism and recreation. They provide a reasonably accurate estimate of wages, employment, number of establishments, and GDP for the ocean economy.

In 1999, the National Ocean Economics Program (NOEP) was launched at the Massachusetts Institute of Technology (MIT). Its mission has been to produce time series data using NIPA on the value of the oceans and coasts to the U.S. economy. The first of its kind, NOEP data provide both snapshots in time as well as trends in ocean-dependent economic activities by county, state, and the nation as a whole. Data are not available at a finer level because Internal Revenue Service (IRS) and Bureau of Labor Statistics (BLS) rules do not allow disclosing information that could compromise competition in the market place.

The NOEP data has revealed that marine transportation and coastal tourism and recreation generate far more jobs and revenue for the U.S. economy than other ocean industries such as fishing, offshore minerals, construction, and ship building. The NOEP also provides marine natural resource production and value estimates, with data on annual production of offshore oil and gas in state and federal waters, and fisheries for more than 50 years (which shine a light on which species have declined and which are stable).

The NOEP data have been cited hundreds of times by government, NGOs, and business, as it has become the premier source for economic statistics on the status of and trends in the ocean and coastal economy. The NOEP produces state-level multipliers, which can be used to determine the overall wage and employment impacts of a dollar spent in an ocean industry after it percolates through the rest of the economy. The NOEP also produces estimates of the size of the coastal economy using shore adjacency criteria. In the U.S. in 2010 the total coastal states' economy generated 83 percent of total U.S. GDP, while only representing about 20 percent of the land mass. Coastal counties produced more than 50 percent of all jobs and wages.

The NOEP not only pioneered the use of ocean and coastal accounts in the U.S., but its methodology has become the international standard, although each nation has customized the basic categories for their own economies. Studies in Canada have been undertaken at the national level (Gardner Pinfold 2009) and provincial levels (Gardner 2005), while ocean economy estimates have been undertaken for the United Kingdom (Pugh 2008), France (Kaladjian 2009, Girard and Kalaydjian 2014), Australia (Allen Consulting Group 2004), New Zealand (Statistics New Zealand 2006) and China (2006–2012) (Song *et al.* 2013). Kildow and McIlgorm (2010) compare some of these international programs. Many Asian and Southeast Asian nations published their ocean accounts in 2009 (Tropical Coasts, 2009), expanding the number of nations to more than 20 that measure the ocean's direct contribution to their economies.

Non-market values are understandably much more difficult to measure than market values. Over many decades, however, economists have refined numerous statistical methods for estimating them. The two primary methodological categories for assessing non-market values can be divided into "revealed preference" techniques and "stated-preference" techniques.

Revealed preference methodologies examine observable human behavior that can be used to estimate values derived from ecosystem services. For example, the Travel Cost Method (TCM) uses data on how much time and money people spend to visit natural resources, such as beaches, lakes, rivers, or national parks, to determine their consumer surplus from the public provision of these goods.

The Hedonic Price Method (HPM) relies on real estate values to separate out from final home prices the incremental contribution of environmental amenities, such as nearby air quality, proximity to natural resources, and scenic views (Kildow 2009), as a means to directly assign monetary values to these ecosystem services (in some ways HPM can be viewed as a "quasi-non-market" valuation method since it relies on market prices to estimate what are normally considered non-market values).

It is worth noting the magnitude of many of these non-market values in the coastal zones and the associated policy implications. The value of an ocean view that is capitalized into a single home can be worth upwards of a million dollars in California (and in this range in many other parts of the world). This means that any attempts to diminish these views with offshore oil rigs could potentially lead to large decreases in property values. In the same manner, proximity to clean beaches with abundant wildlife can be capitalized at hundreds of thousands of dollars per home, and therefore, anything that could potentially despoil the beaches and the surrounding coastal areas could also greatly impact real estate prices (Kildow 2009). There is new research underway that links home prices to the proximity to surf breaks and suggests that this ecosystem service significantly increases nearby property values as well (Scorse *et al.* 2015).

More recently, the enormous damage and loss of life caused by the land fall of unusually intense storms such as Hurricane Sandy in the U.S. and Typhoon Haiyan in the Philippines may lead to plummeting coastal real estate values, both from the risk of loss and unaffordable insurance premiums.

Coastal ecological services generate a relatively stable stream of revenue to local governments through property taxes that can be sustained indefinitely if these resources are protected. Since many communities and states rely on property tax as a major source of overall revenue, it is critically important to understand the extent to which property values are impacted by changes in nearby ecosystem services, for better and worse (Bagchi 2003, *The Economist* 2011).

Stated preference methods are used to estimate the most controversial elements of non-market value: non-use value. Non-use value includes three sources of value—the pure *existence value* of knowing ecosystems (or parts of ecosystems) are being protected, the *option value* of wanting to reserve the right to visit these natural systems sometime in the future, and the *bequest value* of wanting to ensure that future generations have the option of enjoying these ecosystems. Non-use values are most easily understood in the context of a remote area such as the Arctic or the middle of the Brazilian rainforest, which few people are ever likely going to visit, but from which are derived some value from simply knowing that they are being protected and sustained.

The non-market value of ecosystem services for natural processes such as beach nourishment or storm protection can be estimated using "replacement or avoided cost" methods. For example, if a new jetty disrupts natural sand dispersal and reduces the width of down-current beaches, the cost of replacing the sand can be imputed as an added cost of building the jetty—

a non-market impact. Similarly, if mangroves provide storm protection that reduces average annual damages to life and property of $100 million that value can be imputed to the mangroves. In the wake of the 2004 Asian tsunami, analysis of wave impacts in areas where there were relatively intact mangrove systems versus those where they were largely removed in favor of coastal development showed that the former fared much better overall (Dahdouh-Guebas *et al.* 2005, Alongi 2008, Agardy 2014).

Mangroves are also being studied for their carbon sequestration potential. Recent research posits that even at relatively modest CO_2 prices, this "blue carbon" may be extremely valuable (Murray *et al.* 2011, McLeod *et al.* 2011, Ullman *et al.* 2012, Lau 2013). Increasing attention is also being paid to the potential value of rare deep-sea creatures for use in medical research or materials innovation. With deep-sea mining operations expanding exponentially around the world (Haefner 2003, deVogelacere *et al.* 2005), it is critical to document the non-market values of deep-sea life and the threats they face. Even though the profits from deep-sea oil and mineral operations are often great, the costs and risks they impose on marine systems and on society can be great as well in terms of lost opportunities; only a comprehensive accounting can inform policy makers and the public as to whether the net benefits are positive and worthwhile.

An illustrative example of the power of non-market valuation comes from a recent decision by the government of the small island nation of Palau, which is extremely dependent on tourist revenue for its economy. After careful analysis, the government concluded that the value of keeping their sharks alive is orders of magnitude greater than selling them for shark fin soup; research conducted by the Australian Institute of Marine Science indicates that over the course of their lifetime each shark off the coast of Palau provides on average of $1.9 million in tourist revenue (Jolly 2011).

Many of the dozens of nations that have robust whale-watching industries have also estimated the economic benefits these (and other) cetaceans provide, and assessed ways to augment this industry (Cisneros-Montemayor *et al.* 2010). While hunting whales is not allowed in most parts of the world, whale mortality is often due in large part to human causes, whether through ship strikes, pollution, or underwater sonar, and efforts to mitigate these impacts can provide net benefits to regional whale-watching industries.

The social value of ecosystem services

Measuring the social value of ecosystems is considerably more difficult than economic value, not only because social variables are harder to quantify, but because there is no consensus on the definition of social value (see for example Schumpeter 1908, Tool 1977, Schweitzer 1981, Morris and Shin 2002, Lange and Toppel 2006, Lewin and Trumball 2012). In general, social values refer to issues of equity, opportunity, and vulnerability, as well as cultural benefits. For example, a coastal ecosystem can provide social value through its local maritime cultures, the sports and recreational activities that bring the community together, the variety of economic activities that provide for social cohesion and mobility, and the resilience provided against extreme climactic events. Given the complexity in measuring social value and the lack of an agreed-upon definition, it is easiest to illustrate how to think about the concept using some current case studies.

Take offshore oil and mineral development, which is expanding rapidly around the world and tends to concentrate a tremendous amount of wealth in a small number of hands (see Chapters 18 and 20 this Handbook). The world's oil and mining companies exert significant market power, and act much more like oligopolies than firms in a competitive market (Moran 1987, Bolckem 2004, Dore 2006). This means that they often have the ability to set very favorable terms in the areas where they operate, and when they collude they can both move market prices and restrict access (as may be the case with Chinese companies that control much of the world's rare earth minerals).

While the majority of the profits (benefits) of offshore oil and mineral extraction go to the few, the risks of their activities are often spread out over large populations. A major oil spill can decimate the livelihoods of literally millions of inhabitants in nearby coastal areas, and the lost non-use values from degraded marine ecosystems will be felt way beyond the shores where the spill occurred. This asymmetry between concentrated benefits and widespread social costs is endemic in many areas around the world. The situation in the Niger Delta in Nigeria is particularly extreme, where the environmental and social costs of oil production are externalized onto a large and mostly poor population with very few mechanisms for redress (UNDP 2006). But one doesn't have to travel to the less-developed parts of the world to witness such power and equity imbalances; in the U.S. some of the Alaskan indigenous populations are threatened by oil and mineral exploration that undermines traditional ways of life.

From a social value perspective, such unequal power dynamics increase vulnerability for local populations, and can greatly increase inequality and poverty in the event of an accident. They also present a more fundamental problem in that the local people are often at the mercy of the environmental and safety policies of companies that are not held accountable for their misdeeds.

There are ways for governments to address the social inequities inherent in this type of industrial activity, even if it is very difficult to do. First, they can insist on high royalty payments for the energy and minerals, and either funnel this revenue directly back to the communities or use it for social ends, such as education, infrastructure, and healthcare. Governments can also not only require strong environmental and safety protections from firms, but mandate environmental insurance bonds as a precondition for approving offshore operations. These bonds must be high enough to cover clean-up and compensation costs in the event of an accident. The companies get the money returned (with interest) based on environmental and safety performance metrics that must be met not only during operations, but during any decommissioning and remedial phase of the energy and mining projects.

The example of offshore mining presents a case where the diminishment of social value is often imposed on communities through intensive natural resource exploitation in the coastal zones, unless strong government negotiations can tip the scale in favor of greater social benefits and protections. But what if communities want affirmative models for producing strong social values?

The natural sciences have demonstrated that more diverse ecosystems are the most resilient (Chapin *et al.* 2000, Loreau *et al.* 2001), and new social science research points to economic diversity as a key indicator of social resilience as well (Holling 2001, Folke *et al.* 2002, Pickett *et al.* 2004). Communities that derive employment and economic value through a variety of activities are better suited to weather economic crises as well as environmental change. They also provide greater opportunity and mobility to a wider class of the citizenry.

Monterey, California is a good illustration of a community that has evolved from an economy predominantly based on natural resource-intensive industries (fishing and agriculture) to a much more diverse economy. While fishing and agriculture still account for a large share of employment and GDP, the development of very robust tourism, research, and education sectors—based on the very ecological systems that have been under pressure for so long—has produced a much more vibrant economy overall (Kildow and Colgan 2005, Thornberg *et al.* 2011). Monterey is home to many of the premier marine science and policy institutions, which attract highly-educated students and workers, and Monterey Bay Aquarium is world famous. According to the NOEP, which maintains a database on coastal economies at www.ocean economics.org, wildlife viewing and other forms of coastal tourism and coastal recreation are now huge industries, employing more than 13,000 people and generating over $700 million dollars in 2010 in Monterey. The much greater variety of economic activity supports a large service, tech, and construction sector (Thornberg *et al.* 2011).

Tensions remain between the natural resource sectors and the larger community due to continued threats to marine life, and very serious water and water quality issues. But whereas in the past fishermen and farmers controlled most of the levers of power, there are increasing pressures in this more diversified economy to balance needs and promote activities with less of an ecological footprint and more dependence on ecological health.

Florida is an example of a state that has also realized that diversification is a key to economic strength and resilience. Over the past decade, the state has attracted a number of highly prestigious marine and medical research institutions that provide stability, high wages, and an alternative to dependence on tourism, which is volatile given the frequency of extreme weather events in Florida. Scripps Research Institute and Mayo Clinic have been enticed to Florida with generous incentives. The Hubbs Research Institute complements Sea World and its program in Orlando, and numerous other world-class research centers have established satellite operations in Florida. In addition, the government of Florida has increased its own ocean research programs at its universities; marine research budgets in Florida in 2008 were more than $300 million, with only a little more than half of the institutions reporting precise figures (NOEP 2008).

Any discussion of social value from ocean and coastal ecosystems would be incomplete without mentioning the fascinating new neuroscience research that has shown that the proximity to the oceans, with their miles of beaches, dramatic sunsets, and abundant wildlife provide tremendous psychological value to humans. Coastal zones make people happier, healthier, and reduce stress through a number of channels (Volker and Kistemann 2011, White *et al.* 2010). This tremendous social value is not captured by any of the current models, but is potentially very large. The policy implications are large as well. The key, as always, is how to encourage human settlement and use of coastal zones in ways that don't threaten the very ecological systems that make them so attractive; and also to generate diverse streams of economic activity to maximize resilience and opportunity.

Incorporating ecological services into national accounts

Environmental accounts provide a framework for collecting and organizing information on the status, use, and value of a nation's natural resources and environmental assets, as well as on expenditures on environmental protection and resource management (INTOSAI WGEA 2010, and see also Chapter 2 of this Handbook).

Creating "green" national accounts that incorporate changes in ecosystem services and natural capital into current measures of GDP has been an overarching goal of the environmental movement for decades. The logic is simple: by incorporating the full range of environmental values into economic accounts areas can identify where certain industrial activities actually make society worse off, and also the areas where investments in natural capital can provide the greatest returns to society.

The United Nations Conference on the Environment in Rio de Janeiro in 1992 produced Agenda 21, which called for the UN to begin work on a handbook for green accounting. The finished product was based on numerous approaches to environmental accounting, pioneered in a series of workshops by the United Nations Environment Programme (UNEP) in collaboration with the World Bank. Because of the embryonic nature of this work, the discussion of concepts and methods never reached any final conclusions, and the UN handbook and its System of integrated Environmental and Economic Accounting (SEEA) were therefore issued as an interim version of work in progress (UNSD and UNEP 2000). However, the SEEA was subsequently tested in Canada, Colombia, Ghana, Indonesia, Japan, Mexico, Papua New Guinea, the Philippines, the Republic of Korea, Thailand, and the U.S.

In response to the issuance of the UN handbook, the U.S. Bureau of Economic Analysis (BEA) in the Department of Commerce began to develop a system for extending NIPA to include both market and non-market estimates of ecosystem values. Members of Congress were tipped off about this work in 1995 and held hearings. Some in Congress believed that the methods for valuing the environment were still immature and not ready to be institutionalized; they were also responding to pressures from the coal and other extraction industries, who feared that their activities would show up negatively in the new green accounts. Other members felt that it was inappropriate to change an economic accounting system that everyone trusted.

The result of the hearings was that Congress withdrew funding for this BEA experiment, imposed a ban on any additional work until further notice, and asked the National Academy of Sciences National Research Council to review the BEA strategies and report back to them. The resulting report entitled, "Nature's Numbers" (NRC 1999), provided an unequivocal endorsement of green accounts and a call for a comprehensive assessment of market and non-market values of ecosystem services. The authors expressed concern that the U.S. might lag behind other nations if a system of green accounts wasn't developed quickly, and that it was in the best interests for U.S. investors and policy makers to have this information.

In 2006, sufficient progress towards a system of green accounts internationally prompted a major interagency meeting between the U.S. Government Accounting Office (USGAO) and the National Academy of Sciences to once again discuss the topic of environmental accounts (USGAO 2007). This meeting followed the lifting of the Congressionally imposed ban on BEA's activities that had lasted from 1995–2005). In 2010, a report by the U.S. General Accounting Office described the status of environmental accounting around the world, indicating that many nations were now using some form of it and that there was a strong effort to standardize the accounts (INTOSAI WGEA 2010). It is argued that the absence of U.S. participation prevented the U.S. from having a voice in the setting of international green accounting standards.

Since 2010, The European Commission has instituted regulations on green accounts for the entire European Community. Through the UN's work and the European Community's efforts, many nations have now implemented an official system of environmental accounting. However, the U.S. government has yet to follow suit, and there are no indications of any impending plans to do so.

Conclusion

Much of the social and economic value of ocean and coastal resources is obvious to all, as it shows up clearly in market data, for both income and employment. However, the value of the ecosystem services in the ocean and coastal zones is poorly understood as they involve large, distant, and complex processes, the magnitude of which is likely greater than the value of what can be quantified (Moberg and Rönnbäck 2003; Barbier *et al.* 2011). A large portion of this value is conferred to society outside of market mechanisms, and in ways that are often hard to trace. However, new methods for estimating these non-market values are coming online, and must be embraced in order to achieve long-term sustainability that guarantees widespread social benefit and economic prosperity. It is up to natural scientists to establish and communicate the links between ecosystem services, including important services coming from marine and coastal ecosystems, and benefits to humans, and up to economists to determine their monetary values.

In turn, it is the role of decision makers and governing institutions, at all levels, to use this information in promoting sustainable use.

Unfortunately, market failure is the norm, rather than the exception in the environmental realm, meaning that markets alone will almost never take into account the value of ecosystem services without some form of intervention. Governments, NGOs, and businesses, therefore, must take an active role in building institutions and mechanisms to bring the non-market values to light, as well as to incorporate them into decision making. This is hard to do in political and economic climates dominated by short-term thinking, but it is absolutely essential. If this path is not followed, much of what is believed to be economic growth will over the medium- to long-term prove illusory, and the costs that are ignored will increase to the point that they overwhelm the benefits.

There has been significant progress towards an international system of green accounts that attempts to build the full value of ecosystem services into individual national accounts, but this has yet to be fully realized, and U.S. participation has stalled. Even with the advent of national green accounts, much of the most important data on the value of ecosystem services will be needed at finer geographic scales so that cities, states, and provinces can have useful information for planning. The data requirements of such an effort are vast, but the payoffs much larger. The value of marine and coastal ecosystem services will only grow over the coming decades, as population increases put ever greater pressures on natural resources, and humans around the world seek the beauty of the oceans and coasts for the high quality of life they provide. At the same time, as these natural assets and services in coastal areas grow in value, they are increasingly threatened by climate change impacts of Sea Level Rise and Ocean Acidification to name but two. Ignoring the value and importance of natural capital along coasts will be at society's peril.

References

Agardy, T. (2014). Mangrove Ecosystem Services. In van Bochove, J., Sullivan, E., Nakamura, T. (Eds.). *The Importance of Mangroves to People: A Call to Action*. Cambridge UK: United Nations Environment Programme World Conservation Monitoring Centre, pp. 43–68.

Allen Consulting Group. (2004). Economic contribution of Australia's marine industries 1995–96 to 2002. Retrieved from: www.environment.gov.au/resource/economic-contribution-australias-marine-industries-1995–96–2002–03 (accessed July 15, 2015).

Alongi, D. M. (2008). Mangrove forests: Resilience, protection from tsunamis, and responses to global climate change. *Estuarine, Coastal and Shelf Science*, 76(1), 1–13.

Bagchi, S. (2003). Politics and economics of property taxation. *Economic and Political Weekly*, 38(42), 4482–4490.

Barbier, E. B., S. D. Hacker, C. Kennedy, E. W. Koch, A. C. Stier, and B. R. Silliman (2011). The value of estuarine and coastal ecosystem services. *Ecological Monographs*, 81(2), 169–193.

Bolckem, S. (2004). Cartel formation and oligopoly structure: A new assessment of the crude oil market. *Applied Economics*, 36, 1355–1369.

Boyd, J. and S. Banzhaf (2007). What are ecosystem services? The need for standardized environmental accounting units. *Ecological Economics*, 63, 616–626.

Chapin, F. S., E. S. Zavaleta, V. T. Eviner, R. L. Naylor, P. M. Vitousek, H. L. Reynolds, D. U. Hooper, S. Lavorel, O. E. Sala, S. E. Hobbie, M. C. Mack, and S. Diaz (2000). Consequences of changing biodiversity. *Nature*, 405(6783), 234–242.

Cisneros-Montemayor, A. M., U. R. Sumaila, K. Kaschner, D. Pauly (2010). The global potential for whale watching. *Marine Policy*, 34(10), 1273–1278.

Costanza, R., O. Perez-Maqueo, M. Luisa Martinez, P. Sutton, S. J. Anderson, and K. Mulder (2014). Changes in the global value of ecosystem services. *Global Environmental Change*, 26, 152–158.

Dahdouh-Guebas, F., L. P. Jayatissa, D. Di Notto, J. O. Bosire, D. Lo Seen, and N. Koedam (2005). How effective were mangroves as a defense against the recent tsunami? *Current Biology*, 12(12), R443–R447.

DeVogelacre, A. P., E. J. Burton, T. Trejo, C. E. King, D. A. Clague, M. N. Tamburri, G. M. Cailliet, R. E. Kochevar, and W. J. Douros (2005). Deep-sea corals and resource protection at the Davidson Seamont, California, U.S.A. In A. Freiwald and J. M. Roberts (Eds.) *Cold-water Corals and Ecosystems*, Springer-Verlag, Berlin, Heidelberg, pp. 1189–1198.

Dore, M. H. (2006). Mineral taxation in Jamaica: An oligopoly confronts taxes on resource rents—and prevails. *American Journal of Economics and Sociology*, 46(2), 179–203.

Folke, C., S. Carpenter, T. Elmqvist, L. Gunderson, C. S. Holling, and B. Walker (2002). Resilience and sustainable development: Building adaptive capacity in a world of transformations. *AMBIO: A Journal of the Human Environment*, 31(5), 437–440.

Gardner Pinfold Consulting Economists Ltd., and MariNova Consulting Ltd (2005) Economic Value of the Nova Scotia Ocean Sector. Prepared for Government of Canada and Nova Scotia Government. Retrieved from: http://www.dfo-mpo.gc.ca. (accessed 15 September 2015).

Gardner Pinfold (2009). Economic Impact of Marine-Related Activities in Canada. Economic Analysis and Statistics Branch, Fisheries and Oceans Canada, Ottawa, Ontario.

Girard S. and Kalaydjian, R. (2014). French Marine Economic Data 2013. Retrieved from http://dx.doi.org/10.13155/36455 (accessed 15 June 2015).

Haefner, B. (2003). Drugs from the deep: Marine natural products as drug candidates. *Drug Discovery Today*, 8(12), 536–544.

Holling, C. S. (2001). Understanding the complexity of economic, ecological, and social systems. *Ecosystems*, 4(5), 390–405.

INTOSAI Working Group on Environmental Auditing (WGEA) (2010). Environmental accounting: Current status and options for SAIs (pg. 38). Retrieved from: www.environmental-auditing.org/LinkClick.aspx?fileticket=s%2fFCvUzSKsk%3d&tabid=128&mid=568 (accessed July 3, 2015).

Jolly, D. (2011). Priced off the menu? Palau's sharks are worth $1.9 million each, a study says. *The New York Times*, May 2.

Kalaydjian, R., F. Daurès, S. Girard, S. VanIseghem, H. Levrel, and R. Mongruel (2009). French Marine Economic Data, IFREMER.

Kildow, J. T. (2009). The influence of coastal preservation and restoration on coastal real estate values. In L. H. Pendleton (Ed.) *The Economic and Market Value of America's Coasts and Estuaries: What's at Stake*, Coastal Ocean Values Press, Washington DC, pp. 97–114.

Kildow, J. T. and C. S. Colgan. (2005). California's Ocean Economy. National Ocean Economics Program, California State University, Monterey Bay, CA.

Kildow, J. T. and A. McIlgorm (2010). The importance of estimating the contribution of the oceans to national economies. *Marine Policy*, 34(3), 367–374.

Lange, F. and R. Topel (2006). The social value of education and human capital. *Handbook of the Economics of Education*, 1, 459–509.

Lau, W. W. Y. (2013). Beyond carbon: Conceptualizing payments for ecosystem services in blue forests on carbon and other marine and coastal ecosystem services. *Ocean and Coastal Management*, 83, 5–14.

Lewin, J. L., and W. N. Trumbull (2012). The social value of crime?. *International Review of Law and Economics*, 10(3), 271–284.

Loreau, M., S. Naeem, P. Inchausti, J. Bengtsson, J. P. Grime, A. Hector, D. U. Hooper, M. A. Juston, D. Raffaelli, B. Schmid, D. Tilman, D. A. Wardle (2001). Biodiversity and ecosystem functioning: Current knowledge and future challenges. *Science*, 294(5543), 804–808.

Mcleod, E., G. L. Chmura, S. Bouillon, R. Salm, M. Bjork, C. M. Duarte, C. E. Lovelock, W. H. Schlesinger, and B. R. Silliman (2011). A blueprint for blue carbon: Toward an improved understanding of the role of vegetated coastal habitats in sequestering CO_2. *Frontiers in Ecology and the Environment*, 9(10), 552–560.

Millennium Ecosystem Assessment (2005). *Ecosystems and Human Well-Being: Synthesis*. Washington, DC: Island Press.

Moberg, F. and P. Rönnbäck (2003). Ecosystem services of the tropical seascape: Interactions, substitutions and restoration. *Ocean & Coastal Management*, 46(1), 27–46.

Moran, T. H. (1987). Managing an oligopoly of would-be sovereigns: The dynamics of joint control and self-control in the international oil industry past, present, and future. *International Organization*, 41(4), 575–607.

Morris, S. and H. S. Shin (2002). Social value of public information. *The American Economic Review*, 92(5), 1521–1534.

Murray, B. C., L. Pendleton, W. A. Jenkins, and S. Sifleet (2011). Green payments for blue carbon: Economic incentives for protecting threatened coastal habitats. Nicholas Institute Report.

National Ocean Economics Program (2008). Phase II Facts and Figures Florida's Ocean and Coastal Economies Report. Retrieved from: www.oceaneconomics.org (accessed July 6, 2015).

National Research Council (1999). *Nature's Numbers: Expanding the National Economic Accounts to Include the Environment*, W. Nordhaus and E. Kokkelenberg (Eds). Washington, DC: The National Academies Press.

PEMSEA (2009). The marine economy in times of change. *Tropical Coasts* 16(1) 1-80. Retrieved from: www.pemsea.org/sites/default/files/tc-v16nt.pdf (accessed 8th September 2015).

Pickett, S. T., M. L. Cadenasso and J. M. Grove (2004). Resilient cities: meaning, models, and metaphor for integrating the ecological, socio-economic, and planning realms. *Landscape and Urban Planning*, 69(4), 369–384.

Pugh, D. (2008). Socio-economic indicators of marine-related activities in the UK economy, Final Report, Crown Estate, March.

Schumpeter, J. (1908). On the concept of social value. *Quarterly Journal of Economics*, (23), 213–232.

Scorse, J. (2010). *What environmentalists need to know about economics*. New York: Palgrave MacMillan.

Scorse, J., A. Sackett, and F. Reynolds (2015). Impact of surf breaks on home prices in Santa Cruz, CA. *Tourism Economics*, 21(2), 409–418.

Schweitzer, A. (1981). Social values in economics. *Review of Social Economy*, (39)3, 257–278.

Song, W. L., G. H. He and A. McIlgorm (2013). From behind the Great Wall: The development of statistics on the marine economy in China. *Marine Policy*, 39, 120–127.

Statistics New Zealand (2006). New Zealand's Marine Economy 1997–2002. Retrieved from: http://m.stats.govt.nz/browse_for_stats/environment/natural_resources/marine.aspx (accessed July 3, 2015).

Stern, N. (2006). *Stern review on the economics of climate change*. London: HM Treasury.

The Economist (2011). Buttonwood Blog Mar 29 2011. Taxation policy: You can't take land offshore. Retrieved from *The Economist* online: www.economist.com/blogs/buttonwood/2011/03/taxation_policy_0 (accessed July 3, 2015).

Thornberg, T., J. G. Levine, P. Duffy, A. L. Beale, B. Abe, and M. Chow (2011). *2011 Monterey Economic Forecast*. Los Angeles, CA: Beacon Economics, LLC.

Tool, M. R. (1977). A social value theory in neoinstitutional economics. *Journal of Economic Issues*, 823–846.

Ullman, R., V. Bilbao-Bastida, G. Grimsditch (2012). Including blue carbon in climate market mechanisms. *Ocean and Coastal Management* 83,15–18.

United Nations Development Programme (UNDP) (2006) Niger Delta Human Development Report. United Nations Development Programme. New York.

United Nations Statistics Division (UNSD) and United Nations Environment Programme (UNEP) (2000). *Handbook of National Accounting: Integrated environmental and economic accounting—an operational manual*. New York: United Nations. Retrieved from: http://unstats.un.org/unsd/publication/ SeriesF/ SeriesF_78E.pdf (accessed July 6, 2015).

U.S. General Accountability Office (2007). Highlights of a GAO/NAS Forum: Measuring our Nation's Natural Resources and Environmental Sustainability, GAO-08-127SP. Washington, DC: General Accounting Office, 2007. Retrieved from www.gao.go (accessed March 9, 2013).

Volker, S. and T. Kistemann (2011). The impact of blue space on human health and well-being—Salutogenic health effects of inland surface waters: A review. *International Journal of Hygiene and Environmental Health*, 214(6), 449–460.

White, M., A. Smith, K. Humphreys, S. Pahl, D. Snelling, and M. Depledge (2010). Blue space: The importance of water for preference, affect, and restorativeness ratings of natural and built scenes. *Journal of Environmental Psychology*, 30(4), 482–493.

12

STRATEGIC ENVIRONMENTAL ASSESSMENT

Richard Kenchington and Toni Cannard

Introduction

In the 1960s the impacts of human activities on natural resource systems and environmental health became matters of increasing national and international policy concern. The United States National Environmental Policy Act (1969) introduced the concept of Environmental Impact Assessment to address the possible impacts on the environment of proposed projects. During the 1970s many nations drew on this example to develop similar legislation

The 1972 Stockholm Conference on the Human Environment (UNEP, 1993) and subsequent collaboration by IUCN, UNEP, and WWF with UNESCO and FAO addressed broader concepts of 'sustainable development' in the World Conservation Strategy (International Union for the Conservation of Nature and Natural Resources, 1980). This reflected growing agreement that environment and development strategies should be considered together which was further developed by the Brundtland Commission (WCED, 1987) and in Agenda 21 adopted at the Rio Summit (UN, 1992). By the1980s there was increasing criticism of the typically narrow spatial and temporal scope of EIA and discussion of the need for Strategic Environmental Assessment (SEA) to develop national, regional and global policy and legislative initiatives to address environment and development issues on a sustainable basis (Kelly *et al.* 1987). Specific outcomes included Australia's National Strategy for Environmentally Sustainable Development (NSESD, 1992), the European Union Directive 2001/42 (EC, 2001) on the assessment of the effects of certain plans and programmes on the environment (known as the SEA Directive) US SEA and the application of SEA in development assistance programs (Chaker *et al.*, 2006).

Addressing these issues in marine ecosystems has generally involved the transfer of terrestrial practice to marine ecosystems. This presents challenges because many of the terrestrial practices are centred on concepts of title and management responsibility with respect to areas defined by geodesic points on the surface of the earth. These concepts reflect a presumption of separability, limited connectivity ('good fences make good neighbours') because of the ability of substantially different suites of activities to take place on either side of a boundary with little need for cross-boundary management coordination.

The presumption of separability and primacy of use within defined areas is generally reasonable on land where, at any location, the characteristics of soils, exposure, local rainfall, and unidirectional drainage flows down and within catchments are primary determinants of site-associated ecological communities and potential economic uses of land. There can be tricky trans-boundary issues through connectivity of migratory species, catchment water flows or atmospheric interactions particularly where these transfer pollutants but most issues can be addressed within a framework of spatial rights and responsibilities.

In contrast, the active, mobile third dimension of the water column is itself an important functional habitat and a mass transport linkage sustaining complex assemblages of species for all or part of their lifespan. It is a highly active and variable transport medium, providing horizontal and vertical connection between habitats and biodiversity populations through currents, tides and surface wind/wave interactions that typically have little relationship to jurisdictional boundaries. Beyond the reach of sunlight, life on or within the seabed can depend on detritus falling from the upper levels of the water column often brought over distances significantly greater than national boundaries. Halpern *et al.* (2008, 2012) have reported the decline of marine ecosystems and increasing reach of human impacts including pollution and debris to the remotest marine areas.

The issues of linkage and scale present two substantial challenges to strategic assessment and management of coastal and ocean areas. Intertidal and estuarine habitats are key ecological boundary hotspots because of the linkage to events on land to through terrestrial run-off and consequent impacts of changed water flows, water quality, pollution and debris. The costs of social, environmental and economic impacts to a downstream or receiving jurisdiction and ecosystem may exceed the social and economic benefits of the activities in an upstream jurisdiction that causes those impacts. Where the scale of the linkage is greater than national or regional jurisdictions the resolution of such issues is extremely complex. The issue of scale is compounded by the possibility of multiple uses on the sea surface, in the water column, on and beneath the seabed and the cumulative effects of an increasing range and volume of land use and marine activities that have cross-boundary impacts on marine ecosystems and natural resources.

Characteristics of SEA

While the concept of SEA developed from concerns over the narrow spatial and temporal scope of EIA there has been a lack of systematic clarity concerning the differences between SEA and EIA. The critical limitation of EIA is that it is concerned with specific projects and does not apply further to policies, plans or programs (Therivel *et al.* 1992). Noble (2000) sought a better understanding through case studies of 18 SEA projects but saw no consensus for a definition based on the strategic characteristics of SEA noting that Court *et al.* (1994) considered that SEA represented an 'extension of project-based EIA' to higher levels of decision making. This extension of SEA involves moving beyond early single, sectoral, short-term considerations and influence towards broader community/political consensus and greater certainty through long-term planning for ecologically sustainable development.

Noble (2000) developed a list of defining characteristics that distinguish the tactical advanced project/proposal focus of EIA from the broader strategic policy focus of SEA (Table 12.1).

Noble (2000) concluded that 11 of the 18 case studies demonstrated the characteristics of strategic assessment in establishing a strategic approach to minimise potential negative (environmental and social) outcomes by selecting the least negative options for meeting defined sectoral goals/objectives in an economically feasible and environmentally acceptable manner.

In 2005, SEA Theory and Research were addressed in a series of workshops at a special conference on 'International Experience and Perspectives on SEA'. In an editorial paper for a

Table 12.1 Defining characteristics of EIA and SEA

EIA	SEA
Timescale is project dependent	Timescale 25 years min. commonly in Australia
Represents an end • Brings closure to an issue or undertaking	Leads to a strategy for action • A means to an end
Goals and objectives are predetermined • EIA predicts potential outcomes of an already predetermined option	Set in context of broad vision goals and objectives • examines strategies to accomplish particular goals and objectives
Proponent: predominantly by project development proponent	Proponents can be local, state, regional or national governance institution
Asks 'what are the impacts of our project option?' • Addresses project in terms of pre-determined option • Alternatives often limited to issues of alternative design • Theoretically contains 'no action' alternative – a choice not to proceed • Management emphasis on mitigating likely negative outcomes	Asks 'what is the preferred option/s?' • Broad range of alternative options at early stage • Contains a 'no change'. Options to achieve goal could include existing pathway. • No action not an alternative. • Management emphasis on minimising negative outcomes by selecting the least negative alternative at an early stage.
Fine spatial scale, local direct impact area	Wide spatial scale, local and non-local impact areas
Forecasts • Predicts and assesses likely outcomes of specific undertaking	Backcasts then forecasts • Determines a range of options and then forecasts the likely outcomes of each option
Reactive • EIA is designed to react to, or assess, a predetermined option • Definitive – assesses a single undertaking: well defined start: (project proposal) and end (decision to proceed or not)	Proactive • Creates and examines alternatives leading to the preferred option • On demand: can be implemented at any time to assess whether strategic choices are meeting specified visions or objectives, or should new visions, goals of objectives should be developed in order to maximize positive outcomes
Project specific – assesses options for addressing the impacts of a specific project proposal	Not project specific – assesses alternatives and implications of cumulative and interactive impacts for regions and sectors
Narrow focus and highly detailed • Assessment generally technical, often quantitative and highly detailed • Impacts and benefits of primarily one community analysed	Broad focus and low level of detail • Assessment is broad, usually non-technical and qualitative • Focus broadens moving upscale from programs, plans and policies to alternatives. • Impacts on various communities included in the analysis

Source: After Noble (2000).

special issue on those workshops (Wallington *et al.*, 2007) provide a useful synthesis of discussion among practitioners, policy makers and scholars on the apparent and increasing confusion on the particular roles and methodological approaches to SEA. As part of that discussion Bina (2007) identified three lines of argument for SEA:

- strategy – addressing the paucity of environmental considerations in policies, plans and programmes;
- procedure – responding to perceived limitations in practice of EIA;
- purpose – considering contribution to sustainable development.

These lines of argument are relevant in broader considerations of management of human uses and impacts affecting biological diversity, natural resources in marine space.

Ocean and coastal strategic sustainability issues

By the mid-twentieth century, emergence and application of new technologies enabled a rapid increase in the capacity and range of existing maritime industries of shipping and fisheries and of impacts of terrestrial activities affecting marine ecosystems. Research was enabling the development of new uses, including seabed hydrocarbon and mineral production, and demonstrating the paucity of environmental considerations in policies, plans and programmes for management of marine space. These considerations raised the need for a new governance and strategic policy framework reaching beyond the historic concepts of a 3 nautical mile territorial sea and sectoral regulation of activities on the high seas. Decadal discussions between 1958 and 1987 led to the United Nations Law of the Sea Convention (LOSC) which came into effect in 1994.

At the same time, awareness of the need to protect and manage marine environments and natural resources grew following major oil pollution events arising from the wrecks of supertankers Torrey Canyon and Amoco Cadiz and reports by Heyerdahl (1971) of rafts of tar balls and other floating debris on the surface of the Atlantic Ocean far from land. Three needs were identified: to control pollution of the sea; to conserve marine fishery resources; and to protect biological diversity in representative areas of marine environments. Since then experience of terrestrial pollution of marine environments and the effects of extreme storm events and tsunamis has increased the importance of addressing the strategic implications of human activities and development for marine biodiversity, natural resource productivity and delivery of ecosystem services (Scheffers *et al.* 2012).

While the term biological diversity is not used in the LOSC, several provisions (parts V, VII, IX and XII) require States to cooperate bilaterally and regionally in a sustainable manner to protect and preserve the living resources of the sea. The convention explicitly addresses both conservation and sustainable utilisation of living resources in Part V (Articles 61–68), Part VI (Article 77) and Part VII (Articles 116–120). It also addresses what is now called 'biodiversity' in Part XII when it calls for *the protection and conservation of the natural resources of the Area and the prevention of damage to the flora and fauna of the marine environment* (Article 145b) and for action to *protect and preserve rare or fragile ecosystems as well as the habitat of depleted, threatened or endangered species and other forms of marine life* (Article 194.5).

Strategic environmental considerations for marine ecosystems were specifically addressed in chapter 17 of Agenda 21 which identified actions needed in the twenty-first century to address: 'protection of the oceans, all kinds of seas, including enclosed and semi-enclosed areas and the protection, rational use and development of their living resources' (UN, 1992).

The allocation of exclusive economic zones under the LOSC led governments to review research and commission studies to assess and develop the economic potential of uses of marine space. Brief summaries provide two examples illustrating a more general pattern.

In Australia, the Government commissioned a review of the potential for marine industries, science and technology entitled 'Oceans of Wealth' (Australia DITC, 1989) which provided strategic guidance for subsequent research. An Oceans Policy (Tsamenyi and Kenchington, 2012), adopted in 1998 by Australian federal, state and territory governments, resulted in the National Strategy for Ecologically Sustainable Development (NSESDS, 1992) that included consideration of coastal and marine sustainability. By 2006/7 the economic value of maritime industries to Australia was in excess of $45 billion (AIMS, 2012) and the national research agency (CSIRO) had established research programs with objectives of providing innovations in ocean and atmosphere observation and modelling; to create understanding of the role of oceans; and to deliver significant economic, social and environmental benefits for Australia and the region'.

In 2012 the European Commission produced a communication paper 'Blue Growth – opportunities for marine and maritime sustainable growth' that summarised the outlook and potential for gross value and employment generation after two decades of policy, research and management (Figure 12.1).

The EC communication noted the need for sustainability, respect for environmental concerns given the fragile nature of the marine environment and legislative measures that reassure

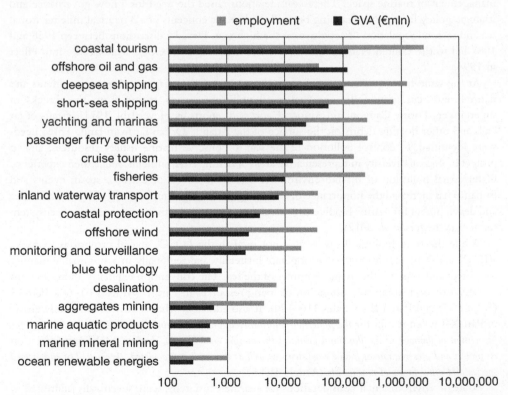

Figure 12.1 European Union: Blue economy values in terms of employment and gross economic value added

Source: European Commission (2012).

investors that there will be no unforeseen delays in planning processes or infrastructure connections.

The tension between proper environmental consideration and timely development decision making is a strategic concern in most jurisdictions. It operates at all scales from single project, through sectoral strategy to integrated management. There are two core elements. The first is the maintenance of the resource of biodiversity and ecosystem processes in their own right and for their productivity of fishery resources and other ecosystem services. The second, particularly for proponents of activities that do not appear to depend directly on ecosystem health is to establish a rational basis for addressing the economic externality of additional project development and operational costs imposed to address reasonable environmental concerns and maximise positive social impacts. In the marine context, this tension is complicated by the scale and connectivity of ecosystems, the legal constraints of allocating and enforcing title in marine space and the high costs of marine research and development.

Procedure

Programmes addressing marine resource and biological diversity conservation and sustainability understandably reflect sectoral objectives and differing views on acceptance of risk, and on the extent, costs and timescales needed for implementation of appropriate management to address risk.

This is reflected in outcomes of international meetings. Thus the World Summit on Sustainable Development (2002) which set a specific fishery target: 'To maintain or restore stocks to levels that can produce the maximum sustainable yield with the aim of achieving these goals for depleted stocks on an urgent basis and where possible not later than 2015.' While in 2004, the 7th Conference of the Parties (CoP) of the CBD, adopted a minimum target of 10 per cent of marine ecosystems to be included in representative networks of MPAs by 2012 (Decision VII/30, Annex II).

The diversity of approaches is apparent in a growing range of terminologies and acronyms for specific but substantially overlapping technical or legislated protocols. A World Bank (2006, pp. 9–12) report listed 32 marine management protocols identifying four groups through a typology based on objectives and extent of the environmental protection offered:

- marine protected area tools, primarily for biodiversity conservation and habitat protection;
- multi-use management tools, primarily for balanced conservation and socio-economic uses;
- sustainable use of marine-resource management tools, primarily for extractive use;
- culture/ecological/social protection reserves, primarily for indigenous and traditional non-indigenous communities.

The list in World Bank (2006) is not exhaustive, and the growing list reflects competing sectoral priorities and technical approaches to addressing overarching issues arising from the increasing range of human uses and impacts affecting marine space. Each approach involves elements of Strategic Environmental Assessment as identified by Noble and Storey (2000) and most apply these through some form of planning and implementation cycle based on a broad goal and more precisely defined objectives consistent with the goal.

The critical issue is the extent to which the principles of strategic environmental assessment are or can be applied effectively within or in association with other marine management procedures.

Purpose

SEA has developed to address the limitations of project-based and often time-critical, pressured, project assessments and decision making. Most of the examples relate to strategic assessment of the environmental constraints on the development of policies or programmes for economic or sectoral activities such as forestry, agriculture, urban development, power supply, transport and service corridors, freshwater supply, location of alternative energy installations and other developing industries such as aquaculture.

To the extent that SEAs are undertaken to address specific sectoral policies, plans or programmes, their capacity to effectively address the likely cumulative impacts of multiple competing sectoral strategic objectives may be limited. This is particularly critical in the context of linkage and cross boundary factors in coastal and marine ecosystems.

Beyond SEA a diverse range of concepts and activities addressing marine conservation and sustainable resource use reflects community and sectoral perspectives on marine ecosystem and resource management.

The first is a broader community cultural perspective, using best available information to determine areas that should be protected and managed to maintain particular environmental, cultural, aesthetic, social or recreational values. Depending on the availability of time and resources the information can range from Delphi consultation with expert scientific, indigenous, local and industry expert knowledge holders to commissioned studies based on remote sensing, seabed mapping and analysis of fishery data.

The second is a fisheries production perspective using commercially confidential data to identify areas of particular importance for fishery production or productivity.

The third is a non-fishery maritime industry perspective relating to minimising construction and operational costs and constraints to activities such as seabed or sub-seabed mineral resources operations; fixed or permanently attached facilities, such as alternative energy, major port or cargo transfer installations; and navigation, pipeline or cable corridors.

All three perspectives should be addressed together to achieve the purpose of Chapter 17 of Agenda 21 (UNEP 1992) or Part XII of LOSC.

The management of the Great Barrier Reef is an example of a management process to address conservation and reasonable use of an iconic marine ecosystem. Specific overarching legislation was passed in 1975 to provide for creation of a multiple use MPA applying processes of an objectives-based, adaptive framework for planning, management, monitoring, outlook reporting. The central element is a zoning system with subsidiary consistent and more detailed area plans as needed. Initial zoning was completed in 1988 and a completed review and revision came into effect in 2004 (Kenchington and Day 2011). The system provides for a wide range of uses including fishing, tourism, shipping and defence operations but is not a complete model for multiple use management because of explicit legislative exclusion of operations for exploration or recovery of minerals except for the purposes of scientific research that reflect a core concern in the history of protection of the iconic nature of the Great Barrier Reef. However, operational history of the Great Barrier Reef Marine Park includes many cross-boundary arrangements with respect to impacts on the Reef from land and water use in jurisdiction of the State of Queensland.

The concept of ecosystem-based management flowed from the Rio Summit of 1992 and projects have been undertaken in many jurisdictions, many with support from the United Nations Environment Programme and Global Environmental Facility. UNEP has drawn on a range of case studies to provide an introductory guide to marine and coastal ecosystem-based management (UNEP, 2011).

Douvere and Ehler (2009) drew on research and case studies of integrated coastal and ocean management to discuss challenges of multi-jurisdictional and multi-sectoral planning and suggest a systematic framework of marine spatial planning for strategic assessment and planning of multiple use.

Discussion

The need to address management of new sectoral uses of marine space, and increasing recognition of the vulnerability of marine ecosystems presented challenges for the established sectoral management systems. For shipping, issues of the increasing length, draft and capacity of vessels, safety of navigation in increasingly crowded shipping lanes, piracy and management of operational and catastrophic pollution may be addressed within protocols under the LOSC but require increasing engagement with other sectors.

From the mid-twentieth century fisheries management has had the strategic challenge of pressured expectation of expanding food production through development of new fisheries and of increasing experience of fish stock collapses (Hueting and Reijnders, 2004). Despite professional appreciation of the medium and longer term resource conservation and economic needs to reduce and better manage effort in many fisheries, the short-term social and economic dependencies present substantial challenges. The development and application of the FAO concept of an ecosystem approach to fishery management provides a basis for addressing the complex sectoral challenges (FAO, 2003; Rice *et al.*, 2012, Kenchington *et al.*, 2014).

The site-related nature of mineral and alternative energy uses are typically addressed by an approach based on project EIA and sectoral SEA to assess and address the risks of establishment; operational and accident contingency management and reasonable provisions for removal and remediation in the event of accident or at project conclusion.

Eales and Sheate (2011), reviewing the effectiveness of policy-level environmental and sustainability assessment, have noted a continuing need to go beyond the 'business-as-usual' approach of simply seeking to balance environmental, social and economic factors. They noted that the most commonly used definition of sustainable development is that of the Brundtland Commission (WCED 1987) but pointed out that its second sentence and key concepts that clearly address the strategic social and environmental imperatives underlying sustainable development are less frequently quoted.

> Sustainable Development is development that meets the needs of the present without compromising the ability of future generations to meet their own needs. It contains within it two key concepts:
>
> – the concept of "needs", in particular the essential needs of the world's poor, to which overriding priority should be given; and
> – the idea of limitations imposed by the state of technology and social organisation on the environment's ability to meet present and future needs.
>
> *(WCED, 1987)*

The references to the essential needs of the poor, and the environment's ability to meet present and future needs introduce strategic considerations of scale in space, connectivity, information and equity that go beyond the narrow spatial and temporal scope of conventional economic development considerations and environmental impact assessment.

Recognition of this need was reflected in Australia's National Strategy for Environmentally Sustainable Development (NSESD, 1992; Commonwealth of Australia, 1992) and Principle 17 of the Rio Declaration (UN 1992):

> Environmental impact assessment, as a notional instrument, shall be undertaken for proposed activities that are likely to have a significant adverse impact on the environment and are subject to a decision of a competent authority.

Agenda 21 of the Rio Summit provided a framework for development of national, regional and global policy and legislative initiatives for strategic consideration of environment and development. Subsequent outcomes include the European Commission's Strategic Environment Assessment Directive (European Commission, 2001) and US SEA and the application of SEA in development assistance programmes (Chaker *et al.*, 2006).

Noble and Storey (2001, pp. 491–502) proposed a seven-phase approach for the conduct of SEA illustrating it with a case study relating to the development of an energy strategy to address expected increase in demand for electricity in Canada (NEB, 1999):

1 Scoping the assessment issues
2 Describing the alternatives
3 Scoping the assessment components
4 Evaluating potential impacts
5 Determining impact significance
6 Comparing the alternatives
7 Identify the best practicable environmental option.

The list suggests that the assessment approach is based on considering the environmental externalities that must be addressed in order to implement a defined and implicitly accepted sectoral policy or programme. In the linked transboundary situation of coastal and oceanic waters many sectors contribute to and should be engaged in design and implementation of comprehensive ecosystem-based management actions to address cumulative and interacting impacts. The objectives of sectors can conflict at scales from local site to regional, they can be shared or mutually supported, as in the case of sanctuaries that also protect significant life cycle sites of commercially or recreationally important species, or they can have no impact on each other (Rice *et al.*, 2012). These issues are ideally addressed through a trusted multi-sectoral process of spatial planning or zoning to determine the purposes and operations for which specific areas may be used, entered or impacted.

Typically such processes are conducted by agencies reporting to fisheries or environment ministries that are perceived or expected to champion the interest of the respective sector. This brings at least a perception of conflict of interest – a decision that favours the objective of the host sector can be perceived as failure to address a legitimate interest of the other sectors, while a decision favouring another sector can be perceived as failure to address the primary sectoral stakeholder interests of the host. Other sectors tend to limit engagement to strong tactical defence of immediate interests and opposition to measures that may limit options for future activities.

Ideally a planning and management regime should be implemented by a trusted non-sectoral operating agency with overarching legislation to engage all sectors. This would deliver a policy, planning and management implementation cycle that is monitored in relation to defined management objectives, evaluated to report on the short and longer term outlook and adapted as necessary to reflect experience and changing circumstances. In such a regime each sector would prepare and engage through a process similar to SEA but decisions relating to impacts of specific proposals and conditions of use would reflect an open process for consideration of cumulative impacts and interactions of multiple human uses.

Acknowledgements

This research is undertaken by the CSIRO Flagship Coastal Collaboration Cluster with funding from the CSIRO Flagship Collaboration Fund. The Coastal Collaboration Cluster is an Australian research program designed to enable more effective dialogue between knowledge-makers and decision-makers in Australia's coastal zone.

References

AIMS. 2012. Valuing the Australian marine industry: discussion paper. Australian Institute for Marine Science. Accessed 3 January 2013 at: www.aims.gov.au/c/document_library/get_file?uuid=25030f5e-02c9–4d8f-9a19-ceca6c6dcba3&groupId=30301.

Bina, O. 2007. A critical review of the dominant lines of argumentation on the need for strategic environmental assessment. *Environmental Impact Assessment Review*, 27 (6), 585–606.

Chaker, A., El Fadl, K., Chamas, L. and Hatjian, B. 2006. A review of strategic environmental assessment in 12 selected countries. *Environmental Impact Assessment Review*, 26 (1), 15–56.

Commonwealth of Australia, Department of Industry, Technology and Commerce. 1989. Oceans of Wealth?: A report by the Review Committee on Marine Industries, Science and Technology (Chairman: K. R. McKinnon). Canberra, Australian Govt. Pub. Service, 1989. xix, 188 pp.

Commonwealth of Australia. 1998. Australia's Oceans Policy, Caring, Understanding, Using Wisely. Accessed 18 July 2015 at: www.environment.gov.au/archive/coasts/oceans-policy/publications/policy-v1.html.

Court, J. D., Wright, C. J. and Guthrie, A. C. 1994. *Assessment of Cumulative Impacts and Strategic Assessment in Environmental Impact Assessment*. A report prepared for the Commonwealth Protection Agency, Australia: J.D. Court and Associates Pty Ltd.

CSIRO. n.d. Accessed 18 July 2015 at: www.csiro.au/en/Research/OandA/About.

DITC. 1989. Oceans of Wealth? Report by Review Committee on Marine Industries, Science and Technology, Dept Industry, Technology and Commerce, Aust Gov. Pub. Service, Canberra. 188 pp.

Douvere, F. and Ehler, C. N. 2009. New perspectives on sea use management: Initial findings from European experience with marine spatial planning. *Journal of environmental management*, 90 (1), 77–88.

Eales, R. P. and Sheate, W. R. 2011. Effectiveness of policy level environmental and sustainability assessment: Challenges and lessons from recent practice. *Journal of Environmental Assessment Policy and Management*, 13 (1), 39–65.

European Commission. 2001. EC. Directive 2001/42/EC of the European Parliament and of the Council on the Assessment of the Effects of Certain Plans and Programmes on the Environment, Luxembourg, 27 June 2001, (PE-CONS 3619/3/01 REV 3). Accessed 4 July 2001 at: http://europa.eu.int/comm/environment/eia/sea-support.htm.

European Commission. 2012. *Blue growth: Opportunities for marine and maritime sustainable growth (text with EEA relevance)*. Communication from the Commission to the European parliament, the Council, the European Economic and Social Committee and the Committee of the Regions. Com (2012) final. 12 pp.

FAO (Food and Agriculture Organisation of the United Nations). 2003. No. 4 Suppl. 2. The Ecosystem Approach to Fisheries. FAO Technical Guidelines for responsible Fisheries. FAO, Rome 112 pp.

Halpern, B. S., S. Walbridge, K. A. Selkoe, C. V. Kappel, F. Micheli, C. D'Agrosa, J. F. Bruno, K. S. Casey, C. Ebert, H. E. Fox, Fujita, R., Heinemann, D., Lenihan, H. S., Madin, E. M. P., Perry, M. T., Selig, E. R., Spalding, M., Steneck, R. and Watson, R. 2008. A global map of human impact on marine ecosystems. *Science*, 319, 948–993.

Halpern, B. S., Longo, C., Hardy, D., McLeod, K. L., Samhouri, J. F., Katona, S. K., Kleisner, K., Lester, S. E., O'Leary, J., Ranelletti, M., Rosenberg, A. A., Scarborough, C., Selig, E. R., Best, B. D., Brumbaugh, D. R., Chapin, F. S., Crowder, L. B., Daly, K. L., Doney, S. C., Elfes, C., Fogarty, M. J., Gaines, S. D., Jacobsen, K. I., Bunce Karrer, U. R., Zeller, D. 2012. An index to assess the health and benefits of the global ocean. *Nature*, 488, 615–620.

Heyerdahl, T. 1971. *The Ra Expeditions*. New York, Doubleday. Various republications.

Hueting R. and Reijnders, L. 2004. Broad sustainability contra sustainability: The proper construction of sustainability indicators. *Ecological Economics*, 50, 249–260.

International Union for the Conservation of Nature and Natural Resources. (1980). *World Conservation Strategy: living resources conservation for sustainable development*. IUCN, Morges. Accessed 18 July 2015 at: cisdl.org/natural-resources/public/docs/wcs.pdf.

Kelly, D., Cote, R. P., Nicholls, B., Ricketts, P. J. 1987. Developing a strategic assessment and planning framework for the marine environment. *Journal of Environmental Management*, 25 (3), 219–230.

Kenchington, R. A. and Day J. C. 2011. Zoning, a fundamental cornerstone of effective Marine Spatial Planning: Lessons learnt from the Great Barrier Reef, Australia. *Journal of Coastal Conservation*, 15 (2), 271–278.

Kenchington, R. A., Vestergaard, O. and Garcia, S. M. (2014). Spatial dimensions of fisheries and biodiversity governance. In S. M. Garcia, J. Rice and A. Charles (eds). *Governance of Marine Fisheries and Biodiversity Conservation*. Wiley Blackwell, Chichester, pp. 110–123.

National Strategy for Ecologically Sustainable Development (NSESD). 1992. Prepared by the Ecologically Sustainable Development Steering Committee, Endorsed by the Council of Australian Governments. Accessed 18 July 2015 at: www.environment.gov.au/about-us/esd/publications/national-esd-strategy.

NEB. 1999. *Canadian Energy: Supply and Demand to 2025*. Calgary, Alberta, Canada, National Energy Board.

Noble, B. F. 2000. Strategic environmental assessment: What is it and what makes it strategic? *Journal of Environmental Assessment Policy and Management*, 2 (2), 203–224.

Noble, B. F. and Storey, K. 2001. Towards a structured approach to strategic environmental assessment. *Journal of Environmental Assessment Policy and Management*, 3 (4), 483–508.

OECD (2012). Strategic Environmental Assessment in Development Practice: A Review of Recent Experience. OECD. Accessed 18 July 2015 at: http://dx.doi.org/10.1787/9789264166745-en.

Rice, J., Moksnsess, E., Attwood. C., Brown, S. K., Dahle, G., Gjerde, K. H., Grefsund, E. S., Kenchington, R., Kleiven, A. R,. McConney, P., Ngoile, M. A. K., Naesje, T. F., Olsen, E., Moland Olsen, E., Sanders, J., Sharma, S., Vestergaard, O. and Westlund, L. 2012. The role of MPAs in reconciling fisheries management with conservation of biological diversity. *Ocean and Coastal Management*, 69, 217–230.

Scheffers, A. M., Scheffers, S. R., Kelletat, D. H. 2012. Coasts at Risk. In *The Coastlines of the World with Google Earth: Understanding our Environment*, pp. 239–286. Coastal Research Library 2, (c) Springer Science+Business Media B.V. 2012.

Therivel, R., Wilson, E., Thompson, S., Heany, D. and Pritchard, D. 1992. *Strategic Environmental Assessment*. London, Earthscan.

Tsamenyi, M. and Kenchington R. 2012. Australian oceans policymaking. *Coastal Management*, 40 (2), 119–132.

United Nations. 1992. UN Conference on Environment and Development. Agenda 21, Chapter 17. Accessed 18 July 2015 at: www.unep.org/Documents.Multilingual/Default.asp?DocumentID=52&ArticleID=65&l=en.

UNCED. 2012. Agenda 21. Accessed 20 September 2015 at: https://sustainabledevelopment.un.org/milestones/unced/agenda21.

UNEP. 1993. Stockholm 1972. Report of the United Nations Conference on the Human Environment. Action Taken by the Conference. Accessed 18 July 2015 at: www.unep.org/documents/default.asp?DocumentID=97.

UNEP. 2011. Taking steps toward Marine and Coastal Ecosystem-Based Management – An Introductory Guide. UNEP, Nairobi. Accessed 18 July 2015 at: www.unep.org/pdf/EBM_Manual_r15_Final.pdf.

Wallington, T., Bina, O. and Thissen, W. 2007. Theorising strategic environmental assessment: Fresh perspectives and future challenges. *Environmental Impact Review*, 27 (6), 569–584.

WCED. 1987. *Our Common Future, World Commission on Environment and Development*. Oxford, Oxford University Press.

World Bank. 2006. *Scaling Up Marine Management: The Role of Marine Protected Areas*. Report No. 36635 – GLB. Washington DC.

WSSD. 2002. World Summit on Sustainable Development Plan of Implementation. Johannesburg.

13

GREENING THE OCEAN ECONOMY

A progress report

Linwood Pendleton, Megan Jungwiwattanaporn,
Yannick Beaudoin, Christian Neumann, Anne Solgaard,
Christina Cavaliere and Elaine Baker

Introduction

The way we manage our global ocean economy continues to evolve. While new research clearly shows the importance of ocean ecosystems to people (Barbier *et al.*, 2011), other evidence clearly depicts an ocean in decline (Pandolfi *et al.*, 2003; Pauly *et al.*, 2005; Worm *et al.*, 2006). In response, the United Nations Environment Program along with organizations including UNDESA, UNDP, IMO, FAO, IUCN, GRID-Arendal and World Fish Center have promoted a new effort in relation to marine management and economic development that applies a green economy approach to the Blue World (UNEP *et al.*, 2012). This approach seeks to change economic and industrial behavior to reduce impacts on the marine environment and in turn increase human welfare by carefully balancing the environmental, economic, and social capital that are required to support a sustainable, ecosystem-based approach to marine economic activity.

What is a green economy?

The green economy offers an alternative framework to the largely unsustainable conditions promoted by current growth and development policies. To date, nearly every ocean and coast on the globe has been impacted by human activity (Lotze *et al.*, 2006; Halpern *et al.*, 2008). This has led to the destruction of 35 percent of the world's mangrove forests and 20 percent of the world's coral reefs, with a further 20 percent of coral reefs considered degraded (MEA, 2005). Over 30 percent of fish stocks are overexploited, depleted, or just recovering from depletion; and over 400 oxygen-poor "dead zones" have been identified throughout the world (Diaz and Rosenberg, 2008). The current trend is pushing the planet's limits and could negatively impact social and economic well-being in the future.

The green economy is an approach that attempts to align economic development with social and environmental goals. A green-economy encourages institutional and policy reforms as well as changes in private and public expenditure, in order to cut carbon emissions, reduce pollution, improve resource efficiencies, improve social equity and prevent biodiversity loss. The green economy relies on a variety of economic and policy tools to promote environmental, social, and economic well-being. The transition to a global green economy will, however, be impossible without considering the planet's heavy reliance on marine and coastal resources.

The three capitals of the green economy

A green economic approach simultaneously pursues economic, social, and environmental goals. It also recognizes the importance of economic, social, and environmental capital in achieving such goals. These three forms of capital are essential to long-term prosperity and together form the foundation of a sustainable "green" economy. The three capitals are closely linked. Environmental capital, such as trees, land, and non-renewable resources, can be transformed into the tools and industry that make up economic capital. Developing economic capital can lead to poverty alleviation and increased standards of living, forming a society's social capital. Ideally, increased productivity and living standards will enable societies to reinvest in their environmental and social capital in order to ensure sustainable growth. Unfortunately, this often does not occur. Even as global GDP increases, poverty rates in many areas are rising as habitat loss and pollution are increasing (UNEP *et al.*, 2012).

The economic value of the marine world

Investing in the long-term health of coastal and marine resources is vital to the success of the global economy. The ocean provides a vast amount of wealth, and yet many of its habitats are deteriorating. Ecosystem services are the benefits humans receive from nature. The marine world offers an abundance of ecosystem services—some of which are currently valued on the market and some of which are not. Current estimates for the value of marine ecosystem services are in the realm of trillions of US dollars per year (Costanza *et al.*, 1997), ranging from the open ocean's value of $491/ha/year to the $352,249/ha/year of coral reefs (de Groot *et al.*, 2012). Yet much of the value of ocean and coastal ecosystems has been lost due to poor management. Fisheries particularly exemplify the potential wealth and loss of the ocean economy. In 2009, over 80 million tonnes of fish were harvested globally with an estimated value exceeding US $100 billion dollars (FAO, 2010). However, overfished stocks mean that fisheries are producing far less value than they could. A World Bank study estimated that overfishing results in lost economic value of $50 billion each year (World Bank, 2009). Proper management of ocean resources would ensure their long-term profit and viability.

Marine values: market and non-market

The seas provide a large array of resources currently valued on the market. Oceans contribute to the market via tourism revenues, improving real estate prices, and through goods sold on the market such as seafood, sand, minerals, and mangrove wood. The market value of these contributions is significant. World travel and tourism currently produce 9 percent of the global GDP, with coastal and marine areas remaining a popular destination (UNEP, 2011c). In 2003, nearly 60 million recreational anglers spent US$40 billion in expenditures (Cisneros-Montemayor and Sumaila, 2010). The 10 million recreational divers and 40 million snorkelers active in the world are estimated to generate over US$5.5 billion each year (Cisneros-Montemayor and Sumaila, 2010). Other sectors, such as fishing, contribute billions of dollars each year to the global market. The ocean is economically important on an international scale; locally, many developing countries are heavily dependent on marine-based revenues.

Many of the services provided by the ocean are not easily captured on the market. Such services include human uses that are not charged for (e.g. recreation and views), natural processes such as nutrient balancing and coastal protection from storms, as well as non–use values that may be rooted in cultural and indigenous values and preferences. Many economists have attempted to capture the economic value of these non-marketed resources through a variety of techniques (Naber *et al.*, 2008; TEEB, 2010; UNEP-WCMC, 2011). Further, markets are being created to capture some of these previously "non-marketed" goods and services through Payments for Ecosystem Services (PES).

Examples of the greener ocean sectors

The following paragraphs highlight trends in ocean uses, wherein marine values are captured with fewer externalities and greater sustainability. This discussion is not meant to be comprehensive but rather exemplary; readers are encouraged to examine sector-specific chapters that provide further detail (especially Chapter 14–17 on fisheries and mariculture; Chapters 18–21 on energy and genetic resources; Chapters 22–24 on shipping, communications, and seapower; and Chapters 26–27 on tourism and marine heritage).

Fishing

The importance of the fisheries sector to food security and poverty alleviation gives it a significant role in the transition to a green economy. Fishers provide food for 500 million people—or 8 percent of the world population (FAO, 2010). There are 120 million people employed by fisheries in the world, 90 percent of which work in small-scale fisheries, mostly in developing countries (World Bank, 2010). Aquaculture is growing, supplying over half of the world's fish; and alone generated $US 98.5 million in 2008 (FAO, 2010). Unfortunately, many of the world's fisheries are being harvested unsustainably—such that 32 percent of global stocks are considered overexploited, depleted or recovering, with a further 50 percent considered fully exploited (FAO, 2010).

Overfishing, especially in small-scale fisheries, could exacerbate poverty levels and affect food security. Already many fishers are finding they must travel farther, and spend more on fuel, in order to find fish (Tyedmers, 2004; World Bank *et al.*, 2010; Suuronen *et al.*, 2012). The fishing sector must also address the effects of agricultural runoff and climate change on fish populations, the increasing number of powerful fishing vessels (Tyedemers *et al.*, 2005), and the pollution produced by aquaculture.

Although the fishing sector faces a variety of issues, its future in the green economy is bright. The industry will need to address the three capitals of the green economy by investing in environmental sustainability via resource efficiency and a reduced carbon footprint, while also considering social equity and the health of small-scale fisheries. This transition to a green economy will likely rely on increased investments in fishing operations and technical innovations, as well as management and governance reforms.

Fortunately, positive examples within the fishing sector already exist. The FAO Code of Conduct for Responsible Fisheries has informed fishery and aquaculture policies around the world. A study of 130 fisheries showed that establishing a system of co-management led to social, economic, and environmental success 70 percent of the time (Gutierrez *et al.*, 2011). The increasing use of eco-labels could lead to increased conservation and the shifting of consumer preferences.

Shipping

The size and international character of the shipping industry make it one of the leading drivers of the global economy. Maritime shipping carries approximately 90 percent of world trade; while freight rates contribute about US$380 billion to the world economy (ICS, 2012a). The shipping industry is furthermore an important employer, giving jobs to 1.5 million seafarers even as it generates many more onshore jobs (ICS, 2010). The industry can thus play an important part in the green economy and has to date recognized this role.

The shipping industry's impacts on the environment can include: pollution, the release of invasive species from ship ballast, sea life collisions, the recycling of old ships, and CO_2 emissions. Such impacts can come at a high economic cost. For example, invasive species can disrupt fisheries, cause fouling, and affect recreation at an estimated cost of $100 billion each year (Chisholm, 2004). Fortunately, the global nature of the shipping industry has long made regulations necessary to ease the flow of trade. This long regulatory history has created the frameworks necessary to implement policies for a green economy.

The main regulatory body for the shipping and cruise line industry is the International Maritime Organization (IMO). Acknowledging the environmental impacts of shipping, the IMO has instituted several conventions, including:

- International Convention for the Prevention of Pollution by Ships (MARPOL, 1973, amended 2010)
- International Management Code for the Safe Operation of Ships and for Pollution Prevention (the ISM Code, 1993)
- International Convention for the Control and Management of Ships' Ballast Water and Sediments (2004)
- International Convention for the Safe and Environmentally Sound Recycling of Ships (2009)
- Ship building standards in the International Convention on the Safety of Life at Sea (SOLAS, 2010).

(IMO, 2012)

Going forward, the main area for improvement in the shipping industry likely lies in reducing CO_2 emissions. The introduction and increasing use of Liquified Natural Gas (LNG)-powered vessels has caused some reduction in CO_2 and in other pollutants such as sulfur. Other sources of power, such as hybridized sail and fuel or solar and fuel ships are in development. However, truly "green" alternative fuel sources are not yet a practical source of power for ship engines. Improvements will most likely come from increasing efficiency across the transport chain, greening supply chains, and building local economies. The shipping industry has begun improving ship performance and expects a 20 percent emission reduction per ton of cargo moved per kilometer by 2020 (ICS, 2012b).

Marine-based renewable energy

The transition to a green economy will require investing in renewable sources of energy that are cleaner and less volatile than the fossil fuels currently in use. Fortunately, coastal and marine environments offer several potential options as research focuses on the ability of wind, tides, ocean currents, salinity gradients, and marine algae to produce energy. Research and development for over 100 different marine-based technologies is currently underway in over

30 countries (IPCC, 2011). Meanwhile, the IPCC's Special Report on Renewable Energy Sources (2011) estimates marine-based renewables could generate 7,400 exajoules (EJ) annually, a number that exceeds today's energy needs.

Currently there are over 85 countries that have established renewable energy targets (UNEP, 2011c). Further, the production of wind energy is increasing as wind power becomes more economically competitive (Mosegaard *et al.*, 2009). The global capacity to generate wind energy increased tenfold from the end of 2000 to June 2011 (WWEA, 2011). Wind energy is cleaner than its non-renewable counterparts; it can also provide certainty to investors as its costs are constant over its lifetime—which helps hedge against the changing prices of fossil fuels (Awerbuch, 2003).

Many marine-based technologies are still undergoing development—tidal, wind, and algae-based energy are not yet economically feasible. Yet research is ongoing to make these energy types more competitive, by reducing upfront capital costs and increasing output. The main challenges to deploying marine-based renewable energy on a global scale will likely revolve around government incentives and policy, continued financing, creating the necessary infra-structure, and gaining social acceptance. However, while challenges do exist, the technologies for renewable energy suggest a positive step forward.

Deep sea mining

There are three main classes of globally occurring deep-sea mineral deposits—manganese nodules, manganese crusts, and seafloor massive sulphides (SMS) (Rona, 2003). Recently, signifi-cant occurrences have been found in the exclusive economic zones of several Pacific Island Countries (PICs) (Glasby, 1982; Hein, *et al.*, 2005); these include SMS deposits containing copper, lead and zinc, gold and silver; and manganese nodules and crusts that contain nickel, copper, cobalt, and rare-earth elements. The refinement of deep-sea mining (DSM) technology, the con-tinued rise in global demand for metals (UNEP, 2011a), the high potential ore grades and increased clarity in the governance of exploration and extraction, have led industry to consider DSM as a viable prospect.

DSM activities have the potential to damage important ecosystem goods and services (e.g., fish habitat, genetic resources, scientific research opportunities). While the mining foot-print at sites (e.g., SMS) is expected to be small in comparison to land-based operations (Scott, 2006), there remain large gaps in our understanding of associated ecosystems, including spatial connectivity and the resilience of the ecosystems (Nautilus, 2008; Van Dover *et al.*, 2011).

Benefits, costs and policy perspectives

The primary potential economic benefit of DSM is linked to the value of metals on the world market. Incidental benefits include advances in technology and advances in scientific under-standing that are difficult to put a price on. Benefits of technological advances fall into two categories: (1) advances that will improve the feasibility and profitability of future DSM, and (2) advances that will benefit other industries.

Key costs of DSM include destruction of the physical habitat of the sea floor and associated biota and accidental release in the water column of contaminated materials during the recovery process. Destruction of ecosystems associated with deep-sea minerals might involve the loss of "existence values," or "bequest values,"[1] or there may be future-use values of which we are unaware. Studies have also shown the link between mining and political instability[2] whereby mineral wealth may increase the risk of conflict in four ways: by affecting a country's performance in other economic sectors; by making government weaker; by giving resource-rich regions incentives to seek autonomy; and by providing financial resources to support political conflicts.

Environmental regulatory regimes that directly address DSM are either new or under development. At a regional scale, Pacific Island Countries are leading the way with the development of a framework for the environmental management of deep sea areas that can be adapted for national implementation. The International Seabed Authority, for its part, recommends the "Dinard Guidelines" for the environmental management of deep-sea ecosystems, which aims to protect natural diversity, ecosystem structure, function and resilience, while enabling rational use (Van Dover *et al.*, 2011).

Biodiversity and pharmaceuticals

The pharmaceutical industry is increasingly engaging in marine bioprospecting in the hopes of discovering new drugs under the sea. Compounds produced by marine plants and animals may hold the secret to new cures and products. Already there are several marine-based drugs on the market. Retrovir (AZT), the first drug licensed for treating HIV, was based on compounds extracted from a sponge (Harbor Branch, 2006a). Prialt (Ziconitide) was created from compounds extracted from sea snails and is used to treat chronic pain in cancer and AIDs patients (Harbor Branch, 2006b). A 2003–2004 marine pharmacology review shows initial results for 166 marine-based chemicals (Mayer *et al.*, 2007). Meanwhile, drug developments from coral reefs are estimated to be over US\$ 6,000 per hectare (OECD, 2005).

The success of this industry is threatened however by biodiversity loss, as pollution, climate change, and other environmental pressures threaten the health of marine populations. A transition to a green economy will be necessary to ensure the continued success of this industry. Bioprospecting itself is not without issues, however. In ensuring a green economy, frameworks will have to be established ensuring the fair distribution of wealth and respect for indigenous knowledge. In terms of environmental effects, the pharmaceutical industry often needs only small samples in which to focus on genetic materials.

Tourism

Tourism is responsible for a significant proportion of world production, trade, employment, and investments (UNEP *et al.*, 2012). It is projected that the number of international tourists will reach the historic one billion mark by December 2012 (UNWTO, 2012). As arguably the largest global industry, tourism is also the largest sector supporting protected areas. The tourism economy represents 9 percent of world GDP and contributes to 6–7 percent of total employment (UNEP *et al.*, 2012). In 150 countries it is one of the five top export earners and in 60 countries it is the first. It is the main source of foreign exchange for one-half of Least Developed Countries (LDCs).

Tourism is growing at more than 4 percent per year; ecotourism is believed to be growing at three times that rate (Milder *et al.* 2010; UNWTO 2012). There is international demand for these services and tourism-related Payment for Ecosystem Services (PES) can be a sustainable financing mechanism for biological and cultural conservation. Globally, coastal tourism is the largest market segment and is growing rapidly (Orams, 1999; Hall, 2001).

Challenges

In a business-as-usual (BAU) scenario, by the year 2050, overall tourism growth will result in increases in energy consumption (111 percent), greenhouse gas emissions (105 percent), water consumption (150 percent), and solid waste disposal (252 percent) (UNEP *et al.*, 2012).

Rapid growth in travel and preferences for further distances, shorter time-periods and energy-intensive activities are resulting in the sector's contribution of 12.5 percent of radiative forcing and 5 percent of anthropogenic emissions of CO_2 (UNEP *et al.*, 2012). Emissions cause coral bleaching, ocean acidification, and sea level rise. Other coastal tourism pressures include water pollution, land conversion, biodiversity, and loss of local and indigenous cultures and built heritage.

Opportunities

Sustainable tourism incorporates positive economic, sociocultural, environmental and climate considerations and impacts, during planning and implementation. Sustainable tourism can serve as a conduit for bio-cultural conservation and has major potential to raise investments for conservation. The green investment scenario is expected to undercut the corresponding afore-mentioned BAU scenario by 18 percent for water consumption, 44 percent for energy supply and demand, and 52 percent for CO_2 emissions (UNEP, 2011c). Efficiency improvements, local hiring, sourcing local products, and safeguarding local culture and environment can reinforce employment potential. On the demand side, more than a third of travellers favor environmentally friendly experiences. Increasing involvement of local communities in the value chain can con-tribute to the development of local economies and poverty reduction and create "green services" in energy, water, and waste management efficiency (UNEP *et al.*, 2012). Meanwhile, a combined Blue Carbon and sustainable tourism strategy can result in conservation and climate change mitigation.

In addition there is a potential for financial mechanisms that result in payments for ecosystems services (Wunder *et al.*, 2008). These services may involve the protection of natural heritage sites, coral reefs, cultural sanctuaries, or traditional livelihoods (Mayrand and Paquin, 2004).

In summary, investment in energy efficiency and improving waste management can save money for tourism businesses, create jobs, and enhance destination aesthetics. Investment requirements in conservation and restoration are small relative to the high value of ecosystem services (ES) that are essential for continued economic activities and human survival (UNEP *et al.*, 2012).

The majority of tourism businesses are small- and medium-sized enterprises (SMEs) and contribute mostly to local livelihoods (UNEP *et al.*, 2012). The use of internationally recognized standards can assist businesses in understanding aspects of sustainable tourism and mobilize invest-ment. Innovative multi-sector partnerships and financing strategies are required and can spread the costs and risks of green investments. Cross-sectoral consultation and Integrated Coastal Zone Management (ICZM) are required for good sustainable tourism, destination planning, and development strategies (UNEP *et al.*, 2012). Tourism planning has to include capacity building, government commitment, enforcement, and climate change considerations. Tourism's impacts on local communities are complex and demand careful planning. Governments can use tax con-cessions and subsidies to encourage investment.

Climate change is a key risk factor for tourism. The information base for effective adaptation remains inadequate for developing nations, particularly SIDS. An efficient instrument to deal with greenhouse gas emissions is to introduce carbon taxes on production and consumption but can be challenging in developing nations (UNEP *et al.*, 2012).

Making tourism businesses more sustainable will foster the industry's growth, create more and better jobs, consolidate higher investment returns, benefit local development and contribute to poverty reduction, while raising awareness and support for the sustainable use of natural resources (UNEP *et al.*, 2012: 107). More research into Payment for Ecosystem Services including markets for landscape beauty are crucial for valuing intact marine and coastal environments.

Incorporating green economic thinking into ocean management

Planning for a green economy

The success of the green economy will largely depend on introducing proper policy frameworks, incentives, and education platforms specific to each sector. In terms of ocean management, many of the green economy success stories thus far have been due to excellent planning. Maritime shipping instituted many green reforms largely because of the international frameworks already in place to regulate shipping. Strides were made towards a greener shipping sector as governments and industry agreed to conventions made through the IMO. In the fisheries sector advances towards a greener economy have been made through co-management plans at the local level and FAO conventions at the international scale. Certifications in Coastal Tourism can engage businesses in sustainable actions and implement internationally recognized standards.

Planning for a green economy can occur at varying scales and a variety of planning tools exist for policy makers. Planning can involve developing the regulatory frameworks necessary at the international, national, or local level. Regulatory bodies can create the guidelines and enforcements that are necessary and unique to each industry. Governments can also create incentive plans to encourage green industry, or use taxes to discourage unsustainable behavior. Education and capacity building can help foster engagement and support for a green economy at a broader level. For example, the use of internationally recognized standards for sustainable tourism is necessary to monitor tourism operations and management. The Global Sustainable Tourism Criteria (GSTC) provides a promising current platform to begin the process of grounding and unifying global standards (UNEP *et al.*, 2012).

Strategic Environmental Assessments (SEAs) are often conducted at the earliest stages of a project in order to evaluate environmental impacts and help decision makers adjust their plans (see Chapter 12 of this Handbook). Such assessments can also include alternative options and ensure projects or policies are aligned with larger national goals.

Ultimately, planning for a green economy will help policy makers achieve strategic decision making, avoid costly mistakes, and strengthen public support for programs that encourage developing all three capitals of a green economy.

Ecosystem based management and marine spatial planning

Marine ecosystems are highly interconnected, with their inhabitants weaving delicate webs of interdependencies. Ecosystems are also spatial units in the oceans, defined by specific characteristics such as productivity, and components both non-living and living.

Classical management of human activities in the marine environment is organized along political boundaries and manages economic sectors independently of each other. This approach does not necessarily align with how natural systems are structured, and so classical management fails to consider critical aspects, including the compatibility of activities with each other and with ecosystems, and the cumulative nature of impacts on species and ecosystems both within and across boundaries. Further, management is often implemented from a human, not an ecosystem perspective, failing to recognize all goods and services valuable in both monetary and non-monetary terms.

Ecosystem based management (EBM) is an approach that explicitly recognizes ecosystem services, builds on ecosystem boundaries and takes account of ecosystems' inherent interactions and dependencies (McLeod and Leslie, 2009; Agardy *et al.*, 2011; UNEP, 2011b). It regards associated human populations as integral parts of the ecosystem (UNEP, 2006). EBM allows for the development of management plans on small and large geographic scales, which can be tailored to meet multiple, defined objectives. It is an approach enabling nature-based socio-economic development.

EBM is not a product or endpoint, but rather an interactive process embracing change and adaptation as objectives are redefined. Through continued stakeholder engagement and monitoring and evaluation, the EBM process allows for management improvements to be made, for example, in response to ecosystem changes (such as climate change).

Building on existing legislation and tools, such as fisheries management or Marine Protected Areas, and building on knowledge that is readily available, EBM can mature from one focused activity, time, or location into an ongoing highly inclusive process. An essential aspect of the process is the continuous integration of management sectors and ecosystem elements. For example, shipping channels can be moved away from cetacean migration corridors, or total allowable fish catches can be linked to maximum bycatch levels of endangered species. EBM shares its integrative nature with other management approaches such as ICZM, Watershed management or Marine Spatial Planning (MSP). In fact, EBM can incorporate these approaches. MSP in particular can be regarded as one of the most widely applied tools of EBM.

Where EBM aims at reconciling human activities with one another on the basis of marine ecosystems' services, MSP focuses on the compatibility of activities. "Marine Spatial Planning is a public process of analyzing and allocating the spatial and temporal distribution of human activities in marine areas to achieve ecological, economic, and social objectives that are usually specified through a political process" (Ehler and Douvere, 2009). Originating in Australia's response to concerns over the need to protect the Great Barrier Reef, it has now been applied by a growing number of countries and regions, including the US, in the UK, the Netherlands, Belgium, Germany, and the Baltic Sea.

Long-established vested interests in the marine environment, such as fishing in certain areas, which competes with growing sectors such as maritime transport and tourism, and new uses of the sea, such as offshore wind farms, have made marine spatial planning a necessary process for conflict resolution. While some human activities are incompatible, such as naval military exercises with small-scale commercial fishing, others may be beneficial to one another, such as small-scale commercial fishing and coastal tourism. In some areas, there may not be perceived user conflict over marine space, and hence no immediate need for an MSP initiative. However, rather than being reactive, MSP allows for future-oriented planning, and for optimizing econ-omic activities.

Usually encompassing the mandates of several management authorities, a clear governance structure should support the application of MSP. These can include new or adapted legislation, but could also be based on inter-ministerial or inter-agency consultations. The objectives of MSP are a matter of definition, and a public process of informed stakeholder consultation should guide MSP from the very beginning. This can ensure the process is meeting both needs and objectives, and should be repeated as spatial plans, maps, and agreed-upon visions are reviewed periodically.

Both EBM and MSP are approaches and tools that can resolve conflicts between our marine activities; and help us achieve cultural, social, and environmental development based on the goods and services marine ecosystems provide us with.

Environmental and ecological impact assessments

Environmental impact assessments (EIAs) have become a standard planning tool that examines the possible positive and negative effects a proposed project may have on the environment and nearby communities. The International Association for Impact Assessment (IAIA) defines EIAs as "The process of identifying, predicting, evaluating and mitigating the biophysical, social, and other relevant effects of development proposals prior to major decisions being taken and commitments made" (IAIA, 1999). Social Impact Assessments (SIAs) too are becoming recognized tools to assess community impacts. SIA and EIA are essentially management tools for policy makers to inform and encourage taking environmental issues into account during their decision making process.

The specific methodology used in an EIA will depend on the industry and project being assessed. Tools can vary from life-cycle analysis to mathematical modeling to rapid rural appraisals (UNEP, 2008). The process, however, generally involves an initial screening and scoping stage, followed by a mitigation stage and monitoring stage, and ending with an audit of the EIA itself. The goal of the EIA is to suggest ways in which a policy or project can mitigate environmental impacts and display a variety of options.

EIAs are currently in use in a variety of countries. The United States was among the first to promote the use of this tool with its National Environmental Policy Act of 1969 (EPA, 2012). The act requires all federal agencies to conduct an EIA on projects receiving federal funding. Many countries have since implemented their own equivalents including Australia, the EU, China, and India (UNEP, 2004).

Conclusions

A greener approach to the marine and coastal economy will require new management approaches and new science. Integrated, indeed transdisciplinary science that combines natural and social sciences will be required to understand how humans affect marine ecosystems and how changes in these ecosystems in turn affect human well-being. Such integrated science needs to be driven by carefully articulated policy and management needs. We will never fully understand the entire marine and coastal ecosystem, but we can begin to understand those key components that are most affected by people and upon which people most critically depend.

Notes

1   Existence value can be defined as the benefit derived from simply knowing something exists even if it is never used. Existence values are often associated with marine biodiversity (Hageman, 1985). Bequest value is the value placed on the knowledge that resources and opportunities will be available to future generations (Beaumont *et al.* 2007).

2   Examples include Professor Michael Ross of UCLA and Professor Paul Collier of Oxford University.

References

Agardy, T., Davis, J. Sherwood, K., and Vestergaard, O. 2011. *Taking Steps Toward Marine and Coastal Ecosystem-Based Management: An introductory guide.* Nairobi: UNEP.

Awerbuch, S. (2003). Determining the real cost: Why renewable power is more cost-competitive than previously believed. *Renewable Energy World.* Retrieved on July 22, 2015 from www.jxj.com/magsandj/rew/2003_02/real_cost.html.

Barbier, E. B., Hacker, S. D., Kennedy, C., Koch, E. W., Stier, A. C., and Silliman, B. R. (2011). The value of estuarine and coastal ecosystem services. *Ecological Monographs,* 81(2), 169–193.

Beaumont, N. J., van Ierland, E., Marboe, A. H., Starkey,D. J., Townsend, M., Zarzycki, T., Austen, M. C., Atkins, J. P., Burdon, D., Degraer, S., Dentinho, T. P., Derous, S., Holm, P., Horton, T. (2007). Identification, definition and quantification of goods and services provided by marine biodiversity: Implications for the ecosystem approach. *Marine Pollution Bulletin,* 54, 219–242.

Chisholm, J. (2004). Initial scoping study to review the global economic impacts of aquatic bio-invasions. Unpublished report, GEF-UNDP-IMO GloBallast Project.

Cisneros-Montemayor, A. M., and Sumaila, U. R. (2010). A global estimate of benefits from ecosystem-based marine recreation: potential impacts and implications for management. *Journal of Bioeconomics*, 12(3), 245–168.

Costanza, R., D'Arge, R., De Groot, R., Farber, S., Grasso, M., Hannon, B., Limburg, K. O'Neill, R., Paruelo, J., Raskin, R. G., Sutton, P. and van den Belt, M. (1997). The value of the world's ecosystem services and natural capital. *Nature*, 387(6630), 253–260.

de Groot, R., Brander, L., van der Ploeg, S., Costanza, R., Bernard, F., Braat, L., Christie, M., Crossman, N., Ghermandi, A., Hein, L., Hussain, S., Kumar, P., McVittie, A., Portela, R., Rodriguez, L.C., ten Brink, P., and van Beukering, P. (2012). Global estimates of the value of ecosystems and their services in monetary units. *Ecosystem Services*, 1(1), 50–61.

Diaz, R. J., and Rosenberg, R. (2008). Spreading dead zones and consequences for marine ecosystems. *Science*, 321(5891), S. 926–929.

Ehler, C., and Douvere, F. (2009). *Marine Spatial Planning: A step-by-step approach toward ecosystem-based management*. Intergovernmental Oceanographic Commission and Man and the Biosphere Programme. IOC Manual and Guide No. 5, ICAM Dossier No. 6. Paris: UNESCO.

EPA. *National Environmental Policy Act*. Retrieved on September 28, 2012 from www.epa.gov/region1/nepa/.

FAO. (2010). *The state of world fisheries and aquaculture*. Rome: FAO.

Glasby, G. P. (1982). Manganese nodules from the South Pacific: An evaluation. *Marine Mining*, 3, 231–270.

Gutiérrez, N., Hilborn, R., and Defeo, O. (2011). Leadership, social capital and incentives promote successful fisheries. *Nature*, 470, 386–389.

Hageman, R., 1985. *Valuing Marine Mammal Populations: Benefit Valuations in a multi-species ecosystem*. National Marine Fisheries Service, Southwest Fisheries Centre, La Jolla, California.

Hall, C. (2001). Trends in ocean and coastal tourism: The end of the last frontier? *Ocean & Coastal Management*, 44(9–10), 601–618.

Halpern, B. S., Walbridge, S., Selkoe, K. A., Kappel, C. V., Micheli, F., D'Agrosa, C. T. M. (2008). A global map of human impact on marine ecosystems. *Science*, 319 (5865), S. 948–952.

Harbor Branch Media Lab (2006a). *The Pipeline and the Finish Line. Marine Biotech in Depth*. Retrieved September 2013 from www.marinebiotech.org/pipeline.html.

Harbor Branch Media Lab (2006b). *Marine Bioprospecting: Mining the untapped potential of living marine Resources*. Retrieved September 2013 from www.marinebiotech.org/biopro.html.

Hein, J., McIntyre, B., and Piper, D. (2005). Marine mineral resources of Pacific Islands: A review of the exclusive economic zones of islands of U.S. affiliation, excluding the State of Hawaii. *U.S. Geological Survey Circular*, 1286, 62.

International Association for Impact Assessments (IAIA). 1999. *Principles of Environmental Impact Assessment Best Practice*. Retrieved September 2012 from www.iaia.org/publicdocuments/special-publications/Principles%20of%20IA_web.pdf.

ICS. (2010). *BIMCO/ISF Manpower 2010 Update*. Retrieved February 23, 2012 from International Chamber of Shipping: www.marisec.org/Manpower%20Study.pdf.

ICS. (2012a). *Shipping Facts: Shipping and world trade*. Retrieved September 28, 2012 from International Chamber of Shipping: www.marisec.org/shippingfacts/ worldtrade/index.php.

ICS. (2012b). *IMO Agreement on CO2 Technical Rules*. Retrieved February 23, 2012 from International Chamber of Shipping: www.shippingandco2.org/imopackage.htm

IMO. (2012). *List of IMO Conventions*. Retrieved February 23, 2012, from International Maritime Organization: www.imo.org/About/Conventions/ListOfConventions/Pages/Default.aspx.

IPCC. (2011). *Renewable Energy Sources and Climate Change Mitigation*. Retrieved July 15, 2015 from Special Report on Renewable Energy Sources and Climate Change Mitigation: http://srren.ipcc-wg3.de/report.

Lotze, H. K., Lenihan, H. S., Bourque, B. J., Bradbury, R. H., Cooke, R. G., Kay, M. C., Jackson, C. H. (2006). Depletion, degradation, and recovery potential of estuaries and coastal seas. *Science*, 312(5781), S. 1806–1809.

McLeod, K., and Leslie, H. (Eds.). (2009). *Ecosystem-Based Management for the Oceans*. Washington DC: Island Press.

Mayer, A. M. S., Rodríguez, A. D., Berlinck, R. G. S., Hamann, M. T. (2007). Marine pharmacology in 2003–4: Marine compounds with anthelmintic antibacterial, anticoagulant, antifungal, anti-inflammatory, antimalarial, antiplatelet, antiprotozoal, antituberculosis, and antiviral activities; affecting the cardiovascular, immune and nervous systems, and other miscellaneous mechanisms of action. In: *Comparative Biochemistry and Physiology*, Part C, 145, 553–581, citation from p. 553.

Mayrand, K., and Paquin, M. (2004). *Payments for Environmental Services: A Survey and Assessment of Current Schemes*. Unisféra International Centre for the Commission for Environmental Cooperation of North America. Retrieved July 22, 2015 from www.cec.org/Storage/56/4894_PES-Unisfera_en.pdf.

MEA. (2005). Current State and Trends Assessment: Coastal Systems, Chapter 19. *Millennium Ecosystem Assessment*. Island Press, Washington. Retrieved September 2012 from www.millenniumassessment.org/en/Conditions.html.

Milder, J. C., Scherr, S. J., and Bracer, C. (2010). Trends and future potential of payment for ecosystem services to alleviate rural poverty in developing countries. *Ecology and Society* 15(2), 4. Retrieved July 22, 2015 from www.ecologyandsociety.org/vol15/iss2/art4/.

Mosegaard, R., Chandler, H., Barons, P., and Bakema, G. (2009). *Economics of Wind Energy part III, Department of Technical University*. Risø DTU: National Laboratory, Technical University of Denmark.

Naber, H., Lange, G-M., Hatziolos, M. (2008). "Valuation of Marine Ecosystem Services: A Gap Analysis". Retrieved July 22, 2015 from www.cbd.int/marine/voluntary-reports/vr-mc-wb-en.pdf.

Nautilus. (2008). *Environmental Impact Statement Solwara 1 Project, Volume A Main Report*. Brisbane: Nautilus Minerals Niugini Limited.

OECD Environment Directorate (2005): The Costs of Inaction with Respect to Biodiversity Loss. Background Paper, EPOC high-level special session on the costs of inaction, 14.04.2005, Paris.

Orams, M. (1999). *Marine Tourism: Development, impacts and management*. London: Routledge.

Pandolfi, J. M., Bradbury, R. H., Sala, E., Hughes, T. P., Bjorndal, K. A., Cooke, R. G., McArdle, D., McClenachan, L., Newman, M. J. H., Paredes, G., Warner, R., and Jackson, J. B. C. (2003). Global trajectories of the long-term decline of coral reef ecosystems. *Science*, 301(5635), 955–958.

Pauly, D., Watson, R., and Alder, J. (2005). Global trends in world fisheries: Impacts on marine ecosystems and food security. *Philosophical Transactions of the Royal Society B: Biological Sciences*, 360(1453), 5–12.

Rona, P. (2003). Resources of the sea floor. *Science* 299, 673–674.

Scott, S. (2006). The dawn of deep ocean mining. *ScienceDaily*. Retrieved July 22, 2015 from www.sciencedaily.com/releases/2006/02/060221090149.htm.

Suuronen, P. F. C., Glass, C., Løkkeborg, S., Matsushita, Y., Queirolo, D., and Rihan, D. (2012). Low impact and fuel efficient fishing: Looking beyond the horizon. *Fisheries Research* 119–120, 135–146.

TEEB (2010). *The Economics of Ecosystems and Biodiversity Ecological and Economic Foundations*. Edited by Pushpam Kumar. London and Washington: Earthscan.

Tyedmers, P. (2004). Fisheries and energy use. *Encyclopaedia of Energy*. Vol. 2, pp. 683–693. Retrieved July 22, 2015 from www.fcrn.org.uk/sites/default/files/Fishing_and_Energy_Use.pdf

Tyedmers, P., Watson, R., and Pauly, D. (2005). Fueling global fishing fleets. *Ambio*, 34 (8), 635–638.

UNEP. (2004). *Environmental Impact Assessment and Strategic Environmental Assessment: Towards an Integrated Approach*. Retrieved on September 28, 2012 from www.unep.ch/etu/publications/textONUbr.pdf.

UNEP. (2006). *Ecosystem-based Management – Markers for Assessing Progress*. Paris: United Nations Environment Programme.

UNEP. (2008). *Desalination Resource and Guidance Manual for Environmental Impact Assessments*. Cairo: United Nations Environment Programme, Regional Office for West Asia, Manama, and World Health Organization, Regional Office for the Eastern Mediterranean.

UNEP. (2011a). *Decoupling Natural Resource Use and Environmental Impacts From Economic Growth*. Paris: United Nations Environment Program – International Resource Panel.

UNEP. (2011b). *Taking Steps towards Marine and Coastal Ecosystem-based Management: An introductory guide*. Paris: United Nations Environment Program.

UNEP. (2011c). *Towards a Green Economy: Pathways to sustainable development and poverty eradication*. Paris: United Nations Environment Program.

UNEP, FAO, IMO, UNDP, IUCN, WorldFish Center, GRIDArendal. (2012). *Green Economy in a Blue World*. Retrieved July 22, 2015 from www.unep.org/greeneconomy and www.unep.org/regionalseas.

UNEP-WCMC. (2011). *Marine and Coastal Ecosystem Services: Valuation methods and their application*. UNEP-WCMC Biodiversity Series No. 33. 46 pp. Cambridge, UK.

UNWTO. (2012) UNWTO Commission For East Asia And The Pacific Unwto Commission For South Asia Twenty-fourth Joint Meeting Chiang Mai, Thailand May 4, 2012.

Van Dover, C., Smith, C., Adron, J., Arnaud, S., and Beaudoin, Y. (2011). Environmental management of deep-sea chemosynthetic ecosystems: Justification of and considerations for a spatially-based approach. *ISA Technical Study: no. 9*, Kingston, Jamaica: International Seabed Authority. p. 90.

World Bank. (2009). *The Sunken Billions: The economic justification for fisheries reform.* Washington DC: The International Bank for Reconstruction and Development/The World Bank.

World Bank. (2010). *The Hidden Harvests: The global contribution of capture fisheries.* Retrieved September 28, 2012 from http://siteresources.worldbank.org/EXTARD/Resources/336681–1224775570533/The HiddenHarvestsConferenceEdition.pdf.

World Bank, FAO and World Fish Centre. (2010). *The Hidden Harvests: The global contribution of capture fisheries.* Washington DC: World Bank.

World Commission on Environment and Development (1987). *Our Common Future.* Retrieved on September 28, 2012 from www.un-documents.net/ocf-02.htm#I.

Worm, B., Barbier, E. B., Beaumont, N., Duffy, J. Emmett, Folke, C., Halpern, B. S., Jackson, J. B. C., Lotze, H. K., Micheli, F., Palumbi, S. R., Sala, E., Selkoe, K. A., Stachowicz, J. J.. and Watson, R., Watson, R. (2006). Impacts of biodiversity loss on ocean ecosystem services. *Science*, 314(5800), 787–790.

Wunder, S., Engel, S., and Pagiola, S. (2008). Taking stock: A comparative analysis of payments for environmental services programs in developed and developing countries. *Ecological Economics*, 65(4), 834–852.

WWEA. (2011). World Wind Energy Annual Report 2011. Retrieved November 29, 2011, from World Wind Energy Association: www.wwindea.org/home/index.php?option=com_frontpage&Itemid=1.

PART 2

The uses of the sea
Living resources

14

GLOBAL FISHERIES

Current situation and challenges

Yimin Ye

Introduction

Fishing for food is an ancient practice in human history, dating back at least 40,000 years. With the development of civilization and the ever-rising human population, the increasing human demand for fish has led to improvement of fishing tools and techniques. Today, large fishing vessels are able to cross oceans to fish anywhere they like, and modern equipment for navigation, fish detection, and handling of heavy machinery have made fishing so efficient that any targeted fish species can be wiped out if fisheries are not well managed.

Although fisheries today still remain an important source of food provision and make a significant contribution to the economy and well-being of human society, commercial fishing vessels have earned a bad reputation because they overfish fish stocks and cause negative impacts on ecosystems. Overexploitation and impacts on ecosystems do not only result in environmental concerns, but also harm the fishing industry itself. In open access fisheries, overexploitation and degradation of ecosystems reduce productivity of fishery resources, while the decline in production in turn makes fishermen fish even harder to stay in business. Without external interruption, such a vicious circle just gets worse over time. In contrast, if proper regulation is exercised, overexploitation can be avoided, overfished stocks can be restored, and the entire fishery ecosystem will become healthier. As a result, fish production will be increased in comparison with the overfished situation. Reconciling fishery production with environmental conservation is thus the goal of sustainable development of fisheries.

This paper first summarizes the social and economic significance of fisheries for food security and nutrition, supporting livelihoods of human society. It then reviews the current situation in global marine fisheries, the status of fish stocks, the state of fishing fleets, and management practices. A diagnosis of the issues and challenges fisheries are facing today follows; finally, this chapter discusses how fisheries can make a successful transition to achieve long-term sustainability.

Significance of fisheries in food security and human well-being

Food security and nutrition

Food security requires 'all people, at all times, have physical, social and economic access to sufficient, safe and nutritious food to meet their dietary needs and food preferences for an active and healthy life' (FAO, 1996). Food security always occupies a high position in national planning, whether in developed or developing countries. Fishing aimed to produce a reliable source of food in ancient times and remains so today in many subsistence fisheries. Although fishing is considered a commercial activity in many fisheries, its contribution to food security and human nutrition needs is still the core concern at national level, particularly in low income, food–deficient countries.

Hunger and malnutrition are still identified as among the most significant problems facing the world's poor, and the major challenge governments and international development communities need to address. Globally, about one billion people – one sixth of humanity – are chronically hungry (FAO, 2010).

Fisheries can contribute to food security in two ways: directly as a source of essential nutrients, and indirectly as a source of income to buy food. In 2010, the global average fish consumption reached 23 kg per person annually. In many small island developing countries, people's dependence on fish for food is much higher than the average, reaching 153.2 kg per person in Maldives and 123.4 kg per person in Palau (FAO, 2007). The role of fish for meeting the food security needs of the poor cannot be overstated.

Fish is a well-known and frequently traded food product in the poorest communities and is therefore a source of income. For some households, selling fish is a means to obtain products or services that household members are not able to produce. Through trading, fishers and aquaculturists thus contribute to better food security, not only for their own households but also for households where members neither capture fish in the wild nor raise it in captivity (Tacon and Metian, 2013).

Fish is highly nutritious, rich in essential micronutrients, minerals, essential fatty acids and proteins, and represents an excellent supplement to nutritionally deficient cereal-based diets (Garcia and Rosenberg, 2010). World wide, fish provide about 17 per cent of the human animal protein intake (FAO, 2012). As can be expected, the contribution of fish as a protein source is particularly important in small island states where, in fact, frequently more than 50 per cent of the animal protein consumed comes from fish. A similar situation prevails in several coastal States in West Africa and in such large countries as Indonesia and Japan. In some lesser developed coastal states – especially in Asia, but also in some parts of Africa – fish proteins are absolutely essential to food security as they comprise a large share of an already relatively low level of animal protein consumption.

Social and economic contributions

Poverty is most often seen in remote rural areas, and so are many small-scale and subsistence fisheries, particularly around lakes, rivers, deltas, floodplains and coastal areas. Globally more than 60 million people are engaged in artisanal capture fisheries, and about 50 per cent of those employed in the sector are women. Therefore, fisheries provide an important source of livelihood. Fisheries and aquaculture directly employed over 34 million people in 2010 (FAO, 2010). Considering secondary industries such as boat building, equipment and maintenance, vessel supplies, fish processing and trade, etc., fisheries support the livelihoods of 660–820 million people, or about 10–12 per cent of world population. The fisheries and aquaculture sector contribution to gross domestic product (GDP) ranges from around 0.5 to 2.5 per cent, but may reach as high as 10 per cent in island countries such as Nauru.

In many parts of the world fisheries form an indispensable source of livelihood, providing food, employment and income to millions of families. For many, fishing is a 'last resort' with an important safety net function: in situations of failing agriculture or job loss, fishing may provide part-time or temporary income and relatively cheap and nutritious food. The social and economic contribution of the small-scale fisheries sector, however, tends to be obscured by national statistics in many countries, because fish landings by the small-scale sector may not be reported, or if data are collected, are typically under-reported. As a result, the role of small-scale capture fisheries in rural livelihood, trade and food security remains critically unrecognized in development and poverty reduction approaches (Garcia and Rosenberg, 2010). However, it is clear that healthy fisheries can contribute to poverty reduction through generation of revenues and wealth creation, operating as a socio-economic 'lift' at the community level and contributing to economic growth at the national level.

Current situation of marine fisheries

Global fish landings

The world's total fish production has increased linearly since 1950, reaching 154 million tonnes in 2011, of which marine capture fisheries contributed 51 per cent. Compared to marine capture fisheries, the largest sector, mariculture (13 per cent), freshwater aquaculture (29 per cent) and inland capture fisheries (7 per cent) contribute relatively less to fisheries production (Figure 14.1). The world's marine fisheries produced only 16.7 million tonnes in 1950, but increased explosively to 87.7 million tonnes in 1986, and remained relatively stable afterwards until 2002. During the last decade, marine fisheries decreased by about 10 per cent from the previous decade, settling at 78.9 million tonnes annually by 2011 (Figure 14.1).

Marine fisheries have experienced different development stages. Rapid development was seen in the late 1950s and 1960s and between 1983 and 1989 (Figure 14.1). The first boom was believed to be caused mainly by post-war ship-building expansion in the 1950s, new technologies such as motor trawlers in the 1960s and the extension of jurisdiction to 12 nautical miles by most coastal states. This is the region that encompassed the ocean's most productive upwelling and continental shelves. The second rapid expansion was associated with the extension of jurisdictions from 12 to 200 nautical miles, which occurred with the establishment of Exclusive Economic Zones (EEZs) under the legal provisions of the UN Convention on Law of the Sea (UNCLOS).

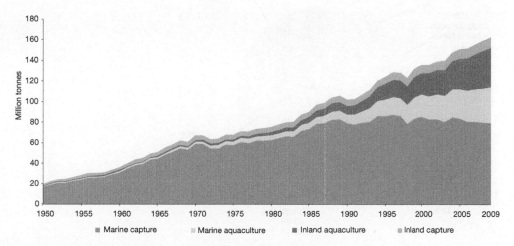

Figure 14.1 World fish production from different sectors of fisheries
Source: Author.

Species composition

Pelagic species dominate global marine catches. Small pelagics (herrings, sardines, anchovies, etc.) contributed about 22 per cent (19.9 million tonnes) of the total catch in 2009 (Figure 14.2). This share is down from 29 per cent in the 1950s and 27 per cent in 1970s. The large pelagics (tunas, bonitos, billfishes and miscellaneous pelagics) accounted for 19 per cent (16.6 million tonnes) of the total catches in 2009. This is an increase in their share from 13 per cent in the 1950s. Demersal fishes (flounders, halibuts, soles, cods, hakes, haddocks and miscellaneous demersals) contributed 12 per cent (10.9 million tonnes), compared with almost 26 per cent in the 1950s and 1970s. Miscellaneous coastal fishes increased slightly to 8 per cent (7.2 million tonnes) from 7 per cent in 2009. Catches of crustaceans (crabs, lobsters, shrimps, prawns, krill, etc.) contributed 6 per cent (5.4 million tonnes) in 2009, slightly lower than the 7 per cent share of marine fisheries in 2002. Molluscs (abalones, conchs, oysters, mussels, scallops, clams, squids, octopus, etc.) increased slightly from 6 per cent in the 1950s and 1970s to 7 per cent (6.2 million tonnes) in 2009.

Species composition varies from area to area around the world. All the major species groups are represented more or less equally in the Northwest Pacific (Area 61) (Figure 14.3). Small pelagics (mostly anchoveta) dominate catches in the Southeast Pacific (Area 87). In the Northeast Atlantic (Area 27), demersal fishes were the most abundant, followed by larger pelagics and small pelagics. In the Western Central Pacific, catches were dominated by larger pelagics, which were also the most abundant group in the Western Indian Ocean (Area 51). Small pelagics were also dominant in the Eastern Central Atlantic (Area 34), Mediterranean and Black Sea (Area 37), Western Central Atlantic (Area 31) and Eastern Central Pacific (Area 77). In contrast, demersal fishes were the dominant species group in the Northeast Pacific (Area 67) and Southwest Pacific (Area 81).

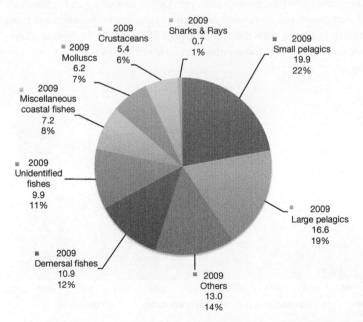

Figure 14.2 World marine catch by main species groups in 2009 (million tonnes and %)
Source: Author.

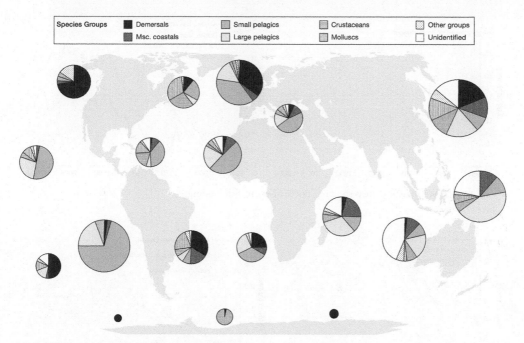

Figure 14.3 Catch species composition by main species groups in FAO statistical areas in 2009
Source: Author.

Regional variations in landings

Based on the average catches in 2005–09, the Northwest Pacific is the largest contributor (25 per cent) to the global catch, followed by the Southeast Pacific (16 per cent), Western Central Pacific (14 per cent), Northeast Atlantic (11 per cent) and Eastern Indian Ocean (7 per cent). All other FAO areas contribute less than 5 per cent of the global total catch (Ye and Cochrane, 2011). World marine fisheries have gone through significant development and changes since 1950 when FAO started collecting fisheries statistics data. Accordingly, the levels of exploitation of fish resources and their landings have also varied over time.

The temporal pattern of landings differs from area to area, depending on the level of urban development and changes that countries surrounding that area have experienced. In general, they can be grouped into three types. The first group are those FAO areas that have demonstrated oscillations in total catch (Figure 14.4). These are the Eastern Central Atlantic, Northeast Pacific, Eastern Central Pacific, Southwest Atlantic, Southeast Pacific and Northwest Pacific. These areas provide about 53.5 per cent of the world's total catch. Some areas in this group may have shown a clear drop in total catch in the last few years, e.g. Northeast Pacific, but, over the longer period, a declining trend is not evident.

The second group consists of areas that have demonstrated a decreasing trend in catch since reaching a peak at some time in the past. This group has contributed 19.9 per cent of global catch on average in the last five years, and includes the Northeast Atlantic, Northwest Atlantic, Western Central Atlantic, Mediterranean and Black Sea, Southwest Pacific and Southeast Atlantic (Figure 14.5). It is interesting and noteworthy that such declines occurred at different times: in the Northwest Atlantic in the late 1960s; in the Northeast and Southeast Atlantic in the mid 1970s; in the Western Central Atlantic and Mediterranean and Black Sea in the mid 1980s; and

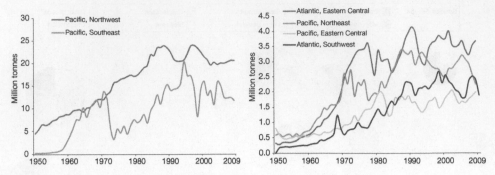

Figure 14.4 FAO statistical areas showing fluctuations in fish landings
Source: Author.

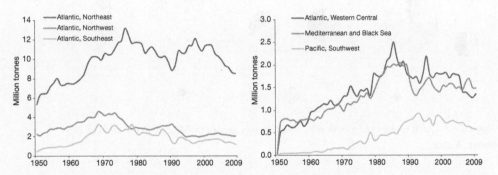

Figure 14.5 FAO statistical areas showing a decreasing trend in fish landings
Source: Author.

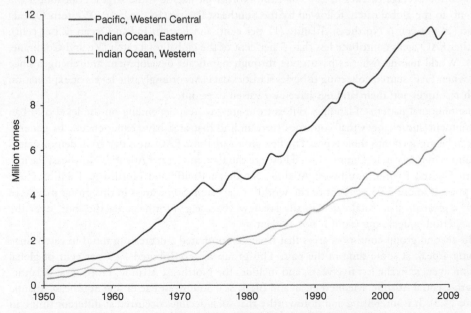

Figure 14.6 FAO statistical areas maintaining an increasing trend in fish landings
Source: Author.

in the Southwest Pacific in the early 1990s (Figure 14.5). This sequence largely reflects the fact that areas surrounded by the most-developed countries experienced the earliest decline in catches. The largest decline was seen in the Northwest Atlantic, where landings dropped by 55 per cent from their peak to 2009. The second-largest drop was in the Western Central Atlantic with 46 per cent, followed by the Southwest Pacific with 37 per cent and the Northeast Atlantic with 35 per cent. The total catches in the Mediterranean and Black Sea dropped by 28 per cent.

The third group comprises the FAO areas that have shown a continual increase in catch since 1950 (Figure 14.6). There are only three areas in this group: Western Central Pacific, Eastern and Western Indian Ocean. They have contributed 26.4 per cent of the total catch on average in the last five years. Minor drops in catch have also been seen in Western Central Pacific and Western Indian Ocean in the last two years. However, considering the uncertainty involved in catch reporting and the natural fluctuations in fish stock abundance, such declines might have been caused by environmental 'white noise' and need to be further monitored.

The state of marine fish stocks

Fish are renewable resources – living beings that replenish their numbers naturally and may be caught, within limits, on a continuous basis without this leading to their elimination. The renewability of a fish stock is determined by its biological parameters and its abundance. When a fish population is large or close to its carrying capacity, fish will grow slowly and mortality caused by natural factors will be high, with surplus reproduction low. In contrast, if the populations of fish stocks are low, fish can grow faster with lower natural mortality, benefiting from a lower competition for food and space among individual fish. In this case the total surplus production of the stock will also be limited to a low level by the low reproductive biomass. Generally speaking, surplus production of a stock is a parabolic function of its biomass. It increases with stock abundance initially, reaches a maximum when the stock grows to a certain level, and then decreases gradually to zero while the stock reaches its carrying capacity.

The maximum annual surplus production is often called maximum sustainable yield in the fishery (MSY). The MSY and its associated stock biomass is the most frequently applied reference point in fisheries management and international instruments. When a fish stock is below the level that can produce MSY, the stock is said to be overfished. Therefore, it is necessary to know the current stock size and the stock level associated with MSY in order to determine whether a stock is overfished. Such information cannot be obtained simply from the landings data alone, but through stock assessment, which requires numerical modelling based on other data such as fishing effort and biological parameters, as well as catch landing data.

The FAO has monitored the state of the world's marine fish stocks since 1974, periodically reviewing 548 stock items. In 2009 the 395 stocks that were assessed represented about 70 per cent of global catch. The assessment methods are described in detail in Ye and Cochrane (2011).

Of the fish stocks assessed, 57.4 per cent were estimated to be fully exploited in 2009. These stocks produced catches that were already at or very close to their maximum sustainable production. These offer no room for further expansion in catch, and even some risk of decline if not properly managed. Among the remaining stocks, 29.9 per cent were overexploited, and 12.7 per cent non-fully exploited in 2009. The overexploited stocks produced lower yields than their biological and ecological potential. They require strict management plans to rebuild their stock abundance to restore full sustainable productivity. The non-fully exploited stocks were under relatively low fishing pressure and have a potential to increase their production.

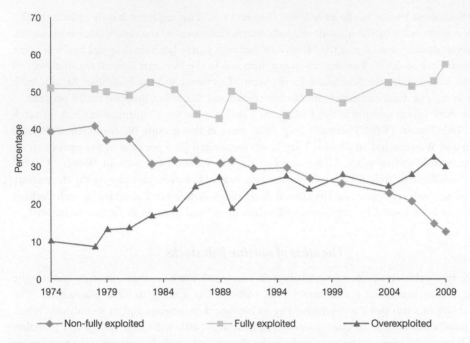

Figure 14.7 Stock status of the world's marine fishery resources from 1974 to 2009
Source: Author.

However, these stocks often do not have high production potential. The potential for increase in catch may be generally limited. Nevertheless, proper management plans should be established before increasing the exploitation rate of these non-fully exploited stocks to avoid following the same track of overfishing.

The proportion of non-fully exploited stocks has decreased continuously since 1974, when the first FAO assessment was accomplished (Figure 14.7). In contrast, the percentage of over-fished stocks has since increased, especially in the late 1970s and 1980s from 10 per cent in 1974 to 26 per cent in 1989. After 1990, the number of overfished stocks continued to increase, but the rate of increase slowed, until the last two assessments, reaching about 30 per cent in 2009. The fraction of fully exploited stocks demonstrated the smallest change over time. The percentage dropped from about 50 per cent at the start of the series to 43 per cent in 1987 and subsequently increased to 57.4 per cent in 2009 (Figure 14.7).

A primary goal of fishery management is to control fishing at a level that allows the fishery to produce a sustained annual yield. This yield should be as close to maximum sustainable yield as allowed by responsible management within the context of an ecosystem approach. At the same time, it allows for increasing exploitation rates on non-fully exploited stocks. This would maximize the sustained contribution of fisheries to global food security and human well-being. The increasing trend in fully exploited stocks after 1990 may indicate the positive impact of fishery management towards maximizing production. However, close attention is required to all fully exploited stocks to ensure that they are not over-exploited in the future. Further, the increase in overfished stocks is cause for concern. Nevertheless, the deceleration in the rate of increase of overfished stocks after 1990 in comparison with the 1980s may indicate some progress in improved management. It suggests that some fish resources have benefited from the management efforts of coastal states and the international community.

The state of the fishing industry

With the development of fisheries, the world's fishing fleets have expanded along with the number of fishers engaged in fishing. The total number of fishing vessels reached 4.36 million in 2010 (FAO, 2012), of which 74 per cent were from marine fisheries. The fleet in Asia was the largest, accounting for 73 per cent of the global fleet, followed by Africa (11 per cent), Latin America and the Caribbean (8 per cent), North America (3 per cent) and Europe (3 per cent). Similarly, total number of fishers in marine fisheries were 12 million in 1970, and reached 34 million in 2008 (FAO, 2012), increasing nearly threefold.

Along with the fleet expansion, fishing technology and equipment has also improved greatly. As a result, fishing effort, defined as the number of decked vessels multiplied by technological coefficients that reflect the increase in fishing efficiency over time, increased linearly ninefold from 1970 to 2008 (Ye *et al.*, 2013).

Today's large fishing effort is not at a level compatible with sustainable production of fisheries. Based on Ye *et al.* (2013), the capacity of global fishing fleets needs to be cut back by 36–43 per cent from the 2008 level to match the level at which the MSY can be produced. Achieving this target requires 12–15 million fishers moving out of marine fisheries worldwide.

Overcapacity is a problem recognized 20 years ago. The Rome Consensus on World Fisheries (FAO, 1995a), the Code of Conduct for Responsible Fisheries (FAO, 1995b), and the Kyoto Declaration and Plan of Action (FAO, 1995c) all noted the need for management of fishing capacity. Since then, the issues of fleet overcapacity and capacity management – especially as a key threat to the long-term viability of exploited fish stocks and the fisheries that depend on them – have become essential elements of work to avoid the degradation of fishery resources, the dissipation of potential food production, and significant economic waste.

In 1999, the FAO Committee on Fisheries adopted the International Plan of Action for the Management of Fishing Capacity (FAO, 1999), which specifies a number of actions to be urgently taken:

(i)   assessment and monitoring of fishing capacity;
(ii) preparation and implementation of national plans;
(iii) international consideration; and,
(iv) immediate actions for major international fisheries requiring urgent attention.

Indeed, the immediate objective of the Plan of Action is for 'States and regional fishery organizations, in the framework of their respective competencies and consistent with international law, to achieve worldwide, preferably by 2003 but no later than 2005, an efficient, equitable and transparent management of fishing capacity'.

Unfortunately, overcapacity of the global fishing fleets remains a serious issue ten years after this deadline. The major constraints and issues include difficulties in finding alternative employment for displaced fishers, pressures imposed by industry (harvesting and processing) not to reduce fleets or catch, difficulties in monitoring, control and surveillance and a lack of institutional capacity to develop and implement capacity management plans as well as to undertake the appropriate research required (e.g. stock and capacity assessments).

The ninefold increase in fishing effort since 1970 has resulted in only 50 per cent increase in fish landings. This clearly suggests that catch rate per unit effort has decreased and so has the direct economic efficiency of fishing operations. The excess fishing effort has also caused fish stocks to be overfished. The double edges of overcapacity made world fisheries lose US$5 billion in 2004 (The World Bank and FAO, 2009).

The state of international trade of fish products

Fish and fishery products are among the most traded food commodities worldwide. Trade plays a major role in the fishery industry as a creator of employment, food supplier, income generator, and contributor to economic growth and development. About 25 per cent of fish production entered international trade in 1976, increasing to about 38 per cent (57 million tonnes) in 2010 (FAO, 2012). This increase reflects the sector's growing degree of openness to, and integration in international trade, as well as sustained demand, trade liberalization policies, globalization of food systems and technological innovations in the international fish trade.

In the period 1976–2010, the world trade in fish and fishery products grew significantly also in value terms, rising from US$8 billion to US$109 billion, with annual growth rates of 8.3 per cent in nominal terms and of 3.9 per cent in real terms (FAO, 2012).

The average trade flows of fish and fishery products for the period 2008–2010 show that Latin America and the Caribbean have solid positive net fishery exports, as is the case for Oceania and the developing countries of Asia. By value, Africa has been a net exporter since 1985, but it is a net importer in quantity terms, reflecting the lower unit value of imports (mainly for small pelagics). Europe and North America are characterized by a fishery trade deficit. In general, developing countries tend to export products at high prices for hard foreign currencies, and developed countries often import fish products to meet their increasing demand for fish.

Through international trade, fisheries provide an important source of foreign currency for developing countries. In addition, trade creates more, often new, employment and enhances labour based entitlements particularly among women. It is now well established that women's employment tends to contribute more to family welfare and food security. It may be concluded that the international trade in fish will improve fishers' income and consequently contribute to food security and livelihoods of the fishing communities.

Moving towards sustainable fisheries

What is a sustainable fishery?

The concept of sustainable fisheries is a derivative of the more general concept – sustainable development. In 1987 the United Nations released the Brundtland Report (WCED, 1987), which included what is now one of the most widely recognized definitions: 'Sustainable development is development that meets the needs of the present without compromising the ability of future generations to meet their own needs.' The United Nations 2005 World Summit Outcome Document (UN, 2005) refers to the 'interdependent and mutually reinforcing pillars' of sustainable development as economic development, social development and environmental protection.

People working on fisheries used the term of sustainable yield long before 1987. Sustainable yield is the ecological yield that can be extracted without reducing the base of the capital resource itself, emphasizing the indefinite sustainability of the yield. The concept of sustainable development also focuses on temporal continuity through balance between generations, but has a broader content. Fishing as an economic activity pursues catch and its derivative profit, and in turn the catch it caught and job opportunities it provides support people's livelihoods and also contribute to human well-being, although fishing may also cause changes to the environment and ecosystems. So, the social, economic and environmental aspects of fishing are consistent with the three pillars of sustainable development. A sustainable fishery must strike a balance among the three pillars that is pragmatic and satisfactory for all the relevant stakeholders of a specific fishery.

The complexity and the diversity of fisheries lead to the promotion of different concepts. For example, maximum sustainable yield (MSY) is a popular concept used in many international instruments: the Law of the Sea (UN, 1982), the United Nations Fish Stocks Agreement (UN, 1995), the 2002 World Summit on Sustainable Development Goals (UN, 2002), the United Nations Millennium Development Goals (UN, 2013), the Convention on Biological Diversity 2010 targets (UNEP, 2010), and the outcome document of the Rio+20 World Summit, 'The Future We Want' (UN, 2012). In contrast, economists believe that maximum economic yield (MEY) should be the goal of sustainable fishery management as MEY supports a higher economic efficiency. In practical fishery management, the USA, the EU and New Zealand have set MSY as their target reference point and Australia sets MEY as its management target reference point.

Although MSY and MEY seem to be working for commercial fisheries, a large number of small-scale fisheries and subsistence fisheries may find both MSY and MEY unsatisfactory because they pay more attention to social contributions of fisheries to local communities. Social benefits are reflected in many different aspects such as food production, employment, and support to upstream and downstream industries. Some of these are proportional to the volume of production, and others in line with the scale of employment and the extent of support to secondary industries. The three metrics of social, economic and ecological sustainability do not always vary proportionally with the development of fishing, and in some circumstances, they may conflict with each other. For example, when fishing effort exceeds the level that can produce MSY, any increase in effort will reduce production from the fishery, but can add more job opportunities. As a result, the impact on upstream industries such as gear and boat maintenance may still be positive, but that on downstream industries for instance marketing and distribution is more likely to be negative as a result of reduced production volume. In addition, social benefits often materialize in various forms of ripple effects in a society, and therefore cannot be easily measured using a single quantitative measure. Further, economic efficiency or rent maximization is achieved even at an effort level lower than that associated with MSY. Difference in emphasis on social, economic and production concerns will lead to different levels of desired development.

Adding to the difficulty in measuring economic and social benefits is environmental sustainability. For pure conservation purposes, the pristine condition of a fish stock may be an ideal situation. However, this is a utopia difficult to justify for fisheries because catching fish for human well-being is the primary purpose of all fisheries. Environmental sustainability is to protect fish stocks so that human well-being derived from these stocks can be maintained forever. In this sense, environmental sustainability serves the purpose of sustainable production.

The single species theory tells us that surplus production maximizes when abundance of a fish stock is fished down to a certain level, and this level is called biomass at maximum sustainable yield (BMSY). Once a stock falls below BSMY, sustainable productivity of the stock will be impaired. The low stock level may also increase the likelihood of causing negative impacts on other components of the ecosystem in which the stock resides. Given the common belief that the higher the stock level, the less the environmental impact, it may be justifiable to set BMSY as the limit level for environmental sustainability.

The environmental impact discussed above is limited to what may occur through the target species. There are also other kinds of impacts that are side effects of fishing activities, for example, bycatches incidentally caught, and damage to benthic habitats by bottom trawling. These kinds of impacts on the environment are often considered separately and dealt with by supplementary regulations.

In summary, a sustainable fishery should be defined by three pillars: social, economic and environmental. The metrics of the three pillars may conflict with each other. Each fishery is different and every fishery operates under different circumstances. There is no universal definition for sustainable fisheries, nor a single measure for sustainability. It is necessary to balance social, economic and environmental objectives based on the social and economic conditions and the characteristics of the fishery through smart design, realigning incentives and involving stakeholders. Nevertheless, it is not difficult to find a limit beyond which each of the three pillars of sustainability shall not fall. A sustainable fishery should not impair its production potential, should not run in economic deficit, and should not cause irrecoverable damage to the ecosystem upon which it relies.

Challenges to achieving sustainability in fisheries

Resource overexploitation, economic loss, and ecological degradation are three critical issues the world fisheries face today (Ye *et al.*, 2013). These certainly fall beyond the limits that define unsustainable fisheries discussed above. How can fisheries solve these problems and move towards long-term sustainability?

These three critical issues are consequences of overfishing, which is an inevitable result of overcapacity of fishing fleets. Why has the world fishery accumulated such an excess capacity in fishing fleets? Poor management and governance are to blame. In the past, fishery management failed to control fishing at a level that can maintain long-term sustainability. This is partly caused by the open-access nature of fish resources and partly by the myriad subsidies provided to fisheries. The former allows free entry of vessels into fishing, and the latter provides economic incentives for fishing vessels to continue fishing even if they operate in the red. Subsidies exist initially for amelioration of the economic difficulties encountered by fishing vessels due to consideration of social concerns, but these often make the situation worse (OECD, 2009), albeit providing temporary relief. If clear ownership of fish resources is established, fishermen will have the incentives to move towards an environmentally sustainable and economically profitable fishery. Therefore, the priority to transform fisheries is to promote rights-based management approaches. A successful transition of the world fisheries to sustainability also requires capacity building in design and implementation of such rights-based approaches at all levels from individual to communities, national and regional.

Rights-based management

The current unsatisfactory situation of the world's fisheries is the tragedy of the commons that lack an appropriate system of property rights. Common pool resources such as fisheries are open to anyone who wants to enter. The benefits of resource exploitation accrue to individuals, each of which is motivated to maximize his or her own use of the resource, while the costs of exploitation are distributed among all those to whom the resource is available. Fishermen have no incentives to protect fish resources and to arrange fishing activities to achieve a long-term social, economic and environmental sustainability.

Rights-based management (RBM) is a tool that creates rules that define not only the right to use and the allocation of fisheries resources, but also the responsibility to protect resources and to sustain the fishery (Scott, 2000). Thus, fishermen, fishing vessels, fishing communities and so forth can be awarded a licence, quota or fishing right to stocks. These rights in turn also stimulate their owners to act accordingly for their own maximum benefits to retain these rights.

There are a large number of different RBM approaches, such as limited non-transferable licensing; community catch quotas; individual non-transferable or transferable effort quotas, individual non-transferable or transferable catch quotas, vessel catch limits or territorial use rights

Table 14.1 Examples of different rights-based approaches

Rights-based approach	Different categories	Examples
Access rights	• TURFs (Territorial Use Rights in Fishing)	• Chilean Loco Fishery rights • Ben Tre Clam Fishery
	• Limited Entry Licenses	• Australian Spencer Gulf Prawn fishery • Kuwait prawn fishery
	• Community-based management	• Negros Island Community Fishery • Zanzibar's Small Pelagic Fishery
Input rights	• Numbers of gear units/fishermen • Numerical rights to use a certain amount of fishing time for a specific gear	• Indonesian Sardine Fishery • Australian Northern Prawn Fishery • Parties to the Nauru Agreement Tuna fishery
Output rights	• ITQs (Individual Transferable Quotas)	• Alaskan Bering Sea Crab Fishery • California Morro Bay Groundfish Fishery
	• TACs (Total Allowable Catches)	• New Zealand Sanford Fishery • Torres Strait Rock Lobster Fishery

Source: Revised from Tindall, 2012.

in fisheries. Table 14.1 gives examples of such RBM approaches used for different fisheries around the world.

For an RBM system to function in line with sustainability, it needs to be applied within a framework that incorporates the three pillars of sustainable development. When determining resource access, rights-based approaches to the management of small-scale fisheries need to take account of their collective nature, as well as the social and cultural dimensions of their activities.

One of the important elements of a rights-based approach to fisheries management is that it gives the fishing industry security. A rights-based system gives fishermen security and a sense of ownership of the resource and therefore significantly increases their willingness to invest and participate in initiatives to protect fishing grounds and fish resources. They know that even though they may not be able to fish in any one year, they will be able to fish in future years and reap the benefits. By changing the incentives, it is within fishermen's interests to help manage the resource for the long term.

By reducing the 'race to fish', management authorities may loosen controls over when to fish and where to fish so that fishermen can make more scientific arrangements that will subsequently benefit the fish supply chain and safety at sea as seen, for example, in the North Pacific halibut fishery (Tindall, 2012). The benefits seen in the transition to rights-based management differ from fishery to fishery (Table 14.2).

However, rights-based management is by no means a panacea and also has social and economic impacts. For example, Individual Transferable Quotas (ITQs) may marginalize the small-scale fishing sector. Moreover, ITQs may also concentrate fishing rights in the hands of a minority of individuals or companies who have the available capital to buy out their competitors' share (Beddington *et al.*, 2007). The price of quotas may increase dramatically over time, making it difficult for new entrants. This is a particular issue for communities that depend on fishing and have limited other employment opportunities. These impacts can be mitigated by tailoring the rights-based management system to the specific characteristics of the fishery. For instance,

Table 14.2 Benefits from rights-based management approaches

	Benefits
Economic	• Improving economic returns and efficiency of the fishery through better capacity management • Reducing fishing costs via removing the 'race for fish' factor • Increasing price by adjusting fishing in accordance with market demand
Social	• Clear rights ownership leading to clearer responsibility and collective incentives of all stakeholders • Improved safety at sea • Fish supply better matched with market demand • Benefits for future generations
Environmental	• Incentives to prevent overfishing • Incentives to protect environment such as reducing bycatch and discards and protecting habitats • Fishermen's willingness to support management for the long term

Source: Tindall (2012).

the North Pacific halibut fishery imposed an ownership cap as well as rules on leasing to ensure it essentially remained a fisher-owned enterprise (Tindall, 2012).

Integrating fisheries management into national development plans

Fisheries are a complex dynamic natural–human system, extending vertically from social–economic–natural ecosystems and horizontally from fish stocks to environmental issues (Ye, 2012). Different parts of a complex system are intertwined and interact to form collective consequences. Therefore, a successful fishery requires coordinated policy and actions in all other relevant sectors. Unfortunately, fisheries are frequently marginalized in the development of national policy and development plans and do not receive sufficient support and attention from other sectors, because in most countries fisheries form a small component of the national economy, 0.5–2.5 per cent of GDP, with the exception of a few small island nations.

A popular topic in many developing countries is the increase of fishery production in order to address national food security concerns. Governments encourage more vessels to become engaged in fishing. Such decisions are often not based on scientific advice and even contradict the need for reducing the excess capacity of fishing fleets. To avoid conflicting sectoral policies, it is necessary to mainstream fisheries into national plans for development, food security and poverty eradication in order to achieve a general appreciation of the value of natural resources and thereby the need for conservation and management, as well as the national coordination of management actions.

Decommissioning the excess fleet capacity requires 12–15 million people moving out of the fishery industry and a large lump investment sum of $96–358 billion in the event of implementation of buyback programmes (Ye *et al.*, 2013). Both the financial support and the creation of alternative employment opportunities can hardly be accomplished by the fishery sector alone, but demand coordinated efforts from different sectors and levels of government under national planning policy frameworks.

Illegal, unreported and unregulated (IUU) fishing is a global threat to sustainable fisheries and to the management and conservation of fisheries resources and marine biodiversity. It is currently estimated that IUU fishing practices are worth around $13 billion each year. IUU can take place in zones of national jurisdiction, within the areas of competence of regional fisheries bodies, and on the high seas. An FAO Conference approved the Agreement on Port State Measures to Prevent, Deter and Eliminate Illegal, Unreported and Unregulated Fishing (PSM) in 2009. However, implementing and enforcing PSM may require national legal framework and coordination that can enable port states to apply PSM to combat IUU fishing.

Capacity building

To achieve a sustainable fishery, it is necessary to collect data, carry out assessment and develop a management plan for the fishery, which specifies the goal and specific regulations to achieve that goal. Without stock assessment, people do not know the productivity of a fish stock, the present rate of fish removal, and the current stock status. What is not measured cannot be managed efficiently (Ye, 2012).

This is not to say that without formal stock assessment, it is not possible to manage fisheries. If the management system is sufficiently responsive and adaptive, it is possible to monitor some simple indicators such as catch rates and total landings, though the effectiveness and efficiency of management may not be comparable to an assessment-based management system. However, such a responsive and adaptive system often requires good infrastructure and an efficient regulatory mechanism, which is often lacking in fisheries that do not have the capacity for stock assessment and implementing regulatory measures. The costs of such non-assessment-based management may also prove high, as indicated by the estimated loss of US$50 billion a year of the world fisheries by the World Bank and FAO (2009).

Unfortunately, worldwide only about 20 per cent of global marine landings have sufficient data and have been quantitatively assessed and only a small fraction of fish stocks have sufficient data and have been quantitatively assessed (Branch *et al.*, 2011). Many fisheries have no management plan in place, especially in developing countries where the majority of fishing takes place. It is easy to come to the conclusion that without scientific management of all the world fisheries, the current gloom regarding the state of the world's fisheries will remain. There is an urgent need to provide transfer of technology and to build capacity for developing country fishing communities.

Sustainable fishery management requires a strong partnership among all stakeholders, fishers, managers, scientists, conservationists and other sectors such as marketing and the seafood industry. People are used to thinking inside the box of their own sector and consider only sectoral objectives. Such segmented approaches focus on only a few dots in the complex fishery system. A partnership approach can connect all the dots in the system, provide a comprehensive solution that balances all the objectives of different sectors, and form a collaborative working arrangement. As discussed above, the decommissioning of excess fishing capacity can be more easily achieved through a collective effort involving all sectors of society. However, this kind of partnership is new and capacity building is needed for all those involved.

Concluding remarks

The world's fisheries face serious problems: overfished stocks, excess fleet capacity, degradation of habitats and environment. As a result, food production potential has been lost, fishing fleets are running in the red, and biodiversity and functioning of marine ecosystems are compromised. Some of these consequences are temporary but others may have implications for many generations to come. It is necessary to act immediately to make changes towards sustainability and the benefits will be tangible.

The international community has launched a series of campaigns to end overfishing. The 2002 World Summit on Sustainable Development and the 2010 Conference of the Parties to the CBD set time-bound goals for fisheries, and the Rio+20 Conference in 2012 reiterated its commitment to achieve these goals. Why has so little obvious progress been made so far? The solutions to overfishing are known and well-documented, but rebuilding overfished stocks for sustainability is complicated by institutional weakness and short-term socio-economic consequences. A partnership approach of all stakeholders may unite different sectors into a whole society task force to tackle the problems that confront the world fisheries.

A global agenda for fisheries must also realize that fisheries in different regions have unique economic and institutional barriers to sustainable fishing. The appropriate policies best able to achieve the goal can vary widely. There is a need for countries to build practical policy- and decision-making capacity for sustainable management of fisheries. Capacity building efforts should typically work in service to developing countries as well as to many fisheries in the developed world. Today, the fisheries that do not have a management plan are still the majority. Without bringing these under management, sustainability of the world fisheries will be difficult to achieve.

References

Beddington, J. R., Agnew, D. J. and Clark, C. D. (2007) 'Current problems in the management of marine fisheries', *Science*, vol. 316, pp. 1713–1716.

Branch, T. A., Jensen, O., Ricard, E., Ye, Y. and Hilborn, R. (2011). 'Contrasting global trends in marine fishery status obtained from catches and from stock assessments', *Conservation Biology*, vol. 25, pp. 777–786.

FAO. (1995a) 'The Rome consensus on world fisheries', www.fao.org/docrep/006/AC441E/AC44 1E00.HTM, accessed 30 August 2013.

FAO. (1995b) *Code of Conduct for Responsible Fisheries*. Rome: Food and Agriculture Organization.

FAO. (1995c) 'Kyoto declaration and plan of action', www.fao.org/docrep/006/ac442e/AC442e3.htm, accessed 30 August 2013.

FAO. (1996) 'World food summit', www.fao.org/wfs/index_en.htm, accessed 30 August 2013.

FAO. (1999) 'International plan of action for the management of fishing capacity', www.fao.org/ docrep/006/x3170e/x3170e04.htm, accessed 30 August 2013.

FAO. (2007) *Fish and Fishery Products: World apparent consumption statistics based on food balance sheets*. FAO Fishery Circular No.821. Rome: Food and Agriculture Organization.

FAO. (2010) *The State of Food Insecurity in the World: Addressing food insecurity in protracted crises*. Rome: Food and Agriculture Organization.

FAO. (2012) *The State of World Fisheries and Aquaculture 2012*. Rome: Food and Agriculture Organization.

Garcia, S. M. and Rosenberg, A. A. (2010) Food security and marine capture fisheries: characteristics, trends, drivers and future perspectives. *Philosophical Transactions of the Royal Society B*, vol. 365, pp. 2869–2880.

OECD. (2009) *Reducing Fishing Capacity: Best practices for decommissioning schemes*. Paris: OECD Publishing.

Scott, A. (2000) 'Introducing property in fishery management', in R. Shotton (eds), *Use of Property Rights in Fisheries Management*, FAO Fisheries and Technical Paper 404/1, Rome: Food and Agriculture Organization.

Tacon, A. G. J. and Metian, M. (2013) 'Fish matters: importance of aquatic foods in human nutrition and global food supply', *Reviews in Fisheries Science*, vol. 21, pp. 22–38.

The World Bank and FAO. (2009) *The Sunken Billions*. Washington DC: World Bank.

Tindall, C. (2012) 'Fisheries in Transition: 50 Interviews with the Fishing Sector', http://pcfisu.org/wp-content/uploads/2012/01/TPC1224-Princes-Charities-case-studies-report_WEB-02.02.pdf, accessed 30 August 2013.

UN (United Nations). (1982) *The Law of the Sea*. New York: The United Nations.

UN. (1995) 'The United Nations fish stocks agreement', www.un.org/depts/los/convention_agreements/ convention_overview_fish_stocks.htm, accessed 30 August 2013.

UN. (2002) 'World summit on sustainable development', Johannesburg, South Africa 26 August–4 September 2002. United Nations A/CONF.199/L.6/Rev.2 www.rrcap.unep.org/wssd/Political%20 declaration_4%20Sep%2002.pdf, accessed 30 August 2013.

UN. (2005) '2005 World Summit outcome document', www.who.int/hiv/universalaccess2010/world summit.pdf, accessed 30 August 2013.

UN. (2012) 'The future we want: The outcome document adopted at the Rio+20 Conference', www.un.org/en/sustainablefuture, accessed 30 August 2013.

UN. (2013) *The Millennium Development Goals Report 2013*. New York: United Nations.

UNEP (The United Nations Environmental Program). (2010). *Report of the 10th Meeting of the Conference of the Parties to the Convention of Biological Diversity*. New York: UNEP.

WECD (World Commission on Environment and Development of the United Nations). (1987) 'Our common future', www.un-documents.net/our-common-future.pdf, accessed 30 August 2013.

Ye, Y. (2012) 'World fisheries: fish stock status, sustainability goal and future perspectives', in K. Soeters (eds), *Sea the Truth: Essays on overfishing, pollution and climate change*. The Netherlands: Nicolaas G. Pierson Foundation, pp. 119–132.

Ye, Y. and Cochrane, K. (2011) 'Global review of marine fishery resources', in FAO, *Review of the State of World Marine Fishery Resources*, FAO Fisheries and Aquaculture Technical Paper 569. Rome: Food and Agriculture Organization.

Ye, Y., Cochrane, K., Bianchi, G., Willmann, R., Majkowski, J., Tandstad, M., Carocci, F. (2013) 'Rebuilding global fisheries: the World Summit Goal, costs and benefits', *Fish and Fisheries*, vol. 14 (2), pp. 174–185.

15

THE HIGH SEAS AND IUU FISHING

*Henrik Österblom, Örjan Bodin, Anthony J. Press
and U. Rashid Sumaila*

The high seas and global trends in fisheries

The high seas are the areas beyond national jurisdiction, offshore from the productive shelf areas, where most fisheries are operating (Figure 15.1). However, decreasing fish stocks in many coastal areas (Jackson *et al.* 2001, Christensen *et al.* 2003) have historically pushed fisheries operations further offshore and to new species. In the 1950s, only 9 per cent of global catches were taken from the high seas, whereas the corresponding number in 2010 was 12 per cent (Sumaila *et al.* 2015). Global and regional patterns of fishing are characterized by catches at increasing depths (Pauly *et al.* 2003, Morato *et al.* 2006, Villasante *et al.* 2012) and further from markets (Swartz *et al.* 2010). The deep sea beyond the continental shelf edges and away from individual nations' jurisdiction is thus becoming an increasingly important area for wild capture fisheries. Many of the species caught in these areas are long-lived, start reproducing late and have limited capacity to sustain commercial catches (e.g. Norse *et al.* 2012). It is also not uncommon that these species aggregate around deep-sea features such as hydrothermal vents or deep-sea corals, with high or unknown biodiversity values making them vulnerable to overfishing.

Although a large number of fisheries are licensed to operate in the high seas and have legitimate quotas, monitoring and enforcement is a problem for much of high seas fisheries (Norse *et al.* 2012). Compounding this problem is the problem of flags of convenience, the use of which is more prevalent on the high seas (Miller and Sumaila 2014).

Challenges to high seas governance

Building on earlier work on fisheries governance, Sumaila (2012) identified a number of reasons why oceans and fisheries governance is challenging. By the very nature of the high seas, the problems identified in that research become even more acute.

A key challenge is illegal, unregulated, unreported (IUU) fishing[1] that occurs in the high seas (and within exclusive economic zones (EEZs), see further down) because this part of the global ocean is, in contrast to the coastal zones, relatively poorly regulated (Sumaila *et al.*, 2006). IUU fishing leads to a failure to achieve both management goals and sustainability of fisheries

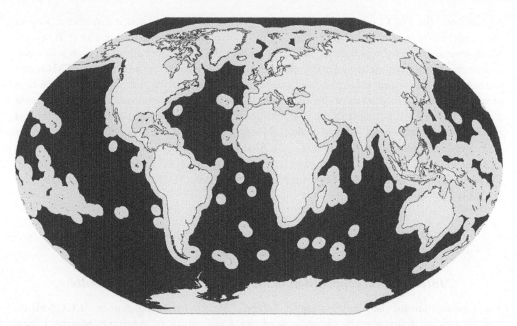

Figure 15.1 Global map of the world's oceanic fishing areas. Light and dark areas indicate the locations of coastal exclusive economic zones (EEZs) and the high seas, respectively
Source: Sumaila *et al.* (2007).

(Pitcher *et al.* 2002). These reasons make it important to eliminate IUU fishing, which is known to be a widespread activity worldwide (Sumaila *et al.* 2006, fig. 1; Agnew *et al.* 2009).

A second challenge is that the common property and/or open access nature of fishery resources, which was identified as a generic challenge to successful resource management decades ago (Gordon 1954; Hardin 1968), is a bigger problem in the high seas than in coastal waters, as ownership is essentially shared by all citizens of the world (UN 1982).

The third challenge relates to the provision of subsidies to the fishing sector which is a key driver of over-capacity and overfishing (Milazzo, 1998). Again, the effect of subsidies on overfishing is likely to be higher with respect to high seas fishing because, all else being equal, it will have a bigger effect on the profitability of fishing companies (Sumaila *et al.*, 2010).

A fourth challenge stems from recent technological progress, which many have argued has been instrumental for the global expansion of fishing towards the high seas. The development of flash freezing technology, for example, has directly contributed to the observed patterns. During the post-war period, the world has also seen a rapid globalization of fish markets (Swartz *et al.* 2010) combined with industry consolidation and increased capital investments in high seas fishing fleets, driven by global commercial actors. It is worth noting that many technological developments (e.g. satellites) have improved the capacity for monitoring and enforcement on the high seas.

Governance and policy frameworks for the high seas

Coastal states can declare an EEZ out to 200 nautical miles, within which they have exclusive right to regulate their fisheries. Much of the high seas areas beyond national jurisdiction (200 nm) are covered in global policy frameworks, including UNFSA (United Nations Fish Stocks

Agreement), the FAO Port state agreement, IPOA-IUU, the Compliance agreement. Implementation of these agreements is the responsibility of member states collaborating in Regional Management Fishery Organizations (RFMOs). These RFMOs cover all of the high seas areas, and develop individual policies for the species and fisheries under their mandate. RFMOs and the member states within them also have a responsibility for ensuring that vessels flying the flags of member states, or nationals of member states flying flags of convenience, comply with these policies. However, an analysis of the performance of RFMOs suggests that many have limited regulation and enforcement (Cullis-Suzuki and Pauly 2010). In the rest of the chapter, we describe an example of high seas management (i.e. CCAMLR, an international institution for managing fisheries in the Southern Ocean) that has been relatively successful in reducing IUU fishing in the high seas. We focus in particular on the factors that made CCAMLR successful where other RFMOs have failed.

CCAMLR – an example of successful reduction of IUU fishing in the high seas

International cooperation in the Southern Ocean yields results

The Commission for the Conservation of Antarctic Marine Living Resources, or CCAMLR, was established following the agreement of the Convention of Antarctic Marine Living Resources, in 1983. Its mandate is to manage finfish and krill stocks in the Southern Ocean, around Antarctica, using an ecosystem approach. The area within which the Commission has responsibility includes vast high seas areas around the Antarctic continent, as well as a number of areas within the national jurisdiction of some member states with Sub-Antarctic islands and associated EEZs[2] (Figure 15.2). The Commission includes some twenty member states that meet annually to negotiate new and revised policy measures, review compliance and take decisions about fishing quotas, based on scientific advice (Constable *et al.* 2000, Miller *et al.* 2010).

As a consequence of depleted fish stocks in other regions, substantial fishing efforts displaced to the Southern Ocean during the 1990s. This overcapacity from elsewhere included, e.g. tuna vessels from adjacent RFMOs. When toothfish fisheries developed in the 1990s, CCAMLR had still to develop important policy measures to regulate this fishery. The international markets for Patagonian toothfish (aka Chilean Sea Bass) developed rapidly, however (e.g. in the USA), and the high price of this fish in combination with low likelihood of detection contributed to a rapid increase in IUU fishing. In the mid 1990s, it became evident that this was a critical challenge for CCAMLR. The scientific committee of CCAMLR concluded that this fishery risked not only leading to the substantial reduction of toothfish stocks, but also to the likely collapse of globally threatened seabird stocks, caught on the baited hooks set to catch toothfish.

However, as a consequence of extensive diplomatic pressure directed at flag and port states associated with IUU fishing, combined with new policy tools (including vessel blacklists), and novel forms of collaboration, IUU fishing has been substantially reduced (see Figure 15.2 – Österblom et al. 2010, 2011, Österblom and Sumaila 2011, Österblom and Bodin 2012, Bodin and Österblom 2013, Österblom and Folke 2013). Many countries in CCAMLR, including for example Great Britain, Australia, France and New Zealand, have devoted substantial resources to monitoring and international cooperation. No new ships have been discovered during the past few years. This reduction of IUU fishing has enabled CCAMLR to raise the levels of licensed quotas (Österblom and Sumaila 2011) and there are now signs that albatross populations (previously caught as bycatch on baited longlines used by IUU operators) are now beginning to recover from illegal and unsustainable fishing (Robertson *et al.* 2014). Detecting

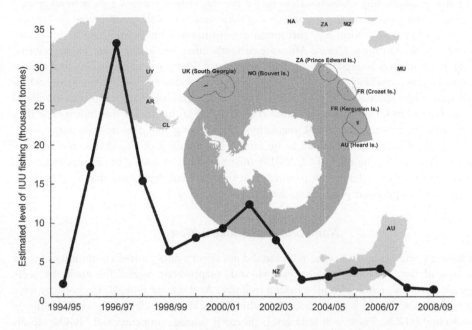

Figure 15.2 A map of the CCAMLR area (shaded), including relevant Sub-Antarctic Islands and adjacent EEZs (or fisheries zones) and the official estimates of IUU fishing in the CCAMLR area
Source: modified from Österblom and Sumaila 2011, Österblom and Folke 2013.

vessels, uncovering information on illegal landings or transport, investigating complex networks of ownerships and securing convicting sentences for globally operating and adaptive illegal fishing operators has required extensive multinational cooperation. The results from CCAMLR show convincingly that there is great potential in international cooperation for marine resources (Österblom and Folke 2013). Although there are still many unresolved questions – for example where will the illegal fishing operators appear next – it is clear that a number of lessons can be drawn from the CCAMLR 'solution' to illegal fishing. Here we present how international political pressure, the inclusion of non-state actors, problem reconceptualization, and the development of tools that facilitate collaboration have contributed to the successful reduction of the IUU fishing in the Southern Ocean.

Political pressure to clean up your own backyard

During the first years of high levels of IUU fishing, most identified IUU vessels were flagged to CCAMLR member states. The shipowners who were pioneering illegal fishing in the region first used mainly Chile and Argentina as flag states. Since these countries are members of CCAMLR, it created a delicate diplomatic situation when it became known that they were associated with IUU activities. These and other countries within CCAMLR that were used as flag states, however, mostly had a good capacity to react when it was discovered that 'their' ships were involved in irregularities, and were therefore able to take appropriate action. Hence, these countries took rapid steps to deregister the identified vessels. Several vessels received substantial fines and some were even seized and scrapped. Owners of vessels that managed to escape enforcement, however, developed increasingly sophisticated techniques to minimize the risk of penalties, including continuous change of the flag state – there is a clear trend among many illegal fishing operators, over time, to flag their vessels to states with lower governance capacity where problems with, e.g. corruption are common (Österblom *et al.* 2010). North Korea and several countries in Central Africa are currently often used as flag states. These countries probably have limited interest and capacity related to fisheries regulations in very distant regions. IUU vessels operating in the CCAMLR are also continuously changing the name, colour and call sign. Over the past fifteen years there has been a gradual change and now no illegal vessels are flagged to CCAMLR (Österblom *et al.* 2010). This has meant that in principle they are also not subject to the CCAMLR regulations. The fishing activities are thus unregulated, rather than illegal, but still considered to be serious by members of CCAMLR and result in blacklisting. However, a number of CCAMLR member states, including the European Union, have also developed complementary national (and international) legislation that is focusing on improving the prospects for prosecuting their nationals (Erceg 2006).

Non-state actors take action

Illegal fishing operators in the Southern Ocean did not affect a disorganized community of weak voices. Instead, they were affecting very valuable and untapped commercial fish stocks that were of large interest to a growing licensed fishing industry. At the same time, IUU operators were killing thousands of charismatic seabirds, which quickly mobilized the community of environmental NGOs. These non-state actors (licensed fishing companies and NGOs) rapidly formed an alliance as they had a shared and strong interest in reducing IUU fishing. Governments and non-state actors, or environmental NGOs and the fishing industry do not always share the same agenda when it comes to practical management of natural resources, but in the case of CCAMLR, they did. States representing emerging licensed industries were quick to support the development of policy measures directed at reducing IUU fishing, but due to consensus mechanisms in CCAMLR, this turned out to be a slow process. Initial attempts were filled with loopholes, which were only corrected several years after the implementation of some policy measures (Miller *et al.* 2010). The US, one of the major markets for toothfish, enforced a unilateral ban on the import of toothfish products from some areas as an emergency measure – although this was not perceived as the way forward for CCAMLR as a forum for international collaboration building on consensus. Initially, the political will was not sufficient to mobilize a strong commitment for developing and enforcing effective policies. However, the licensed fishing industry and the NGO community were able to engage in a different way than CCAMLR member states. For instance, they conducted their own investigations, carried out their own monitoring in the high seas and used this information for 'naming and shaming' suspected offenders. They went public with controversial reports that implicated countries (including CCAMLR member states), but also individual fishing companies and banks implicated as associated with IUU fishing. These reports used information derived from the NGO community and the fishing industry, who were conducting investigations in ways that governments were unable to do. The publication of such material raised the level of public awareness in several countries and hence also the political awareness within CCAMLR (Fallon and Kriwoken 2004, Baird 2006). Over time, member states also increasingly included non-state actors in their national delegations, as they could contribute with complementary functions. There has been an increasing number of NGOs and fishing industry representatives in the Commission, as well as an increasing diversity in national delegations (Österblom and Sumaila 2011). Reasons for the very active engagement by these non-state actors include the high conservation value of the region, combined with the very high value of the toothfish stocks. The fishing industry is relatively consolidated, with only a small number of fishing companies engaged in the licensed fishery from each country. There is thus a very clearly defined system of property rights for the industry, which consequently has a clear stake in reducing IUU fishing – lower levels of IUU fishing mean higher levels of licensed quotas (Österblom and Sumaila 2011).

Conceptualizing the problem in an attractive narrative

Successful policy entrepreneurs (cf. Kingdon 1984) often use narratives to conceptualize problems that need to be addressed (Huitema and Meijerink 2010). One powerful narrative that appears to have changed the perception of many organizations involved in reducing IUU fishing is that of the problem as a form of organized international crime (Österblom *et al.* 2011). This is a powerful analogy and one that appears highly relevant for this fishery (Österblom *et al.* 2011). For instance, both organized crime and this fishery are carried out by loosely organized and adaptive networks, both without any central authority or form of control. Both activities use bribes to intimidate members of the networks and use shell companies to hide the beneficial owners (Griggs and Lugten 2007, Österblom *et al.* 2011). A 'fight' against both types of problems often involves spending large resources on identifying and investigating potential major bosses, while the problems often are complex and involve a diversity of actors operating in loose networks, often without centralized control. Organized crime and illegal fishing in countries' economic zones both represent a new form of security threat – and are both perceived as threatening the sovereign rights of affected states; they are subject to a new form of non-traditional security threat that requires other than traditional military means. Conceptualizing IUU fishing as a form of organized crime probably contributed substantially to generating important political will, within CCAMLR, to protect national borders against perceived foreign perpetrators (Österblom *et al.* 2011).

Facilitate collaboration

Several member states were constrained by their limited jurisdiction in the high seas. Australia, for instance had limited ability to engage in fisheries enforcement outside of their EEZs around the Sub-Antarctic Islands of Heard and McDonald Islands. Effective enforcement in the high seas required well-developed policy measures and international collaboration. Initial policy attempts were filled with loopholes that were easy to circumvent by IUU operators and IUU fishing remained a key challenge for CCAMLR during most of the early 2000s. CCAMLR has subsequently developed more effective policy measures, however, that has contributed substantially to the current and much reduced levels of IUU fishing (Österblom and Sumaila 2011). These tools were often developed as a response to crisis situations within CCAMLR triggered by IUU fishing (Österblom and Sumaila 2011) and include, e.g., an electronic catch documentation scheme (e-CDS, see Agnew 2000) that enables tracking toothfish products from vessels to markets and a black list of all vessels detected to be operating outside the agreed framework. The catch documentation scheme and vessel IUU list have been identified as critical resources for CCAMLR (Österblom and Bodin 2012). Governments, but also NGOs and the licensed fishing industry can report suspected activities and thereby contribute to the monitoring of compliance – at sea and in the market. The CCAMLR secretariat has been empowered with these tools and functions as an important coordinating node in the network by facilitating collaborating (Österblom and Bodin 2012). There is frequent cooperation between and within countries, and between governments and non-state actors – the diversity of actors contributes with important complementary components (Bodin and Österblom 2013). These different organizations have different agendas and priorities – but they also have a common interest in reducing IUU fishing. This common interest, in combination with the tools developed by CCAMLR and the coordinating function provided by the secretariat has made it easier for a diverse set of organizations to collaborate at the global level for sustainable management of Antarctic marine living resources.

An important aspect for the success in CCAMLR is also the substantial investments made in surveillance capacity. Australia and France have invested in a shared system of satellite monitoring over the French Kerguelen Island and the Australian Heard and McDonald Islands. Both countries also have substantial enforcement capacities. For instance, Australia has constructed a purpose built monitoring and enforcement vessel that operates more than 200 days per year in the region. France has substantial monitoring and enforcement capacity, in part by using a converted (and previously confiscated) IUU vessel whose operations are in part funded by the licensed French toothfish industry. France also has substantial military capacity stationed at Reunion, at close range to the French Sub-Antarctic Islands. Australia and France carry out joint training and enforcement in the region (Österblom and Sumaila 2011), while Great Britain and New Zealand have substantial monitoring and enforcement capacities that are partially deployed to the region (including aerial surveillance). This strong and increasing presence of CCAMLR member states in the high seas is producing an important deterrent for IUU operators.

Conclusion

Fishing has been expanding into the high seas in recent decades mainly because of overfishing in coastal waters and the increasing demand for fish due to increasing populations and rising incomes in many parts of the world. We have identified and presented a number of reasons why high seas fisheries governance is challenging. We illustrated, using CCMLR as an example, that it is possible to be successful in governing high seas fishing, given the right conditions. The experiences from CCAMLR are relevant for other regions in the high seas, but also at the global level. For instance, a number of RFMOs are increasingly using tools that have been critical for the success of CCAMLR, including vessel blacklists and catch documentation schemes. There are also some indications that policy makers and practitioners are starting to address the global challenge of IUU fishing in a more consistent matter (Österblom 2014), but the four challenges addressed initially still remain as key for marine sustainability in many areas of the high seas.

Notes

1   Illegal fishing is conducted by national or foreign vessels in waters under the jurisdiction of a state without its permission, or in contravention of its laws and regulations. Unregulated fishing occurs within the high seas under the management jurisdiction of RFMO, by a flagless vessel, or vessel flying the flag of a state or entity not a party to the RFMO, in a manner inconsistent with the conservation and management measures of the RFMO. Unreported Fishing is defined as fishing that has not been reported, or has been misreported, to the relevant national authority or to the relevant RFMO.
2   It should be noted that in addition to EEZ there are also extended continental shelf claims.

References

Agnew, D. J. 2000. The illegal and unregulated fishery for toothfish in the Southern Ocean, and the CCAMLR catch documentation scheme. *Marine Policy* 24: 361–374.
Agnew, D. J., Pearce, J., Pramod, G., Peatman, T., Watson, R., Beddington, J. R. and Pitcher, T. J. 2009. Estimating the worldwide extent of illegal fishing. *PloS ONE* 4: e4570.

Baird, R. J. 2006. *Aspects of Illegal, Unreported and Unregulated Fishing in the Southern Ocean*. Springer, Dordrecht.

Bodin, Ö. and Österblom, H. 2013. International fisheries regime effectiveness: Activities and resources of key actors in the Southern Ocean. *Global Environmental Change* 23: 948–956.

Christensen, V., Guénette, S., Heymans, J. J., Walters, C. J., Watson, R. Zeller D. and Pauly, D. 2003. Hundred year decline of North Atlantic predatory fishes. *Fish and Fisheries* 4(1): 1–24.

Constable, A. J., de la Mare, W. K., Agnew, D. J., Everson, I. and Miller, D. 2000. Managing fisheries to conserve the Antarctic marine ecosystem: practical implementation of the Convention on the Conservation of Antarctic Marine Living Resources (CCAMLR). *ICES Journal of Marine Science* 57: 778–791.

Cullis-Suzuki, S. and Pauly, D. 2010. Failing the high seas: a global evaluation of regional fisheries management organizations. *Marine Policy* 34: 1036–1042.

Erceg, D. 2006. Deterring IUU fishing through state control over nationals. *Marine Policy* 30: 173–179.

Fallon, L. and Kriwoken, L. 2004. International influence of an Australian nongovernment organization in the protection of Patagonian toothfish. *Ocean Development and International Law* 35: 221–266.

Gordon, H. S. 1954. The economic theory of a common property resource: the fishery. *Journal of Political Economy* 62: 124–42.

Griggs, L. and Lugten, G. 2007. Veil over the nets: unravelling corporate liability for IUU fishing offences. *Marine Policy* 31: 159–168.

Hardin, G. 1968. The tragedy of the commons. *Science* 162: 1243–1248.

Huitema, D. and Meijerink, S. 2010. Realizing water transitions: the role of policy entrepreneurs in water policy change. *Ecology and Society* 15(2): 26. Available at: www.ecologyandsociety.org/vol15/iss2/art26/ (accessed 6 June 2015).

Jackson, J. B. C., Kirby, M. X., Berger, W. H., Bjorndal, K. A., Botsford, L. W., Bourque, B. J., Bradbury, R. H., Cooke, R., Erlandson, J., Estes, J. A., Hughes, T. P., Kidwell, S., Lange, C. B., Lenihan, H. S., Pandolfi, J. M., Peterson, C. H., Steneck, R. S., Tegner, M. J. and Warner, R. R. 2001. Historical overfishing and the recent collapse of coastal ecosystems. *Science* 293: 629–637.

Kingdon, J. W. 1984. *Agendas, Alternatives and Public Policies*. HarperCollins College Publishers, New York.

Milazzo, M. 1998. Subsidies in World Fisheries: A Re-examination. World Bank Technical Paper No. 406, World Bank, Washington, p. 86.

Miller, D. D. and Sumaila, U. R. 2014. Flag use behavior and IUU activity within the international fishing fleet: refining definitions and identifying areas of concern. *Marine Policy* 44: 204–211.

Miller, D. G. M., Slicer, N. and Sabourenkov, E. 2010. IUU fishing in Antarctic Waters: Actions and Regulations by CCAMLR, in D. Vidas (ed.) *Law, Technology and Science for Oceans in Globalisation*. Brill Academic Publishers, Leiden, pp. 175–196.

Morato, T., Watson, R., Pitcher, T. J. and Pauly, D. 2006. Fishing down the deep. *Fish and Fisheries* 7: 24–34.

Norse, E. A., Brooke, S., Cheung, W. W. L., Clark, M. R., Ekeland, I., Froese, F., Gjerde, K. M., Haedrich, R. L., Heppell, S. S., Morato, T., Morgan, L. E., Pauly, D., Sumaila, U. R. and Watson, R. 2012. Sustainability of deep-sea fisheries. *Marine Policy* 36: 307–320.

Österblom, H. 2014. Catching up on fisheries crime. *Conservation Biology* 28: 877–879.

Österblom, H. and Bodin, Ö. 2012. Global cooperation among diverse organizations to reduce illegal fishing in the Southern Ocean. *Conservation Biology* 26: 638–648.

Österblom, H. and Folke, C. 2013. Emergence of global adaptive governance for stewardship of regional marine resources. *Ecology and Society* 18(2): 4.

Österblom, H. and Sumaila, U. R. 2011. Toothfish crises, actor diversity and the emergence of compliance mechanisms in the Southern Ocean. *Global Environmental Change* 21: 972–982.

Österblom, H., Constable, A. and Fukumi, S. 2011. Illegal fishing and the organized crime analogy. *Trends in Ecology and Evolution* 26: 261–262.

Österblom, H., Sumaila, U. R., Bodin, Ö., Hentati Sundberg, J. and Press, A. J. 2010. Adapting to regional enforcement: fishing down the governance index. *PLoS ONE* 5: e12832.

Pauly, D., Alder, J., Bennett, E., Christensen, V., Tyedmers, P. and Watson, R. 2003. The future for fisheries. *Science* 302: 1359–1361.

Pitcher, T., Watson, R., Forrest, R., Valtýsson, H. P., Guénette, S. 2002. Estimating illegal and unreported catches from marine ecosystems: a basis for change. *Fish and Fisheries* 3: 317–339.

Robertson, G., Moreno, C., Arata, J. A., Candy, S. G., Lawton, K., Valencia, J., Wienecke, B., Kirkwood, R., Taylor, P., Suazo, C. G. 2014. Black-browed albatross numbers in Chile increase in response to reduced mortality in fisheries. *Biological Conservation* 169: 319–333.

Sumaila, U. R. 2012. Seas, oceans and fisheries: a challenge for good governance. *Round Table* 101: 157–166.

Sumaila, U. R., Alder, J. and Keith, H. 2006. Global scope and economics of illegal fishing. *Marine Policy* 30(6): 696–703. Available at: www.seaaroundus.org/journal/2006/GlobalScopeEconomicsIllegal Fishing.pdf (accessed 6 June 2015).

Sumaila, U. R., Zeller, D., Watson, R., Alder, J. and Pauly, D. 2007. Potential costs and benefits of marine reserves in the high seas. *Marine Ecology Progress Series* 345: 305–310.

Sumaila, U. R., Khan, A., Teh, L., Watson, R., Tyedmers, P., Pauly, D. 2010. Subsidies to high seas bottom trawl fleets and the sustainability of deep-sea demersal fish stocks. *Marine Policy* 34: 495–497.

Sumaila, U. R, Lam, V., Miller, D. D., Teh, L., Cheung, W. W. L., Watson, R., Pauly, D., Zeller, D., Rogers, A. D., Côté, I. M., Roberts, C. 2015. Winners and losers in a world where the high seas is closed to fishing. *Scientific Reports* 5: Article number 8481.

Swartz, W., Sala, E., Tracey, S., Watson, R. and Pauly, D. 2010. The spatial expansion and ecological footprint of fisheries (1950 to present). *PLoS One* 2010 5: e15143.

United Nations (1982). United Nations Convention on the Law of the Sea. 1–202. Division for Ocean Affairs and the Law of the Sea, Office of Legal Affairs, United Nations. Available at: http://www.un.org/depts/los/convention_agreements/texts/unclos/UNCLOS-TOC.htm (accessed 6 June 2015).

Villasante, S., Morato, T., Rodriguez-Gonzalez, D., Antelo, M., Österblom, H., Watling, L., Nouvian, C., Gianni, M. and Macho, G. 2012. Sustainability of deep-sea fish species under the European Union Common Fisheries Policy. *Ocean and Coastal Management* 70, 31–37.

16

RETHINKING SMALL-SCALE FISHERIES GOVERNANCE

Ratana Chuenpagdee and Svein Jentoft

Introduction

Despite the general recognition of their social, cultural and economic importance, small-scale fisheries are still ignored or dismissed as relics of the past. In many countries, this marginalization is shown by inadequate financial, institutional and scientific support for small-scale fisheries (Béné and Friend, 2011; Teh *et al.*, 2011). Further, management decisions, especially regarding rules and regulations, are not sensitive to the conditions and needs of this sector. They do not reflect the roles and contributions that small-scale fishing people make to the society in terms of income generation, food security, poverty alleviation and fisheries sustainability. They also do not respect traditional user and access rights of small-scale fishing people. On the contrary, fisheries management tends to view the small-scale fishing sector as a problem that needs to be solved. For instance, when faced with resource degradation and overexploitation issues, one common solution has been to reduce fishing capacity, which often implies cutting down on the large number of small boats. On the other hand, gear and technological enhancement aimed at improving the fishing efficiency of large-scale fishing operations may be framed as a good thing, when argued from a food provision perspective. But from the justice point of view there are many other factors that need to be considered in the decision about what policy interventions are most suited to address the current and emerging fisheries challenges.

Four key concerns have been identified in fisheries governance, namely, ecosystem health, social justice, livelihoods and food security (Chuenpagdee *et al.*, 2005; Bavinck *et al.*, 2013). They are all related to issues and challenges found in many fisheries around the world, such as unsustainable resource exploitation, poor recognition of fishing rights, poverty, lack of alternative employment and distortion in markets and supply chains (Jentoft and Eide, 2011). These concerns are exacerbated by current global changes in the environment, economics, demography and politics. Despite major efforts expended to address these challenges, fisheries resources continue to dwindle, conflicts continue to rise, poverty persists and livelihood displacement is widespread. For people interested in fisheries, whether from a research, governance or human rights perspective, the challenge is daunting.

It has been suggested that dealing with major concerns in fisheries requires a perspective that explicitly differentiates small-scale from large-scale fisheries (Jentoft and Chuenpagdee, 2009;

Khan and Neis, 2010; Berkes, 2012). After all, very little attention has been paid to the fact that the overwhelming majority of people involved in fisheries are small scale. Alternative governance thinking has been proposed (see for example, Gray, 2005; Kooiman *et al.*, 2005). Transdisciplinary and interdisciplinary research, as well as stakeholder participation and local knowledge integration have been promoted. Still, tools and measures employed in fisheries management nowadays appear to have seen little innovation. Even recent ideas, such as catch shares for resource allocation (Costello *et al.*, 2008) or Marine Stewardship Council seafood certification (Ponte, 2008), are still based on a thinking that considers fisheries as dominated mostly by the large-scale industrial and commercial sector.

We argue in this chapter that any discussion about fisheries should begin with the recognition that: (1) fisheries differ with, but also beyond, scale, and (2) fisheries governance needs therefore to correspond with the specificities of fisheries in order to address the key concerns. This implies that the particular context of fisheries also must be taken into account. For instance, while it may matter less for large-scale fisheries what goes on in fishing communities, the governance of small-scale fisheries must be attuned to their needs. It follows thus that a policy directed at alleviating poverty in small-scale fisheries must also include community welfare services such as health and education. Further, commonly used instruments such as marine protected areas and individual transferrable quotas are not similarly applicable in large- and small-scale fisheries. Rather, a broader range of approaches and tools, including those generated from within communities based on local experience and knowledge, should be considered.

In the following, we present aspects of small-scale fisheries that differentiate them from their large-scale counterparts. We discuss important characteristics and considerations for small-scale fisheries that can help make them viable while serving both their communities and the concerns of the society at large. We conclude with a discussion of relevant governance principles for small-scale fisheries that if implemented could facilitate better governance and sustainable livelihoods.

Fisheries differ beyond scale

At one stage, all fisheries were operated using traditional, small-sized gears in near shore areas. Fishing was an integral part of the livelihood portfolio of coastal communities, along with farming and other food production systems. The 'great fish race' (Butcher, 2004), an era of ocean industrialization (Smith, 2000), and the 'blue revolution' (Bailey, 1988; Bavinck, 2011) in the late nineteenth and early twentieth century all contributed to a paradigm shift in the industry and subsequent changes to the sea- and landscapes. Large, engine-powered vessels, with mechanized towing gears, enhanced storage capacity and refrigeration were introduced and operated alongside the small boats, as well as further from shore. Major investment in infrastructure such as harbours and roads, improved processing facilities and international market expansion were part of the development. For many countries, fish became an important export commodity, bringing in foreign currency and generating national growth. Such expansion resulted in a fivefold increase in the world fisheries production (marine/inland capture and culture) in less than fifty years, from about 20 million tonnes in 1950s to about 90 million tonnes in 1995. Production has continued to grow but at a much slower rate (about 1 per cent annually from 2002 to 2008), with the estimate of 154 million tonnes for 2012 (FAO, 2009a, 2010, 2012a). This increase was accompanied, however, by a shift from fisheries taking place mostly in developed countries to developing and less developing countries, where small-scale fisheries dominate (Chuenpagdee *et al.*, 2006). Today, about 80 per cent of the world's fisheries production comes from developing countries, with the value surpassing 50 per cent (FAO, 2009a).

The changing nature of fisheries, and the consequences for fishing communities, is not trivial. With 20 per cent of the world's population relying on fish as the main source of protein, and at least 135 million people depending on fishing for their livelihoods (FAO, 2009a), it is imperative to have not only a whole portrait of the fisheries, but also an ability to differentiate the multidimensional characteristics of fisheries. The term 'fishery' when used without a qualifier can lead to different interpretations. For instance, what type of fisheries 'are in crisis' (McGoodwin, 1990; Clark, 2006) – marine, inland or brackish, or capture or culture? Similarly, what fisheries 'rhyme with poverty' (Béné, 2003) – small, medium or large-scale? The diversity, complexity, dynamics and scale issues associated with fisheries, which give rise to concerns and challenges in governance, are different depending on the type and nature of fisheries (Kooiman *et al.*, 2005; Jentoft and Chuenpagdee, 2009). This implies that small-scale fisheries cannot be governed as if they are a miniature version of large-scale fisheries or a small business sector that is waiting to grow. Neither should the goal of governance necessarily be to expand and develop all fisheries into large scale. Also in fisheries, the 'small is beautiful' doctrine has merit in ecological as well as in social and economic terms. From a local food security perspective, the role of small-scale fisheries should not be undermined or underestimated, but in reality this has happened. Fisheries development promoted in most countries tends to encourage modernization and industrialization of fisheries sectors and professionalization of fishers, replacing the occupational pluralism that often characterized small-scale fisheries with one that emphasizes specialization (Bavinck, 2011). Those who are not able to adapt, are unwilling to change, or unable to demonstrate financial contribution to the country are marginalized, as a consequence (Cadigan, 2009). This has led to declining fishing populations and displacement in developed countries, as seen in Spain (Pascual-Fernández and De la Cruz Modino, 2011), Newfoundland (Walsh, 2011), Iceland (Einarsson, 2011) and Norway (Jentoft, 2013).

Making a distinction between large and small is not as straightforward as it seems. Physical characteristics of the fishing operation are the first noted aspects since they are the most observable. Based on a global review, Chuenpagdee *et al.* (2006) conclude that most countries define small-scale fisheries using features such as fishing methods, gears employed, boat type and length, engine size, number of crew, fishing location (distance from shore) and time spent at sea. Pauly (2006) and Jacquet and Pauly (2008) add other attributes such as level of fuel consumption, amount of discards and utilization of catch to distinguish small-scale from large-scale fisheries. These aspects are commonly referred to because they can be directly compared with the large-scale sector. They are also easier to measure, monitor and control than social and cultural dimensions commonly associated with small-scale fisheries, such as roles and involvement of family members, including women and children in fishing and post-harvest activities. Rather than defining what small-scale fisheries are, the United Nations Food and Agriculture Organization (FAO) refers to their characteristics as a dynamic and evolving sector that employs labour-intensive harvesting, processing and distribution technologies in their exploitation of aquatic resources (FAO, 2005). Similar to Johnson (2006), they also note the seasonal nature of the fisheries, the use of food for household consumption, and the different commodity chains associated with small-scale fisheries. Whether they are in developed or developing countries, policies for small-scale fisheries have direct consequences on the viability and livelihoods of coastal people who consider fishing not only as an income generating activity, but also as a way of life (Thompson *et al.*, 1983; Kraan, 2011; Marciniak, 2011; Onyango, 2011).

Small-scale fisheries contribute to revenue generation, job creation and employment too, and in many cases, more so than the large-scale fisheries sector. The 'values' of small-scale fisheries can be spelled out in how well the sector performs with regard to fuel efficiency and ecological sustainability, as noted by Thomson (1980), Berkes *et al.* (2001) and Pauly (2006).

The importance of small-scale fisheries goes beyond the quantifiable items, however, with a good portion of catches shared and consumed within fishing households and in the community (Dyer and McGoodwin, 1994). From the social perspective, it can also be argued that kinship and other relationship networks unique to small-scale fisheries are the sources of resilience and safety nets for the communities (Johnson, 2006; Islam, 2011). They represent, for instance, cultural heritage that helps to sustain communities, providing the inhabitants not only with a place to live but also markers of identity, family and kinship that are trans-generational (Carothers *et al.*, 2010).

In the new guidelines for securing sustainable small-scale fisheries coordinated by FAO, they refrain from defining what small-scale fisheries are on the global scale due to their contextual particularities, leaving it to the individual member states to determine what small-scale fisheries are in their own situation (FAO, 2015). In governance terms, this means that policies for sustaining small-scale fisheries must happen at a lower scale than the global. In other words, policies need to be tailored to accommodate the local characteristics of small-scale fisheries as they exist, and where they exist. Small-scale fisheries are simply too diverse for 'one size fits all' policies. Such policies must take into account the contributions of small-scale fisheries to food security, liveli-hoods, community identity and social cohesion (Allison and Ellis, 2001; Acheson and Gardner, 2010; Onyango and Jentoft, 2010), as well as ownership and tenure structure (Korten, 1986; FAO, 2012b) and use of local knowledge (Neis and Felt, 2000; Ruddle, 2000; Bundy and Davis, 2013). Combined, all of these characteristics further distinguish small-scale fisheries from their large-scale counterpart. Because these qualities are harder to capture, not readily quantified, and unique to small-scale fisheries, they are often not well incorporated in the institutional design and decision making about fisheries. It is precisely these aspects of small-scale fisheries, however, not the scale per se or the technological attributes, that lead us to care for them, find them important from a societal perspective and intriguing from a social science research point of view.

Key concerns affecting small-scale fisheries

Four key concerns have been identified in fisheries governance, namely, ecosystem health, social justice, livelihoods and food security (Chuenpagdee *et al.*, 2005; Bavinck *et al.*, 2013). In their deliberation about these four main concerns, the authors do not make a point of distinguishing between large- and small-scale fisheries. It is indisputable, however, that issues related to ecosystem health, social justice, livelihoods and food security affect small-scale fisheries differently than they do large-scale. It can also be said that the ability to address these concerns is not the same in small- and large-scale fishing enterprises. Again, the need to consider the two sectors separately, although not independently from each other, is critical for fisheries policies and governance. In fact, their relationships are in many instances a key determining factor for small-scale fisheries development. As the weaker party, they often lose in the competition for resources and space. It is for these reasons that FAO emphasizes the need to provide secured tenure rights for small-scale fisheries. In reality, the efforts to keep large-scale fishing vessels outside of inshore waters where small-scale fisheries operate are simply not sufficient. Many countries, particularly in the South, lack the institutional capacity to provide such security. Small-scale fisheries' tenure rights therefore end up as words on paper that do not make much difference on a daily basis.

Ecosystem health

Concerns about ecosystem health stem largely from the increasing recognition of impacts that fishing practices generate on aquatic ecosystems (see, for example, Dayton *et al.*, 1995; Watling and Norse, 1998; NRC, 2002). It is well recognized that many fishing practices are harmful to the ecosystem. While not scale-specific, the assessment of fishing gear impacts by Chuenpagdee *et al.* (2003), in terms of by-catch and habitat damage, illustrates that impacts differ among gear types with large-scale industrialized towing gears such as bottom trawls causing the most damage. On the other hand, gears generally employed in a smaller operation such as traps and hand-lines generate a lower level of impact. These findings align well with what Pauly (2006) and others have long argued and called for, which is to consider gears and fleet interaction on the ecosystem in fisheries management policies and decisions (McConney and Charles, 2010). Global efforts are required to prevent destructive gears known to cause high levels of impacts from operating, at least in biologically sensitive locations, and to minimize impacts from other gears, including small-scale ones, through technological innovation, regulations and local cooperation.

The flip side of damage and destruction is conservation and stewardship. Since these concepts are closely linked with community-based management (Conrad and Daoust, 2008) and other civic engagement initiatives (Shandas and Messer, 2008), it is not unreasonable to expect high level of contributions from small-scale fishing people towards these efforts. Examples of these can be found in the establishment and management of several marine protected areas around the world, as well as in other customary practices and voluntary measures, including area and temporal closures during migration or spawning seasons in accordance with conservation and precautionary principles. The potential for successful implementation of these initiatives is higher when fishing communities are involved in the process than when a top-down process is taken (Chuenpagdee *et al.*, 2013). In addition, other features of small-scale fisheries that make eco-system health very relevant to their existence are related to a high level of dependency on fisheries resources and good knowledge about the fisheries and the ecosystem. Given the above, it is highly plausible that successful maintenance and restoration of ecosystem health would rest upon the goodwill and involvement of small-scale fishing people. It is also worth noting that concerns from various global change processes, including those related to climate and markets, on ecosystem health are rising and affecting the viability of fisheries. Despite being highly vulnerable to these changes, small-scale fishing communities may be able to find mechanisms to cope and adapt with their strong social capital and local support network (Islam and Chuenpagdee, 2013). In instances where occupational pluralism exists, fishing communities can rely on other sources of food and income to sustain their families during the stressful periods, and be resilient in case of natural disasters.

Social justice

Given the marginalization, the prevalence of poverty and low political priority, social justice is a major concern for small-scale fisheries, requiring careful consideration in all aspects of governance. The concept is difficult to define, however, and since it often means different things to different people and relates to various aspects, its inclusion in decision making is always contentious. In the case of distributional justice, for instance, allocation rules that seem just to some may seem unfair to others, particularly if they are on the losing end of the deal. Like many other concepts, justice can be contextual (Walzer, 1983) or local (Elster, 1992), with principles that do not always apply in the same way and without a single criterion to evaluate its function (Dahl, 1989). Throughout fisheries development history, many rules and regulations have been made in favour of industrial fishing companies, allowing them to expand their operations and secure their access to resources. In a similar vein, subsidies have often favoured large-scale fisheries, creating over-capacity and subsequent competitions that have negatively affected small-scale fisheries. Furthermore, individual catch quotas, whether transferable or not, employed in many developed countries, have generated uneven opportunities for fishers, disadvantaging those who may have traditional rights but without the financial assets to compete in the market-based, property-rights system. Even in Iceland where the implementation of such a scheme receives high praises, the success is questionable from a social justice perspective with small-scale fishing people being excluded from the programme (Sabau, 2011 and Einarsson, 2011). In Denmark, individual transferable quotas have turned small-scale fisheries into a lease-based fishery, where small-scale fishers lose access to the fisheries resources they relied on (Høst, 2015).

In the context of small-scale fisheries, Jentoft (2013) submits that a good place to start a discussion about justice is human rights. At an international meeting on small-scale fisheries held in Bangkok in 2008, a statement that fishing rights for small-scale fisheries are also human rights was clearly articulated. The event was one of the major steps toward the development of the 'Voluntary Guidelines for Securing Sustainable Small-Scale Fisheries' (FAO, 2015), which included inputs from fisheries stakeholders, for discussion and approval by member states. In the document, small-scale fishing people must be respected in order to secure their human rights, including their right to food, and their right to have secured access and use of fisheries resources. These rights are made explicit as one of the key principles in accord with the FAO Code of Conduct for Responsible Fisheries (FAO, 1995). Small-scale fishers' rights include the rights to organize and to participate in management and governance of fisheries resources, as well as in other aspects affecting their communities. Such involvement is part of procedural justice and the right to be heard, which can encourage people to publicly state their opinions and concerns about allocation rules and other regulations (Perusse Daigle *et al.*, 1996). As well, given that participation in management is time consuming, governments and other relevant agencies need to provide support to enable fishers' involvement. Considerations about mechanisms, formats and avenues for participation are also required for meaningful engagement, as promoted in co-management (Jentoft, 1989; Pinkerton, 1989).

Livelihood viability

The dismal state of the world fisheries today begs for a reiteration of the importance of fisheries to the livelihoods and food security of small-scale fishing people. These two concerns are most prominent for them because of their vulnerability and high dependency on fisheries resources. When fishing is not just an occupation of last resort but in fact a way of life (Onyango, 2011), its meaning with regards to a livelihood goes beyond income and employment. For many small-scale fishing communities, fishing is a part of their heritage, culture, tradition, and has other intrinsic values that cannot be easily replaced. While fishers in an impoverished condition may be willing to move out of the sector for a better life, studies show that many would prefer to stay, given a choice (see examples in Jentoft and Eide, 2011). This strong tie to a fishing lifestyle, an important feature distinguishing small-scale from large-scale fishers, is a part of the overall well-being of small-scale fishers that should not be undermined in fisheries governance. Instead, efforts should be made to understand fisheries' livelihood values and to incorporate them in the formulation of policies and the design of institutions. Principles relevant to livelihoods and well-being of small-scale fisheries, such as inclusion, reflexivity, adaptation, precaution and social justice, should be applied (Johnson, 2013).

Not only do fisheries support small-scale fishing livelihoods, small-scale fishers in turn contribute significantly to the well-being and livelihoods of their communities. This connection is obvious in places where small-scale fishing people are well integrated in the communities. Consequently, small-scale fisheries serve an important niche, as well as contributing to local food security and providing a social safety net along with local employment and development opportunities. In the south of Thailand, for instance, the social and cultural importance of small-scale fisheries to the communities was most evident in the recovery efforts after the Indian Ocean tsunami in 2004 (Chang *et al.*, 2006). This is, in fact, one of the distinguished features of small-scale fisheries. Small-scale fishers fish from communities where they and their families also live – the very places where they have histories and connections. The social values guiding fishing activities are also those guiding community lives (Pálsson, 1991). At the end, it is small-scale fishing people who live, work and spend money in their communities, while large-scale fishing enterprises place their investments elsewhere and fail to contribute to sustaining the local economy.

Food security

Among the key concerns, food security is probably the most recognized by government and inter-governmental bodies, non-government organizations and donors alike. This is also why small-scale fisheries have recently received international interest. They are increasingly valued for their contribution to food provision at the local level and also globally. Food security is indeed one of the Millennium Development Goals, alongside poverty eradication. While its importance in fisheries governance cannot be disputed, achieving food security – ensuring the right of all people to have access to sufficient, safe and nutritious food – is, as Pullin (2013) puts it, the 'wickedest of wicked problems' (p. 87). Indeed, the 'fish chain', which controls both supply and demand through the interconnectivity of the ecosystem, harvest and post-harvest systems, all the way to consumption, is highly complex and dynamic, and is affected by global change processes.

The importance of fish to food security increases with a substantial portion of small-scale fisheries catches consumed within fishing households and distributed to others in the communities, local markets and beyond. Access to markets is, however, often hampered by poor infrastructure such as roads and means of transportation. In many instances, such markets do not exist but must be developed in order to provide equitable and non-discriminatory trade options for small-scale fishers at local, regional and national levels. Rather than employing eco-labelling and certification schemes developed for industrial fisheries, the nature and conditions of small-scale fisheries need to be accommodated more so than has been the case in the past (Ponte, 2008; Jacquet *et al.*, 2010). This may include developing area-specific labelling schemes (FAO, 2009b) and supporting alternative strategies such as the 'Slow Fish' network, which promotes locally caught fish as ingredients in restaurants (Slow Food, 2013).

Securing the contribution of small-scale fisheries to food security would therefore require attention along the entire fish chain, particularly at the post-harvest where in fact the majority of small-scale fishing people, especially women, are employed. This may also involve providing support to fishers and fish workers' organizations, such as cooperatives, as a means of removing bottlenecks in the value chain so small-scale fishing people can take advantage of market segments that would otherwise not be available.

Governance for sustainable small-scale fisheries

Given the characteristics and values of small-scale fisheries, their actual and potential contribution to society at large, and the concerns that confront the world's fisheries nowadays, governance of small-scale fisheries requires deep, as well as new, thinking rooted in key principles relevant to the sector. Governance principles for small-scale fisheries must be different from those associated with large-scale fisheries, but not necessarily unique for small-scale fisheries. Small-scale fisheries are a way of life for the millions of people who depend on them, whose livelihoods and human rights must be secured. In other words, small-scale fisheries and the people employed in them are important in themselves. That requires governance mechanisms that are people-centred, community-oriented, culturally sensitive, and democratically structured. Second, governance must aim to realize the potential of small-scale fisheries as a sector, along with their important contributions to the overall social and economic well-being of the entire society. For this reason, they are too important to fail. Their profile should be elevated. They should not be considered a marginal sector, left to fend for themselves. In both instances, they need collective actions and representations at all scales, i.e. from within the communities as well as at regional, national and international scales. They also need a solid backing, not only by governments but also by civil society organizations. Given the processes that are now occurring within international bodies such as the FAO where countries in many instances seem to underestimate the challenges faced by small-scale fisheries as well as opportunities available to them, their future is far from secured. Appropriate governance mechanisms and suitable platforms need to be fostered to enable small-scale fisheries to speak up against power, indifference and ignorance. For that, they need to be empowered through various individual and collective capacity development programmes.

While environmental organizations have a role to play in fisheries governance, their actions must be critically examined since it cannot be assumed that they will always have the interest of small-scale fisheries in mind. In many instances, they have agendas that go against those of small-scale fishing people, as can sometimes be seen in the case of marine protected areas where small-scale fisheries are seen as part of the problem rather than a solution. There are examples around the world where small-scale fishers have been pushed aside in the name of conservation, thus denying them the opportunities to feed themselves (Mascia *et al.*, 2010; Isaacs, 2011; Rees *et al.*, 2013). Such initiatives go against the human right to food concept. However, there are instances where synergies have been created between marine protected areas and small-scale fisheries to achieve both conservation and sustainable livelihoods goals (see, for instance, Jentoft *et al.*, 2012). Focusing on a single goal without consideration on others often results in mistrust, resentment and conflicts, affecting both the processes and the outcomes of such initiatives. These are, in many respects, the reasons why marine protected areas fail to deliver (Chuenpagdee *et al.*, 2013).

Governance principles for sustainable small-scale fisheries are well articulated in the new small-scale fisheries guidelines (FAO, 2015). Rooted strongly in international human rights standards and existing tenure rights, they include general principles related to human dignity, respect for cultures, non-discrimination practices, equity and equality, meaningful participation, rule of laws, transparency and accountability. In the context of fisheries governance, the human rights based approach represents a new perspective on small-scale fisheries governance that has not been highlighted before. Small-scale fishing people have both individual and collective rights. Securing those rights, for instance through the rule of law that guarantees their customary practices, is essential not only in the context of providing viable livelihoods and food security but also in creating safe working conditions and living environments, free from external encroachment, including human trafficking. Fisheries governance is also about protecting and encouraging their rights to express their opinion, to organize and to be involved in the political process, as essential to social justice and freedom (Sen, 2009; Jentoft, 2013).

The governance principles expressed in the small-scale fisheries guidelines are to be implemented in concert with other guiding principles such as precautionarity and sustainability, using holistic and integrated approaches that promote ecosystem health, social responsibility and economic viability. These two principles, in particular, would involve a broader perspective – one that goes beyond ecological considerations to one that also includes, and emphasizes, the social and cultural aspects of small-scale fishing communities, who are often equally vulnerable as the natural ecosystems. As was originally emphasized for instance by the so-called Brundtland

Commission (United Nations, 1987), the conservation of nature cannot realistically be obtained if the issues of poverty are left unaddressed. People who starve will tend to overfish if fishing is the only way to feed themselves, despite their innate stewardship ethics (Nguyen and Flaaten, 2011). For instance, marine protected areas that do not take poverty issues into account are likely to fail due to lack of compliance (Isaacs, 2011; Onyango, 2011). For the most part, major institutional reform will be required to enable the implementation of these principles, which may not be easy but is necessary if nations are serious about securing rights and sustainability of the small-scale fisheries sector. Resistance may also be expected, particularly if such change invokes conflicts with how things have always been and when only the benefits of large-scale fisheries, and related enterprises, are considered. Some coordination and cooperation will need to be facilitated, along with awareness raising, to garner support from the general public and civil society organizations.

In addition to the principles promoted in the small-scale fisheries guidelines, others have been suggested by social science research in small-scale fisheries. In Jentoft and Eide (2011), for instance, a dexterity principle is called for to deal with the diverse ecological, social and political context associated with small-scale fisheries. Governance based on the dexterity principle is sensitive to details and the local context, while taking into account the importance of fisheries to small-scale fishing people's livelihoods. In support of the dexterity principle, the subsidiarity principle offers legitimate reasoning for stakeholder involvement in fisheries governance (McCay and Jentoft, 1996). Referring to how decision making should take place where the problem is experienced and where many of the solutions may be found, the subsidiarity principle suggests that management authority should be vested at the lowest possible organization. Also, decisions about resource extraction and allocation should reside at the same level (Bavinck and Jentoft, 2011). This suggestion reflects the realization that states are, for the most part, incapable of addressing all the issues pertaining to small-scale fisheries on their own, or even if they could, it may not be sufficiently timely or effective. It should be noted that the subsidiarity principle does not preclude the states from delivering services that they are responsible for and must continue to provide (Jentoft and Eide, 2011). In fact, most of the paragraphs in the small-scale fisheries guidelines start with a sentence, 'State should. . .,' which suggests that the state has a broad responsibility vis-à-vis small-scale fisheries, including the realization of the subsidiarity principle. As a step in this direction, co-management schemes should be promoted by the state, with the provision of enabling legislation as well as the organizational capacity to make this management system implementable and effective.

Concluding thoughts

Given the dismal situation of small-scale fisheries in many parts of the world, the 'urgency principle' is deemed essential. Small-scale fisheries can no longer be ignored. They are simply too big and too important in world fisheries, from ecological, social and cultural perspectives. Similar to other sectors addressed in the Millennium statement, efforts to alleviate their situation cannot be postponed. As argued by Jentoft and Eide (2011), small-scale fisheries, especially those in impoverished conditions, are vulnerable to the changing circumstances taking place around them. But that is not the only reason why good governance thinking and proactive policies that support their sustainability are needed. As previously argued, their contribution to poverty alleviation, food security and ecosystem stewardship needs to be maintained and promoted. Failure to do so will have negative consequences on small-scale fisheries, further marginalizing small-scale fishing people and undermining their prospects and future, including damaging numerous communities around the world that depend on them.

Governability is an analytical lens provided by the interactive governance theory to help address these concerns. A governability assessment involves an examination of the overall quality of the governance system and its ability to deliver what it sets out to do (Kooiman, 2003; Jentoft and Chuenpagdee, 2015). It takes into account the basic characteristics of small-scale fisheries, in terms of diversity, complexity, dynamics and scale associated with their biological, social, economic and political dimensions. The analysis also includes an examination of the fit, responsiveness and performance of chosen institutional arrangements. It helps to understand the extent to which existing institutions and governance systems contribute to facilitating or inhibiting the quality of governance relevant to small-scale fisheries. For example, it asks whether instruments such as individual transferrable quotas promoted to manage large-scale fisheries are really suitable for small-scale. Finally, the governability assessment pays close attention to the meta-order elements of governance, such as the fundamental values, images and principles influencing how people behave and act, arguing that they are the essence of what makes fisheries more or less governable. An analysis of current governance discourse, e.g. what principles and practices underlie key policies and decisions, how they materialize in governing institutions and strategies, as well as how closely they relate to those of small-scale fishing people, fall within the purview of the governability assessment.

Both 'soft' approaches in detecting and understanding values, images and principles of fisheries stakeholders and 'hard' approaches from governments at local, national and global levels are required to achieve the goal of securing sustainable small-scale fisheries. The governability assessment offers a thorough analysis of where the opportunities for intervention in the fish chain as well as in the governing system exist. Such analysis helps determine what needs to be done to address social justice and power imbalance issues, and to design institutions and governance mechanisms that foster the viability of small-scale fisheries. The realization of this depends largely on the commitment of the government to act responsibly and with determination, and in accord with the guidelines and principles suggested in the FAO guidelines for small-scale fisheries. Governments must not only want the goals of sustaining small-scale fisheries, they must also want the means that would be required in order to fulfil them. The voluntary nature of the guidelines raises some concern about how it will be implemented. The experience in the North of what has happened in small-scale fisheries may signal what may happen to those in the South, which is not a positive experience. It will take effort from all parties, including civil society organizations and research institutions, to turn adversity and antagonism into opportunities and synergies that help conserve and protect small-scale fisheries and their future, for their own sake as well as for society at large.

Acknowledgements

We thank the Social Sciences and Humanities Research Council of Canada for funding the Too Big To Ignore partnership, which helps foster our investigation in small-scale fisheries governance.

References

Acheson, J. M. and Gardner, R. (2010) 'The evolution of conservation rules and norms in the Maine lobster industry', *Ocean & Coastal Management*, vol. 53, no. 9, pp. 524–534.

Allison, E. H. and Ellis, F. (2001) 'The livelihoods approach and management of small-scale fisheries', *Marine Policy*, vol. 25, no. 5, pp. 377–388.

Bailey, C. (1988) 'The political economy of marine fisheries development in Indonesia', *Indonesia*, vol. 46, pp. 25–38.

Bavinck, M. (2011) 'The mega-engineering of ocean fisheries: a century of expansion and rapidly closing frontiers', in S. D. Brunn (ed.) *Engineering Earth: The Impacts of Mega-engineering Projects*, Kluwer, Dordrecht, pp. 257–273.

Bavinck, M. and Jentoft, S. (2011) 'Subsidiarity as a Guiding Principle for Small-Scale Fisheries' in R. Chuenpagdee (ed.) *Contemporary Visions for World Small-Scale Fisheries*, Eburon, Delft, pp. 311–320.

Bavinck, M., Chuenpagdee, R., Jentoft, S. and Kooiman, J. (eds) (2013) *Governability in Fisheries and Aquaculture: Theory and Applications*, Springer Publication, Amsterdam.

Béné, C. (2003) 'When fishery rhymes with poverty: a first step beyond the old paradigm on poverty in small-scale fisheries', *World Development*, vol. 31, pp. 949–975.

Béné, C. and Friend, R. M. (2011) 'Poverty in small-scale fisheries: old issue, new analysis', *Progress in Development Studies*, vol. 11, no. 2, pp. 119–144.

Berkes, F. (2012) 'Implementing ecosystem based management: evolution or revolution', *Fish and Fisheries*, vol. 13, no. 4, pp. 465–476.

Berkes, F., Mahon, R., McConney, P., Pollnac, R. and Pomeroy, R. (eds) (2001) *Managing Small-Scale Fisheries: Alternative Directions and Methods*, International Development Research Centre, Ottawa.

Bundy, A. and Davis, A. (2013) 'Knowing in context: an exploration of the interface of marine harvesters' local ecological knowledge with ecosystem approaches to management', *Marine Policy*, vol. 38, pp. 277–286.

Butcher, J. G. (2004) *The Closing of the Frontier: A History of the Marine Fisheries of Southeast Asia c. 1850–2000*, KITLV Press, Leiden.

Cadigan, S. T. (2009) *Newfoundland and Labrador: A History*, University of Toronto Press, Toronto.

Carothers, C., Lewb, D. K. and Sepez, J. (2010) 'Fishing rights and small communities: Alaska halibut IFQ transfer patterns', *Ocean & Coastal Management*, vol. 53, no. 9, pp. 518–523.

Chang, S. E., Adams, B. J., Alder, J., Berke, P. R., Chuenpagdee, R., Ghosh, S. and Wabnitz, C. (2006) 'Coastal ecosystems and tsunami protection after the December 2004 Indian Ocean Tsunami', *Earthquake Spectra*, vol. 22, no. S3, pp. S863–887.

Chuenpagdee, R. and Jentoft, S. (2013) 'Assessing governability: what's next', in M. Bavinck, R. Chuenpagdee, S. Jentoft and J. Kooiman (eds) *Governability in Fisheries and Aquaculture: Theory and Applications*, Springer Publication, Amsterdam, Chapter 18, pp. 335–349.

Chuenpagdee, R., Liguori, L., Palomares, M. L. D. and Pauly, D. (2006) 'Bottom-up, global estimates of small-scale fisheries catches', *Fisheries Centre Research Report*, vol. 14, no. 8, p. 105.

Chuenpagdee, R., Morgan, L. E., Maxwell, S. M., Norse, E. A. and Pauly, D. (2003) 'Shifting gears: assessing collateral impacts of fishing methods in the U.S. waters', *Frontiers in Ecology and the Environment*, vol. 1, no. 10, pp. 517–524

Chuenpagdee, R., Pascual-Fernandez, J. J., Szelianszky, E., Alegret, J. L., Fraga, J. and Jentoft, S. (2013) 'Marine protected areas: Re-thinking their inception', *Marine Policy* vol. 39, pp. 234–240.

Chuenpagdee, R., Degnbol, P., Bavinck, M., Jentoft, S., Johnson, D., Pullin, R. and Williams, S. (2005) 'Challenges and Concerns in Fisheries and Aquaculture', in J. Kooiman, J. M. Bavinck, S. Jentoft and R. S. V. Pullin (eds) *Fish for Life: Interactive Governance for Fisheries*, University of Amsterdam Press, Amsterdam, pp. 25–27.

Clark, C. W. (2006) *The Worldwide Crisis in Fisheries: Economic Models and Human Behavior*, Cambridge University Press, New York.

Conrad, C. T. and Daoust, T. (2008) 'Community-based monitoring frameworks: increasing the effectiveness of environmental stewardship', *Environmental Management*, vol. 41, no. 3, pp. 358–366.

Costello, C., Gaines, S. D. and Lynham, J. (2008) 'Can catch shares prevent fisheries collapse?', *Science*, vol. 321, pp. 1678–1681.

Dahl, R.A. (1989) *Democracy and its Critics*, Yale University Press, New Haven, CT.

Dayton, P. K., Thrush, S. F., Agardy, M. T. and Hofman, R. J. (1995) 'Environmental effects of marine fishing', *Aquatic Conservation: Marine and Freshwater Ecosystems*, vol. 5, pp. 205–232.

Dyer, C. L. and McGoodwin, J. R. (eds) (1994) *Folk Management in the World's Fisheries: Lessons for Modern Fisheries Management*, University Press of Colorado, Niwot, CO.

Einarsson, N. (2011) *Culture, Conflict and Crises in the Icelandic Fisheries: An Anthropological Study of People, Policy and Marine Resources in the North Atlantic Arctic*, Uppsala Publisher, Sweden.

Elster, J. (1992) *Local Justice: How Institutions Allocate Scarce Goods and Necessary Burdens*, Russel Sage Foundation, New York.

FAO. (1995) *Code of Conduct for Responsible Fisheries*, Food and Agriculture Organization, Rome.

FAO. (2005) 'Increasing the contribution of small-scale fisheries to poverty alleviation and food security', *FAO Technical Guidelines for Responsible Fisheries*, Food and Agriculture Organization, Rome.

FAO. (2009a) *The State of World Fisheries and Aquaculture 2008*, Food and Agriculture Organization, Rome.

FAO. (2009b) *Guidelines for the Ecolabelling of Fish and Fishery Products from Marine Capture Fisheries (Revision 1)*, Food and Agriculture Organization, Rome.

FAO. (2010) *The State of World Fisheries and Aquaculture 2010*, Food and Agriculture Organization, Rome.

FAO. (2012a) *The State of World Fisheries and Aquaculture 2012*, Food and Agriculture Organization, Rome.

FAO. (2012b) *Voluntary Guidelines on the Responsible Governance of Tenure of Land, Fisheries and Forests in the Context of National Food Security*, Food and Agriculture Organization, Rome.

FAO. (2015) *Voluntary Guidelines for Securing Sustainable Small-Scale Fisheries in the Context of Food Security and Poverty Eradication*, Food and Agriculture Organization, Rome.

Gray, T. S. (2005). *Participatory Fisheries Governance*, Amsterdam, Springer.

Høst, J. (2015). *Market-Based Fisheries Management. Private Fish and Captains of Finance*. Dordrecht: Springer Science.

Isaacs, M. (2011) 'Creating action space: small-scale fisheries policy reform in South Africa', in S. Jentoft and A. Eide (eds) *Poverty Mosaics: Realities and Prospects in Small-Scale Fisheries*, Springer, Dordrecht, pp. 359–382.

Islam, M. M. (2011) 'Living on the margin: the poverty-vulnerability nexus in the small-scale fisheries of Bangladesh', in S. Jentoft and A. Eide (eds) *Poverty Mosaics: Realities and Prospects in Small-Scale Fisheries*, Springer, Dordrecht, pp. 71–95.

Islam, M. M. and Chuenpagdee, R. (2013) 'Negotiating risk and poverty in the Sundarbans mangrove fishing communities of Bangladesh', *Maritime Studies (MAST)*, vol. 12, no. 7, pp. 3–20.

Jacquet, J. L. and Pauly, D. (2008) 'Funding priorities: big barriers to small-scale fisheries', *Conservation and Policy*, vol. 4, pp. 832–835.

Jacquet, J., Pauly, D., Ainley, D., Holt, S., Dayton, P. and Jackson, J. (2010) 'Seafood stewardship in crisis', *Nature*, vol. 467, pp. 28–29.

Jentoft, S. (1989) 'Fisheries co-management: delegating government responsibility to fishermen's organizations', *Marine Policy*, vol. 13, no. 2, pp. 137–154.

Jentoft, S. (2013) 'Social justice in fisheries: a governability challenge', in M. Bavinck *et al.* (eds) *Governability in Fisheries and Aquaculture: Theory and Applications*, Springer Publication, Amsterdam, pp. 45–65.

Jentoft, S. and Chuenpagdee, R. (2009) 'Fisheries and coastal governance as a wicked problem', *Marine Policy*, vol. 33, pp. 553–560.

Jentoft, S. and Chuenpagdee (eds) (2015). *Interactive Governance for Small-Scale Fisheries: Global Reflections*. Dordrecht: Springer Science.

Jentoft, S. and Eide, A. (eds) (2011). *Poverty Mosaics: Realities and Prospects in Small-Scale Fisheries*, Springer, Dordrecht.

Jentoft, S., Pascual-Fernandez, J. J., De la Cruz Modino, R., Gonzalez-Ramallal, M. and Chuenpagdee, R. (2012) 'What stakeholders think about marine protected areas: case studies from Spain', *Human Ecology*, vol. 40, no. 2, pp. 185–197.

Johnson, D. S. (2006) 'Category, narrative, and value in the governance of small-scale fisheries', *Marine Policy*, vol. 30, no. 6, pp. 747–756.

Johnson, D. S. (2013) 'Livelihoods in fisheries: a governability challenge', in M. Bavinck, R. Chuenpagdee, S. Jentoft and J. Kooiman (eds) *Governability in Fisheries and Aquaculture: Theory and Applications*, Springer Publication, Amsterdam, pp. 67–86.

Khan, A. and Neis, B. (2010) 'The rebuilding imperative in fisheries: clumsy solutions for wicked problems?', *Progress in Oceanography*, vol. 87, pp. 347–356.

Kooiman, J. (2003) *Governing as Governance*, Sage Publications, London.

Kooiman, J., Bavinck, M., Jentoft, S. and Pullin, R. (eds) (2005). *Fish for Life: Interactive Governance for Fisheries*, Amsterdam University Press, Amsterdam.

Korten, D. (ed.) (1986) *Community Management: Asian Experience and Perspectives*, Kumarian Press, West Hartford.

Kraan, M. (2011) 'More than income alone: the Anlo-Ewe Beach Seine Fishery in Ghana', in S. Jentoft and A. Eide (eds) *Poverty Mosaics: Realities and Prospects in Small-Scale Fisheries*, Springer, Dordrecht, pp. 147–172.

McCay, B. J. and Jentoft, S. (1996) 'From the bottom up: participatory issues in fisheries management', *Society and Natural Resources*, vol. 9, pp. 237–250.

McConney, P. and Charles, A. T. (2010) 'Managing small-scale fisheries: moving toward people-centered perspectives', in R. Q. Grafton, R. Hilborn, D. Squires, M. Tait, M. Williams (eds) *Handbook of Marine Fisheries Conservation and Management*, Oxford University Press, New York, pp. 532–545.

McGoodwin, J. R. (1990) *Crisis in the World's Fisheries: Peoples, Problems, and Policies*, Stanford University Press, Stanford, CA.

Marciniak, B. (2011) 'Vanished prosperity: poverty and marginalization in a small Polish fishing community', in S. Jentoft and A. Eide (eds) *Poverty Mosaics: Realities and Prospects in Small-Scale Fisheries*, Springer, Dordrecht, pp. 125–146.

Mascia, M. B., Claus, C. A. and Naidoo, R. (2010) 'Impacts of Marine Protected Areas on Fishing Communities', *Conservation Biology*, vol. 24, pp. 1424–1429.

NRC (National Research Council). 2002. *Effects of Trawling and Dredging on Seafloor Habitat*. National Academy Press, Washington, DC.

Neis, B. and Felt, L. (eds) (2000) *Finding Our Sea Legs: Linking Fishery People and Their Knowledge with Science and Management*, ISER Books, St John's, Canada.

Nguyen, K. A. T. and Flaaten, O. (2011) 'Facilitating change: a Mekong Vietnamese small-scale fishing community', in S. Jentoft and A. Eide (eds) *Poverty Mosaics: Realities and Prospects in Small-Scale Fisheries*, Springer, Dordrecht, pp. 335–357.

Onyango, P. (2011) 'Occupation of last resort? Small-scale fishing in Lake Victoria, Tanzania', in S. Jentoft and A. Eide (eds) *Poverty Mosaics: Realities and Prospects in Small-Scale Fisheries*, Springer, Dordrecht, pp. 97–124.

Onyango, P. and Jentoft, S. (2010) 'Assessing poverty in small-scale fisheries in Lake Victoria, Tanzania', *Fish and Fisheries*, vol. 11, no. 3, pp. 250–263.

Pálsson, G. 1991. *Coastal Economies, Cultural Accounts. Human Ecology and Icelandic Discourse*, Manchester University Press, Manchester.

Pascual-Fernández, J. J. and de la Cruz-Modino, R. (2011) 'Conflicting gears, contested territories: MPAs as a solution?', in R. Chuenpagdee (ed.) *Contemporary Visions for World Small-Scale Fisheries*, Eburon, Delft, pp. 205–220.

Pauly, D. (2006) 'Major trends in small-scale marine fisheries, with emphasis on developing countries, and some implications for the social sciences', *Maritime Studies (MAST)*, vol. 4, no. 2, pp. 7–22.

Perusse Daigle, C., Loomis, D. K. and Ditton, R. B. (1996) 'Procedural justice in fishery resource allocations', *Fisheries*, vol. 21, no. 11, pp. 18–23.

Pinkerton, E. (ed.) (1989) *Co-Operative Management of Local Fisheries: New Directions for Improved Management and Community Development*, The University of British Columbia Press, Canada.

Ponte, S. (2008) 'Greener than thou: the political economy of fish ecolabelling and its local manifestations in South Africa', *World Development*, vol. 36, no. 1, pp. 159–175.

Pullin, R. S. V. (2013) 'Food security, governability and fish chains', in M. Bavinck, R. Chuenpagdee, S. Jentoft and J. Kooiman (eds) *Governability in Fisheries and Aquaculture: Theory and Applications*, Springer Publication, Amsterdam, pp. 87–109.

Rees, S. E., Rodwell, L. D., Searle, S. and Bell, A. (2013) 'Identifying the issues and options for managing the social impacts of Marine Protected Areas on a small fishing community', *Fisheries Research*, vol. 146, pp. 51–58.

Ruddle, K. (2000) 'Systems of knowledge: dialogue, relationships and process', *Environment, Development and Sustainability*, vol. 2, no. 3–4, pp. 277–304.

Sabau, G. (2011) 'Whose fish is it anyway? Iceland's cod fishery rights', in R. Chuenpagdee (ed.) *Contemporary Visions for World Small-Scale Fisheries*, Eburon, Delft, pp. 173–184.

Sen, A. (2009) *The Idea of Justice*, Penguin, London.

Shandas, V. and Messer, W. B. (2008) 'Fostering green communities through civic engagement: community-based environmental stewardship in the Portland area', *Journal of the American Planning Association*, vol. 74, no. 4, pp. 408–418.

Slow Food. (2013) *Slow Fish*, www.slowfood.com/slowfish, accessed 7 June 2013.

Smith, H. D. (2000) 'The industrialization of the world ocean', *Ocean and Coastal Management*, vol. 43, pp. 11–28.

Teh, L. S. L., Teh, L. C. L. and Sumaila, U. R. (2011) 'Quantifying the overlooked socio-economic contribution of small-scale fisheries in Sabah, Malaysia', *Fisheries Research*, vol. 110, no. 3, pp. 450–458.

Thomson, D. (1980) 'Conflict within the fishing industry', *ICLARM Newsletter*, vol. 3, pp.3–4.

Thompson, P., Wailey, T. and Lummis, T. (1983) *Living the Fishing*, Routledge & Kegan Paul, London.

United Nations (1987) 'United Nations World Commission on Environment and Development, Our Common Future' (The Brundtland Report), Oxford University Press, Oxford.

Walsh, D. (2011) 'What restructuring? Whose rationalization? Newfoundland and Labrador's Memorandum of Understanding on its fishing industry', in R. Chuenpagdee (ed.) *Contemporary Visions for World Small-Scale Fisheries*, Eburon, Delft, pp. 81–97.

Walzer, M. (1983) *Spheres of Justice: A Defence of Pluralism & Equality*. Basil Blackwell, Oxford.

Watling, L. and Norse, E. A. (1998) 'Disturbance of the seabed by mobile fishing gear: a comparison with forest clear-cutting', *Conservation Biology*, vol. 12, pp. 1189–1197.

17

MARICULTURE

Aquaculture in the marine environment

Selina Stead

Introduction

This is an exciting period for the global aquaculture industry as one of the fastest-growing food producing sectors. In 2012, the aquaculture sector provided almost half of all fish for human food (FAO, 2014). This chapter provides a broad overview of marine farming as a multi-species and diverse industry that transcends complex biomes, ecosystems, governance arrangements and social acceptability levels, to mention only a few of the current and future issues impacting development.

Mariculture, farming of aquatic organisms in the marine environment, is increasingly being considered by governments and non-governmental organisations as a potential solution to addressing some of the world's common and bigger societal issues, namely, food security and income generation. As a commercial sector, aquaculture is relatively young, 4–5 decades, with the Atlantic salmon, *Salmo salar* L., being the most intensely farmed in seawater (Figures 17.1 and 17.2). Interest in culturing other marine species such as aquatic plants, e.g. seaweed, *Kappaphycus striatum*, in Zanzibar (Figure 17.3), molluscs including oysters, *Crassostrea gigas* in Scotland (Figure 17.4) and various finfish is growing, unlike the availability of appropriate coastal and marine space for farming purposes.

Aquaculture attracts a lot of media attention where arguably it is presented more commonly in a negative context especially in terms of environmental impacts and sourcing of feed ingredients. In contrast, positive benefits such as aquaculture providing an alternative or supplementary form of protein and livelihood option receives less coverage. Aquaculture as a global food production sector could do more to improve awareness about the benefits associated with aquatic farming. This is important because it is harder to communicate benefits of food produced under water which most people will not see compared with its terrestrial counterpart, agriculture which is more visible and better related to by consumers.

This chapter aims to present a broad overview of mariculture. The first objective and focus of the next section is to describe production trends at different geographical scales including global, regional, national and local level examples to help identify emerging patterns in production. The second objective is to identify constraints to development of the sector through reflecting on issues identified in developed countries and lessons learned using a case study from a developing country before the final conclusions section.

Figure 17.1 Photograph of a seawater salmon, *Salmo salar* L., cage on the west coast of Scotland

Figure 17.2 Photograph of a freshly processed salmon, *Salmo salar* L., farmed on the west coast of Scotland

Figure 17.3 Photograph of a seaweed *Kappaphycus striatum* farm in Zanzibar, East Africa

Figure 17.4 Photograph of the non-native Pacific oyster, *Crassostrea gigas* being cultured on the west coast of Scotland. The Pacific oyster is preferred for aquaculture rather than the native oyster *Ostrea edulis*

Production trends

A global overview

Up to the last decade (2000–2010), estimates have shown an overall surplus of meat from cattle, poultry and pork, as sources of protein, in meeting the predicted global human population's needs. Supplies from these terrestrial sources and their ability to meet human demand are reaching a limit, more quickly than predicted in the case of some countries. Many governments are looking at opportunities to expand production to provide sources of aquatic protein to meet any short-falls from availability of meat. With fish remaining the most traded food commodity worldwide (FAO, 2014), access to new markets especially given emerging changes in international trade patterns will require a more integrated socio-economic and environmental evidence base to support effective management and governance efficacy. Human behavioural drivers influencing production trends including tenure security, market access, human rights and traceability in the food supply chain are more widely acknowledged as important determinants to the future growth of aquaculture around the world. Although beyond the scope of this chapter to detail these influences on aquaculture production, more information can be found in the following report, FAO, 2014 and should be considered in any future development planning of the sector.

Approximately 89 per cent of all global aquaculture (freshwater, brackishwater and marine) production, measured by volume of species farmed, occurs in Asia, with China accounting for more than 60 per cent in 2010 (FAO, 2012). The majority (88 per cent) of worldwide production is focused in the following countries, mostly Asian: Bangladesh, India, Indonesia, Japan, Myanmar, Thailand, the Philippines and Vietnam (FAO, 2012). In 2010, aquaculture production worldwide was estimated to be worth US$119 billion, a record level of 60 million tonnes (excluding aquatic plants and non-food products), according to figures reported by FAO (2012). In terms of trends in global consumption, farmed fish consumption, per capita on a global scale increased from 1.1kg in 1980 to 8.7kg in 2010 (7.1 per cent average rate of increase per annum) with demand estimated to continue rising (FAO, 2012). Values available in 2014 show 66.6 million tonnes of food fish were produced with estimates for 2013 expected to reach 70.5 million tonnes (FAO, 2014). Production trends vary significantly around the world depending on: biophysical features of coastal areas that determine the species that are suitable for mariculture, economic conditions, social factors, management efficacy, governance systems in place, political and technological impacts among others.

Around 17 million people (3 per cent of the world's population) are involved in fish farming, with 97 per cent of those engaged in finfish aquaculture living in Asia, followed by Latin America and the Caribbean (1.5 per cent), and Africa (1 per cent), according to FAO (2012). This report showed the number of people engaged in fish farming had increased by 6 per cent per annum compared with 0.8 per cent per year for those in the capture fisheries sector over a five year period (2005–2010). Noteworthy is employment in capture fisheries is stagnating or decreasing while aquaculture is providing increased opportunities (FAO, 2012). However, although the largest decrease in individuals engaged in capture fisheries on a global scale occurred in Europe (2 per cent p.a. between 2000 and 2010) there was little to no increase in people employed in fish farming (FAO, 2012) suggesting minimal transfer of personnel between the fisheries and aquaculture sectors. Recognising this trend in Europe, a conference was organised in Vigo in 2009 specifically to address opportunities for the fisheries sector to invest in developing aquaculture (Stead, 2009). Africa during the same period had a 6 per cent increase in people's involvement in fish farming, followed by Asia (5 per cent), and Latin America and the Caribbean (3 per cent). According to the FAO in 2014, Brazil has improved its global ranking significantly in recent years in terms of food fish production.

In 2010, FAO (2012) reported the least-developed countries (LDCs) have the smallest involvement in aquaculture (4.1 per cent by quantity and 4 per cent by value), thus less access to benefits associated with related food security and income generation. These were mostly in sub-Saharan Africa and in Asia. They include Bangladesh, Myanmar, Uganda, the Lao People's Democratic Republic and Cambodia, developing countries in Asia and the Pacific (Myanmar and Papua New Guinea). In sub-Saharan Africa the list includes Nigeria, Uganda, Kenya, Zambia and Ghana. In South America, Ecuador, Peru and Brazil have been highlighted. All these countries have efforts towards expanding development to become significant aquaculture producers in their regions with varying rates of success. A major challenge is having a comprehensive aquaculture development strategy that has the political will and financial support to optimise opportunities that match market conditions. In contrast, developed industrialised countries have seen a decline in production levels from 22 per cent in 1990 to 7 per cent (4 mt) by quantity and 14 per cent (US$17 billion) by value in 2010. One exception to this trend is Norway, where, Atlantic salmon (*Salmo salar* L.) aquaculture in marine cages increased from 151,000 tonnes in 1990 to over one million tonnes in 2010 (FAO, 2012).

According to the FAO (2012), marine finfish aquaculture only represented 3 per cent or 2 million tonnes (mt) of global aquaculture production with freshwater finfish comprising the majority (56 per cent and 34 mt), followed by molluscs (24 per cent, 14 mt), crustaceans (10 per cent, 6 mt), diadromous fishes (6 per cent, 4mt), and other aquatic animals (1 per cent, 0.8 mt) in 2010. Mariculture accounted for approximately 29 per cent of world aquaculture production by value compared with 62 per cent being comprised of freshwater species. Brackishwater aquaculture makes up 8 per cent in terms of quantity, equal to 13 per cent by value due to the high value of marine shrimps (FAO, 2012).

In summarising worldwide production trends over the last 50 years, growth of the aquaculture sector was slowest during the 1970s, 1980s and 1990s. This was due to early piloting stages for many different species to test their appropriateness for up-scaling to commercial levels of culture, finding the best technologies that were cost-effective and optimising growth performance under varying conditions. Knowledge of species with specific growth rates that returned greater profits for the investor (reached market size in a short time) was a priority objective during this period. Efforts to improve production continue to focus on improving husbandry and feed formulations to achieve: good food conversion rates and growth performance; robust life history strategies that are elastic and can tolerate high stocking densities among many factors; resistance to disease; enhanced desirable genetic traits and good palatability of end products that appeal to a wide range of consumers. Increased awareness about animal welfare including recommendations on stocking densities, how animals are handled and killed have particularly influenced the way in which the aquaculture sector operates its production units. Health and safety of personnel and species grown in aquaculture has also received considerable attention in the last two decades. Retailers have influenced calls for traceability of products demanding, for example, evidence that organisms are fed from sustainable feed sources, which led to various certification schemes being introduced.

Feed is considered to be a major constraint to expansion of global aquaculture development and for some species such as the Atlantic salmon feed can represent nearly 50 per cent of production costs. Improvements in the number of finfish requiring to be fed from artificial sources continues as research in this field is realised and translated to development of the sector with around 33 per cent of fish being fed diets including artificial ingredients in 2010, compared with 50 per cent in 1980 (FAO, 2012). This decrease is due partly to improved diets resulting in better food conversion rates. Furthermore, one-third of the world's cultured food fish harvested was without the use of artificial feed, mainly through production of bivalves and filter-feeding carps (FAO, 2012).

Thus two further major constraints to development of the sector globally are: 1) matching production levels to target market demand to reflect dynamic changes in consumer preferences and perceptions; and 2) applying good governance principles to improve decision making about aquaculture management. The latter constraint includes aquaculture policies needing to better integrate some of the unique characteristics of the sector by providing, for example, incentives for new and smaller businesses to have access to financial and business support to initiate a variety of production models. In developing countries such as Africa, there is a growing consensus that aquaculture should be developed as a commercial activity and private enterprise encouraged – see the FAO (2014) report for more information.

Advancements in research and development on husbandry, feeding, immunology, production technology, processing, packaging and logistics to mention only a few activities important to the value-chain continue to develop; however, the rate of growth for the aquaculture sector in some regions continues to be relatively slow. This is especially the case in Europe which will be discussed later in this chapter.

Regional trends

Brief overviews of aquaculture production levels for different regions are presented in this section using statistics from FAO (2012) and recent observations during fieldwork. The aforementioned report states although there have been improvements in some geographical areas regarding systematic collection of data and analyses, around 30 per cent of 190 countries failed to submit national aquaculture production statistics. Therefore caution should be exercised in using these trends reported herein beyond the purpose for which they are intended; that is, the information is used to give context about regional differences to try and identify constraints influencing production trends. Nonetheless general patterns are described and some gaps are filled using data from a range of sources to provide an indication of relative changes in production levels.

Asia

As indicated earlier, most aquaculture occurs in Asia, home to 89 per cent of worldwide production in 2010 with China accounting for more than 60 per cent (FAO, 2012). The main Asian producing countries are Bangladesh, India, Indonesia, Japan, Myanmar, Thailand, the Philippines and Vietnam, dominated by finfishes (65 per cent), followed by molluscs (24 per cent), crustaceans (10 per cent) and other species (1.5 per cent; FAO, 2012). With the current high levels of interest in vulnerable small-scale fisheries sectors, especially in Asia, non-government organisations such as the Haribon Foundation in the Philippines with whom I continue to collaborate (Lavides *et al.*, 2010) are exploring with fishery-dependent communities the opportunities for introducing aquaculture as an alternative or supplementary livelihood. In the past, many aquaculture livelihood programmes have been introduced in this and other parts of the world without a fuller understanding of the socio-economic variables that contribute to successful adoption of aquaculture in addition to knowledge needed about the environmental considerations (Slater *et al.*, 2013; 2014). This lack of a socio-economic context led to many failed aquaculture projects aimed at alleviating poverty in communities due to human behavioural drivers not being considered as part of introducing livelihood initiatives. In aquaculture livelihood programmes where communities were selected based on their vulnerability to food insecurity, for example, high dependence on depleted fisheries, with the aim being for aquaculture activities to provide food security and/or a source of income, failure of these programmes can compound problems and hardship. Issues might include lack of trust in individuals being unwilling to consider aquaculture at a future date if projects fail. A greater emphasis is now being placed on integrated approaches to introducing community-based aquaculture where equal consideration is given to the social, economic, environmental, policy, legal and technological dimensions of production (Stead, 2013).

Oceania

Oceania makes a minor contribution to global aquaculture production levels compared to other regions previously discussed and farms mostly marine molluscs (64 per cent in 2010, a decrease from 95 per cent in the early 1980s) and finfishes (32 per cent). The major countries involved are Australia and New Zealand which mainly farm Atlantic (*Salmo salar*) and Chinook salmon (*Oncorhynchus tshawytscha*), respectively. Freshwater aquaculture accounts for <5 per cent of the Oceania's production (FAO, 2012). Major constraints to development in this region have been identified as: lack of high quality genetic seed stock and the need for higher levels of expertise in feed technology.

In New Zealand, reasons presented around the debate on whether the country should expand mariculture or not can be partly explained by the historic arguments over uncertainty perceived by some over ownership of the foreshore and seabed. There is also opposition from a range of protesters particularly in relation to expanding salmon aquaculture in New Zealand that include environmental advocates, inshore fishers, recreational anglers, holiday home owners and tourism companies. The uncertainty over sea and ocean tenure including wide ranging debates on how to value natural resources including marine environments is a common debate around coastal parts of the world. Not dealing with this uncertainty is a barrier to many countries in their efforts to attract investors, especially international companies to help build the mariculture sector. In contrast, Tasmania continues to attract investment in developing its aquaculture sector, including for investment for expanding salmon production.

North and South America

In North and South America aquaculture in terms of production volume is mostly finfishes (58 per cent), crustaceans (22 per cent) and molluscs (20 per cent), and according to FAO (2012), South America is expanding its aquaculture development especially in Brazil and Peru. This is in contrast to North America which has maintained production levels in the past few years although this is changing. In North America estimates state that around 84 per cent of sea-food is imported to meet consumer demands as the aquaculture industry is dominated by freshwater farms rearing catfish and carp. There is growing interest in developing mariculture to meet the rising demand for seafood in North America and the National Oceanic Atmospheric Administration developed a marine aquaculture policy in 2011 to aid development of the sector. Furthermore, future changes to trade routes as a result of climate change and possible new routes opening through the Arctic could influence the ratio of imports to exports in North America. In South America, there has been increased interest particularly in investment from international companies to research and develop new species, methods and technologies. Clams, mussels and oysters are being explored especially in Brazil where conflicts surrounding shrimp aquaculture have influenced perceptions and attitudes about future sustainability of aquaculture production.

Europe

Mariculture in Europe has increased from 56 per cent of total aquaculture production in 1990 to 82 per cent in 2010 influenced mostly by expansion of the marine cage culture of Atlantic salmon (*Salmo salar*; FAO, 2012*)*. Finfish aquaculture represents 75 per cent of all European aquaculture production with the remaining 25 per cent largely comprised of farming molluscs (FAO, 2012). European producers have either maintained their current levels of production over recent years or left the sector, especially in the case of the marine bivalve sector, e.g. in Scotland a number of mussel, *Mytilus edulis*, farms ceased trading in 2013 (personal observation). The proportion of bivalves in terms of total production in Europe has continued to decline from 61 per cent in 1980 to 26 per cent in 2010 (FAO, 2012). Mussels were predicted to be one of the fastest growing sub-sectors of mariculture, thus this decrease was unexpected but illustrates some of the negative impacts of natural (unseasonal blooms), economic (decreased price), social (low consumer demand) and technological (production levels not cost-effective with available methods) influences that can be difficult to mitigate.

Africa

Like Europe, Africa is dominated by finfish aquaculture (99.3 per cent by volume in 2010 according to FAO, 2012), with only a small percentage of production arising from marine shrimps and marine molluscs. Overall, African aquaculture production has increased from 1.2 per cent to 2.2 per cent in the last ten years whereas freshwater finfish farming declined from 55 per cent to 22 per cent in the 1990s but was again slowly increasing to 40 per cent in 2010 due to development in sub-Saharan Africa, namely: Nigeria, Uganda, Zambia, Ghana and Kenya (FAO, 2012). To date, there is a lack of success stories in this region for providing evidence to persuade larger scale investment in diversification from finfish with more unsuccessful than successful examples to draw on. To illustrate, in 2011, disease outbreaks were responsible for failure in marine shrimp farming production in Mozambique. In contrast, an international company called Aquapemba was the first to set up a pilot coastal cage farm to grow the marine finfish, Dusky Kob (*Argyrosomus japonicas*) in Pemba Bay in Northern Mozambique, yet the latter received less media coverage and support than the former. Major constraints to development of the mariculture sector include: limited microfinance and small loans especially to purchase feed; lack of capital and investment especially from larger and more experienced investors; natural events causing destruction of equipment, for example, storms damaging cages; risk of theft; training opportunities limited in aquaculture; availability of a regular supply of good quality seed stock; market intelligence; lack of aquaculture-related policies; and negative perceptions of aquaculture. This is not an exhaustive list and constraints will vary according to the size of the operation, intensity of different species farmed and geopolitical characteristics. What is clear is that Africa needs to rethink how best to build its aquaculture sector and, according to the FAO (2014), commercial growth including private investment is the way forward.

In the Western Indian Ocean and islands off East Africa there has been growing interest in invertebrate mariculture, namely sea cucumber (*Holothuria scabra*) farming where community-based projects have been set up in Madagascar and piloted in Zanzibar, a semi-autonomous part of Tanzania. The Seychelles is also exploring the potential to set up commercial production of sea cucumbers in 2014 (personal communication). The next section presents a brief overview of a case study from the Western Indian Ocean to share some of the lessons learned from piloting community-based mariculture (Slater *et al.*, 2013; 2014; Stead, 2013).

Developing country case study: piloting community and private based partnerships to address poverty alleviation using sea cucumber aquaculture in Tanzania

Aim

In 2009, the Leverhulme Trust funded a pilot project with the overarching aim to investigate whether poverty eradication could be addressed through a community-led aquaculture development project in partnership with a University and private business enterprise in Tanzania. This was a three year project that started in 2010, working with Professor Yunus Mgaya (University of Dar es Salaam, Tanzania) and with a postdoctoral researcher, Dr Matthew Slater (Slater *et al.*, 2013; 2014).

Approach and expected outcome

The main approach adopted to address the project aim was participative governance (Stead, 2005a). The philosophy underpinning this approach was to provide a participatory forum so that individuals and groups with an interest and/or who might be impacted and/or could benefit from the expected project outcomes were given the opportunity to be involved in the decision-making process of the project from the agenda-setting stage and throughout until its completion when funding ended.

The expected outcome of the project was to develop and pilot a method that employed good governance principles (e.g. cohesiveness, openness, participation, effectiveness and accountability) to identify social (e.g. perceived interest in aquaculture as a livelihood) and economic (e.g. income generated through farming sea cucumbers) drivers to introducing sea cucumber aquaculture in coastal communities in Tanzania. Local conditions, cultures, practices and political sensitivities were identified and taken into consideration when designing survey instruments and activities.

Objectives

Five broad objectives were used to focus implementation of the research methods to: 1) develop an adaptive learning process for introducing community-led sea cucumber aquaculture using an interactive communication process between researchers, policy makers, private investors/partners and community members; 2) build a stakeholder-led process to support the structure and type of sea cucumber aquaculture operation desired by a local community; 3) construct a user-friendly framework to aid managers and researchers in how to analyse results from a community-led aquaculture development project; 4) identify future constraints to development including guidance on communication of lessons learned; and 5) understand considerations of regulatory and governance frameworks that impact private partners' involvement.

An important part of the community-led project was to set up a pilot sea cucumber hatchery and to attract a private partner. Larval rearing activities commenced in November 2010; however, production of sea cucumber juveniles was not optimised, partly constrained by withdrawal of the private partner and commercial support, limitations of infrastructure (unreliable seawater supply and unstable electricity supplies) and resource constraints. Cage construction of sea cucumbers and pilot farming workshops were undertaken nonetheless to show interested individuals what would be involved if a successful hatchery operation could be realised.

Main project outcomes

Lessons learned in initiating an aquaculture community-based project

Establishing a good working relationship between the main project partners, Newcastle University (UK) and University of Dar Es Salaam (Tanzania) in advance of the project starting was considered a priority to ensure that expectations from all involved were established and taken into consideration in tailoring the fieldwork. In particular, identifying key personnel who could dedicate time to the project and had an established relationship with the communities we worked in helped to facilitate progress.

Attending and presenting at a national workshop on aquaculture in Tanzania in 2009 (Stead, 2009) in advance of the project starting, provided an excellent opportunity to raise awareness about the community-led approach being developed for sea cucumber aquaculture, establish contacts at a local, national and international level, especially with prospective key informants.

The aims and findings were presented in person and orally to village members at the start and before the project finished, taking into account sensitivities around those with a lower educational background. Illustrated reports were supplied to village members, village councils and local authorities. Face-to-face meetings were effective in providing an opportunity to enthuse locals about the project and resulted in providing a forum for successfully recruiting willing local volunteers for the aquaculture pilot studies and extension work during the initial stages.

Findings

The details of how social and economic drivers can influence willingness of individuals to consider sea cucumber aquaculture can be found in Slater *et al.* (2013). The important lesson learned was that when trying to motivate decision makers to consider aquaculture livelihood programmes there needs to be a demonstration of what information individuals use when trying to decide whether to get involved in aquaculture as an alternative or supplementary livelihood. Unless this information is fully understood and used by policy makers then future aquaculture livelihood programmes aimed at tackling poverty will continue to fail.

The main difficulty encountered in this project was achieving the continued support of an investor. A private partner was identified and recruited to the project to help develop the pilot sea cucumber hatchery to supply juveniles for on-growing in sea pens alongside coastal villages; however, the private partner withdrew and ceased hatchery support, thus constraining development of the production facility. There were numerous reasons given, including there needed to be more research to ensure completion of the life cycle under cultured conditions in the pilot hatchery in Tanzania before the company would justify investing more time and resources. The private partner decided to invest in research and development on sea cucumbers back at its main head quarters in South Africa, to address observed constraints to the current hatchery set-up. Constraints identified were: unstable electricity supply, unreliable fresh seawater availability, remoteness of hatchery location and difficulties with obtaining specialised diets for early stages of development post-spawning.

Future research would need to address the set-up of a fully operational hatchery in advance of community involvement unless alternative resources and funding can be found. This is because interviewees expressed that it is important for communities to see a successful hatchery first so that they know there will be a supply of juveniles they can purchase for on-growing (Slater *et al.*, 2013). Many projects on developing aquaculture as a livelihood have failed in the past and the reasons why have not been fully understood; thus the lessons learned from this project may help others working in the sustainable development of the aquaculture field.

In terms of how the five broad objectives were addressed we: 1) used semi-structured and face-to-face interviews, focus groups, workshop meetings and participant observation to show that an adaptive process with two-way communication between researchers, community members, policy makers and private partners was important to maintain from start to finish to facilitate openness and participation; 2) realised learned communities needed to see a successful hatchery in operation before they were willing to commit and get fully involved; 3) identified Bayesian Belief Networks to be useful for analysing socio-economic data (Slater *et al.*, 2013); 4) were made aware of major constraints including a lack of incentives to attract investors and insufficient resources to meet their infrastructure requirements; and 5) highlighted the importance of regulatory and governance frameworks needing to be in place to reduce risk and uncertainty in investment by private partners interested in community/private aquaculture development partnerships.

The participatory governance principles applied to the process and methods used in this project led to a user-friendly framework being designed to aid managers and researchers on how to conduct community-led aquaculture (for further information refer to Slater *et al.*, 2013, 2014).

The framework is generic and may be used for other community-based aquaculture development projects to help optimise the project outcomes. Before ending this brief overview of this project, special thanks must also go to Mr Kithakeni and Dr Lugendo for their technical support. Thanks must also go to all those involved in the project in Tanzania and the students who helped with this work who so kindly gave their time and effort to help identify lessons learned for future development of community-based aquaculture.

Conclusions

In general, most research in aquaculture, especially in developed countries has focused on the science underpinning growth performance and technologies underpinning production. This chapter does not outline advancements in the different technological fields related to aquaculture as these topics were considered beyond the scope of this review and there are some excellent sources of information available on this topic. Nonetheless the patterns described in regional production trends will be significantly influenced in future by the availability of technological expertise and the resources required to implement technological advances in practice. The most successfully farmed marine species to date has been the Atlantic salmon, although part of its life cycle is spent in freshwater. Much interest in expanding mariculture has focused on offshore aquaculture; however the costs to implement some of the technological specifications to deal with extremes of weather conditions, logistics and safety of personnel make this currently an expensive option with low margins for profit based on current seafood market prices. Thus it is more likely inshore mariculture will expand faster than the offshore sector. There are many opportunities to examine how mariculture can be developed, especially as co-location options with other sectors such as the renewable energy sector including the offshore wind industry, in those countries where this is being promoted.

Increasing attention has been paid to aquaculture management where aquaculture is more frequently included as part of plans using integrated coastal zone management or marine spatial planning principles (Stead, 2005a). Although widely recognised that management alone cannot address sustainability goals of aquaculture development as its focus is too narrow, there is growing recognition about the need for a broader governance framework that can take into consideration social, economic, political and market-related issues, to mention only a few (Stead, 2005b; 2009; 2012a; 2012b; 2013; FAO, 2014).

Approaches to enhancing or introducing aquaculture in developing regions have focused more on technical production, with less effort placed on social, economic, political and market related issues (Ahmed and Lorica, 2002; Stead, 2005a; 2005b; 2009; 2013). Growth rates of aquaculture in some developing regions (e.g. Western Indian Ocean – WIO) and developing countries (e.g. Tanzania) have been slowed by insufficient investment combined with lack of knowledge on governance – decision-making processes needed to support management, research and business capacity to strengthen this sector. Furthermore, when aquaculture programmes are trialled, for example, in Africa, there can be a lack of context-specific data or focus on resource availability, environmental, social and market conditions (Brummet and Williams, 2000; Eriksson *et al.*, 2012) at multiple governance levels (local, national and regional). Thus a major constraint to advancing aquaculture in regions such as the WIO is the lack of a regional coordination unit that includes all the aforementioned considerations, leading to interventions tending to be uncoordinated, of an ad hoc nature and in the longer term unsustainable (Charles *et al.*, 1997).

Mariculture if applied at the appropriate scale and context-specific has the potential to offer livelihoods and support sustainable development of marine and coastal environments if an ecosystem's approach to aquaculture (EAA) is adopted (FAO, 2010; Troell *et al.*, 2011). In conclusion, mariculture operations are commonly undertaken in isolation from other marine sectors and aquaculture is not afforded the opportunity to play on a level playing field in terms of access to support required to develop the sector. On a final note, unless there is a stronger political will to promote mariculture with the resources required to implement advances made through research and development, combined with effective communication strategies aimed at increasing demand for farmed sources of seafood then the mariculture sector will not fulfil its full potential in contributing to tackling global food insecurity.

References

Ahmed, M., Lorica, M. H. 2002. Improving developing country food security through aquaculture development: lessons from Asia. *Food Policy*, 27, 125–141.

Brummett, R. E., Williams, M. J. 2000. The evolution of aquaculture in African rural and economic development. *Ecological Economics*, 33, 193–203.

Charles, A. T., Agbayani, R. F., Agbayani, E. C., Agüero, M., Belleza, E. T., González, E., Stomal, B., Weigel, J.-Y. 1997. *Aquaculture Economics in Developing Countries: Regional assessments and an annotated bibliography*. FAO, Rome.

Eriksson, B. H. B., Robinson, G., Slater, M. J., Troell, M. 2012. Sea cucumber aquaculture in the western Indian Ocean: challenges for sustainable livelihood and stock improvement. *AMBIO: A Journal of the Human Environment*, 41(2), 109–121.

FAO. 2010. *Aquaculture Development. 4. Ecosystem approach to aquaculture. FAO technical guidelines for responsible fisheries*. No. 5, Suppl. 4. Rome, Food and Agriculture Organization, 53 pp.

FAO. 2012. *The State of World Fisheries and Aquaculture 2012*. Rome, Food and Agriculture Organization, 209 pp.

FAO. 2014. *The State of World Fisheries and Aquaculture 2014*. Rome, Food and Agriculture Organization, 223 pp.

Lavides, M. N., Polunin, N. V. C., Stead, S. M., Tabaranza, D. G., Comeros, M. T., Dongall, J. R. 2010. Finfish disappearances around Bohol, Philippines inferred from traditional ecological knowledge. *Environmental Conservation*, 36(3), 1–10.

Slater, M. J., Mgaya, Y. D., Mill, A. C., Rushton, S. P., Stead, S. M. 2013. Effect of social and economic drivers on choosing aquaculture as a coastal livelihood. *Ocean and Coastal Management*, 73, 22–30.

Slater, M. J., Mgaya, Y. D., Stead, S. M. 2014. Perceptions of rule-breaking related to marine ecosystem health. *PLOS ONE*, 9(2), e89156.

Stead, S. M. 2005a. A comparative analysis of two forms of stakeholder participation in European aquaculture governance: self-regulation and integrated coastal zone management. In: Gray, T. S., ed. *Participation in Fisheries Governance*. Dordrecht, Springer, pp. 179–192.

Stead, S. M. 2005b. Changes in Scottish coastal fishing communities: understanding socio-economic dynamics to aid management, planning and policy. *Ocean and Coastal Management*, 48, 670–692.

Stead, S. M. 2009. Plenary entitled 'Future of marine aquaculture in a changing fisheries environment. In *Proceedings of AQA Marine Aquaculture – the 21st century opportunity and challenge for world fishing conference*', 18 September, Vigo, pp. 14–17.

Stead, S. M. 2012a. Keynote entitled 'Integrated coastal management and the development of a knowledge-based marine economy', Ningbo, China, September 2012.

Stead, S. M. 2012b. Keynote entitled 'Balancing food security and environmental concerns: challenges of sustainability', British Ecological Society, Birmingham, UK, December 2012.

Stead, S. M. 2013. Keynote entitled 'Future strategic direction for sustainable aquaculture in western Indian Ocean. Community based aquaculture in WIO: challenges and lessons learned', Zanzibar, 9–11 December 2013.

Troell, M., Hecht, T. Beveridge, M. Stead, S. Bryceson, I. Kautsky, N. Ollevier, F. and Mmochi, A. 2011. Mariculture in the Western Indian Ocean region: introduction and some perspectives. In: Troell, M., Hecht, T., Beveridge, M., Stead, S., Bryceson, I., Kautsky, N., Mmochi, A., Ollevier, F. (eds) *Mariculture in the WIO Region: Challenges and prospects*. WIOMSA Book Series No. 11. viii+ pp. 1–5.

Energy and materials

18

OIL AND GAS

Hance D. Smith and Tara Thrupp

Introduction

The development of offshore hydrocarbon resources has a long history, with small beginnings in the closing decades of the nineteenth century, followed by rapidly increasing importance after the Second World War, from the late 1940s onwards. This chapter begins by tracing the stages of development of the offshore industry within the broad context of the global oil industry. This is followed by an outline of the resources concerned, ranging from crude oil to natural gas, as well as their innumerable combinations, taking due account of the key factors that influence the estimation, production and consumption of these resources. The offshore oil and gas industry is associated with the development of advanced technologies in a number of fields, ranging from petroleum engineering on the one hand, to offshore engineering designed to overcome the considerable obstacles to working in the marine environment on the other. This in turn focuses attention not only on the complex interplay of environmental influences and impacts, but also on the economic and social implications of offshore industry operations at various geographical scales. Finally, the governance and management of the industry are discussed.

Development

From the beginning of the use of crude oil in the course of the nineteenth century, the development of hydrocarbon resources beneath the seafloor has proceeded in parallel with the much more extensive development on land. The reason for this is the extensive presence of hydrocarbon-bearing geological formations that straddle present-day postglacial shorelines, often with greater resource potential offshore compared to onshore, so that oil production onshore could relatively readily be extended onto the adjacent continental shelf, at least close to the shore. Particularly notable early developments included offshore Baku in the Caspian Sea, where the first offshore waterproofed wells were sunk 20–30m offshore as early as 1824, with much more extensive developments in the final two decades of the nineteenth century (Patin 1999); the Caspian was also associated with the first modern oil tanker, the *Zoraster*, built in 1878, which made its maiden voyage from Baku to Astrakhan in the same year (LeVine 2007). A second significant development was offshore California in the 1890s (Yergin 1991). Later, in the 1920s, offshore production commenced in Lake Maracaibo in Venezuela (Salas 2009). In the long stage of global development between the 1870s and the 1930s, demand for oil was notably driven by the replacement of whale oil by kerosene; and by naval interests with the adoption of the steam turbine in the years before the First World War: oil was twice as economical as coal for raising steam.

Moreover, oil was four times as economical as coal when used in internal combustion engines, which played a decisive role in the expansion of demand throughout the next long stage of development between the 1940s and the 1990s through its applications in land, sea and air transport, together with developments in electrical power generation and petrochemical industries. Thus much more significant offshore development commenced in the late 1940s adjacent to the United States coast in the Gulf of Mexico in particular. Later, in the 1960s and early 1970s, offshore development commenced in the North Sea, initially associated with gas production in the southern North Sea following the discovery in 1958 of the giant Slochteren onshore gas field in the Netherlands. This precipitated the North Sea continental shelf cases between 1965 and 1971, when the region was subdivided among the littoral states by median lines. Most of the development of production took place in the United Kingdom and Dutch sectors (Hutcheson and Hogg 1975). Other notable continental shelf developments occurred at about the same time in the Persian Gulf and South China Sea (Figure 18.1).

Production of hydrocarbons from the marine environment is almost always more costly than comparable developments on land. Thus a crucial circumstance driving offshore exploitation was the step changes in the value of crude (Figure 18.2). Between the initial steep rise in the crisis of 1973 and the early 1980s, the value of crude in the market had increased approximately tenfold. Thus, towards the end of this second stage advancing technology associated with continuing high prices of hydrocarbons made it economical both to explore and begin exploitation in much deeper waters beyond the continental shelf, extending well down the continental slope (Figure 18.3), as the demand for both oil and gas continued its global expansion. For example, between 1948 and 1973 world oil consumption grew sixfold; it continued to grow thereafter, with gas becoming ever more important, as it was especially suited to national gas grids in key markets such as the United Kingdom; as well as for power generation and in the petrochemical industry.

The industrial structure of the offshore oil industry is part of the wider structure of the global oil industry, but has distinctive and sometimes specialised components derived from the nature of the marine environment, associated technological development and high costs. The basic operations in the handling of oil and gas may be classified into five groups: exploration, production, transportation, processing and distribution, the first three of these being especially maritime-oriented. In the long-term development of offshore provinces such as the Gulf of Mexico and the North Sea – a process that takes place over several decades – there is a distinctive sequence of development, beginning with exploration and the development of the service industry, primarily located in ports accessible to the offshore fields. This is followed by the manufacture and installation of production facilities, which is the primary function of offshore engineering. The next and longest stage is production and maintenance, which is followed by decommissioning of offshore installations as the provinces gradually near the end of their productive lives. For relatively small provinces with few fields, all of these stages may be part of a single development phase. For large provinces, such as those already mentioned, there are several iterations involved that overlap with one another in time. Thus in the North Sea, for example, the gas fields in the south were developed first; followed successively by the oil (and some gas) fields in the central and northern region; and followed yet again by the deep water developments on the edge of the continental shelf and on the continental slope west of Shetland.

The combination of operations and stages led in turn to the emergence of an offshore industry with four major components. First is exploration and the development of production. Exploration begins with airborne aeromagnetic and gravimetric surveys aimed at delineation of possible hydrocarbon formations detectable at the regional geology scale. This leads into seismic surveying at local scales to map out the detailed geology, which is done by seismic survey ships. Once promising geological structures are identified, the next stage is test drilling, using exploration rigs, discussed further below.

The offshore engineering industry is the second major element of the offshore industry. It begins with the building of mobile exploration rigs, including jack-ups and semi-submersibles; and drill ships. Once fields are proved and production is decided upon, production facilities are developed, including the building and installation of production platforms, pipelines and associated equipment. On shore this involves the development of production platform yards, module building yards and pipe-coating yards. In the case of smaller scale steel-based production platforms and module yards, yards may be developed from conventional shipyards and other dockside industrial facilities. Large-scale steel or concrete platforms require dedicated dry dock yards adjacent to deep water such as the sea lochs of western Scotland and the Norwegian fjords. Decommissioning facilities may be located adjacent to service bases – operational or redundant. The offshore engineering industry shades into the wider engineering industry which supplies a wide range of specialised equipment, notably the risers used to transport chemicals, muds, oil and gas between the reservoirs and the exploration and production rigs on the surface. Meanwhile, offshore engineering firms involved in construction of platforms in particular are closely related to or even offshoots of the larger building and civil engineering industry companies.

Exploration is developed in tandem with the build-up of the service industry, which then continues throughout the production and, to a diminished extent, decommissioning stages. A primary requirement is the availability of sea- and airport facilities around the clock at locations that are readily accessible to the exploration and later, production locations offshore. In the North Sea in the 1960s and 1970s, for example, the maximum geographical sphere of influence of service industry ports was strongly determined by the geographical ranges of operation of offshore supply vessels and helicopters respectively. These roughly coincided at around 350km, the convenient overnight operating range of the supply vessels and the maximum operating range of the helicopters then used to ferry personnel to and from the exploration rigs and pro-duction platforms. Ashore, the service industry requires shore bases in port areas in all-weather harbours with ample adjacent storage space for pipes, drilling muds and chemicals, and engineering equipment warehouses. Also needed are engineering facilities and engineering supplies and services, including diving; specialised drilling; and hydrographic, geophysical and engineering surveying firms. Also important are transport services, including specialised oil service vessels and tugs; cargo shipping services; rail transport and heavy road haulage, scheduled and charter air services, and helicopter services. In the accommodation and catering field as well as housing it may be necessary to establish temporary accommodation including work camps; as well as contract catering, laundry services, and other functions.

The fourth and final element is the finance industry. Offshore exploration and production are capital intensive, operating at the cutting edge of a whole range of technologies, with concomitant demands on financing. Even individual field developments may run into billions of US dollars, beyond even the largest oil companies' abilities to undertake the risk involved single-handed. Thus such developments are financed via specialised offshore divisions of large banks, often operating as consortia to spread the risk; as well as dealing with the long time horizons of such developments, which are measurable in decades.

Figure 18.1 Offshore exploration and production
Source: Author.

Defining characteristics of the offshore industry include its complexity and global scale of operations. At the core of the industry are the major international oil companies and state-owned oil companies of some of the major producing countries, notably Russia and major members of the Organisation of Petroleum Exporting Countries (OPEC) such as Saudi Arabia and Venezuela. Here major exploration and production decisions are taken, as well as management of long-term production and decommissioning. The international oil companies are head-quartered in a handful of major cities in North America and Western Europe, with exploration and production divisions located adjacent to major production areas, in cities such as Houston and Aberdeen. Exploration and production components of the industry are necessarily located adjacent to major oil provinces (Figure 18.1). While specialised engineering is widely distributed in the larger engineering industries, offshore engineering is also concentrated geographically near a handful of offshore provinces, but operating at global level. Supreme is the US Gulf Coast, followed by certain areas adjacent to the North Sea such as Scotland, north-east England and southern Norway. Transport operates in a similar way, notably including the specialised shipping divisions of large shipping companies and national-level rail and road haulage organisations.

Resources and reserves

Hydrocarbons under the seabed are the same as hydrocarbons under the land. There exists a huge group of compounds ranging from crude oil to natural gas, with innumerable combinations of the two. Crude oil is refined into a large range of products, including kerosene, used as aircraft fuel; gasoline and diesel; as well as providing feedstock for the vast range of chemicals produced by the petrochemical industry, including plastics, industrial chemicals, dyes and detergents. The most important natural gases include methane, ethane, propane and butanes. Methane is used directly for domestic gas supply through gas grids; ethane is mainly used in the petrochemical industry to produce ethylene, the most important manufactured organic chemical.

Hydrocarbons are mainly located in sedimentary rock formations that include combinations of source rocks, such as shales and coal; reservoir rocks including shales and limestones; and massive cap rocks such as sandstones. The geographical distribution of some of these resources on the world's continental shelves is indicated in Figure 18.1. Migration of hydrocarbons takes place from the source rocks to the reservoir rocks on geological timescales. The cap rocks prevent continued migration upwards from the reservoir rocks, not only by virtue of their lithological characteristics, but also by geological structures such as folds and faults, which influence the dispositions of the three respective categories of rocks in such a way as to seal the reservoirs to produce structural or stratigraphic traps. With the aid of seismic data it is possible to construct reservoir maps in 3D as a first step in estimation of the material reserves present.

The calculation of recoverable reserves is a complex business that continues throughout the life of an individual field as well as that of an offshore province consisting of many fields. To begin with, only a proportion of the resources can be extracted from the reservoir, partly due in the first instance to the nature of the hydrocarbons, together with the lithology and structure of the reservoir itself. This is further complicated by the installation of the production facilities, notably the disposition of wells relative to the pressure patterns in the reservoir, and the presence or absence of drive mechanisms such as water underlying the hydrocarbons that may maintain pressure for a considerable time; drive pressures may be maintained artificially as pressure drops by, for example, injection of water to lengthen the productive life of a field. Computer-based simulation modelling is used to predict the behaviour of fields using various well configurations for example, bearing in mind that the economic lifetime of a field is likely to extend over several decades.

Since the 1970s estimated production lifetimes of individual fields and provinces have tended to increase in the course of production, due to improvements in the science of petroleum geology coupled with advances in the technology of petroleum engineering, so that recoverable reserves have tended to increase from around 30 per cent to around 50 per cent in some cases. Calculation of material reserves for individual reservoirs may employ a number of methods, including comparison with nearby fields, volumetric estimation and material balance. However, the most important calculations are based on the production decline curve, based on monitoring of production throughout the productive life of the field. For offshore provinces – key to national interests – it is possible to simply calculate the resources in place by summing the data of the individual fields present. However, more useful is the use of the proved reserves concept that is used to predict production based on existing economic and technological circumstances at the time of the initial calculation, without taking into account economic and technological changes (van Meurs 1971). As a result, offshore production in areas such as the Gulf of Mexico and North Sea has already lasted much longer than initially assumed (Odell and Rosing 1975).

Beyond the physical and petroleum engineering influences on reserve estimation are economic factors. The starting point is the demand for hydrocarbons as reflected in the price. As the marine environment is associated with high costs of production, exploration and exploitation of offshore hydrocarbons only really took off in the early 1970s when the first major rise in the price of crude took place; the pace of development tended to slow down during periods of relatively low prices, most notably in the later 1980s and 1990s, accelerating again in the first decade of the twenty-first century (Figure 18.2). Bearing in mind the decadal lifetimes of individual fields and the even longer lifetimes of offshore provinces, it is difficult or even impossible from an economic perspective to be precise about reserve estimation more than a few years ahead. The economics of offshore hydrocarbon production is further complicated by the role of the state, in which the ultimate ownership of the resources is invested. This is discussed in the final section of this chapter.

In summary, the resources, and therefore reserves of offshore hydrocarbons are in practice primarily the outcome of the interplay of the geology of the resources; the influence of petroleum

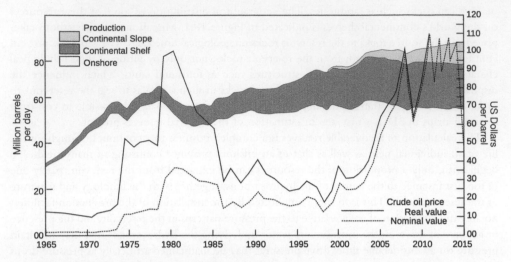

Figure 18.2 Offshore production: continental shelf and slope
Sources: Ferentinos 2013; Jashuah 2012; Market Realist 2014.

engineering factors that come into play once production commences; the relatively long-run role of the economics of the industry expressed in supply and demand (BP annual) mediated through prices and costs; and the role of the state through national policies, which are underlain by political factors (Sandrea and Sandrea 2007).

Technology

While the marine environment influences the nature of the reserves calculations profoundly through its imposition of high costs of production, it most obviously influences the industry through the specialised technology required to overcome obstacles of water depth and the dynamic ocean–atmosphere system. This is subject to large-scale variations on a range of timescales including short-term storms of various kinds, through longer term seasonal variations to even longer term climate changes that affect both atmosphere and ocean. Technology may be usefully discussed in relation to the operations of exploration, production, transportation, processing and distribution already introduced above, with the focus on the sea for the first three of these functions, and the coast for the second two.

For the exploration phase, aeromagnetic and gravimetric surveys are conducted from the air. The marine influence really begins with seismic exploration, which is undertaken using specialist survey vessels. For the ensuing test drilling, a primary requirement is for mobility of the drilling rigs, which can be readily moved from one test well to the next. Thus in relatively shallow waters, less than 100m deep, jack-up rigs are commonly used: the deck can float, and the supporting legs can be raised when the rig is moved and lowered to the seabed when drilling. For deeper parts of the continental shelf and upper parts of the continental slope just beyond 200m depth, exploration is largely effected using semi-submersible rigs that are self-propelled and can be maintained in position using systems of anchors supplemented by thrusters. In the deeper waters down to the seaward limits of the continental slope at depths of several thousand metres, dynamically positioned drill ships are employed.

The decisive influence of water depth on design remains present during the production phase (Chakrabati 2005). Although both jack-ups and semi-submersibles can be modified for the production stage, and piers built out from the shore were often used during the very early stages of production noted at the beginning of this chapter, a range of mainly steel fixed and floating structures are used for production. These include a range of steel jacket platforms piled into the seabed, which are widely used on the continental shelves in the major offshore provinces. In a few areas, such as the northern North Sea there are also between 20 and 30 concrete gravity platforms equipped with storage capacity. Fixed structures in the form of flexible compliant towers and tension leg platforms are used in deeper water on the outer edge of the continental shelf and well down the continental slope. In the deepest waters semi-submersibles, floating production storage and offloading (FPSO) technology, and spar platforms are employed. Subsea completions are used at all depths: in shallower waters these are useful for exploiting small, otherwise uneconomic fields that can be connected to existing pipeline systems. In deep water, these are connected to floating systems such as FPSOs. By 2005 the full current range of water depths had been reached (Office of Ocean Exploration 2010) (Figure 18.3).

In offshore transportation, an initial distinction has to be made between liquid crude oil on the one hand, and natural gas on the other, bearing in mind that it is necessary to separate the two for large-scale transportation. For gas, pipelines are essential, and are linked to coastal shore terminals, which may be either connected directly to land grids or processing facilities, or used for loading liquefied natural gas (LNG) tankers, such as at Braefoot Bay and Moss Morran adjacent to the Firth of Forth in Scotland, which were respectively developed both for tanker loading and siting of an ethylene cracker plant. In the case of crude oil, various configurations of pipelines,

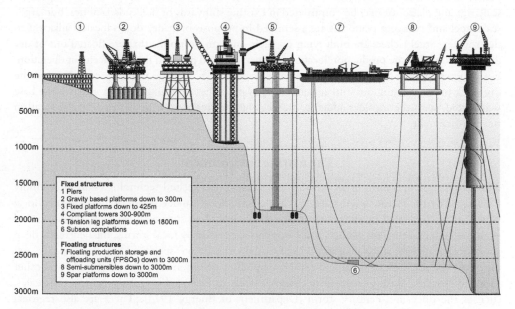

Figure 18.3 Types of offshore oil and gas structures (GOM – Gulf of Mexico). For 3: (deepest: Shell's Bullwinkle, 1991, 412m/1,353ft GOM); For 4: (deepest: Chevron Texaco's Petronius, 1998, 534m/1,754ft GOM); For 5: (deepest: ConocoPhillips Magnolia, 2004, 1,425m/4674ft GOM); For 6: (deepest: Shell's Coulomb tie to NaKika, 2004, 2,307m/7570ft GOM); For 7: (deepest: 2005, 1,345m/4,429ft Brazil); For 8: deepest: Shell's NaKika, 2003, 1,920m/6,300ft GOM); For 9: (deepest: Dominion's Devil's Tower, 2004, 1,710m/5,610ft GOM)

Source: Author. Office of Ocean Exploration, NOAA – oceanexplorer.noaa.gov/explorations/06mexico/background/oi/media/types_06.html; and Chapter 3, Shell Oil and Gas Offshore Production – http://s081.static-shell.com/content/dam/shell/static/usa/downloads/alaska/os301-ch3.pdf, both accessed 10 January 2015.

offshore storage and transport to shore by tankers are used. For small fields not near other fields, and for the initial stages of large field developments dedicated tankers with or without offshore storage are likely to be most economical, for example in the case of the Brent field in the North Sea, where a floating spar was employed during the early phase of development (Owen and Rice 1999), although tanker operations are vulnerable to and often interrupted by adverse weather conditions. For large-scale transport over long timescales, such as large individual fields and groups of fields typical of offshore provinces such as the Gulf of Mexico and the North Sea, pipeline systems have been installed. These are linked to tanker terminals on the coast and may also be directly linked to coastal or inland gas processing plants, refineries and petrochemical plants.

Offshore exploration and production also requires a substantial range of ancillary technologies. These include supply and standby vessels as well as helicopters; crane barges, barges for transporting modules and other equipment, and accommodation barges during platform installation; trenching and pipe-laying barges. Firefighting barges may be needed throughout both stages; and crane barges are needed for both installation and decommissioning, when platforms are dismantled. Deep diving technology and remotely operated vehicles (ROVs) are needed for work on the seabed, for example, periodic inspections of platforms and pipelines. There are many thousands of wellheads on the seabed, comprising many abandoned dry wells or wells that have ceased production. Flare stacks are needed to regulate the flow of gas from reservoirs.

Finally, while the shore side of the offshore industry includes a vast range of technologies, particular interest in the present context attaches to the construction of rigs and platforms. Small-scale rigs and platforms may be constructed in existing shipyards or dockside facilities, but large-scale steel and concrete platforms are assembled in purpose-built dry docks located adjacent to deep water. Steel jackets are built lying on their sides on pontoons that are floated out of dry dock and towed to the production location for upending and piling in the emplacement operation. For concrete gravity platforms, only the base containing the storage facility is completed in the dry dock before floating out into adjacent deep water. The legs are then slip-formed as the base is ballasted, with the topsides added to the legs before towing out to the production location for emplacement.

The marine environment

The design and operation of all offshore installations and related technologies are profoundly influenced by the marine environment. This includes local surface winds; wind-generated local waves; swell generated by distant weather systems including storms, hurricanes, typhoons and cyclones; and currents, including surface currents generated by local storms, tidal currents, deep water ocean currents, and non-storm-related currents that are site-specific. On the seabed pipelines especially are at risk from scour and exposure by shifting sand moved by bottom tidal and other localised currents. Risk assessment in design is likely to be based on relatively long timescales, such as the once-in-a-century storm (Department of Energy 1974). There are also regional contrasts, notably between the temperate regions of the North Atlantic that exhibit large-scale seasonal variations in weather with storms concentrated in the winter months; sub-tropical and tropical regions such as the Gulf of Mexico and South China Sea that are characterised by hurricane and typhoon seasons respectively; and the ecologically fragile Arctic and sub-Arctic seas and coasts. Global climate change may also be a consideration on longer investment decision-making scales (Burkett 2011).

The environmental impacts of offshore hydrocarbon operations (Boesch and Rabalais 1987; Cairns 1990; Holdway 2002; Speight 2015) may be considered in relation to exploration, production, transportation, processing and distribution, together with decommissioning. These may in turn be classified into two broad categories, namely, impacts associated with routine operations on the one hand, and those concerned with accidents of various types on the other. Impacts can also be specifically related to key elements of the environment: the atmosphere, sea surface, water column, seabed and coast with their associated ecosystems. Many, but not all of the impacts relate to pollution by crude or refined oils and associated drilling muds and chemicals that persist to varying degrees in the marine and coastal environment. Much of the information and data on oil in the marine environment relates to large-scale oil spills associated with tanker accidents, most of which are not directly related to offshore operations.

In the field of routine operations, noise generation by seismic surveying impacts cetaceans; and flaring poses a hazard to seabirds. Exploratory rig operation impacts the seabed through anchoring. Service industry shipping and port operations are associated with routine discharges from vessels and port facilities. Offshore, exploration and production drilling gives rise to large quantities of drill cuttings and associated drilling muds deposited on the seabed surrounding wellheads which both smothers the seabed and pollutes the immediate environment to a limited extent, while produced water and minor oil and chemical discharges from the rigs and platforms affects the sea surface and enters the water column (Bakke *et al.* 2013). The building of coastal installations including service bases, platform yards, terminals, pipelines and pipeline landfalls and downstream plants such as refineries has major environmental impacts. These include modification of land use, visual intrusion, noise, pollution, and modifications of ecology and hydrology. Offshore, platforms, pipelines and wellheads impact the seabed directly; while sea use conflicts are especially important in intensely used sea areas, notably in the case of fisheries, navigation associated with commercial shipping and peacetime military operations, and aggregate dredging. Many of these impacts, especially those associated with construction offshore and on the coast, are large scale and long term, although the impacts of routine marine pollution appear to be more limited in both time and location.

By contrast, the environmental impacts of major accidents are spectacular, although again frequently relatively limited with regard to locational extent. The principal outcomes are large oil spills associated either with blowouts or tanker accidents; explosion and fire associated with rig and platform accidents that may or may not involve oil pollution, but are liable to lead to loss of life; and loss of exploration rigs – again associated with loss of life (Alford *et al.* 2014). In the case of blowouts, the effects are variable and unpredictable: key examples include those of Ekofisk in the North Sea in 1977, which did not catch fire and resulted in a large sea surface slick comprised of light oil fractions that dispersed before reaching land; and the Ixtoc I blowout in the Gulf of Mexico that was associated with an explosion and fire. Platform accidents have notably included the *Piper Alpha* explosion and fire in the North Sea in 1988 that resulted in 167 fatalities and the loss of the platform (Cullen 1990); and the 2010 *Deepwater Horizon* blowout, explosion and fire that caused 11 fatalities and massive oil pollution of the water column and adjacent coasts (National Commission 2011). An example of a large-scale pollution incident at a terminal was that of the *Esso Bernicia*, which collided with a jetty at the Sullom Voe Terminal in Shetland (Scotland) in 1978, rupturing the ship's heavy fuel oil tanks. However, the majority of serious tanker accidents are not directly related to offshore operations, such as one of the best documented environmental impacts of a tanker accident, the loss of the *Braer* offshore Shetland in 1993 (Ritchie and O'Sullivan 1994). Between 1955 and 1968 over 20 mobile drilling units had been lost in disasters (Chakrabati 2005), although the worst single accident was the capsizing of the accommodation platform *Alexander Kielland* in the North Sea in 1980, with the loss of 123 lives.

The impact of oil on the ocean takes a wide variety of forms (Speight 2015). Chemical and physical changes include among others, biodegradation, dispersion, emulsification, evaporation, oxidation and weathering, physical transport, sedimentation and spreading. At the level of individual organisms such as seabirds, fish and benthic fauna, implications include ingestion; tainting; lethal toxicity; development, growth and reproductive effects; and smothering. At ecosystem level there may be bioaccumulation in organisms and sediment; changes in ecosystem structure; and degradation of ecosystems. Hazards are particularly acute in coastal intertidal wetlands and in benthic ecosystems on the sea floor.

Economic, social and political implications

The driving forces in development of offshore oil and gas involve the interaction of economics and technology. At local and regional scales, development most obviously creates employment, not only within the specialised sectors of the offshore industry itself noted above, but through multiplier effects in a wide range of industries in the engineering and service sectors, including specialised survey and engineering firms, accommodation, transport and finance. A large proportion of the jobs created are highly skilled specialisms mainly concentrated in urban industrial regions in the developed world: the most important regions are the Gulf Coast of the United States and the littoral states of the North Sea, especially the United Kingdom. Even in such areas, development of offshore engineering and service bases are necessarily often located in rural areas lacking the skilled populations required. Thus development is associated with substantial population movements into these locations from other parts of the country involved and abroad. There is considerable urban expansion. A high level of skills coupled with the time pressures of development lead to high levels of both personal and regional incomes. Regional industrial structures become more diversified, and substantial pressures may be placed on pre-existing industries that find themselves in competition with the oil industry for labour (McNicholl 1977).

It is at local and regional levels that the social impacts of the offshore hydrocarbon industry are most evident (Button 1976; Lyddon 1976; Moore 1982; Wills 1991). The proximate driver is the increase in employment, associated with rises in wages and salaries, which increases personal incomes. This is allied to influxes of population, either temporarily on large construction projects, where workers are housed in purpose-built camps or liners moored alongside construction sites; or permanently in expanded urban settlements. The former creates social tensions; and the latter greatly increases pressure on infrastructure due to the supply of housing and public services. In rural areas in particular, political tensions may arise within communities driven by the desire to increase employment, personal and regional incomes on the one hand; and the resistance to development perceived as undermining other industries and the way of life on the other.

At national level the economic implications of offshore development are frequently significant, even if the offshore industry is only a part of a larger onshore oil industry. In the major industrial centres adjacent to the Gulf of Mexico and the North Sea in particular, the full range of oil and related industries are present (Mackay and Mackay 1975). However, many of the important off-shore developments are in countries and large regions that do not possess the full range of offshore engineering industries, so that the direct exploitation of the resource itself is paramount as, for example, in Alaska, Brazil, Nigeria, Angola, India and Sakhalin (Figure 18.1). Either way, the offshore hydrocarbon industry evolves into a major sector in national economies, and it becomes necessary to develop clear national policies for development. These may vary, an often cited example being the contrasting depletion policies of the United Kingdom on the one hand, favouring rapid development; and Norway on the other, where a relatively slow rate of development was preferred (Earney 1992). Internal national political tensions may also arise regarding the role of government in the promotion of the industry and the use to which large-scale government revenue deriving from the offshore industry may be put, ranging from support for industrial development, through social programmes to creation of sovereign wealth funds (Harvie 1994; Kemp 2011–2012; Smith 2011).

At global level it is notable that the share of world oil production from the marine environment has steadily increased since its inception in the 1960s, first on the continental shelves and latterly on the continental slopes; it is now over 30 per cent, despite the offshore – especially the continental slopes – being a high-cost environment. At the time of writing in early 2015 the vulnerability imposed by high costs is becoming evident, with the postponement or shelving of plans for deep water projects, although several of these continue to progress. Politically, the development of offshore hydrocarbon resources was a driver in the extension of coastal state jurisdiction negotiated for the Law of the Sea Convention between 1973 and 1982 at a time when offshore developments were accelerating worldwide. A comparable mindset is also currently driving national claims to sea areas in the Arctic Ocean as sea ice coverage progressively shrinks. Existing or potential developments are sometimes a catalyst for political instability, for example, in the Niger Delta region of Nigeria; and offshore the Falklands/Malvinas, where Argentina regards current exploration licensed by the Falkland Islands' Government within the exclusive economic zone as illegitimate. Offshore oil operations have also been severely affected by corruption in certain cases, notably in Brazil.

Governance and management

The offshore oil and gas industry has evolved to constitute a large part of the oil industry overall, a truly global industry dominated, as already noted, by a relatively small number of transnational and state-owned oil companies. These lead on the major investment and production decisions, including the never-ending quest for new resources, where the marine environment has come to represent a decisive area of operation. The dominance of hydrocarbons in the energy budget of the global economy is also associated with the large practical role and political influence of states in its governance and management, in the present context coastal states with substantial resources within their maritime jurisdictions. A third group of stakeholders is to be found in civil society, notably the voluntary organisations that campaign on oil-related issues; and communities at local and regional levels that are impacted by offshore developments. The geopolitics and decision making regarding offshore oil are complex, and lie largely beyond the scope of this chapter. Rather the discussion is now focused upon the practical measures of governance and management that mediate the relationships between the offshore industry on the one hand and the marine environment on the other (Kemp 2011–2012; Bridge and Le Billon 2013).

The first task for management is licensing of exploration and production, which occurs before and during the exploration and subsequent production stages. For a coastal state, the continental shelf is first generally divided into blocks for licensing. Exploration licences are issued for relatively short periods to avoid licence holders simply hanging on to the licences without doing any exploration, and use of the licences is governed by sets of conditions. Licences are often issued in a series of rounds: during early stages of exploration, if states wish to have substantial exploration activity under way quickly, this may involve licensing many blocks at the same time, while in a mature province the rate of exploration can be influenced by simply issuing relatively small numbers of blocks for licensing in successive rounds. The selection of blocks is also influenced by the progressive increase in geological knowledge as the offshore province is explored and exploited. Once production decisions are taken by oil companies, governments then begin to issue production licences, also governed by sets of conditions, but necessarily extending over much longer time periods. Overall, the licensing system can be used by government as a tool to encourage exploration and production and to vary the rate at which these operations take place.

The second set of tools that coastal state governments use relates to taxation. Oil companies are subject to normal company taxation, although by virtue of their transnational nature are subject to double taxation arrangements among states, to avoid the companies being taxed more than once. Normal state practice also involves levying royalties on the wellhead value of hydrocarbons produced, which is generally levied at a low rate of not more than the 10–15 per cent range. However, because of the very large scale of revenues associated with offshore production, states also levy additional special taxes on oil companies operating within their marine jurisdictions, such as the petroleum revenue tax and supplementary petroleum duty introduced in the United Kingdom in the mid 1970s and early 1980s respectively (Rowland and Hann 1987). In the United Kingdom's case this resulted in over 80 per cent of the income being generated being paid as tax during the period of peak production in the early 1980s, although big variations in the circumstances of individual field developments resulted in these being individually ring-fenced for tax purposes. Taxation is thus a major influence on investment decisions; during periods of low prices, such as the late 1980s and at the time of writing in early 2015, governments are pressurised to reduce the tax burden to maintain investment in the industry. Taxation is also a major tool in depletion policy.

The third management field concerns offshore safety. As already noted, the industry operates in a hostile environment at the limits of technological development. Early stages of development are thus associated with putting in place increasingly comprehensive safety legislation governing all aspects of working, although this lagged behind events in the North Sea case, for example, where several accidents occurred before legislation was put in place. Offshore industry-specific measures relate to design, construction and operation of all installations and engineering equipment, together with specialised functions such as diving. In common with many industries, although safety was a major consideration it was not generally fully integrated into production systems. This changed with the *Piper Alpha* accident referred to above, after which safety cases have had to accompany offshore developments. The offshore industry is also subject to related legislation, especially that relating to shipping operations. After the loss of the *Braer* noted above, for example, standby tugs were based in northern Scottish waters to cope with emergencies (Department of Transport 1994).

Onshore, coastal developments are also subject to government regulation. Much of this is conventional environmental and planning legislation. However, in cases where large-scale developments are being undertaken, especially if these are likely to be held up by bureaucratic delay, governments may legislate. A striking example relates to Scotland, where the United Kingdom Government enacted The Petroleum Development (Scotland) Act 1975 to clear the way for platform yard development, together with the Zetland County Council Act 1974 and the Orkney County Council Act 1974 on the initiative of the respective local authorities, to similarly expedite the building of tanker terminals.

Conclusion

For both investment decision-making and operational reasons, the offshore oil and gas industry necessarily operates on long timescales of many decades, but it can be profoundly influenced by short-term economic and political circumstances. Of special interest is the transition to the next long stage in global economic development, now under way (Chapter 1). On this timescale the importance of new sources of energy is growing, including marine renewable resources already regionally significant, albeit on a limited scale, although there is no reason to suppose that the dominance of hydrocarbons will be seriously affected at the global scale in the first half of the twenty-first century (Odell 2004; Voudouris 2014). However, on the short timescale of the business cycle, downturns have already led to substantial restructuring of the industry and the introduction of economy measures, together with continuing technological innovation, not least the move into ever deeper water, all being trends likely to extend into the longer term future (Upton 1996; Pinder 2001; Smith 2011, Wood 2014).

References

Alford, J. B., Peterson, M. S., Green, C. C. (eds) (2014) *Impacts of oil spill disasters on marine habitats and fisheries in North America*. Oxford, CRC Press.

Bakke, T., Klungsoyr, J., Sanni, S. (2013) Environmental impacts of produced water and drilling waste discharges from the Norwegian offshore petroleum industry. *Marine Environmental Research* 92, 154–169.

Boesch, D. F., Rabalais, N. N. (eds) (1987) *Long-term environmental effects of offshore oil and gas development*. London and New York, Elsevier Applied Science.

BP (annual) *BP Statistical Review of World Energy*. London, BP.

Bridge, G., Le Billon, P. (2013) *Oil*. Cambridge, Polity Press.

Burkett, V. (2011) Global climate change implications for coastal and offshore oil and gas development. *Energy Policy* 39(12), 7719–7725.

Button, J. (ed.) (1976) *The Shetland way of oil: reactions of a small community to big business*. Sandwick, Thuleprint.

Cairns, W. J. (ed.) (1990) North Sea oil and the environment: developing oil and gas resources – environmental impacts and responses. London, Taylor & Francis.

Chakrabati, S. K. (ed.) (2005) *Handbook of offshore engineering. Volume 1*. Oxford, Elsevier.

Cullen, The Hon. Lord W. Douglas (1990) *The public inquiry into the Piper Alpha disaster*. London, HMSO.

Department of Energy (1974) *Guidance on the design and construction of offshore installations*. London, HMSO.

Department of Transport (1994) *Safer ships, cleaner seas: Report of Lord Donaldson's Inquiry into the Prevention of Pollution from Merchant Shipping*. London, HMSO.

Earney, F. C. F. (1992) The United Kingdom and Norway: offshore development policies and state oil companies. *Ocean & Coastal Management* 18(204), 249–258.

Ferentinos, J. (2013) Global Offshore Oil and Gas Outlook. Gas/Electric Partnership 2013. Infield Systems. Accessed 21 January 2015 from www.gaselectricpartnership.com/HOffshore%20Infield.pdf.

Harvie, C. (1994) *Fool's gold: the story of North Sea oil*. London, Penguin Books.

Holdway, D. A. (2002) The acute and chronic effects of wastes associated with offshore oil and gas products on temperate and tropical ecological processes. *Marine Pollution Bulletin* 44(3), 185–203.

Hutcheson, A. M., Hogg, A. (1975) *Scotland and oil*. 2nd edn. Edinburgh, Oliver & Boyd.

Jashuah (2012) Crude oil prices since 1861.png. Own work by uploader, data from BP workbook of historical data. Accessed on 10 January 2015 from http://en.wikipedia.org/wiki/File:Crude_oil_prices_since_1861.png.

Kemp, A. (2011–2012) *The official history of North Sea oil and gas*. 2 vols. Abingdon, Routledge.

LeVine, S. (2007) *The oil and the glory: the pursuit of empire and fortune in the Caspian Sea*. New York, Random House.

Lyddon, W. D. C. (1976) North Sea oil and its consequences for housing and planning. *Planning and Administration* 3, 71–86.

Mackay D. I., Mackay, G. A. (1975) *The political economy of North Sea oil*. London, Martin Robertson.

McNicholl, I. H. (1977) The impact of supply bases on the economy of Shetland. *Maritime Policy & Management* 4(4), 215–226.

Market Realist. (2014) Year in Review – The Curtains Fall On 2014. Accessed on 18 January 2015 from http://marketrealist.com/2015/01/year-review-curtains-fall-2014/.

Moore, R. S. (1982) *The social impact of oil: the case of Peterhead*. London, Routledge & Kegan Paul.

National Commission on the BP Deepwater Horizon Oil Spill and Offshore Drilling (2011) *Deep water: the Gulf oil disaster and the future of offshore drilling: Report to the President*. Washington DC, US Government.

Odell, P. R. (2004) Why carbon fuels will dominate the 21st century's global energy economy. Brentwood, Multi-science Publishing.

Odell, P. R., Rosing, K. E. (1975) *North Sea Oil Province: an attempt to simulate its development and exploitation, 1969–2029*. Littlehampton, Littlehampton Book Services.

Office of Ocean Exploration, National Oceanographic and Atmospheric Administration (NOAA), U.S. Department of Commerce (2010) *Ocean Explorer Gallery*. Accessed 10 January 2015 from http://ocean explorer.noaa.gov/explorations/06mexico/background/oil/media/types_600.html.

Owen, P., Rice, T. (1999) *The decommissioning of Brent Spar*. London, E & F Spon.

Patin, S. (1999) *Environmental impact of the offshore oil and gas industry*. East Northport, EcoMonitor Publishing.

Pinder, D. (2001) Offshore oil and gas: global resource knowledge and technological change. *Ocean & Coastal Management* 44(9–10), 579–600.

Ritchie, W., O'Sullivan, M. (eds) (1994) *The environmental impact of the wreck of the 'Braer'*. Edinburgh, The Scottish Office.

Rowland, C., Hann, D., (1987) *The economics of North Sea oil taxation*. London, Macmillan.

Salas, M. T. (2009) *The enduring legacy: oil, culture and society in Venezuela*. Durham and London, Duke University Press.

Sandrea, I., Sandrea, R. (2007) Global offshore oil: geological setting of producing provinces, E&P trends, URR, and medium term supply outlook. *Oil and Gas Journal*, 5 and 12 March.

Smith, N. J. (2011) *The sea of lost opportunity: North Sea oil and gas, British industry and the Offshore Supplies Office*. Oxford, Elsevier (Handbook of Petroleum Exploration and Production, 7).

Speight, J. G. (2015) *Handbook of offshore oil and gas operations*. Oxford, Elsevier.

Upton, D. (1996) *Waves of fortune: the past, present and future of the United Kingdom offshore oil and gas industries*. Chichester, Wiley.

Van Meurs, A. P. H. (1971) *Petroleum economics and offshore mining legislation*. Oxford, Elsevier Science.

Voudouris, V. (ed.) (2014) Special section: oil and gas perspectives in the 21st century. *Energy Policy* 64, 1–174.

Wills, J. (1991) *A place in the sun: Shetland and oil*. Edinburgh, Mainstream.

Wood, Sir Ian (2014) *UKCS Maximising Recovery Review: Final Report*. London, HM Government.

Yergin, D. (1991) *The Prize: the epic quest for oil, money and power*. New York, Simon & Schuster.

19

RENEWABLES

An ocean of energy

Sean O'Neill, Carolyn Elefant and Tundi Agardy

Introduction

Some look to the ocean and take in seascapes that calm the mind and soothe the senses, while others see a bounty of living resources and biodiversity. The oceans have supported great societies and civilizations, and have been the setting for innumerable historical events. But more and more, people are looking to the sea for something the land is increasingly unable to provide at the levels we demand: energy. Oceans offer a vast array of energy options, whether conventional sources of oil and gas, renewable energy such as wind, wave and tidal, thermal, or radical new forms of energy such as algal-based biofuels. As energy demands continue to grow and our conventional sources dwindle or become inaccessible, the oceans will be looked to more and more to meet energy needs.

The oceans span nearly three quarters of the earth's surface, and directly support 70 percent of the planet's photosynthesis. There is thus a huge resource base and vast amounts of space in which to derive and exploit energy. Solar, mechanical (wave and tidal), thermal, and wind energy can all be generated at sea, to supplement conventional sources of non-renewables such as oil and gas. And the oceans can support biofuel production as well.

It is true that whether we are at or near peak oil is a matter of great controversy. What is indisputable is that new supplies and even new forms of energy are needed to meet ever-growing demands, reduce the risks of continuing to emit high levels of greenhouse gases and other pollutants, and allow countries to develop energy-independence. Besides emissions, fossil fuels are subject to price spikes and can foment political conflict. This has led inventors and entrepreneurs to develop myriad ways to tap renewable energy from the sea. The ocean of energy is there for the taking, and if we use it wisely and carefully, ocean energy may well be central to supporting human life on the planet for many centuries to come.

Ocean-based renewable energy

Fossil fuels such as petroleum or other hydrocarbon resources are considered nonrenewable, since it takes millions of years to convert organic matter into these energy resources. As oil and gas become increasingly difficult to recover, as geopolitics complicate the access to both the resources and the markets for such energy products, and as existing supplies in accessible areas diminish, interest in renewables is on the increase (IEARETD 2012). Additionally, recent attention on carbon emissions from hydrocarbon use and its role in global warming and other climate change means that renewable energy resources are fast becoming a preferred alternative. However, renewable energy resources are still, at this point in time, generally more expensive than conventional non-renewables. Much technology remains in the research and development domain, and few marketable technologies have come to scale. Nonetheless, developing, deploying, and marketing ocean renewables has become a big business, and investments from both public and private sectors will propel marine renewables to center stage in energy development.

More than one hundred ocean renewable technologies are being investigated worldwide with approximately 40 different technologies being researched in the United States. Marine renewables have great potential as a sustainable source of energy for numerous reasons. For one, water is approximately 800 times more dense than air, providing greater power and more accurate predictability than most other renewable energy sources. Additionally, ocean renewable energy is easily accessed from the most densely populated areas on major rivers and near coasts—minimizing the need to construct thousands of miles of new transmission lines.

It is estimated that in the U.S., ocean wave and in-stream tidal hydrokinetic energy resources could soon provide 7–10 percent of present electricity consumption—roughly equal to the entire generating capacity provided by conventional hydropower. Ocean renewables support energy independence by utilizing an abundant domestic resource for electricity generation while creating jobs, supporting economic development, and diversifying the electric generating portfolio. This diversification of electric generating resources is the foundation of a reliable electrical system providing consumers with protection from over reliance on any single source of electric supply, financial and other risks.

There are five major classes of renewable energy currently available at sea: 1) kinetic energy derived in a way that is unique to the ocean, such as energy provided by waves and tides, 2) renewable energy at sea that is not unique to the ocean, such as offshore wind and solar energy, 3) thermal energy, such as that produced by the temperature differential of surface and deep ocean waters, 4) salinity gradient or osmotic power, and 5) marine biofuels, such as those derived from algae. These are discussed separately in the following sections.

Kinetic energy: wave, tide, and current devices

Attempts have been made to harness the enormous energy potential of moving ocean water for decades (Ocean Energy Systems 2012). As far back as the middle of the eleventh century, people were making the logical extension from exploiting energy in running rivers, streams, and canals (an ancient technology that probably predates even waterwheels for grinding flour) to trying to harness that same mechanical energy contained in waves and tides. The first commercial scale wave energy plant was commissioned for the Isle of Islay (Scotland) in 2000; at about the same time, the Japan Marine Science and Technology Center created a large-scale experimental wave energy platform.

Early harnessing of the kinetic energy contained in moving seawater was focused on estuaries, where both river hydrology and tides influence the movement of seawater or brackish water. But wave energy can be harnessed, in theory, anywhere where there are predictable waves, including in offshore areas (Shields and Payne 2014). Much current effort is focused on identifying criteria and developing algorithms that can prioritize areas suitable for marine renewable energy extraction (see for instance Davies and Pratt 2014 for a description of strategic planning of renewable energy locations in Scotland). In addition, this information is increasingly driving comprehensive spatial planning and ocean zoning in many parts of the world (Agardy 2010).

The first commercial scale wave power station was established in Scotland at the beginning of the century. The designs for harnessing this water movement vary, but all systems use turbines or attenuators that convert kinetic energy by using the water movement to propel turbines that drive an electrical generator. Some of these devices use water movement to move air across turbine blades, others use the movement of water directly to generate kinetic energy. In recently developed devices, wave energy is converted to electrical energy using floating hinged devices known as attenuators that operate on the surface parallel to the waves.

Oscillating Wave Surge Converters sit on the floor of the ocean and capture power using a pendulum mounted on a pivoted joint. Scotland's Aquamarine Power has developed their Oyster 800 device, which successfully operated through three winters at the European Marine Energy Centre (EMEC). Oscillating Water Column (OWC) devices are partially submerged hollow structures that capture the ebb and flow of waves by having the rise and fall of water compress and decompress air that in turn drives a two directional turbine to create electricity. After a successful pilot in 1991, WaveGen developed the first commercial OWC in 2001, called the LIMPET, on the Scottish Island of Islay to produce electricity for sale on the grid. Engineering giant Voith subsequently bought WaveGen and developed 16 units in Mutriku, Spain. Another well-known OWC developer is Ocean Energy, Ltd of Ireland which uses an oscillating water column device floating on a barge or boat that was successfully tested in Galway Bay, Ireland.

Other marine kinetic devices rely on what is known as heaving buoy technology, which captures the kinetic energy in the orbital motion of surface waves. Point absorbers capture power from waves as a heaving buoy moored or connected to the floor of the ocean, for example the Wave Energy Technology–New Zealand device developed by Power Projects Limited and Industrial Research Limited that was successfully tested at the open water test center operated by the Northwest National Marine Renewable Test Center in Oregon (U.S.A.).

Similarly, overtopping devices capture water as waves break into a storage reservoir and the captured water is returned to the ocean through a conventional turbine. AWS of the U.K. is the best-known developer of overtopping devices.

There are also hybrids combining point absorbers and attenuators such as Columbia Power Technologies device and Nualgi Nanobiotech's Rock n Roll Wave Energy device in India, or OWCs combined with wind turbines developed by Floating Power Plant AS of Denmark. Columbia Power Technologies began with a direct drive point absorber that evolved to a proprietary rotary hybrid design. Oscilla Power, Inc. of Seattle, Washington has developed a wave energy converter based on magnetorestrictive materials or SRI International's electroactive polymer artificial muscle (EMPAMT) technology. With so many renewable wave energy technologies being investigated and the technology evolving so rapidly, it is difficult to describe them, and some technologies may well be obsolete (while unimagined new ones arrive on the scene) by the time this Handbook of the Oceans is published.

Tidal and current energy technology development is running ahead of wave development, although advances in wave energy are sure to bring these two sectors closer. Most tidal energy plants use a dam, known as a barrage, that spans a narrow bay or inlet. Sluice gates on the barrage allow the tidal basin to fill on the incoming high tide and empty through the turbine system on the outgoing, or ebb tide. As in wave energy systems, there are units that generate electricity on both the incoming and outgoing tides (Charlier 2003). Like wave energy, tidal technologies depend on the more esoteric principles of physics including vortex-induced vibrations, as well as swarm technologies used to define and capture the strongest tidal or current flows (Lyatkher 2014).

Some tide energy technologies use the rise and fall of tides to generate power while others are designed to capture the power of the tidal currents, even the geographically large and powerful currents such as the Gulf Stream that flows past the North American coast then veers off toward Europe in the North Atlantic. Like wind power, tidal energy devices include horizontal and vertical axis turbines, and in each of these classes there are subsets depending on blade styles and configuration. For example, Marine Current Turbines (MCT) of the U.K. uses two double bladed turbines sharing one large metal pole for installation, while Verdant Power of the U.S. uses three bladed horizontal axis turbines mounted each on their own pole, very much like onshore wind turbines. Ocean Renewable Power Company of the U.S. uses a horizontal axis helical turbine that looks very much like an old push lawn mower.

Other tidal technologies include the Oscillating Hydrofoil that harnesses tidal energy as it creates lift on either side of a foil to raise and lower an hydraulic system. Venturi devices concentrate tidal flow through a turbine using a funnel-like collecting device and are favored by the French engineering giant Alstom Hydro; this technology is also incorporated in devices developed by Open Hydro. Archimedes Screw is a helical corkscrew device championed by Flumill of Norway. Tidal Kites allow energy development in areas of relatively slow current or nominal tide range, the prototype kites show how they fly in a figure eight motion while tethered to the seafloor to attenuate water movement eightfold in order to efficiently generate kinetic energy. Early developments in kite technology were undertaken by UEK in the U.S., but Minesto of Sweden is now the leader in this field. Minesto was recently the winner of the Navigator Prize of the 2014 Ocean Exchange (see www.oceanexchange.org), competing against innovations in energy, food production and resource use, waste management, shipping, water management, and other technology challenges.

Marine wind and solar energy

Wind and solar energy generation are of course not unique to oceans. Yet oceans not only provide vast amounts of space and sufficient sunlight and wind—they also provide these as a commons property that can in theory be more easily accessed than private property to meet the public good.

Offshore windfarms are common in some parts of the world, such as Northern Europe. Oceanic wind is a preferred alternative to other forms of energy generation in areas where land is in short supply, and where coastal winds are sustained and strong. As wind installations on land have created visual impact and driven concerns about noise and light pollution, energy investors and utilities have increasingly looked offshore (Shields and Payne 2014). Denmark initially led the effort in harnessing sea wind, and constructed the first offshore wind farm in 1991 off the Port of Vineby. The U.K. opened its first offshore wind farm in 2000 in Northumberland, and is following Denmark's lead with expanded wind farms and feasibility studies for siting in new areas. Offshore wind developments have been prolific elsewhere in northern Europe and the support of the European Commission for carbon-free electricity is tremendous.

Offshore wind energy benefits from long stretches of flat seas allowing wind to gain speed, as well as proximity to population centers (Lynn 2012). The siting of large-scale offshore wind farms has generated public controversy, but this renewable sector has also catalyzed much systematic marine spatial planning, at the local (state) and national level. For instance, the spectre of wind farm development propelled the state of New Jersey (U.S.A.) to develop a comprehensive marine spatial plan for state waters to three nautical miles offshore; similarly, the planned Cape Wind development off Cape Cod, MA and the Block Island Wind Farm in RI drove state authorities to engage in some of the most comprehensive participatory planning processes ever seen in the US. One area with huge potential for growth in the renewables sector is combined systems, as described in a review of dual wave and wind systems (Pérez-Collazo *et al.* 2014).

The oceans are also the world's largest solar collector: one square mile contains more energy potential than 7,000 barrels of oil (DiChristina 2007). Solar arrays with unfettered access to sunlight can be installed in virtually any coastal area sheltered from excessive wind or waves. Currently most offshore solar plants are used to power oil platforms and *in situ* research equipment. Solar energy can be placed on many types of structures allowing for multiple uses co-locating with other human uses.

Thermal energy: ocean thermal energy conversion (OTEC)

Marine areas—or at least some areas—can also be harnessed for energy by using the temperature differential of surface and deep waters to drive energy generation. The differential exists because the sun warms the surface layers of the ocean, especially in the tropics, while deep waters stay cool. In order for the technology to be able to capture the thermal energy, this temperature differential must be more than 25 degrees Celsius.

Using the temperature of seawater to make energy actually dates back to 1881, when a French Engineer by the name of Jacques D'Arsonval first thought of using ocean thermal energy gradients. His student, Georges Claude, built the first OTEC plant in Cuba in 1930, producing 22 kilowatts of electricity with a low-pressure turbine. OTEC (Ocean Thermal Energy Conversion) is shorthand for all such thermal energy technologies, but it also refers specifically to the best-known and largest scale pilot effort to harness ocean thermal energy, initiated in Hawaii in 1974.

Three types of systems are used to convert ocean thermal energy to electrical energy. Closed cycle systems use the warm surface water to vaporize a low-boiling point fluid such as ammonia. As the vapor boils and expands, it drives a turbine, which then activates a generator to produce electricity. Open cycle systems operate at low pressure and actually boil the seawater, which produces steam to drive the turbine/generator. Hybrid systems use elements of each, in an attempt to improve conversion efficiencies.

Although the temperature differential between surface waters and the deep ocean is significant in almost all parts of the globe, there are constraints to being able to harness this potential energy. Main among them is having deep cold water in close proximity to warm surface waters. Tropical island nations in the Pacific Ocean are particularly suited. According to NASA, some 98 tropical countries could benefit from the technology. OTEC also has spin-off benefits, including air conditioning, chilled-soil agriculture, aquaculture, and desalination.

Thermal energy conversion has great potential, but enormous challenges remain. The technology is still very inefficient and piping large volumes across great depths of ocean (a kilometer or more) is a major engineering feat. Yet some energy experts believe OTEC could produce billions of watts of electrical power if it could be made cost-competitive with conventional power technologies.

Salinity gradient power

Salinity gradient or osmotic power is generated by the difference in the salt concentration between *seawater* and *river water* that exists in estuaries around the world. Salinity gradient power relies on *osmosis* with ion-specific *membranes* to generate electricity, through either reverse electro-dialysis (RED) or *pressure retarded osmosis* (PRO). First invented in Israel by Prof. Sidney Loeb, osmotic power operating plants have been developed by the Norwegian utility Statkraft, and REDstack in the Netherlands. The technology is still in its infancy and has yet to come to scale, but salinity gradients have potential as sustainable sources of energy because the osmotic gradient is available everywhere rivers meet the sea, and because the only waste product is brackish water (Jones and Finley 2003).

Marine biofuels

The last two decades have witnessed a flurry of interest in alternative fuels, especially biofuels. Terrestrial sources of biofuels are now a major agricultural commodity and are the foundation for profitable businesses in developed and developing countries alike. Biofuels can be derived from agricultural and forestry residues, energy crops, landfill gas, and the biodegradable components of municipal and industrial wastes. Such fuels can be used for transportation fuel, to provide heat, or to generate electricity. Biomass residues have been burned to create power since at least the middle of the nineteenth century, but inefficiencies tended to be extremely high until research and development efforts became focused on making biofuels economically viable.

Corn and switchgrass have received most of the attention as biofuel sources, but there is no reason why marine plants cannot provide the same cellulose for fuel conversion. This emerging technology is being tested in various venues, including research and development of marine algae as biofuels being undertaken by the U.S. National Space Agency (NASA). Marine algae produce lipids and photosynthesize very rapidly, making them highly suitable for biofuels. Marine biofuels have several advantages over terrestrially derived biofuels—notably that land conversion away from food crops is not necessary, and nutrient supply in most waters is sufficient to support algal growth.

The potential global market for marine renewables

According to Ocean Energy Systems, "by 2030 ocean energy will have created 160,000 direct jobs and saved 5.2 billion tonnes of CO_2 emissions" (Huckerby *et al.* 2011). As new countries enter the race to commercialize ocean renewables, the industry has shown unprecedented cooperation as competitors acknowledge that no country has substantial market share at this time, and that all boats rise with the tide. Some areas of international collaboration include U.S. participation in the Ocean Energy Systems (OES) initiative, which brings together 22 countries and the International Electrotechnical Commission's (IEC) Technical Committee 114 (TC-114) for marine energy, with 14 participating countries and nine observing countries.

At this point in the development of this emerging and potentially highly profitable industry, issues such as deployment, moorings, and power take-off systems continue to be the primary focus of technology developers. However, for those companies involved in project development —the necessary steps for getting these technologies into open waters—there are numerous challenges in the areas of permitting and regulation, potential environmental impacts, funding, the need for adaptive management, emerging marine spatial planning (MSP—see Chapter 31 on MPAs and MSP, this volume), and public acceptance of these developments. The main constraints are detailed below, with a special focus on the dynamic and rapidly evolving situation in the United States.

Current U.S. regulatory schemes for marine renewables

There are a number of components that make up the marine renewable energy regulatory process: legislation or regulations that govern the consent or approval process (including any special processes for demonstration projects), procedure for obtaining a lease or rights to use lands for the project, review of project impacts, including environmental, navigation, fishing and recreational use, and grid access.

In the United States, the Federal Power Act (FPA), 16 U.S.C. 791 et seq. governs licensing of marine renewables projects. Under the FPA, Federal Energy Regulatory Commission (FERC) may issue preliminary permits and licenses for marine renewables. A preliminary permit enables a developer to study a site for three years and maintain priority to apply for a license over competing applicants but does not authorize construction of a project (Federal Power Act, 16 U.S.C. sec. 800). As a result, a preliminary permit does not provide any opportunity to test projects in real world conditions. A FERC license, by contrast, allows a developer to construct and operate a project, generally for a term of up to 50 years. But the process for obtaining a license is lengthy (as long as three to seven years) and requires data on a project's potential impacts, which are often unknown until a project is deployed and observed.

Recognizing the limited options for demonstration projects, FERC developed two alternatives. The first alternative, known as "the Verdant exception" allows a developer to deploy and operate a small (less than 5 MW) project for 18 months or less to gather data to support a license application, so long as the developer agrees not to sell power to the grid during the test period. The second alternative is the FERC created "pilot license process" for new technologies in 2007. A pilot license has a five-year term, a processing time of one year, limited study requirements up-front but rigorous post-deployment monitoring requirements. At the end of the five-year pilot license term, a developer has the option of removing the project or applying for a long-term license at the site. Two United States developers—Verdant Power and Ocean Renewable Power Corporation were awarded pilot licenses by FERC in 2012.

The FERC process authorizes project operation but does not confer property rights for constructing the project. For projects located on "state submerged lands"—that is, lands up to three miles off shore (with the exception of Texas and the West Coast of Florida where states own lands up to ten miles offshore)—a developer will typically obtain a land lease or rights of usage from the state. Projects beyond these limits are located on the Outer Continental Shelf, where a developer must obtain a lease from the Minerals Management Service (MMS). In April 2009, MMS issued rules for grant of leases and also entered into a Memorandum of Understanding (MOU) with BoEM to coordinate the BoEM leasing process with the FERC licensing process. Under the MOU between FERC and MMS, a developer must secure a lease from MMS, before it can receive a FERC license.

For projects that connect to the interstate grid, FERC has power, under the Federal Power Act and FERC's own regulations, to oversee interconnection. FERC established a straight-forward protocol that developers must follow to obtain grid access; the rules for smaller generators are not complicated and the process is relatively quick. As marine renewables projects expand in size, they will impose greater demands on the grid. Marine renewables projects

Table 19.1 List of U.S. Federal Authorities and legislation requiring consideration in offshore renewable energy development

- National Environmental Policy Act
- Endangered Species Act
- Marine Mammal Protection Act
- Magnuson Stevens Fishery Conservation and Management Act
- Marine Protection, Research and Sanctuaries Act
- Executive Order 13186 (Migratory Birds)
- Coastal Zone Management Act
- Clean Air Act
- Clean Water Act
- Marking of Obstructions
- Executive Order 13547 (Stewardship of the Oceans, Our Coasts and Great Lakes)
- Ports and Waterways Safety Act
- Rivers and Harbors Appropriation Act
- Resource Conservation and Recovery Act
- National Historic Preservation Act
- Archaeological and Historical Preservation Act
- American Indian Religious Act
- Federal Aviation Act
- Federal Power Act
- Executive Order 13007 (Indian Sacred Sites)

may face longer "queues" for access, as the utility or the regional transmission system operator evaluates how to incorporate large amounts of new and variable power into the system.

In the United States, federal agencies that issue a license must prepare an environmental analysis to assess the impacts of a project on the surrounding environment and other uses. The FPA also requires FERC to review the effect of a project on navigation and to consider whether it makes best use of the waterway (FPA, Section 803). Projects must also comply with a variety of federal environmental laws, such as the Endangered Species Act (protects endangered species), the Coastal Zone Management Act (CZMA—ensures that the project is consistent with state plans for use of coastal areas), the Clean Water Act (protects water quality), while abiding by state environmental regulations as well. In addition to the FERC license and a land lease, developers must also obtain authorizations from the agencies that administer these federal statutes. There is no process for coordinating issuance of a FERC license and issuance of a CZMA authorization (issued by the state) or a water quality certificate and, as a result, the license process is quite lengthy.

The list of U.S. Federal Authorities and legislation in Table 19.1, though by no means complete, pertains to activities on the continental shelf and therefore must as a minimum be considered in marine renewable development.

Some of the implications of this legislative context are discussed in the following section on potential environmental impacts of ocean renewables.

Potential environmental impacts

Potential environmental effects of marine renewable energy development are being carefully studied, and appear to be similar for many of these technologies. Effects on biological resources could include alteration of the behavior of animals, damage and mortality to individual plants and animals, and potentially larger, longer-term changes to plant and animal populations and communities (Gill *et al.* 2014; Shields and Payne 2014). These impacts can occur at any stage during the development of marine renewables, including during technology testing, site characterization, device installation, operation and maintenance, and/or decommissioning. Many installation and decommissioning impacts are analogous to those from conventional energy industries (e.g., marine oil and gas) and appear to be short term in nature.

Ocean renewable energy systems are becoming both more efficient and more economically viable (IEARETD 2012). But these energy systems are not without cost. First, there are the prospective ecological impacts. Constructing and operating facilities will undoubtedly have environmental costs, as will diverting, moving, or variously treating large volumes of seawater. Facilities will be generating their own pollution and wastes, including light pollution. Wind turbines and underwater turbines generate noise, which is a growing concern of marine conservationists. And removal of nonrenewable resources such as methane hydrates and renewable ones such as algae may alter both the geology or oceanography and the ecology of some marine areas.

The specific environmental effects of marine hydrokinetics are similar for many of the ocean renewable technologies. Assessments have identified a number of potential environmental effects; Shields and Payne (2014) detail both direct and indirect ecological effects that would result from extensive installation of offshore renewable energy developments. These include:

- alteration of currents and waves;
- alteration of substrates, sediment transport, and deposition;
- alteration of habitats for benthic organisms;
- noise during construction and operation;
- emission of electromagnetic fields;
- toxicity of paints, lubricants, and antifouling coatings;
- interference with animal movements and migrations; and
- strike by rotor blades or other moving parts.

It should be noted (and is stressed by Shields and Payne 2014) that not all environmental effects are negative impacts. In some cases marine renewable installations boost biodiversity and productivity, in part because the installations provide new habitat or habitat heterogeneity, and in part because other more damaging uses of ocean space and resources are restricted at such sites. Furthermore, as Shields and Payne (2014) and other authors have pointed out, the environmental costs of marine renewable industries must be evaluated against the benefits that these technologies provide—not least is the lessening of climate change-driving greenhouse emissions.

Potential environmental impacts remain a concern, however. In the U.S., the Department of Energy and other agencies, including the Bureau of Ocean Energy Management, the Federal Energy Regulatory Commission, and the National Oceanic and Atmospheric Administration (NOAA) have funded numerous studies since 2005 when ocean renewable energy was included in the federal definition of renewable energy for the first time. The studies have been conducted by national laboratories, the Electric Power Research Institute and private companies such as HT Harvey Associates, Alden Research labs, and Re-Vision. Add to these broader studies the site specific, *in situ* studies required by state and federal agencies in the permitting of the early projects. While the U.S. has been amassing data from these early projects, there has been international cooperation and a common thirst for data, as embodied in Annex IV of OES.

Annex IV is a collaborative project Ocean Energy Systems-Implementing Agreement (OES-IA) to examine environmental effects of ocean energy devices and projects. There is currently a wide range of ocean energy technologies and devices in development around the world; the few data that exist on environmental effects of these technologies are dispersed among different countries and developers.

Annex IV member countries are currently collaborating to create a searchable, publicly available database of research and monitoring information to evaluate environmental effects; the U.S. currently leads the effort. The database, publicly available via Tethys, includes data from ocean energy projects and research studies and case study reports compiled as part of this effort. Annex IV will address wave, tidal, and ocean current energy development, but not ocean thermal energy conversion (OTEC) or energy derived from salinity gradients. The construction of the data-base, or knowledge management system, will be followed by a comprehensive report with a worldwide focus on monitoring and mitigation methods including findings from the database, the results of an experts' workshop, and lessons learned from the project (Copping *et al.* 2014; Polagye 2011).

The quest for data unites the ocean stakeholder community more than any other dynamic. IBM, as part of its Smart Planet Initiative supported the launch of the Irish Marine Weather Buoy Network that provides weather and wave data at six locations around the Irish coast. Real-time wave height and direction data are available on the website—updated hourly. Commercial fishers can identify where the fish are, scientists are provided with real-time acoustic information, surfers can tell when the surf's up, and restaurants can identify what the catch of the day will be—all because of IBM's *in situ*, real-time data gathering.

Funding for ocean renewables

A recent study from the University of Edinburgh comparing U.K. research and development funding with that of the U.S. revealed that a total approaching £160 million of public money has been targeted at the ocean energy sector in the U.K., and approaching £60 million in the U.S. These figures do not include production incentives such as the U.K.'s Renewables Obligation Certificates, or other price regulation systems that encourage renewable energy production in countries throughout Europe (Amar and Suarez 2011).

Consistent funding for research and development and testing infrastructure marks the situation in Europe. Production-based incentives for marine and hydrokinetic energy are far higher than in the U.S. For example, where the U.S. Production Tax Credit is 2.2 cents (U.S.) per kilowatt hour for wind power and 1.1 cents for ocean renewables, in Italy Production Based Incentives are over 32 Eurocents per kilowatt hour followed by Scotland at just over 27 Eurocents, with Portugal, Ireland, England, Wales, and Northern Ireland, France, and Denmark providing production-based incentives greater than fivefold the amount of U.S. Production Tax Credits.

Incentivizing marine renewables is also common elsewhere. In 2010, South Korea passed a nationwide Renewable Portfolio Standard requiring utility companies to generate a certain portion of energy from renewable resources (about 2 percent or 1,474 MW by 2012 and 8 percent or 6,648 MW by 2020. In 2008, the New Zealand Government launched the NZ$ 8 million Marine Energy Deployment Fund (MEDF) to promote the deployment of prototype projects in NZ waters. The Fund was a matching fund—developers had to provide at least 60 percent of the proposed cost of the project. Four competitive rounds were conducted from 2008 to 2011 and six awards were made. These awards went to both wave and tidal energy projects but required developers to meet strict contractual targets. Providing incentives for marine renewables is a priority in many coastal countries, and innovative mechanisms for financing R&D and capital construction are springing up as demand for energy only continues to increase. Marine spatial planning underway in many of these countries is also acting to streamline permitting and lends confidence to investors.

Ocean governance – the global context for energy development

Renewable energy has been welcomed into the many discussions on ocean governance happening throughout the world and this Oceans Handbook has several in-depth articles covering topics including Marine Spatial Planning, Strategic Environmental Assessments, and Adaptive Management that are all central to developing good and effective ocean governance.

The cornerstone of global cooperation to effectively govern and protect ocean uses and the ocean itself is the U.N. Convention on Law of the Sea (UNCLOS). The Law of the Sea Treaty defines the rights and responsibilities of nations in their use of the world's oceans, establishing guidelines for businesses, the environment, and the management of marine natural resources. The treaty also formally recognizes 12-mile territorial sea boundaries and 200-mile exclusive economic zones (EEZs).

The United States has not ratified the Law of the Sea. Opponents are concerned that doing so could compromise U.S. sovereignty or potentially subject users to added taxes. Support comes from the oil and gas industry, environmental non-governmental groups, and the past five Secretaries of State—both Democrat and Republican—have endorsed U.S. participation. President Obama's Executive Order 13547—Stewardship of the Ocean, Our Coasts, and the Great Lakes—which laid the foundation for MSP in the United States, also calls for the U.S. to sign the Law of the Sea. Despite the fact that offshore oil and gas interests, along with environmental non-governmental organizations, agree that the U.S. should become a signatory to UNCLOS, the prospects for the U.S. signing and ratifying the treaty remain bleak. In the absence of a unifying convention to guide marine renewable energy development, ocean governance remains fragmented and unable to set global standards for sustainability and equity.

Conclusions

The potential of ocean renewables to meet growing energy needs was elegantly summed up by the International Energy Agency over a decade ago:

> There is a growing awareness of economic, energy security and environmental values of renewables and of its critical role to sustainable development. This is leading to political initiatives to promote their development, such as the EU Directive that establishes the target of increasing the 1997 6% share of renewables to 12% in 2010, and recent approvals and ratifications of the Kyoto Protocol. Renewables are also high on the agenda of developing countries, and expanded renewable energy deployment is one of the key goals of the World Bank.

The U.S., Canada, Mexico, the U.K., Ireland, Sweden, Norway, Denmark, Belgium, France, Australia, New Zealand, Japan, China and India are now all invested in ocean energy development.

For too long the people of the developed world have taken energy for granted; it is only in times of high energy costs (particularly rising costs at the fuel pump or on home heating bills) that the public is even conscious of the fact that supplying energy is a costly, and sometimes unpredictable, endeavor. The sudden surge of interest in the effects of global warming, and increasing geopolitical tensions between oil supplying and oil consuming countries, has opened many people's minds to considerations of new sources of energy, as well as to issues of energy conservation.

Ocean renewable energy is now big business. In July of 2011, the United Kingdom's Carbon Trust estimated that the worldwide tradable market for marine energy devices accessible to U.K.-based business was valued at £340 billion. Other countries increasingly rely on marine renewables to supplement energy demand currently served mainly by fossil fuels. With fossil fuel-based sources of electricity subject to price changes much like a variable rate mortgage, renewables represent a fixed rate mortgage and diversity of supply. These features of renewable energy support sustainability, reliability and affordability. To make sound energy choices and meet future energy demands, new generating capacity needs to be installed and operating as quickly as possible, the world over.

Acknowledgment

Some introductory text is taken from Agardy, T. 2007: *An ocean of energy, there for the taking.* Published in World Ocean Observatory (www.w2o.net) and used with permission.

References

Agardy, T. S. 2010. *Ocean Zoning: Making Marine Management More Effective.* Earthscan, London.

Amar, E. and J. Suarez. 2011. Seasteading engineering report: floating breakwaters and wave power generators. Available online: http://seasteading.wpengine.netdna-cdn.com/wp-content/uploads/2012/02/Wave_Energy_and_Break_Water_Survey_Research.pdf (accessed July 22, 2015).

Charlier, R. H. 2003. Sustainable co-generation from the tides: bibliography. *Renewable and Sustainable Energy Reviews* 7, 215–247.

Copping, A., H. Battey, J. Brown-Saracino, M. Massaua, and C. Smith. 2014. An international assessment of the environmental effects of marine energy development. *Ocean and Coastal Management* 99, 3–13.

Davies, I. M. and D. Pratt. 2014. Strategic sectoral planning for offshore renewable energy in Scotland. In M. A. Shields and A. I. L. Payne (Eds.) *Marine Renewable Energy Technology and Environmental Interactions.* Springer Verlaag, Berlin, pp. 141–152.

DiChristina, M. 2007. Solar Power. Ocean Planet Smithsonian. Popular Science. Available online: http://seawifs.gsfc.nasa.gov/OCEAN_PLANET/HTML/ps_power.html (accessed July 22, 2015).

Gill, A. B., I. Gloyne-Philips, J. Kimber and P. Sigray. 2014. Marine Renewable Energy, Electromagnetic (EM) Fields and EM-Sensitive Animals. In M. A. Shields and A. I. L. Payne (Eds.) *Marine Renewable Energy Technology and Environmental Interactions.* Springer Verlaag, Berlin, pp. 61–80.

Huckerby, J., H. Jeffrey, and B. Jay. 2011. *An International Vision for Ocean Energy.* Ocean Energy Systems, A Technology Initiative of the International Energy Agency. Available online: www.ocean renewable.com/wp-content/uploads/2011/05/oes_vision_brochure_2011.pdf (accessed July 22, 2015).

IEA. 2001. Agreement on Ocean Energy Systems. Available online: www.iea.org/topics/oceanenergy/ (accessed July 22, 2015).

IEARETD (International Energy Authority Renewable Energy Technology Deployment). 2012. *Offshore Renewable Energy: Accelerating the Deployment of Offshore Wind, Tidal, and Wave Technologies.* Routledge, London.

Jones, A. T. and W. Finley. 2003. Recent developments in salinity gradient power. Available online: www.waderllc.com/2284–2287.pdf (accessed July 22, 2015).

Lyatkher, V. 2014. *Tidal Power: Harnessing Energy from Water Currents.* Wiley & Sons, New York.

Lynn, P. A. 2012. *Onshore and Offshore Wind Energy: An Introduction.* Wiley & Sons, Chichester, New York.

Pérez-Collazo, C., D. Greaves, and G. Iglesias. 2014. A review of combined wave and offshore wind energy. *Renewable and Sustainable Energy Reviews* 42, 141–153.

Polagye, B., B. Van Cleve, A. Copping, and K. Kirkendall (Eds.) 2011. Environmental Effects of Tidal Energy Development. NOAA Technical Memorandum NMFS F/SPO-116.

Shields, M. A. and A. I. L. Payne. (Eds.) 2014. *Marine Renewable Energy Technology and Environmental Interactions*. Springer Verlaag, Dordrecht, Heidleberg, New York, London.

20

OCEAN MINERALS

James R. Hein and Kira Mizell

Introduction

Nearly 71 percent of the Earth is covered by ocean, yet during the entire history of societies, the mineral resources essential for nation building have been acquired solely from the continents. As we will discuss, this is changing for many reasons. There is a rapidly increasing need for rare metals required for emerging technologies, such as cobalt, nickel, copper, zinc, gallium, germanium, zirconium, niobium, molybdenum, platinum, indium, tellurium, tungsten, silver, gold, bismuth, and the rare-earth elements (REEs). These metals are concentrated in a variety of deep-ocean mineral deposits. This new search for resources is greatly expanding investigations of the deep-ocean environment.

Deep-ocean minerals were discovered over a century ago during the Challenger expedition of 1873–1876, but only relatively recently did programs develop to determine their origin, distribution, and resource potential. Modern scientific studies began in the 1970s with manganese nodules. Mining of manganese nodules in the northeast Pacific Clarion-Clipperton Zone (CCZ) was expected to take place in the late 1970s or early 1980s but did not occur. The study of seafloor massive sulphides (SMS) began soon thereafter in the late 1970s, and ferromanganese (Fe-Mn) crusts as a potential resource for cobalt commenced in the early 1980s. Investigations of deep-ocean minerals continued for the remainder of the twentieth century, but due to low global metal prices, there was little incentive to mine the deep ocean, until the turn of the century.

Today, vast areas of deep-ocean floor have been contracted for exploration and several deep-ocean mines are scheduled to start operations within the next few years. The primary economic interests have been in nickel, copper, and manganese from nodules, cobalt, nickel, and manganese from crusts, and copper, zinc, gold, and silver from SMS. Research undertaken over the past decade has identified additional rare metals and REEs that are potential by-products of mining of those focus metals (Hein *et al.*, 2010, 2013). These metals are essential for a wide variety of high-tech, green-tech, emerging-tech, and energy applications (Table 20.1). Over the past decade, global consumption of many rare metals has increased, but supplies can be unreliable due to the limited number of major producers. Increased competition for metal resources from rapidly expanding economies may lead to shortages. Deep-ocean mineral deposits will not replace land-based mining but will offer an additional source of raw materials to meet increasing demand.

Table 20.1 Example rare metals for emerging and next generation technologies

Metal	Application
Tellurium	Photovoltaic solar cells; computer chips; thermal cooling devices
Cobalt	Hybrid and electric car batteries, solar energy storage, magnetic recording media, super-alloys, supermagnets, cell phones
Bismuth	Liquid Pb-Bi coolant for nuclear reactors; bi-metal polymer bullets, superconductors, computer chips
Tungsten	Negative thermal expansion devices, superalloys, X-ray photo imaging
Niobium	Superalloys, next generation capacitors, superconducting resonators
Platinum	Hydrogen fuel cells, chemical sensors, cancer drugs, flat-panel displays, electronics
Yttrium	Compact fluorescent lamps, LEDs, flat-screen TVs, medical applications, ceramics
Neodymium	Hard disk drives, medical applications, portable electronics and small motors, high-strength permanent magnets, wind turbines
Praseodymium	Flat screen TVs, portable electronics and small motors, hard disk drives, magnets, lasers, pigments, cryogenic refrigerant
Cerium	Catalysts, metal alloys, radiation shielding, phosphors for flat screen TVs
Gadolinium	Magnetic resonance imaging contrast agent, memory chips
Europium	Liquid crystal displays, fluorescent lighting, LEDs, red and blue phosphors for flat screen TVs, small motors
Terbium	Green phosphor for flat screen TVs, lasers, fluorescent lamps, optical computer memories, medical applications

Mining raw materials from the ocean is not new. Shallow-water, continental-margin deposits have been mined for decades, particularly aggregate (sand and gravel) deposits used for construction, and tin placer deposits have been extracted from offshore Southeast Asia and elsewhere. Continental shelf phosphorite deposits also have an economic potential and may be the next deposit type to be mined offshore. Here, we emphasize deep-ocean mineral deposits, but will briefly describe these continental margin deposits that are presently being mined and those that may be mined in the near future, including extraction from seawater.

Continental margin marine mineral deposits

Continental margin marine mineral deposits include aggregate, sand, placer minerals, and phosphorite. Aggregate, sand, and placers are detrital minerals that were transported and deposited on the shelf, whereas phosphorite is a chemical sedimentary deposit that formed in place from chemical reactions in the near-surface sediment.

Aggregate deposits are composed of quartz and lesser amounts of other minerals such as feldspar, as well as rock fragments and shells. These deposits form by sorting that occurs when waves and currents remove small particles and the larger sand and gravel particles are then concentrated into mineable aggregate. Continental margin aggregate typically formed during the fall and subsequent rise of sea level during Quaternary glaciations. Shell aggregate forms when marine shells are broken up in high-energy marine environments and concentrated into shell banks. Shells respond uniquely to wave and current action because of their differing sizes and shapes and therefore can be easily segregated from the other sediment.

Aggregate is used primarily in the construction industry, and by volume it is the most important offshore hard mineral deposit being extracted today. Most aggregate is used to manufacture concrete (sand, gravel) and cement (shell). Other construction uses include roads, fillers, railway ballasts, mortar, and glass. Another important use of offshore sand is to replenish or nourish eroding beaches, which requires high quality white sand judged suitable for beaches. Beaches are essential to the tourist economy of many nations, and many are undergoing active replenishment, including such famous beaches as Waikiki in Hawaii. Aggregate is currently being mined in shallow water in many places worldwide, most notable in Japan, the U.K., Canada, and the U.S.A. The U.K. acquires approximately 25 percent of its aggregate from offshore sources. Japan is the largest producer of offshore aggregate, with production ongoing for more than 50 years. Like the U.K., Japan mines about 20–25 percent of its aggregate offshore.

Placer deposits are concentrations of gemstones or metallic minerals transported to the ocean by rivers, and concentrated into mineable deposits by sorting, except for placer diamond deposits where sorting does not occur. Placer minerals are the resistant products (refractory) of the breakdown of rock on the continents. All known placer deposits occur within the 200 nautical mile (360 kilometers (km)) exclusive economic zone (EEZ) of coastal nations.

Placer minerals are classified according to their specific gravity. The heavy-heavy minerals (specific gravity (SG) 6.8–21) include precious metal deposits such as platinum and gold and the important tin placer mineral, cassiterite; the light-heavy minerals (SG 4.2–5.3) include the refractory accessory minerals of igneous rock, such as zircon (source of zirconium), monazite (source of REEs), and ilmenite and rutile (both a source of titanium). The most important placer deposits currently being mined are diamond placers (SG 2.9–4.1) mined off Namibia. The water depth of placer mining varies from the beach to about 200 meters (m); placer diamond mines are the deepest water operations today.

Phosphorite is classified as a rock composed predominantly of phosphate minerals. The most common phosphate mineral in marine deposits is carbonate fluorapatite (CFA). Phosphate minerals replace and cement carbonate sediments; precipitate as oolites, granules, or pellets that concentrate by sorting into sand deposits; and form slabs and nodules in the sediment that can later be exposed at the seabed by erosion.

Phosphorite deposits occur predominantly along the west coast of continents on the continental shelf and upper slope at water depths of less than 1000 m, and at the margins of the oxygen minimum zone (OMZ). Marine phosphorite forms in low-to-middle latitude regions, and is linked to divergence, coastal upwelling, and high primary productivity.

Phosphorite is used almost exclusively in the agricultural industry as fertilizers and also is the source for production of phosphoric acid found in all soda drinks. Potential by-products of phosphorite mining include vanadium, uranium, fluorine, and REEs.

A large submarine phosphorite deposit also occurs on Chatham Rise southeast of New Zealand. The Chatham Rise deposit spans an area of 150 by 1000 km and consists of phosphorite gravel. Chatham Rock Ltd. has an exploration contract granted by New Zealand for an area estimated to contain 100 million tonnes of rock phosphate. The only other currently permitted area for phosphate extraction belongs to Namibian Marine Phosphates, Ltd. for an area about 60 km off Namibia covering 7000 km^2. The United States EEZ is estimated to contain approximately 6000 million tonnes of phosphate rock; Mexico's deposits occur over a wide belt covering about 13,000 km^2. Due to the importance of phosphate in agriculture, rapidly decreasing supplies from land-based mines that currently supply all phosphorite, the lack of substitutes, and the increasing global population, demand is expected to increase significantly.

Mining techniques: Most aggregate mining is accomplished by large dredging operations using hydraulic or mechanical systems at water depths of less than 40 m. The types of operations are dependent on local environmental conditions and are therefore specific to location and deposit type. For example, strong winds, waves, and currents have a high impact and areas protected from these forces are favored. Environmental restrictions can also affect the type of dredging employed; the main consideration, besides removal of benthic organisms directly in the path of the miner, is creation of a sediment plume that can affect benthic biology outside of the immediate impact area. Also, the sand component of seafloor aggregate may limit the source and therefore supply of sand available for the natural maintenance of adjacent beaches. Other concerns include multiple use issues such as recreation, fishing, aquaculture, navigation routes, seabed pipes and cables, among others.

Like aggregate, metallic placers are mined using both hydraulic and mechanical dredging operations. Diamond mining uses traditional placer mining techniques as well as more sophisticated vertical and horizontal suction dredging. Vertical mining entails drilling into the seabed with a 6 to 7 m diameter drill head and sucking up the diamond bearing material. Horizontal mining employs the use of Seabed Crawlers (remotely controlled, caterpillar-tracked underwater mining vehicles) that move across the seafloor pumping gravel up to vessels where it is separated and sorted. Limits to current ocean placer mining systems include extreme susceptibility of metal components to corrosion, surface wind and wave impacts on platforms, and the creation of sediment plumes.

Mining seawater

Seawater makes up 98.8 percent of the world's surface water and contains every element in the periodic table, mostly in trace concentrations. Currently, only sodium, magnesium, calcium, potassium, and bromine are extracted for profit, but there is interest in the extraction of lithium, uranium, and deuterium.

Sodium chloride (common table salt) makes up 71 percent of the dissolved solids in seawater and is immensely important to society. Most sea salt is produced via fractional solar evaporation in which brines are allowed to evaporate for various periods of time in multiple ponds. Magnesium and potassium are then recovered using electrolytic processes. India produces over half of the world's sea salt, and other coastal areas with a combination of low precipitation and high evaporation such as Mexico, France, Spain, and Italy, make up the majority of remaining production. Bromine occurs in seawater at a low concentration of about 65 parts per million (ppm = grams per tonne) but can be produced as a co-product of sea salt extraction.

Lithium is used virtually in all electronic devices that require a rechargeable battery. It occurs in seawater at 0.18 ppm and Japan and Korea have development programs for its extraction from seawater using ion-exchange, adsorbent, and polymeric membrane technologies. Lithium from seawater may become economically viable within the next few years. The U.S., Japan, France, South Korea, Sweden, Taiwan, U.K., Canada, India, and Italy have research programs to extract and concentrate uranium from seawater, which occurs at 0.003 ppm. The discovery of high-grade, land-based uranium deposits in Australia and Canada has curbed investment in ocean uranium extraction technology, although research programs continue. Deuterium is of interest due to its potential as a fuel for nuclear fusion energy. The oceans are estimated to contain approximately 46 million tonnes of deuterium; however, extraction processes have yet to be perfected.

Deep-ocean mineral deposits: Ferromanganese (Fe-Mn) crusts

Fe-Mn crusts (Figure 20.1) are found on rock surfaces of seamounts, ridges, and plateaus as pavements and coatings on talus in areas that remain sediment-free for millions of years. Fe-Mn crusts precipitate from cold ambient seawater and are thus classified as hydrogenetic, also called hydrogenous. Crusts are found at water depths of 400–7000 m, with the thickest and most metal-rich crusts occurring at depths of about 800–2500 m. The distribution of crusts and

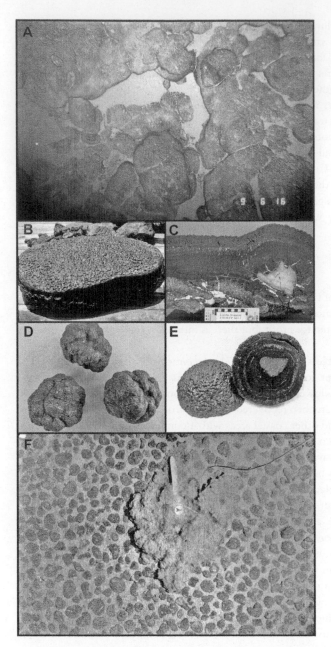

Figure 20.1 Representative photographs of seafloor and samples of Fe-Mn crusts and nodules: (A) seafloor paved with Fe-Mn crust from Horizon Seamount in the Johnston Island EEZ, central Pacific, 2000 m water depth; (B) Fe-Mn crust collected from a seamount in the Marshall Islands EEZ, NW Pacific, water depth 1780 m; (C) cross-section of a 18 centimeter (cm) thick Fe-Mn crust showing multiple growth layers and substrate rock, from seamount in the Marshall Islands EEZ, 1800 m water depth; (D) diagenetic-hydrogenetic manganese nodules from the CCZ, NE Pacific Ocean; (E) cross-section of a 13.6 cm diameter hydrogenetic manganese nodule from Lomilik Seamount, Marshall Islands EEZ, showing concentric accretion layers; (F) seafloor photograph of 3 x 4 m area of a nodule field in the CCZ and the sediment plume created by impact of the weight

Source: Author.

characteristics of seamounts indicate that mining operations will likely take place at water depths from about 1500 to 2500 m (Hein *et al.*, 2009).

Fe-Mn crusts vary in thickness from <1 to 260 millimeters (mm), with thicker crusts occurring on older seamounts. Crusts are thickest in the NW Pacific where the ocean floor is the oldest, more than 145 million years old (Jurassic age). NW equatorial Pacific Fe-Mn crusts also typically have the highest concentrations of rare metals and for these reasons the NW-central equatorial Pacific is considered the prime zone for crust extraction (PCZ; Figure 20.2). The Atlantic and Indian Oceans have less Fe-Mn crusts than the Pacific because there are fewer and generally younger sediment-free seamounts and ridges.

Fe-Mn crusts adsorb large quantities of metals from seawater because of their very high porosity (mean 60 percent), extremely high specific-surface areas (mean 325m^2 per gram), and remarkably slow growth rates of 1–5 mm per million years (Hein *et al.*, 2000). Seamounts and ridges have unique characteristics that aid in the development of Fe-Mn crusts and the acquisition of metals. Obstructional upwelling is created by the impingement of deep-water currents along the flanks of seamounts and ridges, which produces turbulent mixing that keeps them sediment free; the upwelling also supplies nutrients to surface waters that promotes primary productivity. Organic matter generated from primary productivity sinks and oxidizes in the water column creating an OMZ that is a reservoir for dissolved manganese and its associated metals; the OMZ also decreases Fe-Mn crust growth rates, thereby allowing considerable time for acquisition of metals from seawater.

Hydrogenetic Fe-Mn crusts form by accretion of hydrated manganese oxide (MnO_2) and iron oxyhydroxide ($FeO(OH)$) colloids, which acquire trace metals by surface sorption. An electrochemical model describes a first-order process for sorption of metals from seawater, with positively charged ions sorbed onto the negatively charged surface of MnO_2 and negatively charged and neutral ions in seawater sorbed onto the slightly positive charged surface of $FeO(OH)$ (e.g., Hein *et al.*, 2013; Hein and Koschinsky, 2014). During formation of surface metal complexes, both electrostatic bonding and chemical bonding play roles in the accumulation of trace metals. Second-order processes include surface oxidation (cobalt, platinum, cerium, tellurium, thallium) and substitution.

Fe-Mn crusts are composed of δ-MnO_2 (vernadite) and X-ray amorphous feroxyhyte. Todorokite is rare, found in only 2 percent of open-ocean Pacific crusts (Hein *et al.*, 2000). CFA can make up more than 20 percent of the older layers (pre-middle Miocene) of thick crusts. Minor quartz, feldspar, and other detrital minerals in the crusts are delivered by winds and by seafloor weathering.

Iron and manganese occur in subequal amounts in crusts with manganese generally higher in open-ocean Pacific crusts and iron generally somewhat higher in continental-margin crusts around the Pacific and in Atlantic and Indian Ocean crusts. Cobalt and nickel have been the metals of greatest economic interest in Fe-Mn crusts, and mean concentrations for large areas of the global ocean range from 0.30 percent to 0.67 percent and 0.23 percent to 0.46 percent, respectively. Smaller areas that would compose a 20-year mine site (Hein *et al.*, 2009) can average about 0.8 percent cobalt and 0.5 percent nickel. Another metal enriched in Fe-Mn crusts and of great interest to the photovoltaic solar cell industry is tellurium, which globally averages about 50 ppm in crusts, with a maximum of 205 ppm. Total REEs average about 0.16 percent to 0.25 percent over large regions of the global ocean. However, localized areas can yield contents as high as 0.7 percent and individual samples over 1 percent total REEs. Platinum has also received attention because concentrations of this precious metal as high as 3 ppm have been reported. However, for most locations, even small areas, platinum does not average more than about 0.7 ppm, which is still significant compared to land-based deposits if extractive metallurgy can

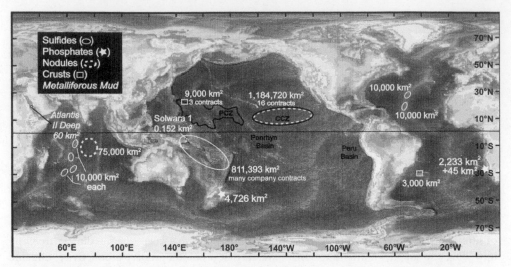

Figure 20.2 Map of deep-ocean exploration contracts as of January 2013, which also marks the only extraction contract area, for SMS at Solwara 1. Note that some marked areas comprise contracts for multiple countries/agencies; size of ovals not to scale. See text for description of contracts, pending contracts and plans of work. About half of the 1,940,000 of contracted areas are in EEZs and the other half in areas beyond national jurisdictions (The Area)

Source: Author.

be optimized. Other platinum-group metals (PGMs) are much less enriched in crusts. Mean contents of other metals that are potential by-products of cobalt–nickel–manganese mining in the PCZ crusts are bismuth (42 ppm), molybdenum (463 ppm), niobium (54 ppm), titanium (1.2 percent), tungsten (89 ppm), and zirconium (559 ppm) (Hein *et al.*, 2013).

Manganese nodules

Manganese nodules occur throughout the global ocean on the surface of sediment-covered abyssal plains at water depths of 3500 to 6500 m. The most extensive deposits have been found in the Pacific Ocean, especially within the CCZ, the Peru Basin, and Penrhyn Basin (Figure 20.2). A large nodule field also occurs in the Central Indian Ocean basin. Other nodule fields may also occur in the Argentine Basin in the SW Atlantic and in the Arctic Ocean, although those areas are poorly explored. The CCZ is the area of greatest economic interest for nodule mining because of high Ni and Cu contents in the nodules and high nodule abundances.

Nodules are most abundant in abyssal areas with oxygenated bottom waters, low sedimentation rates (less than 10 cm per kiloannum), and where sources for abundant nuclei occur. High-grade (>2 percent Cu+Ni) nodules occur in areas of moderate primary productivity in surface waters; nodule grade is influenced by location relative to the calcite compensation depth (CCD), which is controlled by primary productivity in surface waters. Above the CCD, deposition of biogenic calcite (shells of plankton) increases sedimentation rates and dilutes the organic carbon necessary for the chemical reactions in the sediment (called diagenetic) that release Ni and Cu (Verlaan *et al.*, 2004; Cronan, 2006). The highest-grade nodules form near but generally below the CCD.

Nodules grow by accumulation of manganese and iron hydroxides around a nucleus. Unlike Fe-Mn crusts, nodules acquire metals from two sources, seawater (hydrogenetic) and sediment pore fluids (diagenetic). Nodules occur that are solely hydrogenetic (seamounts) or solely diagenetic (Peru Basin), but most sequester metals from both sources. The pore fluids are the predominant source of nickel and copper, whereas seawater is the dominant source of cobalt. Pore fluid metals are derived from oxidation/reduction (redox) reactions and organic matter decomposition in upper sediment layers and are subsequently incorporated into the manganese nodules forming at the seabed. The diagenetic component of nodules increases their growth rates by a factor of 2 to 50 compared to Fe-Mn crusts, with measured rates of up to 250 mm per megaannum. The greater the diagenetic input, the faster the growth rate. Growth rates of hydrogenetic end-member nodules converge with those of Fe-Mn crusts. Typical nodules are 1–5 cm long. Diagenetic nodules in the Peru Basin can be up to 20 cm long.

Manganese nodules are composed predominantly of hydrogenetic δ-MnO_2 (vernadite), diagenetic 10Å manganate (todorokite, buserite, asbolan), lesser amounts of feroxyhyte, and, less commonly, diagenetic birnessite. Minor amounts of detrital aluminosilicate minerals and authigenic minerals are commonly present. Key physical properties of Fe-Mn nodules are the sheet and tunnel structures of the diagenetic manganese minerals, which allow for the acquisition of large amounts of nickel, copper, and other elements that stabilize the mineral structure.

Manganese nodules typically have three to six times more manganese than iron, distinguishing them from Fe-Mn crusts where Fe and Mn contents are similar. Nickel and copper have been the metals of greatest economic interest and have a mean concentration in the CCZ nodules of 1.3 percent and 1.1 percent, in Peru Basin nodules 1.3 percent and 0.60 percent, and in Central Indian Basin nodules 1.1 percent and 1.0 percent, respectively. These abundances vary somewhat in mine-site-size areas within these major nodule fields. Nodules contain more nickel, copper, and lithium than crusts, and crusts are more enriched in the other rare metals and REEs. Lithium in CCZ nodules averages 131 ppm and is especially high in diagenetic nodules, averaging 311 ppm in Peru Basin nodules (Hein and Koschinsky, 2014). REEs in nodules are also of economic interest but are generally two to six times lower than they are in Fe-Mn crusts, with maximum total REEs plus yttrium in CCZ nodules of 0.08 percent. Other metals of interest as potential by-products of nickel–copper–manganese mining in the CCZ include cobalt (0.21 percent), molybdenum (590 ppm), and zirconium (307 ppm).

Seafloor massive sulphides (SMS)

SMS deposits are metal-bearing sulfide mineral-rich deposits (Figure 20.3) precipitated from hydrothermal fluids on and below the seabed. In areas of volcanic activity, cold seawater moves through cracks into the seafloor down to depths of several kilometers and is heated up to 410°C. Heated seawater reacts with the surrounding rocks during its descent and leaches their contained metals. These chemical reactions produce a fluid that is hot, slightly acidic, reduced, and enriched in dissolved metals and sulphur. Due to the lower density of the hot fluid, it rises rapidly to the seafloor and vents as focused flow into the water column and produces black-and-white smoker chimneys. The dissolved metals precipitate as sulfides on contact with cold oxygenated seawater. Much of the metal carried to the seafloor is deposited as fallout from the hydrothermal particle plume, as well as forming the chimney structures. The ascending hydrothermal fluid is replaced by descending cold seawater thereby forming a hydrothermal circulation cell.

The most abundant minerals in SMS deposits are pyrite (FeS_2) and other iron sulphides. The minerals of economic interest are chalcopyrite (copper sulphide) and sphalerite (zinc sulphide), and their contained precious metals gold and silver (Petersen and Hein, 2013). Non-sulphide minerals such as sulphates, amorphous silica and silicates also occur in SMS deposits.

The contents of metals are variable and not all are of commercial interest. Copper and zinc are commonly less than 25 percent each. Gold and silver are highly enriched in some deposits, up to several tens of ppm for gold and several hundreds of ppm for silver. Other trace elements

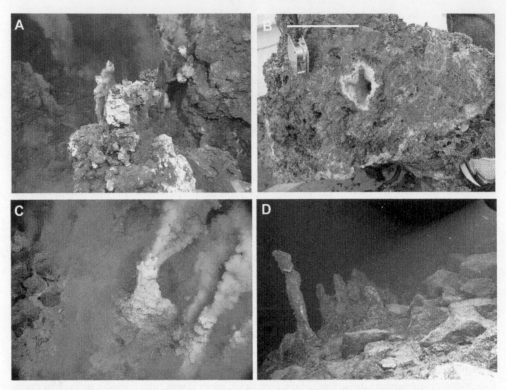

Figure 20.3 Representative photographs of seafloor and a sample collected of SMS from the Mariana volcanic arc, West Pacific: (A) active sulphide chimney emitting high-temperate fluids, East Diamante Caldera, 354 m water depth, JAMSTEC cruise NT10–12; (B) cross-section through a large chimney; the yellow conduit lining is silica colored with a trace of elemental sulphur; red-brown barite surrounds the zinc-sulphide-rich chimney collected from 377 m water depth, JAMSTEC cruise NT10–12; digital scale bar represents 10 cm; (C) white smoker chimneys from Champagne vent site, EW Eifuku volcano, 1610 m water depth; NOAA cruise TN167; (D) cluster of dead black smoker zinc sulphide chimneys and host blocky dacite rocks, East Diamante Caldera, 348 m water depth, JAMSTEC cruise NT09–08
Source: Author.

(bismuth, cadmium, gallium, germanium, antimony, tellurium, thallium, indium) that usually occur in SMS deposits in low concentrations (ppm level) can be up to several tens to hundreds of ppm in some deposits, especially those that form in volcanic arc settings. These rare metals are important in high-tech and green-tech applications. Weathering of old SMS exposed at the seabed may increase the copper contents in the deposit due to formation of secondary copper-rich sulphides.

Hydrothermal convection cells are found in all oceans and depend only on a heat source, a fluid to transport heat and chemical elements to the seafloor, and pathways for that fluid to penetrate and then exit the seafloor. Submarine volcanic activity is most extensive along the divergent plate boundaries characterized by mid-ocean ridges but is also common along convergent plate boundaries where volcanic arcs and back-arc spreading centers develop through subduction of oceanic crust. The divergent/convergent zones in the oceans have a combined length of 89,000 km, including 64,000 km of oceanic spreading centers and 25,000 km of submarine volcanic arcs and back-arc-basin spreading ridges. The abundance of massive sulphide deposits is related to the magmatic activity along those plate boundaries. Most high-temperature hydrothermal systems at mid-ocean ridges occur in the axial zones of the spreading center and are associated with basaltic volcanism; however, a number of large hydrothermal systems also develop at off-axis volcanoes. A new class of SMS deposit occurs along slow spreading ridges where detachment faults direct fluid flow away from the ridge axis (Cherkashov *et al.*, 2010). These detachment faults expose ultramafic mantle rocks at the seafloor and hydrothermal circulation can produce copper- and gold-rich deposits.

Hydrothermal circulation cells in subduction-related environments are similar to those at mid-ocean ridges, but the geology and tectonic setting influence the mineralogy and chemistry of the deposits. The most influential differences are host-rock compositions, the input of magmatic fluids, volatiles, and metals into the hydrothermal fluids and venting in shallow-water environments. Active hydrothermal systems and sulphide deposits occur in the craters and calderas of large arc volcanoes but most high-temperature vents and the largest sulphide deposits in the western Pacific occur in the back-arc spreading centers (e.g., Lau Basin, North Fiji Basin, Mariana Trough).

Besides these focused-flow hydrothermal systems, diffuse-flow systems form when the rising metal-rich fluid mixes with downward moving seawater, which results in the precipitation of mostly metal oxides over large areas of seabed. This type of hydrothermal deposit does not usually contain metals of economic interest.

Drivers for confronting the challenges of mining deep-ocean minerals

The global population passed seven billion people in November 2011. More than 2.5 billion live in countries with expanding economies and a rapidly growing middle class. An ever-growing number of people live in developing countries that need to build the infrastructure and acquire resources necessary for a sustainable energy future. The mineral resources required to sustain that growth and to support green- and emerging-technologies can no longer be supplied economically solely from land-based sources. Many rare metals required for these green- and high-technology applications are abundant in deep-ocean mineral deposits. Existing sources for many critical metals are limited; China is the major producer for 30 critical metals, including the 14 REEs and yttrium. China's exports of these metals are decreasing as internal use increases.

The average grades being mined in land-based mines are continually decreasing. For example, average copper ore mined in 1900 contained 4 percent copper, whereas now it is close to 0.5 percent copper (Mudd, 2009). Consequently, more ore must be processed to yield the same amount of metal. In comparison, SMS deposit grades vary from about 1 percent to 12 percent copper, although the tonnage reserves are unknown. Nodules also contain more than 1.0 percent copper and occur in high tonnage deposits. Terrestrial mines must remove an increasing amount of overburden to reach ore bodies, with super-deep open-pit and mega-underground mines planned. In contrast, marine Fe-Mn crust, manganese nodule, and SMS deposits have little or no overburden.

REEs found in Fe-Mn crusts and nodules have several distinctions from those produced from terrestrial mines. The marine deposits have significantly lower grades, but the tonnages of total REE metals are comparable in the CCZ nodules and PCZ crusts with the largest land-based carbonatite-hosted deposits, Bayan Obo in China and Mountain Pass in the USA. More important is the much larger complement of heavy REEs (HREEs) relative to light REEs in the marine deposits. The large terrestrial REE deposits have <1 percent HREEs, whereas the PCZ crusts and CCZ nodules have respectively 18 percent and 26 percent HREE complements of the total REEs (Hein, 2012). This is key because the HREEs have the greatest economic value (Hein *et al.*, 2013).

Current activities and major players in development of deep-ocean mineral deposits

As of January 2015, exploration contracts have been signed or are pending signature for over 2 million km² of seabed (Figure 20.2), about half by coastal States for operations within their respective EEZs, and the remainder by the International Seabed Authority (ISA) in areas beyond national jurisdictions, known as The Area. About 40 percent of the contracted area is for SMS deposits, most within EEZs of SW Pacific States, and only 60,000 km² for SMS in The Area. Most of the remaining 60 percent of the contracts are for Fe-Mn nodules, all in The Area, with two small contract areas for phosphorite, one off New Zealand, one off Namibia, and one very small contract area for metalliferous mud in the Red Sea. In July 2012, the Council and Assembly of the ISA passed regulations for the exploration for Fe-Mn crusts and soon after received three plans of work for contract areas (9000 km² total) in the West Pacific, which were approved in 2014. Brazil submitted a plan of work for Fe-Mn crusts on Rio Grande Rise, Atlantic Ocean, which has been approved by the Council of Nations; this 3000 km² is included in the total contract area.

The following States through their Federal agencies have signed contracts for Fe-Mn nodule areas of about 75,000 km² each: China, France, Germany, India, Japan, Korea, Russia, and a group of States under the name Inter-Ocean Metals (Bulgaria, Cuba, the Czech Republic, Poland, Russia, Slovakia); in addition, eight companies have nine contracts or pending contracts of 75,000 km² areas for nodules, one for 58,620 km². China, France, Germany, India, Korea, and Russia have signed or have pending contracts for SMS of 10,000 km² each. Japan, China, Russia, and Brazil have plans of work approved by ISA Council for Fe-Mn crusts of 3000 km² each.

Comparisons with land-based reserves

To get a measure of how the metal tonnages of deep-ocean deposits compare with land-based deposits, we use the best-studied area for nodules, the CCZ, and the best-studied area for crusts, the PCZ. Comparisons of total contained tonnages of metals from those two marine areas are made with the global terrestrial reserves (U.S. Geological Survey, 2012) and the global terrestrial reserve base (TRB; U.S. Geological Survey, 2009), which includes resources that are currently economic (reserves), marginally economic, and subeconomic. A conservative estimate of the dry tonnage of nodules in the CCZ is 21,100 million tonnes (ISA, 2008) and for crusts in the PCZ is 7533 million tonnes (Hein and Koschinsky, 2014).

Nodules in the CCZ have 6000 times more thallium, 1.2 times more manganese, 1.6 times more tellurium, 1.8 times more nickel, 3.4 times more cobalt and 4 times more yttrium than the entire TRB for those metals. Metals in CCZ nodules as a percent of the TRB are arsenic 88 percent, molybdenum 63 percent, thorium 27 percent, vanadium 25 percent, copper 23 percent, tungsten 21 percent, lithium 20 percent, niobium 15 percent, total REEs as oxides (TREO) 10 percent, titanium 7 percent and PGM 4 percent.

With only 36 percent the tonnage of CCZ nodules, Fe-Mn crusts in the PCZ have just over 1700 times more thallium, 9 times more tellurium, 3.8 times more cobalt, 3.4 times more yttrium, and 1.8 times more arsenic than the TRB. Metals in PCZ crusts as a percent of the total TRB are bismuth 46 percent, manganese 33 percent, nickel 21 percent, molybdenum 18 percent, vanadium 13 percent, niobium 13 percent, TREO 11 percent, tungsten 11 percent, titanium 10 percent, thorium 8 percent, PGMs 5 percent, and copper 0.7 percent. These calculations illustrate that significant tonnages of metals occur in the marine deposits, however, it should be kept in mind that not all the nodules in the CCZ or crusts in the PCZ would be recovered, as is also true for the metals that make up the TRB.

Unique characteristics for recovery of deep-ocean minerals

Land-based mines leave a substantial footprint of roads, building complexes, and open pits, as well as impacted waterways and millions of tonnes of waste rock; marine-based mine sites will have no roads, seafloor ore-transport systems or buildings, and little seafloor infrastructure. Fe-Mn crusts and nodules are essentially two-dimensional deposits and exposed at the seabed. SMS deposits have a third dimension of several tens of meters, but have little or no overburden to be removed. In land-based mines, removal of overburden to reach an ore body can be as much as 75 percent of the material moved. Since the deep-ocean mining platform will be a ship, it can be moved to new locations so that small high-grade deposits can be selectively mined. In addition to the higher grades of the marine deposits, three or more metals can be recovered from a single mine site for each of the three main types of deep-ocean mineral deposits. Deep-ocean marine operations will not impact indigenous or native human populations; an increasing concern with land-based mine sites. Further, no personnel will be in harm's way at the mine site in deep-ocean operations.

Technical challenges

There is a large body of literature on potential technologies for the exploration and exploitation of deep-ocean mineral deposits (see Hein *et al.*, 2000 for review). However, the first-generation technologies for exploitation are currently available or are presently being built for mining Fe-Mn nodules and SMS, but not Fe-Mn crusts. Extraction technology for SMS has been adapted from that used in deep-ocean petroleum operations, such as seabed pipe trenching operations, and from offshore placer diamond mining, the latter of which is being adapted from shelf-depth operations to deep-water operations (see section above on placer diamonds).

Fe-Mn crusts have two major technological hurdles to overcome, one for exploration and mine-site characterization and one for extraction. It is essential that an exploration tool be developed that is deep-towed, or Autonomous Underwater Vehicle (AUV)-mounted, and will measure crust thicknesses (tonnage) in place in real time. The best avenues may be the development of a multispectral seismic instrument or a gamma-radiation detector, although issues with attenuation of the gamma-ray signal in seawater must be overcome. A second issue is the development of a mining tool that will be able to separate the Fe-Mn crust from the substrate rock on which it is attached without collecting the substrate rock, which would significantly dilute the ore grade. This removal must take place on an irregular and often rough seabed at 1500–2500 m water depths, and with crusts attached with variable degrees of adherence depending on the type of substrate rock, which will require substantial technological advancements.

Environmental considerations

Disturbance of the Earth's surface whether land-based (farming, logging, mining, cities, roads, wind farms, solar farms, etc.) or deep-ocean based (mining, trawling, wind farms, etc.) will disturb or even destroy habitat, so it is essential that the most environmentally sound practices be developed and employed for all Earth-surface activities. Besides the potential environmental advantages of deep-ocean mining listed above, deep-ocean ecosystems generally show low population densities compared to areas of many land-based mines, with the exception of deep-ocean active hydrothermal black smoker vent fields. However, those vigorously venting systems will not likely be exploited in the near term, rather extinct systems are currently the focus of industry. Ecosystem compositions will need to be characterized at each potential mine site, the degree of endemism determined, and an environmental assessment made. The International Marine Minerals Society produced an Environmental Code that is posted on their website (www.immsoc.org). A real-world example of an environmental impact statement is provided by Nautilus Minerals Inc. for its Solwara 1 mine site and posted on their website (www.cares. nautilusminerals.com).

The potential affects of Fe-Mn nodule mining have been studied during several international programs that were based on extensive field studies, as well as theoretical and experimental laboratory studies. It is clear that there will be damage of habitat in the path of a mining vehicle and that a sediment plume of unknown extent will be created in the bottom-water layer. The ISA's Kaplan Project (ISA, 2008) concluded that it is difficult to predict the threat of nodule mining to biodiversity and the risk of species decline because of limited knowledge of the numbers and geographic distributions of species. Approaches to ameliorate these problems include development of proper mining equipment, use of un-mined corridors within the ore field so that impacted species can repopulate the area, and mining down current from reference sites with equivalent ecosystems so that larvae would be transported to the mined area.

Fe-Mn crust mining would not significantly involve sediment suspension and would cover an area much smaller than that needed for nodule mining (Hein *et al.*, 2009). Crust particles can be an effective scavenger of trace metals, which might lead to depletion of some micronutrients, whereas a release of particles within an OMZ around seamounts could lead to release of metals. The generalization that seamounts are island habitats with highly endemic faunas that comprise unique communities distinct in species composition from other deep-sea habitats is not supported by a recent compilation of existing data (Rowden *et al.*, 2010).

Deep-ocean ores will be transported to land-based processing plants. Once the ores are transported to existing or newly built processing plants, the same environmental issues that presently exist at such plants will continue; however, newly built plants may be more efficient and employ advancements in green technologies. Processing on board ship will likely be limited to dewatering of the ore, which would be pumped back down to the water depth of the mine site.

References

Cherkashov, G., Poroshina, I., Stepanova, T., Ivanov, V., Bel'tenev, V., Lazareva, L., Rozhdestvenskaya, I., Samovarov, M., Shilov, V., Glasby, G. P., Fouquet, Y., and Kuznetsov, V. (2010) "Seafloor massive sulfides from the Northern Equatorial Mid-Atlantic Ridge: New discoveries and perspectives," *Marine Georesources and Geotechnology*, vol. 28, pp. 222–239.

Cronan, D. S. (2006) "Processes in the formation of central Pacific manganese nodule deposits," *Journal of Marine Science and the Environment*, vol. C4, pp. 41–48.

Hein, J. R. (2012) "Prospects for rare earth elements from marine minerals" Briefing Paper 02/12, International Seabed Authority.

Hein, J. R. and Koschinsky, A. (2014) "Deep-ocean ferromanganese crusts and nodules," in Holland, H. D. and Turekian, K. K. (Eds.) *Treatise on Geochemistry*, 2nd ed., vol. 13, Elsevier, Oxford, Chapter 11, pp. 273–291.

Hein, J. R., Conrad, T. A., and Dunham, R. E. (2009) "Seamount characteristics and mine-site model applied to exploration- and mining-lease-block selection for cobalt-rich ferromanganese crusts," *Marine Georesources and Geotechnology*, vol. 27, pp.160–176.

Hein, J. R., Conrad, T. A., and Staudigel, H. (2010) "Seamount mineral deposits, a source of rare metals for high-technology industries," *Oceanography*, vol. 23, pp. 184–189.

Hein, J. R., Mizell, K., Koschinsky, A., and Conrad, T. A. (2013) "Deep-ocean mineral deposits as a source of critical metals for high- and green-technology applications: Comparison with land-based resources," *Ore Geology Reviews*, vol. 51, pp. 1–14.

Hein, J. R., Koschinsky, A., Bau, M., Manheim, F. T., Kang, J. K., and Roberts, L. (2000) "Cobalt-rich ferromanganese crusts in the Pacific," in Cronan, D. S. (Ed.) *Handbook of Marine Mineral Deposits*, CRC Press, Boca Raton, FL, pp. 239–279.

ISA, International Seabed Authority (2008) "Biodiversity, species ranges, and gene flow in the abyssal Pacific nodule province: predicting and managing the impacts of deep seabed mining," ISA Technical Study, No. 3, Kingston, Jamaica

Mudd, G. (2009) "The Sustainability of Mining in Australia: Key production trends and their environmental implications for the future," Research Report, Department of Civil Engineering, Monash University and Mineral Policy Institute, Australia, pp. 1–277.

Petersen, S. and Hein, J. R., 2013. The geology of sea-floor massive sulphides, in Baker, E. and Beaudoin, Y. (Eds.), Secretariat of the Pacific Community, Deep Sea Minerals 1A: Sea-floor massive sulphides, a physical, biological, environmental, and technical review, v. 1A, SPC, Suva, Fiji, pp. 7–18 (http://grida.no/publications/deep-sea-minerals/, accessed June 13, 2015).

Rowden, A. A., Dower, J. F., Schlacher, T. A., Consalvey, M., and Clark, M. R. (2010) "Paradigms in seamount ecology: fact, fiction and future," *Marine Ecology*, vol. 31, pp. 226–241.

U.S. Geological Survey (2009) *Mineral commodity summaries 2009*, U.S. Geological Survey, Washington D.C.

U.S. Geological Survey (2012) *Mineral Commodity Summaries 2011*, U.S. Geological Survey, Washington D.C.

Verlaan, P. A., Cronan, D. S., and Morgan, C. L. (2004) "A comparative analysis of compositional variations in and between marine ferromanganese nodules and crusts in the South Pacific and their environmental controls," *Progress in Oceanography*, vol. 63, pp. 125–158.

21

MAKING PROGRESS WITH MARINE GENETIC RESOURCES

Salvatore Aricò

Marine genetic resources: an introduction

On the basis of the definition provided in article 2 of the Convention on Biological Diversity (CBD), which deals with use of terms, marine genetic resources can be described as material to be found in the marine environment of any living origin (plant, animal, microbial, other) containing functional units of heredity of potential or actual value (SCBD, 1992; Aricò and Salpin, 2005). The 2010 Nagoya Protocol on Access to Genetic Resources and the Fair and Equitable Sharing of Benefits Arising from their Utilization to the Convention on Biological Diversity expands the definition of genetic resources in the text of the CBD to also encompass marine 'derivatives'; these, in article 2 of the Nagoya Protocol, on use of terms, are defined as naturally occurring biochemical compounds resulting from the genetic expression or metabolism of biological or genetic resources, even if they do not contain functional units of heredity (SCBD, 2011). These definitions de facto imply that, from the standpoint of the CBD and Nagoya Protocol regime, all material that contains functional units of heredity and all naturally occurring biochemical compounds to be found in the marine environment potentially fall within the category of marine genetic resources.

Article 3 of the Nagoya Protocol, which deals with access to genetic resources, specifies that the Protocol applies to genetic resources within the scope of article 15 of the CBD (access to genetic resources), that is to say, with a strong focus on the sovereign rights of States over their natural resources in determining access to genetic resources. But article 15 also contains provisions related to creating the conditions to facilitate access to genetic resources for environmentally sound uses by other Contracting Parties;[1] the full involvement in scientific research based on genetic resources of the Contracting Parties providing such resources;[2] and for sharing in a fair and equitable way the results of research and development and the benefits arising from the commercial and other utilization of genetic resources with the Contracting Party providing such resources.[3]

Article 11 of the Nagoya Protocol, which deals with transboundary cooperation, states:

> In instances where the same genetic resources are found in situ within the territory of more than one Party, those Parties shall endeavour to cooperate, as appropriate,

with the involvement of indigenous and local communities concerned, where applicable, with a view to implementing this Protocol.

Moreover, the global vocation of the Nagoya Protocol is well reflected in its article 10, which deals with a global multilateral benefit-sharing mechanism, also in relation to benefits derived from the utilization of genetic resources that occur in transboundary situations. The article states:

> Parties shall consider the need for and modalities of a global multilateral benefit-sharing mechanism to address the fair and equitable sharing of benefits derived from the utilization of genetic resources and traditional knowledge associated with genetic resources that occur in transboundary situations or for which it is not possible to grant or obtain prior informed consent. The benefits shared by users of genetic resources and traditional knowledge associated with genetic resources through this mechanism shall be used to support the conservation of biological diversity and the sustainable use of its components globally.

The definitions and provisions in the articles of the CBD and the Nagoya Protocol referred to above, albeit cognisant of sovereignty issues in relation to resources to be found in areas within national jurisdiction, seem to reflect the underlying assumption that, in the context of the CBD and Nagoya Protocol regime, genetic resources are considered to be of potential importance to humankind as a whole.

The collaborative provisions envisaged by the Nagoya Protocol in relation to transboundary cooperation and to a global multilateral benefit-sharing mechanism can greatly assist in promoting cooperation in regards not only to shared marine genetic resources or to genetic resources the origin of which cannot be determined, but also to marine genetic resources that clearly originate from areas beyond national jurisdiction, in support of implementation of the general provisions on Marine Scientific Research under the UN Convention on the Law of the Sea (UNCLOS).

Discriminating between marine genetic resources within areas of national justification from shared marine genetic resources and from those to be found in areas beyond national jurisdiction, albeit legally relevant, is not very meaningful from a scientific perspective:

- First, due to the very broad scope of the definition of genetic resources under the CBD and Nagoya Protocol regime, in many cases it might be difficult, not to say impossible, to define exact geographic boundaries when dealing with material of any living origin containing functional units of heredity or naturally occurring biochemical compounds resulting from the genetic expression or metabolism of biological or genetic resources.
- Second, marine species are distributed according to both taxonomic as well as physiognomic features (habitat, water characteristics, seabed topography, ecological processes); these features, together with those of the life cycles of species (for example, several sedentary marine species have pelagic stages), determine patterns of species dispersal, isolation and evolution (Agostini *et al.*, 2008). Hence, from a scientific perspective, for genetic resources derived from several marine species, it may be difficult to associate them with a specific geographic location.
- Third, the application of metagenomic analysis, which has shifted the biodiversity paradigm from species and ecosystem centred to gene centred,[4] in the marine environment has indicated clearly that the diversity of genes even in 'poor' parts of the world oceans such as the Sargasso Sea[5] is vast (Venter *et al.*, 2004). Furthermore, metagenomic analysis has showed that some genes are clearly ubiquitous (Aziz *et al.*, 2010); at the same time, research

showed that some genetic sequences of marine samples in the water column were unique in almost every sample collected (Rush *et al.*, 2007).

In essence, our scientific knowledge of how genetic diversity shapes itself on Earth, including in the marine environment, is in its infancy, and it seems difficult to accommodate these findings in the theoretical framework of conventional ecological sciences, to which legal instruments often refer.

From a policy perspective, in the context of the current legal framework on the law of the sea, it has proven very difficult even to agree on some basic principles on how to deal with access to and the sharing of benefits deriving from the utilization of marine genetic resources. The debate on the legal status of marine genetic resources in areas beyond national jurisdiction has become increasingly heated since the first observations by Glowka in 1996 (Glowka, 1996) on the applicability of the freedom of use principle versus the principle of common heritage of mankind to these resources to more recent observations on the complexity of the debate, which is still to be resolved (Scovazzi, 2015).

Treves (2010) has called the legal regime of genetic resources in the seabed beyond the limits of national jurisdiction 'the most "fashionable" law-of-the-sea problem of the present'. This can be justified in light of two main types of reasons.

The first is the general perception that applications of findings of research on marine genetic resources might be economically prosperous and societally beneficial. This is potentially true, as illustrated by the number of, and trends in, patents filed on the basis of these resources (Arnaud-Haond *et al.*, 2011); as well as successful examples of applications of scientific discoveries based on these resources in sectors such as health, industry and cosmetics (Leary *et al.*, 2009; Leary, 2011).

Defining the regime according to which access to marine genetic resources in areas beyond national jurisdiction and the sharing of benefits deriving from their utilization would be regulated has a direct impact on who would benefit from these resources. In the absence of clarity with regard to the legal status of, and due to the high costs involved in operations related to research on these resources, access to them and the derived benefits have been a prerogative of a lucky few (Aricò and Salpin, 2005; Ruth, 2006; Leary, 2007; Arnaud-Haond *et al.*, 2011).

The second type of reason why the debate on marine genetic resources in areas beyond national jurisdiction has attracted so much attention is its complexity. Issues related to these resources are multiple and multifold, as they span many aspects of a scientific, technical, technological, policy, social, economic and political nature. These entail inter alia issues related to intellectual property rights, including modalities related to public–private partnerships and the licensing to the private sector of patents filed on the basis of research conducted with public funding (Aricò and Salpin, 2005; Salpin and Germani, 2007; Chiarolla, 2013); who owns the technology needed to explore the extreme environments (in terms of depth, temperature, toxicity and remoteness) where marine genetic resources of particular interest are found, and issues related to technology transfer and capacity building in general; ensuring disclosure with regard to where marine genetic resources have been sampled (Vierros *et al.*, 2015); and issues related to the conservation of vulnerable species and ecosystems and legal and technical difficulties related to the establishment of marine protected areas beyond areas of national jurisdiction (Salpin and Germani, 2010).

The debate on marine genetic resources, as part of the consideration by the United Nations General Assembly of the conservation and sustainable use of marine biodiversity beyond areas of national jurisdiction since 2004, is largely responsible for triggering a process on the possible elaboration of a new implementing agreement on the conservation and sustainable use of marine biological diversity of areas beyond national jurisdiction under UNCLOS, with the aim of strengthening the current law of the sea framework. Discussions on a possible new implementing agreement have been polarized as some States favour focusing on the implementation of existing instruments, while others insist that the current law of the sea regime should be strengthened, including by taking in the emerging and unresolved issues related to marine genetic resources, including questions on the sharing of benefits, as well as the need for a global mechanism related to area-based management tools, environmental impact assessments and capacity building and technology transfer. In Rio in 2012, high-level State representatives, building on the work of the Ad Hoc Open-ended Informal Working Group to study issues relating to the conservation and sustainable use of marine biological diversity beyond areas of national jurisdiction and before the end of the sixty-ninth session of the UN General Assembly, committed to address, on an urgent basis, the issue of the conservation and sustainable use of marine biological diversity of areas beyond national jurisdiction, including by taking a decision on the development of an international instrument under UNCLOS.[6] In my opinion this process can be seen as a de facto reopening of the way in which the UNCLOS regime is currently arranged.

Much has been written on the multiple issues that the debate on marine genetic resources of areas beyond national jurisdiction has entailed; hence this chapter will not repeat those considerations. Moreover, despite the complexity of the multiple dimensions of the problem, there is a need to demystify and to unbundle the current debate on marine genetic resources.[7]

Further to the introduction provided thus far, this chapter recollects the history of the policy debate on marine genetic resources of areas beyond national jurisdiction from its inception in 1995 up to the present, which may provide a useful lens though which to analyse and better understand current discussions and deliberations under the auspices of the UN General Assembly. The chapter focuses on those directions that seem to be more promising so as to deal with issues related to marine genetic resources in an effective manner, namely technology transfer and capacity building. It refers to the latest scientific developments in relation to the study of genetic resources – a field that evolves at a rapidity much more significant than the pace of the policy discussions. It also attempts to provide examples of practical experiences and lessons learned in dealing with these resources at multiple scales, as well as best practices by the scientific community and the private sector. Finally, the chapter suggests future prospects to make progress with the current debate on marine genetic resources of areas beyond national jurisdiction, so as to move from discourse to action on their conservation and sustainable and equitable use.

The primacy of the UN Convention on the Law of the Sea and the important backstopping role of the Convention on Biological Diversity

In 1995, the report on access and benefit-sharing in relation to genetic resources prepared by the Secretariat of the CBD for consideration by the second meeting of the Conference of the Parties (COP) to the Convention noted:

> The genetic resource provisions of the Convention on Biological Diversity do not apply to genetic resources in areas outside national jurisdiction, such as the high seas and the deep seabed. Genetic resources in these areas may, however, have major value for humanity. UNCLOS did not anticipate this value and it is unclear whether or how the common heritage principle applies to the living resources of the deep seabed.[8]

The Secretariat suggested that an in-depth analysis be commissioned by the COP to the Convention's Subsidiary Body on Scientific, Technical and Technological Advice (SBSTTA);

the study was to treat the relationship between the CBD and UNCLOS in relation to how to address the use of genetic resources outside national jurisdiction and how UNCLOS and the Convention on Biological Diversity could be mutually reinforcing with regard to access to marine genetic resources under national jurisdiction.

The legal interpretation advanced by the CBD Secretariat with regard to the non-applicability of the provisions of the CBD to genetic resources in areas outside national jurisdiction is correct in light of Article 4 of the CBD, which deals with the jurisdictional scope of the Convention (SCBD, 1992), and which in its paragraph (a) states: '[the provisions of this Convention apply, in relation to each Contracting Party] [i]n the case of components of biological diversity, in areas within the limits of its national jurisdiction'. It can be questioned, however, that this interpretation albeit correct is only partial in that paragraph (b) of Article 4 of the CBD states:

> [the provisions of this Convention apply, in relation to each Contracting Party] [i]n the case of processes and activities, regardless of where their effects occur, carried out under its jurisdiction or control, within the area of its national jurisdiction or beyond the limits of national jurisdiction.

In the latter regard, Salpin and Aricò (2005) effectively make the argument that bioprospecting of genetic resources from areas beyond national jurisdiction would be covered by the CBD in so far as it is an activity that may have impacts on the biodiversity in those areas, not necessarily from an access and benefit-sharing perspective.

As Aricò and Salpin (2005) demonstrate extensively, 'processes and activities' also entail a number of steps related to seabed bioprospecting of marine genetic resources, which the authors loosely define as the search for, and exploitation of, valuable compounds from genetic resources of the seabed beyond national jurisdiction.

Regardless, the suggestion proactively put forward by the CBD Secretariat triggered a debate at the second meeting of the COP, which led to the adoption of decision II/10, in which:

> The Conference of the Parties, [r]equests the Executive Secretary, in consultation with the United Nations Office for Ocean Affairs and the Law of the Sea, to undertake a study of the relationship between the Convention on Biological Diversity and the United Nations Convention on the Law of the Sea with regard to the conservation and sustainable use of genetic resources on the deep seabed, with a view to enabling the Subsidiary Body on Scientific, Technical and Technological Advice to address at future meetings, as appropriate, the scientific, technical, and technological issues relating to bio-prospecting of genetic resources on the deep seabed.[9]

In 1996, the CBD Secretariat prepared document SBSTTA/2/15 on bioprospecting of genetic resources of the deep seabed with a view to assisting SBSTTA in its work and recommendations on the subject.[10] This document was never considered, as the provisional agenda submitted to the second meeting of SBSTTA meeting was orally amended by the Executive Secretary of the Convention, who reported:

> the Bureau had considered it inappropriate to include for discussion an item on bioprospecting of genetic resources of the deep seabed, since the Secretariat had been unable to consult with the United Nations Convention on the Law of the Sea (UNCLOS) in time to coordinate their input.[11]

The position of the COP Bureau was reflective of a certain degree of nervousness on behalf of some Contracting Parties to the Convention to tackle the potentially sensitive issue of if and how the CBD should deal with an issue that many perceived as a legal gap under UNCLOS (CBD Secretariat, pers. comm.).

When joining the CBD Secretariat in 1998 as the Head of its Marine and Coastal Unit, which was in charge of coordinating the development and monitoring the implementation of the Convention's first thematic programme of work dealing with marine and coastal biological diversity, the author of the current chapter found it of urgency to follow-up to decision II/10, paragraph 12, so that progress could be made in relation to clarifying the legal status of marine genetic resources in areas beyond national jurisdiction. This process involved many interactions between the CBD Secretariat and the UN Division for Ocean Affairs and the Law of the Sea (DOALOS) as well as iterations by CBD's SBSTTA and COP.

The 2003 study of the relationship between CBD and UNCLOS with regard to genetic resources of the deep seabed jointly prepared by the CBD Secretariat and DOALOS concluded that a legal lacuna existed in relation to these resources and that both CBD and UNCLOS could offer potential solutions and tools to deal with issues related to marine genetic resources in areas beyond national jurisdiction.[12]

Consideration of the outcomes of the joint study by SBSTTA and subsequently by the COP at its fourth meeting in 2004 led to the invitation by the COP to the UN General Assembly to further coordinate work relating to conservation and sustainable use of genetic resources of the deep seabed beyond the limits of national jurisdiction.[13] Indeed, the dominant position of countries, not only in the context of the CBD process but also in relation to annual debates and deliberations of the UN General Assembly on ocean affairs and the law of the sea, was that UNCLOS was the framework within which issues related to marine biodiversity in areas beyond national jurisdiction ought to be tackled and resolved.

At the same time, the COP invited Parties and other States to identify activities and processes under their jurisdiction or control that may have a significant adverse impact on deep seabed ecosystems and species beyond the limits of national jurisdiction,[14] thus reaffirming its competence in relation to processes and activities affecting biodiversity, even in the event that those took place in areas beyond national jurisdiction, consistently with Article 4 of the Convention.

The implications of the facts recollected above are the following:

- first, the CBD has demonstrated its ability to contribute to elucidating issues related to a complex and potentially controversial topic such as that of marine genetic resources in areas beyond national jurisdiction while respecting its scope, which focuses on issues within national jurisdiction;
- second, the debate initiated within the CBD, and the explicit recognition of the primacy of UNCLOS when dealing with marine areas beyond national jurisdiction, has triggered an interlinked but also self-determining process under the UN General Assembly in relation to UNCLOS and marine biodiversity in areas beyond national jurisdiction.

In its resolution 59/24 of 17 November 2004, the UN General Assembly decided to establish an Ad Hoc Open-ended Informal Working Group to study issues relating to the conservation and sustainable use of marine biological diversity beyond areas of national jurisdiction (hereinafter referred to as 'the Working Group').[15]

The Working Group has met six times to date, in 2006, 2008, 2010, 2011, 2012 and 2013. Since 2004, where it was tasked with, inter alia, examining the scientific, technical, economic, legal, environmental, socio-economic and other aspects and indicating possible options and approaches to promote international cooperation and coordination for the conservation and sustainable use of marine biological diversity beyond areas of national jurisdiction, its mandate has evolved into undertaking a process with a view to ensuring that the legal framework for such biodiversity effectively addresses conservation and sustainable use by identifying gaps and ways forward, including through the implementation of existing instruments and the possible development of a multilateral agreement under UNCLOS. At its sixth meeting in August 2013, the Working Group recommended that the UN General Assembly establish a process within the Working Group to prepare for the decision to be taken at its sixty-ninth session on the development of an international instrument under UNCLOS.[16]

The increased frequency of the meetings of the Working Group and the fact that it became tasked with making recommendations to the UN General Assembly show the importance that States attach to the issue.

Transfer of marine technology: from discourse to reality?

The history of technology transfer in the context of the law of the sea is summarized in the IOC[17] Criteria and Guidelines on the Transfer of Marine Technology (IOC, 2005).[18] The account recalls the genesis of the debate on marine technology transfer in the context of the III UN Conference on the Law of the Sea in 1970 in relation to the proposal put forward in 1967 by Ambassador Arvid Pardo of Malta to declare the seabed beyond national jurisdiction as common heritage of mankind. This proposal was subsequently reflected in the 1970 UN General Assembly Declaration of Principles Governing the Sea-Bed and the Ocean Floor, and the Subsoil Thereof, beyond the Limits of National Jurisdiction.[19]

Issues related to technology transfer were dealt with in the context of the III UN Conference on the Law of the Sea (1973–1982) under the agenda item dealing with 'Development and Transfer of Technology' at the Conference's first session in 1974. At the third session of the Conference in 1975, specific provisions on technology transfer in a treaty language were introduced. At the eighth session of the Conference in 1978, substantive negotiations on Part XIV of UNCLOS, which deals with the development and transfer of marine technology, were concluded and subsequently adopted at the eleventh session of the Conference in 1982 as part of the final text of UNCLOS.

Despite the fact that the origin of the debate on technology transfer relates to non-living resources in areas beyond national jurisdiction and that Part XIV of UNCLOS does not only refer to technology for the exploitation of resources but also to research technology and other technology, provisions in Part XIV of UNCLOS can be applied to marine resources in general (a possible notable exception relates to articles 273 and 274 of UNCLOS, which refer specifically to activities in the Area and the role of the International Seabed Authority.[20]

Article 271 of UNCLOS deals with the promotion by States, directly or through competent international organizations, of the establishment of generally accepted guidelines, criteria and standards for the transfer of marine technology on a bilateral basis or within the framework of international organizations and other fora, taking into account, in particular, the interests and needs of developing States.

IOC is recognized by UNCLOS as a competent international organization in relation to marine scientific research.[21] It is also mentioned among the organizations with which the Commission on the Limits of the Continental Shelf may cooperate in order to exchange scientific and technical information that might be of assistance in discharging the Commission's responsibilities.[22]

The explicit recognition of IOC as a competent international organization by UNCLOS, which entered into force in 1994, stimulated the establishment by the IOC Assembly in 1997 of the IOC Advisory Body of Experts of the Law of the Sea (IOC/ABE–LOS). The group's work has so far focused on the practice of States in the application of part XIII, dealing with marine scientific research, and XIV, dealing with the transfer of marine technology, of UNCLOS; the procedure for the application of Article 247 of the Convention in relation to marine scientific research projects undertaken by or under the auspices of international organizations; and the legal framework, within the context of UNCLOS, which is applicable to the collection of oceanographic data.

In 2003, the XXII Session of the IOC Assembly adopted the IOC Criteria and Guidelines on the Transfer of Marine Technology that had been developed by IOC/ABE-LOS following an invitation by the UN General Assembly.[23]

The scope of the Criteria and Guidelines reflects the related provisions of UNCLOS on the transfer of marine technology (part XIV). In the Criteria and Guidelines, marine technology is defined as encompassing information and data; manuals and guides; standards and reference materials; sampling and methodology equipment; observation facilities and equipment; equipment for *in situ* and laboratory observations, analysis and experimentation; computer hardware and software, including modelling techniques; and expertise, knowledge and skills in relation to scientific, technical and legal know-how related to marine scientific research and observations.

Against the assumption that the transfer of marine technology should enable all parties concerned to benefit, on an equitable basis, from developments in the area of marine sciences, the criteria presented involve: the requirement to develop specific legal, scientific and financial schemes at national, sub-regional and regional levels; the need to conduct transfer of technology on a free-of-charge basis or at a reduced rate for the benefit of the recipient country; the specification to take into account the needs and interests of developing and land-locked countries; the rights and duties of holders, suppliers and recipients of marine technology; the importance of the transfer of environmentally sound technologies; and the need to rely on existing or new cooperation schemes, including joint ventures and partnerships among States, intergovernmental organizations, governmental and non-governmental organizations and private entities.

Guidelines for the application of the above-mentioned criteria involve the development of a clearing-house mechanism on the transfer of marine technology. This mechanism should list donor parties; provide scientific and technological information and data; inform on marine research institutes involved in training, cruises and universities and other organizations that provide study grants, and access to workshops and seminars; list experts who could provide assistance in the field of marine technology; and provide networking opportunities between these organizations. The Guidelines also deal with the inclusion of technology transfer into the national strategic plans of IOC Member States; the establishment of regional and sub-regional focal points for marine technology, possibly with IOC's regional commissions; and the establishment of a voluntary cooperation fund for promoting and facilitating the transfer of marine technology. The Guidelines also foresee a Transfer of Marine Technology Application (TMTA) and the provision of assistance by IOC in the identification of donors of marine technology and in the development of possible cooperative schemes, including joint ventures and partnerships. The Commission would also promote the participation of the recipient country and facilitate the organization of expert missions, the delivery of technical training and the assessment of the results of the projects implemented.

Issues related to the technological capacity of States to deal with marine genetic resources in areas beyond national jurisdiction would in many instances apply also to resources from areas within national jurisdiction of those States where the scientific, technical and technological capacity is still lacking. Hence, the discrimination between marine genetic resources within areas of national jurisdiction from those in areas beyond national jurisdiction appears artificial from the point of view of scientific cooperation, technology transfer and capacity building on marine genetic resources.

The current legal framework in relation to technology transfer provided by UNCLOS is comprehensive enough to operationalize existing provisions on technology transfer to deal with marine genetic resources. Because of its neutral nature, technology transfer can provide a constructive contribution to solving tensions between developed and developing countries when dealing with access to and the sharing of benefits arising from the utilization of marine resources. The Nagoya Protocol expressly refers to technology transfer as an element of benefit-sharing, and Salpin (2013) demonstrates that possible synergies between the Protocol and UNCLOS to foster coordination at national level and international cooperation largely overcome the challenge of implementing two legal instruments that have been conceived thirty years apart.

Not surprisingly, on several instances, technology transfer has been referred to by many countries as a most promising way forward to deal with many of the issues related to conservation and sustainable use of marine biodiversity in areas beyond national jurisdiction, including marine genetic resources:

- At the first meeting of the Working group in 2006, delegations recognized the need for transfer of technology to developing countries in relation to the conservation and sustainable use of marine biodiversity in areas beyond national jurisdiction. They also noted that the relevant provisions of UNCLOS, which call for the transfer of technology on fair and reasonable terms and conditions, were not effectively implemented. Developed countries, relevant international organizations and financial institutions were invited to support deep-sea scientific research in developing countries through bilateral, regional or global cooperation programmes and partnerships, and developing countries were encouraged to compile a list of relevant experts.[24]
- At the second meeting of the Working Group in 2008, it was stressed that capacity-building and technology transfer were key efforts to address implementation gaps in UNCLOS in relation to marine biodiversity in areas beyond national jurisdiction.[25]
- At its third meeting in 2010, the Working Group reiterated the need to promote capacity building and the transfer of marine technology, including South–South cooperation, in favour of marine biodiversity in areas beyond national jurisdiction. Competent organizations, in cooperation with States, were invited to develop capacity building programmes and workshops for the purpose of sharing information, knowledge and skills on these resources. The application by States of the IOC Criteria and Guidelines on the Transfer of Marine Technology was recommended.[26]
- The fourth meeting of the Working Group in 2011 saw the adoption of the historical recommendation according to which

  > a process be initiated, by the General Assembly, with a view to ensuring that the legal framework for the conservation and sustainable use of marine biodiversity in areas beyond national jurisdiction effectively addresses those issues by identify-ing gaps and ways forward, including through the implementation of existing instruments and the possible development of a multilateral agreement under the

United Nations Convention on the Law of the Sea; [t]his process would address the conservation and sustainable use of marine biodiversity in areas beyond national jurisdiction, in particular, together and as a whole, marine genetic resources, including questions on the sharing of benefits, measures such as area-based management tools, including marine protected areas, and environmental impact assessments, capacity building and the transfer of marine technology.[27]

- At the fifth meeting of the Working Group in 2012, several delegations highlighted the need for increased transfer of marine technology noting that this was the most significant gap in the implementation of UNCLOS.[28]
- At the sixth meeting of the Working Group in 2013, the representative of the Group of 77 and China reiterated that technology transfer constituted the greatest implementation gap in UNCLOS in relation to marine biodiversity from areas beyond national jurisdiction.[29]

In light of the controversial nature of the debate on genetic resources, technology transfer measures provide not only a 'quick win' approach to deal, in an effective way, with issues related to access and benefit-sharing in relation to these resources as of now, until a possible international instrument under UNCLOS is completed; they may also constitute the only approach currently at our disposal.

At the substantive level, it would be useful to produce a comprehensive up-to-date assessment of the technology capacity as well as needs of developing countries to assist in the way technology transfer can further assist the current discussions on marine genetic resources in areas beyond national jurisdiction and in general, which is currently lacking.

Recent developments related to scientific and technological aspects of marine genetic resources

In addition to oceanographic vessels, the technology involved needed to reach deep sea environments includes manned as well as remotely operated vehicles (ROVs), sampling devices, pressured aquaria, devices for culturing of microorganisms under pressure conditions equivalent to the depth from which samples were collected and conventional biological oceanography laboratory equipment. These equipments and operations, especially those related to manned vehicles and ROVs, are very sophisticated and costly, and only a few countries have access to them.[30]

In the case of genes or compounds of potential interest that are isolated, conventional molecular biology techniques such as those employed for sequence analysis, gene expression and protein structure are used. More recently, shotgun sequencing methods developed in the context of the Human Genome Project have allowed gene sequencing to be enhanced significantly (Venter *et al.*, 2001).

The application of IT to organize and study the data collected through molecular biology analyses (bioinformatics) and computer-based modelling of genetic information and biological systems (computational biology) are seen as approaches that could considerably shorten the time of study and discovery, as well as the life cycle that goes from discovery to application (Leary, 2007; Glowka, 2010). Synthetic biology is an emerging discipline that aims to design and engineer biologically based parts, novel devices and systems as well as redesigning existing, natural biological systems (Royal Academy of Engineering, 2009). All these areas of application of IT to biology are potentially relevant to R&D on marine genetic resources.

The computer technology to perform bioinformatics and computational biology is becoming widely accessible. However, the knowledge associated with their application is not yet widespread. Moreover, issues related to interoperability of databases, and the challenge to deal with today's 'data deluge' indicate that bioinformatics will still require efforts to promote its very basic methods before this tool can be applied routinely and on a large scale.

On the other hand, techniques to conduct rapid sequencing that can assist in the discovery of new gene systems are rather low-priced (Royal Academy of Engineering, 2009), and these are likely to become more and more diffused.

Shimmield (2013) reviews the extent and types of research related to microbial marine genetic resources to be found in open waters below the depth of 200 m (which may correspond to areas beyond national jurisdiction, depending on sea floor topography features). These include metagenomic (gene sequencing of generic samples); advanced cell sorting techniques followed by massive DNA amplification; traditional culture methods; and culture collections of marine microbes, many of which are associated with the World Federation for Culture Collection.[31] The author recalls the distinction between sequence-based metagenomics and function-based metagenomics; the latter allows ecosystem functions to be identified such as production of antibiotics, which in turn allows the genes that code for that particular function to be tracked back. This approach may find a good response in those conducting research and development (R&D) on marine genetic resources in the near future.

Selected examples of successful experiences to deal with the management of marine genetic resources

The multiple theoretical aspects of issues related to marine genetic resources have been dealt with in length and have largely been clarified from an academic perspective already. Because solutions to deal with issues related to marine genetic resources ought to be based on practical approaches that reflect successful experiences on the ground; and because information on experiences by individual countries and best practices by the scientific community and the private sector is generally poor, there is a need to start filling this significant knowledge gap in relation to case studies on successful examples to deal with the management of marine genetic resources, especially those to be found in areas beyond national jurisdiction.

The selection of case studies presented below exemplifies the type of information that is needed to match the needs of multiple stakeholders concerned by marine genetic resources with relevant experiences and best practices. Successful experiences may be replicated in the context of a possible new implementing agreement under UNCLOS dealing with the conservation and sustainable use of marine biodiversity in areas beyond national jurisdiction, including issues related to marine genetic resources.

Examples of global, regional and national model regulations and experiences dealing with genetic resources

The intersessional workshop on marine genetic resources convened under the auspices of the Working Group held in New York in May 2013 depicts well the general state of current reflections on marine genetic resources in relation to issues such as meaning and scope; types and extent of research; applications and uses; impacts and conservation challenges; technological aspects; access and benefit-sharing; intellectual property rights; relevant legal and policy regimes; and international scientific cooperation and technology transfer. In addition, the workshop represented an important step towards the identification of successful experiences to deal with marine genetic resources.[32]

The World Federation of Culture Collections (WFCC) relies on the Microorganisms Sustainable Use and Access Regulation International Code of Conduct (MOSAICC) and the World Data Centre for Microorganisms for promoting the exchange of microbial genetic resources for research purposes, including marine genetic resources from areas beyond national jurisdiction, and for recording their geographical origin. Broggiato refers to the work of WFCC as contributing to operationalizing the notion of 'common pools of resources'.[33]

Broggiato recalls that in the 1990s, the CBD sparked the development of regional regimes on genetic resources, which provided common frameworks with minimum legal requirements for access and benefit-sharing in relation to genetic resources. The European Culture Collection Organization (ECCO) has developed a core material transfer agreement for the transfer of material for research purposes between entities using the same licensing conditions.[34] Through the Micro B3 project (microbial biodiversity, bioinformatics and biotechnology), the European Union plans to develop a standard access and benefit-sharing agreement for ocean sampling activities that will discriminate between R&D for the public domain and corporate R&D.[35] Broggiato stresses the innovative character of this standard in providing a 'come-back clause' to renegotiate monetary benefit-sharing if the product developed reaches commercialization.

The Mediterranean Science Commission has developed a draft charter on access and benefit-sharing – a voluntary scheme for scientists that applies to sampling in marine areas within and beyond national jurisdiction. The contents of the draft charter are organized according to the following 'core values': equity and fairness; certainty of property rights; legality/liability; transparency; traceability; reciprocal relations; concerted handling of commons; nature conservation (environmental respect); efficiency.[36]

A few States are quite active in the field of R&D on marine genetic resources, while many are in need of assistance to develop adequate individual and institutional capacity to deal with these resources. Yet comprehensive information in relation to national capabilities and needs is lacking.

Gabrielsen considers that biological banks and repositories of sampled organisms play a significant role in promoting research and applications on marine genetic resources, due to the high costs involved in research on these resources. Norway has developed a legal regime to regulate access to and use of genetic resources in areas of national jurisdiction, which includes biogeographic information of samples, a Material Transfer Agreement (MTA) when applicable, ownership, intellectual property rights and benefit-sharing arrangements.[37] In 2001, Norway had proposed a plan of action for Marine Scientific Research in areas under national jurisdiction, which reflected Norway's legislation in relation to the provisions under part XIII of UNCLOS.[38]

Japan's practical experience with research of marine genetic resources from the deep seabed is recognized. The Japan Agency for Marine-Earth Science and Technology (JAMSTEC)[39] has a longstanding experience, in particular, with the technology involved in the study of these resources. This technology includes unconventional oceanographic vessels such as manned research submersibles and ROVs, which are based on very sophisticated and expensive technology, often comparable to space technology. Moreover, JAMSTEC is in the lead also in relation to the sampling techniques and the sample laboratory analysis techniques involved. As a result of these activities, JAMSTEC has filed several patents in relation to technological inventions as well as scientific discoveries based on deep seabed genetic resources.[40]

An analogous organization very active in deep sea research technology and research is the Woods Hole Oceanographic Institution (WHOI)[41] in the USA, which unlike JAMSTEC is a private and non-profit organization. WHOI has collaborated with the company Diversa in the sampling of organisms that has ultimately led to the development of the commercial enzyme for starch processing Ultra-Thin®.[42]

Codes of conduct for scientists

Van Dover (2012) recalls that scientists are responsible stakeholders in relation to the conservation and sustainable use of deep sea ecosystems, as codified in the InterRidge statement of commitment to responsible research practices at deep-sea hydrothermal vents.[43] InterRidge is a non-profit organization that provides a platform for coordinating research focusing on mid-ocean ridge systems and for the sharing of best practices. It contains guidelines for research practices at hydrothermal vents, which include avoiding activities having a deleterious impact; avoiding activities leading to long-lasting alteration; avoiding collections that are not essential; avoiding transplanting biota between sites; avoiding adverse impacts of one's research with research activities conducted by others by familiarizing researchers with the status of current and planned research; facilitating the fullest possible use of collected organisms by the global community of scientists through collaborations and cooperation.

Shimmield (2013) reports that several collections are in the process of aligning their practices with the provisions of the Nagoya Protocol through codes of conduct reflecting inter alia the following principles: sustainability; an ecosystem-based approach; integrated management; the precautionary principle; stakeholder participation; transparency for all stakeholders; and compatibility with international and domestic legislation.

Private institutions and corporations

In 2005, the US-based private company Diversa recognized the ethical, technological and politico-sociocultural challenges of bioprospecting.[44] Diversa's framework model for ethical and successful bioprospecting collaboration aimed at engaging in ethical bioprospecting, in which participating stakeholders (countries, institutions and corporations) can benefit from the research and development applications associated with bioprospecting. In addition to securing legal access through Prior Informed Consent (PIC) and provisions for the sharing of royalties, low-environmental impact was also sought during the operations. Diversa, which then became Verenium, has always been very transparent about having developed commercial products based on bioprospecting of marine genetic resources in areas beyond national jurisdiction. Unfortunately this initiative does not seem to have been replicated and applied widely by other corporations dealing with marine genetic resources.

Open access data management and exchange systems

The Ocean Biogeographic Information System (OBIS) is a global database of geo-referenced records of marine species that stemmed from the ten-year (2001–2010) Census of Marine Life Programme. Currently managed by IOC, OBIS allows searching records per species name, region and bathymetry. Currently 450 institutions from fifty-six countries provide data for OBIS.[45]

More specifically on marine genetic resources, and also contributing to OBIS, the MICROBIS database provides geo-referenced and environmental data associated with environmental sequencing surveys.[46]

Monitoring patents is a difficult endeavour. Specifically related to marine genetic resources is the Bioprospector project, a database of bioprospecting information resources in the Arctic, Antarctic, Pacific and marine (other regions) areas maintained by the United Nations University. As the database focuses on research on, but also commercialized products deriving from, marine genetic resources, and due to the fact that the information available in the public patent domain often lacks transparency, the information presented in the Bioprospector database is patchy, and this database deserves to be populated further.[47,48]

Information on patents filed according to different national patent systems is increasingly available in the public domain, although details on the origin of sampling are often lacking due to the fact that there is no requirement for disclosure of origin under the current patent requirements. Useful resources to search patents are PATENTSCOPE,[49] which is managed by the World Intellectual Property Organization (WIPO), and the Patent Lens initiative.[50]

Efforts centred on the recognition of relevant indigenous and local knowledge and stakeholder participation mechanisms

Vierros (Vierros *et al.* 2010) illustrates the importance of traditional knowledge and associated management practices in dealing with marine genetic resources in the Pacific region. Traditional knowledge has benefited those wishing to commercialize products derived from marine genetic resources; successful experiences of partnerships between indigenous and local communities and R&D endeavours in the region reflect the need for clear royalty policies, regional schemes for revenue-sharing due to the transboundary nature of marine organisms that provide the source of active compounds and capacity-building in marine chemistry and biosystematics at the local level.

Slattery, on the basis of evidence collected in countries in the Indo-Pacific and the Caribbean regions, stresses the importance of education on the fundamental research partnerships that can help countries, through collaborative action, move effectively on a path from discovery to development.[51]

A useful experiment in bringing together multiple stakeholders is the European Commission-funded PharmaSea project, which focuses on bioprospecting of genetic resources from marine extremophiles and sees the participation of partners from academia, industry and non-governmental organizations.[52]

Future prospects of the policy debate: towards a recognition of the contribution of marine genetic resources to the post-2015 development

The debate on marine genetic resources boils down to the benefits of applications of discoveries about the diversity of ways in which life in the marine environment manifests itself (*who should enjoy those benefits?*) but also to the responsibilities entailed by some of those discoveries (*shouldn't we conserve by all possible means such diversity, for the benefit of current and future generations, especially in the case of vulnerable systems and organisms?*). It is time to make significant progress with the conservation and the sustainable use of, and sharing of benefits from, marine genetic resources for the benefits of all.

Aricò (2008) advocates that marine genetic resources have the potential to contribute to meeting several of the Millennium Development Goals as far as health, diseases and possibly also food security are concerned.

Heads of States and Government and high-level representatives attending the UN Conference on Sustainable Development in Rio de Janeiro in 2012 acknowledged 'the role of access and benefit-sharing arising from the utilization of genetic resources in contributing to the conservation and sustainable use of biological diversity, poverty eradication and environmental sustainability'.[53] Indeed, marine genetic resources hold a potential in all of these areas, as well as in relation to improving human health.

The Pacific Small Island Developing States have proposed a Sustainable Development Goal (SDG) on Oceans focusing on healthy, productive and resilient oceans and sustainable energy.[54] It would be important that, if endorsed, such SDG encompass targets related to the conservation and sustainable use of, and sharing of benefits from, marine biodiversity in areas beyond national jurisdiction.

Such strong policy-enabling framework would assist in carrying out much needed actions to improve the status of marine biodiversity in these areas and to prevent its degradation. For example, several ecosystems where organisms with metabolic paths of interest to (often novel) science and industry are to be found in deep sea areas such as hydrothermal vents which are considered as vulnerable systems that are still insufficiently represented in systems of designated marine protected areas, such as in the case of the European Natura 2000 network (Olsen *et al.*, 2013). Moreover, the recognition of the contribution of marine genetic resources to attaining the SDGs is highly desirable if one is to boost international scientific cooperation and technology transfer in the area of marine genetic resources.

Up to the present, issues related to access and benefit-sharing in relation to marine genetic resources from areas beyond national jurisdiction have been dealt with as part of a package agreed on by the Working Group in 2011 and comprised of marine genetic resources, including questions on the sharing of benefits, measures such as area-based management tools, including marine protected areas, and environmental impact assessments, capacity-building and the transfer of marine technology.[55]

The notion of a package of issues was adopted mainly to make progress with negotiations related to a possible new implementing agreement under UNCLOS. At this stage of negotiations, in light of the lack of consensus on the need for a new agreement, there is a need to start unbundling the above-mentioned issues that, albeit interrelated, also present challenges specific to each of them.

Dealing with the multiple issues related to marine genetic resources, including questions on the sharing of benefits, requires a pragmatic approach to make progress on these issues and promote the conservation and sustainable use of, and the sharing of benefits from, these resources. Technology transfer could become an important element to implement such an approach.

Yet, a meaningful programme on technology transfer specifically dealing with marine genetic resources at the intergovernmental level is lacking. The main constituents of such a programme could reflect the current IOC Criteria and Guidelines on the Transfer of Marine Technology. The programme could be developed under the framework of UNCLOS, agreed upon by the UN General Assembly, and benefit from technical backstopping from UN bodies such as the IOC (on modalities for technology transfer and capacity-building) and WIPO (in relation to intellectual property rights issues), and possibly the United Nations Industrial Development Organization (UNIDO) and the World Health Organization (WHO). The involvement of UNIDO would reflect the organization's mandate to promote and accelerate sustainable industrial development in developing countries and economies in transition, while WHO's involvement and, in particular, that of its Department of Public Health, Innovation, Intellectual Property and Trade would be justified in light of the demonstrated contribution of marine genetic resources to health issues.

A process to prepare for a decision on the development of a new implementing agreement under UNCLOS, focusing on the conservation and sustainable use of marine biological diversity of areas beyond national jurisdiction is in train. This decision will be made at the sixty-ninth session of the UN General Assembly on whether or not to have such an implementing agreement.

Some have already compared this process to the III UN Conference on the Law of the Sea process (undisclosed, pers. comm.). We cannot and should not wait any longer to promote the conservation and sustainable use of, and the sharing of benefits from, marine biodiversity in areas beyond national jurisdiction. In this regard, the transfer of technologies represents the most obvious low-hanging fruit to achieve speedy progress with these issues, in a spirit of true partnerships among members of the United Nations and on the basis of the legitimate interpretation of the various debates on marine genetic resources that have taken place since 1995 – that these resources, including those to be found in areas beyond national jurisdiction, are the heritage of mankind as a whole.

Notes

1 Article 15 of the CBD, paragraph 2.
2 Article 15 of the CBD, paragraph 6.
3 Article 15 of the CBD, paragraph 7.
4 Cf. the section of this chapter on recent developments related to scientific and technological aspects of marine genetic resources.
5 The Sargasso Sea is an 'oligothrophic' sea, that is to say, poor in nutrients.
6 Paragraph 162 of 'The Future We Want'. Cf. also next section of this chapter.
7 In addition to investigations by established academia, the subject of marine genetic resources has become increasingly popular with graduate students in the past new years. See for example Appiott, J. (2011) 'Breaking the Stalemate: Analyzing State Preferences in the Global Debates on Marine Biodiversity Beyond National Jurisdiction' (Masters Thesis); and Fedder, B. (2013) *Marine Genetic Resources, Access and Benefit Sharing: Legal and Biological Perspectives*, Routledge, Oxon and New York.
8 Document UNEP/CBD/COP/2/13, available at www.cbd.int/doc/meetings/cop/cop-02/official/cop-02-13-en.pdf (last accessed on 19 August 2015).
9 Paragraph 12 of decision II/10 of the CBD COP, available at www.cbd.int/decision/cop/default.shtml?id=7083 (last accessed on 19 August 2015).
10 Available at www.cbd.int/doc/meetings/sbstta/sbstta-02/official/sbstta-02-15-en.pdf (last accessed 20 September 2015).
11 Report of the Subsidiary Body on Scientific, Technical and Technological Advice on the work of its second meeting, available at www.cbd.int/doc/meetings/cop/cop-03/official/cop-03-03-en.pdf (last accessed 20 September 2015).
12 Study of the relationship between the Convention on Biological Diversity and the United Nations Convention on the Law of the Sea with regard to the conservation and sustainable use of genetic resources on the deep seabed (decision II/10 of the Conference of the Parties to the Convention on Biological Diversity), available at www.cbd.int/doc/meetings/sbstta/sbstta-08/information/sbstta-08-inf-03-rev1-en.pdf (last accessed on 19 August 2015).
13 Paragraph 55 of decision VII/5 of the CBD COP, available at www.cbd.int/decision/cop/default.shtml?id=7742 (last accessed on 19 August 2015).
14 Paragraph 56 of decision VII/5 of the CBD COP.
15 Paragraph 73 of UN General Assembly resolution 59/24, available at www.unga-regular-process.org/images/Documents/un%20a-res-59-24.pdf (last accessed 20 September 2015).
16 Summary of the discussions at the sixth meeting of the Ad Hoc Open-ended Informal Working Group to study issues relating to the conservation and sustainable use of marine biological diversity beyond areas of national jurisdiction (New York, August 2013) by the International Institute for Sustainable Development, available at www.iisd.ca/oceans/marinebiodiv6/brief/brief_marinebiodiv6e.pdf (last accessed on 19 August 2015).
17 Intergovernmental Oceanographic Commission of UNESCO.
18 See also the IOC web pages on the transfer of marine technology at http://ioc-unesco.org/index.php?option=com_content&view=article&id=315&Itemid=100029 (last accessed on 19 August 2015).
19 UN General Assembly resolution 2749 (XXV), available at http://daccess-dds-ny.un.org/doc/RESOLUTION/GEN/NR0/350/14/IMG/NR035014.pdf?OpenElement (last accessed 20 September 2015).
20 Article 1 of UNCLOS defines the Area as 'the seabed and ocean floor and subsoil thereof, beyond the limits of national jurisdiction'.
21 UNCLOS, annex VIII, article 2.2.
22 UNCLOS, annex II, article 3.2.

23 Paragraph 23 of UN General Assembly resolution 56/12, available at http://daccess-dds-ny.un.org/doc/UNDOC/GEN/N01/475/82/PDF/N0147582.pdf?OpenElement (last accessed 20 September 2015).

24 Report of the Ad Hoc Open-ended Informal Working Group to study issues relating to the conservation and sustainable use of marine biological diversity beyond areas of national jurisdiction, document A/61/65, available at www.un.org/depts/los/biodiversityworkinggroup/biodiversityworkinggroup.htm (last accessed on 19 August 2015).

25 Report of the Ad Hoc Open-ended Informal Working Group to study issues relating to the conservation and sustainable use of marine biological diversity beyond areas of national jurisdiction, document A/63/79 and A/63/79/Corr.1, www.un.org/depts/los/biodiversityworkinggroup/biodiversityworkinggroup.htm (last accessed on 19 August 2015).

26 Report of the Ad Hoc Open-ended Informal Working Group to study issues relating to the conservation and sustainable use of marine biological diversity beyond areas of national jurisdiction, document A/65/68, www.un.org/depts/los/biodiversityworkinggroup/biodiversityworkinggroup (last accessed on 19 August 2015).

27 Paragraphs 1(a) and (b) of the report of the Ad Hoc Open-ended Informal Working Group to study issues relating to the conservation and sustainable use of marine biological diversity beyond areas of national jurisdiction, document A/66/119, available at www.un.org/depts/los/biodiversityworkinggroup/biodiversityworkinggroup (last accessed on 19 August 2015).

28 Report of the Ad Hoc Open-ended Informal Working Group to study issues relating to the conservation and sustainable use of marine biological diversity beyond areas of national jurisdiction, document A/67/95, www.un.org/depts/los/biodiversityworkinggroup/biodiversityworkinggroup (last accessed on 19 August 2015).

29 Summary of the discussions at the sixth meeting of the Ad Hoc Open-ended Informal Working Group to study issues relating to the conservation and sustainable use of marine biological diversity beyond areas of national jurisdiction (New York, August 2013) by the International Institute for Sustainable Development, available at www.iisd.ca/oceans/marinebiodiv6/brief/brief_marinebiodiv6e.pdf (last accessed on 19 August 2015).

30 See slides of the presentation by Sophie Arnaud-Haond at the intersessional workshop on marine genetic resources (New York, May 2013), available at www.un.org/Depts/los/biodiversityworkinggroup/workshop1_arnaud.pdf (last accessed 20 September 2015).

31 See www.wfcc.info (last accessed on 19 August 2015).

32 See Intersessional workshops aimed at improving understanding of the issues and clarifying key questions as an input to the work of the Working Group in accordance with the terms of reference annexed to General Assembly resolution 67/78: Summary of proceedings prepared by the Co-Chairs of the Working Group, document A/AC. 276/6, available at www.un.org/depts/los/biodiversityworkinggroup/biodiversityworkinggroup.htm. See also the abstracts of the presentations at the Working Group workshop on marine genetic resources (New York, May 2013), available at www.un.org/Depts/los/biodiversityworkinggroup/workshop1_abstracts_website.pdf (last accessed on 19 August 2015).

33 See the abstracts of the Working Group workshop on marine genetic resources (New York, May 2013), presentation by Arianna Broggiato, available at www.un.org/Depts/los/biodiversityworkinggroup/workshop1_abstracts_website.pdf. See also copy of the presentation's slides at www.un.org/Depts/los/biodiversityworkinggroup/workshop1_broggiato.pdf (last accessed on 19 August 2015).

34 See www.eccosite.org/ (last accessed on 19 August 2015).

35 See www.microb3.eu/ (last accessed on 19 August 2015).

36 See www.ciesm.org/forums/index.php?post/2013/03/14/CIESM-Charter-on-ABS (last accessed on 19 August 2015).

37 See the abstracts of the Working Group workshop on marine genetic resources (New York, May 2013), presentation by Kjersti Lie Gabrielsen, available at www.un.org/Depts/los/biodiversityworkinggroup/workshop1_abstracts_website.pdf (last accessed on 19 August 2015).

38 Document A/AC.259/4 on 'Marine science and the development and transfer of marine technology, including capacity-building' submitted by the Delegation of Norway to the Second Meeting of the United Nations Informal Consultative Process on Oceans and the Law of the Sea, New York, May 2001, available at http://daccess-dds-ny.un.org/doc/UNDOC/GEN/N01/328/92/IMG/N0132892.pdf?OpenElement (last accessed 20 September 2015).

39 See www.jamstec.go.jp/e/index.html (last accessed on 19 August 2015).

40 See slides of the presentation by Kazuhiro Kitazawa at the intersessional workshop on marine genetic resources (New York, May 2013), available at www.un.org/Depts/los/biodiversityworkinggroup/workshop1_kitazawa.pdf (last accessed on 19 August 2015).

41 See www.whoi.edu/ (last accessed on 19 August 2015).
42 This information is reported in Christoffersen, L. P. and Mathur, E. (2005) 'Bioprospecting ethics and benefits: A model for effective benefit-sharing', *Industrial Biotechnology*, vol. 1, no. 4, pp. 255–259.
43 The statement is available at www.interridge.org/IRStatement (last accessed on 19 August 2015).
44 Cf. Christoffersen, L. P. and Mathur, E. (2005) 'Bioprospecting ethics and benefits: A model for effective benefit-sharing', *Industrial Biotechnology*, vol. 1, no. 4, pp. 255–259.
45 See www.iobis.org/ (last accessed on 19 August 2015).
46 See http://icomm.mbl.edu/microbis/ (last accessed on 19 August 2015).
47 See http://bioprospector.net (last accessed 20 September 2015).
48 See the abstracts of the Working Group workshop on marine genetic resources (New York, May 2013), presentation by Paul Oldham, available at www.un.org/Depts/los/biodiversityworkinggroup/workshop1_abstracts_website.pdf (last accessed on 19 August 2015). See also copy of the presentation's slides at www.un.org/Depts/los/biodiversityworkinggroup/workshop1_oldham.pdf (last accessed on 19 August 2015).
49 See http://patentscope.wipo.int/search/en/search.jsf (last accessed on 19 August 2015).
50 See www.patentlens.net (last accessed on 19 August 2015).
51 See the abstracts of the intersessional workshop on marine genetic resources (New York, May 2013), presentation by Marc Slattery, available at www.un.org/Depts/los/biodiversityworkinggroup/workshop1_abstracts_website.pdf (last accessed on 19 August 2015).
52 See www.pharma-sea.eu/pharmasea.html (last accessed on 19 August 2015).
53 Paragraph 199 of 'The Future We Want', the outcome document of the UN Conference on Sustainable Development, endorsed through and annexed to UN General Assembly resolution 66/288, available at www.un.org/en/ga/search/view_doc.asp?symbol=%20A/RES/66/288 (last accessed 20 September 2015).
54 Outcome document of the Pacific SIDS Regional Preparatory Meeting to the 2014 Third International Conference on Small Island Developing States, available at https://sustainabledevelopment.un.org/content/documents/5249233Pacific%20Outcome%20Chairs%20Revised%20Final%20Version.pdf (last accessed 20 September 2015).

References

Agostini, V., Aricò, S., Escobar Briones, E., Clark, M., Cresswell, I., Gjerde, K., Niewijk, D. J. A., Polacheck, A., Raymond, B., Rice, J., Roff, J., Scanlon, K. M., Spalding, M. Tong, E., Vierros, M. and Watling, L. (2008) Global Open Oceans and Deep Seabed (GOODS) Biogeographic Classification, UNESCO-IOC, Paris. Available at http://unesdoc.unesco.org/images/0018/001824/182451e.pdf (last accessed on 26 July 2015).

Aricò, S. (2008) 'Deep sea genetic resources: What is their potential?', in *Proceedings of the Fifth Trondheim Conference on Biological Diversity*, Trondheim, Norway, October 2007, pp. 169–174.

Aricò, S. and Salpin, C. (2005) Bioprospecting of Genetic Resources of the Deep Seabed: Scientific, Technical and Policy Aspects, UNU, Tokyo. Available at www.ias.unu.edu/binaries2/Deep Seabed.pdf (last accessed on 26 July 2015).

Arnaud-Haond, S., Arrieta, J. M. and Duarte C. M. (2011) 'Marine biodiversity and gene patents', *Science*, vol. 25, 1521–1522.

Aziz, R. K., Breitbart, M. and Edwards, R. A. (2010) 'Transposases are the most abundant, most ubiquitous genes in nature', *Nucleid Acids Research*, vol. 38, no. 13, pp. 4207–4217.

Chiarolla, C. (2013) 'Intellectual Property Rights issues', in IUCN Information Papers for the Intersessional Workshop on Marine Genetic Resources (New York, 2–3 May 2013), pp. 37–45. Available at www.un.org/depts/los/biodiversityworkinggroup/documents/IUCN%20Information%20Papers%20for%20BBNJ%20Intersessional%20Workshop%20on%20MGR.pdf (last accessed on 20 September 2015).

Glowka, L. (1996) 'The deepest of ironies: Genetic resources, Marine Scientific Research, and the Area', *Ocean Yearbook*, vol. 12, pp. 154–178.

Glowka, L. (2010) 'Evolving perspectives on the international seabed area's genetic resources: Fifteen years after the "Deepest of Ironies"', in D. Vidas (ed.) *Law, Technology and Science for Oceans in Globalisation: IUU Fishing, Oil Pollution, Bioprospecting, Outer Continental Shelf*, Martinus Nijhoff Publishers, Leiden/Boston, MA, pp. 397–420.

Intergovernmental Oceanographic Commission (2005) IOC Criteria and Guidelines on the Transfer of Marine Technology, UNESCO, Paris. Available at http://unesdoc.unesco.org/images/0013/001391/139193m.pdf (last accessed on 26 July 2015).

Leary, D. K. (ed.) (2007) *International Law and the Genetic Resources of the Deep Sea*, Martinus Nijhoff Publishers, Leiden.

Leary, D. K. (2011) 'Marine genetic resources: The patentability of living organisms and biodiversity conservation', in P. Jacquet, R. K. Pachauri and L. Tubiana (eds) *Oceans: The New Frontier*, TERI Press, Delhi, India, pp. 183–193.

Leary, D. K., Vierros, M., Hamon, G. Aricò, S. and Monagle, C. (2009) 'Marine genetic resources: A review of scientific and commercial interest', *Marine Policy*, vol. 33, pp. 183–194.

Olsen, E. M., Johnson, D., Weaver, P., Goñi, R., Ribeiro, M. C. *et al.* (2013) *Achieving Ecologically Coherent MPA Networks in Europe: Science Needs and Priorities*, Marine Board Position Paper 18, K. E. Larkin and N. McDonough (eds), European Marine Board, Ostend.

Royal Academy of Engineering (2009) *Synthetic Biology: Scope, Applications and Implications*, Royal Academy of Engineering, London.

Rush, D. B., Halpern, A. L., Sutton, G., Heidelberg, K. B., Williamson, S. *et al.* (2007) 'The *Sorcerer II* global ocean sampling expedition: Northwest Atlantic through Eastern Tropical Pacific', *PLOS Biology*, vol. 5, pp. 398–431.

Ruth, L. (2006) 'Gambling, in the deep sea', EMBO Reports, The European Molecular Biology Organization, vol. 7, no. 1, pp. 17–21.

Salpin, C. (2013) 'The Law of the Sea: A before and an after Nagoya?', in E. Morgera, M. Buck and E. Tsioumani (eds) *The 2010 Nagoya Protocol on Access and Benefit-sharing in Perspective: Implications for International Law of Implementation Challenges*, Martinus Nijhoff Publishers, Leiden/Boston, MA, pp. 149–183.

Salpin, C. and Germani, V. (2007) 'Patenting of research results related to genetic resources from areas beyond national jurisdiction: The crossroads of the law of the sea and intellectual property law', *Review of European Community & International Environmental Law*, vol. 16, no. 1, pp. 12–23.

Salpin, C. and Germani, V. (2010) 'Marine protected areas beyond areas of national jurisdiction: What's mine is mine and what you think is yours is also mine', *Review of European Community & International Environmental Law*, vol. 19, no. 2, pp. 174–184.

Scovazzi, T. (2015), 'The assumption that the United Nations Convention on the Law of the Sea is the legal framework for all activities taking place in the sea', in S. Aricò (ed.) *Ocean Sustainability in the Twenty-First Century*, Cambridge University Press and UNESCO Publishing, Cambridge and Paris, pp, 232–248.

Secretariat of the Convention on Biological Diversity (SCBD) (1992) *Convention on Biological Diversity: Text and Annexes*, United Nations, New York. Available at www.cbd.int/doc/legal/cbd-en.pdf (last accessed on 26 July 2015).

Secretariat of the Convention on Biological Diversity (SCBD) (2011) Nagoya Protocol on Access to Genetic Resources and the Fair and Equitable Sharing of Benefits Arising from their Utilization to the Convention on Biological Diversity: Text and Annex, Convention on Biological Diversity, Montreal. Available at www.cbd.int/abs/text/default.shtml (last accessed on 26 July 2015).

Shimmield, G. (2013) 'Extend and types of research, uses and applications', in IUCN Information Papers for the Intersessional Workshop on Marine Genetic Resources (New York, 2–3 May 2013), pp. 7–14. Available at www.un.org/depts/los/biodiversityworkinggroup/documents/IUCN%20Information%20 Papers%20for%20BBNJ%20Intersessional%20Workshop%20on%20MGR.pdf (last accessed on 20 September 2015).

Treves, T. (2010) 'The development of the law of the sea since the adoption of the UN Convention on the Law of the Sea: Achievements and challenges for the future', in D. Vidas (ed.) *Law, Technology and Science for Oceans in Globalisation: IUU Fishing, Oil Pollution, Bioprospecting, Outer Continental Shelf*, Martinus Nijhoff Publishers, Leiden/Boston, MA, pp. 41–58.

Van Dover, C. L. (2012) 'Ocean policy: Hydrothermal vent ecosystems and conservation', *Oceanography*, vol. 25, no. 1, pp. 313–316. Available at http://tos.org/oceanography/archive/25–1_van_dover.pdf (last accessed on 20 September 2015).

Venter, J. C., Adams, M. D., Myers, E. W., Li, P. W., Mural, R. J. *et al.* (2001) 'The sequence of the human genome', *Science*, vol. 291, pp. 1304–1351 (+ corrigendum). Available at www.sciencemag.org/content/291/5507/1304.full.pdf (last accessed on 26 July 2015).

Venter, J. C., Remington, K., Heidelberg, J. F., Halpern, A. L., Rusch, D. *et al.* (2004) 'Environmental genome shotgun sequencing of the Sargasso Sea', *Science*, vol. 304, pp. 66–74.

Vierros, M., Tawake, A., Hickey, F., Tiraa, A. and Noa, R. (2010) *Traditional Marine Management Areas of the Pacific in the Context of National and International Law and Policy*, United Nations University–Traditional Knowledge Initiative, Darwin, Australia.

Vierros, M., Salpin, C., Chiarolla, C. and Aricò, S. (2015) 'Emerging and unresolved issues: The example of seabed and open ocean genetic resources in areas beyond national jurisdiction', in S. Aricò (ed.) *Sustainable Oceans in the Twenty-First Century*, Cambridge University Press and UNESCO Publishing, Cambridge and Paris, pp. 198–231.

Ocean space

22

SHIPPING AND NAVIGATION

Jeanette Reis and Kyriaki Mitroussi

Introduction

Shipping was the world's first truly global industry. The transportation of people and cargoes from one place to another by sea pioneered global trade and has since been regarded as a useful barometer of global economic development (Ng and Wilmsmeier 2012). To some ship owners, the industry is very personal – providing the basis for a rich culture, steeped in tradition. Yet as a business, it is specialised, capital intensive and subject to considerable variations in profits. Approximately 90 per cent of all transport takes place by sea, primarily as it is considered to be the most cost-effective mode of long-distance transport (IMO 2009a). Consequently, current seaborne trade amounts to approximately 8.7 billion tonnes per annum and continues to grow (UNCTAD 2012).

The shipping industry has undergone several major transformations since its inception, many of which continue to influence its current operations. This chapter will examine historical, current and future global shipping trends and consider these within their evolving operational and management framework. It will then go on to consider challenges and opportunities for the industry that include mitigation of, and adaptation to, climate change, technological innovation and geographical shifts in global economic activity. Together, these factors set the course for the future development of the industry.

Shipping routes and commodities

Evidence of transport of cargoes by sea has been traced back nearly 10,000 years to Neolithic times (Sidell and Haughey 2007). At that time, modest cargoes of locally sourced products such as livestock and wood were transported on simple wooden rafts alongshore between settlements. As skills in boatbuilding improved, so wooden boats became stronger and larger, oars were used instead of poles and vessels became more stable in the water, therefore permitting travel further offshore. The introduction of the use of sails, combined with continued enhancement of vessel construction skills, permitted vessels to travel considerably further afield. A long list of seafaring communities including the Phoenicians, Egyptians, Greeks, Carthaginians, Chinese, Vikings, Omanis, Spaniards, Portuguese, Italians, British, French, Dutch, Polynesians and Celts rapidly developed their skills in boatbuilding and navigation, and over time, each took their turn to dominate the world's oceans as part of their quest for new resources. Goods such as spices, silks and minerals were particularly valuable to the early coastal communities and as regular exchange of these precious commodities developed, so the first sea trade routes became established.

Over time, the distribution of raw materials such as coal, from extraction sites to processing and manufacturing centres and on to consumer markets began to shape the local and then the global economy. This laid the foundation of the first global industrial revolution. From the late 1800s onwards ships sailed from ports located near the vast mines of the UK, Australia, South Africa and North America to the manufacturing centres of Europe and later, the Far East. Coal was king for almost seventy years; however, following a relatively long period of exponential growth in manufacturing lasting almost seventy years, the 1950s onwards saw oil supplement and then almost entirely replace coal as the preferred primary source of fuel for manufacturing and transportation. The demand for coal declined rapidly and trade patterns in oil became the new indicator of global economic development. Oil has been transported predominantly from sources in the Middle East, West Africa, South America and the Caribbean to the manufacturing centres of Europe, North America and Asia. In 2011, an unprecedented 22,734 billion ton-miles of oil was transported globally, compared to just 3,763 billion ton-miles of coal (UNCTAD 2012). Coal is no longer king.

Iron ore, another base resource of the industrial revolution was also transported great distances from the 1800s onwards, primarily from extraction sites in South America and Australia to the massive metal manufacturing and processing factories of Europe and East Asia. As populations centralized to support the rapidly emerging manufacturing industries, industrial scale food production and transportation, particularly grain, was required to feed the growing hungry workforce. This led to further increases in seaborne trade and in 2011 grain accounted for 1,940 billion ton-miles (UNCTAD 2012) or 4 per cent of all goods transported by sea. The main trade routes for grain currently operate between North and South America to Asia and Africa. Interestingly, as demand for grain increased, so the demand for specialized vessels that could carry greater quantities and travel faster also increased. This was to enable food products to be delivered as freshly as possible to their destinations. Consequently, specialized vessels began to emerge in the 1980s that were larger, easier to load and unload and that could travel at greater speeds.

Technological developments have not solely related to ship design, but also to the way in which goods are carried. The containerization of goods has led to a significant increase in the volume of goods transported, a decrease in loading times and improved protection for the products inside containers. This is particularly important for high-value goods such as mobile phones and televisions which are becoming more widely used. Consequently, the last thirty years have witnessed a significant growth in the transportation of containerized goods from the high-tech manufacturing centres of the People's Republic of China, Japan and Southeast Asia to the consumer markets of the western world (UNCTAD 2011). In 2011, high-value goods were transported 6,599 billion ton-miles and in the near future, this figure is expected to continue to increase.

In addition to historical trade developments, it is useful to consider the more recent medium-term trade of commodities. The total quantity of cargo transported globally has more than tripled over the last forty years and in that time, the proportions of major cargo types have changed considerably. Figure 22.1 illustrates key global trade routes of all cargo types and highlights the top fifteen busiest ports.

For example, in 1970, oil accounted for 56.2 per cent of all shipped commodities, while main bulks which include liquids, ore, grain, coal, bauxite and phosphate accounted for 17.46 per cent and other dry bulks accounted for 26.34 per cent (UNCTAD 2011). In contrast, by 2010, oil accounted for just 28.33 per cent (2,752 million tonnes) of all shipped commodities,

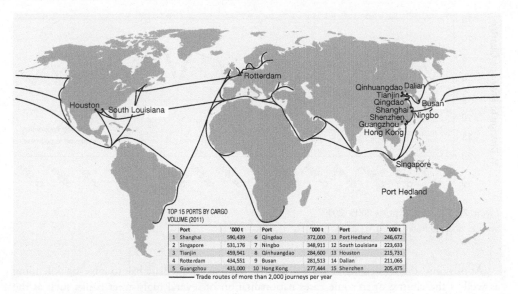

TOP 15 PORTS BY CARGO
VOLUME (2011)

Port	'000 t	Port	'000 t	Port	'000 t
1 Shanghai	590,439	6 Qingdao	372,000	11 Port Hedland	246,672
2 Singapore	531,176	7 Ningbo	348,911	12 South Louisiana	223,633
3 Tianjin	459,941	8 Qinhuangdao	284,600	13 Houston	215,731
4 Rotterdam	434,551	9 Busan	281,513	14 Dalian	211,065
5 Guangzhou	431,000	10 Hong Kong	277,444	15 Shenzhen	205,475

————— Trade routes of more than 2,000 journeys per year

Figure 22.1 Key global trade routes
Source: Adapted from UNCTAD 2012.

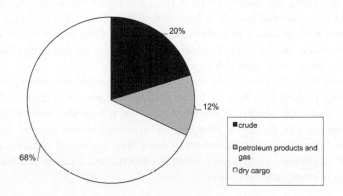

Figure 22.2 Global exports by commodity type 2011
Source: Based on UNCTAD Review of Maritime Transport (UNCTAD 2012).

while main bulks increased to 38.79 per cent (2,333 million tonnes) and other dry cargo increased to 32.88 per cent (3,323 million tonnes). Changes in the types of cargoes transported are a useful indicator of mass production and consumption patterns, and in particular reflect the more recent reduction in the use of oil to produce energy. The most recent figures for global commodity exports are illustrated in Figure 22.2 and reflect the predominance in transportation of dry cargo, followed by crude oil and petroleum products.

The early years of the noughties decade were not good ones for the shipping industry. Technological developments related to advancement of nuclear and renewable energy production meant that fewer commodities were transported. In addition, global trade in oil and petroleum-based products went into serious decline following the 2008 global economic crash, as illustrated in Figure 22.3. This directly led to a dramatic and unprecedented decrease in global manu-facturing, consumption and trade activity.

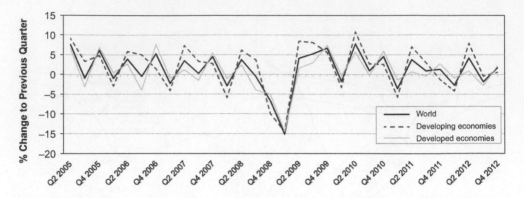

Figure 22.3 Global exports 2005–2012
Source: Adapted from UNCTAD 2013.

At this time, the United States sub-prime crisis led to an immediate halt to inter-bank lending as well as the closure or in some cases nationalisation of several high street banks such as the Royal Bank of Scotland – a leading shipping bank. A deep and long recession followed, but overall, global long-term demands for commodities have remained steady. This is primarily a reflection of the activities of developing countries such as India and China that were relatively unaffected by the crisis and have invested heavily in large-scale infrastructure and energy generation projects that support their growing manufacturing industries. Coal and iron ore for these schemes are being exported mainly from Australia and Brazil, who are benefiting from their neighbours' advancements in these areas (Mitropoulos 2011).

Although there are debates about the production and consumption pendulum swinging from west to east and the shipping yards of East Asia are busier than ever, at present it appears that the pattern of commodity trade will remain as described here for at least the next decade. However, there are clear signals that things are changing and it would be prudent for those involved in the shipping industry to consider how best to meet these upcoming challenges – particularly those relating to adaptation or relocation of existing infrastructure, location of training centres and reconsideration of compliance measures.

The economic structure of bulk and liner sectors

The 'service' of the merchant shipping industry can be considered to be the provision of transport. Shipping has a highly complex economic structure that can primarily be attributed to two apparently contradictory characteristics – its international character and its fragmentation. Its main assets, vessels, are very capital intensive, of diverse size and type and highly dependent on technological advances in materials and processes. Vessels can be financed, owned, built, flagged, operated, managed, crewed, maintained and regulated by entirely different stakeholders. In addition, they can be new or second-hand and can have broadly different life spans, depending on the individual operating, regulatory and market conditions. Shipping therefore is not a homogeneous activity and should be analysed according to its different activity sectors. The main divisions in freight shipping sectors relate to liner and bulk markets. These are considered in more detail below.

'Bulk' is considered to be any cargo with a parcel size large enough to occupy an entire hold, while all other cargoes with smaller parcel sizes are considered to be 'general cargoes'.

The size of the parcels defines the type of shipping services needed and this results in small parcels being served by liner shipping and larger parcels being served by bulk shipping. Liner shipping calls at regular ports, at regular times, offers fixed prices for each commodity and has an obligation to accept cargo from all customers. Dry bulk shipping develops when dry cargo sizes become large enough to be carried in shiploads, and the benefits of economies of scale become apparent. The bulk sector is segmented by type of commodity in the wet bulk market and includes the transportation of crude oil, oil products and orange juice while the dry bulk market includes transportation of products such as iron ore, coal, grain, phosphates and bauxite.

The high level aim of the bulk shipping sector is to provide transport for large volumes of goods in the most cost efficient way (Lun and Quaddus 2009). In many ways, the bulk shipping industry's operation closely resembles an industry under perfect competition conditions (Clarksons Research Services 2004). There are a large number of shipping companies that own ships offering similar services, without entry or exit barriers and with freely available trade information. The transport of bulk cargoes is conducted primarily in a 'tramp' manner, whereby vessels do not offer a scheduled service but are sent to different destinations according to demand. Although the bulk and liner shipping sectors have distinct pricing mechanisms, one is more like a wholesale operation and the other more like a retail operation, respectively. These distinctions determine the design and size of vessels, freight rates and life cycles, which will be examined in more detail in the following sections.

Design and size of vessels

Both bulk and liner shipping are subject to the same dynamics of economy and trade. In the last decade both sectors have benefited from strong economic growth and an increase in international trade, despite the 2008 financial crisis and its wider repercussions. During its most recent heyday, the dry bulk sector enjoyed particularly high freight rates over a relatively long period of time that extended between 2003 and 2008. By mid 2008 dry bulk freight rates had increased by an unprecedented 300 per cent against their 2003 level. China and India's burgeoning iron ore and coal demands were the main drivers behind this increase, and as industrial development continues to expand in the East, so this trend is likely to continue.

In much the same way, liner container trade, which carries 70 per cent of the international seaborne trade value, saw increases at times of strong economic growth. For instance, in 2007 it experienced an increase of more than 10 per cent over 2006 levels but a dramatic collapse in the last part of 2008 saw rates at the end of the year plummet to 60–70 per cent lower than at its beginning (Platou 2010). In 2009 demand in the container ship market responded to the rapid trade decrease and also plummeted by 9 per cent while supply continued to see positive growth of 5.1 per cent resulting in an overcapacity of vessels of a staggering 14.1 per cent. Consequently, liner carriers recorded significant losses, with the reported collective loss for 2009 estimated to be over $20 billion (UNCTAD 2009). Remediation measures adopted to deal with decreased revenues included cutting back on the number of services offered, idling ships, scrapping vessels, cancelling orders, non-delivery of goods and slow steaming at half speed of around 13 knots.

The liner sector is also characterized by significant market concentration. The top twenty liner companies transport almost two-thirds of all containers, carried as TEUs or Twenty Equivalent Units. A number of additional mergers and acquisitions in July 2008 meant there were on average 7.7 per cent fewer companies providing services per country than in July 2004. This has raised concerns about oligopolistic market structures, particularly for countries with low connectivity. For example, of the top fifty well-connected countries, as identified through the Liner Shipping Connectivity Index, only two, Egypt and Morocco, are African countries. Islands in the Caribbean area are near the bottom of the index, with Montenegro and Albania ranked as the least well-connected countries. These countries are more reliant on a limited number of services and are also more likely to pay higher prices for those services that are accessible. This has implications for the resilience of such countries and should be addressed as soon as possible (UNCTAD 2012).

In 2010 the resurrection of manufacturing activity and global trade in containerized goods led to a recovery of demand for liner shipping services, although recovery has been slow and erratic. Multiple risks threaten to undermine the prospects of a sustained recovery and a stable world economy, including sovereign debt problems in many European countries, fiscal austerity, natural disasters in Asia, political unrest in the Middle East, and rising volatile energy and commodity prices, among other factors (UNCTAD 2011).

Freight rates

It is accepted that the demand for shipping service is a derived demand, that is, the demand exists for the commodities transported rather than for the shipping service itself. This derived demand, however, is also dependent on a number of internal and external factors, such as the status of the global economy and generic transport costs. The world merchant fleet reflects the capacity of sea transportation that is available at any one time, but this static stock is in turn affected by additions or withdrawals from the fleet as well as operational factors that determine fleet productivity. For instance, loading and unloading times, vessel speed, number of loaded days at sea and number of days taken to clear customs are all factors that affect the total transport capacity of a vessel within any given time period (Stopford 2009).

Due to the international character of the industry and its increased diversification, a number of factors have continued to influence the delicate supply and demand balance of shipping. A concise, although not exhaustive list of factors is shown in Figure 22.4.

Freight rates respond to the interaction between supply and demand of shipping services and represent the point of revenue agreement between shipowners and charterers. The shipping industry is inherently characterized by capital intensiveness, volatile freight rates, seasonality, strong business cycles and exposure to fluctuations in regional and global economies (Kavussanos and Visvikis 2006; Xu *et al.* 2011). The volatility of freight rates, however, is experienced differently throughout the industry. For example, larger vessels generally have less resilience to fluctuating freight rates compared to smaller ones and this appears to be true both for dry bulk vessels (Kavussanos 1996) as well as tankers (Glen and Rogers 1997). This is potentially because smaller vessels can be engaged in a broader range of trades than larger vessels, and also because smaller ships have fewer draft restrictions at ports. In recent years freight rate volatility has been felt in an acute manner not only by shipowners but also by other market participants including charterers, shipping investors, shipyard owners and financiers. The knock-on effect of freight rate fluctuations percolates throughout the entire industry.

In addition, physical factors such as port and strait congestions such as those found at the Bosporus bottleneck, natural disasters such as the 2012 USA hurricane Katrina and the 2012 Australian floods have impacted on the delicate supply and demand balance. Regional disputes in the Middle East, North Africa and around North Korea have also significantly influenced the market. Reduced productivity arising from such occurrences puts additional strain on the potential supply of vessels, which in turn can increase freight rates. Avoidance of transit through dispute areas can also add significantly to transport times and costs. If a vessel sailing from Asia to northern Europe needed to avoid the Suez Canal, it would need to sail an additional seven

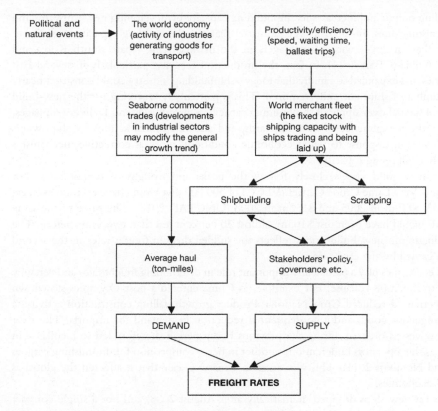

Figure 22.4 Overview of shipping supply and demand factors
Source: Adapted from Stopford 2009.

days around the coast of Africa. With losses of up to $60,000 a day for very large crude carriers, this is an expense shipping companies are keen to avoid. Interestingly, political interventions, such as the prohibition of single-hull tankers, which had two major cut-off dates in close succession – 2005 and again in 2010 – have also significantly affected availability of permitted vessels within restriction zones and thus global freight rates for those vessels has rocketed.

Vessel life cycle

The point at which it becomes economically viable to construct a new vessel, buy a used vessel, or decommission an existing vessel is primarily determined by freight rates. The following sections discuss the main characteristics of each of these stages in vessel life cycle and examine the factors that lead to the transition from one stage to the next.

Cradle-construction of new vessels

During the mid 2000s, shipowners invested heavily in new-build vessels at historically high levels. For example, between 1996 and 2002, 2 million tonnes per annum of new-builds were constructed (Clarksons Shipping Intelligence 2012). By 2007, this had increased dramatically to 7 million tonnes per annum. The new-building market responds to the augmented demand for shipbuilding output and increases its shipbuilding capacity by developing new shipyards and expanding existing ones. In 2010 it was estimated that the shipbuilding capacity was 27 per cent greater than in 2004, with main players in the market being Japan, South Korea and increasingly, China (CESA 2011). In fact, the production capacity, particularly in several East Asian countries, has expanded so rapidly that today's shipbuilding industry could construct nearly twice the number of ships currently required. This is particularly worrying for the new-build market as a substantial oversupply of shipbuilding capacity is already evident. Delivery slippages, partly due to the inexperience of new shipyards, and also intentional actions by shipowners deferring delivery in response to wider economic conditions, work as corrective mechanisms for the supply of shipping capacity.

Prices for new-build vessels closely follow the peaks and troughs of freight rates. For example, a new Very Large Crude Carrier (VLCC) of 300,000 dwt would have cost, on average, $67m in 2003, $119m in 2005 and $153m in 2008 (UNCTAD 2010). The price of the same vessel in 2010 would have been just $103m, almost 30 per cent less than two years before. The dramatic reduction in new-build prices reflects the sudden drop in freight rates in the second half of 2008 caused by the credit crisis.

Regional economics play a particularly important role in determining freight rates and certainly contributed to the 2008 collapse. For example, as China entered a post-Olympics slowdown phase, it experienced reduced Gross National Product growth, falling construction activity, a return to indigenous goods and a consequential reduction in demand for imports. The over supply of new vessels ordered during the previous boom period therefore led to a collapse in freight rates, as ship operators undercut one another in fierce competition for diminishing cargoes (Alizadeh and Nomikos 2011). This was on such a massive scale that it affected the global as well as local economies.

Contracts for new ships dropped dramatically after August 2008, and not a single contract was received by worldwide ship builders during May 2009 (Xu *et al.* 2011). Ship investors and ship manufacturers found themselves in a difficult position. In many cases, freight income fell sharply and was insufficient to cover vessel operating costs and loan repayments. This had serious knock-on implications for the second-hand and decommissioning markets, which, like a line of dominos, also collapsed.

Sale of second-hand vessels

In the mid 2000s shipowners rushed to increase their shipping capacity in order to take advantage of highly profitable freight rates by turning their attention to the sale and purchase markets. Between 2003 and 2008 prices for second-hand dry bulk carriers increased more than threefold and for tankers more than twofold in all ship size categories (UNCTAD 2010). The price of second-hand vessels is related to the age of the vessel as well as technical characteristics of the ship, such as size, equipment, the yard of build, etc. However, prices are critically affected – over and above any other factor – by freight rates. There is a time lag of a couple of months between corresponding peaks and troughs of freight rates and second-hand vessel prices, but the relationship is clearly identifiable. Shipowners can favour the option of a second-hand buy at the peak of a cycle, as this is the fastest way, along with time chartering, to expand shipping capacity to take advantage of booming short-term freight rates. The great demand for second-hand vessels tends to be reflected in the level of sale and purchase activity and the price of vessels, which may even be above the costs of a new-build at times. Consequently, high prices of second-hand vessels can trigger investment in new-builds which are more efficient, more technologically advanced and have an extended economic life.

However, the glory days of the second-hand market peaks rarely last. Between October and November 2008, dry bulk freight rates fell by more than 95 per cent with immediate effects on the values of new and second-hand vessels. In February 2009, many vessels were being sold at half the value of the previous month. For example, in February 2009, the single hull 263,097 dwt tanker *Grand Pacific* was sold at $14m. Just one month prior, it was being negotiated for $28m (Moundreas 2009).

Grave-decommissioning vessels

The decision to decommission a vessel comes only with the gloomiest shipowners' expectations about future freight rates and after contingency reserves for the maintenance of vessels have been exhausted. The high level of freight rates in the period 2003–2007 allowed the oldest and least efficient ships to continue trading at a profit and this led to an increase in the average age of decommission for all ship types. Decommission activity during the booming years was minimal and in 2007, just 5.7m dwt was demolished, the lowest activity since the early 1990s (UNCTAD 2011). Average scrap prices were above $500 per ton for the period 2003–2007, reaching $700 per ton at the beginning of 2008 (Platou 2009), reflecting the higher prices for scrap when supply was lowest.

Individual vessel and freight sectors may exhibit differentiated economic behaviour in the short term, but as they are all affected by common external factors, such as global economic shifts, international governance and technological developments, they are intrinsically linked and whatever happens in one industry sector eventually ripples through to the others.

Management of the shipping industry

Although markets determine the numbers of vessels, the geographical distribution of vessels and the types of cargoes transported at any one moment in time, longer term operations are more heavily influenced by local, national, regional and international management measures. The most important focus of shipping management relates to ensuring the safety of all human and non-human life and property, including the natural environment (IMO 1974). Although the majority of measures still relate to preservation of life and property, in recent years, measures have been extended to include climate change mitigation measures, contingency planning and compensation for pollution.

Despite a complex and disjointed history, modern shipping is the safest and most environmentally friendly form of commercial transport (Bloor *et al.* 2013). Commitment to safety and protection of the marine environment have consistently been central considerations for the majority of those involved in the industry and, consequently, shipping was one of the first industries to adopt safety standards at international scale. This has ensured that vessels travelling between one country and another are subject to a fairly consistent set of regulations, although there remain some inconsistencies in the interpretation of those regulations.

The earliest international safety standards were agreed in 1863 when a treaty signed by thirty countries introduced basic navigational procedures for ships to follow when encountering another vessel at sea, so as to avoid collision. The seas at this time were incredibly congested, with vessels transporting the coal and ore required to sustain the global demands for the first mass manufactured goods. Although the 1863 treaty helped to reduce the number of accidents, it failed to address the specific issues of emergency rescue, communications and mandatory provision of lifeboats. In the winter of 1912, the infamous loss of the *Titanic* led to the first International Conference on Safety at Sea in 1913, where these issues were discussed. The second and third conferences followed in 1929 and 1948, respectively. Although it took many years, the latter conference recognised that a single international organisation was required to implement marine safety measures consistently, and consequently it established the Intergovernmental Maritime Consultative Organisation, now known as International Maritime Organization or IMO. Today the IMO takes the lead on international marine safety matters.

Organisations and management measures

Since then, many other organisations have been formed at international right through to local level, to ensure the safety of people, cargoes, vessels and the marine environment. These include, in addition to IMO, international organisations representing shipowners and operators, charterers, classification societies, brokers, protection and indemnity clubs, and navigation aid organisations. At national and local level, organisations include foreign offices, ministries of defence, marine pollution control units, marine accident investigation branches, hydrographic offices, maritime and coastguard agencies, health and safety executives, port and harbour authorities, lighthouse authorities, and in some countries lifeboat institutions. Each of the above organisation types plays a unique role in marine safety, although works closely with other organisations to deliver a range of management measures. These are illustrated in Figure 22.5.

Of all the organisations involved, it is the IMO that leads the way in marine safety at international level. IMO is a 'Specialized Agency' of the United Nations (UN) that provides a co-ordinated international approach to all maritime matters (IMO 2013). In particular, IMO works towards improving international shipping procedures, raising safety standards and reducing marine pollution. Modern shipping management is guided by a series of international conventions that are administered by IMO. The organisation is therefore a non-regulatory one that makes

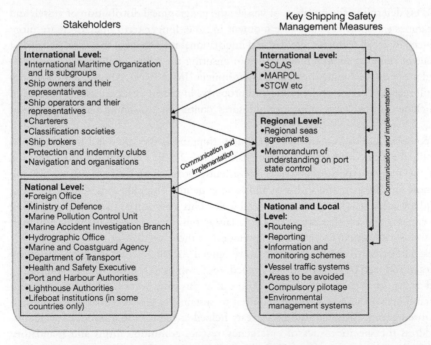

Figure 22.5 Stakeholders and shipping safety management measures
Source: Author.

recommendations based on consensus of its members, which are subsequently enforced by the national laws of member states. Consequently, national implementation of internationally agreed standards does not take place at a uniform rate and can be subject to a broad, sometimes problematic range of interpretations (CEC 1993).

Today, a plethora of IMO Conventions relating to security, pollution and liability determine levels of ship safety. However, three are of particular relevance to global shipping safety – SOLAS (International Convention for the Safety of Life at Sea), MARPOL (International Convention for the Prevention of Pollution from Ships) and STCW (International Convention on Standards of Training, Certification and Watchkeeping for Seafarers). These will be discussed in more detail below.

Safety regulations

The SOLAS Convention was first adopted in 1974 and has since, in its successive forms been generally regarded as the most important of all international treaties concerning the safety of merchant ships (IMO 1974). The main objective of the SOLAS Convention is to specify minimum standards for the provision, construction and operation of equipment on board ships. Flag States are responsible for ensuring that ships under their flag comply with its requirements.

Despite the widespread implementation of SOLAS, recent analyses report that globally, there has been a general increase in the number of ships above 100GT lost at sea. In 2006, IMO reported 120 losses, but by 2010 this had increased to 172 vessels (IMO 2012). Although the increase is quite marked, the overall figures remain relatively low when compared with the total number of vessels operating, and are certainly much lower than the pre-IMO period.

Thankfully, the number of lives lost has not corresponded with this trend and IHS Fairplay reports a decrease in loss of life per year (not including piracy) from 1825 in 2006 to just 250 in 2010 (IMO 2012). This suggests that safety measures such as SOLAS are becoming increasingly effective.

Pollution regulations

MARPOL 73/78 was, as its name suggests, adopted initially in 1973, although a number of additional annexes have since entered into force, so as to support other Conventions being introduced (IMO 1973). MARPOL applies to all types of vessels and concerns technical aspects of pollution, with the exception of the disposal of waste into the sea by dumping. There have been many amendments to the original Convention, some of the most important concerning the introduction of double hulls in 1992, the organisation of inspection timetables to assist operators in 1994, the submission of pollution reporting forms in 1996, and the identification of 'Special Areas', where pollution discharges are prohibited, in 1999.

Oil spills have remained one of the most visible and emotive forms of pollution and have been closely monitored at a global scale for many decades. Reports suggest that the number of large spills (>700 tonnes) has reduced in recent years. According to the International Tanker Owners Pollution Federation, between 2000 and 2009, on average, there were 3.3 large spills per year, but between 2010 and 2012, there were on average just 1.7 large spills a year, a significant reduction (ITOPF 2013). Again, this emphasises the effectiveness of the design and implementation of international shipping measures such as MARPOL.

Marine litter is also a highly contentious issue that has been the subject of considerable debate in recent years (IMO 2012). The decisions relating to when and where to dispose of waste produced during a ship's voyage are regulated through MARPOL Annex v Garbage.

The requirements are much stricter in 'Special Areas' but perhaps the most important feature of the Annex is the complete ban imposed on the dumping of all forms of plastic, which can have a degradation rate of several decades. Despite actions taken nationally and internationally, quantities of marine litter continue to increase and it is likely to take many decades before the results of these measures become evident.

Although there are no recent or certain figures on the global quantities or composition of marine litter, a 1997 study by the US Academy of Sciences estimated the total input of marine litter into the oceans, worldwide, to be approximately 6.4 million tonnes per year (IMO 2012). If this study was to be repeated today, it is expected that figures would be substantially higher. It is estimated in the 1997 study that up to 70 per cent of the marine litter that enters the sea ends up on the seabed, whereas about 15 per cent is washed up on beaches and 15 per cent floats on the water surface. It's sobering to think that plastics washed up on beaches after storms represent just 15 per cent of total plastic pollution and that another 85 per cent is out of sight. To worsen the situation, identification of polluters is notoriously difficult, especially further offshore. Without accurate real-time monitoring, it is likely that the dumping of such waste will continue.

Although oil and litter pollution cause major environmental impacts, a number of less visible pollution issues have become apparent in recent years. Among other things, these occur as a result of ballast discharge operations and also from the use of anti-fouling paint. It is estimated that current shipping activities transfer approximately 3–5 billion tonnes of ballast water internationally each year and host at least 7,000 different species of plants and animals (GLOBALLAST 2013). Ballast water plays an important role on board vessels and is essential for maintaining stability; however, it also presents risks to the marine environment, particularly where invasive species are inadvertently released into new environments. The port of Cardiff in South Wales, UK is currently experiencing prolific zebra mussel inundation, resulting in the rapid decline of the much smaller local native populations. This is directly related to ballast water discharge. In response to this risk, IMO has developed the *International Convention for the Control and Management of Ships' Ballast Water and Sediments* which requires that all ships carry out ballast water management procedures. This was adopted in 2004, but has only recently been ratified by the required thirty States that will permit its entry into force. Guidelines for sampling and environmental management schemes to assist implementation of the Convention have been adopted, therefore it is expected that ballast water risks will reduce in the near future.

Positive developments in the regulation of the use of chemical hull coatings have also been successful in the last few years. Vessel hulls must be kept free from marine growth to ensure streamlined movement through the water, thus maintaining transit speeds and reducing fuel costs. Until recently, tributyltin- (TBT) based hull coatings were used to prevent marine growth. Recent research has found that as well as causing deformities in some shellfish species, TBT coatings accumulate up the food chain, meaning that larger species of fish, seals and dolphins build up quantities of the toxic chemical as they feed on contaminated smaller species. The reputation of TBT is poor and it has been described as one of the most harmful substances to be intentionally introduced to the marine environment. Recognition of the harmful effects of TBT led to the introduction of IMO's *International Convention on the Control of Harmful Anti-fouling Systems on Ships*, which came into force in 2008. Fortunately, many responsible shipowners recognised the damage TBT did and had already been abiding by the Convention's requirements since it was first consolidated in 2003. Copper-based antifoulants have predominantly replaced TBT, although these are also harmful to the marine environment. At present, copper-based antifoulants present the best practicable environmental option.

Training regulations

The 1978 STCW Convention was the first to establish basic requirements for training, certification and watchkeeping of seafarers at an international level (IMO 1978). The Manila amendments to the STCW Convention and Code were adopted in 2010 and entered into force in 2012. These included requirements for hours of rest, alcohol and drug abuse security, use of new technologies, environmental awareness and teamwork. These reflect the emerging demands of the shipping industry, particularly the need to develop capacity to keep up with the rapidly growing technological developments as well as protection of crew's basic human rights will be protected (Couper *et al.* 1999).

Regional, national and local level maritime management measures

In addition to international agreements, many States are party to a range of regional agreements. For example, the UK is a signatory of the Paris Memorandum of Understanding on Port State Control, the Bonn Agreement, the Mancheplan, the Norbrit Agreement and the UK/Ireland Agreement. Regional agreements primarily concern survey of vessels using ports, cooperation during oil spill clean-up, contingency planning and search and rescue operations. These are particularly beneficial for marine operations because it is recognised that the shipping industry is international in nature and therefore issues that affect one country easily extend to neighbours.

At national and local levels, initiatives such as routeing and reporting are commonplace. These are particularly prevalent where traffic is congested, where sea conditions are regularly severe, where the environment is sensitive, or where there have been previous marine incidents. The number of national level initiatives has increased in recent years, mainly in response to the political need to respond to major incidents. Most states now have electronic charts that provide up-to-date details of such initiatives, as well as pilotage services that assist vessels to safely transit complex areas. For example, deep water routes for vessels have been implemented in the Bay of Fundy, Canada (CNTM 2014) and a traffic separation scheme has been established in the Dover Strait UK, one of the busiest shipping routes in the world (IMO 2014).

Discussion

The relationships between international, regional, national and local maritime stakeholders, as illustrated in Figure 22.5 depends on good levels of communication and coordination, clear reporting frameworks and local inspection regimes that are consistent and robust. Although the current framework for marine safety management at first appears to be straightforward, there are inevitable accidental and intentional breakdowns in communications and implementation. One of the most serious issues to affect global shipping industry relates to rogue traders – those that fail to pay fair wages, adequately consider the health and safety of their staff and who regularly flout the law. Fortunately, these tend to operate on a limited scale, generally on an ad hoc basis and remote from the main shipping routes. However, another and more serious development, particularly in parts of the world where bribery and corruption are rife, is the once almost forgotten, but now returning issue of piracy. Sea areas around the coasts of Nigeria and Somalia are increasingly at risk and consequently international military operations are almost constantly seeking to ensure safe passage of vessels in these areas. In 2013, a total of 234 piracy incidents were reported to the International Maritime Bureau, with twelve actual hijackings (ICC 2014). Considering that a typical ransom fee for an oil tanker could be $2–$3 million with up to $9.5 million for larger vessels (CNBC 2013), it is unsurprising that this is a tempting profession for cash-strapped coastal communities.

Some might argue that the existing marine management framework is limited in its ability to be able to tackle such issues and although it is recognised that a change is needed, this is unlikely in the near future. The current approach to setting international standards for shipping tends to be reactive, slow, and based on industry-driven compromises. An example of this relates to the phase-out of single-hulled oil tankers. It was only after the single-hulled *Exxon Valdez* (NOAA 1989) sank off the coast of Alaska in 1989 that the USA introduced a mandatory phase-out of these unsafe tankers within its waters. It took the sinking of the single-hulled *Erika* (CPEM 2000) ten years later off the coast of France for the member states of the IMO to accelerate the global phase-out to match that of the USA. Even then, the provisional target date for the phase-out of all single-hulled vessels was 2015, more than twenty-six years after the initial identification of the problem. One wonders whether the equivalent rate of change would be considered to be acceptable in the automotive or aeronautical industry? Most likely not.

Infiltrating at an even slower pace, and less visible than the pollution and safety issues described above, is a global phenomenon that has the potential to radically shift the way in which the global shipping industry operates – climate change. Climate change is now a widely accepted concept that is likely to have significant environmental, economic and social implications. A changing climate presents a number of risks and opportunities for the shipping industry, many of which are likely to influence the pace and direction of the global economy. Governments and industries have accepted the link between carbon dioxide (CO_2) and global mean temperature increase and have agreed to reduce emissions. Changes in climate impact both positively and negatively on the shipping industry. For example, on the one hand, increased storminess may lead to an increase in the number of accidents, with associated loss of life, cargoes, ships and pollution; however, on the other hand, new routes could become available as ice melts and opens up the polar routes all year round, saving time and money for ship companies and facilitating rapid delivery of goods for customers. In addition, emerging markets, potentially influenced by mass migrations will require significant infrastructure that will require vast quantities of raw materials that would primarily be transported by sea.

The potential risks posed by climate change are so great that stakeholders have agreed to take steps to minimise CO_2 and other greenhouse gas emissions. In June 2009, an IMO Study on greenhouse gas emissions from ships (IMO 2009b) presented a comprehensive assessment of the contribution made by international shipping to climate change. According to the study, international shipping was estimated to have emitted 870 million tonnes, or about 2.7 per cent of all global emissions of CO_2 in 2007. By the year 2050, in the absence of further regulation, ship emissions could increase by up to 300 per cent compared with 2007 levels, primarily as a consequence of the expected growth in global sea trade.

Although high levels of greenhouse gas emissions are considered to be harmful at a global scale, local impacts are just as important. A 2013 study carried out by the Seafarers International Research Centre concluded that continual high levels of atmospheric pollution were recorded at several ports in North West Europe (Bloor *et al.* 2013). Health impacts such as breathing difficulties were reported to be associated with locally high levels of atmospheric pollution, caused by ships in port areas. The report concluded that this was especially the case for larger and busier ports (EPA 2013).

The shipping industry is a relatively minor contributor to the total volume of atmospheric emissions compared to road vehicles and public utilities, such as power stations (IMO 2012); however, it still emits as much as the aviation industry (Bloor *et al.* 2013). Atmospheric pollution from ships has reduced in the last decade primarily as a result of improvements in engine efficiency, improved hull design and the ability to carry larger cargo quantities. A modern container ship uses only a quarter of the energy per cargo unit of container ships in the 1970s, and the Korean-built *Aniara*, one of the world's largest car and truck carriers, is considered to be the most environmentally friendly vessel of its type.

In July 2011, IMO's Marine Environment Protection Committee (MEPC), at its 62nd session, adopted a new chapter to MARPOL Annex VI that includes a package of mandatory technical and operational measures to reduce greenhouse gas emissions from international shipping activities, with the aim of improving the energy efficiency for new ships through improved design and propulsion technologies and for all ships, both new and existing, through improved operational practices. The measures came into force on 1 January 2013.

This is a significant achievement for IMO as for the first time in history it has established a global mandatory greenhouse gas emission reduction regime for an entire economic sector and is the first legally binding climate deal with global coverage since the Kyoto Protocol. The measures could see carbon dioxide emissions reduced by between 45 and 50 million tonnes a year by 2020 as from 2013 all ships will be required to implement an energy efficiency management plan, including monitoring of fuel consumption, and all new ships built from 2013 and onwards will be required to meet specific energy targets relating to grams of CO_2 per ton-mile, which will be reviewed and decreased every five years.

Progressive reductions in sulphur dioxide emissions from marine engines were also agreed at MEPC62, and vessels constructed on or after 1 January 2016, operating in Emission Control Areas (ECA) will be subject to the new sulphur dioxide limits. The Baltic and North Sea areas and the East and the West Coast of the United States and Canada have already been designated as ECAs and it is expected that more will follow. The revised measures are expected to reduce contributions to climate change and improve human health, particularly for coastal communities.

The 2009 Greenhouse Gas Study referred to earlier (IMO 2009b), identifies a significant potential reduction in greenhouse gas emissions through a range of technical and operational measures. If implemented, these measures could reduce emission rates to 25–75 per cent below current levels. As well as environmental benefits, such measures provide cost savings, therefore are widely supported by industry (Eide *et al.* 2011). Innovative propulsion systems such as 'Aquarius' and 'Energysail' (Maritime CEO 2013), are becoming more commonplace. These incorporate rigid sails, solar panels and energy storage modules that allow ships to harness the power provided by the wind and sun, saving up to 40 per cent fuel costs. It is recognised, however, that the recent global economic decline has hindered the short-term expansion of this market.

Conclusions

Following a very difficult period, rays of hope are beginning to shine through for the shipping industry. Picking itself up from the 2008/9 economic collapse, where most of the major industrialised economies recorded negative growth, has not been easy. Fortunately, the innovative approaches adopted by all those involved in the industry have allowed them to sustain themselves until the present time, as moderate global economic growth is beginning to return. However, the global recovery is uneven, much slower than the recoveries that followed previous recessions, and is challenged by fragile economic, political and environmental conditions. Multiple risks continue to undermine the prospects of a sustained recovery, including debt problems, natural disasters, political unrest, and volatile energy and commodity prices (UNCTAD 2011).

Although the causes of these risks cannot be entirely addressed through management practices, many of their consequences can. Emerging regulation around safety, pollution and crew continues to respond to systemic changes in the industry and there are more international regulations now than there ever before. Although there are short-term costs associated with implementing these regulations, for those that can adopt a long-term approach, success will follow. For instance, regulation concerning the training and well-being of crew may have significant cost implications in the short term, but in the longer term, leads to a more productive, resilient workforce. Similarly, regulations requiring reduced levels of emissions have initial cost implications for the shipping industry, yet, as they also enhance efficiency, will improve long-term stability. Such environmental regulation is also expected to put pressure on ageing vessels, making them potentially unviable and pushing them towards decommissioning, thus making way for newer, energy efficient and safer vessels. When one considers the industry through a systems lense, it becomes apparent just how closely the different elements interrelate.

Perhaps in recognition of global economy risks, an emerging concept that is beginning to gain traction is that of a 'green economy' (UNEP 2013). Dwindling natural and economic resources have led many countries including China, Japan and New Zealand to reconsider how their economies might develop in the future. Green economies focus on developing short supply chain, low carbon, high value extraction, processing and consumption models. Although it may sound idealistic, serious consideration is being given to the concept and it is likely to become more widespread in the near future, as countries rely less on international trade and more on national and regional scale economics. If this occurs, there could be widespread implications for global shipping as smaller coastal vessels transporting a variety of products over short distances becomes the norm. The changes may result in higher risks of accidents and pollution, which may in turn require further evolution of management measures.

Finally, it is recognised that there is an overarching need to balance economic, social and environmental demands in order to maintain a sustainable shipping industry. The industry as a whole has successfully innovated in the past and will undoubledly continue to adapt in the future. As described in this chapter, evolving vessel design, loading and unloading practices and freight rates are an intrinsic part of this dynamic industry. Albeit with a time lag, management advancements in maintaining safety levels, reducing pollution and improving conditions for crew have supported and are expected to continue to support the evolution of the industry as a whole. For now, the global economy barometer continues to rise.

References

Alizadeh, A. and Nomikos, N. 2011. Dynamics of the Term Structure and Volatility of Shipping Freight Rates. *Journal of Transport Economics and Policy*, 45(1) pp. 105–128.

Bloor, M. Baker, S. Sampson, H. and Dahlgren, K. 2013. *Effectiveness of International Regulation of Pollution Controls: The Case of the Governance of Ship Emissions*. Seafarers International Research Centre, Cardiff University, Cardiff.

CEC Commission of the European Communities 1993. A Common Policy on Safe Seas. Communication from the Commission COM (93)66. Brussels, 24 February 1993.

CESA 2011. *Annual Report 2010–2011* Community of European Shipyards' Association.

Clarksons Research Services 2004. *The Tramp Shipping Market*. Clarkson Research Services, London.

Clarksons Shipping Intelligence Network 2012. Downloaded from www.clarksons.net/sin2010/ on 23 January 2012.

CNBC 2013. Pirates Release Tanker and 26 Crew Seized Last Year. *CNBC*. Downloaded from www.cnbc.com/id/100543040 on 13 January 2014.

CNTM 2014. *Canadian Annual Notice to Mariners*. Downloaded from www.notmar.gc.ca/eng/services/annual/section-a/notice-10.pdf on 13 January 2014.

Couper, A., Walsh, C., Stanberry, B. and Boerne, G. 1999. *Voyages of Abuse: Seafarers, Human Rights and International Shipping*. Pluto Press, London.

CPEM 2000. *National Report into the Sinking of the Erika off the Coasts of Brittany.* The Permanent Commission of Enquiry into Accidents at Sea.

Eide, M., Longva, T., Hoffmann, P. Endresen, O. and Dalsøren, S. 2011. Future Cost Scenarios for Reduction of Ship CO_2 Emissions. *Maritime Policy and Management – The Flagship Journal of International Shipping and Port Research*, 38(1), pp. 11–37.

EPA 2013. *The Plain English Guide to the Clean Air Act.* Downloaded from www.epa.gov/airquality/peg_caa/concern.html on 16 June 2013.

Glen, D. and Rogers, P. 1997. Does Weight Matter? A Statistical Analysis of the SSY Capesize Index. *Maritime Policy and Management*, 24, pp. 351–64.

GLOBALLAST 2013. *Project Overview.* Downloaded from http://globallast.imo.org/index.asp?page=gef_interw_project.htm on 16 June 2013.

ICC 2014. International Chamber of Commerce Piracy Statistics. Downloaded from www.icc-ccs.org/piracy-reporting-centre/piracynewsafigures on13 January 2014.

IMO 1973. MARPOL. Downloaded from www.imo.org/About/Conventions/ListOfConventions/Pages/International-Convention-for-the-Prevention-of-Pollution-from-Ships-%28MARPOL%29.aspx on 15 November 2012.

IMO 1974. SOLAS. Downloaded from www.imo.org/About/Conventions/ListOfConventions/Pages/International-Convention-for-the-Safety-of-Life-at-Sea-%28SOLAS%29,-1974.aspx on 15 November 2012.

IMO 1978. STCW. Downloaded from www.imo.org/About/Conventions/ListOfConventions/Pages/International-Convention-on-Standards-of-Training,-Certification-and-Watchkeeping-for-Seafarers-%28STCW%29.aspx on 15 November 2012.

IMO 2009a. *International Shipping and World Trade Facts and Figures.* Maritime Knowledge Centre, London.

IMO 2009b. *Prevention of Air Pollution from Ships. Second IMO GHG Study 2009 Update of the 2000 IMO GHG Study.* MEPC 59/4/7. Downloaded from www.imo.org/blast/blastDataHelper.asp?data_id=26046&filename=4–7.pdf on 25 January 2014.

IMO 2012. *International Shipping Facts and Figures – Information Resources Trade, Safety, Security, Environment.* Downloaded from www.imo.org/KnowledgeCentre/ShipsAndShippingFactsAndFigures/TheRoleand ImportanceofInternationalShipping/Documents/International%20Shipping%20%20Facts%20and%20 Figures.pdf on 21 February 2013.

IMO 2013. *IMO – What It Is.* International Maritime Organization, London. Downloaded from www.imo.org/en/About/Documents/What%20it%20is%20Oct%202013_Web.pdf on 19 July 2015.

IMO 2014. *Ships Routeing.* Downloaded from www.imo.org/OurWork/Safety/Navigation/Pages/Ships Routeing.aspx on 13 January 2014.

ITOPF 2013. *Oil Tanker Spill Statistics 2013.* Downloaded from www.itopf.com/information-services/data-and-statistics/statistics/documents/OilSpillstats_2013.pdf on 21 February 2013.

Kavussanos, M. 1996. Comparisons of Volatility in the Dry-cargo Ship Sector: Spot Versus Time Charters, and Smaller Versus Larger Vessels. *Journal of Transport Economics and Policy*, 30, pp. 67–82.

Kavussanos, M. and Visvikis, I. 2006. Shipping Freight Derivatives: A Survey of Recent Evidence. *Maritime Policy and Management*, 33(3), pp. 233–255.

Lun, Y. and Quaddus, M. 2009. An Empirical Model of the Bulk Shipping Market. *International Journal of Shipping and Transport Logistics*, 1(1), pp. 37–54.

Maritime CEO 2013. *Eco Marine Power- Sustainable Shipping Technology.* Downloaded from www.maritime-ceo.com/news_content.php?fid=3w3c109&goback=.gde_44626_member_245419857 on 30 May 2013.

Mitropoulos, E. 2011. Next Generation Shipping – What's next, IMO? *Nor-Shipping Conference.* 24 May 2011, Oslo, Norway.

Moundreas, G. 2009. Sale–Purchase Reports, 25 February 2009. Downloaded from www.gmoundreas.gr/moundreas/reports/sale_purchase_reports on 28 February 2009.

Ng, A. and Wilmsmeier, G. 2012. The Geography of Maritime Transportation: Space as a Perspective in Maritime Transport Research. *Maritime Policy and Management: The Flagship Journal of International Shipping and Port Research*, 39(2), pp. 127–132.

NOAA 1989. *The Exxon Valdez Oil Spill. A Report to the President.* Prepared by the National Response Team. National Oceanographic and Atmospheric Administration.

Platou, R. 2009. The Platou Report 2009. RS Platou, Oslo.

Platou, R. 2010. The Platou Report 2010. RS Platou, Oslo.

Sidell, J. and Haughey, F. 2007. *Neolithic Archaeology in the Intertidal Zone.* Oxbow, Oxford.

Stopford, M. 2009. *Maritime Economics*, 3rd edn. Routledge, Oxford.

UNCTAD 2009. *Review of Maritime Transport, 2009*. UN Conference on Trade and Development, Geneva.

UNCTAD 2010. *Review of Maritime Transport, 2010*. UN Conference on Trade and Development, Geneva.

UNCTAD 2011. *Review of Maritime Transport, 2011*. Downloaded from http://unctad.org/en/docs/rmt2011_en.pdf on 25 January 2014.

UNCTAD 2012. *Review of Maritime Transport, 2012*. Downloaded from http://unctad.org/en/Publications Library/rmt2012_en.pdf on 16 June 2013.

UNCTAD 2013. UNCTAD Statistics. Downloaded from http://unctadstat.unctad.org/TableViewer/tableView.aspx?ReportId=99 on 19 July 2015.

United Nations Environment Programme 2013. *Green Economy and Trade – Trends, Challenges and Opportunities*. Prepared by the Trade, Policy and Planning Unit of the United Nations Environment Programme.

Xu, J. J., Yip, T. L. and Marlow, P. B. 2011. The Dynamics between Freight Volatility and Fleet Size Growth in Dry Bulk Shipping Markets, *Transportation Research Part E*, 47(6), pp. 983–991.

23

SUBSEA TELECOMMUNICATIONS

Lionel Carter and Douglas R. Burnett

Introduction

Send an email overseas, download a video, search the internet or make an airline booking – these everyday actions will most likely involve subsea fibre-optic cables. Over 95 per cent of international communications and data transfer are via the global subsea cable network – a general collective term used here to encompass the many cable systems that are owned and operated by independent commercial entities. That dominance of cables reflects their ability to reliably and rapidly transmit large volumes of data and voice traffic in a secure and economical manner (Carter *et al.*, 2009; Burnett *et al.*, 2013). Although satellites carry <5 per cent of international traffic, they are still suited for providing worldwide coverage and television broadcasts as well as bringing communications to remote areas not linked to cables and to regions prone to natural disasters.

Such is the reliance of the world's economy, security and social framework on subsea cables that they are now regarded as *critical infrastructure* and hence are worthy of the best possible protection (e.g. Lacroix *et al.*, 2002; ACMA, 2015). To emphasise this point, the SWIFT (Society for Worldwide Interbank Financial Telecommunication) provides a service that transmits financial data between 208 countries via subsea cables. In 2004, up to $US7.4 trillion were transferred or traded on a daily basis. Thus a failure of the subsea network, no matter how brief, invites large financial repercussions (Rauscher, 2010). Accordingly, cable protection has become a high priority especially at this time of a rapidly increasing human presence offshore. The last two decades have witnessed a marked expansion of shipping, offshore renewable energy generation, hydrocarbon and mineral exploration, industrial fishing and marine research, all of which are taking place against a backdrop of ocean change forced by a climate under the influence of anthropogenic greenhouse gases (e.g. Halpern *et al.*, 2008; UNEP-WCMC, 2009; Smith *et al.*, 2010; IPCC, 2013).

Subsea cables

Around 1.5 million kilometres of fibre-optic telecommunications cables have been laid on the ocean floor since 1988 when the first trans-oceanic system, *TAT-8*, was installed to link the USA, UK and France (Figure 23.1). For water depths exceeding *ca.* 2000 m – an approximate limit for bottom trawl fishing, which is a major cause of cable damage (e.g. Mole *et al.*, 1997;

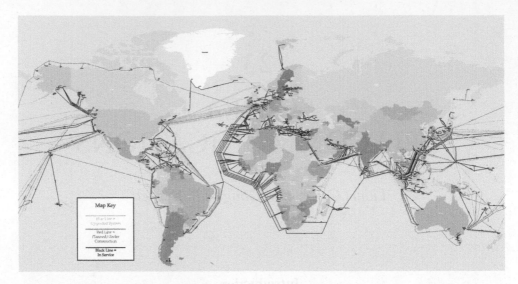

Figure 23.1 Submarine cables of the world
Source: Courtesy of Submarine Telecoms Forum.

Kordahi and Shapiro, 2004) – fibre-optic cables are typically laid on the seabed surface and are about the size of a domestic garden hose, i.e. between 17 and 22 mm diameter. This is the deployment made for most of the global network as water depths >2000 m account for 84 per cent of the ocean. In depths <2000 m, cables are up to 50 mm diameter due to the addition of steel wire armour for protection especially in continental shelf and slope waters shallower than 200 m (Kordahi *et al.*, 2007). These shallow-water cables are also commonly buried below the seabed for additional protection.

Deep-water cables consist of (from inside to outside; Figure 23.2) (i) hair-like glass fibres encased in a steel tube filled with a thixotropic medium, (ii) a covering of steel wire strands to provide strength, (iii) a copper-based composite conductor to carry electrical power and (iv) a protective insulating sheath of polyethylene (Hagadorn, 2009). For water depths <2000 m, various layers of galvanized steel wire armour are applied according to the nature of the risk. These armoured types are finally wrapped in a hard-wearing polyethylene sheath. Because light signals passing along the optical fibres periodically require amplification, repeaters are installed at intervals approaching 100 km along a cable route. Powered via the copper-based conductor, modern repeaters now use optical amplifiers that are essentially glass fibres containing the rare earth element, erbium. When energised by lasers, these erbium-doped fibres amplify the light signals sending them on their way to the next repeater.

The rapidly evolving fibre-optic technology is the latest evolutionary phase in undersea communications that began symbolically in 1850 with the laying of a telegraph link between Dover and Calais. No more than a copper wire insulated by the natural polymer, *gutta percha*, the cable could not withstand the strong waves and currents of the English Channel and failed after a few messages (Carter *et al.*, 2009). A strengthened version was deployed a year later and survived for a decade, which was long enough to encourage installation of other short-haul cables around Europe. In 1858, the first trans-Atlantic telegraphic cable was laid between Ireland and Newfoundland (Gordon, 2002; Ash 2013). Although it operated for only 26 days, it was a start. Following other failed attempts that encouraged improvements in cable design and laying techniques, a reliable operating system was installed in 1866 from the famed cable ship, *Great*

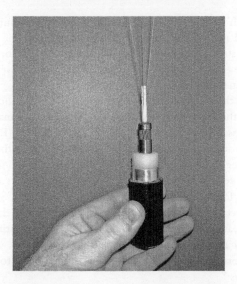

Figure 23.2 A section of lightweight cable designed for deployment on the seabed surface, nominally in water depths exceeding 2000 m, which is the approximate depth limit of deep trawl fishing, a major cause of cable faults
Source: L. Carter.

Eastern. Advances in cable design and construction improved reliability and transmission speeds, which increased from 12 words/minute for the first cables to 200 words/minute by the 1920s. In the 1930s experiments with polyethylene encased coaxial cables were underway and, along with the development of repeaters, set the scene for cables to carry multiple voice channels. In 1955–56 this became a reality with the laying of the coaxial system, TAT-1, between Scotland and Newfoundland. The *telephonic era* was born and the *telegraphic era* became history. On the first day of operation in 1956, 707 telephone calls were made between the USA and UK – a major improvement on telegraphic links but not enough to match developing satellite communications in the late 1970s and 1980s. Even though trans-Atlantic cables achieved capacities of up to 4000 telephone channels, they were mainly viable on major traffic routes. Satellites dominated global communications through the 1980s. However, the laying of the first fibre-optic, trans-oceanic link in 1988 heralded a fundamental shift in communications. The carrying capacity of fibre-optic cables was much larger than that of their coaxial counterparts and so the transition to the *fibre-optic era* began. Today, a cable link, such as that between the USA and Japan, can accommodate 23 million simultaneous voice calls or around 1.9 million simultaneous transfers of 1 Mb files (PC Landing, 2013). Equally important, the rapid evolution of fibre-optic cables coincided with the development of the internet. It was a fortuitous timing. Cables could efficiently and rapidly transfer large volumes of information and data around the world whereas the internet made that material available to a wide range of users with an equally wide range of applications.

International law and submarine cables

The International Convention for the Protection of Cables[1] is the foundation of modern international law for submarine cables as contained in the United Nations Convention on the Law of the Sea (1982) ('UNCLOS') (Burnett *et al.*, 2013). UNCLOS treats cables the same based on their purpose. If a cable is used for dual purposes, telecommunications and science or telecommunications and natural resources, it will in addition to the UNCLOS legal regime governing telecommunication cables be governed by the legal regime for marine scientific research or natural resources as the case may be. UNCLOS expressly provides for the fundamental freedom to lay and maintain submarine cables in ten articles.[2]

UNCLOS establishes the rights and duties of all States, balancing the interests of coastal States in offshore zones with the interests of all States in using the oceans. Coastal States exercise sovereign rights and jurisdiction in the exclusive economic zone (EEZ) and upon the legal continental shelf (LCS) for the purpose of exploring and exploiting their natural resources, but other States enjoy the freedom to lay and maintain submarine cables in the EEZ and upon the LCS. In archipelagic waters and in the territorial sea, coastal States exercise sovereignty and may establish conditions for cables or pipelines entering these zones. Such conditions on cables should generally be effective only within the territorial and archipelagic seas and not extend into the EEZ or high seas.[3] The laying and maintenance of submarine cables is considered a reasonable use of the sea and coastal States benefit from them.

Outside of the territorial sea, the core legal principles applying to international cables can be summarised as follows:[4]

- the freedoms to lay, maintain and repair cables outside of territorial seas, including cable route surveys incident to cable laying[5] (Nordquist *et al.*, 1993);
- the requirement that parties apply domestic laws to prosecute persons who endanger or damage cables wilfully or through culpable negligence;[6]
- the requirement that vessels, unless saving lives or ships, avoid actions likely to injure cables;
- the requirement that vessels must sacrifice their anchors or fishing gear to avoid injury to cables;
- the requirement that cable owners must indemnify vessel owners for lawful sacrifices of their anchors or fishing gear;
- the requirement that the owner of a cable or pipeline, who in laying or repairing that cable or pipeline causes injury to a prior laid cable or pipeline, indemnify the owner of the first laid cable or pipeline for the repair costs;
- the requirement that coastal States along with pipeline and cable owners shall not take actions that prejudice the repair and maintenance of existing cables.

Careful route planning helps to avoid damage to cables (Wagner, 1995). With respect to potential adverse impacts caused by submarine cables, UNCLOS indirectly takes into account their potential environmental impact by distinguishing cables from submarine pipelines; that is, on the continental shelf it allows a coastal State to delineate a route for a pipeline but not for a cable.[7] The reason for this distinction is that there is a need to prevent, reduce and control any pollution that may result from pipeline damage. By comparison, damage to an ocean cable does not involve pollution (Nordquist *et al.*, 1993), but may significantly disrupt international communications or electrical power distribution.

States treat international cables in national maritime zones as critical infrastructure that deserves strong protection to complement traditional international cable law. Australia, consistent with international law, has legislated to protect its vital cable links by creating seabed protection zones that extend out to 2000 m water depth.[8] Bottom trawling and other potentially destructive fishing practices, as well as anchoring, are prohibited inside these zones. New Zealand has enacted legislation that established no fishing and anchoring zones around cables.[9] The trend is expected to continue because most nations depend upon cables for participating in the global economy and for national security. These developments go hand in hand with conservation, as restrictions on trawling to prevent cable damage can also provide direct benefits for biodiversity by protecting vulnerable ecosystems and species such as corals and sponges (Carter *et al.*, 2009).

Since UNCLOS, the parties to the UNESCO Convention on Underwater Cultural Heritage (2001) agreed to exempt cables from that treaty because of the specific provisions of UNCLOS and the agreement of the parties that cable laying and maintenance posed no threat to underwater cultural heritage.

The laying and maintenance of cables is a reasonable use of the sea, and in 166 years of use, there has been no irreversible environmental impact from them. UNCLOS and State practice have provided adequate governance for international cables outside national waters, and State practice increasingly recognises the importance of protecting cables from activities that could damage them.

Protecting the network in a busy ocean

About 150 to 200 cable faults occur annually around the world. Analyses of fault records show that *ca.* 70 per cent and more of all faults result from human activities, notably fishing and ships' anchoring (Kordahi and Shapiro, 2004; Kordahi *et al.*, 2007; Wood and Carter, 2008). Damage caused by natural phenomena such as submarine landslides tends to be <10 per cent but may be locally higher in hazard-prone regions, in particular the seismically active rim of the Pacific Ocean (see Natural hazards). Faults resulting from component failure have a long-term average of 7 per cent but over the past two decades it has dropped below 5 per cent, reflecting the improved reliability of cables (Kordahi *et al.*, 2007). Around 20 per cent of faults are classified as 'unknown', that is there is no conclusive evidence regarding the cause of the fault such as the presence of trawl gear or furrows in the seabed produced by a dragged anchor.

Commercial fishing appears to be the prime cause of cable faults (Drew, 2009). Bottom trawl fishing is especially hazardous because it is a widespread and repetitive practice. The hazard relates to heavy trawl doors or *otter boards* ploughing the seabed and damaging any cable in their path. Several different scenarios are possible; (i) the otter boards pass over the cable and scrape the sheathing without causing significant damage, (ii) the boards break the protective and insulating sheathing to allow seawater to make contact with the live conductor and produce a *shunt fault*, (iii) the cable is dragged and bent sufficiently to damage the glass fibres to form an *optical fault* and (iv) the cable is severed. Fishing operations using an anchor or grapnel also pose a significant risk. Grapnels may be towed by fishing vessels to recover lost gear. If a grapnel snags a cable on or under the seabed, the momentum of the towing vessel may bend, stretch or ultimately break a cable depending on its diameter and amount of wire armouring. A non-armoured or light-weight cable may break under a few tonnes strain whereas a double armoured cable breaks at 40 tonnes strain or more. Anchors are used to install static fishing devices ranging from lobster pots to large fish-trapping systems. The hazard arises when large devices are indiscriminately installed over subsea cables. If these static devices are poorly anchored and eventually drift with the currents than anchors may be dragged onto cables.

The anchoring of ships is the other major cause of faults. Incorrectly stowed anchors are known to break free while a ship is underway. As a result, anchors can be towed long distances over the seabed unbeknown to the vessel operators. To emphasise the hazard, Shapiro *et al.* (1997) noted that a 4 tonne anchor on a 5000 tonne ship could penetrate 5 m into a soft muddy seabed posing a threat to any cables along a vessel's course. Such a case appears to have occurred in 2008 when a ship dragged its anchor across the Mediterranean seabed between Tunisia and Sicily and broke three cables (Orange, 2008). At least 14 nations suffered loss of fibre-optic connectivity. Particularly hard-hit were internet-based businesses such as out-sourcing and call centres in India and Pakistan. Another example of anchoring impacts followed the economic downturn of 2008–2009. International trade declined and cargo-less vessels were laid up around major ports. Many vessels anchored outside port limits (OPL) to avoid pilot and port charges (West of England, 2009). Off Singapore and Malaysia, OPL anchorages became congested. The problem was exacerbated by strong tidal currents that heightened the risk of vessel collision and cable damage as ships dragged their anchors or swung on their moorings. Resultant anchor damage to cables amounted to almost $US4.5 million (Lloyds List, 2009). To resolve the problem, Malaysia, Singapore and Indonesia approached the International Maritime Organization to direct ships to designated anchorages where vessels were better regulated and cables were either absent or well protected.

Protection of subsea fibre-optic cables is a multifaceted process that involves some combination of physical protection, legal protection, active preventative measures and ongoing communication with other seabed users including the public. Physical protection typically centres on the armouring of a cable, and where possible, burying it beneath the seabed to a depth that protects it from a known hazard. Where cables cannot be buried, for example in rocky areas, they may be protected by concrete mats or rock armour, or inserted in iron pipes. For environmentally sensitive coasts, cables may be placed under the littoral zone seabed via directional drilling from shore (e.g. Austin *et al.*, 2004).

Cables within the EEZ (200 nautical mile (370 km) limit) or Territorial Sea (12 nautical mile (22 km) limit) may be afforded legal protection through the creation of cable protection zones (see International Law and Submarine Cables). In the case of Australia, protection zones exclude activities considered hazardous to cables, and departures from the law are punishable by fines and/or ten years imprisonment (ACMA, 2015). Of course a protection zone is only successful through effective policing and education of seabed users (e.g. Transpower and Ministry of Transport, 2013). Policing may involve periodic over-flights, stationing of permanent observers and/or public notification of hazardous activities. Development of new technologies, in particular the Automatic Identification System (AIS) and Vessel Monitoring System (VMS) now permit monitoring of vessels in real time (Drew, 2013). AIS is required for ships over 299 gross tonnes and provides frequent updates via VHF radio of a vessel's name, number, position, speed, direction and other data. Software plots a ship's location and heading and thus provides a warning should a vessel pose a risk; those data are also archived to provide evidence should a vessel be charged with damaging a cable.

Education is an ongoing process and occurs on several fronts; (i) provision of plain-language information on cables and their importance for policy makers, the public and other seabed users (ICPC, 2013; Burnett *et al.*, 2013), (ii) provision of cable locations on navigational charts and official notices to mariners (e.g. ACMA, 2015) and (iii) direct communication and collaboration between the cable industry representatives and other seabed users such as the fishing and offshore wind-farm industries. Those initiatives result in the sharing of knowledge about the operations of the involved parties and have produced guidelines to improve the safety of seabed users and infrastructure (Drew and Hopper, 1996; OFCC, 2007; The Crown Estate, 2012).

Environmental aspects of cables

In traversing the ocean, cables encounter a suite of environments that range from the highly dynamic, wave-dominated surf zone to extreme depths of 5000 m and more where water temperatures are <2°C, pressures are 500 times that at sea level, and currents, if present, are typically slow although they may be subject to strong periodic perturbations (e.g. Hollister and

McCave, 1984). For practical purposes, cables can be separated into (i) those laid on the seabed usually in water depths >*ca*. 2000 m beyond the present limit of bottom trawl fishing, but also in shallower water where cable burial is not possible or needed due to unsuitable seabed conditions, the presence of a cable protection zone or an absence of other seabed users whose activities may pose a risk and (ii) cables buried under the seabed for protection from such activities that occur mainly in water depths <200 m but can extend to *ca*. 2000 m.

Cables laid on the seabed

Once deployed on the seabed, cables are exposed to physical and biological forces (Carter *et al*., 2009). In the shallow waters (*ca*. <30 m) of the inner continental shelf, tides in association with storm-forced ocean currents and waves move seabed sediment that can abrade, bury, expose or undermine cables. The last phenomenon can produce suspensions that sway under current/wave action and eventually induce cable fatigue (e.g. Carter *et al*., 1991; Kogan *et al*., 2006). In depths down to *ca*. 130 m – the global average depth limit of the continental shelf – effects of storm-forced currents and waves decrease but can nonetheless disturb the seabed under extreme storms. If the continental shelf has a strong supply of sediment, as is the case for shelves of the circum-Pacific rim (e.g. Milliman and Syvitski, 1992), the rapid accumulation of sediment (up to 1 cm/year) can bury a cable.

In the deep ocean (>2000 m), the deposition of sediment is considerably slower (e.g. 0.001–0.004 cm/year). However, higher rates occur locally where land-derived sediment is discharged to depth via (i) submarine canyons linked to large rivers and (ii) submarine landslides (e.g. Heezen and Ewing, 1952; Carter *et al*., 2012). These settings can be found off major rivers such as the Zaire (Droz *et al*., 1996), but the prime sites are off earthquake- and storm-prone margins such as those bordering the Pacific Ocean. There, small rivers discharge a disproportionally large volume of sediment because of unstable landscapes and pronounced rainfalls. In such instances cables may receive rapid influxes of mud and sand from submarine landslides and mud-laden flows or *turbidity currents* with sufficient power to break cables (see Natural Hazards). Elsewhere, cables may be subject to locally intensified, deep-ocean currents that flow along the western boundaries of the major oceans (Hogg, 2001) and are known to erode and deposit sediment especially during periods when the flow is reinforced by powerful eddies (Hollister and McCave, 1984; Carter and McCave, 1997).

Any interaction of surface-laid cables with the marine biota appears to be negligible to minor judging by several cable-based studies. Using sediment cores and video footage collected by a remotely operated vehicle along a subsea cable off California, Kogan *et al*. (2006) showed that animals living in or on the seabed 1m and 100 m from the cable were not statistically different. In a similar vein, Andrulewicz *et al*. (2003) likewise reported no change in the abundance, composition and biomass of organisms before and one year after a subsea power cable deployment. Grannis (2001) reached similar conclusions regarding the biota along a cable off the northeastern USA. Where cables remain uncovered by sediment they can act as substrates for encrusting organisms as long as the cable is within the depth range of an organism. As a consequence, recovered cables have been a source of specimens from parts of the ocean seldom visited by researchers (e.g. Levings and McDaniel, 1974).

Surface laid cables are exposed to fish, in particular sharks and marine mammals. Deployment of the first trans-ocean fibre-optic cable was disrupted by shark attacks in 1985–1987 off the Canary Islands (Marra, 1989). Bites from the deep-dwelling *crocodile shark*, identified from its teeth embedded in the cable sheath, were sufficient to damage the cable and force repairs. Why the fish attacks took place remains unclear, but they prompted a redesign of deep-water cables that improved protection against fish bites. The interaction with marine mammals was highlighted by Heezen (1957) who reported on a series of cable faults caused mainly by sperm whales whose remains were entangled in the recovered cables. Analysis of fault records revealed the entanglements were confined to telegraphic era cables and were located mainly on the continental shelf and upper continental slope where cables had been previously repaired – a factor that suggests the whales may have been caught in coils of cable formed from the repair. Cable fault databases show a cessation of whale entanglements with the introduction of coaxial and fibre-optic systems (Wood and Carter, 2008). This marked change resulted from improved design and laying techniques that prevented coiling, and cable burial beneath the seabed.

Comprised of high density, ultra-violet resistant polyethylene sheathing without antifouling agents, modern cables are chemically stable. When samples of different cable types – some with cut ends sealed and others exposed – were immersed in 5 litre containers of natural seawater and tested for leached metals. Only zinc was detected and this was from the galvanised wire-armoured samples with exposed ends (Collins, 2007). Up to 11 parts per million Zn were measured initially whereas iron and copper were at natural background levels, i.e. no leaching was detected. Bearing in mind that (i) in the open ocean Zn concentrations would be much lower due to dilution, (ii) the rates of Zn leaching declined markedly after ten days' exposure in the seawater containers and (iii) Zn is a common trace element in seawater that is essential for marine biological processes (Morel and Price, 2003), the amount of leachate recorded in the laboratory tests is small.

Cables buried under the seabed

The protective burial of cables on the continental shelf and slope usually involves mechanical ploughing. A plough, towed by a cable ship, opens a furrow in the seabed into which the cable is inserted. In soft sediments, the furrow wall collapses to encase the cable whereas more con-solidated materials may only partially collapse. In both instances, burial is facilitated by natural sediment accumulation. Ploughing disturbs the seabed, the extent of which varies with substrate conditions, plough type and the depth of burial, which is dictated by the nature of the expected hazard (e.g. Rapp *et al.*, 2004). For heavily fished areas where the trawl doors penetrate *ca.* 0.5 m into the seabed, a cable may be buried 1m deep in which case the plough share will have a disturbance strip *ca.* 0.3 m wide. In addition, the skids that support the plough share can disturb the immediate seabed surface over a 2 m to 8 m wide strip.

The other main burial technique is jetting. High pressure water jets, commonly incorporated on a remotely operated vehicle (ROV), liquefy seabed sediments allowing the cable to sink to a required depth. The technique is used for substrates that are unsuitable for mechanical ploughing such as steep slopes, very soft muddy sediments and water depths *ca.* >1000 m (Hoshina and Featherstone, 2001). Jetting is also used to bury repaired sections of cable. Again, there is some disturbance relating to the width of the jetted furrow and dispersal of turbid water whose extent and effect on biota will depend upon local oceanographic conditions.

Burial disturbance should be viewed in context. Burial is a restricted and non-repetitive activity in the 20–25 year design life of a cable unless a repair is required. In contrast, bottom trawl fishing has a wider footprint and is a repetitive process (e.g. Puig *et al.*, 2012). Another consideration is the ability of the seabed to recover naturally from burial-related disturbance. For sheltered coasts, recovery from ploughing of mangrove swamps and salt marshes ranges from 2 to 7 months and 1 to 5 years respectively (Dernie *et al.*, 2003; Ecoplan, 2003). On exposed coasts down to *ca.* 30 m water depth, wave and current action shift sediment on a daily to annual basis resulting in amelioration or removal any burial scars (Carter and Lewis, 1995; CEE, 2006). The benthic fauna are also adapted to the frequently mobile sediments and appear to be unaffected by cable deployment as evinced by Andrulewicz *et al.* (2003). For the remainder of the continental shelf out to *ca.* 130 m depth, the influence of waves and wind-driven currents declines with depth although tides are omnipresent and can instigate daily movement of sediment in tide-dominated regions such as the English Channel and Cook Strait, New Zealand (Grochowski *et al.*, 1993; Carter and Lewis, 1995). Plow scars on shelves with a substantial supply of sediment may be infilled by the natural deposition of mud that can locally reach 1 cm/year or more (e.g. Huh *et al.*, 2009). At the shelf edge, seabed recovery may be facilitated by tidal and ocean currents that are intensified against the edge topography. For shelf areas with a limited sediment supply and weak or infrequent current/wave action, recovery is slower (e.g. NOAA, 2005; California Coastal Commission, 2007).

Cable life cycle

A cradle-to-grave analysis of fibre-optic cables was made by Donovan (2009) to assess the environmental effects – both positive and negative – of cable manufacture, operation and recovery. Potential environmental effects are associated mainly with the electrical power required to operate cable terminal stations and with fuel used by cable ships for laying and maintenance. Taking those and other factors such as cable manufacture and recycling into account, it was estimated that 7 g of carbon dioxide equivalents were released for every 10,000 gigabit km, i.e. the transmission of 1 gigabit of data over 10,000 km of cable. The relevance of this carbon dioxide equivalent emission becomes apparent when a cable-based teleconference between Stockholm and New York is compared to an equivalent face-to-face meeting (Donovan, 2009). Because face-to-face meetings involve air and other travel, the carbon dioxide equivalent emissions were 1920 kg compared to 5.7 kg for the teleconference.

Natural hazards

Although natural hazards cause <10 per cent of all cable faults (Kordahi *et al.*, 2007) they become more prominent in depths >1000–2000 m, where human operations are markedly reduced. Furthermore, a major hazard such as a submarine landslide can damage multiple cables to cause a widespread reduction or even loss of internet and communication traffic. This was the case for earthquake-triggered submarine landslides and turbidity currents off Algeria 2003 (Dan *et al.*, 2003; Cattaneo *et al.*, 2012), southern Taiwan, 2006 (Hsu *et al.*, 2009) and northern Japan, 2011 (studies underway).

Hazards from the coast to abyssal ocean

On the continental shelf, cables are exposed to mobile sand and gravel (Allan, 2000) that can (i) abrade, (ii) bury or (iii) undermine cables to produce suspensions that may result in cable fatigue (e.g. Kogan *et al.*, 2006). However, improved design, construction and deployment have produced robust shallow-water cables.

Severe storms are a major threat to coastal and shelf cable infrastructure. Winds increase wave and current action to enhance sediment mobility. During Hurricane Iwa (1982), wind-forced waves and currents set off submarine landslides and turbidity currents that swept down the continental slope off Oahu, Hawaii to break six cables (Dengler *et al.*, 1984). Winds and changes in barometric pressure also set up storm surges as occurred during Typhoon Nargis (2008) when a 4 m high storm surge passed over the Irrawaddy Delta to damage a cable station (Ko, 2011). Hurricane Sandy (2012) also generated a 4 m high surge, which together with 180 mm of rain, flooded lower Manhattan (NASA, 2012; USGS, 2014). At least one fibre-optic link was damaged, but a key impact related to the flooding of several major cable-fed datacenters (Cowie, 2012).

River floods are also a hazard especially where a river is linked directly to a submarine canyon across which cables pass. Typhoon Morakot (2009) was the wettest tropical cyclone on record for Taiwan. Almost 3 m of rain generated exceptional floods that caused the Gaoping River to discharge an estimated 150 million tonnes of sediment in just six days (Carter *et al.*, 2012). River discharge eventually transformed into two turbidity currents that swept down the submarine Gaoping Canyon to damage eight cables along a 370 km long pathway in water depths down to 4200 m.

Submarine landslides and turbidity currents are also triggered by earthquakes, so it is unsurprising that such occurrences are relatively common in seismically active regions especially where tectonic plates collide as is the case for the circum-Pacific Ocean and parts of the Mediterranean region. However, the textbook example is from the comparatively stable Grand Banks, Newfoundland, which was shaken by a magnitude M7.2 earthquake in 1929. The shock set off subsea landslides that immediately caused 12 faults in Atlantic telegraphic cables (Heezen and Ewing, 1952; Piper *et al.*, 1985). Landslide debris contributed to a major turbidity current that broke a further 16 faults as it flowed 650 km over the seabed into water depths of over 5000 m. Current speeds of up to 65 km/hour were achieved en route. Since this pioneering set of observations, earthquakes have been implicated in a range of cable-damaging events, e.g. Orleansville (Heezen and Ewing, 1955), Papua-New Guinea (Krause *et al.*, 1970) with one of the more recent being the 2006 Hengchun earthquake (M7.0) off southern Taiwan (Hsu *et al.*, 2009). For the latter event the main shock was also accompanied by the near-instantaneous cable failures (three). Several turbidity currents followed with some possibly triggered by aftershocks. These debris-choked currents flowed down Gaoping Canyon and into the Manila Trench to cause a total of 19 cable faults. Hengchun was followed three years later by Typhoon Morakot (2009) and in 2010 by another turbidity current that broke nine cables. Clearly the ocean floor off southern Taiwan, which is a major cable corridor, is a highly hazardous region.

Offshore earthquakes can also generate tsunami that pose a risk to coastal and continental shelf telecommunications. The 2004 Indonesian mega-earthquake (M9.1–9.3) and tsunami damaged terrestrial telecommunications and possibly a subsea cable off South Africa (M. Green, BT, personal communication). Another tsunami, this time formed by the 2011 Tohoku mega-earthquake (M9.0) off northern Japan, extended up to 5 km inland and reached surge heights of up to 15 m (Tanaka *et al.*, 2012). Such a devastating inflow of water severely damaged at least one cable landing station as well as fixed line and mobile phone infrastructure. In addition, submarine telecommunications were damaged (BBC, 2011), but the full extent and specific cause(s) of the damage (landslide, turbidity current, tsunami) have yet to be determined.

Damage from active submarine volcanoes and icebergs/sea ice is rare although the latter hazard may come to the fore as remote Arctic areas are connected to the global network (see Hazards under climate change). Submarine volcanic activity is widespread especially along mid-ocean ridges where tectonic plates diverge and around the Pacific rim. Volcanoes pose potential hazards through eruptions, landslides, hot water vents and other phenomena including tsunami such as occurred during the eruption of Krakatau (Krakatoa) in 1883 (Winchester, 2003). Following a main blast, a tsunami that locally reached 35 m height, radiated southwards across the Indian Ocean severing the local subsea telegraphic link with the rest of the world. However, direct impacts of submarine volcanism can be minimised because volcanic structures have distinctive geological and geophysical features and are avoided by cable route planners.

Hazards under climate change

Coastal and ocean environments are responding to the present phase of climate change as documented by the IPCC (2013) and a plethora of ongoing research.

Meltwater from glaciers and ice sheets, together with thermal expansion of the ocean have produced a global average rise in sea level of 3.2 mm/year (University of Colorado, 2015). Tectonic plate movements, ocean currents, gravitational effects among others also locally affect sea level; hence the need to determine local conditions when assessing the risk posed to coastal cable infrastructure. As sea level rises, coasts become more prone to erosion and flooding as demonstrated by data from Australia that suggest the frequency of extreme high sea level events increased by a factor of 3 in the twentieth century (Ocean Climate Change, 2012).

Ocean and atmospheric warming are likely to lead to more intense and/or frequent storms (IPCC, 2013) that will strengthen wave and current activity at the coast and adjacent continental shelf. In addition, stronger winds and associated drops in barometric pressure will enhance storm surges of which Hurricane Sandy may be a harbinger judging by a recorded increase of extreme climatic events in the northeast USA (NOAA, 2012).

Warming is likely to affect rainfall patterns and by association river discharge as exemplified by Typhoon Morakot (Carter *et al.*, 2012). However, although Morakot was the wettest cyclone on Taiwanese records it cannot be confidently attributed to modern climate change although its characteristics are consistent with climate projections.

The core of strong westerly winds has moved towards the poles resulting in changes in wave and current regimes (Toggweiler *et al.*, 2006; Thompson *et al.*, 2011). The Southern Hemisphere, for example, is witnessing stronger wave activity and strengthened ocean currents (e.g. Böning *et al.*, 2008) – responses that will have a bearing on coastal and shelf settings as well as cable laying and maintenance operations.

In addition to direct environmental effects, climate change is altering other seabed activities and hence is changing the risk such activities pose to cables. Wind turbine farms are expanding as nations seek to reduce greenhouse emissions, meet increasing demand for electrical power and establish more secure energy supplies. The growth of wind farms and other renewable energy schemes, as well as plans to create new submarine energy grids (e.g. IEEE Spectrum, 2010), will restrict the choice of viable telecommunication cable routes and will impact upon laying and maintenance operations. Industrial fisheries may also be responding to climate change (e.g. Frost *et al.*, 2012). Ocean warming in the Northern Hemisphere has encouraged southern fish species to migrate north, for example, previously Mediterranean-dwelling anchovies now occur in commercial quantities off the UK. Furthermore, deep-dwelling fish are increasing their preferred depth by 3.6 m/decade. Such trends are likely to alter the style and depth range of fishing practices.

Ocean/climate change is also bringing new opportunities for cables. The Arctic Ocean has lost much of its summer sea ice and in September, 2012, reached its minimum extent since 1978 when satellite monitoring began (NSIDC, 2012). Ice loss reflects a warmer ocean and increased storminess, and if the current trend continues, the summer Arctic could be ice-free within a decade. Such a marked environmental shift has fostered plans to install fibre-optic links with remote Arctic communities and the rest of the world. The Russian Optical Trans-Arctic Submarine Cable System (ROTACS), for instance, is planned to link Tokyo with the Russian Arctic and London (*New Scientist*, 2012) with branches to South Korea and China.

Telecommunications in an evolving seascape

Humanity's increasing presence in and on the ocean has placed pressures on the environment especially in relation to the extraction of living and non-living resources (e.g. UNEP, 2006). One response has been the creation of Marine Protected Areas (MPA), which presently encompass *ca.* 3.2 per cent of the global ocean (Marine Reserves Coalition, 2012). The nature of that protection varies among nations. In the case of Australia, whose MPAs extend over 3.1 million km² of ocean and seabed (Australian Government, 2015), protection is afforded at different levels ranging from MPAs where all commercial activities are prohibited to those where limited activities are permitted except those that are damaging to the environment.

The laying and maintenance of submarine telecommunications cables are generally permitted activities in multi-purpose protected areas, especially in light of their designation as *critical infrastructure*, their low environmental impact (OSPAR, 2008; Carter *et al.*, 2009; Burnett *et al.*, 2013) and their special status under UNCLOS (see International law and cables).

Marine research, especially that related to climate/ocean change, natural hazards and resource assessment, has been bolstered by the development of ocean observatories. These fibre-optic and power cable-based systems have been designed for long-term (20–25 years) monitoring and *in situ* experiments, the data from which are available in near-real time for the public and science community (Carter and Soons, 2013). Although subsea communications cables have been used to measure currents and thermal structure of the ocean since the 1980s (Baringer and Larsen, 2001; Howe, 2004), it has only been in the last five years that large observatories capable of conducting multidisciplinary research, have come to the fore. One of the first cabled observatories is the Monterey Accelerated Research System (MARS) situated in Monterey Bay, California (MARS, 2015). It began operation in 2008 and is based on a *node* located on the seabed at 891 m depth. Shaped like a truncated pyramid and weighing several tonnes, the node is a trawler-resistant submarine housing into which various sensors and experiments can be plugged. It supplies communications and power, and is connected to a shore-based receiving centre by a 52 km long fibre-optic/power cable.

On a larger scale is the North East Pacific Time-series Undersea Network Experiments (NEPTUNE) observatory, which began operation in 2009. NEPTUNE is presently the largest observatory with an 812 km long fibre-optic/power cable that interconnects five nodes distributed from the continental shelf to abyssal plain at 2660 m depth (Ocean Networks Canada, 2015). In that configuration, NEPTUNE covers a range of key marine environments with experiments tailored for a specific setting. For example, the node in the submarine Barkley Canyon (400–1000 m depth) is the hub for research into (i) the movement of water and sediment along the canyon, (ii) the composition and change of canyon ecosystems, (iii) gas hydrates – mixtures of methane gas and ice that are a potential source of hydrocarbons and (iv) the impacts of earthquakes and tsunami. Such information is relevant to the cable industry by virtue of the hazards posed by landslides and turbidity currents generated by unstable sediments on submarine slopes and canyons, and by the potential mining of gas hydrates as an energy source.

Whatever the activity, the increasing human presence offshore has prompted regulatory regimes such as Marine Spatial Planning (MSP). These regimes provide frameworks to address coastal and marine issues concerned with environmental conservation and sustainability, commercial and recreational activities as well as scientific research. Implementation of MSP is underway in Europe, North America and Oceania among other regions (e.g. DEFRA, 2009; Ministry for the Environment, 2012). The USA, for example, created a National Ocean Council to implement policy concerning stewardship of the Great Lakes, coasts and oceans (National Ocean Council, 2012). Policy aims are wide ranging; from protecting and restoring ocean biodiversity to bettering the public's knowledge of its offshore estate. Of relevance to submarine telecommunications are; (i) the support of sustainable, safe, secure and productive access to, and uses of the ocean and (ii) the exercise of rights and jurisdiction in accordance with international law that involves respect for and preservation of navigational rights and freedoms. The latter point is critical in that while not a signatory to UNCLOS, the USA recognises the importance of that convention as 'the bedrock legal instrument governing activities on, over and under the world's oceans'. There is also recognition of cables as critical infrastructure, which DEFRA (2011) describes as 'socially and economically crucial to the UK'. Both the legal and critical infrastructural aspects, along with acknowledgement of the nil to low environmental impact of submarine cables (e.g. DEFRA, 2011), should not disadvantage submarine cables under MSP.

Notes

1   The Cable Convention continues to be widely used in the cable industry. While its essential terms are included in UNCLOS, the Cable Convention remains the only treaty that provides the detailed procedures necessary to implement them. *See* Art. 5 special lights and day shapes displayed by cable ships; minimum distances ships are required to be from cable ships; Art. 6 minimum distance ships are required to be from cable buoys; Art 7 procedures for sacrificed anchor and gear claims, Art. 8 competency of national courts for infractions; and Art. 10, procedures for boarding vessels suspected of injuring cables and obtaining evidence of infractions. Article 311(2) of UNCLOS recognises the continued use of these provisions, which are compatible and supplement UNCLOS.
2   Articles 21, 51, 58, 79, 87, 112–115 and 297.
3   Article 79(4).
4   Articles 51, 58, 79, 87, 112–115, and 297.(1)(a).
5   The term laying refers to new cables while the term maintaining relates to both new and existing cables and includes repair. Nordquist, *United Nations Convention on the Law of the Sea 1982: A Commentary*, Vol II (1993) at p. 915.
6   The origin of the term 'culpable negligence' is found in Renault, Louis, The Protection of Submarine Telegraphs and the Paris Conference (October–November 1882) at p. 8 where reference is made to two early English cases *Submarine Cable Company v. Dixon, The Law Times*, Reports, Vol. X, N.S. at 32 (Mar. 5, 1864) and *The Clara Killian*, Vol. III L.R. Adm. and Eccl. at 161 (1870). These cases hold that culpable negligence involves a failure to use ordinary nautical skill which would have been used by a prudent seaman facing the situation that caused the cable fault. Since the term 'culpable negligence' was adopted in UNCLOS without discussion, it is reasonable to assume that the same standard applies under UNCLOS.
7   Article 79(3).
8   Telecommunications Act of 1997, including amendments up to Act No. 169 of 2012.
9   Submarine Cable and Pipeline Protection Act (16 May 1966).

References

ACMA (Australian Communications and Media Authority), 2015. ACMA's Role in Protection of Sub-marine Cables. www.acma.gov.au/Industry/Telco/Infrastructure/Submarine-cabling-and-protection-zones/submarine-telecommunications-cables-submarine-cable-zones-i-acma (accessed 8 July 2015).

Allan, P., 2000. Cable security in sandwaves. Paper ICPC Plenary 2000, Copenhagen.

Andrulewicz, E., Napierska, D. and Otremba, Z., 2003. The environmental effects of the installation and functioning of the submarine SwePol Link HVDC transmission line: a case study of the Polish Marine Area of the Baltic Sea. *Journal of Sea Research* 49, 337–345.

Ash, S., 2013. The development of submarine cables, in Burnett, D. R., Beckman, R. C. and Davenport, T. M., eds. *Submarine Cables: The Handbook of Law and Policy*. Martinus Nijhof Publishers, Leiden, pp. 19–40.

Austin, S., Wyllie-Echeverria, S., Groom, M. J., 2004. A comparative analysis of submarine cable installation methods in Northern Puget Sound, Washington. *Journal of Marine Environmental Engineering* 7, 173–183.

Australian Government, 2015. *Commonwealth Marine Reserves*. Department of the Environment. www.environment.gov.au/topics/marine/marine-reserves (accessed 15 July 2015).

Baringer, M. O. and Larsen, J. C., 2001. Sixteen years of Florida Current Transport at 270 N. *Geophysical Research Letters* 28, 3179–3182.

BBC, 2011. Japan to repair damaged undersea cables. www.bbc.co.uk/news/technology-12777785 (accessed 15 July 2015).

Böning, C. W., Dispert, A., Visbeck, M., Rintoul, S. R., Schwarzkopf, F., 2008. Response of the Antarctic Circumpolar Current to recent Climate Change. *Nature Geoscience* 1, 864–869.

Burnett, D. R., Beckman, R., Davenport, T. M., 2013. *Submarine Cables: The Handbook of Law and Policy 2013*. Martinus Nijhof Publishers, Leiden.

California Coastal Commission, 2007. Coastal Permit Development Amendment and Modified Consistency Certification E-98–029-A2 and E-00–0004-A1. http://documents.coastal.ca.gov/reports/2007/11/Th8a-s-11–2007.pdf (accessed 8 July 2015).

Carter, L. and Lewis, K., 1995. Variability of the modern sand cover on a tide and storm driven inner shelf, south Wellington, New Zealand. *New Zealand Journal of Geology and Geophysics* 38, 451–470.

Carter, L. and McCave, I. N., 1997. The sedimentary regime beneath the deep western boundary current inflow to the Southwest Pacific Ocean. *Journal of Sedimentary Research* 67, 1005–1017.

Carter, L. and Soons, A. H. A., 2013. Marine scientific research cables, in Burnett, D. R., Beckman, R. C. and Davenport, T. M., eds. *Submarine Cables: The Handbook of Law and Policy*. Martinus Nijhof Publishers, Leiden, pp. 323–337.

Carter, L., Wright, I. C., Collins, N., Mitchell, J. S., Win, G., 1991. Seafloor stability along the Cook Strait power cable corridor. *Proceedings of the 10th Australasian Conference on Coastal and Ocean Engineering 1991*, pp. 565–570.

Carter, L., Milliman, J. D., Talling, P. J., Gavey, R., Wynn, R. B., 2012. Near-synchronous and delayed initiation of long run-out submarine sediment flows from a record-breaking river flood, offshore Taiwan. *Geophysical Research Letters* 39, L12603, 5 pp.

Carter, L., Burnett, D., Drew, S., Hagadorn, L., Marle, G., Bartlett-McNeil, D., Irvine, N., 2009. *Submarine Cables and the Oceans - connecting the world*. UNEP-WCMC Biodiversity Series 31. ICPC/UNEP/UNEP-WCMC, 64 pp.

Cattaneo, A., Babonneau, N., Ratzov, G., Dan-Unterseh, G., Yelles, K., Bracène, R., Mercier de Lèpinay, B., Boudiaf. A, Déverchère, J., 2012. Searching for the seafloor signature of the 21 May 2003 Boumerdès earthquake offshore central Algeria. *Natural Hazards Earth System Science* 12, 2159–2172.

CEE, 2006. Basslink Project Marine Biological Monitoring, McGauran's Beach. Report to Enesar Consulting. www.dpiw.tas.gov.au/inter.nsf/Attachments/CMNL-6T52TV/$FILE/CEE_monit_report_1_final.pdf (accessed22 December 2013).

Collins, K., 2007. Isle of Man Cable Study – preliminary material environmental impact studies. Preliminary Report, University of Southampton.

Cowie, J., 2012. Hurricane Sandy: outage animation. www.renesys.com/blog/2012/10/hurricane-sandy-outage-animati.shtml (accessed 8 July 2015).

Dan, G., Sultan, N., Catteano, A., Déverchère, J., Yelles K., 2003. Mass transport deposits on the Algerian margin (Algiers area): morphology, lithology and sedimentary processes. *Advances in Natural and Technological Hazards Research* 28, 527–539.

DEFRA, 2009. Marine Coastal and Access Act 2009. www.defra.gov.uk/environment/marine/mca/ (accessed 8 July 2015).

DEFRA, 2011. Marine Planning System (2011) www.defra.gov.uk/environment/marine/protect/planning (accessed 15 July 1015).

Dengler, A. T., Wilde, P., Noda, E. K., Normark, W. R., 1984. Turbidity currents generated by Hurricane IWA. *Geomarine Letters* 4, 5–11.

Dernie, K. M., Kaiser, M. J., Richardson E. A., Warwick R. M., 2003. Recovery of soft sediment communities and habitats following physical disturbance. *Journal of Experimental Marine Biology and Ecology* 285–286, 415–434.

Donovan, C., 2009. Twenty thousand leagues under the sea: a life cycle assessment of fibre optic submarine cable systems. Degree Project SoM EX2009–40 KTH Department of Urban Planning and Environment, Stockholm. www.kth.se/polopoly_fs/1.190775!/Menu/general/column-content/attachment/MScThesDonovan09.pdf (accessed 8 June 2015).

Drew, S., 2009. Submarine cables and other maritime activities, in Carter, L., Burnett, D., Drew, S., Hagadorn, L., Marle, G., Bartlett-McNeil, D. and Irvine, N., eds. *Submarine Cables and the Oceans: Connecting the World*. UNEP-WCMC Biodiversity Series 31. ICPC/UNEP/UNEP-WCMC, pp. 43–48.

Drew, S., 2013. External aggression: global risks, local solutions. *Submarine Telcomms Forum* 67, 31–36. www.subtelforum.com/articles/wp-content/STF-67.pdf (accessed 8 July 2015).

Drew, S. and Hopper, A. G., 1996. *Fishing and Submarine Cables – Working Together.* International Cable Protection Committee publication, 48 pp. www.iscpc.org/publications/18.3Fishing_Booklet.pdf (accessed 8 July 2015).

Droz, L., Rigaut, F., Cochonat, P., Tofani R., 1996. Morphology and recent evolution of the Zaire turbidite system, Gulf of Guinea. *Bulletin Geological Society of America* 108, 253–269.

Ecoplan, 2003. *Monitoring der Salz-wiesen Vegetation an der Bautrasse im Ostheller von Norderney 1997–2002.* Ecoplan report for Deutsche Telekom.

Frost, M., Baxter, J. M., Buckley, P. J., Cox, M., Dye, S. R., and Harvey, N. W., 2012. Impacts of climate change on fish, fisheries and aquaculture. *Aquatic Conservation: Marine and Freshwater Ecosystems.* http://onlinelibrary.wiley.com/journal/10.1002/(ISSN)1099–0755 (accessed 8 July 2015).

Gordon, J. S., 2002. *A Thread across the Ocean.* Simon & Schuster, London, 239 pp.

Grannis, B. M., 2001. Impacts of mobile fishing gear and a buried fiber-optic cable on soft-sediment benthic community structure. MSc thesis, University of Maine, 100 pp.

Grochowski, N. T. L., Collins, M. B., Boxall, S. R., Salomin, J. C., 1993. Sediment transport predictions for the English Channel, using numerical models. *Journal of the Geological Society* 150, 683–695.

Hagadorn, L., 2009. Inside submarine cables., in Carter, L., Burnett, D., Drew, S., Hagadorn, L., Marle, G., Bartlett-McNeil, D. and Irvine, N., eds. *Submarine Cables and the Oceans: Connecting the World.* UNEP-WCMC Biodiversity Series 31. ICPC/UNEP/UNEP-WCMC, pp. 17–20.

Halpern, B. S., Walbridge, S., Selkoe, K. A., Kappel, C. V., Micheli, F., D'Agrosa, C., Bruno, J. F., Casey, K. S., Ebert, C., Fox, H. E., Fujita, R., Heinemann, D., Lenihan, H. S., Madin, E. M. P., Perry, M. T., Selig, E. R., Spalding, M., Steneck, R. and Watson, R., 2008. A global map of human impact on marine ecosystems. *Science* 319, 948–952.

Heezen, B. C., 1957. Whales entangled in deep sea cables. *Deep-Sea Research* 4, 105–115.

Heezen, B. C. and Ewing, M., 1952. Turbidity currents and submarine slumps, and the 1929 Grand Banks earthquake. *American Journal of Science* 250, 849–873.

Heezen, B. C. and Ewing, M., 1955. Orleansville earthquake and turbidity currents. *American Association of Petroleum Geologists Bulletin* 39, 2505–2514.

Hogg, N. G., 2001. Quantification of the deep circulation, in Siedler, G., Church, J. and Gould, J., eds. *Ocean Circulation and Climate.* Academic Press, London, pp. 259–270.

Hollister, C. D. and McCave, I. N., 1984. Sedimentation under deep-sea storms. *Nature* 309, 220–225.

Hoshina, R. and Featherstone, J., 2001. Improvements in submarine cable system protection. Proceedings SubOptic 2001, Kyoto; paper P6.7, 4 pp. www.scig.net/ (accessed 8 July 2015).

Howe, B. M., 2004. Expanding the ATOC/NPAL North Pacific Array using the TPC-4 submarine cable. Proceedings. Earthquake Research Institute of Tokyo, 8–9 November 2004.

Hsu, S-K., Kuo, J., Lo, C-L., Tsai, C-H., Doo, W-B., Ku, C-Y., Sibuet, J-C., 2009. Turbidity currents, submarine landslides and the 2006 Pingtung earthquake off SW Taiwan. *Terrestrial. Atmospheric and Oceanic Science* 19, 767–772.

Huh, C-A., Lin, H-L., Lin, S. and Huang, Y-W., 2009. Modern accumulation rates and a budget of sediment off the Gaoping (Kaoping) River, SW Taiwan: a tidal and flood dominated depositional environment around a submarine canyon. *Journal Marine Systems* 76, 405–416.

IEEE Spectrum, 2010. *Europe Plans a North Sea Grid.* http://spectrum.ieee.org/energy/the-smarter-grid/europe-plans-a-north-sea-grid (accessed 8 July 2015).

Intergovernmental Panel on Climate Change, 2013. Climate Change 2013: The Physical Science Basis. Working Group I Contribution to the IPCC 5th Assessment Report – Changes to the Under-lying Scientific/Technical Assessment (IPCC-XXVI/Doc.4). www.ipcc.ch/report/ar5/wg1/#. Uq5tSxaWfHg (accessed 8 July 2015).

International Cable Protection Committee, 2013. Publications. www.iscpc.org (accessed 8 July 2015).

Ko, L. L., 2011. Experience of Nargis Storm in Myanmar and emergency communications. Presentation Ministry of Communications, Posts and Telegraphs. www.itu.int/ITU-D/asp/CMS/Events/2011/disastercomm/S4C-Myanmar.pdf (accessed16 July 2015).

Kogan, I., Paull, C., Kuhnz, L., Burton, E., Von Thun, S., Greene, H. G., Barry, J., 2006. ATOC/Pioneer Seamount cable after 8 years on the seafloor: Observations, environmental impact. *Continental Shelf Research* 26, 771–787.

Kordahi, M. E. and Shapiro, S., 2004. Worldwide trends in submarine cable systems. Proceedings SubOptic 2004, Monaco; paper We A2.5, 3 pp., www.suboptic.org/conference-archives/?y=2004&q= Kordahi&o=date|DESC (accessed 8 July 2015).

Kordahi, M. E., Shapiro, S., Lucas, G., 2007. Trends in submarine cable system faults. Proceedings SubOptic 2007, Baltimore, MD. 4 pp., www.scig.net/ (accessed 8 July 2015).

Krause, D. C., White, W. C., Piper, D. J. W., Heezen, B. C., 1970. Turbidity currents and cable breaks in the western New Britain Trench. *Geological Society of America Bulletin* 81, 2153–2160.

Lacroix, F. W., Button, R. W., Johnson, S. E., Wise, J. R., 2002. A Concept of Operations for a New Deep-Diving Submarine. Appendix 1, Submarine Cable Infra structure pp 139–149, in Rand Corporation Monograph/Report Series MR1395, 182 pp., www.rand.org/pubs/monograph_reports/MR1395.html (accessed 8 July 2015).

Levings, C. D. and McDaniel, N. G., 1974. A unique collection of baseline biological data: benthic invertebrates from an underwater cable across the Strait of Georgia. Fisheries Research Board of Canada. Technical Report 441, 19 pp.

Lloyds List, 2009. Singapore and Malaysia take hard line on anchoring. www.lloydslist.com/ll/incoming/article714.ece (accessed 8 July 2015).

Marine Reserves Coalition, 2012. See Marine Protected Areas. www.marinereservescoalition.org/ (accessed 8 July 2015).

Marra, L. J., 1989. Shark bite on the SL submarine light wave cable system: history, causes and resolution. *IEEE Journal Oceanic Engineering* 14, 230–237.

MARS, 2015. The Monterey Accelerated Research System (MARS). www.mbari.org/mars/(accessed15 July 2015).

Milliman, J. D. and Syvitski, J. P. M., 1992. Geomorphic tectonic control of sediment discharge to the ocean: The importance of small, mountainous rivers. *Journal of Geology* 100, 525–544.

Ministry for the Environment, 2012. Managing Our Oceans. Discussion Document 112pp. www.mfe.govt.nz/publications/marine/managing-our-oceans-discussion-document-regulations-proposed-under-exclusive (accessed 8 July 2015).

Mole, P., Featherstone, J., Winter, S., 1997. Cable protection – solutions through new installation and burial approaches. Proceedings SubOptic 1997, San Francisco, CA, pp. 750–757.

Morel, F. M. M. and Price, N. M., 2003. The biogeochemical cycles of trace metals in the oceans, *Science* 300(5621), 944–947.

NASA, 2012. Hurricane Sandy Atlantic Ocean. www.nasa.gov/mission_pages/hurricanes/archives/2012/h2012_Sandy.html (accessed 8 July 2015).

National Ocean Council, 2012. National Ocean Council. www.whitehouse.gov/administration/eop/oceans (accessed 5 July 2015).

New Scientist, 2012. Fibre optics to connect Japan to the UK via the Arctic. www.newscientist.com/article/mg21328566.000-fibre-optics-to-connect-japan-to-the-uk—via-the-arctic.html (accessed 8 July 2015).

NOAA, 2005. Final environmental analysis of remediation alternatives for the Pacific Crossing-1 North and East submarine fiber-optic cables in the Olympic Coast National Marine Sanctuary. National Oceanographic and Atmospheric Administration, 77 pp. + appendix. http://sanctuaries.noaa.gov/library/alldocs.html (accessed 8 July 2015).

NOAA, 2012. US Climate Extremes Index; NE USA. National Climate Data Center, NOAA. www.ncdc.noaa.gov/extremes/cei/graph/ne/cei/01–12 (accessed 8 July 2015).

Nordquist, M. H., Grandy, N. R., Nandan, S. N., Rosenne, S., 1993. *United Nations Convention on the Law of the Sea 1982: A Commentary*, Vol II. Brill Academic Publishers, the Netherlands, 1,088 pp.

NSIDC, 2012. Arctic Sea Ice News and Analysis. National Snow and Ice Data Center. http://nsidc.org/arcticseaicenews/(accessed 8 July 2015).

Ocean Climate Change, 2012. Marine Climate Change Impacts and Adaptation Report Card 2012. www.oceanclimatechange.org.au/content/index.php/2012/report_card/ (accessed 8 July 2015).

Ocean Networks Canada, 2015. NEPTUNE in the NE PACIFIC. 2015. www.oceannetworks.ca/installations/observatories/neptune-ne-pacific (accessed 5 July 2015).

OFCC, 2007. Oregon Fishermen's Cable Committee. www.ofcc.com (accessed8 July 2015).

Orange, 2008. Three undersea cables cut: traffic greatly disturbed between Europe and Asia/Near East zone. www.orange.com/en_EN/press/press_releases/cp081219en.html (accessed 8 July 2015).

OSPAR Commission, 2008. *Background Document on Potential Problems Associated with Power Cables Other Than Those for Oil and Gas Activities*. Biodiversity Series 370/2008, 50 pp.

PC Landing, 2013. PC Landing Corp. comments on draft environmental assessment. Federal Energy Regulatory Commission P-12690–005. heraldnet.com/assets/pdf/DH132226219.pdf (accessed 8 July 2015).

Piper, D. J., Shor, A. N., Farre, J. A., O'Connell, S., Jacobi, R., 1985. Sediment slides and turbidity currents on the Laurentian fan: side-scan sonar investigations near the epicentre of the 1929 Grand Banks earthquake. *Geology* 13, 538–541.

Puig, P., Canals, M., Company, J. B., Martín, J., Amblas, D., Lastras, G., Palanques, A., 2012. Ploughing the deep sea floor. *Nature* 489, 286–289.

Rapp, R., Gaitch, I., Lucas, G., Kuwabara, T., 2004. Marine installation operations: expectations, specifications, value and performance. *Proceedings SubOptic 2004, Monaco*. Poster We 12.5, 3 pp.

Rauscher, K. F., 2010. ROGUCCI (Reliability of Global Undersea Cable Communications Infrastructure), *The Report, Issue 1*. IEEE Communications Society, 150pp + Appendices.

Shapiro, S., Murray, J. G., Gleason, R. F., Barnes, S. R., Eales, B. A., Woodward, P. R., 1997. Threats to submarine cables. *Proceedings SubOptic 1997, San Francisco*, pp. 742–749. Available at Submarine Cable Improvement Group, l.carter@vuw.ac.nz (accessed 12 April 2013).

Smith, J., Wigginton, N., Ash, C., Fahrenkamp-Uppenbrink, J., Pennisi, E., 2010. Changing Oceans. *Science* 328 (5985), 1497.

Tanaka, H., Nguyen X. T., Makoto, U., Ryutaro, H., Eko, P., Akira, M. and Keiko, U., 2012. Coastal and estuarine morphology changes induced by the 2011 great east Japan earthquake tsunami. *Coastal Engineering Journal* 54(1). www.worldscientific.com/worldscinet/cej (accessed 15 June 2015).

The Crown Estate, 2012. Publication of offshore proximity guidance. www.thecrownestate.co.uk/news-media/news/2012/collaboration-between-submarine-cable-and-offshore-wind-industries-ushered-in-with-publication-of-proximity-guidance/ (accessed 8 July 2015).

Thompson D. W. J., Grise K. M., Solomon S., Kushner, P. J., England, M. H., Grise, K. M., Karoly, D. J., 2011. Signatures of the Antarctic ozone hole in Southern Hemisphere surface climate change. *Nature Geoscience* 4, 741–749.

Toggweiler, J. R., Russell, J. L., Carson, S. R., 2006. Mid-latitude westerlies, atmospheric CO_2, and climate change. *Paleoceanography* 21, PA2005, www.fisica.edu.uy/~barreiro/papers/rtoggweiler1.pdf (accessed 15 July 2015).

UNEP, 2006. *Ecosystems and Biodiversity in Deep Waters and High Seas*. UNEP Regional Seas Report and Studies 178, 58 pp.

Transpower and Ministry of Transport, 2013. Cook Strait Cable Booklet 16pp. www.transpower.co.nz/resources/cook-strait-cable-booklet (accessed 8 July 2015).

UNEP-WCMC, 2009. United Nations Environment Programme and World Conservation Monitoring Centre datasets. www.unep-wcmc.org/oneocean/ datasets.aspx (accessed 8 July 2015).

University of Colorado, 2015. Global Mean Sea Level Time-Series. http://sealevel.colorado.edu/results.php (accessed 8 June 2015).

USGS, 2014. Hurricane Sandy Storm Tide Mapper. http://54.243.149.253/home/webmap/viewer.html?webmap=c07fae08c20c4117bdb8e92e3239837e (accessed 15 July 2015).

Wagner, E., 1995. Submarine cables and protections provided by the law of the sea. *Marine Policy* 19, 2, 127–136.

West of England, 2009. Singapore: Risk of Contact Damage When Anchoring. www.westpandi.com/Publications/News/Archive/Singapore-Risk-of-Contact-Damage-When-Anchoring-Outside-Port-Limits/ (accessed 8 July 2015).

Winchester, S., 2003. *Krakatoa: The Day the World Exploded*. Viking, New York.

Wood, M. P. and Carter, L., 2008. Whale entanglements with submarine telecommunication cables. *IEEE Journal of Oceanic Engineering* 33, 445–450.

24

SEA-POWER

Steven Haines

Undoubtedly, the institutions most closely associated with the term 'sea-power' are navies – those military forces belonging to states that are capable, ultimately, of applying destructive and lethal force at or from the sea. In fact, such capabilities are not the exclusive preserve of navies. A maritime force may also include elements of air and land forces; indeed, will frequently do so in the context of contemporary military operations. Yet other maritime agencies are also capable of deploying force, in relation to constabulary tasking (see below). When we talk of 'navies' in general terms, therefore, we also imply civilian manned coastguards and similar seagoing agencies with law enforcement responsibilities. Nevertheless, it is navies that are the main subject of this chapter and it is principally on them and their functions that the following discussion is focused.

Power does not exist in a vacuum; the environmental context is important. The precise roles and functions of sea-power – of navies – evolve over time in response to changes to the maritime strategic environment. We should even be open to the possibility that substantial change in the strategic environment could lead to a transformation in the general utility of sea-power. It is particularly important to make this point because the maritime environment has been notably subject to fundamental, intensive and wide-ranging change in recent decades. For good reason, therefore, this chapter focuses significant attention on the strategic environment – which needs to be understood if the functions of navies are to make any sense.

Ultimately, we focus our attention and our conclusions on the present and immediate future of sea-power. Nevertheless, it will be illuminating to reflect on the past, for two reasons. First, it is almost invariably the case that an awareness of history leads to a better understanding of the present. Second, traditional assumptions can be difficult to challenge and those involved in the business of sea-power tend to be conservative in their thinking. Looking at things through a traditional lens may help to explain new realities in a convincing manner.

This chapter, then, is about naval forces, the environment in which they operate, what they have been used for in the past, what their utility is today and what functions they may have to perform into the future. It is also about the governance of the oceans viewed from a naval perspective, the emphasis being on sea-power's contribution to good order and stability at sea.

Applications of sea-power

In the mid 1990s, as military strategists considered the consequences of the end of the Cold War, the British Naval Staff set about articulating the Royal Navy's maritime strategic doctrine (Naval Staff, 1995, 1999). It analysed the historic functions of navies as a precursor to its analysis of the contemporary attributes of sea-power and, in so doing, categorised naval tasks in a way that provided a framework for the understanding of sea-power in all eras. No better way of doing this has been devised since then; it is an especially appropriate way to look at the legal dimension of naval tasking, and it is the framework we employ in this chapter.

Operationally, navies are instruments for the application of sea-power in three distinct ways, each of which provides a means of categorising naval tasks:

- *military application*: military tasks rely on the ability of navies to apply lethal or destructive force in the traditional war-fighting sense (Naval Staff, 1999, pp. 52–62). While, ultimately, such tasking may include all the requirements of combat, individual military tasks do not necessarily require the actual application of force. Maintaining a naval presence as a form of conventional strategic deterrence is a good example of such a task, the successful achievement of which will lead to the avoidance of conflict but will depend on navies possessing a demonstrable capability to apply force if it does prove necessary. When force is applied for military purpose, its international legitimacy at the point of application has until now been measured against the Laws of War (these days more usually referred to as the Law of Armed Conflict or International Humanitarian Law);
- *constabulary application*: constabulary tasks are to do with the enforcement of law (Naval Staff, 1999, pp. 62–65). This can be either domestic law (especially within the jurisdictional zones of coastal states) or international law (especially on the high seas). Enforcement tasks also rely, ultimately, on the ability of navies to apply force when required – although the hope is that law enforcement proceeds successfully without the need to apply potentially lethal force. Indeed, when applying force for constabulary purposes, the guiding principle is to use the minimum force necessary to enforce the law, complying with what is frequently referred to these days as the 'law enforcement paradigm';
- *benign application*: finally, benign tasks are those for which the application of potentially lethal or destructive force is not a consideration. They include such things as humanitarian operations, search and rescue, disaster response operations and hydrographic surveying (Naval Staff, 1999, pp. 65–67).

In distinguishing between these different applications, one can usefully regard them as enduring over time, even though the range of individual tasks categorised under their headings will certainly change in response to environmental fluctuations. They serve to explain the past, the present and the immediate future uses of sea-power. Here, for the sake of brevity, however, we will consider only the military and the constabulary applications. While benign uses of sea-power make an important contribution to the stability and security of the maritime environment, it is military and constabulary capabilities that influence the development of navies over time and, indeed, serve to define them.

As already noted, we need to discuss these applications in relation to their strategic environmental context. There are many ways of analysing strategic and environmental complexities. One good one is to break the maritime strategic environment down into various dimensions, analysing each in turn and then determining how each affects the others (Ministry of Defence, 2001, pp. 21–23; Gray, 2010, pp. 38–41). The political, economic and technological dimensions are especially significant as far as sea-power is concerned. They have also certainly combined to have a profound effect on the normative. We say something about each of these in what follows.

In relation to the past we examine the environment that shaped sea-power in the 'Grotian Era'. This is taken to be the period from the middle of the seventeenth century to the middle of the twentieth. As far as the present is concerned, we describe the changes that have taken place since the end of the Second World War and how navies have coped so far with them. Finally, we very briefly comment on the future.

The maritime environment in the Grotian Era

Although they would certainly not have realised it at the time, Europeans in the first half of the seventeenth century were living through a relatively brief period of profound geopolitical and international legal significance. We trace the modern international political system from the Peace of Westphalia in 1648, frequently referring to what we have today as the 'Westphalian state system' (Jackson and Owens, 2005, p. 53). The philosophical thinking underpinning this notion is principally to do with the nature of territorial sovereignty on land, on the central importance of states to the international system, and about the conduct of relations between them. It has traditionally had little to do with notions of sovereignty over the oceans that cover over 70 per cent of the earth's surface.

Only a tiny proportion of those waters fall under the legitimate territorial jurisdiction of states – essentially, the narrow strip of internal and territorial waters and such features as historic bays. The total area of high seas – effectively the area beyond territorial jurisdiction – well exceeds 300 million square kilometres. This has arguably remained an anarchic space, possessed and governed by nobody and, for much of the last three centuries, largely unregulated.

Although he died three years before the Peace of Westphalia, the Dutch lawyer, Hugo de Groot (Grotius) said important things that helped to define significant features of the emerging international political system. Frequently referred to as the 'father of international law', he wrote one of the earliest international legal texts (*De Jure Belli ac Pacis* published in 1625), but it is a much shorter piece entitled *Mare Liberum* (The Freedom of the Seas) published in 1609 that has particular relevance to ocean governance. *Mare Liberum* was a part of a case Grotius made in defence of Dutch maritime commercial interests (Neff, 2012, pp. xv–xxi). They favoured free trade and the free use of the oceans to give meaning to it. Others opposed this idea at the time, in particular the English who wished to control trade, especially that entering English ports. The dispute was so serious it resulted in war (Fulton, 1911, pp. 9–11). Ironically, however, British imperial ambitions were eventually realised through her maritime reach and naval supremacy and the British Empire became the champion of the very high seas freedom it had first so strongly resisted.

There evolved a general acceptance that the high seas were free for all to use, that no state could exercise general jurisdiction over the oceans or claim them for itself, and that high seas freedom was to be guaranteed by the major naval powers – with the British Empire playing a pre-eminent role for much of the next three centuries. Coastal states could only exercise their jurisdiction to the outer limit of what became a 3 mile territorial sea around their coasts (Colombos, 1967, p. 94).

The freedom of the high seas and the strictly limited extent of the territorial sea prevailed until after the Second World War. The idea and the reality of high seas freedom was sustained over three centuries with relative ease for two reasons directly related to the economic value of the oceans. The oceans have economic potential in two senses. First they are an essential medium for the transportation of goods (about 90 per cent of global trade is shipped by sea). Second they are a source of both living and non-living resources – fish and minerals.

From the eighteenth century onwards, the major maritime powers were prepared to insist on free passage for their vessels and had the naval wherewithal to enforce it. In a mercantilist trading environment, navies and trading fleets existed within an almost symbiotic relationship. Those states with large navies also had large merchant fleets – and those with large merchant fleets also had large navies. The maritime power of the state consisted of a combination of those two institutions, with each depending on the other. Navies protected merchant fleets and were themselves funded through the wealth generated by imperial and trading activities.

Until relatively recently, the resources of the ocean had little or no economic value. The mineral deposits on or under the seabed were largely unknown and certainly unrecoverable. Fish, on the other hand, were so plentiful that they had no value prior to being caught. There was no sense of 'ownership' of the fish stocks in the sea, no incentive to convert them into 'property' and the generality of stocks in the high seas required no protection. With mineral resources either undiscovered or physically unexploitable, and fish stocks not generally under threat, the scarcity or management of the ocean's resources were not incentives for the regulation of activity on the high seas.

The high seas became free for all to use and for all to exploit in any way they wished, consistent with the need to respect others' use of them. Four fundamental normative rules emerged that governed activity on the oceans and gave substance to high seas freedom:

- The regulation of the high seas was to be of a bare minimum, with rules only established if high seas freedoms could not be preserved without them (O'Connell, 1982, pp. 9–20).
- States at war (which were free to use the high seas as a place to do battle with their naval forces) were to allow free trade to continue and, in particular, to respect the rights of neutral states to use the ocean even while they were themselves engaged in mortal combat upon it (O'Connell, 1984, pp. 1101–1102).
- Piracy was outlawed as a measure aimed at providing security to ships, exercising their rights freely to navigate the oceans (O'Connell, 1984, pp. 966–983).
- The acceptance of exclusive flag state jurisdiction on the high seas meant that vessels remained the exclusive responsibility of their state of registration. Warships – the agents of the sovereign – could demand the right to visit vessels registered in their own states but had no general right to visit those registered in others. There were certain circumstances in which warships could insist on stopping, visiting and searching foreign flagged vessels but they were very much exceptional (including, for example, the exercise of belligerent rights of visit and search to ensure vessels' neutrality and confirm the legitimacy of trade in time of war, and if a vessel were suspected of involvement in piracy) (O'Connell, 1984, pp. 794 and 801–810).

These norms were a vital feature of the maritime strategic environment that characterised the Grotian Era. They also had a significant influence on the utility of navies in that time. While the vestiges of these four norms survive to this day, they have been under considerable pressure since the Second World War. Indeed, the first and most fundamental – by which regulation was to be of a bare minimum – has seemingly been abandoned altogether. That is something to which we shall return, after discussing sea-power during the Grotian Era.

Sea-power in the Grotian Era

So, what were the military and constabulary tasks that navies could perform consistent with those four norms? Briefly, they were as follows.

Military application

The priority military task for all navies in war is to establish sufficient control of the sea to allow them to conduct those other operations that are their essential belligerent purpose. During the Revolutionary and Napoleonic Wars, for example, Britain needed to establish sufficient control over the seas around the coast of France to conduct effective blockade operations (a form of economic warfare) off French ports. Similarly, it needed to achieve control of the seas off the coasts of Spain and Portugal to allow the Royal Navy to provide essential support to Wellington's campaign ashore in the Peninsular (an example of power projection). Significant major fleet actions of the Revolutionary and Napoleonic wars were frequently to do with sea control, although by their nature they were slowly building Britain's command of the sea which, following Trafalgar in 1805, was to be an assumed attribute of the Royal Navy for the rest of the nineteenth century. Indeed, with the British maritime dominance in mind, early theorists of naval warfare focused on the need to 'command the sea' which, as the term suggests, meant a degree of control verging on total dominance of the oceans. Admirable though that would be, it would be likely to prove extremely difficult to achieve. It was also not essential. A sufficient but limited control of the sea was actually consistent and very much in keeping with one of the principles of war – namely 'economy of effort'(Gooch, 1989, pp. 27–45).

The correlative or corollary of sea control is sea denial. In their struggles with each other, navies are engaged in achieving control of the sea for their own purposes while denying its use to their opponents. In the later decades of the Grotian Era, technology greatly enhanced sea denial capabilities. Both mines and submarines became notable means of denying large sea areas to opposing surface naval forces. They rendered sea command noticeably more difficult, making the less expansive concept of sea control more appropriate.

Following their achievement of sea control, navies during the Grotian Era had two belligerent functions to perform: to project power from sea to land; and to conduct economic warfare. Projecting power ashore or supporting land forces has always been an important war fighting role for navies – Wellington's successful Peninsular campaign would have been impossible without the support of the Royal Navy, for example (Rodger, 2004, p. 561). Nevertheless, during the Grotian Era the principal means by which navies applied strategic pressure on an enemy was through attacks on its commerce and trade on the high seas.

High seas interdiction (or *guerre de course*) operations and the mounting of blockade operations off the coasts and ports of an enemy, were both intended severely to disrupt trading activities and apply economic pressure to weaken the enemy's resolve. Through the age of sail and into the subsequent era of steam propulsion, such operations had potentially profound effect. The great maritime imperial powers were heavily reliant on trade, so attacks aimed at disrupting it were an important method of warfare. This was especially so in general great power wars such as the Revolutionary and Napoleonic Wars and the First and Second World Wars.

In the two world wars, particular attention became focused on the interdiction of shipping by submarines. The decisions by Germany in both wars to resort to unrestricted submarine attacks on shipping were controversial and contrary to the Laws of War at Sea (Halpern, 1994, pp. 64–100; MacIntyre, 1971, pp. 11–15). An entirely lawful *guerre de course* operation, however, was that conducted by the German pocket battleship *Admiral Graf Spee* in the closing months of 1939. The warship intercepted British flag merchant ships in the southern Indian and Atlantic oceans, boarded them, removed their crews to safety and then sank them (Miller, 2013, pp. 83–105). This is often referred to as an example of a legitimate *guerre de course* operation. Nothing of that nature has occurred since 1945, however, although the theoretical legitimacy of such attacks was reaffirmed by an influential group of naval and legal experts as recently as 1995 (Doswald-Beck, 1995).

Constabulary application

The constabulary use of navies was by no means their defining purpose in the Grotian Era. The one exceptional requirement on the high seas was, of course, in relation to the suppression of piracy (and latterly also, the suppression of the slave trade). Within the very narrow coastal jurisdiction of coastal states, there was a limited requirement for coastal protection, customs and excise revenue collection, the enforcement of quarantine regulations, and the protection of local fishing interests. Constabulary functions were, however, far from defining tasks for major navies, which would tend to conduct law enforcement operations in the margins of their principal, military, activities.

The purpose of navies in the Grotian Era

Great power navies during the Grotian Era were concerned with establishing their own control (or command) of the ocean as a precursor to substantial engagement in economic warfare and the projection of power ashore. Great power war was a frequent, almost constant, feature of an international system in which imperial, expansionist ambitions resulted in war being regarded as a routine means of conducting relations with rival powers. While war was always regarded as the ultimate means by which states could impose their will on rivals, it was by no means regarded as the 'last resort' measure in the way it tends to be regarded today. Navies were the instruments of imperial and great power ambition and, while that might be merely to do with maintaining equilibrium as between the great powers, territorial expansion was also seen as a legitimate purpose. War fighting was the *raison d'être* for navies; it is what they were expected to do and what their officers and men expected to be engaged in. Navies existed and were conditioned for military purpose. The constabulary function consisted of very little, although with the scourge of piracy and the threat it posed to free navigation and trade, it was certainly a feature. The lack of constabulary tasks was due entirely to the minimalist approach to the regulation of the high seas, coupled with the very narrow extent of the territorial sea falling within the jurisdiction of the coastal state. To put it very simply, activity at sea was hardly regulated at all and that obviated the need for substantial law enforcement effort from navies.

The maritime environment transformed

What we have described so far was the situation up to 1945. Significantly, this no longer adequately accounts for either the current state of the maritime strategic environment or the function of navies within it. The last six decades have witnessed a substantial transformation of the maritime environment, in particular in economic, political, technological and normative terms. Its political and technological dimensions have, in important respects, been the principal drivers of change, although economic factors have added considerably to the mix. While the process of developing the law of the sea, especially through a series of UN conferences culminating in the 1982 UN Convention on the Law of the Sea, has certainly had important consequences for ocean governance to date, it may have fallen short of establishing appropriate governance for the oceans as a whole. We certainly need to understand the change that has occurred so far before we can sensibly embark on a discussion of navies and their current and future roles.

By the end of the twentieth century, the age of empires appeared at an end. It was destroyed by the long period of global great power war and ideological conflict that encompassed the two world wars and the cold war that followed, as well as by the force of nationalism. The evidence seems to suggest that imperial great power rivalry and open conflict between the great powers is unlikely to break out. The reasons for this include: the actual disappearance of territorially expansionist empires by the late twentieth century; a profoundly significant moral and ethical shift in attitudes over the last century towards war and its legitimate purpose; the work of the UN and its effect on great power relations and conflict containment; a substantial increase in other international organisations and an increasingly sophisticated diplomatic milieu; and the possession of nuclear weapons by the great powers (which has seemingly discouraged them from using war as an instrument of policy in their relations with each other).

Nonetheless, even though there has been no great power conflict for seventy years (the longest period without such wars ever) it would be irresponsible to assume that conflict between the great powers is now confined to history; the crisis over the Ukraine and Chinese policy in the Far East are indicative of residual tendencies towards traditional great power behaviour. As we examine developments in the strategic environment that appear to signal profound – and positive – systemic change, we need to be conscious of the remaining potential for serious instability in the international system. For this reason, strategic deterrence remains a vitally important component in the mechanism of international security. The role of navies in that remains significant.

Despite that caveat, it is manifestly the case that the maritime strategic environment is qualitatively different from that in 1945. Arguably, the nature of the strategic environment in the 1940s was closer to that of two hundred years before than to that of today. The end of empires, the role of the UN, the expansion of international society, the increasing sophistication of diplomatic processes, the existence of nuclear weapons – all of these have already been mentioned, but it is worth pointing to some other significant maritime developments that will have had an influence on the utility of sea-power:

- International law regulating the conduct of international relations has become more influential and the nature of international diplomacy has changed with a particular emphasis on the avoidance and resolution of conflict.
- The law of the sea has undergone profound change, in particular during the Third UN Conference of 1973–82, and conventional law has proliferated, largely negating the traditional understanding that the seas would be subject to the barest minimum of regulation.
- Politically, the oceans are now the focus of attention of four times as many states as previously, most of which have a direct economic interest in the recently established and extensive jurisdictional zones adjacent to their coasts.
- That jurisdiction has been greatly extended – literally by a factor of over a hundred if we consider the potential outer limit of the continental shelf; its juridical status, with a number of new forms of maritime zone to consider, is now substantially more complex than that of the 3 mile territorial sea of 1945.
- It is far from obvious that the traditional view of the high seas as a free space for warring states to fight their naval battles remains as acceptable as once it was.
- Technology has massively increased the exploitation of the living resources of the oceans, threatening the existence of global fish stocks. The total global fish catch today is over five times that in the 1940s. This has generated a need for the detailed regulation of fishing in all regions of the globe, including on the high seas.
- Technology has also revealed the mineral resources of the oceans and rendered them exploitable, thereby generating significant industrial activity at sea. Currently, mineral exploitation is concentrated on the continental shelf, but deep sea-bed mining may well become viable in the near future.

- The global economy has settled into a phase of rapid, intense and sustained globalisation, with trade (90 per cent of which is carried in ships) having increased fourfold since the 1960s alone.
- The global shipping industry has changed beyond recognition, with a multitude of different interests engaged in virtually all ocean-going vessels and a clear state-related identity often impossible to discern. There is no longer an essential symbiotic relationship between major powers' naval forces and their merchant marines, with the bulk of global shipping now registered in states with little naval capability.
- A substantial proportion of trade is carried in containers that cannot be accessed at sea and require specialist container handling facilities in a limited number of port terminals to load and unload. This renders the conduct of traditional economic warfare at sea impossible.

We live today in a profoundly different era of ocean governance from that which was familiar to those engaged in maritime affairs even less than a century ago. The four essential norms that governed the ways the oceans were regarded in the Grotian Era have barely survived the change that has already occurred. Clearly, regulation of the oceans is proliferating at a significant rate – because the increased use of the oceans renders it essential. The absence of naval war on the high seas, the change in global attitudes to inter-state war, and structural change to the international shipping and trading industries, all lead to serious questions about the military purpose of navies. The recent experience of counter-piracy operations in the Indian Ocean and Gulf of Aden have revealed legal complexities to do with the treatment and prosecution of offenders – and their rights – that are substantially different from the permissive norm that governed the treatment of pirates in the age of sail. Finally, although it is still largely respected, the notion of exclusive flag state jurisdiction on the high seas is increasingly out of kilter with the realities of ship registration and flag state responsibilities for both the regulation of shipping activity at sea and the security of trade.

Sea power in the post-Grotian Age

How have the roles and purposes of naval forces changed in response to the substantial changes wrought within the maritime strategic environment? They are certainly not unaffected. As before, we will look at both military and constabulary applications, before drawing some conclusions.

Military application

Sea control operations are always essential; naval forces need to ensure their own security. The characteristics of sea control do, of course, change over time as technology advances, sensors become more effective and weapons systems achieve greater accuracy and range. Fundamentally, however, the nature of sea control operations remains consistent over time – it is always about navies ensuring the use of the sea for themselves and, if necessary, denying it to others. The major navies of the world all fashion their forces with a clear eye on the potential need to conduct direct combat operations against other naval (including maritime air) forces. Major surface and sub-surface combatants, the bulk of naval weapons and sensors, and investment in their development are predicated on the requirements of sea control.

Since the end of the Second World War, conflicts with notable naval dimensions have included the Korean War in the early 1950s, Suez in 1956, the war in Vietnam in the 1960s and early 1970s, the Indo-Pakistani War of 1971, the Battle of the Paracels between the Chinese and Vietnamese navies in 1974, the Iran–Iraq War between 1980 and 1988, the Falklands/Malvinas War of 1982, and the two Gulf wars of 1991 and 2003. Intriguingly, however, given the complete absence of major naval war, in which opposing sides with roughly symmetrical capabilities are pitted against each other for a sustained period, there has been very little real experience of actual and intense combat operations for sea control since 1945. The closest to that was, arguably, the conflict between Britain and Argentina in 1982. That lasted but a few weeks, however, and was fought between two ill-matched adversaries, one a fully professional and highly sophisticated medium maritime power with considerable naval tradition and combat experience to draw on (the world's third most capable navy after those of the United States and the Soviet Union), the other a third tier conscript navy with no combat experience, limited exercise experience and no substantial naval history for inspiration. Fundamentally, of course, the British navy prevailed in 1982 because it was fully prepared for major sea control operations against a highly capable adversary (the Soviet Navy and its Warsaw Pact allies); the Argentine navy was no real match for it. In particular, the Royal Navy's principal sea denial capability (its nuclear powered hunter-killer submarines) proved decisive.

Sea control requirements aside, in the period since the end of the Cold War, the major Western navies have focused a great deal of their attention on the provision of support to military forces operating onshore as part of a joint campaign. Indeed, so-called 'littoral' operations rapidly became the principal focus, as the major Cold War confrontation between NATO and Warsaw Pact naval forces in the North Atlantic and Pacific gave way to operations against very different forces for very different reasons, in theatres such as the Balkans and the Gulf. Naval doctrine certainly began to reflect the new characteristics of the strategic environment in that respect. Even the doctrinal justification for new aircraft carriers, written to support the naval case in Britain's Strategic Defence Review in 1997/98, relied heavily on land forces' doctrine to identify the utility of carrier air power in relation to power projection (Haines, 1997). Power projection has become the principal war fighting focus of major navies. Whether it is the insertion of land forces into an operational theatre, strategic targeting using ship and submarine launched cruise missiles, or tactical fire support (including by carrier-launched aircraft against land targets), power projection is what navies are notable for being able to deliver today.

The contemporary importance of power projection contrasts starkly with the apparent redundancy of offensive economic warfare. The mounting of belligerent blockade and high seas *guerre de course* operations seem to have lost their utility, their legitimacy and their appeal. Until the end of the 1980s, it had seemed that attacks on shipping would become a feature of any war between the Soviet Union and its Warsaw Pact allies, on the one hand, and the North Atlantic and other Western allies on the other. NATO maintained a sophisticated organisation for naval control of shipping (NCS) to facilitate the protection of merchant ships, principally through convoying. Fortunately, that 'hot' war never materialised and, almost as soon as the Soviet empire disintegrated, NATO's extensive NCS organisation began to wind down. There remains an ability to manage shipping for defensive purposes but this is used today for such things as trade security in regions such as the Gulf and for routeing ships as part of the counter-piracy effort in the Indian Ocean. While defensive arrangements remain, offensive economic warfare has virtually disappeared. An examination of the Royal Navy's doctrinal statements from the 1990s and early 2000s, for example, reveals no mention of traditional offensive economic warfare operations against enemy merchant shipping (Naval Staff, 1995, 1999, 2004).

Following the disintegration of the Soviet Union (arguably the last of the traditional 'empires'), there are currently no expansionist great powers with sufficient capacity and naval capability to wage sustained war at sea in the manner of the great power conflicts of the Grotian Era – although some might argue that the likes of China and India may well achieve that. Even if aggressive great powers with major naval capabilities did emerge, however, economic warfare as waged in the past would not be feasible today.

The changed characteristics of the global shipping industry have rendered traditional economic warfare campaigns impossible to mount. The basic belligerent right of visit and search is, for example, impossible to exercise in relation to the vessels engaged in the bulk of global trade. Put quite simply, containers cannot be inspected at sea. The shipping industry today is considerably more international in its make-up. Historically, cargo vessels were registered in ports and wore the flag of the state in which their ownership was located. Those serving at sea tended to be nationals of the flag state. The ships in which they served tended to carry cargos either destined for their own state of registry or being exported from it, or in some way related to trade within an imperial framework. Today, vast numbers of merchant vessels are registered in states whose ports the vessels will never visit, their crews will be multinational with few, if any, nationals of the flag state, and their cargoes may have nothing whatever to do with commercial interests in the flag state.

In these circumstances, if the notion of exclusive flag state jurisdiction did not already exist, nobody would suggest establishing it as a fundamentally important norm at the heart of arrangements for ocean governance. For it to work effectively, states need to be able to both regulate and protect their merchant fleets wherever in the world they may be. None of the ten states with the largest registered merchant fleets has this capability. Those states are Panama, Liberia, the Marshall Islands, Hong Kong, The Bahamas, Singapore, Greece, Malta, China and Cyprus. Only one of those – China – has a reasonable claim to be regarded as a major naval power. The fact that none of these states has the capacity to meet the global obligations of exclusive flag state jurisdiction raises serious questions about its viability as a basis for regulating and protecting world trade. Indeed, a great many ships are flagged out to convenience registries deliberately to reduce the possibility of them having to comply with inconvenient regulations imposed by the state in which their ownership is located. Those with a criminal turn of mind may even use a convenience flag to avoid being subject to the law enforcement processes in the flag state. The current arrangement for the registration of vessels is open to serious abuse.

All of these new features of maritime trade serve to negate a substantial proportion of the laws of war at sea which, despite an informal review in the mid 1990s, have not been updated since the early twentieth century. Nevertheless, one cannot dismiss entirely the need for some measure of economic warfare capacity. Trade is vulnerable to attack and navies should today focus their attention on threats from non-state armed groups determined to upset global trading processes. Navies have a continuing need to maintain a capability to defend and protect shipping at sea. The likely threat seems not to be from a rival naval power's offensive economic warfare activities, however. Of greatest current concern is the threat from non-state groups, either politically or criminally motivated. This leads us on neatly to constabulary functions.

Constabulary application

Responding to criminal activity at sea is about far more than suppressing piracy, although that obviously remains a problem. It is also about illegal, unregulated and unreported fishing that is destroying fish stocks through the taking of more fish than the maximum sustainable yield for important stocks around the world. It is about suppressing the illicit trade in people, in narcotics, in arms and in a wide range of other goods. There is even, for example, a booming trade in illicit oil out of Nigeria. It is about preventing ships from being sunk or having their identities changed to engineer maritime fraud. It is about having legal arrangements in place for dealing with the full range of maritime security demands, including those resulting from terrorist threats.

It is also about more than those issues with a hard security edge to them. Generally, the oceans are still remarkably empty places but they are becoming less so every year. The areas within two or three hundred miles of land are becoming especially busy in terms of transportation and navigation, of ocean resource exploitation and leisure, and even power generation through wind farms. While the areas of deep ocean most distant from land remain largely free of human activity today, we are perhaps about to witness the beginning of deep sea bed mining, an industry that has been threatening to develop for almost half a century. As global population increases yet further, demand for fish protein will increase and the effective regulation of fishing, both within coastal state jurisdiction and on the high seas, will become an even greater priority than it is today. Environmental protection requires regulation that will need much more than simply a willingness by states to enforce standards on vessels visiting their ports. Evidence suggests that the migration of people will become an increasing activity on the oceans and that this could become especially problematic as coastal communities are threatened by the effects of rising sea levels.

It is simply no longer possible for the oceans to be subject to the barest minimum of regulation. That much is already clear, given the substantial proliferation of maritime regulation in the period since the Second World War. For a large proportion of the regulations existing today, effective enforcement requires policing action at sea. That policing action is the growth activity for naval forces (including, of course, those civilian manned seagoing forces whose designated primary role is law enforcement).

Within the maritime domains of coastal states (the collections of jurisdictional zones that states may claim in accordance with the 1982 UN Convention on the Law of the Sea), the law enforcement responsibility falls to the coastal state's authorities. Since the substantial extensions of jurisdiction (many dating from the 1950s but especially those dating from the 1970s and 1980s) the effectiveness of coastal state policing of maritime domains has been less than entirely satisfactory, with many states simply not having the wherewithal to conduct enforcement operations due to lack of expertise and funding. An effective enforcement effort requires ships, aircraft and well trained personnel to operate at sea. It also requires relevant and appropriate domestic legislation to give effect to regulation at sea. Although not yet fully effective for all states, the situation is improving globally. Both navies and civilian manned constabulary forces are beginning to impose good governance on maritime zones and the activities within them. A total of 177 states are currently listed as having naval forces of some description, many of which are effectively coastguards rather than navies intended for traditional military applications (Wertheim, 2013).

Beyond coastal state limits of jurisdiction, in the more than 300 million square kilometres of ocean designated as high seas, there is no force responsible for law enforcement, except through the mechanism of exclusive flag state jurisdiction, under which all warships have the right in international law to visit and search merchant vessels registered in their own states. As we have noted already, however, the vast majority of registered merchant vessels today are not subject to effective flag state jurisdiction. For the moment, there is no movement to terminate exclusive flag state jurisdiction. Indeed, it is jealously guarded, especially by the major maritime powers. Ways are sought periodically to circumnavigate the restrictive nature of jurisdiction and various international diplomatic and conventional solutions are admittedly arrived at. Nevertheless, as activities on the oceans continue to increase in scope and intensity, and as increasing requirements for regulation are met through legal developments, the retention of exclusive flag state jurisdiction will look increasingly strange – especially as there is often only a very tenuous link between a ship and its state of registration. Whatever the solution, enforcement of regulations will require the active participation of naval forces in constabulary roles. The need for the constabulary application of force can only increase.

Concluding comments

As the various dimensions of the maritime strategic environment have undergone important change since 1945, so the main focus of naval activity has shifted. By far the most import-ant shift, occasioned by the proliferation of regulations and increased activity on the oceans, has been towards constabulary functions. These have undoubtedly multiplied in number, especially through the entirely post-Second World War phenomenon of coast state jurisdiction beyond 3 miles. Constabulary functions can only increase into the future.

At the moment, at least, it appears as though military functions are on the wane, especially notable being the apparent demise of the offensive economic warfare role for navies in wartime. This suggestion will be vigorously resisted by those responsible for the development of naval forces. Traditional naval powers, whose force development remains wedded to the procure-ment of war-fighting capabilities, will be extremely cautious in the face of such suggestions. Importantly, it is no purpose of this assessment to undermine substantially the war-fighting credentials of the world's major navies. It would be profoundly irresponsible to read too much into the available evidence and abandon the future development of war-fighting capabilities altogether. With emerging great powers such as India and China posing some measure of competition to others, abandoning the ability to counteract future naval expansionism would go against the responsible collective instincts of the existing major maritime powers. The rivalries that are developing centred on the South China Sea, for example, certainly have the potential to produce conflict at sea (Kaplan, 2014). There is a very real need still to maintain a posture of preparedness as a form of strategic deterrence. While it is no longer generally acceptable to use navies (or any other military forces) for blatant aggrandisement and expansionism, it is an imperative that states remain capable of responding if others seem likely to develop aggressive intent. For very powerful reasons, therefore, it is simply not sensible to jump to the conclusion that naval war-fighting is a thing of the past and navies in future will be focused almost exclusively on constabulary tasking – even if that is the direction in which the evidence seems to be pointing.

While the vast majority of the 177 navies in existence today will be focused almost exclusively on the security of their own maritime domains, with an emphasis on constabulary tasking, the navies of the major maritime powers and those aspiring to that status will continue to maintain an appropriate presence in the oceans in which they have interests. For the United States' Navy, that means a continuing global presence in all oceans. A range of other sophisticated navies, capable of deploying and operating effectively beyond their own regions (as examples, those of Britain, France, Japan, Russia and China), will play an important role in maintaining stability in the absence of an acknowledged global 'authority' with the wherewithal to impose good governance on the oceans. Naval diplomacy will be a continuing feature of such naval presence – and there will clearly be political rivalry between states with deployable naval capability. Nonetheless, there will also need to be a flowering of naval cooperation for stability at sea. The accumulation of naval forces in the Indian Ocean and Gulf of Aden in response to the threat from Somalian-based piracy is indicative of the sort of cooperation that will be needed. While much was achieved by both European Union and NATO naval deployments in that context, naval forces from other non-European states were also involved and integrated into the constabulary effort. This augers well for the future. Naval forces, especially since the Second World War, have naturally cooperated and operated together in task forces for various purposes. They can integrate and create multinational forces far more readily than land forces. It seems more likely that effective sea-power for the future will be about cooperative ventures for ocean security and good governance rather than about great power rivalry descending into open armed conflict.

Sea-power is in a state of flux at present. It is somewhat reluctantly changing its characteristics as a consequence of fundamental change in the maritime strategic environment. It is difficult for the more traditional great power navies to come to terms with the end of the Grotian Era and to embrace a post-Grotian world of law enforcement and cooperation for stability. On a positive note, however, their caution demonstrates their wisdom. The transition from Grotian great power conflict to post-Grotian cooperation for stability and good governance is by no means entirely a given. Until it is, we should be grateful for the continuing existence of sophisticated great power navies with the wherewithal to deliver lethal and destructive force wherever necessary and if required in extremis.

Note

1   Steven Haines is Professor of Public International Law at the University of Greenwich and formerly Professor of Strategy and the Law of Military Operations at Royal Holloway College in the University of London. For over thirty years he was an officer in the Royal Navy, serving at sea and ashore on a wide range of operations, including UN maritime embargo operations, maritime counter-terrorism, coastal policing and fishery protection. His final operational deployments were to the British led NATO Multinational Brigade (Centre) in Pristina, Kosovo and to the British Joint Task Force in Freetown towards the end of the Sierra Leone civil war. As a Ministry of Defence staff officer, he wrote the second edition of the Royal Navy's military strategic doctrine (*British Maritime Doctrine*, 1999) and chaired the Editorial Board of the UK's official *Manual of the Law of Armed Conflict* (Oxford University Press, 2004), being also joint author of its chapter dealing with Maritime Warfare.

References

Colombos, C. (1967) *The International Law of the Sea*, 6th edn (London: Longman).

Doswald-Beck, L. (Ed.) (1995) *San Remo Manual on International Law Applicable to Armed Conflicts at Sea* (Cambridge: Cambridge University Press).

Fulton, T. W. (1911) *The Sovereignty of the Sea* (Edinburgh: Blackwood & Sons).

Gray, C. (2010) *The Strategy Bridge: Theory for Practice* (Oxford: Oxford University Press).

Gooch, J. (1989) 'Maritime Command: Mahan and Corbett', in C. Gray and R. Barnett (eds), *Seapower and Strategy* (London: Tri-Service Press, pp. 27–46).

Haines, S. (1997) Directorate of Naval Staff Duties, 'The Utility of the Future Aircraft Carrier [CV(F)]', UK Ministry of Defence, Ref: D/DNSD/8/36/8b of 12 August 1997 (Staff Author: Cdr S. Haines RN). Unclassified paper in the author's possession.

Halpern, P. G. (1994) *A Naval History of World War I* (London: UCL Press).

Jackson, R. H. and Owens, P. (2005) 'The Evolution of World Society', in J. Baylis and S. Smith (eds). *The Globalization of World Politics: An Introduction to International Relations*, 3rd edn (Oxford: Oxford University Press, pp. 45–62).

Kaplan, R. (2014) *Asia's Cauldron: The South China Sea and the End of a Stable Pacific* (New York: Random House).

MacIntyre, D. (1971) *The Naval War Against Hitler* (London: Batsford).

Miller, D. (2013) *Langsdorff and the Battle of the River Plate* (Barnsley: Pen & Sword).

Ministry of Defence (2001) *Joint Warfare Publication 0–01: British Defence Doctrine* (Shrivenham: JDCC).

Naval Staff (1995) *BR1806: The Fundamentals of British Maritime Doctrine* (London: HMSO).

Naval Staff (1999) *BR1806: British Maritime Doctrine*, 2nd edn (London: The Stationery Office).

Naval Staff (2004) *BR1806: British Maritime Doctrine*, 3rd edn (London: TSO).

Neff, S. (Ed.) (2012) *Hugo Grotius on the Law of War and Peace* (Cambridge: Cambridge University Press).

O'Connell, D. P. (1982) *The International Law of the Sea: Volume I* (ed.: I. A. Shearer) (Oxford: Clarendon Press).

O'Connell, D. P. (1984) *The International Law of the Sea: Volume II* (ed.: I. A. Shearer) (Oxford: Clarendon Press).

Rodger, N. A. M. (2004) *The Command of the Ocean: A Naval History of Britain 1649–1815* (London: Allen Lane).

Wertheim, E. (2013) *The Naval Institute Guide to Combat Fleets of the World* (Annapolis: Naval Institute Press).

The marine environment

The marine environment

25

WASTE DISPOSAL AND OCEAN POLLUTION

Michael O. Angelidis

Introduction

The marine environment

Waste is the result of human activities and its quantity is continuously increasing following the rapid growth of human population and the prevalence of waste-producing practices in our modern society. Waste is being transported to the marine environment through many pathways, including river discharge, atmospheric deposition or direct discharge/disposal of gas and liquid emissions, solid wastes, radioactivity, thermal energy and noise. In this article, we will briefly present the major categories of waste that are reaching the marine environment and we will attempt to highlight the major environmental concerns, as well as the existing relevant policy and management frameworks.

Sources and classification of marine waste

One approach to categorizing waste is to distinguish between sources, which is also a way of understanding the major pressures in relation to different pollutions in the framework of the DPSIR Approach (Drivers, Pressures, State, Impact and Response). On the other hand, a categorization per substance can also be made, in an attempt to group substances/waste with similar chemical characteristics. In the present article, the different categories of waste will be briefly presented, including their environmental impact and their relationships with economic development. Then, we will briefly present the relevant global and regional policy and management frameworks for controlling pollutants emissions and protecting the quality of the marine environment.

Sewage/ Organic matter and nutrients

Major sources of biodegradable organic matter include sewage from urban areas and wastewater generated from industrial and animal breeding activities. Under certain conditions (important load of organic matter, stratification of the water column, low mixing rate of the water masses) the impacts on the receiving waters include eutrophication phenomena and a decrease in the dissolved oxygen concentration near the bottom and in the sediment, which leads to serious degradation of the ecosystem and, in some cases, to massive killing of organisms. Municipal sewage is also a major source of pathogens, which may have a serious health-related impact on humans through contact (bathing water and sand) or consumption of contaminated seafood (shellfish).

Coastal populations, living within 200 km from the coastline, represent almost 50 per cent of the total Earth population (7 billion in 2011). In addition, urban populations are growing fast, already representing half of the global population with a tendency to increase. Already 14 of the world's 17 largest cities are located along coasts (UNEP and IOC-UNESCO 2009). Unfortunately, in many cases the growth of sanitary infrastructures in cities does not keep up with the increase in the coastal urban population. As a result, an important part of the effluents generated in cities located directly on the coast or on river banks is discharged into the aquatic environment with no adequate treatment.

To cope with the wastewater problem in cities, centralized systems are set for the collection, transportation and treatment of municipal effluents. However, wastewater treatment plants, which usually apply a secondary (biological) treatment process, remove the organic load but do not substantially reduce the nitrogen and phosphorus load. In cases that the receiving water body has the capacity to absorb the nutrient load without negative consequences, this treatment is enough. On the other hand, in areas where the receiving body is sensitive to eutrophication because of its morphology and water circulation, this treatment does not prevent the degradation of the marine environment. In those areas, excess concentrations of nutrients in the water results in excessive growth of phytoplankton beyond natural levels (eutrophication) and to changes in the structure and functioning of the ecosystem, reducing its stability. Direct results of eutrophication are increase in water turbidity, depletion of dissolved oxygen in bottom water (hypoxia) due to the decay of organic matter, and occasionally the occurrence of Harmful Algal Blooms (HABs). HABs are extremely dangerous because of the production of toxins that are poisonous to marine life, and to humans through the consumption of seafood.

Eutrophication is a problem that was recognized as early as 1960 and solutions have been proposed for controlling the loads of organic matter, nitrogen and phosphorus discharged from point and diffuse sources. Urban and industrial effluents rich in organic matter, nitrates and phosphates are the main point-sources of nutrients. Because of the rapid urbanization of the world's population, nutrients from effluents are considered a major issue of concern for the health of the marine coastal environment. Over the next 25 years the annual growth rate in urban areas is predicted to be twice as high as that projected for the total population (1.8 per cent versus almost 1 per cent). As early as 2030, 4.9 billion people, roughly 60 per cent of the world's population, will be urban dwellers (UNDESA 2006). Most of the rapid expansion in urbanization is taking place in relatively small and medium-sized cities with populations of less than 500,000 (UNFPA, 2007). Growth is often unplanned and attracting government and private investment to infrastructure development is difficult (UNEP-GEO5, 2012). In addition, nutrients emissions from diffuse sources (leaching of fertilizers from agricultural land, urban runoff, and atmospheric deposition) are also an issue of concern, since they are globally not efficiently controlled. Anthropogenic contribution to global nutrient cycles is continuously increasing and in some areas it is of the same order of magnitude as natural processes. It is estimated that the global river discharge of nutrients has increased by 15 per cent since 1970, with South Asia accounting for half of the increase (Seitzinger *et al.* 2010). Furthermore, low dissolved oxygen concentrations are reported from many estuaries and coastal areas around the world, while 169 areas are considered hypoxic, especially in Southeast Asia, Europe and eastern North America (UNEP-GEO5, 2012).

Hazardous substances

Hazardous substances are compounds (or groups) that are toxic, persistent and liable to bioaccumulation, or give rise to an equivalent level of concern through synergistic effects or degradation into hazardous substances (EC Directive 2000/60/EC). Hazardous substances include trace metals (such as cadmium, lead and mercury) and metalloids (arsenic), pesticides and their by-products, industrial chemicals and combustion by-products. Also, some of the petroleum compounds and combustion by-products (PAHs) are also included in this category.

Chemical industries continuously produce new compounds for the needs of our society, and as of 2012 it is estimated that more than 295,000 chemical substances are being produced and marketed worldwide (CAS 2012, Figure 25.1). The new globalized economy has resulted in a gradual shifting of the production and use of the chemical compounds towards the emerging economies. Since 1970, chemical production and consumption of chemicals in the countries of the Organization for Economic Cooperation and Development (OECD) has decreased by 9 per cent to the benefit of developing economies, such as China, India, Russia and Brazil. It is estimated that by 2020 chemical consumption in the developing countries could account for a third of global consumption (UNEP-GEO5, 2012). Increasing production and use of chemical compounds also leads to increasing generation of relative waste (air, liquid or solid emissions) containing these compounds. Among the chemicals produced and used around the world, there are many that have a negative impact on the marine environment and may be considered hazardous. Chemical pollutants enter the marine environment mainly (80 per cent) from land-based sources (UNEP 2011), but maritime sources (shipping, activities for the exploitation of marine resources) also play an important role. Pollutants may be released directly to the sea or follow a more indirect pathway through rivers and atmospheric deposition.

The identification of substances that are considered as potentially harmful in the different national and regional legislative systems has been based on specific criteria. Such criteria may include: octanol/water partition coefficient (Kow), acute toxicity, persistence, production volume and use of the chemical compound. Based on such criteria, the Group of Experts on the Scientific Aspects of Marine Environmental Protection (GESAMP, 1990) produced a list of potentially harmful substances, consisting mainly of low molecular weight (C1–C3),

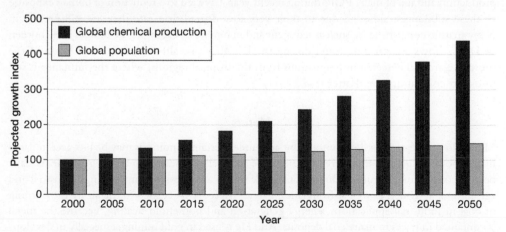

Figure 25.1 Global chemical production is projected to grow at a rate of 3 per cent per year, outpacing the global population growth

Source: Wilson and Schwarzman, 2009, in ESF-Marine Board, 2011.

chlorinated alcanes (chlorinated methane) and medium molecular weight compounds (chlorinated benzenes, phenols and toluenes, PCBs, PCDDs/Fs). Toxicity, persistence and bio-accumulation are the main selection criteria for characterizing a substance as potentially harmful in many regional marine Conventions (UN-ECE POP (Europe, USA, Canada), EU Water Framework Directive (Europe), Arctic Monitoring and Assessment Programme – AMAP, Mediterranean Action Plan – Barcelona Convention (Mediterranean), OSPAR (North Sea), HELCOM (Baltic Sea), BSC (Black Sea).

Persistent organic pollutants (POPs)

POPs are a group of chemicals that are persistent, toxic and bioaccumulative. They are found in different compartments of almost all water bodies, even in remote areas such as the Arctic (AMAP, 2004). The organisms can accumulate POPs from water, sediment and food, resulting in high concentrations of contaminants in the tissues of organisms. For example, organochlorine compounds such as polychlorinated biphenyl (PCB) or organochlorine pesticides (such as DDT) concentrate in fatty tissues of the organisms, remain for long periods in their body and biomagnify up the food chain, with the highest concentrations found in top predators, including humans. Already 22 POPs are included in the list of the Stockholm Convention (2001), which aims at the protection of human health and the environment from POPs. The 22 priority POPs are: aldrin; chlordane; DDT; dieldrin; endrin; heptachlor; hexachlorobenzene; mirex; polychlorinated biphenyls (PCBs); polychlorinated dibenzodioxins (PCDDs); polychlorinated dibenzofuranes (PCDFs); toxaphene; alpha hexachlorocyclohexane; beta hexachlorocyclohexane; chlordecone; hexabromodiphenyl; hexabromodiphenyl ether and heptachlorodiphenyl ether (commercial octabromodiphenyl ether); gamma-hexachlorocyclohexane (lindane); pentachlorobenzene; perfluorooctane sulfonic acid, its salts and perfluorooctane sylfonyl fluoride; tetrabromodiphenyl ether and pentabromodiphenyl ether; and Endosulfan.

POPs may reach the marine environment from direct discharge of industrial effluents if proper treatment is not applied, but also through runoff from contaminated soil and industrial/urban areas, or through atmospheric deposition. Due to the ban on the use of certain substances by international agreements, regional regulations and national legislations, the decrease in the production and use of many POPs during recent years have led to a reduction of human exposure to these substances, although due to their very low degradation rate they are still present in many marine compartments, such as sediment and in biota. For example, the human body burden of DDT shows a decreasing trend during the last 40 years, although important levels of the substance are still detected in populations from the tropical regions, where the substance is still in use to combat malaria (Ritter *et al.*, 2011).

Metals

Metals do not degrade in the environment and under certain conditions may be bio-accumulated in the body tissue and transferred through the food chain. Serious poisoning problems have been recorded for mercury, cadmium and lead, which are released from mining, industrial and agricultural activities, as well as from landfill areas. Mercury is released to the air by the burning of coal in industrial applications, energy generation and household heating, because the metal is contained in traces in many coal deposits. Also Hg is used in gold mining, especially in developing countries (Bose-O'Reilly *et al.*, 2008), as well as in many industrial processes, including production of electronic equipment. Cadmium is used in the production of pigments and electroplating and in the manufacturing of NI-Cd batteries, and is contained in Cd-rich phosphate deposits used for the production of phosphate fertilizers used in agriculture. Leaded gasoline is still in use in many parts of the world, emitting volatile Pb compounds (tetraethyl lead) to the atmosphere, which are then transported to remote places and deposited in the soil and the sea. Additional lead sources include the production of lead paints, batteries and electronic equipment (lead in solder). Due to the atmospheric transportation of many metals (in particles or/and in gaseous form) metal pollution is a global problem, and elevated concentrations of many toxic metals are found in remote areas such the Arctic (AMAP, 2004) and in deep marine sediments (Angelidis *et al.*, 2011).

In general, toxic metals continue to be a serious threat to human health and the marine environment and this is one of the reasons why these compounds continue to be included in all national and regional monitoring schemes.

Emerging issues

Pollution of the marine environment by toxic elements and chemical compounds has been a concern since the mid 1950s and monitoring programmes have been gradually organized to provide necessary data at national and later at regional levels, in order to assess the impact of chemical pollution on marine organisms and human health. Heavy metals and pesticides (DDT) were the first chemicals to be monitored in the early 1970s, followed by PCBs and later by other compounds, such as PAHs, Lindane, TBT and Dioxins (ESF – Marine Board, 2011). As understanding on the effects of commercialized compounds progressed, the uses of conventional toxic pollutants were better regulated in many industrialized countries. However, as new chemicals were commercialized, new concerns were raised on their effects on the environment. Although commercializing new compounds requires toxicological testing by the manufacturers, there is often not enough time to assess all potential toxicological or ecotoxicological effects of these chemicals. As a result, new issues of concern are emerging, requiring additional research and monitoring efforts.

a) Endocrine disruptors

These compounds interfere and alter hormonal signals of organisms and the most widely studied case is that of Tributyl Tin (TBT) and its effects on shellfish. TBT was used as an active agent in antifouling paints in shipping and high concentrations accumulated in the water of small harbours leading to effects on the local oyster population (Alzieu, 2000). Furthermore, studies on marine gastropods (*Nucella lapillus*) exposed to TBT revealed the condition of imposex (superimposition of male features in females, sterilization) (Mathiessen and Gibbs, 1998). In addition to TBT, other chemicals have the potential to disrupt the endocrine system of organisms, such as: i) environmental oestrogens (ex. bisphenol a), ii) environmental anti-oestrogens (ex. dioxin, endosulfan), iii) environmental anti-andogens (ex. DDE, procymidone), iv) chemical reducing steroid hormone levels (ex. fenarimol), v) chemicals affecting reproduction through effects on the central nervous system (ex. dithiocarbamate pesticides), vi) chemicals with multiple mechanism of endocrine action (ex. Phthalates, TBT) (EFS – Marine Board, 2011). Unfortunately, there is widespread use of chemicals that may have endocrine disruption properties and therefore this is a group of contaminants that needs much attention in the coming years.

b) Organohalogenated compounds

In addition to the well-known and monitored organochlorinated compounds, substances such as the polybrominated flame retardants (ex. brominated diphenyl ethers – BDEs) are also of concern because they show increased concentrations and persistence in sediment and marine biota (Tanabe, 2008). The loads of these compounds discharged from urban and industrial areas to the marine environment (directly or through atmospheric transport) are important (Law *et al.*, 2006). Other organohalogenated compounds considered as priority substances are the short-chain chlorinated paraffins (SCCPs).

c) Perfluorinated compounds

Perfluorinated compounds (PFCs – ex. Perfluorooctane sulphonate – PFOS) have been manufactured for many decades and are used as firefighting foams and surface coatings for paper and carpets. Also, the growth of electronic waste due to the high turnover of equipment in the information and telecommunication industry is another source of brominated flame retardants. PFCs are already found in all natural compartments around the world (Paul *et al.*, 2009) and they have recently been included in the list of Persistent Organic Pollutants of the Stockholm Convention.

d) Antifouling paint booster biocides

After the ban on the use of TBT-containing antifouling paints (2003), organic booster biocides were used as alternative compounds in antifouling paints to protect ships' hulls from fouling. These products usually contain copper oxides and organic biocides (ex. Irgarol 1051, diuron) and they are toxic for marine life. It has to be noted that the use of these compounds poses a problem that is difficult to solve, because by definition an antifouling paint must contain a toxic element to keep the hulls of ships free from fouling organisms. Therefore the challenge is to develop efficient biocides with low persistency in the marine environment to avoid accumulation in marine organisms.

e) Pharmaceuticals

A great range of pharmaceutical products have been used for many decades for human and veterinary medicine, and through effluent discharge they are entering the marine environment where their residence time depends on their persistence to degradation. There is relatively little information on the impact of pharmaceuticals on marine ecosystems, but some studies demonstrate such a relationship; for example Näslund *et al.* (2008) found that antibiotics affect the bacterial community in marine sediments, affecting among others nutrient recycling and degradation of organic pollutants.

f) Personal care products

This is a category of pollutants that has been studied relatively recently (late 1990s). Personal care products that are found in the marine environment include synthetic musk of the fragrance industry (ex. nitromusks, polycyclic musks) which may bioaccumulate in fish and invertebrates (Gatermann *et al.*, 2002). Other such products detected in the marine environment include organic UV filters, which are also accumulating in marine mammals and seabirds (Nakata *et al.*, 2007). The long-term risks to aquatic organisms and humans are largely unknown, but the bioaccumulation potential of some of these products in the food chain indicates that they might have biological effects (ESF – Marine Board, 2011).

Nanoparticles involve materials with a size from 1 to 100 nm that are either carbon-based (fullerenes, carbon nanotubes) or inorganic (metal oxides such as TiO_2, metals and quantum dots such as cadmium sulfide) (EFS – Marine Board, 2011). These materials are used in nanotechnological application in electronics, engineering, medicine, personal care products, marine technology and pollution remediation. There is not much information on the impact of such materials on the marine biota, although due to their size they could interfere in processes at cell level (Moore *et al.*, 2009). Also, because of their surface properties, they can also act as vehicles to transport toxic chemicals to the cells.

Litter and microplastics

Litter is being produced in increasing quantities by all human activities, and it is generated from a variety of land-based and sea-based sources. All human activities are somehow connected with litter production and the establishment of a consumer society in many countries since the Second World War has led to a sharp increase of the amount of litter generated per capita, initially in the developed countries and gradually also in the developing world. It is estimated that in North America the yearly use of plastic materials is 100kg per capita, while in the rapidly developing Asian countries the actual yearly use of 20kg/capita is expected to increase to 36kg by 2015 (European Plastics Converters *et al.*, 2009). The sea is the final destination of a great part of the generated litter, if it is not properly managed and/or recycled. The vectors that transport litter to the sea are rivers, drains, sewage outlets, storm water outfalls, run-off from roads and the wind that blows lighter litter items. Land-based sources include all possible human activities (urban, industrial, agriculture, tourism, transportation), while sea-based sources include all merchant and leisure shipping, offshore installations for extraction of marine resources and aquaculture (JRC, 2009). Based on assessments conducted in the North Sea (OSPAR, 2009) and the Mediterranean (UNEP/MAP-MED POL/WHO, 2008) the number of plastic items collected on the beach represents 75 per cent of items in the North Sea and 83 per cent of items in the Mediterranean (JRC, 2009). The 'plastics' category is not uniform but includes a variety of items such as packaging, fishing nets and small pieces of unidentified plastics.

Plastics production began in the middle of the twentieth century and has increased almost continuously since then, reaching 230 million tonnes in 2009 (Plastics Europe, 2010). The most commonly used plastics are: polyethylene, polypropylene, polyethylene tetraphthalate (PET), polyvinyl chloride (PVC), polystyrene and polyamide (Nylon). Due to their low degradation rate, plastics that reach the marine environment may break down to smaller particles but remain in the sea for long periods, floating on the surface or sinking to the seafloor, depending on the buoyancy of the plastic compound. Floating plastics in the oceans include small floating particles (named microplastics – size from 1 μm to 5 mm in diameter), the result of 'weathering' of the initial items, as well as larger plastic items (such as abandoned fishing nets, called 'ghost nets', or bigger floating plastic debris) that are drifting in the oceans following the sea currents. As a consequence, buoyant plastics accumulate in subtropical convergence zones (gyres) where microplastics concentration may reach 200,000 pieces per square kilometre (North Atlantic gyre, Law, K. I. *et al.*, 2010). Floating plastics may remain in the gyres for many years, and there is no evidence yet on trends in their concentration over time. Some may sink to the bottom due to change in buoyancy (caused by biofouling), others may escape from the gyres and end up on the coasts of mid-oceanic islands. Also, the breaking down of microplastics to even smaller pieces may lead to particle sizes undetectable to the applied monitoring methods for floating litter.

Floating plastics such as 'ghost nets', plastic bags or other relatively big items physically affect marine life through entanglement and continuous 'fishing'. Also ingestion of different sized plastic items is another physical process affecting marine life. Monitoring of the stomach content of beached fulmars in the North East Atlantic area revealed on average up to 0.6 g of plastics per bird (1995–1999) which was recently (2004–2008) reduced to 0.3 g (Van Franeker *et al.*, 2010). Entangling and ingestion of marine litter directly affects the life span of individual organisms, through suffocation, obstruction of damage of the gut lining and poor nutrition, but it is not clear to what extend they also affect species and population levels. On the other hand, ingested plastics may additionally lead to bioaccumulation of persistent toxic compounds and acute toxicity to the organism, due to the release of chemicals contained in the original formulation of the plastic or to the desorption of adsorbed pollutants on the surface of plastic particles in the stomach of the organism. Many plastics used in the industry (such as nonyphenols, phthalates, bisphenol A (BPA), styrene monomers and other POPs) have potentially harmful effects including endocrine disruption, which affects reproduction and increases mutagenicity and carcinogenicity in marine organisms (UNEP, 2011).

Plastics accumulation is more difficult to monitor, but available data indicate a widespread distribution of such items in deep-water canyons (Galgani *et al.*, 2000) and abyssal depths (Galil *et al.*, 1995). Their impact on the deep-sea marine organisms is not yet well understood, but their life-span is expected to be much longer than that of floating plastics because of the lack of UV radiation and the cooler temperatures prevailing in this environment.

Oil

Oil spills from normal operations or accidents, from shipping and other oil-related activities (drilling, refining) continue to be an issue of concern for the marine ecosystem, although relevant legislation, international and national regulations and technological tools have substantially improved the control of oily emissions to the sea. The number of oil spills due to transportation (tankers) has significantly reduced since the 1970s (ITOPF, 2010) but serious spills occurred recently in offshore oil and gas exploration activities. In the Deepwater Horizon oil spill in the Gulf of Mexico (2010), 4.9 million barrels of crude oil were released to the marine environment in an accident that is considered to be the largest in the history of oil spills. The full extent of impact on marine life and habitats, as well as on the fishing industry and tourism, has not yet been fully assessed. According to the National Commission on the BP Deepwater Horizon Oil Spill and Offshore Drilling (2011), tar balls continued to wash ashore and wetlands marsh grass continued to foul and die many months after the accident. More recent recorded accidents in the North Sea were the release of at least 1,300 barrels of crude oil from a SHELL platform in the North Sea (August 2011) and a gas leak from a TOTAL exploration platform (March 2012). Because of increasing demands in oil and gas, more offshore exploration facilities are expected to be established in the years to come, increasing the risk of oil spills. Furthermore, because of global warming and the expected melting of Arctic ice, many countries are considering establishing offshore drilling platforms in an area that may contain 20 per cent of the world's undiscovered oil and gas resources (AMAP, 2007; UNEP-GEO5, 2012). This is a potentially dangerous development, because the Arctic environment is particularly vulnerable to oil pollution due to the low rate of oil degradation in low temperatures and the presence of a large number of marine mammals.

Policy and governance frameworks at international or regional level

A number of international initiatives have been established to protect the marine environment from pollution, including hazardous substances and wastes, and marine litter from land-based and sea-based pollution sources.

- The United Nations Convention on the Law of the Sea (UNCLOS 1982), entered in force on 1994, and represents a unified approach towards shared use of the oceans and their resources, addressing navigation, economic rights, pollution, marine conservation and scientific exploration. Following a decision of the World Summit on Sustainable Development (Johannesburg 2002), UNCLOS provides the Secretariat of the UN Regular Process for global reporting and assessment of the state of the marine environment, including socio-economic aspects (the 'Regular Process'). The Regular Process is a regular repeat of assessments, as an integral part of adaptive management in response to changing conditions, and its First Global Integrated Marine Assessment is expected to be completed by 2014.
- The United Nations Environment Programme (UNEP) is the environmental component of the UN system and its mission is to provide leadership and encourage partnership in caring for the environment by inspiring, informing and enabling nations and people to improve their quality of life without compromising that of future generations. The Regional Seas Programme of UNEP, which includes 18 Regional Seas and more than 140 coastal states and territories, is an action-oriented programme focusing on the causes, mitigation and elimination of the consequences of environmental degradation in the marine environment.

The major global multilateral conventions dealing with hazardous substances and wastes in relation to the marine environment are listed here:

- The International Convention for the prevention of pollution from ships (MARPOL) (1973 and 1978) aims at the prevention of marine pollution caused by operational or accidental shipping activities. MARPOL contains annexes dealing with different forms of pollution from ships: oil, noxious liquid substances carried in bulk, harmful substances carried in packaged form, sewage, garbage and air pollution.
- The Stockholm Convention (2001) is a global treaty signed by 152 governments, aiming at the protection of human health and the environment from persistent organic pollutants (POPs). Initially the Stockholm Convention identified 12 priority POPs: aldrin; chlordane; DDT; dieldrin; endrin; heptachlor; hexachlorobenzene; mirex; polychlorinated biphenyls (PCBs); polychlorinated dibenzodioxins (PCDDs); polychlorinated dibenzofuranes (PCDFs) and toxaphene. In 2009 nine additional POPs were added to the list of the Convention: alpha hexachlorocyclohexane; beta hexachlorocyclohexane; chlordecone; hexabromodiphenyl; hexabromodiphenyl ether and heptachlorodiphenyl ether (commercial octabromodiphenyl ether); gamma-hexachlorocyclohexane (lindane); pentachlorobenzene; perfluorooctane sulfonic acid, its salts and perfluorooctane sylfonyl fluoride; tetrabromodiphenyl ether and penta-bromodiphenyl ether. The agreement entered into force in 2010. Endosulfan was also added to the list in 2011. Today 22 POPs are included in the list of the Stockholm convention and the proposed indicators for their monitoring are: a) trends in levels of selected POPs in human tissue; and b) trends in atmospheric levels of selected POPs (such as PCBs and Endosulfan).

- The Rotterdam Convention promotes 'shared responsibility and cooperative efforts among Parties in the international trade of certain hazardous chemicals in order to protect human health and the environment from potential harm and to contribute to their environmental sound use, by facilitating information exchange about their characteristics'. The Convention entered into force in 2004.
- The Basel Convention on the control of transboundary movement of hazardous wastes and their disposal aims at the protection of human health and the environment against adverse effects resulting from the generation, management, transboundary movements and disposal of hazardous and other wastes (entered into force in 1992).

At regional level, several frameworks are in place addressing, among others, issues related to pollution from chemical substances and wastes.

The Regional Seas Programme (RSP) of UNEP covers 18 regions of the world, making it one of the most globally comprehensive initiatives for the protection of marine and coastal environments. The RSP includes the following areas: Antarctic, Arctic, Baltic, Black Sea, Caspian, Eastern Africa, East Asian Seas, Mediterranean, North-East Atlantic, North-East Pacific, North-West Pacific, Pacific, Red Sea and Gulf of Aden, ROPME Sea Area, South Asian Seas, South-East Pacific, Western Africa and the Wider Caribbean. The major role of UNEP/RSP is to assist the Regional Seas Programmes to fulfil their responsibilities towards the priorities identified in relevant UNEP Governing Council Decisions, to contribute to reaching the relevant targets of Agenda 21, the Plan of Implementation of the World Strategy for Sustainable Development and the Millennium Development Goals, and in reconciling global conservation priorities with the realities of implementation at the regional level. Fourteen of the Regional Seas Programmes have adopted legally binding Conventions that express the commitment and political will of governments to tackle their common environmental issues through joint coordinated activities. Most Conventions have adopted protocols, legal agreements addressing specific issues. For example, the Barcelona Convention–Mediterranean Action Plan (Mediterranean Sea), consists of a legislative framework of the Convention for the protection of the Marine Environment and the Coastal Region of the Mediterranean Sea and seven specialized Protocols: i) Protocol for the prevention and elimination of pollution of the Mediterranean Sea by dumping from ships and aircraft; ii) Protocol for the protection of the Mediterranean Sea against pollution from land-based sources and activities (LBS Protocol, UNEP/MAP-MED POL, 2011); iii) Protocol concerning specially protected areas and biological diversity in the Mediterranean; iv) Protocol for the protection of the Mediterranean Sea against pollution resulting from the exploration and exploitation of the continental shelf and the seabed and its subsoil (Offshore Protocol); v) Protocol on the prevention of pollution of the Mediterranean Sea by transboundary movements of hazardous wastes and their disposal; vi) Protocol concerning the cooperation in preventing pollution from ships and, in case of emergency, combating pollution of the Mediterranean Sea; vii) Protocol of Integrated Coastal Zone Management in the Mediterranean (UNEP/MAP 2011).

The European Union legislative framework to protect the marine environment from pollution includes the Dangerous Substances Directive (Directive 76/64/EC) to control their emission and its daughter Directives that set up emission limit values and quality objectives for specific pollutants (82/176/EEC, 84/156/EEC, 85/513/EEC, 84/149/EEC, 86/280/EEC). Legislation to protect the environment from urban effluents and nutrients from agricultural activities is the Urban Waste Water Treatment Directive (91/271/EEC) and the Nitrates Directive (91/676/EEC) and further control of hazardous pollutants was made through the Drinking Water Directive (98/83/EC) and the Directive for Integrated Pollution and Prevention Control (2008/1/EC), which addresses pollution from large industrial plants. Also the EU Regulation

REACH (Registration, Evaluation, Authorization and Restriction of Chemicals) (EC/1907/2006) industries have to assess and manage the risk of chemicals and provide appropriate safety information to users. In order to address water policy issues in a more coherent and integrated way, the EU adopted the Water Framework Directive (2000/60/EC) in order to 'contribute to the progressive reduction of emission of hazardous substances to water' in order to achieve concentrations of hazardous substances in groundwater, surface water and coastal water, near background values. Furthermore, the EU adopted the Marine Strategy Framework Directive (2008/56/EC; Law, R. *et al.*, 2010) that installs an integrated EU policy on marine protection in order to achieve a 'good environmental status' (GES) of marine ecosystems by 2020. In order to assess the achievement of GES, 11 Descriptors have been identified including concentrations and effects of contaminants in the marine environment and seafood (Descriptors 8 and 9), eutrophication (Descriptor 5) and marine litter (Descriptor 10).

Management of hazardous substances and waste

Management of hazardous substances and wastes is not the same around the world: in developed countries there are usually comprehensive systems for such management, while in developing counties this is not usually the case. Furthermore, many developing countries have not ratified or transposed into their national legislation the multilateral environmental agreements on chemicals and wastes.

On a global level, the three Conventions of hazardous substances (Stockholm, Basel and Rotterdam Conventions), as well as the establishment of the Strategic Approach to International Chemicals Management (SAICAM), provide a basis to address the entire life cycle of chemicals and to better understand its relationship with the generation and processing of waste. Since 2011 the three Convention Secretariats have worked under a joint Executive Secretary, in order to proceed to a more holistic approach to the management of chemicals and waste. Under the SAICAM Global Plan of Action, 17 multilateral agreements are in place and over 300 activities have been conducted for the sound management of chemicals (UNEP-GEO5, 2012). However, there is a lot more to be done for the environmentally friendly management of chemicals and waste, not only on the improvement of the national legislation but also on its implementation. Important information gaps still exist especially in the life cycle of the chemicals. Also, because of the lack of financial resources in the developing world, proper management of chemicals, including the remediation of contaminated sites, is a major obstacle to achieving the goals set for 2020 (Agenda 21 and the Johannesburg Plan of Implementation by 2020). As a result, the poorest part of the society is at higher risk of exposure, because of working conditions, poor living conditions often near the vicinity of polluting activities, lack of access to clean water and food, and lack of knowledge on the effects of pollution on their health. Also, there are several stocks of obsolete pesticides around the world that are stored in inappropriate conditions. It is estimated (FAO, 2012) that 290,668 tons of obsolete pesticides are stored in Eastern Europe, Africa, Latin America and the Caribbean, the Near East and Asia.

UNEP has set six goals for the 'sound management of chemicals and waste in order to protect human health and the environment, while ensuring resources efficiency' (from UNEP-GEO5, 2012):

- sound management of chemicals throughout their life cycle, including persistent organic pollutants, heavy metals, and of waste;
- control of the transboundary movement of hazardous wastes as well as responsible trade in hazardous chemicals;

- transparent science-based risk assessment and risk management procedures, as well as monitoring systems at the national, regional and global level;
- support for countries to strengthen their capacity for the sound management of chemicals and wastes;
- protection and preservation of the marine environment from all sources of pollution;
- safe radioactive and nuclear waste management.

Furthermore, in relation to the protection of the marine environment from land-based pollution sources, the Global Programme of Action (GPA), adopted in 1995, was designed to guide national and regional authorities in undertaking sustained action to prevent, reduce and/or eliminate marine degradation from land-based activities. Current priorities are sewage, nutrients and litter.

On a regional level, Regional Seas Conventions (RSC) and other action plans, use integrated management approaches to the protection of the marine environment, including hazardous substances and wastes.

A case study: Barcelona Convention – Mediterranean Action Plan (MAP)

In order to improve the management of chemicals and wastes in the Mediterranean region, MAP adopted (1997) a Strategic Action Programme (SAP MED) to address pollution from land-based activities, as the operational instrument of the Protocol for the Protection of the Mediterranean Sea against Pollution from Land-Based Sources and Activities (the LBS Protocol). The SAP describes the main regional land-based pollution problems, identifies possible control measures, with an estimate of their cost, and sets targets and deadlines for their achievement, through National Action Plans (NAPs). The key land-based activities addressed in the SAP are linked to the urban environment (particularly municipal wastewater treatment and disposal, urban solid waste disposal and activities contributing to air pollution from mobile sources) and to industrial activities, focusing on those responsible for the release into the marine environment of substances that are toxic, persistent and liable to bioaccumulation (TPB), with special attention being paid to persistent organic pollutants (POPs). In addition to activities directly addressing pollution, the SAP also envisages the implementation of capacity-building programmes, provision of external support in accordance with the available sources, implementation of the participatory principle and procedures for monitoring, evaluation, reporting, and regular updating.

The relevant activities of SAP at the regional level predominantly relate to: a) the preparation of the respective guidelines for environmentally sound disposal or management; b) the establishment of Environmental Quality Criteria, emission standards, etc.; c) the development and implementation of technical programmes for the exchange of experience and the provision of information; d) the implementation of research programmes for the validation of technologies; e) the preparation of guidelines in relation to BATs, BEP and clean technologies; and f) participation in selected activities implemented by other international organizations (FAO, OECD, etc.). Also, activities to be implemented at the national level include a) the establishment or updating of the respective national regulations, in accordance with the LBS Protocol and the common measures adopted by the Contracting Parties for the control of pollution; b) the development of specific national (sectoral) plans; c) a percentage reduction or the total elimination of the respective types of discharges, pollutants and polluting activities; d) the promotion of selected and or specific measures and procedures (public transport, the use of environmentally sound fuels and petrol, the participation of selected stakeholders, etc.); e) the preparation of inventories of discharges/emission sources/polluting industries in selected areas (hot spots, areas of concern, critical habitats); and f) the implementation of environmental audits and the application of BATs and BEP. Activities accompanying the implementation of the SAP include monitoring, capacity building, public participation and reporting (UNEP/MAP-MED POL, 2011).

In the framework of SAP, seven (7) Regional Plans (RP) have been agreed by the Contracting Parties for the Mediterranean region: i) RP on the reduction of BOD_5 from urban waste-water; ii) RP for the reduction of BOD_5 in the food sector; iii) RP on the phasing out of aldrin, chlordane, dieldrin, endrin, heptachlor, mirex and toxaphene; iv) RP on the phasing out of DDT; v) RPs on the reduction of inputs of nine chemicals (alpha hexachlorocyclohexane; beta hexachlorocyclohexane; hexabromobiphenyl; chlordecone; pentachlorobenzene; tetra-bromodiphenyl ether and pentabromodiphenyl ether; hexabromodiphenyl ether and heptabromo-diphenyl ether; lindane; endosulfan; perfluorooctane sulfonic acid, its salts and perfluorooactane sulfonyl fluoride), vi) RP for the reduction of mercury vii) RP for the management of marine litter.

In order to plan for an integrated management of human activities for the protection and restoration of the marine environment, the Contracting Parties to the Barcelona Convention have already decided (Decision IG. 17/6 of the 15th COP Barcelona Convention, 2008) to gradually apply an Ecosystem Approach to the management of human activities in the Mediterranean. Following the completion of an initial assessment of the status of the marine environment using the DPSIR approach and the agreement on 11 Ecological Objectives for the region (17th COP, 2012) the countries adopted definitions of Good Environmental Status (GES) and the set targets, (18 COP, 2013). An integrated monitoring programme (including parameters on hazardous substances, nutrients and marine litter) is under development and measures will be drafted (including the control of hazardous chemicals and waste) in order to achieve GES in the marine water bodies of the Mediterranean Sea.

Conclusions

Most probably, we cannot avoid producing waste or using hazardous substances. However, our challenge is to find ways to reduce the loads entering the marine environment, by using less toxic substances, applying cleaner production practices in industry and good practices in agriculture, by recycling materials and by treating wastewater and air emissions. Although we are now in a more advanced stage of predicting potential harmful effects of new chemicals, we have to keep analysing the new chemical compounds and studying their effect on the marine environment in order to reveal potential impacts that were not clearly defined in the initial stages of their introduction in the market. This is a continuous process, because new chemicals are continuously commercialized to fill the needs in our everyday life. On the other hand we already have evidence of the negative impact of many widely used chemicals on the marine ecosystem and human health, and we have to keep trying to regulate their use in all countries, through international and regional regulatory frameworks. International Conventions for hazardous substance control and Regional Seas Conventions for the protection of the marine environment can play a key role in the implementation of the necessary measures at regional and national levels, through coordinated mitigation measures replacing well-known hazardous substances with less harmful substitutes, reducing the plastics and packaging material used, and increasing the recycling potential of products through design and use of appropriate materials.

Furthermore, it has to be recognized that there is no way to get out of this spiral of continuous increase of waste generation and chemical pollution without adopting an individual attitude to reduce waste, as informed, responsible and actively involved citizens of the world.

References

Alzieu, C. 2000. Environmental impact of TBT: the French experience. *The Science of Total Environment*, 258 (1–2), 99–102.

AMAP. 2004. AMAP Assessment 2002: Persistent Organic Pollutants (POPs) in the Arctic. Arctic Monitoring and Assessment Programme, Oslo, Norway, Xvi+310 pp.

AMAP. 2007. Arctic Oil and Gas 2007: Overview Report of the Assessment of Oil and Gas Activities in the Arctic. Arctic Monitoring and Assessment Programme, Oslo. www.amap.no/oga/(accessed 22 July 2015).

Angelidis, M. O., Radakovitch, O., Veron, A., Aloupi, M., Heussner, S. and Price, B. 2011. Athropogenic metal contamination and sapropel imprints in deep Mediterranean sediments. *Marine Pollution Bulletin*, 62, 1041–1052.

Bose-O'Reiley, S. B., Lettmeier, R. M., Gothe, R. M., Beinhoff, C., Siebert, U. and Drasch, G. 2008. Mercury as a serious hazard for children in gold mining areas. *Environmental Research*, 107 (1), 89–97.

CAS. 2012. Chemical Abstract Services, www.cas.org (accessed 22 July 2015).

European Plastics Converters, European Association of Plastics Recycling and Recovery, European Plastics Recyclers and PlasticsEurope, 2009. The compeling facts about plastics 2009: an analysis of European plastics productivity, demand and recovery for 2008. Plastics Europe, Brussels, Belgium.

ESF-Marine Board. 2011. Monitoring Chemical Pollution in Europe's Seas, Programmes, Practices and Priorities of Research, Marine Board-ESF Position Paper 16.

FAO. 2012. *Prevention and Disposal of Obsolete Pesticides*. Food and Agriculture Organization of the United Nations, Rome.

Galgani, F., Leaute, J. P., Moguedet, P., Souplet, A., Verin, Y., Carpentier, A., Goraguer, H., Latrouite, D., Anreal, B., Cadiou Y., Mahe, J. C., Poulard, J. C. and Nerisson, P. 2000. Litter on the sea floor along European coasts. *Marine Pollution Bulletin*, 40 (6), 516–527.

Galil, B. S., Golik, A. and Turkay, M. 1995. Litter in the bottom of the sea: a sea bed survey in the Eastern Mediterranean. *Marine Pollution Bulletin*, 30 (1), 22–24.

Gaterman, R., Biselli, S., Huhnerfuss, H., Rimkus, G. G., Hecker, M. and Karbe, L. 2002. Synthetic musks in the environment. Part 1: Species-dependant bioaccumulation of polycyclic and nitro musk fragrance in freshwater fish and mussels. *Archives of Environmental Contamination and Toxicology*, 42 (4), 437–446.

GESAMP. 1990. Review of potentially harmful substances. Choosing priority organochlorines for marine hazard assessment. FAO, GESAMP Reports and Studies No. 42, 7 pp.

JRC. 2009. Marine Litter – Technical Recommendations for the Implementation of MSFD Requirements. MSFD Technical Subgroup on marine litter. ISSN 1018–5593.

ITOPF. 2010. Oil tanker spill statistics. International Tanker Owners Pollution Federation Ltd. www.itopf.com/information-services/data-and-statistics/statistics/index.html (accessed 22 July 2015).

Law, K. I., Moret-Ferguson, S., Maximenko, N. A., Proskurowski, G., Peakock, E. E., Hafner, J. and Reddy, C. M. 2010. Plastic accumulation in the North Atlantic Subtropical Gyre. *Science*, 329 (5996), 1185–1188.

Law, R. J., Allchin, C. R., de Boer, J., Covaci, A., Herzke, D., Lepom, P., Morris, S., Tronczynski, J. and de Witt, C. A. 2006. Levels and trends of brominated flame retardants in the European environment. *Chemosphere*, 64, 187–208.

Law, R. J., Hanke, G., Angelidis, M., Batty, J., Bignert, A., Dachs, J., Davies, I., Denga, Y., Duffek, A., Herut, B., Hylland, K., Lepom, P., Leonards, P., Mehtonen, J., Piha, H., Roose, P., Tronczynski, J., Velikova V., and Vethaak, D. 2010. Marine Strategy Framework Directive. Task Group 8 Report Contaminants and pollution effects. http://ec.europa.eu/environment/marine/pdf/7-Task-Group-8.pdf (accessed 22 July 2015).

Mathiessen, P. and Gibbs, P. E. 1998. Critical appraisal of the evidence for tributyltin-mediated endocrine disruption in molluscs. *Environmental Toxicology and Chemistry*, 17, 37–43.

Moore, M. N., Readman, J. A., Readman, J. W., Lowe, D. M., Frickers, P. E. and Beesley, A. 2009. Lysosomal cytotoxicity of carbon nanoparticles in cells of the molluscan immune system: an in vitro study. *Nanotoxicology*, 3, 40–45.

Nakata, H., Sasaki, H., Takemura, A., Yoshioka, M., Tanabe, S. and Kannan, K. 2007. Bioaccumulation, temporal trend and geographical distribution of synthetic musks in the marine environment. *Environmental Science and Technology*, 41, 2216–2222.

Näslund, J., Hedman, J. E. and Agestrand, C. 2008. Effects of the antibiotic ciprofloxacin on the bacterial community structure and degradation of pyrene in marine sediment. *Aquatic Toxicology*, 90, 223–227.

National Commission on the BP Deepwater Horizon Oil Spill and Offshore Drilling. 2011. Deep Water: The Gulf Oil Disaster and the Future of Offshore Drilling. Report to the President, United States of America.

OSPAR. 2009. *Marine litter in the North-East Atlantic Region: Assessment and priorities for response.* OSPAR, London, 127 pp.

Paul, A. G., Jones, K. C. and Sweetman, A. J. 2009. A first global production, emission and environmental inventory for Perfluorooctane Sulphonate. *Environmental Science and Technology*, 43 (2), 386–392.

Plastics Europe 2010. Plastics: the facts – An analysis of European plastics production, demand and recovery for 2009. www.plasticseurope.org/document/plastics—-the-facts-2010.aspx?FolID=2 (accessed 22 July 2015).

Ritter, R., Scheringer, M., MacLeod, M. and Hungerbuhler, K. 2011. Assessment of nonoccupational exposure to DDT in the tropics and the north: relevance of uptake via inhalation from indoor residual spray. *Environmental Health Perspectives*, 119, 707–712.

Seitzinger, S. P., Mayorga, E., Bouwman, A. F., Kroeze, C., Beusen, A. H. W., Billen, G., Van Drecht, G., Dumont, E., Fekete, B. M., Garnier, J. and Harrison, J. A. 2010. Global river nutrient export: a scenario analysis of past and future trends. *Global Biogeochemical Cycles*, 24, GBOA08 doi:10.1029/2009GB003587 (accessed 22 July 2015).

Tanabe, S. 2008. Temporal trends of brominated flame retardants in coastal waters of Japan and South China: retrospective monitoring study using archived samples from es-bank, Ehime University Japan. *Marine Pollution Bulletin*, 57, 267–274.

UNDESA. 2006. World Urbanization Prospects: The 2005 Revision. www.un.org/esa/population/publications/WUP2005/2005WUP,Highlights_Final_Report.pdf (accessed 22 July 2015).

UNEP. 2011. *UNEP Yearbook 2011: Emerging issues in Our Global Environment.* UNEP, Nairobi.

UNEP and IOC-UNESCO. 2009. An Assessment of Assessments, Findings of the Group of Experts, Start-up Phase of a Regular Process for Global Assessment of the State of the Marine Environment including Socio-economic Aspects.

UNEP-GEO5. 2012. Global Environment Outlook GEO5, www.unep.org (accessed 22 July 2015).

UNEP/MAP. 2011. Convention for the Protection of the Marine Environment and the Coastal Region of the Mediterranean and Its Protocols. UNEP Mediterranean Action Plan – Barcelona Convention, Athens, 143 pp.

UNEP/MAP-MED POL. 2011. Planning, designing and putting into effect a regional policy to control land-based pollution: lessons from the implementation of the Mediterranean LBS Protocol. Published and accessible on the UNEP/MAP-MED POL website, listed as MED WG 357 7 eng.doc.

UNEP/MAP-MEDPOL/WHO.2008. Results of the assessment of the status of marine litter in the Mediterranean. Athens 2008, UNEP/MAP.

UNFPA. 2007. State of the world population. Chapter 2 People in Cities: Hope Countering Desolation. www.unfpa.org/swp/2007/english/chapter_2/slums.html (accessed 22 July 2015).

Van Franeker, J. A. Meijboom, A., de Jong, M., Verdaat, H. 2010. Fulmar litter EcoQO monitoring in the Netherlands 1979–2007 in relation to the EU Directive 200/59/Econ Port reception facilities. Report C032/09. IMARES, Wageningen UR.

Wilson M. P. and Schwarzman M. R. 2009. Toward a new U.S. chemicals policy: rebuilding the foundation to advance new science, green chemistry, and environmental health. *Environmental Health Perspectives*, 117, 1202–1209. http://dx.doi.org/10.1289/ehp.0800404 (accessed 22 July 2015).

26

MARINE LEISURE
AND TOURISM

Michael Lück

Introduction

With approximately two thirds of our planet covered with water, it is no surprise that the coastal and marine environments have always played a major role in human life. Throughout history the sea has been used by humans for a variety of functions, such as food gathering, transport and waste disposal. Only in the twentieth century have health, leisure and tourism become a major player in the use of the coasts and seas. Early explorers used the oceans for their endeavours, from the Egyptian use of sails to Chinese junks and Viking boats (Lück, 2007; Orams, 1999). With the development of the world's population and improvements in technologies, the importance of the coasts and oceans has become even more important. For example, today in Southeast Asia more than 350 million people live within 50km of the coast (Burke *et al.*, 2002).

The coastal and marine systems are complex and can be subdivided into the littoral, the neritic and the oceanic provinces. The littoral province refers to the area between high and low water (intertidal zone). Adjacent to the littoral province, the neritic province includes the shallower waters above the continental shelf. This zone is usually nutrient rich, and may include ecosystems, such as estuaries, coral reefs and saltwater wetlands. Finally, the oceanic province includes the waters beyond the continental shelf and the ocean floors that lie below them (Dobson, 2008; Lück, 2007). It is the relative ease of access that makes the littoral and the neritic provinces the basis for most human activities, including leisure, recreation and tourism.

Defining marine tourism

The recreational use of the marine environment can be traced back to the Roman era, where upper-class Romans retreated to the summer homes at the sea, in order to avoid the heat of the cities (Jennings, 2007). A major breakthrough in the recreational use of the coasts followed the industrial revolution, where an increase in leisure time, combined with a higher disposable income led to a rapid increase of holidays at the British seaside resorts. Jennings (2007, p. 3) contends that 'toward the end of the eighteenth century in England, spa towns, previously the realm of the upper class, were inundated with lower-middle class populaces'. Today, the recreational activities in coastal and marine environments are extremely diverse, and range from the passive use (such as the scenic coasts as a backdrop) to very involved activities (such as scuba diving). Thus, it is a challenging task to appropriately define coastal and marine tourism. Orams

(1999) cautions that neither a too strict, nor a too liberal definition would be very useful. For example, are cyclists on a scenic shoreline cycle trail actually marine tourists? What about visitors to maritime museums, aquaria and marine parks, which are often hundreds of miles away from the coasts, but have the marine life as their focus? In the search for a meaningful definition, Orams (1999, p. 9) provides the following:

> Marine tourism includes those recreational activities that involve travel away from one's place of residence and which have as their host or focus the marine environment (where the marine environment is defined as those waters which are saline and tide-affected).

This definition is suitable and fitting, because it includes both the tourism component ('travel away from one's place of residence') and the reference to the marine environment. It is important to note that this definition uses the terms 'tourism' and 'recreation' interchangeably, and for this chapter the same approach is taken. This is due to the fact that most activities in the coastal and marine environments are not distinctively 'tourism' or 'recreational' activities, but performed by tourists and recreationists alike.

The marine tourism industry

The marine tourism industry is as diverse as are marine tourists. It ranges from small family-owned ecotour operators to multi billion dollar corporations in the cruise industry. Equally, marine tourists range from very specialized dragonfly tourists (Lemelin, 2013) to the traditional sun, sand and sea mass tourists (Bramwell, 2004). Most forms of coastal and marine tourism have experienced rapid growth over the past decades. In fact, many activities are growing at a much faster rate than tourism in general. The following examples illustrate this growth.

The cruise industry

For more than three decades, cruise lines and their organizations have been repeating the message that cruising is one of the fastest growing segments of the tourism industry (Vogel and Oschmann, 2012). The Cruise Lines International Association (CLIA) represents approximately 80 per cent of the global cruise capacity (CLIA, 2014). Their member lines carried almost 15 million passengers in 2010 (Table 26.1), and it is estimated that in 2014 this figure passed the 18 million passenger mark (CLIA, 2013). In total, i.e. on ships of CLIA members and non-members, cruise passenger numbers will exceed 21.7 million in 2014.

Table 26.1 Annual cruise passenger numbers (CLIA members)

Year	Passenger (million)	Growth (% over 2 years)
2000	7.214	–
2002	8.648	19.88
2004	10.46	20.95
2006	12.01	14.82
2008	13.01	8.33
2010	14.82	13.91
2012	16.95	14.37
2013	17.60	–

Source: CLIA, 2012, 2013,

SCUBA diving

The acronym SCUBA stands for Self Contained Underwater Breathing Apparatus, and refers to equipment that provides an independent source of air and thus allows divers to stay under water for extended amounts of time (dependent, for example on tank size and diver experience). The development of the scuba equipment is an improvement on the 'aqualung', and attributed to Jacques Cousteau and Emile Gagnan in 1943 (Martinez, 2008). Scuba diving has developed in serious terms since 1967 and today represents a multibillion dollar industry worldwide (Musa and Dimmock, 2012). Thus, it is no surprise, that scuba diving has also been proclaimed as 'one of the world's fastest growing recreational sports' (Musa and Dimmock, 2012, p. 1). Indeed, Table 26.2 illustrates that the annual number of certified scuba divers by the world's largest scuba diving organization, PADI, reached almost one million in 2013, and the accumulated total number of certified divers exceeded 22 million (PADI, 2014).

Table 26.2 PADI scuba certifications

Year	Certifications per year	Cumulative certifications
1970	23,736	47,572
1980	107,404	732,266
1990	440,418	3,381,254
1995	680,263	6,237,185
2000	852,702	10,127,054
2005	927,529	14,720,502
2010	923,571	19,382,866
2013	936,149	22,195,063

Source: PADI, 2014.

Whale watching

Commercial whale watching dates back to the 1950s and was developed in Baja California/ Mexico and in Hawaii (Tilt, 1987). According to Hoyt (2001, p. 3) whale watching can be defined as: 'tours by boat, air or from land, formal or informal, with at least some commercial aspect, to see, swim with, and/or listen to any of the some 83 species of whales, dolphins and porpoises.' It is important to note that according to this definition, the term 'whale watching' includes the viewing of other cetaceans, i.e. dolphins and porpoises.

Since its first modest developments in the 1950s, the activity of whale watching has grown significantly in most parts of the world. Indeed, it was estimated that in in 2008, a total of 13 million people took whale watching tours in 119 countries worldwide (Lück, 2012; O'Connor *et al.*, 2009). These figures include a range from a modest 301,616 whale watchers in Central America and the Caribbean, to a whopping 6,256,277 in North America (Table 26.3).

These three cases are examples exemplifying the rapid development of coastal and marine tourism. The industry overall includes a range of activities, such as water sports (surfing, kite surfing, wind surfing, water skiing, jet skiing, sea kayaking, paddle boarding), the cruise industry (including freighter travel and ferries), submersibles and semi–subs, motor boating and yachting, maritime events and regattas, swimming and snorkelling, scuba and snuba diving, fishing, beach resorts and activities, visits to maritime museums and aquaria/marine parks, seafood tourism and observing marine wildlife (fish, pelagic birds, polar bears, penguins, pinnipeds, sharks, manatees and many more).

Table 26.3 Whale watcher numbers and growth rates by region

Region	Whale watchers		Average annual growth (%)
	1998	*2008*	
Africa and Middle East	1,552,250	1,361,330	−1.3
Europe	418,332	828,115	7.1
Asia	215,465	1,055,781	17.2
Oceania, Pacific Islands and Antarctica	976,063	2,477,200	9.8
North America	5,500,654	6,256,277	1.3
Central America and Caribbean	90,720	301,616	12.8
South America	266,712	696,900	10.1
Total	9,020,196	12,977,218	3.7

Source: O'Connor *et al.*, 2009.

Technological advances in marine tourism

Many of the above-mentioned activities experienced significant growth rates because technological developments made access to the marine environment easier, more comfortable and cheaper. It is important to remember that humans are not inherently 'marine creatures', and thus not designed to survive in, under or on the water. Thus, they are heavily reliant on equipment that facilitates most marine activities. Orams (1999) discusses a range of activities that have been heavily influenced by technology, including scuba diving, accommodation, passenger-carrying vessels, recreational vessels, surfboards and sailboards, submersibles and navigational aids. These examples represent the plethora of developments across the coastal and marine industries that tourism can benefit from. Miller (2007) contends that over the last century, the design and development of ocean-going passenger vessels greatly benefited from new building materials, innovative design and engineering, and even new financing arrangements. After the 'Golden Age' of ocean liners between the early 1900s and World War II, ocean liners became increasingly redundant, mostly due to the rapid development of passenger air travel. The first super ship was introduced by Norwegian Cruise Lines (NCL) in 1980. NCL bought the former ocean liner *S.S. France* in 1979 and spent $80 million converting it into a cruise ship about 150 per cent the size of its competition. The renamed *Norway* could carry 2,181 passengers and offered a wide range of entertainment options (Delp, 2010). She was finally retired from service and sold to be scrapped in India in 2006. Advanced technological capabilities enabled a race to the competing cruise corporations commissioning even bigger and better cruise ships. Following the super ship, the first mega ship was Royal Caribbean's *Sovereign of the Seas*, built in 1987 and offering space for 2,276 passengers at a size of 73,129 GRT (Miller, 2007). Further milestones in the cruise ship development include Carnival Cruise Lines' *Carnival Destiny* and Royal Caribbean's *Freedom of the Seas* which were the first cruise ships to exceed 100,000 GRT and 150,000 GRT, respectively. Again, it was Royal Caribbean with ships that exceeded the 200,000 GRT mark with the *Oasis of the Seas* in 2009 (225,282 GRT). The rapid growth of ships is illustrated in Table 26.4, showing cruise ships that exceed 100,000 GRT.

Among the mega developments, at this stage as concepts, are the *America World City* and the *Freedom Ship*. Both concepts are of a magnitude that dwarfs current cruise ships (America World City, n.d.; Freedom Ship International, 2009). They are also largely geared towards retired passengers who will buy an apartment on the vessel and permanently live on board. *America World City (AWC)* is planned to feature 1,800 guest rooms and is too large to dock in most ports. Four 400-passenger vessels are docked inside the belly of the *AWC* and will shuttle

Table 26.4 Cruise ships over 100,000 GRT

Ship name	Cruise line	Year built	Tonnage (GRT)	Class
*To be named	Royal Caribbean	2018	tba	Oasis Class
*To be named	Royal Caribbean	2016	227,700	Oasis Class
Allure of the Seas	Royal Caribbean	2010	225,282	Oasis Class
Oasis of the Seas	Royal Caribbean	2009	225,282	Oasis Class
*Anthem of the Seas	Royal Caribbean	2015	167,800	Quantum Class
*Ovation of the Seas	Royal Caribbean	2016	167,800	Quantum Class
Quantum of the Seas	Royal Caribbean	2014	167,800	Quantum Class
Freedom of the Seas	Royal Caribbean	2006	154,407	Freedom Class
Independence of the Seas	Royal Caribbean	2008	154,407	Freedom Class
Liberty of the Seas	Royal Caribbean	2007	154,407	Freedom Class
Queen Mary 2	Cunard	2004	148,528	Express Class
*To be named	Princess Cruises	2017	143,000	Royal Class
Royal Princess	Princess Cruises	2013	142,714	Royal Class
Regal Princess	Princess Cruises	2014	142,229	Royal Class
Mariner of the Seas	Royal Caribbean	2003	138,279	Voyager Class
Navigator of the Seas	Royal Caribbean	2002	138,279	Voyager Class
Explorer of the Seas	Royal Caribbean	2000	137,308	Voyager Class
Adventure of the Seas	Royal Caribbean	2001	137,276	Voyager Class
Voyager of the Seas	Royal Caribbean	1999	137,276	Voyager Class
*Carnival Vista	Carnival Cruise Lines	2016	135,000	Vista Class
Carnival Breeze	Carnival Cruise Lines	2012	128,500	Dream Class
Carnival Magic	Carnival Cruise Lines	2011	128,500	Dream Class
Carnival Dream	Carnival Cruise Lines	2009	128,250	Dream Class
Diamond Princess	Princess Cruises	2004	115,875	Grand Class
Sapphire Princess	Princess Cruises	2004	115,875	Grand Class
Crown Princess	Princess Cruises	2006	113,561	Grand Class
Emerald Princess	Princess Cruises	2007	113,561	Grand Class
Ruby Princess	Princess Cruises	2008	113,561	Grand Class
Carnival Splendor	Carnival Cruise Lines	2008	113,323	Splendor Class
Caribbean Princess	Princess Cruises	2004	112,894	Grand Class
Carnival Conquest	Carnival Cruise Lines	2002	110,239	Conquest Class
Carnival Freedom	Carnival Cruise Lines	2007	110,320	Conquest Class
Carnival Liberty	Carnival Cruise Lines	2005	110,320	Conquest Class
Carnival Glory	Carnival Cruise Lines	2003	110,239	Conquest Class
Carnival Valor	Carnival Cruise Lines	2004	110,239	Conquest Class
Star Princess	Princess Cruises	2002	108,977	Grand Class
Golden Princess	Princess Cruises	2001	108,865	Grand Class
Grand Princess	Princess Cruises	1998	107,517	Grand Class
Carnival Sunshine (launched as *Carnival Destiny*)	Carnival Cruise Lines	1996	102,853	Destiny Class
Carnival Triumph	Carnival Cruise Lines	1999	101,509	Triumph Class
Carnival Victory	Carnival Cruise Lines	2000	101,509	Triumph Class

* Planned ships/ships under construction.

passengers to and from shore. The *Freedom Ship* will be significantly larger, and features its own airfield on the top deck, accepting fixed-wing aircraft carrying up to 20 passengers. It will have space for a population of 100,000, including 40,000 full-time residents, 30,000 daily visitors, 10,000 hotel guests and 20,000 full-time crew. Like *AWC*, it is planned to continuously sail around the globe in two-year trips. While these projects are far away from reality, and may never be built, the concept of apartment ships is not new: ResidenSea has operated the apartment ship *The World* since 2002. It offers 110 luxury apartments and many amenities known to regular cruise ships, such as swimming pools, restaurants, shops, tennis court, disco, minigolf and much more (Dowling, 2006).

However, technological advances are not always of the size displayed by the cruise ship industry. Many developments are small in size, but have significant impacts on the recreational and touristic opportunities. Probably the most significant invention in this field was the scuba gear. Cousteau and Gagnon's development made it possible for millions of recreationists to enjoy an underwater world that would have otherwise been the domain of just a few very specialized people. In addition to the traditional scuba gear, Garrod and Gössling (2008) discuss a number of more recent, related inventions. For example, there is equipment using so-called rebreathers that allow the air to be recycled, rather than being exhaled into the water. This provides the possibility of staying under water for longer periods of time. For the same reason, snuba has become popular. This is a cross between snorkelling and scuba diving, in which divers are attached to an air hose that in turn is attached to the vessel above. This system enables divers to stay under water without having a tank strapped to their back, and provides more freedom while diving. An additional advantage, especially for inexperienced tourists, is that this system does not require formal training and certification (Garrod and Gössling, 2008). Other inventions include the scooter and the sled. The former is a diver-driven propulsion vehicle (DPV) and the latter is pulled by a vessel along the surface, with divers hanging onto it, allowing them to travel faster, with less effort (Garrod and Gössling, 2008).

In addition to these examples, there are myriads of technological advances that provide better, safer, faster, easier, more reliable and cheaper coastal and marine recreation. For example, new building materials for vessels and surfboards, such as fibreglass, aluminium, polyethylene and carbon fibres made equipment lighter, faster and more durable. Mass production processes keep costs down, and make equipment affordable to a much wider population (Orams, 1999). The constant improvements of navigation aids, including Global Positioning Systems (GPS) at relatively low cost now allow sailors to navigate more accurately and safely, even in adverse weather conditions.

Management issues

The marine environment is so vast and complex, that it poses significant challenges for managers. The sheer size and the variety of users make it a difficult task to develop management strategies that are comprehensive and fair to all users. Such strategies can be very diverse in scope, and range from no management to very tight regulation (Orams, 1995). In order to better understand the plethora of strategies, Orams (1999) divided these into four distinct categories: regulatory, physical, economic and educational (Table 26.5).

It is widely accepted that a mix of some of the techniques shown in Table 26.5 commonly provides the best benefits for nature, wildlife, tourists and residents. Management approaches, such as the implementation of zoning or marine protected areas (MPAs) attempt to set defined zones for specific uses only (or no use whatsoever), or for use at specific times or seasons. Marine spatial plans outlining such zones are akin to land-use plans and accommodate different uses in

Table 26.5 Management strategies and techniques

Regulatory	Physical	Economic	Educational
Limit visitor numbers	Site hardening	Differential fees	Printed material
Prohibit certain activities	Facility placement	Damage bond	Low-power radio
Close areas to activities/use	Facility design	Fines	Signs
Separate activities	Sacrifice areas	Rewards	Visitor centres
Require minimum skill level	Remove/alter attraction		Guided walks/talks
	Rehabilitation		Activities
			Personal contact

Source: After Orams, 1999.

different zones. Zones are displayed on zoning maps available to all users (Agardy, 2010). The challenge with zoning is that the majority of the marine environments are multi-user areas, i.e. users range from recreationists and tourists to commercial users (tourism, primary industries, transport). Thus, Agardy (2010, p. 11) argues that:

> planners must recognize connections, including the connections between different elements in an ecosystem, between land and sea, between humans and nature, and between uses of ocean resources or ocean space and the ability of ecosystems to deliver important goods and services.

Based on the same recognition of the interface between human activities, the terrestrial and the marine environments, Integrated Coastal Management has been developed as a means to manage the coastal environment. This is important because any activities in the coastal and marine environments cause a range of impacts. These include effects users have on the environment, but also on other users (Cicin-Sain and Knecht, 1998). These effects can be in coexistence, and not cause any major concern. However, they can also be conflicting, and be to the detriment of other users and/or the environment. While these management approaches are very complex and multidisciplinary, they often lack a focus on coastal and marine recreation and tourism, and are more ecology focused. In order to address recreation and tourism in marine settings, Orams (1999) offers a planning and management tool, based on the Recreation Opportunity Spectrum (ROS): The Spectrum of Marine Recreation Opportunities (SMARO).

The Spectrum of Marine Recreation Opportunities (SMARO)

ROS has its origins in the management of American national parks, forests and other natural areas. It is based on the specific setting of the particular area in question, and recognizes the diversity of users, and the associated challenges. ROS has been adopted by many natural area management agencies, such as the US National Park Service and the Department of Conservation in New Zealand. However, it is a land-based planning and management tool, and Orams adapted it for the recreational use of the marine environment. He contends that SMARO is based on the understanding that the intensity of use is related to the distance from shore, i.e. the closer to shore the more intense the use (Orams and Lück, 2012). As shown in Table 26.6, SMARO is based on the classification of five different zones (Classes 1–5), and three specific characteristics (experiences, environment, location).

While SMARO identifies these five classes, it is important to note that some activities may reach across a number of classes, such as recreational fishing. This activity can occur in Classes

Table 26.6 The Spectrum of Marine Recreation Opportunities (SMARO) (after Orams, 1999)

Characteristics	Class 1 *Easily accessible*	Class 2 *Accessible*	Class 3 *Less accessible*	Class 4 *Semi-remote*	Class 5 *Remote*
Experiences	Much social interaction. High level of services. Usually crowded. Noisy. Lots of activity.	Frequent contact with others. Some spaces/times to escape from others.	Some contact with others. Locations/times when no other people present. Quieter.	Infrequent contact with others. Peace and quiet. Close to nature.	Virtually no contact with others. Solitude. Tranquillity. Self-sufficiency.
Environment	Many human influences. Highly modified. Lower-quality environment.	Human structures and influences visible and close by. Environment quality variable.	Few human structures close by but some may be visible. Higher environmental quality.	Limited evidence of human activity/structures. High quality environment.	Isolated. Little to no evidence of human activity. Pristine environment.
Locations	Close to or in urban areas. Beaches, docks, piers, urban coastal parks and walkways/cycleways. Close to road, parking and mass-transport options.	Intertidal zone and areas up to 100 meters offshore. Often requires walk of several hundred meters from car parking, mass-transport.	100 m to 1km offshore. More than 20 km from major urban area. More remote beaches, islands and coastal areas. Often requires boat for access or walk >500 m.	1–50 km offshore. Isolated coasts/islands/reefs difficult to access. Boats or overnight hiking needed to access.	>50 km offshore. Coastal areas >100 km from any significant human habitation.
Examples of activities	Sunbathing/baking. People-watching. Playing games. Eating. Social gatherings. Sightseeing. Special events (e.g. concerts)	Swimming. Fishing. Boating. Surfing. Windsurfing/kite-boarding.	Boating. Fishing. Snorkelling/SCUBA. Nature-study. Surfing. Sailing. Sea-kayaking.	Coastal sailing. Remote coast hiking and camping. Live-aboard vessel (fishing, diving, surfing etc.)	Offshore bluewater sailing. Live-aboard offshore vessels (for fishing, diving etc.). Remote coast/reef sea-kayaking and surfing.

Source: Orams and Lück, 2012, p. 174.

1–4, and in some cases even in Class 5. It is equally important to note that over time some activities may move across classes, due to the improvements in technologies (as discussed above) and due to seasonal changes (for example, Dusky dolphins in Kaikoura, New Zealand, can be found closer to shore during summer, but further offshore during the winter period. Subsequently, dolphin watching might take place in Classes 2 and 3 during summer, and in Classes 3 and 4 during winter). In essence, the SMARO model is designed to understand general activity patterns, and thus inform and support any management approaches.

Governance and policy

One of the key components of tourism planning and development is policy (Wong, 2009), and coastal and marine tourism is no exception. When pursuing the goal of sustainable tourism development, to the benefit of all stakeholders, extensive consultation and planning, as well as policy development and implementation need to be at the forefront of any endeavour. In contrast to terrestrial policy, the marine environment faces two major challenges: there are few policies governing international waters, and it is significantly more difficult to police and enforce policies at sea than it is on land. While some countries have only relatively small stretches of coastline and adjacent marine environments, other countries feature extensive coasts and vast marine environments. For centuries, seafarers roamed the seas in search of fishing and whaling, transport and discovery without acknowledging any particular ownership of the seas. After some early declarations of local and regional maritime laws, the United Nations convened the first UN Conference on the Law of the Sea in 1958, establishing the principle of the 12-mile con-tiguous zone, giving countries the right to regulate and enforce customs, sanitary and fiscal regulations within this zone (Pernetta, 2004). In the 1970s and 1980s, the United Nations con-vened further conferences and multiple meetings, culminating in a treaty (adopted in 1982 and implemented for all parties in 1994), outlining the 12-mile zone as territorial seas. In addition, countries have the option to declare a 200-mile exclusive economic zone (EEZ) with the right to exploit all living and non-living resources within this zone. However, they do not have territorial rights beyond the 12-mile zone (Pernetta, 2004). The waters beyond the 200-mile EEZ are referred to as the High Seas.

It is also well established that the coastal and marine environments are greatly affected by developments and usage on land. For example, agricultural runoffs, sewage and other waste originate on land, but often pollute the seas. In fact, impacts generated on land can be more of a threat to marine environments than those created by tourism and recreation activities. Harriott (2004) surveyed experts (academics, researchers, postgraduate students and environmental managers), as well as the general public about the perceived main threats to Australia's Great Barrier Reef. While the general public perceived tourism as the main threat to the reef, the experts ranked tourism and recreation-related activities and developments lower than other impacts, such as global warming or agricultural runoffs (Table 26.7) (Harriott, 2004).

The example of the Great Barrier Reef illustrates the limited effectiveness of marine policies, because land-based activities are not covered by these. Equally difficult is the regulation of activities at sea, especially beyond the 12-mile zone. For example, the International Convention for the Prevention of Marine Pollution from Ships (MARPOL), governed by the UN offshoot Inter-national Maritime Organization (IMO), regulates the pollution from ships beyond the 12-mile zone. It came into effect in 1973 and was modified and amended several times afterwards (Lück, 2010). With the exception of some designated Special Areas (Mediterranean Sea, Baltic Sea, Black Sea, Red Sea, Persian Gulf, North Sea, Antarctica, Caribbean) these regulations are relatively lax: treated sewage may be discharged at only four miles from the coast, and untreated

Table 26.7 Perceptions of threats to the Great Barrier Reef by a group of coral reef professionals

Rank	Potential threat	Mean (n=66)	SD
1	Global warming	4.55	0.75
2	Coral bleaching	4.24	0.98
3	Agricultural runoff	4.02	1.00
4	Trawling	3.83	0.91
5	Coastal development	3.76	1.00
6	Commercial line and net fishing	3.62	0.95
7	Pollution (industrial and urban)	3.48	0.97
8	Oil spills	3.24	1.08
9	Crown of thorns starfish	3.14	1.01
10	Introduced species, pests	3.14	1.02
11	Sewage discharge	2.98	1.01
12	Recreational fishing	2.86	0.97
13	Collection of marine life (e.g. shells, coral)	2.55	0.86
14	Boating, anchoring	2.54	0.76
15	Tourism infrastructure	2.46	0.86
16	Ship groundings	2.45	0.97
17	Aquaculture	2.44	0.80
18	Tourist activities	2.41	0.79
19	Rubbish, litter	2.34	0.88
20	Natural disasters (e.g. cyclones, floods)	2.18	0.99
21	Recreational activities by residents	2.10	0.74
22	Scuba diving and snorkelling	1.67	0.65

Note: Means are the average response on a scale from 1 to 5, with 1 representing no threat and 5 representing a very serious threat.

Source: After Harriott, 2004.

sewage at 12 miles and beyond. Bilge water may be discharged beyond 12 miles if filtered, and ships may discharge to within 12 miles of the nearest coast all garbage (except plastics), and even beyond only three miles if these materials are ground (Lück, 2010; Sheppard, 2008). This gives cruise ships the opportunity to discharge most of their liquid and solid wastes, without even breaching the law. Not only that, but many cruise lines have a long history of breaching even these relatively lax laws, dumping within the 12-mile zone (even in harbours) or discharging oil (Klein, 2002).

In addition to MARPOL, there are a variety of international conventions and treaties regulating users of the marine environment. They are not specifically tourism related, but apply to all users of the seas, and include the International Convention for the Safety of Life at Sea 1974 (SOLAS), the International Convention of Standards of Training, Certification and Watchkeeping for Seafarers (STCW Convention), the International Convention on the Control of Harmful Anti-fouling Systems on Ships (Anti-fouling Convention), the Convention on the Conservation of Antarctic Marine Living Resources (CCAMLR), and the Convention on Wetlands of Importance (Ramsar). There are also a multitude of conventions and regulations that are not specifically marine related, but are equally applicable to marine tourism, such as the Convention Concerning the Protection of the World's Cultural and Natural Heritage (for example, Australia's Great Barrier Reef is a World Heritage Site under this convention), the Convention on Biological Diversity, the Convention on the Conservation of Migratory Species of Wild Animals (CMS), or the Convention on International Trade in Endangered Species (CITES).

Looking at these regulations, it becomes clear that they have not been created to specifically address tourism and recreational activities and developments, but they are just as important for these as they are for any other businesses and developments. This phenomenon underlines the vast interconnectivity of (coastal and marine) tourism and recreation with a plethora of other sectors.

Conclusion

The oceans are host to vast marine ecosystems and tourists/recreationists are taking advantage of these environments more than ever before. Supported by significant developments in technology, many marine tourism activities are among the fastest growing tourism sectors. However, this rapid growth and heavy use of the coastal and marine environments brings with it a number of challenges. The impacts of the industry are difficult to manage, and even with policies and regulations in place, it is difficult to enforce these. In addition, the interconnectivity of tourism and other industries, and the interconnectivity of marine activities with terrestrial activities, make coastal and marine tourism and recreation a largely diverse sector that is inherently difficult to manage. When pursuing the goal of sustainable tourism development, to the benefit of all stakeholders, extensive consultation and planning, as well as policy development and implementation need to be at the forefront of any endeavour. In the pursuit of such sustainable management options, a variety of management tools and approaches have been developed. Introducing the Spectrum of Marine Recreation Opportunities (SMARO), Orams (1999) offers a tool that is designed specifically for the understanding and management of coastal and marine tourism and recreation.

References

Agardy, T. (2010). *Ocean Zoning: Making Marine Management More Effective*. London: Earthscan.

America World City. (n.d.). *America World City* [website]. Retrieved 30 December 2012, from http://americanflagship.com/about.

Bramwell, B. (ed.). (2004). *Coastal Mass Tourism: Diversification and Sustainable Development in Southern Europe*. Clevedon, OH: Channel View Publications.

Burke, L., Selig, L., and Spalding, M. (2002). *Reefs at Risk in Southeast Asia* [webpage]. Retrieved 10 May 2002, from www.wri.org/reefsatrisk/reefsatriskseaisa.html.

Cicin-Sain, B., and Knecht, R. W. (1998). *Integrated Coastal and Ocean Management: Concepts and Practices*. Washington, DC: Island Press.

Cruise Lines International Association. (2012). CLIA Industry Update. Arlington, VA: CLIA.

Cruise Lines International Association. (2013). CLIA Passenger Carrying Report (Q4). Retrieved 24 November 2014, from http://cruising.org/sites/default/files/misc/2013_Q4_Cum_All.pdf.

Cruise Lines International Association. (2014). 2014 CLIA Annual State of the Industry Press Conference & Media Market Place. Retrieved 24 November 2014, from www.cruising.org/sites/default/files/pressroom/PressConferencePresentation.pdf.

Delp, L. (2010). *Cruise Ship Firsts Through History* [website]. Retrieved 30 December 2012, from http://news.travel.aol.com/2010/03/05/cruise-ship-firsts-through-history/.

Dobson, J. (2008). Littoral. In M. Lück (ed.), *The Encyclopedia of Tourism and Recreation in Marine Environments* (p. 274). Wallingford: CABI.

Dowling, R. K. (2006). Looking Ahead: The Future of Cruising. In R. K. Dowling (ed.), *Cruise Ship Tourism* (pp. 414–434). Wallingford: CABI.

Freedom Ship International. (2009). *Freedom Ship International* [Website]. Retrieved 30 December 2012, from www.freedomship.com/freedomship/overview/overview.shtml.

Garrod, B., and Gössling, S. (2008). Introduction. In B. Garrod and S. Gössling (eds), *New Frontiers in Marine Tourism: Diving Experiences, Sustainability, Management* (pp. 3–28). Amsterdam: Elsevier.

Harriott, V. J. (2004). Marine Tourism Impacts on the Great Barrier Reef. *Tourism in Marine Environments*, *1*(1), 29–40.

Hoyt, E. (2001). *Whalewatching 2000: Worldwide Numbers, Expenditures, and Expanding Socioeconomic Benefits* [webpage]. Retrieved 24 August 2001, from www.ifaw.org/press/whalewatching2000.html.

Jennings, G. (2007). Water-Based Tourism, Sport, Leisure, and Recreation Experiences. In G. Jennings (ed.), *Water-Based Tourism, Sport, Leisure, and Recreation Experiences* (pp. 1–20). Amsterdam: Elsevier Butterworth-Heinemann.

Klein, R. A. (2002). *Cruise Ship Blues: The Underside of the Cruise Industry*. Gabriola Island, BC: New Society Publishers.

Lemelin, R. H. (ed.). (2013). *The Management of Insects in Recreation and Tourism*. Cambridge: Cambridge University Press.

Lück, M. (2007). Nautical Tourism Development: Opportunities and Threats. In M. Lück (ed.), *Nautical Tourism: Concepts and Issues* (pp. 3–13). Elmsford, NY: Cognizant Communication.

Lück, M. (2010). Environmental Impacts of Polar Cruises. In M. Lück, P. T. Maher, and E. Stewart (eds), *Cruise Tourism in the Polar Regions: Promoting Environmental and Social Sustainability?* (pp. 109–131). London: Earthscan.

Lück, M. (2012). Exploring the Depths: Whale Watching Around the World (textbox). In E. C. M. Parsons (ed.), *An Introduction to Marine Mammal Biology and Conservation* (pp. 292–293). Burlington, MA: Jones & Bartlett Learning.

Martinez, E. (2008). Cousteau, Jacques-Yves. In M. Lück (ed.), *The Encyclopedia of Tourism and Recreation in Marine Environments* (pp. 120–121). Wallingford: CABI.

Miller, M. L. (2007). Remarks on Innovation in Marine Tourism Systems. In M. Lück (ed.), *Nautical Tourism: Concepts and Issues* (pp. 37–58). Elmsford, NY: Cognizant Communication.

Musa, G., and Dimmock, K. (2012). Scuba Diving Tourism: Introduction the Special Issue. *Tourism in Marine Environments*, *8*(1/2, Special Issue), 1–5.

O'Connor, S., Campbell, R., Cortez, H., and Knowles, T. (2009). *Whale Watching Worldwide: Tourism Numbers, Expenditures and Expanding Economic Benefits*. Yarmouth, MA: International Fund for Animal Welfare (IFAW).

Orams, M. (1995). A Conceptual Model of Tourist–Wildlife Interaction: The Case for Education as a Management Strategy. *Australian Geographer*, *27*(1), 39–51.

Orams, M. (1999). *Marine Tourism: Development, Impacts and Management*. London, New York: Routledge.

Orams, M., and Lück, M. (2012). Marine Systems and Tourism. In A. Holden and D. A. Fennell (eds), *The Routledge Handbook of Tourism and the Environment* (pp. 170–182). London: Routledge.

PADI. (2014). *Worldwide corporate statistics 2014 (data for 2008–2013)*. Retrieved 24 November 2014, from www.padi.com/scuba-diving/about-padi/statistics/.

Pernetta, J. (2004). *Guide to the Oceans*. Buffalo, NY: Firefly Books.

Sheppard, V. (2008). International Convention for the Prevention of Marine Pollution from Ships (MARPOL). In M. Lück (ed.), *The Encyclopedia of Tourism and Recreation in Marine Environments* (pp. 236–237). Wallingford: CAB International.

Tilt, W. C. (1987). From Whaling to Whalewatching. Symposium conducted at the meeting of the 52nd North American Wildlife and Natural Resources Conference.

Vogel, M., and Oschmann, C. (2012). The Demand for Ocean Cruises: Three Perspectives. In M. Vogel, A. Papathanassis, and B. Wolber (eds), *The Business and Management of Ocean Cruises* (pp. 3–18). Wallingford: CABI.

Wong, P. P. (2009). Policy and Planning Coastal Tourism in Southeast Asia. In R. Dowling and C. Pforr (eds.), *Coastal Tourism Development* (pp. 103–119). Elmsford, NY: Cognizant Communication.

27

MARITIME HERITAGE CONSERVATION

Juan-Luis Alegret and Eliseu Carbonell

Heritage and maritime heritage

Only fifty years ago the word "heritage" was not used as it is today. At that time, when talking about heritage, reference was mainly to monuments, or what is now called real property heritage, which may include buildings, monuments, and historical sites with a certain historical, artistic, or cultural value.

The protection of monuments, or heritage, began to be institutionalized internationally in the period between the two World Wars. In 1931 in Athens, the League of Nations organized the first International Congress of architects and technicians of historical monuments that gave rise to the famous Athens Charter, which was followed by the Venice Charter in 1964.

One decade later, when UNESCO adopted the Convention Concerning the Protection of the World Cultural and Natural Heritage in 1972, the concept of heritage began to be widespread throughout the world. But the expansion of this concept was not only geographical but also typological, expanding the original meaning of heritage as buildings, monuments, or sites to include the intangible or immaterial heritage. This immaterial or intangible heritage must be attached directly to cultural heritage, which until then had not been taken into account as a kind of heritage.

With the general expansion of the concept of heritage, multiple areas of expertise and application appeared, one of which was the maritime specification of heritage. Maritime heritage is a recent concept linked to the emergence of other types of heritage, such as industrial heritage.

An initial descriptive definition of maritime heritage could be the set of tangible or intangible items linked to human activities that relate to the marine environment and resources, whether developed in the past or in the present. This set of elements can also be called maritime culture as part of cultural heritage. The main feature of elements in this set we generically call maritime heritage is that they are recognized by social groups and are organized according to different geographical or cultural areas.

Maritime heritage in the international context: emergence and evolution

The process of creation and management of theoretical and practical maritime heritage followed different paths over time and at different institutional levels. It received very different social supports from different agents and groups. The three main manifestations of maritime heritage are those now known as underwater heritage, floating heritage and maritime cultural heritage. Today, the concept and practice of maritime heritage should be seen as an amalgam of these and other manifestations.

Many legal instruments and policies exist that, at international, regional, national or local levels, try to safeguard the heritage of marine or aquatic ecosystems, especially the maritime cultural, archeological or floating heritages. At the international level, the United Nations and the European Union have played an important role in the construction of a normative framework for the safeguarding and regulation of cultural heritage in general, and maritime heritage in particular.

The idea of protecting cultural heritage has evolved during the last two centuries. But it was only in the twentieth century that the international community created the legal basis for the safeguarding and protection of cultural heritage as a whole, or maritime cultural heritage as an example.

After the consequences of World War I, the Treaty for the Protection of Artistic and Scientific institutions and Historical Monuments known as the "Roerich Pact" was signed in New York in 1935. This Pact, still in force for ten states, can be considered the first conventional instrument, albeit one made at the regional level of the Americas, that addresses in a specific way the issue of agreed upon respect for and protection of the goods and cultural values and their defense in all situations of peace or war. The Pact established the foundations for the later regulatory standards proclaiming that the preservation of cultural heritage is of great importance for all peoples of the world and that it is important that this heritage should receive international protection.

The United Nations Education, Scientific and Cultural Organization (UNESCO) is the institution in charge of the legal protection of cultural heritage, through international cooperation among 195 Member States and eight Associate Members. Its mandate has been developed through many conventions, declarations, and recommendations, including the following:

- Convention for the Protection of Cultural Property in the Event of Armed Conflict (1954). First Protocol of 1954 and Second Protocol of 1999;
- Convention on the Means of Prohibiting and Preventing the Illicit Import, Export and Transfer of Cultural Property (1970);
- Convention Concerning the Protection of the Word Cultural and Natural Heritage (1972);
- Recommendation Concerning the Safeguarding and Contemporary Role of Historic Areas (1976);
- Convention on the Protection of the Underwater Cultural Heritage (2001);
- Convention for the Safeguard of the Intangible Cultural Heritage (2003).

Specific mention of the maritime dimension of cultural heritage only appeared in all these legal instruments in 2001 with the convention concerning underwater cultural heritage, which initiated the process of recognizing and protecting this form of heritage.

At a regional level, legal instruments adopted for the protection of cultural heritage include *The Convention for the Protection of Archeological, Historical and Artistic Heritage of the American Nations*, adopted by the Organization of American States in 1976 (OAS 1976) and the *European Convention on Offences Related to Cultural Property*, adopted by the Council of Europe in 1985.

From a more specific maritime perspective, in 1982 a set of legal and policy means were created. the Law of the Sea Convention (UNCLOS). The most comprehensive international legal and policy regime related to maritime affairs ever made, UNCLOS represents something like the "Constitution of the Oceans" (UN 1982).

Given that the core of this "great agreement" was centered on trade, fishing, commercial exploitation of natural resources, and environmental protection, UNCLOS considered maritime cultural heritage protection only in terms of underwater cultural heritage and only in two articles: Article 149:

> All objects of an archeological and historical nature found in the area shall be preserved or disposed of for benefit of mankind as a whole, particular regard being paid to the preferential rights of the State or country of origin, or the Sate of cultural origin, or the State of historical and archeological origin.

Article 303:

> 1   States have the duty to protect objects of an archeological and historical nature found at the sea and shall cooperate for this purpose.
> 2   In order to control traffic in such objects, the coastal State may . . . presume that their removal from the seabed in the zone. . . without its approval would result in an infringement within its territory or territorial sea of the laws and practices with respect to cultural exchanges.

In spite of the specific reference to the cultural heritage in those two articles, they refer only to the submerged maritime material cultural heritage. It would be two years later that the intangible or immaterial heritage was taken into account through the 2003 UNESCO *Convention for the Safeguarding of the Intangible Cultural Heritage* (UNESCO 2003).

Another way to protect maritime heritage was, at regional level, the European Convention on the Protection of the Archeological Heritage (Valetta Convention) concluded by the 1992 Council of Europe (EU 1992). Two articles of this Convention state that:

> Art. 1. "Archeological heritage shall include structures, constructions, groups of buildings, developed sites, moveable objects, monuments of other kinds as well as their context, whether situated on land or *under water*".

> Art. 2. ". . . the creation of archeological reserves, even where there are no visible remains on the ground or *under water*." (Emphasis added)

Those two articles indicate that underwater heritage was gaining recognition and it was beginning to be understood as part of the archeological heritage and as including the context in which it was found.

In 1996, the ICOMOS General Assembly ratified the Charter for the Protection and Management of the Underwater Cultural Heritage (ICOMOS 1996). This Charter, a supplement of the Charter of 1990, gives a definition of underwater cultural heritage not present in the previous Charter:

> Underwater cultural heritage is understood to mean the archeological heritage which is in, or has been removed from, an underwater environment. It includes submerged sites and structures, wreck sites, and wreckage and their archeological and natural context.

From this moment, underwater cultural heritage was understood in many countries as a key element in the process of strengthening national identity, as well as a non-renewable resource, and as fundamental to the promotion of recreation and tourism.

The Parliamentary Assembly of the Council of Europe recognized the interest regarding maritime and fluvial heritage in 2000, and recommended to the Committee of Ministries of the Council the need to generate and strengthen the measures to achieve the safeguarding of this heritage, through the creation of experts networks and the support of government entities, museums, and research centers.

This recommendation marked a significant change in the conceptualization of the concept of maritime cultural heritage that now includes both the material and immaterial heritage, not only of the seas but also of other aquatic contexts. The expanded concept of maritime cultural heritage:

> will recognize that the maritime and fluvial heritage comprises much more than submerged sites, be they fixed or movable. The maritime and fluvial heritage is not confined to that which existed in the past, was lost and can be recovered. It extends to artifacts which are neither submerged nor lost in any other way, but *which are in danger of being lost unless active steps are taken to preserve them.* These include defunct dock and harbor installations, coastal defenses such as estuary forts, lighthouses, dykes and tidal mills, fish traps and fishing stations, or vessels whose natural working life may be over but can be kept as nearly as possible in their original condition, or even in working order, for commercial gain, for private or public pleasure, or for educational or training purposes. The maritime and fluvial heritage extends to associated traditions, be they technical, such as techniques of boatbuilding or of handling vessels and their cargoes, or artistic, such as decorative features of ships or equipment, maritime lore or folk music, including, but not confined to, sea shanties.

Through this recommendation, the Parliamentary Assembly proposed the creation of a convention for the analysis and protection of maritime heritage broadly defined, in addition to the establishment of the necessary national and international legal and political means to achieve its goals. The proposition was addressed to UNESCO, which, in its General Conference of 2001, adopted what is considered the only legal tool related to underwater cultural heritage: *The Convention on the Protection of Underwater Cultural Heritage*, which is still in the ratification process (UNESCO 2001).

At the European regional level, maritime heritage was treated in a different way. In 2006, the European Union (EU) decided to launch a debate about a future Maritime Policy that treated the oceans and seas in a holistic way. To this end the EU circulated a work document named *Towards a Future Maritime Policy for the Union: A European Vision for the Oceans and Seas* (EU 2006), known as the Green Paper, that sought to strike the right balance between the economic, social, and environmental dimensions of sustainable development. The Green Paper addressed, among other issues, an essential aspect of European maritime life, namely, the cultural dimension of maritime heritage, which today needs to be reclaimed to reacquaint Europeans with their maritime past and to give them a sense of their present-day maritime identity.

In the Green Paper, Europe's maritime cultural heritage needs to be understood in a broader sense than in the UNESCO context, to include historic and industrial buildings in coastal cities, works of art inspired by the sea, the development of marine flora and fauna, seafood cookery, traditional fishing techniques and different types of boats and navigation techniques.

This cultural heritage of the past is considered now as an integral part of Europe's maritime identity. This means that it can help European citizens become more aware of the importance of the sea in their lives, and enable those involved with the sea themselves to better assess the value of their activity in a European Union turned towards the seas. This heritage is being used as a new political instrument, as the cement for building a common maritime identity for all Europeans. When all maritime stakeholders have a sense of common identity they will want to participate in common decision making and activity planning processes, making implementation of these processes more effective. To reclaim this heritage, the Green Paper recommends a multiplication of cultural links between different maritime sectors; for example, industrial sponsorship of museums or the organization of award ceremonies for best practice. It also proposes making European funds available to help coastal regions to preserve their heritage and attaching importance to educational activities focusing on this cultural heritage.

In response to the strong support shown by many stakeholders during the consultation process for the new policy to promote Europe's maritime culture and heritage, the Commission focused in 2007 on raising the visibility of Maritime Europe as one of the key objectives of a holistic, integrated maritime policy. The way to pursue this goal was not through maritime cultural heritage in the abstract, but through a number of concrete and simple proposals that included the decision to celebrate a European Maritime Day and the creation of a European Atlas of the Seas.

The political and ideological shift from the Green paper was evident. The use of maritime cultural heritage changed from an essentialist consideration to being the cement for building a common maritime identity for all Europeans as was expressed in the Green Paper; to be a much more practical instrument to raise the visibility of the maritime sectors and build upon best practice to support the further development of an integrated approach to maritime affairs as expressed in the document *An Integrated Maritime Policy for the European Union* of 2007.

The most recent appearance of maritime cultural heritage in the institutional development of the EU is through the European Maritime and Fisheries Fund (EMFF), created in 2011 with the aim of achieving the objectives of the reformed Common Fisheries Policy (CFP) and of Integrated Maritime Policy (IMP) through the support given to the implementation of local development strategies by, for example, promoting social well-being and cultural heritage, including maritime cultural heritage, in fishery areas.

At this point, one conclusion can be stated: the long journey to arrive at the definition and protection of maritime heritage as a whole was made across three convergent and simultaneous paths: the underwater heritage, the floating heritage and the maritime cultural heritage.

Underwater maritime heritage

United Nations Convention on the Law of the Sea (UNCLOS) of 1982 is an important reference text for underwater maritime heritage. Although it was drafted with a view to offering general provisions on the law of the sea, it includes two provisions that refer specifically to archaeological and historical objects. Such explicit reference not only confirms the specificity of these objects, differentiating them from "ordinary" objects, but the content of these provisions establishes an obligation for States to protect such objects.

Like UNCLOS, the 2001 UNESCO Convention represents an international regulation specific to underwater cultural heritage, although its contents differ from those of UNCLOS. As with any treaty, the Convention and this specific regulation are effective only among States Parties.

The general principles of the Convention can be summarized as follows:

1   Underwater cultural heritage means all traces of human existence having a cultural, historical or archaeological character that have been partially or totally under water, periodically or continuously, for at least 100 years.
2   The preservation *in situ* of underwater cultural heritage (i.e. in its current location on the seabed) is considered as the first option before allowing or engaging in any activities directed at this heritage. The preference given to *in situ* preservation as the first option stresses the importance of, and the respect for, the historical context of the cultural object and its scientific significance and recognizes that such heritage is, under normal circumstances, preserved underwater owing to the low deterioration rate and lack of oxygen and therefore not necessarily in danger per se.
3   States Parties shall preserve underwater cultural heritage for the benefit of humanity, and take action individually or jointly to achieve this end.

The 2001 Convention does not directly regulate the delicate issue of ownership of any given cultural property between the various States concerned (generally flag states and coastal states); it does, however, establish clear provisions for the States concerned and for international cooperation.

The principle that underwater cultural heritage shall not be commercially exploited for trade or speculation or irretrievably dispersed is not to be understood as preventing professional archaeology, or the deposition of heritage recovered in the course of a research project in conformity with the Convention or salvage activities or actions by finders as long as the requirements of the Convention are fulfilled.

Depending on the current location of the underwater cultural heritage, specific regimes for cooperation between coastal and flag states are applicable: States Parties have the exclusive right to regulate activities in their internal and archipelagic waters and their Territorial Sea; within their Contiguous Zone States Parties may regulate and authorize activities directed at underwater cultural heritage; and within the exclusive economic zone, or the continental shelf and within the Area (i.e., the waters outside national jurisdiction), a specific international cooperation regime encompassing notifications, consultations, and coordination in the implementation of protective measures is established.

The 2001 Convention focuses on the protection of underwater cultural heritage and ensures respect for all human remains located in maritime water and focuses on promoting training in underwater archaeology, the transfer of technologies and information sharing, and trying to raise public awareness concerning the value and significance of the underwater cultural heritage.

Floating heritage

The principles for this kind of heritage are basically two: the vessel as an object that needs to be protected, recovered and conserved; and the use of the vessel as individual or collective cultural reality that implies knowledge, skills, and behaviors developed by the users.

The Athens Charter of 1931 contributed to the development of an extensive international movement that has assumed a concrete form in the work of ICOM and UNESCO and in the establishment by the latter of the International Centre for the Study of the Preservation and the Restoration of Cultural Property. The Venice Charter was created in 1964 as a statement of principles for the conservation and restoration of monuments and sites. Both Charters focus on monuments and sites ashore, but maritime heritage is not covered by them despite its close affinity. Therefore the 4th European Maritime Heritage Congress in Barcelona in 2001, resolved to adapt the *Venice Charter* for maritime heritage in Europe, to be known as the *Barcelona Charter*.

As the Barcelona Charter states, the new concept of floating maritime heritage embraces the single traditional ship in which is found the evidence of a particular civilization or significant development as well as traditional sailing, seamanship, and maritime workmanship. This applies both to larger ships and to more modest craft of the past that have acquired cultural significance with the passing of time, that is, have become maritime heritage. The Charter also states that the preservation, restoration, and operation of traditional ships must have recourse to all the sciences, techniques, and facilities that can contribute to the study and safeguarding of the floating maritime heritage. But perhaps the more important aspect of the Barcelona Charter lies in its aim: "The intention in preserving and restoring traditional ships in operation is to safeguard them whether as works of art, as historical evidence or for perpetuating traditional skills."

Public Administrations of different European states, particularly those with an Atlantic tradition, have already found the necessary instruments to develop a wider and more responsible policy related to the preservation of floating maritime heritage. These policies, which include the essential legal instruments, permit the development of a safeguarding policy for tangible and intangible traces of maritime activities and, in turn, promote long-lasting craft activity and the birth of new cultural and tourist activities related to this heritage.

The endorsement of some European states in 2000 to the "Memorandum of Understanding on the mutual recognition of certificates for the safe operation of traditional ships in European waters and of certificates of competency for crews on traditional ships", can be seen as an important step towards the acknowledgement of traditional vessels as a important part of the Maritime Cultural Heritage and the promotion of their use as such.

Not only states but also individuals and groups have also mobilized in defense of this heritage and have constituted associations with the aim of recovering traditional maritime knowledge by means of restoration and preservation of historical and traditional vessels. Examples include the Association of European Maritime Heritage (2002), the Maritime Heritage Association of Australia, the Scottish Maritime Heritage Association and the magazine *Chasse-Marée* in France, Héritage Maritime Canada in Québec, among many others around the world.

The Barcelona Charter of 2002 established, for the first time, the criteria for the recovery and safeguarding of active traditional vessels seen now as maritime cultural heritage, and acts as a code of best practices and minimum standards for conservation and restoration of vessels to be recognized as traditional ships in operation. In fact, the most successful way to preserve the floating heritage, particularly its associated intangible heritage, as specified in the principles and spirit of the UNESCO Convention on Intangible Heritage of 2003, is to keep vessels in service, with all the problems that this proposal means.

Some of the best examples around the world of this work of recovering heritage are the Sail Training International (STI), created in 2003 in the UK, which attracts up to 4 million visitors annually to its events involving between seventy and 135 sailing vessels, or the Temps de Fête sur Duarnenez in France. Elsewhere there are hundreds of smaller events around the world that attract more modest crowds but bring pleasure to millions of people, whether as spectators or participants.

Another important and more recent example of work on floating heritage is the advisory group named International Historic & Traditional Vessels Panel. This group, put together in 2011 in Rotterdam with 17 specialists from twelve maritime organizations of Australia, Europe and North America, proposed discussions on how historic and traditional vessels can be operated in compliance with modern safety regulations without adverse impact on the very features that provide their historic appeal.

A conclusion we can draw is that the most successful way to preserve the floating heritage, particularly with respect to its intangible dimensions, is to keep vessels in operation. In this sense, it is very important to promote recovery and preservation measures for floating heritage in order to avoid its loss due to scrapping or loss of its ability to sail.

Maritime cultural heritage

Underwater and floating heritages, two specific forms of maritime heritage, were complemented recently with a new perspective that already existed but previously played another role: maritime cultural heritage.

The term cultural heritage has changed content considerably in recent decades, partially owing to the instruments developed by UNESCO. Cultural heritage does not end at monuments and collections of objects, it also includes traditions or living expressions inherited from our ancestors and passed on to our descendants, such as oral traditions, performing arts, social practices, rituals, festive events, knowledge and practices concerning nature and the universe and the knowledge and skills to produce and use traditional crafts.

While fragile, intangible cultural heritage is an important factor in maintaining cultural diversity in the face of growing globalization, an understanding of the intangible cultural heritage of different communities helps with intercultural dialogue, and encourages mutual respect for other ways of life.

The importance of intangible cultural heritage is not the cultural manifestation itself but rather the wealth of knowledge and skills that is transmitted through it from one generation to the next. The social and economic value of this transmission of knowledge is relevant for minority groups and for mainstream social groups within a State, and is as important for developing States as for developed ones.

The starting point is 1972 when the Convention for the Protection of World Cultural and Natural Heritage was adopted by UNESCO (1972). In 1989 the Recommendation on the Safeguarding of Traditional Culture and Folklore defines the cultural heritage henceforth called "immaterial" as:

> All the creations of a cultural community founded on tradition, expressed by a group or individuals and recognized as reflecting the expectations of the community in terms of expression of their cultural identity and social norms and values. Its forms are, among others, language, literature, music, dance, games, mythology, rituals, customs, handicrafts, architecture and other arts.

When we apply the concept of cultural heritage to the maritime context, we find many different perspectives but a common view links to the inseparable material–immaterial distinction.

Maritime cultural heritage can be understood as all those cultural material goods (in water and on land) and cultural immaterial goods (such as representations, perceptions, discourses, practices, material culture, customs, traditions, imageries, cultural landscapes) that are expressions of the maritime culture, of the maritimity, of the maritime differential fact and of the relation among man and the sea, whether possessing a cultural, emotional, or use value.

The maritime cultural heritage extends to associated traditions, be they technical, such as the techniques of boatbuilding or of handling vessels and their cargoes, or artistic, such as the decorative features of ships or equipment, maritime lore or folk music, including, but not confined

Table 27.1 Material maritime heritage and intangible maritime heritage examples

Material maritime heritage	Intangible maritime heritage
Coastal settlements sites (on land or submerged): urban, sea shell mounds, country properties, and their material culture	Languages, oral expressions and names of places
Coastal infrastructures (on land or submerged) with a specific history or acting actively in the maritime landscape, for example, ports, shipbuilding sites, arsenals, coastal fortifications, customs houses, pilot stations, tidal mills, specific industrial facilities, bond stores and other port facilities and navigational aids including lighthouses and maritime signals, ovens, docks, dykes	Techniques and practices preserved on the school vessels and in the national navies
Ritual or Ceremonial Coastal Sites and their material culture	Elaboration of objects, instruments, wardrobes, constructions and corporal ornamentations
Crafts (on land or submerged), comprising all kinds of ships, jetties or wharves of any type or size, and their material culture. Preserved or rebuilt vessels symbolizing a type of techniques, or a kind of disappeared vessel	Knowledge and practices on nature and the universe: related with activities such as fishing and elaboration of objects of the material culture and of ornament
Emblematic animals of the marine habitat such as whales or vegetables exploited at the littoral such as seaweed, or salt marsh	Culinary or gastronomic knowledge and practice, on board or ashore
Maritime and coastal visual representations of the activities related with the sea, of the elements with which these activities are carried out and of the territorial delimitation with regard to the sea. These can be represented in any form, including petroglyph, drawing, stamp, painting, sculpture, photograph, film, video, or infography	The diverse social actors and representations regarding the maritime environments and the social, economic, and political aspects related to them
Historical documents: all written testimony, primary or secondary sources (including literary ones) on the seas and maritime environments, practices, history, discourses or imageries	Dance and musical expressions related to the sea
Material culture of the daily life (on land or submerged): all type of objects that have made part of the daily discursive practices of coastal communities	Ritual, scenic and ceremonial expressions, festivals, games, and sport
	Traditional forms of social, legal, and political organization
	Traditional medicine related to the sea

to, sea shanties, the underwater cultural heritage, paleontological remains and historic landscapes, among others.

The presentation of maritime heritage: selected examples

The experiences of recovery and presentation of maritime heritage are almost inexhaustible. In the world there is a huge number of associations at international, national and local levels working for this purpose; for example, to name just two, the European Maritime Heritage, a non-governmental organization for private owners of traditional ships, as well as for maritime museums, and the Council of American Maritime Museums, an organization dedicated to preserving North America's maritime history, with a network of over 80 maritime museums, institutions, and individuals across North America and beyond.

Nowadays the importance of maritime heritage is evident in almost all the maritime nations in the world, as can be seen across the considerable number of maritime museums spread along the coasts of the five continents. Certainly, the number of maritime museums is relative to the total amount of museums in each country. But in the sphere of museums of History and Ethnology, maritime museums are one of the most numerous. The most visited maritime museums in the world are the National Maritime Museum in London (UK), the Vasa Museum in Stockholm (Sweden) and the USS *Arizona* Memorial at Pearl Harbor in Honolulu (USA), each with over a million visitors per year.

Maritime museums can be organized according to their status: first, those supported by the national Government either directly or through Departments such as the Ministries of Defense; second, those supported by regional or local government; finally, those supported by charitable trusts, volunteer associations, public or private companies, and not-for-profit educational institutions.

In the first group we can highlight the following museums: the Australian National Maritime Museum, established in 1991 and located in Darling Harbour, Sydney; the Het Scheepvaart-museum in Amsterdam (the Netherlands), renovated in 2011, with the eighteenth century ship replica *Amsterdam* moored outside the museum; the Viking Ship Museum in Roskilde (Denmark) built in 1969, which deals with seafaring and shipbuilding in the medieval period; the Istanbul Naval Museum (Turkey), established in 1897 with an important collection of artifacts pertaining to the Ottoman Navy; the Museum of the History of Riga and Navigation (Latvia), originated in 1773, one of the oldest museums in the world; as well as the National Maritime Museum of London (UK), established in 1937, which is not only one of the most visited maritime museums but also probably the largest museum of its kind in the world.

In the second group we can highlight the following museum: the Museu Marítim of Barcelona (Spain), situated in a shipyard that dates back to the thirteenth century; the Galata Museo del Mare (Italy), belonging to the Municipality of Genoa, which includes, along with the Galata, three other major museums, the Museo Navale, the Commerda di Prè and the Castello d'Albertis; the Kobe Maritime Museum (Japan), built to look like a sailing ship, which was opened in 1987 to commemorate the 120th anniversary of the opening of Kobe Port; the Museu Nacional do Mar in São Francisco do Sul (Brasil), created in 1993 by the State of Santa Catalina to host the large diversity of Brazilian vessels; and, finally, the Port-musée of Douarnenez (France) and the Fishing Museum of Palamós (Spain), which will be discussed later.

In the third group we can highlight the following museums: the South African Naval Museum, situated in Simon's Town (South Africa) and a project by the South African Naval Heritage Trust, whose main exhibit is a submarine, the SAS *Assegaai*; the Nao Victoria Museum, located in Punta Arenas (Chile) and opened in 2011 by a private initiative, which aims to show replicas of ships that have contributed to the discovery of this region; the Voyager New Zealand Maritime Museum in Auckland, run by the NZ National Maritime Trust Board, which houses exhibitions spanning New Zealand's maritime history from the first Polynesian explorers to modern-day triumphs at the America's Cup; and, finally, the Museum of America and the Sea, established in 1929 in Mystic, Connecticut (USA), which is a private, not-for-profit educational institution notable both for its collection of sailing ships and for the re-creation of the crafts and daily life of an entire nineteenth century seafaring village, consisting of more than 60 original historic buildings that were moved to the site.

But we should not forget that the recovery, preservation, and presentation of maritime heritage are tasks performed by countless cultural organizations, universities, researchers, big or small museums, and individuals working with enthusiasm in this field. It would be impossible to give an overview of the whole universe of the preservation of maritime heritage in the world, but in this last section we will mention only some significant examples.

The San Juan, a Basque ship sunk in Labrador waters, Canada

In the summer of 1565 on the Labrador coast a Basque ship loaded with melted whale fat sank when it was ready to sail back to Europe. Its discovery and excavation in the cold waters of the Red Bay have become a symbol of the underwater heritage, to the extent that the International Scientific Committee for Underwater Archaeology (UNESCO) decided to adopt as a symbol the silhouette of the *San Juan*.

In the second half of the sixteenth century Red Bay was a port that housed a large population of Basque whalers who settled there during the hunting season. It is estimated that there could be up to nine or ten whaling ships like the *San Juan* moored in the Red Bay at that time. Whale oil was a rare and highly prized commodity. It burned brighter than the more common vegetable oils, and was used in the manufacture of soap, treatment of fabrics and in pharmaceutical products.

The *San Juan* had been built in the port of Pasajes (Guipúzcoa, Spain). It was anchored in the Red Bay when it sank for unknown reasons. The excavations didn't uncover any human remains so probably the entire crew escaped the sinking. Selma Barkham's research located the wreck in 1978, when excavations began that would last until 1985. The research uncovered some very interesting things: it was a vessel of 25 meters long and 7.5 wide, with a crew of 60 people. At the time of sinking, it transported between 800 and 1000 barrels of whale fat, corresponding to the fat from six to nine whales, valued of about $9 million in current value. The results of the excavation uncovered a puzzle of over 3000 pieces of wood, took over 22,000 photographs, and recovered thousands of objects and fragments of objects that give much information about daily life aboard ship, from sailors' shoes to ceramic vessels, some nearly intact. Among other important discoveries there were navigational instruments: an hourglass, a compass and an astrolabe.

Excavation of the *San Juan* has transformed the Red Bay and the Basque ship into a point of reference for all researchers dedicated to underwater archaeology, as well as scholars working in the fields of the history of navigation and of ship construction, as well as the study of the life of ancient communities of seafarers who spent much time living hundreds of miles from their homes.

The recovery of the Melanesian maritime heritage in Vanuatu

The Vanuatu Cultural Centre in the Republic of Vanuatu, Melanesia, organizes an annual training workshop on ethnographic methods for field workers, with the aim of training young people to carry out recovery work to document knowledge and oral traditions in their communities. The collection conducted by these researchers becomes part of the documentation of the Vanuatu Cultural Centre and is embodied in publications, seminars, and exhibitions of the National Museum of Vanuatu.

In the Vanuatu archipelago there is a wide diversity of styles of canoes. Three different types of sails from before the arrival of Europeans have been documented. The canoes were the key element of trade and social relations between the islands. But most of the boats disappeared in the mid-nineteenth century. Sail canoes became early victims of the sandalwood traders and even the blackbirders, some of whom would ram and destroy canoes to steal people to be traded. A couple of generations later, these impacts, coupled with the declining population of the islands, led to the decline of many cultural elements, including their maritime heritage.

The Vanuatu Cultural Centre's work has allowed the recovery of traditional knowledge on shipbuilding, local winds, tides and currents, and navigation skills. According to the island's newspaper, the *Vanuatu Daily Post* (06/08/2011), the island of Aneityum was the first in the 1990s to revitalize their navigation techniques and to produce a traditional vegetable sail from pandanus tree that had not been produced for generations. Currently, sail canoes are used for fishing and small cruise ships. The Vanuatu Cultural Centre also organizes workshops for schoolchildren and workshops with elders to recuperate knowledge about weather reading skills and security in the open sea sailing with canoes. Another remarkable initiative developed by the Vanuatu Cultural Centre is the "Traditional Marine Tenure Project" aimed at the recovery and documentation of knowledge from the elderly about ecology and the various methods employed to harvest and manage marine resources and their associated custom stories.

The floating heritage in Duarnenez, France

Douarnenez, a small town of about 15,000 inhabitants on the coast of Finistère (Brittany, France), contains the largest collection of floating heritage of France. This collection is hosted by the Port-Museum of Douarnenez, which opened in 1993 and is based on the old Musée du bateau, which opened in 1985 at the initiative of a local cultural association established in 1979. More than thirty years of work for the recovery of the maritime heritage is represented by this major museum, which is divided into two sections: the floating museum and the onshore museum.

The onshore museum is hosted in a former factory where sardines were salted and canned. It has a large permanent collection of old boats, both pleasure boats and fishing boats, and other related objects. In addition, the Port-Museum of Douarnenez presents various temporary exhibitions on themes related to navigation and the sea.

The floating museum, open from April to October, is the main floating museum in France. Currently there are five ships moored at the estuary of Port-rhu, near the island of Tristan, open to visitors: the *Dieu Protège*, a Breton vessel from 1951; the *Notre-Dame de Rocamadour*, a Breton lobster boat from 1959; the *Northdown*, an English sailing barge from the Thames from 1924; the *Saint-Denis*, a British steam tug from 1929; and the *Notre-Dame des Vocations*, a Breton tuna fishing boat from 1962. The museum has other boats waiting for restoration before being included among those open for visits by the public.

Douarnenez is also a referent in the field of maritime heritage because of the publication, beginning in 1981, of the magazine *Chasse-Marée*, which, over time, has become a symbol of the defence and protection of maritime heritage in France and elsewhere. Since 1986, Douarnenez has also become well known for the organization of the meeting of traditional boats, which has become a major maritime festival in the world. Given the success of this meeting, it has been organized since 1992 jointly with the neighbouring port of Brest.

"The Fish Space" (Espai del Peix) of Palamós, Spain

More than twenty years ago, Palamós, a small town of about 18,000 inhabitants on the Spanish Catalan Mediterranean coast, began working on its connection with the sea from a cultural heritage perspective. The first cultural facility was created in 2001, the Fisheries Museum of Palamós. It is the only one specifically devoted to fisheries in the entire Mediterranean. Then the Documentation Centre of the Sea and Fisheries (DOCUMARE) and the Chair of Marine Studies (CEM), attached to the University of Girona, were created with the mission of promoting and disseminating maritime knowledge and collaborating and supporting institutions such as the traditional fishermen associations. Other highly successful cultural activities on maritime heritage carried out by the Museum are the "Tavern Talks" (*Converses de taverna*) and "Pictures that Generate Talk" (*Imatges que fan parlar*). Both activities are based on the participation of the townspeople, as well as visitors, related to sea activities and trades.

The most recent, and most versatile and innovative, maritime heritage facility in Palamós is "The Fish Space" (*Espai del Peix*), designed to complement the other facilities, shows all the maritime heritage linked to the capture, marketing, preparation, and consumption of local fish. It can be considered a neo-museum. The exhibition begins with a view of the fish market with the display of the live auction, which serves as a preamble to the guided tour explaining the activities of the fishing fleet, the type of species caught, the physiological and nutritional characteristics of these species, the different culinary traditions, and the different gastronomic traditions related to these species. Visits always culminate in a tasting of some of these species prepared by a fisherman–chef using traditional recipes.

Despite the fact that Palamós has a long culinary and gastronomic tradition linked to seafood, especially the red shrimp (*aristeus antennatus*), it was thought desirable for the promotion of local fishing activities not to focus on all species, particularly not those already well known. From the perspective of local development and social economy, it was thought that the cultural activities of the Fish Space should focus specifically on the promotion of those species that had little commercial value and were less known generally. Although these species are well known to all fishermen and their families, most people do not even know their names, how they taste or how to prepare them. In fact, disseminating knowledge about the maritime heritage is the main task of the Fish Space of Palamós.

The Zuiderzee Museum in Enkhuizen, the Netherlands

One of the greatest examples in the world of conservation of Maritime Heritage is without doubt the Zuiderzee Museum located in Enkhuizen, the Netherlands. This museum is divided into two parts: an outdoor museum and a traditional one. The open-air section covers 15 acres and accommodates authentic buildings from the region, such as a church, a smokehouse for curing fish, a mill, a cheese warehouse, shops and dwelling houses moved from the surrounding fishing villages. The indoor museum includes a magnificent collection of wooden ships from the Netherlands and several rooms devoted to temporary exhibitions on traditional craftsmanship, local history, photography, etc.

At the indoor museum, the permanent exhibition in the Ship's Hall shows the largest collection of ships of all Dutch museums, for example an *'ijsvlet'* (ice boat), a transport ship or a Frisian hunting boat, as well as an audio–visual installation that shows the life of the Zuiderzee fishermen. The outdoor museum offers to the visitors the possibility of experiencing the life of the old inhabitants in the region, such as visiting the interior of real fishermen houses and comparing it with the houses of the upper classes. Staff and volunteers demonstrate historical crafts from everyday life at the beginning of the previous century, such as the smoking of fish, basket weaving or roping making. The outdoor museum offers a lot of activities for children, as for example painting clogs or making a clog boat as did the children who lived in the fishing villages. Children can also dress in traditional clothes and participate in many activities, for example, via an interactive game and film program in which they can take the identity of a curator and gain insight into the way in which presentations and layouts are realized in the Museum.

The origin of the Zuiderzee Museum dates back to the decade of the 1930s, when the Zuiderzee was cut off from the North Sea, becoming an inland sea, and people were concerned that the culture of the former Zuiderzee region would disappear completely. To preserve the local maritime heritage the indoor museum was inaugurated in 1948. At that moment an "Association of Friends of the Zuiderzee Museum" was established and promoted the creation of the open-air section of the Zuiderzee Museum. To this purpose, 130 historical buildings of the region were donated by several municipalities, transported and allocated in a peninsula created by spraying up sand from the seabed on the outside of the seawall separating Enkhuizen from the water on its east side. In order to produce a picture that was as lifelike as possible, some replicas were produced; for instance, a lifeboat house to accommodate a beautiful model of a lifeboat with all the old equipment.

References

Association of European Maritime Heritage (2000) *Memorandum of Understanding on the Mutual Recognition of Certificates for the Safe Operation of Traditional Ships in European Waters and of Certificates of Competency for Crews on Traditional Ships.* Available at www.european-maritime-heritage.org/docs/sc/NewMOU South.pdf (last accessed September 29, 2015).

Basque Whaling Boat. Available at www.pc.gc.ca/lhn-nhs/nl/redbay/natcul/basque.aspx (last accessed September 29, 2015).

Espai del Peix. Palamós. Available at www.espaidelpeix.org/index.php/en.html (last accessed September 29, 2015).

EU (1992) *European Convention on the Protection of the Archaeological Heritage (Revised-Valetta Convention).* Available at www.conventions.coe.int/Treaty/en/Treaties/Html/143.htm (last accessed September 29, 2015).

EU (2006) *Towards a Future Maritime Policy for the Union: a European Vision for the Oceans and Seas* [COM (2006) 275 final]. Available at www.europa.eu/legislation_summaries/maritime_affairs_and_fisheries/maritime_affairs/l66029_en.htm (last accessed September 29, 2015).

ICOMOS (1996) *International Charter on the Protection and Management of Underwater Cultural Heritage Sofia.* Available at www.unesco.org/csi/pub/source/alex7.htm#A%20n%20n%20e%20x (last accessed September 29, 2015).

OAS (1976) *The Convention for the Protection of Archeological, Historical and Artistic Heritage of the American Nations.* Available at www.oas.org/juridico/english/treaties/c-16.html (last accessed September 29, 2015).

Port-musée Duarnenez. Available at www.port-musee.org/ (last accessed September 29, 2015).

UN (1982) *United Nations Convention on the Law of the Sea (UNCLOS).* Available at www.un.org/Depts/los/index.htm (last accessed September 29, 2015).

UNESCO (1972) *Convention Concerning the Protection of the World Cultural and Natural Heritage.* Available at www.whc.unesco.org/en/conventiontext/ (last accessed September 29, 2015).

UNESCO (2001) *Convention on the Protection of the Underwater Cultural Heritage.* Available at www.portal.unesco.org/en/ev.php-URL_ID=13520&URL_DO=DO_TOPIC&URL_SECTION=201.html (last accessed September 29, 2015).

UNESCO (2003) *Convention for the Safeguarding of the Intangible Cultural Heritage.* Available at www.portal.unesco.org/en/ev.php-URL_ID=17716&URL_DO=DO_TOPIC&URL_SECTION=201.html (last accessed September 29, 2015).

Vanuatu National Museum. Available at www.vanuatu.travel/index.php/en/australian-agents1/133-things-to-do/vanuatu-cultural-center/424-vanuatu-cultural-center-and-national-museum (last accessed September 29, 2015).

Zuiderzee Museum. Available at www.zuiderzeemuseum.nl/ (last accessed September 29, 2015).

PART 3

The geography of the sea
Spatial organisation

28

STATE MARITIME BOUNDARIES

Chris M. Carleton

The requirement for maritime boundaries

Ever since mankind formed states there was a potential for the state to extend seaward if a coastline existed. However, in the early civilizations the sea was regarded as free to anybody who wished to use it. This concept began to change following the 1494 Bull of Pope Alexander VI. He declared that all the oceans to the east of a meridian of longitude drawn through Brazil were Portuguese, while those to the west remained Spanish. This attempt at controlling the seas by two early colonial powers was challenged in a treatise by Grotius entitled *Mare liberum*. Grotius stated that there was a right of the freedom of the seas to enable a right of trade and was published as a direct challenge to the Portuguese claim to the eastern seas, an area that the Netherlands had begun to colonise. The subsequent debate between the freedoms of the ocean and the right of a coastal state to claim jurisdiction over sea areas adjacent to its coast is well documented by Professor O'Connell (O'Connell, 1982). The freedom of navigation has been broadly sustained beyond a coastal state's territorial sea, and even within the territorial sea a right of innocent passage has been maintained for all vessels in international law. The requirement for maritime boundaries between states started very modestly. Having established that a state had a right to claim a band of sea off its coastline by the end of the nineteenth century, customary international law developed a consensus that this band of sea should generally be 3 nautical miles (nm) by the first quarter of the twentieth century. This in turn established a requirement for adjacent coastal states to extend their land boundaries out to this 3 nm limit. This modest extension was often achieved by simply extending the final leg of the land boundary out to 3 nautical miles beyond the low-water line in the same direction as the last leg of the land boundary. This was achieved largely by bilateral agreement between states.

The first maritime boundary to go before an international tribunal for settlement was between Norway and Sweden in 1919. Known as the Grisbadarna Case (Scott, 1916) it concerned fishing rights beyond the boundary already established through the Ide fjord out to the limit of the claimed 4 nm territorial sea of the two states. A *Compromis* was signed on 14 March 1908 to submit the case to an ad hoc arbitral tribunal to delimit the boundary seaward from the point established by a joint commission that had been charged with determining the boundary in accordance with the 1661 Treaty of Copenhagen (Scott, 1916). The result was surprisingly forward-thinking and considered by many to be well ahead of its time; the outcome was an

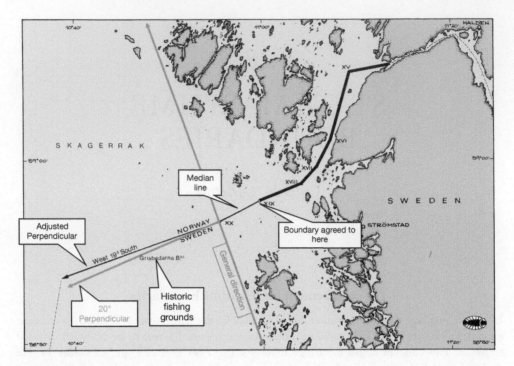

Figure 28.1 The Grisbadarna judgement
Source: Adapted from Francalanci *et al.*, 1994.

adjusted line perpendicular to the general direction of the relevant coasts, as illustrated in Figure 28.1.

This case was considered before either codified law or even state practice was available to the court to help it with its deliberations. Many experts, both legal and technical, have considered this decision a sensible compromise between the two claims. Judge David Anderson published an account of this case, concluding that 'the perpendicular to the general direction of the coast should be retained as a permissible method of delimiting the territorial sea and other maritime areas between adjacent coastal States, for use in appropriate places such as Grisbadarna' (Anderson, 1996).

This case was unusual in that it dealt with a boundary delimitation problem in a way that can be readily recognised today, producing an equitable result with a robust technical input. The way maritime boundaries should be delimited was not really considered in depth until the International Law Commission included this topic in its considerations on the codification of the regime of the high seas and the territorial sea between 1949 and 1956, prior to the Geneva Conference. It was clear that states were beginning to recognise that the oceans off their coasts contained valuable resources that required some certainty of ownership. This was initially the living resources of the oceans but increasingly during the middle of the twentieth century the non-living resources of oil and gas under the continental shelf. The true concept of the right of a coastal state to the seabed and subsoil to seaward of its land territory beyond the limits of the territorial sea was first formulated by the US in the Truman Proclamation of 28 September 1945 (USA, 1945). However, the first maritime boundary to be delimited beyond a claimed territorial sea limit was agreed between the UK and Venezuela in 1942 (Charney and Alexander,

1991a). This relied upon the notion of acquisition by occupation of the seabed; it thus differed considerably from the 'right' proclaimed by Truman. However, the concept of the continental shelf was established and was to be further developed by the International Law Commission in its deliberations leading up to the Geneva Conventions of 1958. It was these four Conventions, opened for signature in 1958 that codified international maritime law for the first time. The Continental Shelf Convention enabled those states that had ratified it the possibility of exploiting the non-living resources on and under the seabed of the continental shelf with a certainty of the right. What was then needed was a certainty of ownership and that meant a requirement for delimitation of the continental shelf beyond the territorial sea.

The development of the legal and technical issues concerning delimitation

The first legal guidance concerning maritime delimitation was to be found in Article 12 of the 1958 Geneva Convention on the Territorial Sea and Contiguous Zone. This article stated that the territorial sea boundary should generally be the median line every point of which is equidistant from the nearest points on the baselines from which the territorial sea is measured. However, this provision did not apply if historic title or other special circumstances required the states to delimit this boundary in another way.

Similarly, in Article 6 of the 1958 Geneva Convention on the Continental Shelf mention is also made of the median or equidistance line; however, there is a requirement for coastal states to determine their continental shelf boundary by agreement and the median/equidistance line is only used in the absence of such an agreement.

The development of maritime boundaries following the 1958 Conventions was generally a requirement of the offshore oil and gas industry, which required certainty on the issue of the sovereignty of the resource being exploited. For the first decade following the coming into force of the continental shelf Convention, these boundaries were determined graphically and were generally based on equidistance. This was the only means by which a maritime boundary could be delimited prior to the advent of the computer.

The first UK continental shelf boundary to be delimited at this time was with Norway (Treaty Series No. 71, 1965). The oil industry required a swift result so that highly prospective areas close to the probable boundary could be exploited. The bilateral negotiations only took about 12 months, a quite remarkable achievement for several reasons. Not only was it the first continental shelf boundary to be delimited in the North Sea, but also both parties were fully aware of the potential wealth that was being divided up. What the parties did not know at this time was the bonanza that would be realised during the next decade. The agreement was important in that it became the model for subsequent UK boundaries in the North Sea. Equidistance was still the prime principle for delimitation at this time and this agreement was no different. It was agreed at a very early stage that the northern limit of the boundary would terminate at or near the 100 fathom isobath, and the Norwegian Trench would be ignored. This latter point caused a few eyebrows to be raised in the UK, but was based on the principles of the 1958 Geneva Convention on the Continental Shelf, Article 1, which states:

> For the purpose of these articles, the term 'Continental Shelf' is used as referring (a) to the seabed of the submarine areas adjacent to the coast but outside the area of the territorial sea, to a depth of 200 metres or, beyond that limit, to where the depth of the superjacent waters admits of the exploration of the natural resources of the said areas; (b) to the seabed and subsoil of similar areas adjacent to the coasts of islands.

The British delegation at Geneva had also publicly accepted a scientific assessment by UNESCO to the effect that this depression in the bed of the North Sea formed part of the continental shelf for legal purposes on account of the existence of a sill at its northern end. This view was shared and acted upon by Norway, Sweden and Denmark. Indeed, since boundary agreements have been agreed in the area ignoring the Trench, Norway has successfully exploited the region.

Having decided that an equidistance line produced a satisfactory division of the continental shelf, a number of technical decisions had to be made, including the use of a common datum and the definition of the lines joining the turning points of the boundary. The only common geodetic datum available for use in the North Sea was European Datum (First Adjustment 1950) (ED 50), and this was the datum used to define the geographic coordinates of the boundary. The median line was determined using graphical methods drawn on specially constructed charts using an equal area projection. This method gave an accuracy of ±100 feet and was the best that could be achieved at this time. The arc of a great circle was the line used to define the boundary between the turning points. All these factors were subsequently used in the later agreements in this area, until the advent of computers in the early 1970s enabled median/equidistance lines to be calculated with greater precision.

Other similar boundaries in different parts of the world were not so technically robust. Indeed, the Bahrain/Saudi Arabian boundary that entered into force on 26 February 1958 (Charney and Alexander, 1991b) does not even list the turning points of the boundary because it was considered that the charts covering this area at the time were not accurate enough to accurately delimit these points. Other continental shelf boundary agreements at this time did use charts to both determine their boundary and officially illustrate the line as part of the treaty document. An example of this approach is contained within the continental shelf boundary agreement between Iran and Qatar, which was signed on 20 September 1969 and entered into force on 10 May 1970 (Charney and Alexander, 1991c). Admiralty chart No. 2837 forms an integral part of the agreement. The difficulty today is that the chart concerned is at a relatively small scale, and no geodetic datum is specified for the chart. In fact, the chart, which covers a fairly large area of the Gulf, is probably made up of several unspecified datums. It is therefore impossible to define this boundary in modern World Geodetic Datum terms. The use of a chart to define a boundary, as opposed to simply illustrating the line, is not recommended; however, in the 1960s and early 1970s there was sometimes no other option. Another example of a bilateral maritime boundary agreement between the adjacent states of the United States and Mexico was signed on 23 November 1970 and entered into force on 18 April 1972. The agreement in the Pacific is illustrated in Figure 28.2.

The first legal challenge to the equidistance/median line principle was by the then Federal Republic of Germany in 1967 against first the Netherlands and then in the following year Denmark at the International Court of Justice (ICJ) in The Hague. These two cases were joined into one case by the Court and became the infamous North Sea Cases. This was the first maritime delimitation case brought before the ICJ and although the Parties did not require the Court to actually delimit the two boundaries, rather to deliver guidance as to how the two boundaries should be delimited by the Parties, the judgment was eagerly awaited. The judgment (ICJ 1969), handed down on 20 February 1969, had a profound effect on the way continental shelf boundaries were to be delimited in the future. The ICJ rejected the contention of Denmark and the Netherlands to the effect that the delimitations in question had to be carried out in accordance with the principle of equidistance as defined in Article 6 of the 1958 Geneva Convention on the Continental Shelf, holding that Germany, which had not ratified the Convention, was not legally bound by the provisions of Article 6; and that the equidistance principle was not a necessary consequence of the general concept of continental shelf rights, and was not a rule of customary

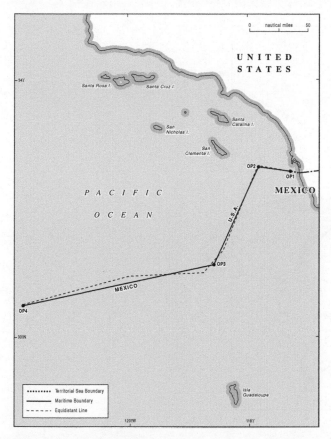

Figure 28.2 United States–Mexico Pacific agreement of 1970
Source: Author.

international law. The ICJ also rejected the contentions of Germany in so far as these sought acceptance of the principle of an apportionment of the continental shelf into just and equitable shares. It held that each party had a right to those areas of the continental shelf that constituted the natural prolongation of its land territory into and under the sea. It was not a question of apportioning or sharing out these areas, but of delimiting them. The ICJ found that the boundary lines in question were to be drawn by agreement between the parties and in accordance with equitable principles, and it indicated certain factors to be taken into consideration for that purpose. The parties agreed to negotiate on this basis. The resulting negotiation between the Parties is illustrated in Figure 28.3.

The strange alterations in the direction of the two boundaries is explained by the fact that oil and gas exploitation was already taking place and there was no desire to place operating installations into different state control.

While this case was proceeding, the UK, a major player in maritime boundary delimitation at this time, was reluctant to continue boundary negotiations in the North Sea. However, once the judgment had been made, a very busy period of boundary making ensued. Germany entered into bilateral negotiations with both Denmark and the Netherlands, as required by the ICJ judgment. At the same time, all three countries held talks with the UK. The outcome was five boundary agreements. Denmark/Germany and Germany/Netherlands agreed in accordance with

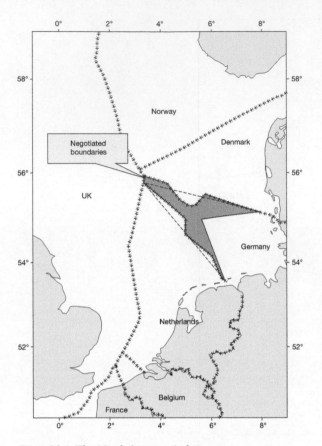

Figure 28.3 The North Sea cases judgement
Source: IBRU Centre for Borders Research, Durham University (Carleton and Schofield, 2002).

the guidelines laid down by the ICJ, and three boundaries with the UK were subsequently delimited: continental shelf boundaries with Germany (Treaty Series No. 7, 1973) and Denmark (Treaty Series No. 6, 1973), and a Protocol with the Netherlands (Treaty Series No. 130, 1972) altering the boundary agreed in 1965 by amending the northern point of the boundary to coincide with the new tripoint with Germany. All these boundaries were ratified in 1972. Despite the ICJ judgment, all three UK boundaries were median lines drawn in a similar way to the earlier boundary agreed with Norway. However, the ICJ judgment definitely sowed a seed of doubt concerning the primacy of a median-line solution for continental shelf boundaries. Two major principles had emerged from this judgment: that of equitable principles; and the fact that a median-line solution was not considered to be a rule of customary international law, even though most of the continental shelf boundaries delimited at that time had been based on equidistance.

During the 1970s the UK continued to hold talks with Norway to extend the boundary agreed in 1965 northwards to the tripoint with the Faeroe Islands. North Sea hydrocarbon exploration and exploitation continued apace and an extension of the boundary was considered essential by the industry. Talks were also continuing with the Republic of Ireland concerning continental shelf boundaries in the Irish Sea and to the west of Scotland.

The next legal challenge concerning maritime delimitation concerned the boundary between the United Kingdom and France. The talks with France did not progress well and both sides

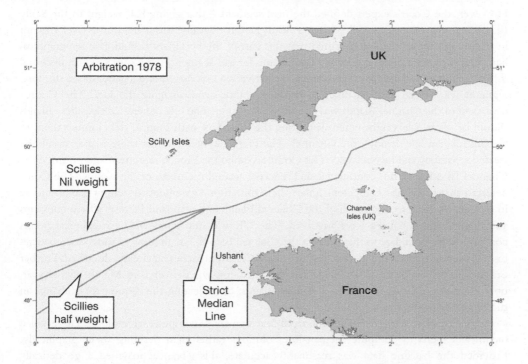

Scillies Nil weight

Scilly Isles

Scillies half weight

Arbitration 1978

UK

Channel Isles (UK)

Ushant

Strict Median Line

France

Figure 28.4 The France/UK Channel case. Ushant is the English name for Oussant
Source: Author.

agreed that the delimitation question should be submitted to an ad hoc Court of Arbitration. Agreement was reached on this procedure in July 1975. The proceedings consisted of three sets of written pleadings, followed by oral hearings in Geneva in January, February and May 1977, all of which required a considerable amount of work. The Court's decision was delivered in July 1977 (Miscellaneous No. 15, 1978). The Court delimited the continental shelf in the Channel by a median line, giving full effect for all islands, including the Eddystone, which the French had accepted as a basepoint for determining fishery limits. In the South West Approaches the Court delimited the boundary giving only half effect to the Scilly Islands as illustrated in Figure 28.4.

This was a disappointing result for the UK. Although the Scilly islands lie some 20 nm to the west of Cornwall, it was considered that the French island of Ouessant, lying some 10 nm to the west of the Brittany Peninsula, matched the geographical effect these islands had on the equidistance line. The Court's technical expert derived the half-effect line for the final segment of the boundary by dropping a line from the westernmost point of the Scilly Islands to the island of Ouessant and producing a perpendicular from this line in a westerly direction. He followed this by dropping a similar line from the western limit of the Cornish Peninsula to the island of Ouessant, and again produced a perpendicular from this line westward. He then bisected these two lines to produce the final segment of the boundary. Although an interesting method to produce half effect, it was accepted by the parties. The boundary through the Channel was simplified and the turning points were joined by loxodromes. Although the UK would have preferred the lines joining the boundary turning points to have been geodesics, it was accepted that the difference between a loxodrome and a geodesic on these short segments was small.

431

However, the Court's expert defined the final segment – that giving half weight to the Scilly Islands – as a loxodrome as well. This was clearly an error because this segment was over 100 nm in length and the Court had specified that this part of the boundary should give geographical half weight to the islands. By using a loxodrome for this segment, the Court's expert distorted this geographical requirement. The difference between a loxodrome and a geodesic on this long segment was several nautical miles to the north, thus disadvantaging the UK. The Court's treatment of the Channel Islands was also very interesting. The UK contended that these islands should be given full weight when delimiting the boundary with France, thus joining them to the full median line through the Channel. The French contended that these islands should be treated separately and be enclaved. The Court favoured the French argument and awarded the Channel Islands a 12 nm continental shelf enclave around the north of the islands. The Court declared that it was not competent under the Arbitration Agreement to delimit the boundary in the narrow belt east and south of the Channel Islands and mainland France, where questions affecting the territorial sea were involved. The UK was not happy on two technical points concerning the boundary to the north of the Channel Islands, which showed some discrepancies in the basepoints used, and the line in the South West Approaches, as already discussed. Further oral hearings were heard in 1978 and the final decision was delivered in March of that year, upholding the UK's query on the Channel Islands, but rejecting the request for a change in the South West Approaches.

A major change in the 1970s was the rapid development of computer technology. This enabled technical experts to compute median-line solutions rather than deriving them graphically. Provided the baseline data was reasonably accurate, this advance provided a geodetically robust solution with far greater accuracy. This is dramatically illustrated by the second continental shelf boundary agreement between the UK and Norway. The boundary was calculated using the same parameters as in 1965, but was computed rather than derived graphically; a difference of 331 metres was found between the northern limit of the 1965 boundary and the computed position of the same point that was the southern point of the extension to the north, see Figure 28.5.

This caused much discussion, and resulted in a linking line 331 metres in length along the parallel of latitude $61°44'12''.00N$ joining the two boundary lines. The protocol supplementary to the agreement of 1965, extending the boundary to the tripoint with the Faeroe Islands, was signed in December 1978 (Treaty Series No. 31, 1980).

The advancement of the delimitation of maritime boundaries following the completion of the Third UN Conference on the Law of the Sea and the opening for signature of the Convention has not helped coastal states to any great extent regarding the techniques to be used in maritime delimitation. Article 15 deals with the territorial sea between states and is written as follows:

> Where the coasts of two states are opposite or adjacent to each other, neither of the two states is entitled, failing agreement between them to the contrary, to extend its territorial sea beyond the median line every point of which is equidistant from the nearest points on the baselines from which the breadth of the territorial sea of each of the two states is measured. The above provision does not apply, however, where it is necessary by reason of historic title or other special circumstances to delimit the territorial seas of the two states in a way which is at variance therewith.

This appears to imply that, in the case of territorial sea boundaries, the median line is still the first point of call when delimiting the boundary.

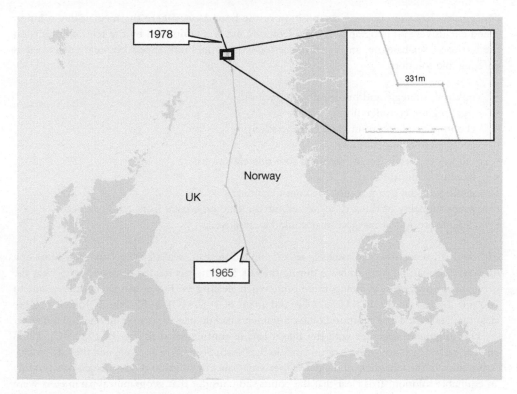

Figure 28.5 The UK/Norway 331 metre discrepancy
Source: Author.

The delimitation of the exclusive economic zone (EEZ) and continental shelf is laid down in Articles 74 and 83. The wording is essentially the same in both. Article 74 states:

1   The delimitation of the exclusive economic zone between States with opposite or adjacent coasts shall be effected by agreement on the basis of international law, as referred to in Article 38 of the Statute of the International Court of Justice, in order to achieve an equitable result.
2   If no agreement can be reached within a reasonable period of time, the States concerned shall resort to the procedure provided for in Part XV. [*This is the provision for third party settlement.*]
3   Pending agreement as provided for in paragraph 1, the States concerned, in a spirit of understanding and co-operation, shall make every effort to enter into provisional arrangements of a practical nature and, during this transitional period, not to jeopardise or hamper the reaching of the final agreement. Such arrangements shall be without prejudice to the final delimitation.
4   Where there is an agreement in force between the States concerned, questions relating to the delimitation of the exclusive economic zone shall be determined in accordance with the provisions of that agreement.

It will be noticed that no mention of an equidistance line is present. This has enabled boundary-makers engaged in bilateral negotiations to use any number of possible circumstances that could conceivably have an effect on the position of the boundary. Where the median-line solution relied exclusively on geographical considerations and was controlled by the relevant points on the territorial sea baseline, any or all of the following could influence a boundary delimited as an equitable solution:

- political, strategic and historical considerations;
- legal regime considerations;
- economic and environmental considerations;
- geographic considerations;
- the use of islands, rocks, reefs and low-tide elevations;
- baseline considerations;
- geological and geomorphologic considerations;
- proportionality of the area to be delimited including coastal front considerations;
- different technical methods that could be employed.

Although all these areas are available to a delimitation team within the relative freedoms of a bilateral negotiation, jurisprudence during the last 30 years has tended to continue to treat the geographic parameters as being paramount when dealing with a maritime boundary out to 200 nm. Indeed, in the ICJ *Malta/Libya* judgment of 1985 (ICJ, 1985), the ICJ made it quite clear that geological and geomorphologic arguments had no part to play within the 200 nm zone.

This judgment, together with the *Libya/Tunisia* continental shelf case of 1982 (ICJ, 1982) and the earlier cases concerning boundaries in the North Sea (*North Sea* case) and the UK/France Arbitration in the Channel, all suggested that a median-line solution did not necessarily provide an equitable solution. It is clear that the courts did consider that geographical parameters were the cornerstone from which the boundary solutions were derived. However, the final result clearly shows that the various geographical parameters were adjusted, sometimes quite dramatically, to produce the equitable solution required.

This change to a more pragmatic approach to the delimitation of maritime boundaries is reflected in the state practice of bilateral agreements within North-west Europe during the decades from the 1970s to the end of the century. The stalled talks between the UK and the Republic of Ireland began again in earnest in the spring of 1986. Both sides began by reviewing their positions, as it had been ten years since the last round of negotiations, in the light of state practice, the provisions contained within UNCLOS and the various international court cases that had occurred during the intervening period. One fact was immediately apparent: neither side could hope to sustain its previous position in the light of developments in international law and state practice. Going to arbitration was still an option, but reservations were expressed over both the probable cost of such action and the fear that the Court would not be prepared to delimit a line when third parties could be involved. This point was prompted by the judgment in the *Malta/Libya* case. It was considered that the best way forward would be to attempt to negotiate a bilateral agreement or, if that failed, a partial agreement, keeping arbitration as the ultimate fallback position. The area to be delimited naturally fell into three sections: the Irish Sea, the South West Approaches and the North West Approaches. They were tackled in that order, but as part of an overall settlement.

Meetings took place alternately between London and Dublin approximately every six weeks for two and a half years. It was agreed at a very early stage that during the negotiations both sides would use Admiralty charts, but this was only for presentation and to give the negotiating teams a general appreciation of the geographic circumstances involved. From time to time specially prepared computer-drawn maps by both Departments of Energy supplemented these charts to illustrate particular points. It was also agreed fairly early on that the boundary would be stepped, complementing the designated area blocks of both countries. Towards the end of the negotiations it was decided to refer the final geographical coordinates of the boundary to the World Geodetic System 1984 (WGS 84), reflecting a growing desire by the oil industry at that time for boundaries to be referred to a modern, worldwide geodetic datum used in satellite navigation systems. It is fair to say that, at some stage during these long and complex negotiations, every method or device that has ever been used in delimitations – and some that have not – was discussed and studied at length in order to produce a result that could be accepted by both sides. Agreement was finally achieved in 1988, to the relief and delight of both parties. The agreement was signed in Dublin on 7 November and ratified on 11 January 1990 (Treaty Series No. 20, 1990). Clive Symmons has written a useful commentary on this boundary in his book, *Ireland and the Law of the Sea* (Symmons, 1993). This agreement is one of the longest continental shelf boundaries delimited to date. The section of the boundary through the Irish Sea and the South West Approaches is some 502 nm in length and the North West some 634 nm, a total of approximately 1,136 nm. The agreement has allowed both countries to extend their respective designated areas up to the boundary. However, although the oil industry has shown some interest in the area, no significant hydrocarbon discoveries have so far been made in the boundary region.

Following the UK's extension of the territorial sea in 1987, a revised boundary agreement was required in the Straits of Dover to transform what had previously been a territorial sea/continental shelf boundary between France and the UK into a pure territorial sea boundary between the two countries. Negotiations began in 1988 to determine the basically technical matter of the two end points of the respective territorial sea limits, as the boundary already in place was not to be altered. Negotiations were straightforward and the new agreement was duly signed on 2 November 1988 and came into force on 6 April 1989 (Treaty Series No. 26, 1989) Of more significance was the joint declaration by the two countries concerning the right of transit passage through the Straits. This is a hugely important freedom of navigation provision, contained within UNCLOS. The terms of the declaration were as follows:

> On the occasion of the signature of the Agreement relating to the Delimitation of the territorial sea in the Straits of Dover, the two Governments agreed on the following declaration: 'The existence of a specific regime of navigation in straits is generally accepted in the current state of international law. The need for such a regime is particularly clear in straits, such as the Straits of Dover, used for international navigation and linking two parts of the high seas or economic zones in the absence of any other route of similar convenience with respect to navigation.'

In consequence, the two Governments recognise rights of unimpeded transit passage for merchant vessels, state vessels, in particular, warships following their normal modes of navigation, as well as the right of overflight of aircraft, in the Straits of Dover. It is understood that, in accordance with the principles governing this regime under the rules of international law, such passage will be exercised in a continuous and expeditious manner. The two Governments will continue to cooperate closely, both bilaterally and through the International Maritime Organization in the interests of ensuring the safety of navigation in the Straits of Dover, as well as in the Southern North Sea and the Channel. In particular, the traffic separation scheme in the Straits of Dover will not be affected by the entry into force of the Agreement. With due regard to the interests of the coastal States, the two Governments will also take, in accordance with international agreements in force and generally accepted rules and regulations, measures necessary in order to prevent, reduce and contain pollution of the marine environment by vessels.

HE Judge David Anderson, formally Second Legal Adviser at the Foreign and Common-wealth Office, expanded on this theme in an excellent paper, 'The Right of Transit Passage and the Straits of Dover', presented to a conference in 1989 (30 March–1 April) at the Centre for Oceans Law and Policy in Washington.

Meanwhile, developments in the southern North Sea were moving apace. France and Belgium had entered into bilateral negotiations concerning their territorial sea and continental shelf boundaries. It will be remembered that the extension of the UK/France continental shelf boundary, agreed in 1982, had been stopped short in the eastern Channel because the France/Belgium boundary had not been delimited at that time. The territorial sea and continental shelf boundaries that had to be delimited between France and Belgium were relatively short, at 12 nm and 18 nm respectively. However, they are calculated using different criteria. The territorial sea boundary is a single straight segment taking account of the low-tide elevations close to the French and Belgian coasts. The low-tide elevations in question were Trapegeer, some 1.3 nm off the Belgian coast west of Nieuwpoort, and Banc Breedt and Banc Smal, 2.5 nm and 1.6 nm off the French coast north of Dunkerque. The difficulty confronting the negotiating teams was not whether these elevations should be used for the calculation of the boundary, but whether they existed as low-tide elevations at all. Both countries use different tidal calculations to determine chart datum, the vertical datum from which all charted depths are measured. The Belgians use the mean lower low-water spring tide, measured over the tidal cycle of 18 and two-third years, to calculate chart datum. The French use the lowest astronomical tide – i.e. the lowest tide level that can be predicted to occur under average meteorological conditions (International Hydrographic Organization, 1990); this is the vertical datum that is recommended by the International Hydrographic Organization and the chart datum used on Admiralty charts. This level is some 0.3 metres lower than Belgian chart datum. This difference of chart datum has a profound effect on the charted existence of low-tide elevations. Banc Breedt, for instance, is shown as a low-tide elevation on both French and British charts but not on Belgian charts. The result of this was that each country calculated the territorial sea median line using different basepoints; the Belgian line did not use Banc Breedt. The solution was to divide the area of difference between the two median lines equally. This compromise solution has, in effect, given half weight to Banc Breedt, without having chosen one vertical datum over the other. It can, however, be described as basically a median-line solution with some adjustment, and fully conforms with the guidelines, such as they are, contained within UNCLOS for the delimitation of territorial sea boundaries.

The second boundary between France and Belgium concerned the continental shelf. Article 2 of the agreement states:

> The points defined above have been arrived at after an attempt to find an equitable solution on the basis mainly of a compromise between two assumptions, one taking into account low-tide elevations at the approaches to the French and Belgian coasts, and one taking into account the low-water line of the coast.

France maintained that the use of low-tide elevations was relevant in the delimitation of a continental shelf boundary, as they had agreed in the UK/France agreement through the eastern Channel. However, Belgium contended that the low-water line of the mainland coast should be decisive and low-tide elevations ignored. The solution is seen as a pragmatic compromise producing an equitable result for both states, thus meeting the requirements of Article 83 of UNCLOS. Analysis of the line indicates that Belgium did not insist that low-tide elevations should be totally ignored, and that France accepted less than full weight for Banc Breedt. It would appear that the result of this compromise was to give full weight to the Belgian low-tide elevation of Trapegeer, which is charted as a low-tide elevation on both French and Belgian charts, but only one-fifth weight to Banc Breedt at the landward side of this boundary. However, at the seaward end of this boundary, full weight has been given to Banc Breedt against the UK low-tide elevation of Long Sand Head in the outer approaches to the Thames Estuary. This seaward end point of the boundary forms a UK/France/Belgium tripoint and is equidistant between Banc Breedt (France), Trapegeer (Belgium) and Long Sand Head (UK), all low-tide elevations and all given full weight for this tripoint. The agreement was signed on 8 October 1990 and entered into force on 7 April 1993 (Charney and Alexander, 1991d).

Once the France/Belgium agreement had been concluded and the UK/France/Belgium tripoint was agreed by all three parties, the way was open for the UK and Belgium to re-open their bilateral negotiations, last conducted in the mid 1960s.

At that time both sides had agreed on the principle of equidistance based on points taken from the territorial sea baselines of the two states; this was the normal practice at that time. Low-tide elevations were used by both sides out to a maximum distance of 3nm from the mainland low-water lines. Both states claimed a 3nm territorial sea limit at that time. At the beginning of the negotiations in the early 1960s the two end points of the boundary were undetermined, as neither the UK nor Belgium had concluded boundary agreements with the Netherlands and France. Both sides were willing to proceed with each end of the line remaining undecided if necessary. In the event, the UK and the Netherlands concluded their agreement in 1965, producing a tripoint between the UK, the Netherlands and Belgium, which was accepted as technically correct by the Belgian authorities. This solved the problem of the end point in the north of the UK/Belgian boundary. Although considerable progress was made in the bilateral negotiations that took place during the next few years, no agreement was actually signed. This was largely due to the fact that the Belgians had not agreed their lateral boundary with France.

Once the Belgians had agreed their boundary with France, the way was open for the UK/Belgian talks to resume, after a gap of some 18 years. The prospective UK/Belgium continental shelf boundary now had both a start and end point in the agreed tripoints. Since the early 1970s both countries had extended their territorial sea limits to 12 nm, which although it made no difference to the Belgian territorial sea basepoints, made considerable differences to the UK basepoints. UK low-tide elevations could now be used in the outer Thames Estuary where they were within 12 nm of the mainland low-water line. This included the low-tide elevation on Long Sand Head and another low-tide elevation on a bank named Shipwash, off the port of Harwich. The equidistance line was thus recalculated using these new basepoints. The result was a line some 4 nm closer to the Belgian coast than the line calculated in 1972. The UK/Belgium/France tripoint gave full weight to the low-tide elevation on Long Sand Head and so coincided with this new line. However, the UK/Belgium/Netherlands tripoint did not use these low-tide elevations, because they were beyond the 3 nm territorial sea limit in force in 1965. On the Belgian side, the same basepoints were still valid from 1972, except for the new outer breakwaters off Zeebrugge that had been completed in 1990. This mismatch of the two lines produced a coffin-shaped area of some 209 square kilometres. It was agreed at an early stage that the boundary should start and end at the two tripoints, with the joining line based loosely on some general method of equidistance. However, it was also agreed that the boundary should take account of the current rules in international law on boundaries, in order to achieve an equitable result. This was consistent with the Belgian/French agreement and the more pragmatic approach adopted by the UK in recent maritime boundary agreements.

During the negotiations a new routine survey was carried out in the approaches to Harwich that showed that the low-tide elevation on the Shipwash had been eroded to the extent that it no longer dried at chart datum. It could therefore no longer be used as a legitimate low-tide elevation in the calculation of a median line. The equidistance line was therefore recalculated using territorial sea basepoints on the mainland low-water line in the vicinity of Orford Ness. This only affected the northernmost turning point of the line. The majority of the line was controlled by the low-tide elevation on Long Sand Head, only just short of 12nm from the mainland low-water line, against baseline points on the Belgian mainland low-water line. The Belgian low-tide elevation on Trapegeer affected only the UK/Belgium/France tripoint, which already gave full weight to Trapegeer and Long Sand Head. With this imbalance of basepoints in mind, it was agreed, in the interests of an equitable solution, to only give about one-third weight to Long Sand Head overall. In effect, the southern part of the line gives full weight to Long Sand Head, and the northern end of the line gives nil effect to this offshore low-tide elevation. This enables the line to join up with the agreed UK/Netherlands line without an unnatural kink in the line. In between the two end tripoints, the line only has one turning point. The position of this point was agreed pragmatically and gave effect to the approximately one-third/two-thirds split of the disputed area. The agreement was signed on 29 May 1991 and ratified on 14 May 1993 (Treaty Series No. 20, 1994).

The agreement was accompanied by a letter from the Belgian foreign minister taking account of the fact that no boundary agreement had been signed between Belgium and the Netherlands delimiting the territorial sea and the continental shelf. Belgium undertook in the letter not to claim across the line agreed by the Netherlands and the UK in 1965, while keeping open the possibility that a future boundary between Belgium and the Netherlands would reach the Netherlands/UK boundary north of the agreed northern point of the present UK/Belgium boundary.

This has indeed proved to be the case. The Netherlands and Belgium finally reached agreement on their territorial sea and continental shelf boundary on 18 December 1996 and it entered into force on 1 January 2000 (Oude Elferink, 1997). The new tripoint with the UK lies some 4.75 nm to the north of the agreed point of 1965. Once again, this is the result of a pragmatic solution to the effect of low-tide elevations, further reinforcing the importance of reaching an equitable solution, particularly for EEZ and continental shelf boundaries.

The result of the Netherlands/Belgium continental shelf agreement required adjustments to both the UK/Netherlands agreement and the UK/Belgium agreement to bring them into line with the new tripoint. These were achieved in 2004 (Colson and Smith, 2011d) and 2005 (Colson and Smith, 2011e) respectively.

The UK Channel Islands boundaries between the Bailiwicks of Guernsey and Jersey had not been completed by the Channel case arbitration in the late 1970s. The court left it to the UK and France to delimit the boundaries between the islands and mainland France. Although France has claimed a 12 nm territorial sea since 1971, the Bailiwick of Guernsey only claims a 3 nm territorial sea at present, hence the 1992 agreed boundary between the islands and mainland France only concerns fisheries (Treaty Series No. 66, 1993). The Bailiwick of Jersey however has claimed 12 nm territorial sea since 1997. Agreement was finally reached on 4 July 2000 (Charney and Smith 2002) after a protracted series of bilateral negotiations that not only concerned the territorial sea boundary, but also the more complicated issues concerning a new fisheries regime in the Granville Bay area.

The delimitation of the territorial sea boundary was first raised during the UK/France Channel Arbitration, but the Court declined to tackle this difficult area because it considered that it did not have jurisdiction to delimit a territorial sea boundary, having been asked to delimit the continental shelf between the two countries. The bilateral negotiations centred on the effect of the large number of small rock features, some of which were above water at high tide and thus technically islands, and many rocks that were low-tide elevations. The tidal range in this area is very large – in the region of 10 metres – thus the geography of the area changes dramatically between low and high water. One of the problems facing the negotiators was the sovereignty of many of these low-tide elevation features. In the event, definitive sovereignty was not possible and a pragmatic solution was the only answer. Another major issue was the effect of the two main groups of islands off the mainland of Jersey, namely Les Écrehous and the Plateau des Minquiers. These groups were not given full weight because their effect, close to the French mainland coast, would have caused an unacceptable result for France. The agreed boundary is a mixture of equidistance, simplification and pragmatism – a truly classic example of flexible negotiation producing an equitable result.

To the north of the UK one further long continental shelf boundary remained outstanding at the beginning of the 1990s. This was the boundary between the UK and the Faeroe Islands. The north-eastern limit of this boundary had been settled in 1978 with the agreed tripoint between the UK, Norway and the Faeroe Islands. The limit to the west was more difficult. A decision had to be made whether to attempt to delimit the boundary to the limit of the continental shelf beyond 200 nm, or only to attempt to delimit the boundary to the 200 nm limit. It soon became clear that it would be very difficult to attempt to delimit this boundary beyond the 200 nm limit, using the continental margin provisions contained within Article 76 of UNCLOS. It was therefore decided that the negotiations concerning the boundary would concentrate on the area within 200 nm only. This ensured that the major considerations would be the geographical aspects of the delimitation, and the difficult questions of the geology and geomorphology would not have to be answered at that time.

The negotiations concerning this boundary had first started in 1978, but at that time there was no pressure from the hydrocarbon industry to progress this boundary; fishery issues were far more important in this area, particularly for the Faeroes. Both countries had enacted 200 nm fishery zones, which produced an overlapping area caused by a different interpretation of the way a median line should be drawn. The UK 200 nm fisheries limit used all territorial sea basepoints on the UK side, including the small islands of Rockall, St Kilda, the Flannan Islands, Sule Skerry, Rona and Sula Sgeir, but ignored the Faeroe Islands' straight-baseline system. The Faeroe Islands, on the other hand, ignored all these small UK offshore islands and included their straight baselines. Although the overlapping area was quite large – some 13,000 square kilometres – the fishery was 'managed' satisfactorily by the two states. Negotiations continued intermittently throughout the 1980s, but it was only in the early 1990s when oil was discovered to the west of the Shetland Islands on the UK continental shelf, close to the western limit of the UK-designed area, that increased interest in the delimitation of this boundary became apparent.

Talks began in earnest, with Denmark leading for the Faeroe Islands, but the difficulties encountered were considerable. The area to be delimited was geographically complicated. The only agreed point on the boundary at the start of this phase of the negotiations was the agreed tripoint with Norway in the north. The majority of the boundary was of an opposite aspect, with the Faeroe Islands facing the west coast of the Shetland Islands, the Orkney Islands and the mainland coast of northern Scotland. The small offshore islands added a further complication to this aspect. The boundary then took on an adjacent aspect as it moved out to the west to the 200 nm limit. Again, Scottish offshore islands were the dominating features, as the Outer Hebrides, the Flannan Islands and St Kilda could all potentially play a part in the delimitation. The UK did not use the small island of Rockall as a basepoint having any effect on this boundary. Once the UK had acceded to UNCLOS in 1997, the UK's 200 nm fishery limit was amended to redraw the limit measured from St Kilda rather than Rockall, in recognition of the fact that Article 121(3) of the Convention applied to this small offshore feature. It is interesting to note that, to date, the UK is the only party to the Convention to have pulled back from a 200 nm zone measured from an island that cannot sustain human life or have an economic life of its own.

The interests of the potential prospectivity for hydrocarbons in the area to be delimited added an extra consideration to be taken into account by both sides, and probably both helped and hindered the progress of the negotiations at different stages. It was quite clear that the oil industry was very keen to work in the boundary area but was reluctant to commit resources until an agreement had been reached. This encouraged both sides to make progress as rapidly as possible but at the same time, to work to produce an agreement that was considered equitable to both parties.

Fishing interests were also considered to be important, particularly by the Faeroe Islands, and negotiations also covered this aspect towards the latter part of the discussions. The final agreement signed in Tørshavn on 18 May 1999 and entering into force on 21 July 1999 (Treaty Series No. 76, 1999) covered both the continental shelf and fisheries. The continental shelf boundary appears to be a mixture of equidistance and pragmatism, while the fisheries boundary follows a median-line solution in the north and west, but introduces a Special Area between the north coast of Scotland and the Faeroe Islands. This Special Area is similar in shape and size

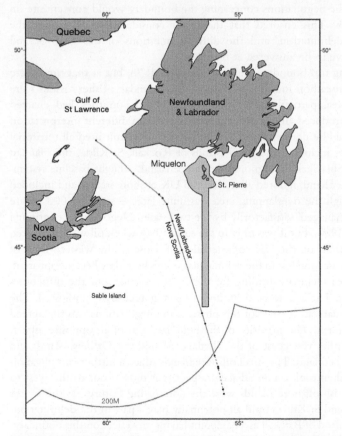

Figure 28.6 France (Saint Pierre and Miquelon)/Canada arbitration 1992 and the Newfoundland/Nova Scotia judgment 2002
Source: Author.

to the old overlapping zone in this area. Once again, there is a similarity, in the way that the agreement has been reached, between this boundary and earlier boundary delimitations since the 1980s.

Has anything changed during the first decade or so of the twenty-first century in the way maritime boundaries have been delimited? I think it is fair to say that bilateral negotiations and their resulting delimitations have not substantially changed in either their practice or resulting delimitations, apart from the substantial change in computer technology and the resulting availability of powerful delimitation and maritime limit programs. The jurisprudence of the last 13 years has seen a change from a rather uncertain prediction of the outcome of a boundary case in the 1980s to far more certainty at the present time. Today the ICJ and international tribunals all agree that the starting point of any maritime delimitation is the median/equidistance line from which a decision is made on whether that line provides an equitable result or whether an adjustment of some kind is required. A classic way of illustrating this change of approach is to compare the results of the boundaries delimiting the French islands of Saint Pierre and Miquelon and Canada (Charney and Alexander, 1997), by a court of Arbitration in 1992 and the inter-State boundary between Newfoundland and Nova Scotia by a Canadian appointed Arbitral Tribunal acting under international maritime law in 2002 (Newfoundland Labrador, 2002) as illustrated in Figure 28.6.

The former could be described as equitable principles taken a step too far, while the latter follows the median line followed by an adjustment principal. Similarly, the ICJ considered that a median line produced an equitable result in the maritime section of the complex sovereignty and delimitation case between Cameroon and Nigeria in their judgment handed down in 2002 (Colson and Smith, 2005) as illustrated in Figure 28.7.

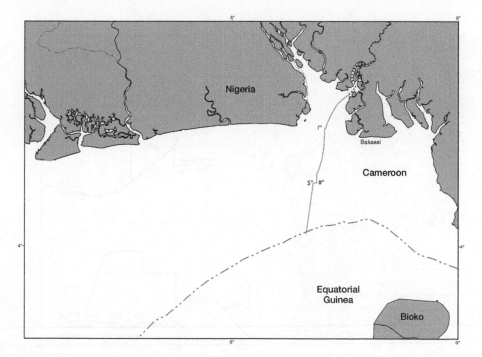

Figure 28.7 Nigeria/Cameroon judgment 2002
Source: Author.

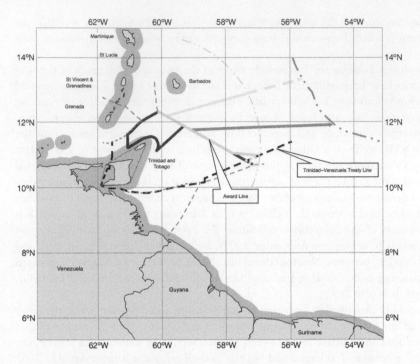

Figure 28.8 Barbados/Trinidad & Tobago judgment
Source: Author.

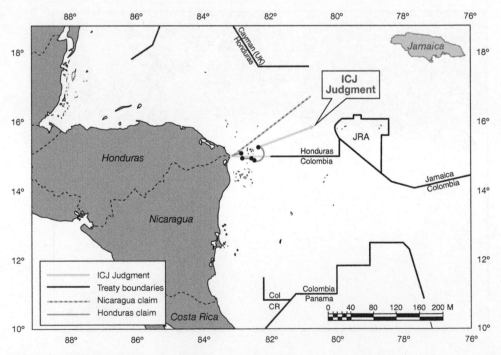

Figure 28.9 Nicaragua/Honduras judgment
Source: Author.

The maritime part of this case hinged on the court's acceptance of the validity of the earlier bilateral agreements of 1971, the Yaoundé II Declaration and the Maroua Declaration of 1975, and then rejecting Cameroon's argument that an adjustment was required to take account of Cameroon's considerable length of coastline. The Arbitral Tribunal in the 2006 judgment of the Trinidad & Tobago–Barbados case (Colson and Smith, 2011b) also only made a slight adjustment of the median line towards the outer end of the boundary up to the 200 nm limit where the boundary terminated as illustrated in Figure 28.8.

The 2007 Nicaragua–Honduras case (Colson and Smith 2011a) was more complex because of the geography concerned. The ICJ had to consider how to delimit a maritime boundary beginning at an unstable land boundary terminal point at the seaward end of the River Coco and a coastline configuration that if used to construct a median line would have meant that the entire boundary was controlled by only two basepoints that were very close together, resulting in an unstable line. Their solution was to construct a bisector between the generalisations of the two coasts, with semi enclaves around some very small Honduran cays towards the beginning of the boundary as illustrated in Figure 28.9.

A bisector constructed in this way is in effect a simplified equidistance line. The ICJ slightly muddied the waters in their 2009 judgment of the Ukraine–Romania maritime boundary (Colson and Smith 2011c). Although the court stated that they had constructed a provisional median

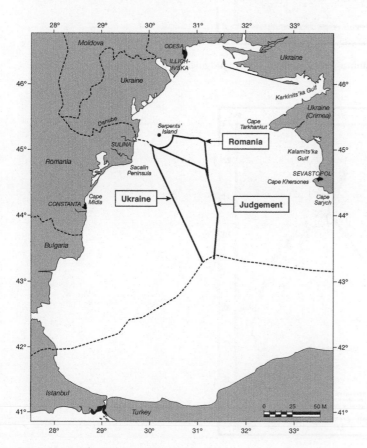

Figure 28.10 Ukraine/Romania judgment
Source: Author.

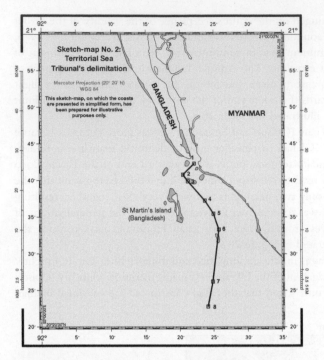

Figure 28.11 Bangladesh/Myanmar territorial sea boundary judgment
Source: ITLOS, 2012, p. 57.

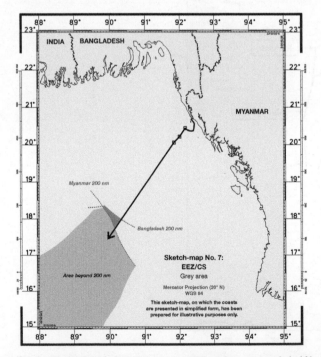

Figure 28.12 Bangladesh/Myanmar EEZ and continental shelf boundary judgment
Source: ITLOS, 2012, p. 138.

line boundary to ascertain whether this provided an equitable result, they did not use basepoints on Serpents' Island, part of the Ukrainian territorial sea baseline, nor points on Sulina Dyke, part of Romania's territorial sea baseline. The line thus constructed is not a true median line but an adjusted line. It would have been clearer if the court had constructed a correct median line and then in a second stage discounted both Serpents' Island and Sulina Dyke, giving their reasons. The end result does provide an equitable result, as illustrated in Figure 28.10, but those states considering bilateral maritime boundary negotiations, following jurisprudence, in the future may be tempted to discount those basepoints they do not like and still insist that this provides a provisional equidistance/median line.

The final two cases to consider are the first maritime boundary delimitation brought before the International Tribunal for the Law of the Sea (ITLOS) by Bangladesh and Myanmar (Burma) and the Nicaragua–Colombia case brought before the ICJ. The judgment for both cases was handed down in 2012. ITLOS was required to delimit the territorial sea, EEZ and continental shelf beyond 200 nm between Bangladesh and Myanmar. The court decided that a median line, giving full weight to the Bangladesh island of St Martin's was the correct line for the territorial sea. This is illustrated in Figure 28.11.

When progressing to the delimitation of the EEZ the court correctly proceeded to construct a provisional equidistance line, thence considering whether this line required adjusting to provide an equitable result. However, the court's provisional equidistance line discounted points on St Martin's Island – it is thus not an equidistance line but an adjusted equidistance line. The correct procedure should be to construct an equidistance line giving full weight to all territorial sea basepoints on each coast, then adjusting as required to produce an equitable result giving a full explanation of why points are discounted and/or why the line has been moved. The court's

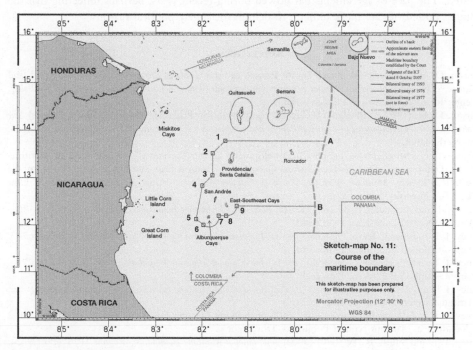

Figure 28.13 Nicaragua/Colombia judgment

Source: Adapted from ICJ (2012). Original full-colour version available online at www.icj-cij.org/docket/files/124/17164.pdf (p. 94, accessed 17 July 2015).

final line is indeed an adjusted equidistance line that has been further adjusted to take account of the concavity of the Bangladesh coast as illustrated in Figure 28.12.

The Nicaragua–Colombia case was a complex combination of the long mainland coast of Nicaragua facing the small remote Colombian islands of Providencia and San Andrés and several uninhabited cays to the north and south. The ICJ in its judgment drew a median line between the Colombian islands and the Nicaraguan coast and then adjusted this line by semi-enclosing the islands with arcs in a ratio of 3:1 in Nicaragua's favour up to stated limits north and south to enable Nicaragua to extend its EEZ to the east of the islands. It then simplifies this line and draws parallels of latitude to the north and south of the islands to extend Nicaragua's EEZ out to 200 nm as illustrated in Figure 28.13.

This elegant solution provides an equitable result taking into account the large disparity of the coastal lengths, the cut-off effect of the Colombian islands to Nicaragua's eastward projection, while maintaining the Colombian islands' entitlement to their own EEZ particularly to the east.

Conclusion

The delimitation of state maritime boundaries is essentially a twentieth century phenomenon, only becoming politically essential once living and non living resources became owned by states with the development of international maritime law through the conventions of 1958 and 1982 and state practice. Boundaries were generally median or equidistance lines up to the early 1980s when equitable principals took precedence. However, within the last decade median/ equidistance has returned as the first step in the delimitation of maritime boundaries adjusted as necessary to produce an equitable result. There are still some maritime boundaries to be delimited. The pace of agreements has slowed during recent years, perhaps reflecting that the remaining boundaries are more difficult to delimit for largely political reasons.

References

Anderson, D. H. (1989) The Right of Transit Passage and the Straits of Dover, in Nordquist, M. H. and Moore, J. N. (eds) *Thirteenth Annual Seminar: Contemporary Issues in United States Law of the Sea Policy.* Center for Oceans Law and Policy, Virginia.

Anderson, D. H. (1996) Grisbadarna Revisited, pp. 155–166 in Platzöder, R. and Verlaan, P. (eds) *The Baltic Sea: New Developments in National Policies and International Co-operation.* Kluwer Law International, the Netherlands.

Carleton, C. and Schofield, C. H. (2002) Developments in the Technical Determination of Maritime Space: Delimitation, Dispute Resolution, Geographical Information Systems and the Role of the Technical Expert. *Maritime Briefing Volume 3 Number 4*, International Boundaries Research Unit (IBRU), University of Durham.

Charney, J. I. and Alexander, L. M. (eds) (1991a) *International Maritime Boundaries, Volume I*, Report 2–13(1). Martinus Nijhoff Publishers, Dordrecht.

Charney, J. I. and Alexander, L. M. (eds) (1991b) *International Maritime Boundaries, Volume II*, Report 7–3. Martinus Nijhoff Publishers, Dordrecht.

Charney, J. I. and Alexander, L. M. (eds) (1991c) *International Maritime Boundaries, Volume II*, Report 7–6. Martinus Nijhoff Publishers, Dordrecht.

Charney, J. I. and Alexander, L. M. (eds) (1991d) *International Maritime Boundaries, Volume II*, Report 9–16. Martinus Nijhoff Publishers, Dordrecht.

Charney, J. I. and Alexander, L. M. (eds) (1997) *International Maritime Boundaries, Volume III*, Report 1–2 Addendum 2. Martinus Nijhoff Publishers, Dordrecht.

Charney, J. I. and Smith, R. W. (eds) (2002) *International Maritime Boundaries, Volume IV*, Report 9–24. Martinus Nijhoff Publishers, Dordrecht.

Colson, D. A. and Smith, R. W. (eds) (2005) *International Maritime Boundaries, Volume V*, Report 4–1 (Add. 2). Martinus Nijhoff Publishers, Dordrecht.

Colson, D. A. and Smith, R. W. (eds) (2011a) *International Maritime Boundaries, Volume VI*, Report 2–24 (Add. 1). Martinus Nijhoff Publishers, Dordrecht.

Colson, D. A. and Smith, R. W. (eds) (2011b) *International Maritime Boundaries, Volume VI*, Report 2–26 (Add. 1). Martinus Nijhoff Publishers, Dordrecht.

Colson, D. A. and Smith, R. W. (eds) (2011c) *International Maritime Boundaries, Volume VI*, Report 8–18. Martinus Nijhoff Publishers, Dordrecht.

Colson, D. A. and Smith, R. W. (eds) (2011d) *International Maritime Boundaries, Volume VI*, Report 9–13 (2). Martinus Nijhoff Publishers, Dordrecht.

Colson, D. A. and Smith, R. W. (eds) (2011e) *International Maritime Boundaries, Volume VI*, Report 9-17 (2). Martinus Nijhoff, Dordrecht.

Francalanci, G., Scovazzi, T. and Romano, D. (1994) *Lines in the Sea*. Martinus Nijhoff Publishers, Dordrecht.

International Court of Justice (ICJ) (1969) North Sea Continental Shelf Cases, ICJ Reports, The Hague.

International Court of Justice (ICJ) (1982) Libya/Tunisia Continental Shelf: Case Concerning the Continental Shelf (Libyan Arab Jamahiriya/Tunisia). The Hague.

International Court of Justice (ICJ) (1985) Malta/Libya Continental Shelf: Case Concerning the Continental Shelf (Libyan Arab Jamahiriya/Malta). The Hague.

International Court of Justice (ICJ) (2012) Territorial and Maritime Dispute (*Nicaragua v. Colombia*) Judgment of 19 November 2012.

International Hydrographic Organization (1990) *Hydrographic Dictionary* – SP No. 32, Monaco.

International Tribunal for the Law of the Sea (ITLOS) (2012) Delimitation of the maritime boundary in the Bay of Bengal (Bangladesh/Myanmar), Judgment, available at: www.itlos.org/fileadmin/itlos/documents/cases/case_no_16/C16_Judgment_14_03_2012_rev.pdf, accessed 17 July 2015.

Miscellaneous No. 15 (1978) Arbitration between the United Kingdom of Great Britain and Northern Ireland and the French Republic on the Delimitation of the Continental Shelf. HMSO, London; and International Maritime Boundaries, Three Volumes, Volume II, Report 9–3. Editors J. I. Charney and L. M. Alexander, Martinus Nijhoff, Dordrecht, 1991.

Newfoundland Labrador (2002) Government reacts to decision in Newfoundland and Labrador/Nova Scotia Offshore Boundary Dispute, available at: www.releases.gov.nl.ca/releases/2002/mines&en/0402n01.htm, accessed 17 January 2014.

O'Connell, D. P. (auth.), Shearer, I. A. (ed.) (1982) *The International Law of the Sea – Vol I*. The Clarendon Press, Oxford.

Oude Elferink, A. G. (1997) Belgium/The Netherlands (Delimitation of Maritime Zones), *International Journal of Marine and Coastal Law*. Volume 12 No. 4, November 1997, p. 548, Kluwer Law International, Dordrecht.

Scott, J. B. (1916) *Hague Court Reports*, available at: https://archive.org/details/haguecourtreport028086mbp, accessed 17 July 2015.

Symmons, C. (1993) *Ireland and the Law of the Sea*, pp. 184–190. The Round Hall Press, Dublin.

Treaty Series No. 71 (1965) Agreement between the Government of the United Kingdom of Great Britain and Northern Ireland and the Government of the Kingdom of Norway relating to the Delimitation of the Continental Shelf between the two Countries, London, March 1965. HMSO, London; and International Maritime Boundaries, Three Volumes, Volume II, Report 9–15. Editors J. I. Charney and L. M. Alexander, Martinus Nijhoff, Dordrecht, 1991.

Treaty Series No. 7 (1973) Agreement between the United Kingdom of Great Britain and Northern Ireland and the Federal Republic of Germany relating to the Delimitation of the Continental Shelf under the North Sea between the two Countries, London, November 1971. HMSO, London; and International Maritime Boundaries, Three Volumes, Volume II, Report 9–12. Editors J. I. Charney and L. M. Alexander, Martinus Nijhoff, Dordrecht, 1991.

Treaty Series No. 6 (1973) Agreement between the Government of the United Kingdom of Great Britain and Northern Ireland and the Government of the Kingdom of Denmark relating to the Delimitation of the Continental Shelf between the two Countries, London, November 1971. HMSO, London; and International Maritime Boundaries, Three Volumes, Volume II, Report 9–10. Editors J. I. Charney and L. M. Alexander, Martinus Nijhoff, Dordrecht, 1991.

Treaty Series No. 130 (1972) Protocol between the Government of the United Kingdom of Great Britain and Northern Ireland and the Government of the Kingdom of the Netherlands relating to the Delimitation of the Continental Shelf under the North Sea between the two Countries, London, November 1971. HMSO, London; and International Maritime Boundaries, Three Volumes, Volume II, Report 9–13. Editors J. I. Charney and L. M. Alexander, Martinus Nijhoff, Dordrecht, 1991.

Treaty Series No. 31 (1980) Protocol Supplementary to the Agreement of 10 March 1965 between the Government of the United Kingdom of Great Britain and Northern Ireland and the Government of the Kingdom of Norway relating to the Delimitation of the Continental Shelf between the two Countries, Oslo, December 1978. HMSO, London; and International Maritime Boundaries, Three Volumes, Volume II, Report 9–15(2). Editors J. I. Charney and L. M. Alexander, Martinus Nijhoff, Dordrecht, 1991.

Treaty Series No. 26 (1989) Agreement between the Government of the United Kingdom of Great Britain and Northern Ireland and the Government of the French Republic relating to the Delimitation of the Territorial Sea in the Straits of Dover, Paris, November 1988. HMSO, London; and International Maritime Boundaries, Three Volumes, Vol. II, Report 9–3. Editors J. I. Charney and L. M. Alexander, Martinus Nijhoff, Dordrecht, 1991.

Treaty Series No. 20 (1990) Agreement between the Government of the United Kingdom of Great Britain and Northern Ireland and the Government of the Republic of Ireland concerning the Delimitation of Areas of the Continental Shelf between the two Counties, Dublin, 7 November 1988, Cm 990. HMSO, London; and International Maritime Boundaries, Three Volumes, Vol. II, Report 9–5. Editors J. I. Charney and L. M. Alexander, Martinus Nijhoff, Dordrecht, 1991.

Treaty Series No. 20 (1994) Agreement Between the Government of the United Kingdom of Great Britain and Northern Ireland and the Government of the Kingdom of Belgium relating to the Delimitation of the Continental Shelf between the two Countries. HMSO, London; and International Maritime Boundaries, Three Volumes, Volume II, Report 9–17. Editors J. I. Charney and L. M. Alexander, Martinus Nijhoff, Dordrecht, 1991.

Treaty Series No. 66 (1993) Exchange of Notes between the Government of the United Kingdom of Great Britain and Northern Ireland and the Government of the French Republic concerning the Activities of Fishermen in the Vicinity of the Channel Islands and the French Coast of the Cotentin Peninsula and, in Particular, on the Schole Bank, Paris 10 July 1992. HMSO, London.

Treaty Series No. 76 (1999) Agreement between the Government of the United Kingdom of Great Britain and Northern Ireland, on the one hand, and the Government of the Kingdom of Denmark together with the Home Government of the Faeroe Islands, on the other hand, relating to the Maritime Delimitation in the area between the United Kingdom and the Faeroe Islands, Tørøshavn, 18 May 1999, Cm 4514. Stationery Office, London; and International Maritime Boundaries, Volume IV, Report 9–23. Editors J. I. Charney and R. W. Smith, Martinus Nijhoff, Dordrecht, 2002.

United States of America (1945) Proclamation No. 2667, 10 Federal Regulation 12303, Washington, DC.

29

THE DEEP SEABED

Legal and political challenges

Tullio Scovazzi

The principle of common heritage of mankind

Under Art. 136 of the United Nations Convention on the Law of the Sea,[1] the 'Area', that is the seabed and ocean floor and subsoil thereof beyond the limits of national jurisdiction, and its resources, are the common heritage of mankind. This is the main innovating aspect of the UNCLOS with respect to the previous law of the sea regime. It is based on a new concept that is completely different from the traditional regimes of sovereignty, which applies in the territorial sea, and of freedom, which applies on the high seas.

The principle of common heritage of mankind was launched in a memorable speech made on 1 November 1967 at the United Nations General Assembly by the representative of Malta, Mr Arvid Pardo. The opportunity for proposing it came from the technological developments that were expected to lead in a relatively short time to the commercial exploitation of poly-metallic nodules lying on the surface of the deep seabed and containing various minerals, such as manganese, nickel, cobalt and copper. The application of a regime of sovereignty was likely to lead to a series of competitive extensions by coastal States of the limits of national jurisdiction on the sea bed. The application of the regime of freedom was likely to lead to a rush towards the exploitation of economically and strategically valuable minerals under a first-come-first-served criterion. According to Mr Pardo, the consequences of both possible scenarios would have been equally undesirable. They would have encompassed political tension, economic injustice and risks of pollution. In a few words, 'the strong would get stronger, the rich richer'.[2]

The basic elements of the regime of common heritage of mankind,[3] applying to the seabed beyond the limits of national jurisdiction, are the prohibition of national appropriation, the destination of the Area for peaceful purposes, the use of the Area and its resources for the benefit of mankind as a whole with particular consideration for the interests and needs of developing countries, as well as the establishment of an international organization entitled to act on behalf of mankind in the exercise of rights over the resources.[4]

The proposal by Malta led to Resolution 2749 (XXV), adopted on 17 December 1970, whereby the United Nations General Assembly solemnly declared that 'the sea-bed and the ocean floor, and the subsoil thereof, beyond the limits of national jurisdiction . . ., as well as the resources of the area, are the common heritage of mankind' (Art. 1). All the basic elements of the concept of common heritage of mankind can be found in Part XI of the UNCLOS. The Area and its resources are the common heritage of mankind (Art. 136).[5] No State can claim or exercise sovereignty over any part of the Area, nor can any State or natural or juridical person appropriate any part thereof (Art. 137, para. 1). The Area can be used exclusively for peaceful purposes (Art. 141). All rights over the resources of the Area are vested in mankind as a whole, on whose behalf an international organization, that is the International Sea-Bed Authority,[6] is entitled to act. Activities in the Area are carried out for the benefit of mankind as a whole, irrespective of the geographical location of States, whether coastal or land-locked, and taking into particular consideration the interests and needs of developing States (Art. 140, para. 1). The ISBA provides for the equitable sharing of financial and other economic benefits derived from activities in the Area through an appropriate mechanism (Art. 140, para. 2).

For the first time in the historical development of international law of the sea a world regime based on the management of resources by an international organization was included in a treaty of codification. The common heritage of mankind is a third conceptual option (*tertium genus*) that applies to a particular kind of resources located in a specific marine space. It does not eliminate the traditional notions of freedom or sovereignty applying in the other marine spaces. But it provides for a different and much more equitable approach.

As is well known, the text of the UNCLOS was submitted to vote after all efforts to reach consensus had been exhausted. It received 130 votes in favour, 4 against and 17 abstentions. Many developed States were among those that cast a negative vote or abstained. The main criticisms were addressed to the regime of the Area. According to the developed States, it would have discouraged mining activities by individual States and private concerns, would have unduly favoured the monopoly of activities by the ISBA, would have burdened the contractors with excessive financial and other obligations relating also to the field of transfer of technology and would have disregarded the interests of industrialized countries in the decision-making procedures of the Council, the executive organ of the ISBA.

In 1994 it was clear that the UNCLOS was expected to formally enter into force without the participation of many developed countries, that is without the participation of the States having the command of the high technological and financial capability required to engage in deep seabed mining activities. To avoid this danger the United Nations promoted a new negotiation on Part XI of the UNCLOS. It resulted in the Agreement Relating to the Implementation of Part XI of the UNCLOS, which was annexed to Resolution 48/263, adopted by the General Assembly on 17 August 1994. This resolution, while reaffirming that the Area and its resources are the common heritage of mankind, recognizes that 'political and economic changes, including in particular a growing reliance on market principles, have necessitated the re-evaluation of some aspects of the regime for the Area and its resources'.

The provisions of the Implementation Agreement and those of Part XI of the UNCLOS 'shall be interpreted and applied together as a single instrument' (Art. 2). However, in the event of any inconsistency between the Implementation Agreement and Part XI, the provisions of the former shall prevail. In fact, the label of 'implementation agreement' is a diplomatic device that covers the evident reality that the UNCLOS was amended and several aspects of the original concept of common heritage of mankind were changed.

Following the adoption of the Implementation Agreement, the UNCLOS has today achieved a broad participation (with some notable exceptions, such as the United States). Although modified, the original spirit of the UNCLOS is not betrayed. The principle of common heritage of mankind is still there and remains a major source of inspiration for a treaty that achieves the codification and the progressive development of international law.

For several years the ISBA has been working on its mandate. In 2000 the ISBA Assembly approved the Regulations on prospecting and exploration for polymetallic nodules and in 2010 the Regulations on prospecting and exploration for polymetallic sulphides.[7] The approval of a third set of regulations, relating to prospecting and exploration for cobalt-rich ferromanganese crusts took place in 2012.[8] Several plans of work for exploration for polymetallic nodules and for polymetallic sulphides have been approved by the ISBA Council. On 1 February 2011 the Seabed Disputes Chamber of the International Tribunal for the Law of the Sea rendered an advisory opinion on *Responsibilities and Obligations of States Sponsoring Persons and Entities with Respect to Activities in the Area*, as requested by the ISBA Council, which provides important clarifications on several aspects of the mining regime.

However, the prospects coming from the mineral resources in the Area remain uncertain, as a number of factors have a negative impact on progress towards their commercial exploitation. These factors include the great depths at which deposits occur, the high costs involved in research and development of mining technology and the fact that, under current economic conditions, deep seabed mining remains uncompetitive if compared to land-based mining.

The genetic resources beyond national jurisdiction

While the prospects for commercial mining in the deep seabed are uncertain, the exploitation of genetic resources found beyond the limits of national jurisdiction may in the near future become a commercially profitable activity.

The deep seabed is not a desert, despite extreme conditions of cold, complete darkness and high pressure. It is the habitat of diverse forms of life associated with typical features, such as hydrothermal vents, cold water seeps, seamounts or deep water coral reefs. In particular, it supports forms of life that present unique genetic characteristics. For instance, animal communities of micro-organisms, fish, crustaceans, polychaetes, echinoderms, coelenterates and molluscs live in the complete absence of sunlight where warm water springs from tectonically active areas (so-called hydrothermal vents).[9] These communities, which do not depend on plant photosynthesis for their survival, rely on specially adapted micro-organisms able to synthesize organic compounds from the hydrothermal fluid of the vents (chemosynthesis).[10] The ability of some deep seabed organisms to survive extreme temperatures (thermophiles and hyperthermophiles), high pressure (barophiles) and other extreme conditions (extremophiles) makes their genes of great interest to science and industry. But marine genetic resources are related not only to organisms from the seabed and its subsoil, but also to those present in the water column.

A difficult question to be addressed is about the international regime applying to genetic resources in areas beyond national jurisdiction.[11] Neither the UNCLOS nor the 1992 Convention on Biological Diversity[12] provides any specific legal framework in this regard. The factual implications of the question are pointed out in a document issued in 2005 by the Subsidiary Body on Scientific, Technical and Technological Advice (SBSTTA) established under the CBD.[13]

Only a few States and private entities have access to the financial means and sophisticated technologies needed to reach the deep seabed:

> Reaching deep seabed extreme environments and maintaining alive the sampled organisms, as well as culturing them, requires sophisticated and expensive technologies. ... Typically, the technology associated with research on deep seabed genetic resources involves: oceanographic vessels equipped with sonar technology, manned or unmanned submersible vehicles; *in situ* sampling tools; technology related to culture methods; molecular biology technology and techniques; and technology associated with the different steps of the commercialization process of derivates of deep seabed genetic resources. With the exception of basic molecular biology techniques, most of the

technology necessary for accessing the deep seabed and studying and isolating its organisms is owned by research institutions, both public and private. To date, only very few countries have access to these technologies.[14]

The prospects for commercial applications of bioprospecting activities, that is, what is currently understood as the search for commercially valuable genetic resources, seem promising:

Deep seabed resources hold enormous potential for many types of commercial applications, including in the health sector, for industrial processes or bioremediation. A brief search of Patent Office Databases revealed that compounds from deep seabed organisms have been used as basis for potent cancer fighting drugs, commercial skin protection products providing higher resistance to ultraviolet and heat exposure, and for preventing skin inflammation, detoxification agents for snake venom, anti-viral compounds, anti-allergy agents and anti-coagulant agents, as well as industrial applications for reducing viscosity.[15]

A further important remark is that the patent legislation of several States does not compel the applicant to disclose the origin of the genetic materials used:

Assessing the types and levels of current uses of genetic resources from the deep seabed proves relatively difficult for several reasons. First, patents do not necessarily provide detailed information about practical applications, though they do indicate potential uses. Moreover, information regarding the origin of the samples used is not always included in patent descriptions.[16]

The 2011 report of the United Nations Secretary-General on 'Oceans and the law of the sea' provides the following information on the relevant commercial developments:

Recent work has focused in discerning the degree to which genetic resources from areas beyond national jurisdiction have contributed to commercial developments, such as patents applied for and granted. To date, it appears that a very small number of patents have originated from the seabed beyond national jurisdiction (generally related to deep-sea bacteria), while a great number have been used on genetic resources from the high seas (primarily micro-organisms, floating sargassum weed, fish and krill). Of concern are applications with potentially large environmental consequences, such as the proposed use of sargassum weed for biofuels.[17]

Common heritage of mankind vs. freedom of the high seas

In 2006 the subject of the international regime for the genetic resources in the deep seabed began to be discussed within the Ad Hoc Open-ended Informal Working Group to Study Issues Relating to the Conservation and Sustainable Use of Marine Biological Diversity beyond Areas of National Jurisdiction,[18] established under United Nations General Assembly Resolution 60/30 of 29 November 2005. Opposite views were put forward by the States concerned.

Some States took the position that the UNCLOS principle of common heritage of mankind applied to marine genetic resources and that the mandate of the ISBA also covered such resources:

Several delegations reiterated their understanding that the marine genetic resources beyond areas of national jurisdiction constituted the common heritage of mankind and recalled article 140 of the Convention, which provides that the activities in the Area shall be carried out for the benefit of mankind and that particular consideration should be given to the interest and needs of developing States, including the need for these resources to be used for the benefit of present generations and to be preserved for future generations. . . . A number of delegations mentioned that the International Seabed Authority constituted an existing mechanism in this area and that consideration should accordingly be given to the possibility of broadening its mandate.[19]

Other States relied on the principle of freedom of the high seas, which would imply a right of freedom of access to, and unrestricted exploitation of, deep seabed genetic resources:

Other delegations reiterated that any measures that may be taken in relation to genetic resources in areas beyond national jurisdiction must be consistent with international law, including freedom of navigation. In their view, these resources were covered by the regime of the high seas, which provided the legal framework for all activities relating to them, in particular marine scientific research. These delegations did not agree that there was a need for a new regime to address the exploitation of marine genetic resources in areas beyond national jurisdiction or to expand the mandate of the International Seabed Authority.[20]

Again, in the 2008 meeting of the Working Group, different views were expressed as regards the regime to be applied to marine genetic resources, repeating what had already taken place in 2006.[21] The same occurred during the 2010 meeting.[22]

This basic disagreement on the international regime of genetic resources leaves a sentiment of dissatisfaction. In fact, both the divergent positions move from the same starting point: 'The United Nations Convention on the Law of the Sea was recognized as the legal framework for all activities in the oceans and seas, including in respect of genetic resources beyond areas of national jurisdiction'.[23] Why do two groups of States, moving from the same assumption, namely that the UNCLOS is the legal framework for all activities taking place in the sea, reach two completely opposite conclusions as regards the matter in question? A possible answer to the question is that the very starting point may not be completely true.[24]

The UNCLOS cannot stop the passing of time

There is no doubt that the UNCLOS is a cornerstone in the field of codification of international law. It has been rightly qualified as a 'constitution for the oceans', 'a monumental achievement in the international community', 'the first comprehensive treaty dealing with practically every aspect of the uses and resources of the seas and the oceans', an instrument that 'has successfully accommodated the competing interests of all nations'.[25]

Nevertheless, the UNCLOS, as any legal text, is linked to the time when it was negotiated and adopted (from 1973 to 1982). Being itself a product of time, the UNCLOS cannot stop the passing of time. While it provides a solid basis for the regulation of many matters, it would be illusory to think that the UNCLOS is the end of legal regulation. International law of the sea is subject to a process of evolution and progressive development that also involves the UNCLOS.

In particular, the UNCLOS cannot make the miracle of regulating those activities that were not foreseeable in the period when it was being negotiated. At this time, very little was known about the genetic qualities of deep seabed organisms. For evident chronological reasons, the potential economic value of the units of heredity of this kind of organism was not taken into consideration by the UNCLOS negotiators. When dealing with the special regime of the Area and its resources, they had only mineral resources in mind. This is fully evident from the plain text of the UNCLOS. While the term 'activities' in the Area is defined as 'all activities of exploration for, and exploitation of the resources of the Area' (Art. 1, para. 1), the 'resources' of the Area are defined as 'all solid, liquid or gaseous mineral resources *in situ* in the Area at or beneath the seabed, including polymetallic nodules' (Art. 133, *a*).[26] The UNCLOS regime of common heritage of mankind does not include the non-mineral resources of the Area.

For the same chronological reasons, the regime of freedom of the high seas cannot apply to genetic resources either. The UNCLOS does not provide any specific regime for the exploitation of marine genetic resources. The words 'genetic resources' or 'bioprospecting' do not appear anywhere in the UNCLOS. A legal gap exists in this regard. Sooner or later it should be filled (better sooner than later) through a regime which, to be consistent, would probably encompass under the same legal framework the genetic resources of both the Area and the superjacent waters.

However, not all of the UNCLOS should be left aside when envisaging a future regime for marine genetic resources beyond national jurisdiction. The scope of the regime of the Area is already broader than it may be believed at first sight. Under the UNCLOS, the legal condition of the Area also has an influence on the regulation of activities that, although different from mining activities, are also located in that space.[27] The regime of the Area already encompasses subjects that are more or less directly related to mining activities, such as marine scientific research,[28] the preservation of the marine environment[29] and the protection of underwater cultural heritage.[30] As far as the first two subjects are concerned, it is difficult to draw a clear-cut distinction between what takes place on the seabed and what in the superjacent waters.

While a specific regime for genetic resources is lacking, the aim of sharing the benefits among all States, which was one of the main aspects of the seminal proposal made by Mr Pardo, can still be seen as a basic objective embodied in a treaty designed to:

> contribute to the realization of a just and equitable international economic order which takes into account the interests and needs of mankind as a whole and, in particular the special interests and needs of developing countries, whether coastal or land-locked.
>
> *(UNCLOS preamble)*

Also in the field of genetic resources, the application of the principle of freedom of the sea (that is the 'first-come-first-served' approach) leads to inequitable and hardly acceptable consequences.[31] New cooperative schemes, based on provisions on access and sharing of benefits, should be envisaged within a relevant future regime.[32] This would also be in full conformity with the principle of fair and equitable sharing of the benefits arising out of the utilization of genetic resources set forth by Art. 1 of the CBD and by Art. 10 of the Protocol on Access to Genetic Resources and the Fair and Equitable Sharing of Benefits Arising from their Utilization, adopted in Nagoya in 2010.[33]

Possible future developments

At the 2008 meeting of the Working Group, the member States of the European Union suggested not to insist on an unproductive dispute about the principles of common heritage of mankind or freedom of the sea, but to envisage practical measures to get out of the deadlock. In this regard, the member States of the European Union stated that the determination of options for the sharing benefits was among the practical measures in question:

> although . . . to date very few products based on MGR [marine genetic resources] from ABNJ [areas beyond national jurisdiction] have been commercialised, the EU holds the view that states should seriously discuss options for facilitating access to samples of MGR that have been collected from ABNJ as well as for sharing in a fair and equitable way benefits that may arise in this regard.[34]

To establish a specific regime, a third UNCLOS implementation agreement was envisaged:

> Several delegations considered that an implementation agreement under the United Nations Convention on the Law of the Sea was the most effective way to establish an integrated regime and address the multiplicity of challenges facing the protection and sustainable use of marine biodiversity in areas beyond national jurisdiction. These delegations suggested that such an instrument was necessary to fill the governance and regulatory gaps that prevented the international community from adequately protecting marine biodiversity in the areas beyond national jurisdiction. It was proposed that such an instrument would address currently unregulated activities, ensure consistent application of modern ocean governance principles in sectoral management regimes and provide for enhanced international cooperation.[35]

At the 2011 meeting of the Working Group, a notable number of States, both developed and developing, proposed the commencement of a negotiation process towards a new implementation agreement of the UNCLOS that could fill the gaps in the present regime of conservation and sustainable use of marine biological diversity in areas beyond national jurisdiction, including the question of the sharing of benefits from genetic resources. However, this suggestion was not shared by some developed States, including the United States and the Russian Federation, which were only in favour of a better implementation of the already existing agreements. In its recommendations, the 2011 meeting of the Working Group made reference to both possibilities (in fact, they should not be seen as one excluding the other):

a) A process initiated by the General Assembly, with a view to ensure that the legal framework for the conservation and sustainable use of marine biodiversity in areas beyond national jurisdiction effectively addresses those issues by identifying gaps and ways forward, including through the implementation of existing instruments and the possible development of a multilateral agreement under UNCLOS.

b) This process would address the conservation and sustainable use of marine biodiversity in areas beyond national jurisdiction, in particular, together and as a whole, marine genetic resources, including questions on the sharing of benefits, measures such as area-based management tools, including marine protected areas, and environmental impact assessments, capacity-building and transfer of marine technology.

c) This process would take place: (i) in the existing Working Group; and (ii) in the format of intersessional workshops aimed at improving understanding of the issues and clarifying key questions as an input to the work of the Working Group.[36]

By Resolution 66/231 of 24 December 2011 on 'Oceans and the Law of the Sea' (para. 167 and annex), the United Nations General Assembly followed the recommendations by the Working Group and decided to start a process to be developed in the Working Group and through intersessional workshops.

During the discussions held at the 2012 meeting of the Working Group on the question of marine genetic resources, different views were expressed on the crucial issue of sharing of benefits:

> Discussions on the types of benefits envisioned, as well as examples of sharing of those benefits were called for. In relation to the sharing of benefits arising from the use of marine genetic resources from areas beyond national jurisdiction, a suggestion was made to consider information sharing and assess whether benefit sharing was desirable and, if so, to what extent and how this could be best achieved. Some delegations were of the view that the experience gained from the implementation of the Nagoya Protocol, along with other instruments such as the International Treaty on Plant Genetic Resources for Food and Agriculture of the Food and Agriculture Organization of the United Nations, could usefully be considered. Concern was expressed that a new legal regime for benefit sharing would impede research and development in that regard. The view was expressed that the greatest benefits from these resources would come from the availability of the products that were made and the contributions of these products to public health, food security and science. Marine scientific research related to marine genetic resources was thus important and should be promoted.
>
> It was suggested that practical mechanisms and options for benefit sharing also include addressing monetary and non-monetary benefits for equitable distribution; fostering effective participation of developing countries in strategic alliances between public sector scientific institutions and private sector biotechnology companies; and establishing research chains beginning in universities and culminating in industry. The view was expressed that the benefits from research and prospecting could be shared in an equitable manner, consistent with the goals of the Convention.[37]

In fact, what is needed for the time being is the consolidation of a general understanding on a number of 'commonalities' that could become the key elements in the 'package' for a future broad regime for the conservation and sustainable use of marine biodiversity in areas beyond national jurisdiction, encompassing a network of marine protected areas, environmental impact assessment, marine genetic resources, including access to and sharing of benefits from them, as well as capacity building and transfer of technology. Again, several States, although not all, were of the view that a new UNCLOS implementation agreement should be developed to cover such a broad range of issues:

> Some delegations emphasized that the conservation and sustainable use of marine biodiversity beyond areas of national jurisdiction was the overall objective of developing an implementing agreement under the Convention. In their view this agreement should cover building blocks such as marine genetic resources, including questions on the sharing of benefits, measures such as area-based management tools, including marine protected areas and environmental impact assessments, capacity-building and the transfer of marine technology.[38]

As an input to its work, the Working Group requested the Secretary-General to convene two intersessional workshops[39] on the topics of, respectively, 'marine genetic resources'[40] and 'conservation and management tools, including area-based management and environmental impact assessment'. The workshops were held in May 2013.

By Resolution 68/70, adopted on 9 December 2013, the UN General Assembly requested, inter alia, the Working Group to make recommendations to the Assembly 'on the scope, parameters and feasibility of an international instrument under the Convention' (para. 198), in order to prepare the decision to be taken at the General Assembly sixty-ninth session (starting in September 2014).

In conclusion, it is today uncertain whether the principle of common heritage of mankind, which is a formidable tool in promoting a just and equitable international economic order, will in the future apply not only to the mineral resources, but also to the genetic resources in marine areas beyond national jurisdiction. The proposal to establish a new regime to ensure benefit sharing from such resources, which would encompass one of the main aspects of the principle of common heritage of mankind, is today opposed by a number of developed States. The discussions held on this topic within the Working Group have taken a slow pace that casts some doubts about an effective outcome in the near future.[41] This uncertain situation is reflected in the outcome declaration ('The Future We Want') adopted in 2012 by the United Nations Conference on Sustainable Development Rio + 20, which envisaged the deadline of the sixty-ninth session of the United Nations General Assembly for the decision on the development of a new agreement:

> We recognize the importance of the conservation and sustainable use of marine biodiversity beyond areas of national jurisdiction. We note the ongoing work under the General Assembly of an ad hoc open-ended informal working group to study issues relating to the conservation and sustainable use of marine biological diversity beyond areas of national jurisdiction. Building on the work of the ad hoc working group and before the end of the sixty-ninth session of the General Assembly we commit to address, on an urgent basis, the issue of the conservation and sustainable use of marine biological diversity of areas beyond national jurisdiction, including by taking a decision on the development of an international instrument under the United Nations Convention on the Law of the Sea.[42]

Post scriptum

On 19 June 2015, following the recommendations by the Working Group, the UN General Assembly adopted without vote Resolution 69/292, on the development of an international legally binding instrument under the United Nations Convention on the Law of the Sea on the conservation and sustainable use of marine biological diversity of areas beyond national jurisdiction. According to the resolution, a preparatory committee will be established and will convene in 2016 and 2017, 'to make substantive recommendations to the General Assembly on the elements of a draft text of an international legally binding instrument under the Convention'. Before the end of its seventy-second session (September 2018) and taking into account the report and the recommendations of the preparatory committee, the General assembly will decide on the convening and on the starting date of an intergovernmental conference under the auspices of the United Nations to elaborate the text of an international legally binding instrument under the Convention. The negotiations:

> shall address the topics identified in the package agreed in 2011, namely the conservation and sustainable use of marine biological diversity of areas beyond national

jurisdiction, in particular, together and as a whole, marine genetic resources, including questions on the sharing of benefits, measures such as area-based management tools, including marine protected areas, environmental impact assessments and capacity-building and the transfer of marine technology.

Notes

1 Hereinafter: UNCLOS.

2　The known resources of the seabed and of the ocean floor are far greater than the resources known to exist on dry land. The seabed and ocean floor are also of vital and increasing strategic importance. Present and clearly foreseeable technology also permits their effective exploration for military or economic purposes. Some countries may therefore be tempted to use their technical competence to achieve near-unbreakable world dominance through predominant control over the seabed and the ocean floor. This, even more than the search for wealth, will impel countries with the requisite technical competence competitively to extend their jurisdiction over selected areas of the ocean floor. The process has already started and will lead to a competitive scramble for sovereign rights over the land underlying the world's seas and oceans, surpassing in magnitude and in its implications last century's colonial scramble for territory in Asia and Africa. The consequences will be very grave: at the very least a dramatic escalation of the arms race and sharply increasing world tensions, also caused by the intolerable injustice that would reserve the plurality of the world's resources for the exclusive benefit of less than a handful of nations. The strong would get stronger, the rich richer, and among the rich themselves there would arise an increasing and insuperable differentiation between two or three and the remainder. Between the very few dominant powers, suspicions and tensions would reach unprecedented levels. Traditional activities on the high seas would be curtailed and, at the same time, the world would face the growing danger of permanent damage to the marine environment through radioactive and other pollution: this is a virtually inevitable consequence of the present situation.

(Pardo, 1975, p. 31)

3 The word 'heritage' itself, which renders the idea of the sound management of a resource to be transmitted to the heritors, was preferred to the word 'property', as the latter could have recalled the *jus utendi et abutendi* (right to use and misuse) that private Roman law gave to the owner: see Mann Borgese 1975, p. x.

4 A fifth element is the protection and preservation of the marine environment, which however applies to any kind of marine spaces.

5 The Area is limited to the seabed and does not include the superjacent waters, which fall under the regime of the high seas. Arts. from 116 to 120 apply to management and conservation of the living resources of the high seas.

6 Hereinafter: ISBA.

7 Prospecting and exploration are the first phases of mining activities. They should be followed by exploitation.

8 Unlike polymetallic nodules, which are found partially buried in areas of the deep seabed, sulphides and crusts are localized in their deposits.

9 Hydrothermal vents may be found both in the Area and on the seabed falling within the limits of national jurisdiction, according to the definition of continental shelf given by Art. 76 UNCLOS.

10 The discovery of hydrothermal vent ecosystems has given rise to a new theory as to how life began on earth. It could have originated and evolved in association with hydrothermal vents in the primeval ocean during the early Archaean period (about 4,000 million years ago).

11 See Glowka 1996; Scovazzi 2004; Arico and Salpin 2005; Leary 2006; Oude Elferink 2007; Millicay 2007; De la Fayette 2009; Armas-Pfirter 2009; Ridgeway 2009; Barnes 2010; Scovazzi 2010.

12 Hereinafter: CBD.

13 Status and Trends of, and Threats to, Deep Seabed Genetic Resources beyond National Jurisdiction, and Identification of Technical Options for their Conservation and Sustainable Use, doc. UNEP/CBD/SBSTTA/11/11 of 22 July 2005.

14 Doc. UNEP/CBD/SBSTTA/11/11 of 22 July 2005, paras. 12 and 13. 'A limited number of institutions worldwide own or operate vehicles that are able to reach areas deeper than 1,000 meters below the oceans' surface, and can therefore be actively involved in deep seabed research' (para. 16).

15 Doc. UNEP/CBD/SBSTTA/11/11 of 22 July 2005, para. 21.

16 Doc. UNEP/CBD/SBSTTA/11/11 of 22 July 2005, para. 22.

17 Doc. A/66/70 of 22 March 2011, para. 63.

18 Hereinafter: the Working Group.

19 Doc. A/61/65 of 20 March 2006, para. 71.

20 Doc. A/61/65 of 20 March 2006, para. 72.

21 Doc. A/63/79 of 16 May 2008, para. 32.

22 Doc. A/65/68 of 17 March 2010, paras. 70–72.

23 Doc. A/63/79 of 16 May 2008, para. 36.

24 Scovazzi 2009.

25 Koh 1983.

26 In so providing, the UNCLOS narrows the term 'resources' that was used in a more abstract and broad sense in Art. 1 of the already mentioned General Assembly Resolution 2749 (XXV).

27   The principle of common heritage in its substantive aspect is, like any norm of international law, capable of being applied in a decentralised manner by states. Even in the absence of *ad hoc* institutions every state is under an obligation to respect and fulfil the principle of the common heritage by ensuring that subjects within its jurisdiction do not act contrary to its object and purpose. This would be the case if a state authorised or negligently failed to prevent biotechnological activities in common spaces that had the effect of causing severe and irreversible damage to the unique biodiversity of that space. Similarly, a state would fail the common heritage if it authorised exclusive appropriation of genetic resources without requiring equitable sharing of pertinent scientific knowledge and without ensuring that a fair portion of economic benefits accruing from their exploitation be devoted to the conservation and sustainable development of such common resources.

(Francioni 2007, p. 14)

28 Art. 143, para. 1, UNCLOS, provides that 'marine scientific research in the Area shall be carried out exclusively for peaceful purposes and for the benefit of the mankind as a whole'. Also bioprospecting, which is a form of marine scientific research, can be considered as covered by this provision.

29 Art. 145 UNCLOS.

30 Art. 149 UNCLOS.

31 See *supra*, under the heading 'The principle of common heritage of mankind'.

32 Art. 241 UNCLOS is also relevant in a discussion on the legal condition of the genetic resources of the deep seabed. It provides that 'marine scientific research activities shall not constitute the legal basis for any claim to any part of the marine environment or its resources'.

33 While the Nagoya Protocol does not apply to areas beyond national jurisdiction, it could become a source of inspiration for a future regime in this regard. Another source of inspiration could be the International Treaty on Plant Genetic Resources for Food and Agriculture, concluded in 2001 under the auspices of the Food and Agriculture Organization (FAO).

34 Statement made on 30 April 2008 on behalf of the member States of the European Union by Amb. Čičerov of Slovenia and distributed at the 2008 session of the Working Group. According to the statement,

a reference point in this regard, in our view, is the International Treaty on Plant Genetic Resources for Food and Agriculture. This Treaty has established a so called 'multilateral system' that includes a negotiated selection of 64 plant genetic resources that are considered most important for world food security and the conservation and sustainable use of plant genetic resources for food and agriculture. Material held in the multilateral system is effectively put into the public domain. It can easily be accessed, provided that the recipient of material commits to comply with pre-determined conditions for the fair and equitable sharing of benefits. Recipients can choose between freely sharing any new developments with others for further research or, if they want to keep the developments to themselves, to pay a percentage of any commercial benefit derived from their research into a common fund. These funds are used for the additional benefit sharing mechanisms of building capacity, for access to and transfer of technology as well as for information sharing. The FAO ITPGRFA illustrates that the international community has been able to develop a functional multilateral system for handling plant genetic resources in the public domain. As such it may provide us with promising avenues for dealing with MGR in ABNJ.

35 Doc. A/63/79 of 16 May 2008, para. 47.
36 Doc. A/66/119 of 30 June 2011, para. 1.
37 Doc A/67/95 of 13 June 2012, paras. 18 and 19 of the Appendix.
38 Doc A/67/95 of 13 June 2012, para. 12 of the Appendix.
39 Doc A/67/95 of 13 June 2012, para. 1 of the Annex.
40 The workshop have addressed the following subjects: meaning and scope; extent and types of research, uses and applications; technological, environmental, social and economic aspects; access-related issues; types of benefits and benefit sharing; intellectual property rights issues; global and regional regimes on genetic resources experiences and best practices; impacts and challenges to marine biodiversity beyond areas of national jurisdiction; exchange of information on research programmes regarding marine bio-diversity beyond areas of national jurisdiction.
41 However, the two above-mentioned workshops ensured a wide participation of scientists, industry and non-governmental organizations and resulted in a well-informed presentation and discussion of the two topics addressed. See doc. A.AC/276/6 of 10 June 2013.
42 Doc. A/Conf.216/L.1 of 19 June 2012, para. 162.

References

Arico, S. and Salpin, C. (2005) *Bioprospecting of Genetic Resources in the Deep Seabed: Scientific, Legal and Policy Aspects*, United Nations University, Yokohama.

Armas-Pfirter, F. (2009) 'How Can Life in the Deep Seabed Be Protected?', 24 *International Journal of Marine and Coastal Law*, p. 281–307.

Barnes, R. (2010) 'Entitlement to Marine Living Resources in Areas beyond National Jurisdiction', in A. Oude Elferink and E. J. Molenaar (eds), *The International Legal Regime of Areas beyond National Jurisdiction: Current and Future Developments*, Martinus Nijhoff Publishers, Leiden, p. 81–141.

De la Fayette, L. (2009) 'A New Regime for the Conservation and Sustainable Use of Marine Biodiversity and Genetic Resources beyond the Limits of National Jurisdiction', 24 *International Journal of Marine and Coastal Law*, p. 221.

Francioni, F. (2007) 'Genetic Resources, Biotechnology and Human Rights: The International Legal Framework', in F. Francioni (ed.), *Biotechnologies and International Human Rights*, Hart Publishing, Oxford, p. 14.

Glowka, L. (1996) 'The Deepest of Ironies: Genetic Resources, Marine Scientific Research, and the Area', 10 *Ocean Yearbook*, p. 156.

Koh, T. (1983) 'A Constitution for the Oceans', in *The Law of the Sea* – Official Text of the United Nations Convention on the Law of the Sea with Annexes and Index, United Nations, New York, p. xxiii.

Leary, D. K. (2006) *International Law and Genetic Resources of the Deep Sea*, Martinus Nijhoff Publishers, Leiden.

Mann Borgese, E. (1975) Introduction, in A. Pardo, *The Common Heritage – Selected Papers on Oceans and World Order*, International Ocean Institute, Valletta. p. x.

Millicay, F. (2007) 'A Legal Regime for the Biodiversity in the Area', in M. Nordquist, R. Long, T. Heider and J. N. Moore (eds), *Law, Science and Ocean Management*, Martinus Nijhoff Publishers, Leiden, p. 739.

Oude Elferink, A. (2007) 'The Regime of the Area: Delineating the Scope of Application of the Common Heritage Principle and Freedom of the High Seas', 22 *International Journal of Marine and Coastal Law*, p. 143.

Pardo, A. (1975) *The Common Heritage – Selected Papers on Oceans and World Order*, International Ocean Institute, Valletta.

Ridgeway, L. (2009) 'Marine Genetic Resources: Outcomes of the United Nations Informal Consultative Process', 24 *International Law of Marine and Coastal Law*, p. 309.

Scovazzi, T. (2004) 'Mining, Protection of the Environment, Scientific Research and Bioprospecting: Some Considerations on the Role of the International Sea-Bed Authority', 19 *International Journal of Marine and Coastal Law*, p. 383.

Scovazzi T. (2009) 'Is the UN Convention on the Law of the Sea the Legal Framework for All Activities in the Sea? The Case of Bioprospecting', in D. Vidas (ed.), *Law, Technology and Science for Oceans in Globalisation*, Martinus Nijhoff Publishers, Leiden, p. 309.

Scovazzi, T. (2010) 'The Seabed beyond the Limits of National Jurisdiction: General and Institutional Aspects', in A. Oude Elferink and E. J. Molenaar (eds), *The International Legal Regime of Areas beyond National Jurisdiction: Current and Future Developments*, Martinus Nijhoff Publishers, Leiden, p. 43.

30

SURVEYING THE SEA

Robert Wilson

Introduction

The title of this chapter is 'Surveying the Sea', the process of which is more commonly known as hydrography, defined by the International Hydrographic Organization (IHO) as:

> that branch of applied sciences which deals with the measurement and description of the physical features of oceans, seas, coastal areas, lakes and rivers, as well as with the prediction of their change over time, for the primary purpose of safety of navigation and in support of all other marine activities, including economic development, security and defence, scientific research, and environmental protection.
>
> *(IHO Hydrographic Dictionary, 2012a)*

Hydrographic surveying has for centuries provided the primary data for nautical charts and publications. The requirement for these vital documents originated from our ancestors' need to navigate trading vessels safely along the world's steadily extending network of sea routes. Indeed, nothing has been more important to the foundation and expansion of the world's maritime trade than the production of nautical charts and publications from the results of the labours of the hydrographic surveyor (Southern, 1938, p. 1). However, modern hydrographic data – depth of water, configuration and nature of the seafloor, identification of submerged hazards, nature of the tidal regime and other marine environmental observations – is the base data from which an ever widening range of products and information services, not necessarily associated with the safety of navigation, are produced. Regrettably, and as acknowledged by the International Hydrographic Organization, despite centuries of hydrographic survey effort we have more and better data to describe the surface of the Moon and Mars than for most of the Earth's seas and oceans. (Decision 17, XVIIIth International Hydrographic Conference, Monaco, 2012a).

Chart evolution

The mariner has required and has been provided with hydrographic information from the earliest days of coastal voyaging. Asia Minor is credited as being the cradle of geography with the first geographical map emerging through the work of Anaximander of Miletus in 568 BCE.

Corinthian maritime forces conquered their way westwards through the Mediterranean probably aided by Phoenician hydrographic information, while the Athenian traveller, Scillax, journeyed beyond the Strait of Gibraltar and later published his discoveries in a periplus, a manuscript document listing ports and coastal landmarks with approximate distances between them, in the third or fourth century BCE (Dawson, 1885, Part 1, pp. 1–2).

Books of directions allowing the mariner to navigate his vessel safely along a coast and variously known as periplus in Classical times, 'routier' or 'Rutter' in the sixteenth century and as Sailing Directions today, slowly relinquished primacy to the nautical chart, a natural development of the geographic map, and designed specifically for the mariner. The advent of the use of the compass at sea in the twelfth century gave rise to longer voyages out of sight of land, thereby driving the need for the specialist nautical chart.

Evolution of the nautical chart has been continuous for centuries. Early charts did not take account of the shape of the earth, with cartographers depicting land masses in approximate relation to each other (Admiralty Manual, 1938, pp. 1–2). By the fifteenth century Portuguese and Spanish portolan charts enabled Mediterranean-based merchant ships to trade with southern England and northern France (IHO, 2005a, p. 2). A major step in the development of both charts and maps came with Mercator's invention of a mathematical projection and the publication of his Universal Map in the late 1560s. Scientific survey methods, better chart projections and more efficient printing methods steadily improved the accuracy and availability of charts.

The modern nautical chart

The paper chart remained the primary means of depicting marine navigational information until the late twentieth century brought the introduction of the first raster charts, in essence a relatively simple electronic facsimile of a paper chart, and later the full Electronic Navigational Chart or ENC. The ENC is not however a chart, but rather an electronic database for use in an Electronic Chart Display and Information System (ECDIS). An ECDIS not only allows the mariner to view a customised chart-like display of the data and view the ship's position, course and speed on it, but also enables its software to analyse the chart data in relation to the ship's position and automatically to alert the mariner to potential hazards in the vessel's vicinity.

The land map and the nautical chart have many similarities in their construction and character; however, the nautical chart, unlike its topographic counterpart is a living document requiring continuous maintenance. The chart by necessity portrays only essential information, therefore it needs to be constructed to highlight dangers and hazards to navigation above all else. As well as the natural, ever changing evolution of the seafloor due to such causes as erosion, moving sandbanks and the growth of reefs, man's influence on the seafloor and the coastal zone adds yet more change with new or modified coastal construction and inadvertent accidents casting wreckage on the seafloor. Short-term changes in the marine environment occur through the stress of weather on aids to navigation such as buoys and beacons, with more permanent changes occurring through such things as the establishment of traffic lanes for the safe routeing of ships. The list of factors causing changes to the nautical chart is extensive, yet the chart must always be kept up to date if navigational hazards are to be notified to the mariner and the safety of life and property at sea assured. Any and all new and relevant information affecting the safety of navigation at sea is known collectively as maritime safety information or MSI. Urgent, safety critical MSI is disseminated via radio navigational and meteorological warnings.

Early charts showed little depth information and what was shown was generally based on imprecise data; accurate survey methods had yet to be developed. However, the early charts still share one thing in common with their modern day successors; they are only as good as the data upon which they are based. This is problematic when many of the changes occur unexpectedly or unseen below the sea surface – and therefore are largely out of sight of the nautical cartographer. If the chart does not show all the relevant changes then the mariner cannot place complete confidence in the chart and its usefulness is diminished.

Surveys may be repeated when new and better technology becomes available, although this is not always the impetus. The latter half of the twentieth century saw a huge change in the type of vessels sailing the world's oceans and seas. Oil tankers built in 1945 with a length of 170 metres and a deadweight tonnage of 18,300 have given way to Ultra-Large Crude Carriers of up to 470 metres in length with a deadweight tonnage in excess of 500,000 tonnes. The increase in size and consequently draft of modern ships is not confined to oil and gas carriers but bulk carriers, container ships and, increasingly, cruise ships. Perhaps the most dramatic change has been that brought about by developments in the offshore oil and gas industries. Offshore oil and gas fields, once the preserve of shallow seas, have moved into deeper water as reserves are exhausted and demand and price increases. Detailed surveys for navigational safety are usually determined by the maximum draft of the ships in service; however, even the largest of today's ships drawing up to 25 metres are small in comparison with offshore gas and oil structures with drafts of over 90 metres. As the use of the maritime environment steadily increases so too do the pressures on its sustainability and preservation and the need for ever more detailed and accurate hydrographic data and services.

Surveying technology

A scientific approach to hydrographic surveying based on sound mathematical principles only arose in the late seventeenth century supported by the development of specific observation and plotting instruments. Towards the end of the eighteenth century the invention of the station pointer, a device with which a vessel's position could be precisely plotted by the observation of two horizontal angles between three fixed marks onshore, was a major technical advance that revolutionised sea surveying and was still in active use for certain applications up until the early 1980s. The station-pointer was followed by Harrison's chronometer. As well as allowing longitude to be calculated at sea, use of the chronometer enabled surveyors to determine the geographic position of significant navigational features such as headlands, capes and shoals as well as ports and harbours.

While developments were made in the ability of a surveyor to determine his position on the earth's surface, similar developments in the determination of depth came much later. The sounding lead, in its various guises, was to remain the primary means of depth determination until the advent of the echo sounder in the 1930s that, for the first time, gave a 'continuous' depth reading directly beneath the ship. The Second World War brought the next great advances with sonar and radio position finding. The advances in anti-submarine sonar searching technology were turned to peaceful use by hydrographers in the location of wrecks and other seafloor hazards; this enabled the hydrographer to 'look between' lines of depth sounding.

By far the greatest advances in survey methods have taken place in the last three decades. Radio-based position fixing systems were relatively short lived, being superseded by high-accuracy satellite positioning techniques – commonly known as GPS. Single beam echo sounders have also given way to swath sounding systems that can provide full seafloor mapping – using sound from ships, boats and underwater vehicles, and lasers from aircraft. This technology leaves less chance of seabed obstructions lying undetected, and can provide an accurate estimate of the depth over detected obstructions. Table 30.1 summarises hydrographic survey development.

Table 30.1 Survey methods table

Date	Sounding method	Position fixing method	Remarks
Pre–1865	Lead line	Visual angles to local landmarks	Surveys were mainly concerned with recording previously undiscovered land with more attention given to positioning the coast than to providing soundings. Soundings, if present, tended to be sparse, with irregular gaps between them.
1865			Steam replaced sail in survey ships and regular lines of soundings begin to appear. Where boats were used instead of ships, oars remained the method of propulsion, and sounding lines continued to be irregular.
1905			Steam replaced oars as the propulsion method for survey boats, allowing regular lines of soundings to be extended to all areas of the survey. The scale of the survey gives an indication for the first time of the expected density of soundings.
1930	SBES		Development of the single beam echo sounder (SBES) allowed the continuous collection of soundings along a ship or boat's track permitting far more detailed surveys to be undertaken.
1950	SBES and sonar	Electronic position fixing	Greater accuracy and consistency of position fixing extending farther offshore than was possible with observing visual angles to shore marks. Visual survey methods still in use for ports and inshore waters. Hull-mounted search sonars, developed during the Second World War, are now used to locate wrecks and obstructions between lines of sounding.
1975	SBES and SSS		Sidescan sonar (SSS) allows the surveyor to locate hazards that exist between lines of soundings. For the first time, surveyors will have examined the entire sea floor. Electronic positioning fixing now used for all surveying.
1990		GPS	Introduction of satellite positioning allows the surveyor to position accurately his ship anywhere in the world to a common horizontal datum.
2000	MBES and ALB		Ship and boat mounted multi beam swathe (MBES) sounding systems replace single beam echo sounders and sidescan sonar. Airborne laser bathymetry (ALB) developed and able to survey both topography and bathymetry simultaneously. Swathe systems permit the surveyor to detect obstructions between survey lines and to gather depths over them giving rise to complete seafloor mapping.
Today	SDB		The current development of satellite derived bathymetry (SDB) techniques pushes the boundaries of surveying through the ability to survey very large areas of land and sea quickly, efficiently and cheaply.

Source. Adapted and updated from NP100 *The Mariner's Handbook*, p. 7.

Our ability to locate our position on the Earth's surface, once a specialised surveying task, is now available to everyone through satellite-based technology even via mobile phones, but it is not only horizontal position that can be derived from satellite positioning data. The position supplied is three-dimensional, allowing the observer to determine height in relation to a fixed vertical datum; this third-dimension permits the real-time observation of offshore tidal heights that may be considered as the largest remaining error in the hydrographer's observations.

Each advance in technology (single beam echo sounding – sonar – GPS – swath sounding) has brought with it the need to repeat previous work and improve nautical charting in step with technological developments. The first distinctive change to charts came with the almost universal metrication of depths on charts from fathoms and feet and the adoption of IHO standard symbols and colours. Satellite positioning became available to mariners as well as hydrographers and so nautical charts had to be converted to a common horizontal datum to allow satellite derived positions to be plotted directly onto a chart without positional error. Today, traditional visual navigation methods and the paper chart are rapidly giving way to GPS positioning and ECDIS or similar electronic charting systems, whereby the ship's position is shown in real-time on a computer controlled display.

Where once the mariner accepted the 'two-dimensional' nautical products direct from the hydrographic office without question, the relationship between the mariner, the hydrographic office and the chart in the modern age is far more complex. An ENC used in an ECDIS is not so much a chart as a digital database that allows various chart-like displays to be created at will by the mariner; it is no longer a 'take it or leave it' two-dimensional paper product but a multi-layered digital version with the mariner given greater control of the navigational picture seen on the bridge. However, although the ENC is to a standard format, ECDIS is not. The user interface varies by manufacturer so although mariners may be using standard ENCs, they need to operate each ECDIS system differently, making the movement of mariners from ship to ship more problematic.

While there are clear advantages in the new digital age there are also potential problems. In the early days of radar, a 'radar assisted collision' was an early feature of its operation at sea until the technology was understood and experience gained in its use. So too, there are problems with ECDIS, whereby the mariner tends to assume that what is seen on a digital display is modern and fit for purpose, oblivious to the fact that the underlying information is often based on old and less reliable data.

National hydrographic services

National hydrographic services emerged in the early eighteenth century. In the colonial period the various imperial powers undertook surveying and charting of their dominions as part of supporting their maritime trade interests, trade routes and command of the sea. The French Hydrographic Office was the first to be established in 1720, the Danish office in 1784 and the British in 1795, although the Honourable East India Company had already been operating its own hydrographic office for over a century by then. These offices and others now in operation have played a significant part in improving the safety and efficiency of seaborne trade but, like other government-supplied utilities, they are frequently taken for granted – until things go wrong.

The decline of empires following the end of the Second World War significantly altered the hydrographic role undertaken by the imperial powers. A mostly unacknowledged consequence of independence for most new maritime States was the implicit transfer of responsibility for national hydrography to the independent States. Unfortunately, many of these now independent States remain largely unaware of this change. So while there has been an ever growing expansion of maritime interests and global dependence on the sea, the surveys to support the existing charts are ageing and increasingly becoming unfit for purpose. There is now a growing concern that too many maritime States are failing in their support for hydrography, despite the increasingly well publicised fact that there are many areas of the world that still lack adequate surveys, up-to-date nautical charts and supporting services.

The IHO has reported that worldwide the number of government survey vessels has declined by 35 per cent in the last 30 years – contract surveys, improved equipment capability and other options have not filled the gap (IHO XVIIIth Conference, 2012b). This situation poses a very real threat to the safety of life at sea and to the marine environment as well as to the sustainable management and development of the seas and oceans as a significant national and global economic resource. At the same time, it must be acknowledged that emerging States have many other national priorities: hydrography is invariably towards the bottom of the list, if on the list at all.

The IHO

International cooperation has long been a feature of hydrography. After the end of the First World War many nations took a renewed interest in hydrography and its international dimension. Such was the interest that the national Hydrographers of France and the United Kingdom proposed the first international hydrographic conference which was held in London in 1919. The conference's crucial resolution established the International Hydrographic Bureau (IHB) 'to ensure effective and continuous cooperation between Hydrographic Offices'. Prince Albert I of Monaco, an eminent marine scientist and explorer, provided the International Hydrographic Bureau's headquarters in Monaco. The IHB was formally founded on 21 June 1921 with 18 Member States (IHO, 2005b). In 1970, an intergovernmental Convention entered into force changing the IHB's name and legal status, creating the International Hydrographic Organization (IHO), with its headquarters and secretariat still based in Monaco. The IHO is recognised by the UN as its competent authority in the fields of hydrography and nautical charting. UN instruments usually refer to the IHO standards and guidelines when it comes to such matters.

The mission of the IHO is to create a global environment in which States provide adequate and timely hydrographic data, products and services and ensure their widest possible use (IHO M-2, 2015, p. 33). The IHO currently (2012) comprises 85 Member States and has two principal programmes: a technical programme devoted to the publication of the international standards and guidelines upon which hydrographic and nautical charting services are provided around the world, and a regional coordination and capacity building programme.

The IHO has always sought to encourage and develop the improvement of surveying and charting standards but, while standards are fine, they are useless unless surveys are actually being conducted and modern up-to-date charts are being produced. The IHO established a Capacity Building Committee in 2003 to 'assess and assist in the sustainable development and improvement of the States, to meet the objectives of the IHO and the Hydrography, Cartography and Maritime Safety obligations and recommendations described in UNCLOS, SOLAS V and other international instruments'. Capacity Building (CB) is considered by the IHO as a strategic objective (IHO Website, n.d. Capacity Building Concept and Strategy).

There are three levels of achievement within the IHO CB programme: the provision of Maritime Safety Information (MSI), the conduct of hydrographic surveying, and the production and maintenance of nautical chart and associated publications. The IHO considers that all States, regardless of development should, and must, be able to achieve the first level – MSI. This level

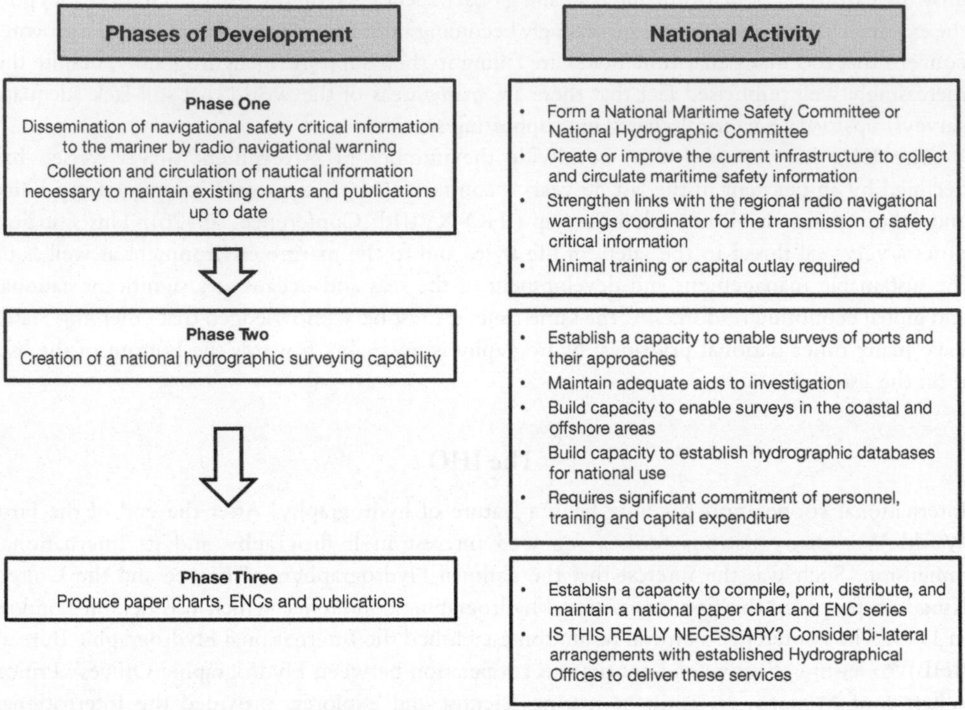

Phases of Development	National Activity
Phase One Dissemination of navigational safety critical information to the mariner by radio navigational warning Collection and circulation of nautical information necessary to maintain existing charts and publications up to date	• Form a National Maritime Safety Committee or National Hydrographic Committee • Create or improve the current infrastructure to collect and circulate maritime safety information • Strengthen links with the regional radio navigational warnings coordinator for the transmission of safety critical information • Minimal training or capital outlay required
Phase Two Creation of a national hydrographic surveying capability	• Establish a capacity to enable surveys of ports and their approaches • Maintain adequate aids to investigation • Build capacity to enable surveys in the coastal and offshore areas • Build capacity to establish hydrographic databases for national use • Requires significant commitment of personnel, training and capital expenditure
Phase Three Produce paper charts, ENCs and publications	• Establish a capacity to compile, print, distribute, and maintain a national paper chart and ENC series • IS THIS REALLY NECESSARY? Consider bi-lateral arrangements with established Hydrographical Offices to deliver these services

Figure 30.1 Capacity building flow diagram
Source: IHO Website, n.d., Capacity Building Concept and Strategy.

requires States to collect information relevant to the safety of the mariner and the maintenance of charts and to promulgate this information through a system of radio navigational warnings and to enable amendments to be made to existing charts and nautical publications. While minimal resources are required to achieve level 1, many States fail to achieve it. Level 2, Hydrographic Surveying, is far more demanding; it requires significant capital expenditure on equipment and the training and employment of surveying staff. Many States can attain this level; however, it is remarkable that there are many coastal States with the financial capability to achieve this level who have not done so. Equally remarkably, there are conscientious and enlightened States that have achieved this level with apparently limited resources. The highest level, Nautical Charting, is the most demanding of all, and carries with it requirements for advanced and sustained technical infrastructure. For this level it is frequently more efficient and realistic for States to form bilateral partnerships with other States that have an existing national Hydrographic Office and to authorise them to produce, maintain and distribute charts and publications on their behalf. There are a number of coastal States that do this. Figure 30.1 illustrates the development phases and the national activity required to achieve the various CB levels.

Regional hydrographic coordination

National, international and global hydrographic programmes are coordinated through a series of regional groupings and through the organisational structure of the IHO. The administration of the IHO, as an inter-governmental body, is carried out by the small secretariat staff at the

IHB in Monaco. The work programme activities of the IHO are carried out through a series of committees and working groups made up of representatives of Member States together with Observer organisations and relevant industry participants. Work is also coordinated at the regional level through 15 Regional Hydrographic Commissions (RHCs) each individually constituted and independent of the IHO. Made up primarily of IHO Member States, RHCs work to help further the ideals and programme of the IHO set down by its Member States. They meet at regular intervals to discuss such matters as mutual hydrographic and chart production problems, joint survey operations, and to resolve schemes for medium and large-scale International Chart coverage in their regions.

The RHC's work is in turn coordinated by one of the two main committees of the IHO, the Inter Regional Coordination Committee (IRCC). The other committee is the Hydrographic Standards and Services Committee (HSSC) that oversees the IHO technical programme through which the various international standards and guidelines governing hydrography and nautical charting services are developed and maintained.

The IRCC's principal objectives as defined by the IHO are to:

> Establish, coordinate and enhance cooperation in hydrographic activities amongst States on a regional basis, and between regions; establish co-operation to enhance the delivery of capacity building programs; monitor the work of specified IHO inter-organizational bodies engaged in activities that require inter-regional cooperation and coordination; promote co-operation between pertinent regional organizations and review and implement the IHO Capacity Building Strategy, promoting Capacity Building initiatives.
>
> *(IHO Website, Inter-Regional Coordination Committee n.d.)*

An element of the monitoring work conducted by each RHC is oversight of the general state of hydrographic surveys and charting within the RHC region; details of each Member State's hydrographic situation are published in IHO publication C-55 – *Status of Hydrographic Surveying and Nautical Charting Worldwide* (IHO, 2012c).

The IHO, through its regional coordination programme has been responsible for one particularly important charting concept, the International Chart Series. The aim of the International Chart Series is to ensure world coverage of standardised paper charts suitable for the navigational requirements of international shipping. This concept also allows IHO Member States who provide, or wish to provide, charts outside their own national waters, to print by facsimile reproduction selected international charts under the terms of a bilateral arrangement between the Member States. Put forward at the IHO Conference in 1967, the concept is now very much a reality (IHO International Charts, 2012d, pp. 1–2).

Status of hydrographic surveys and charting worldwide

While some coastal States may have an understanding of the state of hydrographic surveying within their seas, many do not, and a global picture is difficult to assess. In 1970 the UN raised this question and began a process to evaluate the state of hydrographic surveys, surveying and nautical charting worldwide. The IHO was invited to undertake this task and publishes the current known state in a continuously revised publication C-55 *Status of Hydrographic Surveying and Nautical Charting Worldwide* (IHO, 2012c). While improvements in the world situation have been noted, due in part to the IMO stipulating that traffic routeing measures adopted internationally must be supported by surveys conducted according to IHO standards and through improved regional cooperation to provide modern surveying and charting coverage along shipping routes, the third edition of C-55 still concludes that:

- many governments have still to put in place an effective organisation for the promulgation of information of importance to safe navigation and the protection of the environment, either as navigational warnings or as inputs to those hydrographic offices with responsibility for charting;
- action is needed to implement the Global Maritime Distress and Safety System (GMDSS) in a number of areas, notably in Central America and the Caribbean, most of Africa, and the oceanic areas;
- many coastal States lack the capacity to plan and implement a prioritised survey programme, including top priority routine re-survey of unstable areas along shipping routes and in the approaches to ports;
- failure to apply IHO S-44 (survey) criteria in Marine Scientific Research and offshore industrial surveys leads to lost opportunity data for SOLAS charting purposes.

The passage of time has failed to improve the hydrographic situation in a number of critical areas. Large gaps remain around major international shipping routes in the Indian Ocean, South China Sea, West Pacific and adjacent waters. In the Caribbean, some coastal waters of Africa, Australasia, Oceania and the Polar Regions, modern surveys, the modernization of nautical charts and making charts compatible with satellite-based positioning systems are all urgent requirements in locations that are now frequented by cruise ships (IHO C-55, 2012c).

Data exchange and sharing

To maintain and improve charting there has been a centuries old tradition of data exchange between national hydrographic offices (NHOs). Indeed, the IHO was formed to help foster such exchange. The exchange is made to ensure charts contain the best information available and because all NHOs are, in effect, providing a safety and information infrastructure service that has all the characteristics of a 'public good' service. In other words, it is impossible to recover the full cost of collecting, processing and maintaining the service through sales (Coochey, 1993). However, by the early 1990s a feeling was growing among some IHO Member States that the value of the data that they were providing to other States for inclusion in charts was not being properly recognised. As a result, IHO Technical Resolution 3.4 recognised that copyright or royalty payments should be considered as part of any data exchange arrangement between NHOs. Since this decision, bilateral agreements and cooperation arrangements between not only NHOs but NHOs and others, including port authorities, and the commercial sector have proliferated. The United Kingdom Hydrographic Office, one of the largest NHOs, has agreements with 83 other NHOs, and within the United Kingdom with some 98 port and harbour authorities. The UKHO currently pays in excess of six million pounds per annum in royalties. Other NHOs have been able to defray at least a small part of their operating costs through royalties and payments in kind.

Non charting uses for hydrographic data

Hydrographic data is a scarce and valuable resource. Of the Earth's surface, 71 per cent is covered with water and the IHO estimates that only about 10 per cent of the seafloor has been surveyed by echo sounders at a resolution of 1 minute or better. In 1990 the United States National

Geographic Data Centre established, on behalf of the IHO, the IHO Data Centre for Digital Bathymetry which collects and quality checks oceanic soundings acquired by hydrographic, oceanographic and other vessels. IHO Member States also contribute shallow water bathymetric data and all of these data contribute to the production of more accurate and comprehensive bathymetric maps and grids (IHO World Bathymetry, 2012e).

Hydrographic data is valuable not simply because it is expensive to collect, but because it is of great use to many, emphasising the adage 'gather once – use many times'. An increasing number of governments are beginning to appreciate that good-quality, well-managed spatial data is an essential component of economic and commercial development, and environmental protection – and this includes hydrography. For this reason, many nations have established national spatial data infrastructures (SDI) encompassing such elements as topography, geodesy, geophysics and meteorology. Latterly, and with the availability of high-density data sets obtained from modern equipment, bathymetric and hydrographic parameters are being included either within existing SDIs or as stand-alone Maritime Spatial Data Infrastructures (MSDIs) (IHO M-2, 2015). One area where MSDIs are especially valuable is as a tool for coastal zone management.

There are many more requirements for hydrographic data than just the safety of navigation. Oil, gas and mineral exploration and exploitation require a detailed knowledge of both the seafloor and sub seafloor, coastal facilities for this and other marine activities require careful coastal zone management supported by a detailed knowledge of all the parameters in the coastal zone. The coastal zone itself is used for fishing, recreation and also by those less well aligned to the coastal State's well-being, thereby requiring its maritime defence forces to work safely and efficiently in national waters. Locating safe routes for the submarine cables that carry 93 per cent of the world's intercontinental digital data and telephone traffic requires hydrographic surveys. Marine science is particularly important in monitoring our planet's condition and survival and depends largely on hydrographic information as a fundamental backdrop to help make sense of the science. (IHO M-2, 2015).

National hydrographic data collected in coastal zones provides essential data to MSDIs which are increasingly being used for better overall management and decision making in relation to conflicting requirements and developments within the coastal region. The users of hydrographic data extend beyond the traditional mariner to include government agencies, coastal managers, engineers, scientists, environmentalists and others. (IHO M-2, 2015, p. 14) Coastal zone management has many facets: port design and construction and the development and maintenance of existing ones including dredging operations are probably the most obvious. However, anything built out into the sea or altering the natural flow of tides and currents will alter the hydrography and topography of the coastal zone somewhere. The monitoring and control of coastal erosion, land reclamation from the sea, the establishment and monitoring of dumping grounds for industrial waste, extraction of mineral deposits, developing aquaculture activities come under the broad heading of coastal zone management. High accuracy large-scale surveys provide the primary data essential for coastal zone projects and here the combined knowledge, current and archival, held within a national SDI or MSDI is invaluable.

Global bathymetric data is fundamental to the mapping of our oceans and thereby gaining an understanding of how the earth's global geomorphic, oceanographic and meteorological systems interact. The shape of the ocean basins, ridges and mountains influence the flow of sea water carrying heat, salt, nutrients and pollutants while also influencing the propagation of energy resulting from undersea seismic events, producing potential disasters such as tsunamis. National hydrographic data is routinely published for national waters and ocean passages on standard nautical charts and held in national SDIs; however, the collation and publication of international data in the oceanic region for scientific purposes is a separate issue.

In 1899, at the Seventh International Geographic Congress, a Commission on sub-oceanic nomenclature was formed that, probably more importantly, had the responsibility for the publication of a general bathymetric chart series of the oceans or GEBCO as it is referred to today. Twenty-four sheets covering the world, the Carte Générale Bathymétrique des Océans, were published in Paris in 1905. Advances in bathymetric data gathering from the mid twentieth century onwards increased hugely the amount of data gathered and responsibility for the charts was passed to the then IHB (now IHO) before a joint IHB/Intergovernmental Oceanographic Commission (IOC) of UNESCO Guiding Committee was established in 1974. The GEBCO Guiding Committee was able to offer marine geoscientists across the world the opportunity of publishing their data in a high-quality chart series; world coverage on the original scale of 1:10 million was completed and published by 1982. Paper charts were enhanced in 1994 with the publication of the GEBCO Digital Atlas using bathymetric contours, coastline and ships' tracks from the printed sheets, representing the first seamless, high-quality, digital bathymetric contour chart of the world's oceans (IOC, 2012).

The marine cadastre

In the mid twentieth century the long-held principle of limiting national rights to a narrow belt of sea bordering a nation's coast with the remainder of the world's seas and oceans 'proclaimed to be free to all and belonging to no-one' began to break down. National claims to ever greater sea areas to establish rights over marine and sub seafloor resources and military tensions in sea areas threatened to engulf the world's oceans in conflict and instability. The rapid growth in technology and an insatiable demand for natural resources led to ocean exploitation on a scale undreamt of before the Second World War. Warning shots for action came in the UN in 1967 with a speech by Malta's Ambassador that triggered a 15 year process that, although starting as an exercise to regulate the seabed, developed into a global diplomatic effort to regulate all the world's oceanic areas and the use of our seas and their resources. The UN Convention on the Law of the Sea (UNCLOS), described by the UN Secretary-General as 'possibly the most significant legal instrument of this century' was adopted in 1982 and came into force in November 1994 (UNDOA, 2012). Delimiting all the maritime boundaries described in UNCLOS requires hydrographic data.

National and international obligations

The *Titanic* disaster of 1912 led to the first convention for the safety of life at sea two years later; the current convention, the UN Convention on the Safety of Life at Sea (SOLAS) is the fifth version since 1914. The SOLAS Convention includes Articles setting out general obligations, amendment procedure and so on, followed by an Annex divided into 12 chapters covering minimum standards for the construction, equipment and operation of ships compatible with their safety. Chapter V, applying solely to the safety of navigation, is the key chapter for hydrographic surveying and nautical charting issues. This chapter identifies certain navigation safety services that Contracting Governments undertake to provide and also sets out provisions of an operational nature applicable in general to all ships on all voyages. This chapter applies to the safety of navigation in general in contrast to the remainder of the Convention which only applies to certain classes of ship engaged on international voyages.

Regrettably, another tragedy heralded the need for the establishment of a standardised worldwide system to warn mariners urgently of navigational hazards. In 1971, the tanker *Texaco Caribbean* was struck by the freighter *Paracas* in the Dover Strait; the *Texaco Caribbean* exploded

472

and sank with the loss of eight lives. Despite the wreck being marked by buoys, the freighter *Brandenburg* hit the wreck the following day and sank with the loss of a further 21 lives. Although a light ship and five more light buoys were added, this did not prevent the *Nikki*, which ignored all warnings, from colliding with the submerged wrecks; the *Nikki* went down with her entire crew. Even the addition of a second lightship and about ten more buoys did not prevent a further 16 ships over the following two months ignoring the hazard markers, all fortunately without additional incident or casualties (CEDRE 2012). These tragic events gave rise to the establishment of the Worldwide Navigational Warning Service (WWNWS), replacing a rather patchy set of national services, and the establishment of a standardised international navigational buoyage system. The establishment of the WWNWS system was decided at the Eleventh IHO Conference in 1972 with an IHO Commission established the following year. This service is now well established and plays a significant role in maintaining the safety of navigation at sea (IHO, 2005b, p. 42).

Regulation 4 of Chapter V of SOLAS requires every State that is a Party to SOLAS (162 States in 2012) to ensure that they 'take all steps necessary to ensure that, when intelligence of any dangers is received from whatever reliable source, it shall be promptly brought to the knowledge of those concerned and communicated to other interested Governments'.

As explained earlier in this chapter, these obligations can be met by the establishment of a national MSI infrastructure within each maritime State. This infrastructure may be considered in two parts: safety critical information that is assessed and transmitted by radio and nautical chart and publication corrections that can be published by a variety of means within weeks of the information being received and assessed.

SOLAS Chapter V Regulation 9 requires that States 'undertake to arrange for the collection and compilation of hydrographic data and the publication, dissemination and keeping up to date of all nautical information necessary for safe navigation'. This Regulation obliges all States to ensure that an appropriate level of hydrographic and nautical charting services is available for their waters. There are various ways that this can be achieved and these are discussed later.

UN Resolution A53/32 (1998) also supports the aims of SOLAS and hydrography more generally by stating that:

> The General Assembly invites States to cooperate in carrying out hydrographic surveys and providing nautical services for the purpose of ensuring safe navigation as well as to ensure the greatest uniformity in charts and nautical publications and to coordinate their activities so that hydrographic and nautical information is made available on a worldwide scale.

It is clear that a maritime State that is a Party to SOLAS has a clear treaty obligation for the provision of suitable hydrographic and nautical charting services in support of safety of navigation for the waters under its jurisdiction. This can also be linked directly to UNCLOS where this responsibility also covers the marine environment and its many delicate aspects and the sustained efficient use and management of that environment. This hydrographic obligation is not a 'nice to have'; in the twenty-first century it is as important as education or air traffic control. Limited national education programmes are likely to result in a country's poor economic achievement; insufficient air traffic control can lead to air disasters costing many lives and have a severe effect on a nation's economic development. Simply because there appears to be peace and tranquillity in the seas around a nation's coast does not mean that all is well. A maritime disaster can be hugely expensive in terms of lives and livelihoods lost, and the destruction of delicate marine environments. In the longer term, the lack of hydrographic information is a very serious impediment to developing the seas and oceans for the benefit of a State – the so-called 'blue economy'.

Many maritime States will have, in one form or another, a national maritime policy. This policy will be made up of a number of elements, some of which will be in place to support international obligations, with others to maintain and develop its national infrastructure such as port and maritime authorities, fishing, tourism, etc. Many elements are commonly considered, one is commonly ignored – hydrography.

Too many maritime States fail to understand their liability under SOLAS and UNCLOS within their often proudly publicised sea areas. If anything is placed in, on or under the sea and is known to be there by national authorities who then fail to inform the mariner, through whatever means, then any resulting marine accident and its consequences will be the liability of that maritime State.

Challenges for a national hydrographic office

Any national hydrographic programme or policy requires an overarching public authority to deliver and manage the provision of hydrographic services required by the State or which the State is required to provide. That authority may or may not have the capability to provide MSI, hydrographic surveying and nautical charting services, but it must at least have the power to ensure that these services are provided. The establishment and maintenance of a national hydrographic authority is not an easy task. The maintenance of a highly specialised public organisation is always problematic: staff need to be trained, motivated and retained for almost all their working lives in an organisation where promotion and career advancement is often limited; for hydrographic surveyors with IHO recognised qualifications the draw to move from the public to the private sector for significantly higher remuneration is enormous. For small, and some medium-sized States, the establishment and maintenance of a national hydrographic authority can prove to be a challenge too far. It is common for well-established organisations reliant on maybe fewer than ten staff, to collapse within months and to take years to be re-established. Financial provision within government budgets for equipment and training is rarely made and young and small organisations struggle hard to survive.

National hydrographic needs and obligations can be met in a variety of ways. The first and most obvious is by establishing a national organisation; however, this may not be practical or cost-effective for some countries, and here bilateral arrangements with other States for surveying and or charting may be more appropriate. At the national level there are options for delivering hydrographic services through in-house or commercially contracted services, civil and military services or a mix of all of these. Several government departments will have a need for hydrographic data and to guard against duplication of effort and expenditure and ensure that the maximum value is derived from the national hydrographic effort, every maritime State is strongly advised by the IHO to establish a national hydrographic committee or commission.

Despite, or more likely because of, the advances in hydrographic, positioning, digital and marine technologies, a national hydrographic authority and the hydrographic surveyor both face significant challenges. The growth of international shipping in both the number and size of vessels produces ever greater demand for the modernisation of charts and surveys. The disparity between high positional accuracy of satellite navigation systems and the generally poorer charted data increases the risk of an incident, when the mariner trusts implicitly his position, but does not understand that the chart, and the data from which it is produced, is from an inferior age. Although great strides have been made in the standardisation of hydrographic data collection, access to and identification of all the relevant data for the modernisation of charts is not always forthcoming; some nations even actively prevent the release of data regardless of international agreements or protocols.

And finally, government priorities, particularly in an age of austerity, do not favour hydrography. The state of schools and hospitals are readily apparent and are constantly brought to the attention of politicians; poor charting, the resulting potential risks and missed economic opportunities are not. Several studies by IHO Member States indicate that the cost to benefit ratio of hydrography is at least 1:10 for nations with a significant dependence on maritime trade or interests. Perhaps it is time to spotlight the issue by posing the reverse question: What would the economic, social and environmental implications be if there were *no* hydrographic services?

References

Admiralty Hydrographic Department. 1938. *Admiralty Manual of Hydrographic Surveying*, London: His Majesty's Stationery Office.

Centre of Documentation, Research and Experimentation on Accidental Water Pollution (CEDRE). 2012. Available at www.cedre.fr/en/spill/texaco_caribbean/texaco_caribbean.php (Accessed: 20 July 2012).

Coochey, J. 1993. An Economic Analysis of the Benefits of the RAN Hydrographic Programme, *International Hydrographic Review*, vol. II pp. 22–1–22–2. Available at www.iho.int/iho_pubs/misc/M_2_Suppldocs/1992-Economic_analysis_RAN_Hydro_Prog_Australia.pdf (Accessed: 12 Jul 2015).

Dawson, L. S. 1885. *Memoirs of Hydrography*, Reprinted 1969, London: Cornmarket Press.

International Hydrographic Organization. 2005a. *C-13 Manual on Hydrography*. Available at http://88.208.211.37/iho_pubs/CB/C13_Index.htm (Accessed: 20 July 2012).

International Hydrographic Organization. 2005b. *The History of the International Hydrographic Bureau*. Monaco: IHO.

International Hydrographic Organization. 2012a. *SP-32 Hydrographic Dictionary*. Available at http://88.208.211.37/iho_pubs/standard/S-32/S-32-eng.pdf (Accessed: 26 July 2012).

International Hydrographic Organization. 2012b. XVIIIth IH Conference Papers, April 2012, Proceedings Vol. 1, p.82. Available at www.iho.int/srv1/index.php?option=com_content&view=article&id=388&Itemid=306&lang=en (Accessed: 12 Jul 2015).

International Hydrographic Organization. 2012c. *C-55 Status of Hydrographic Surveying and Nautical Charting Worldwide*. Available at http://88.208.211.37/iho_pubs/CB/C-55/C-55_Eng.htm (Accessed: 21 October 2012).

International Hydrographic Organization. 2012d. *S-4 Regulations of the IHO For International (Int.) Charts and Chart Specifications of the IHO*. Available at http://88.208.211.37/iho_pubs/standard/S-4/S-4_e4.3.0_EN_Aug12.pdf (Accessed: 21 October 2012).

International Hydrographic Organization. 2012e. *World Bathymetry*. Available at www.iho.int/srv1/index.php?option=com_content&view=article&id=300&Itemid=744 (Accessed 29 October 2012).

International Hydrographic Organization. 2015. *M-2 The Need for Hydrographic Services*. Available at www.iho.int/iho_pubs/IHO_Download.htm (Accessed 17 July 2015).

International Hydrographic Organization Website. n.d. Capacity Building Concept and Strategy, available at www.iho.int/srv1/index.php?option=com_content&view=article&id=532&Itemid=407&lang=en (Accessed: 12 July 2015).

International Hydrographic Organization Website. n.d. Inter-Regional Coordination Committee, available at www.iho.int/srv1/index.php?option=com_content&view=article&id=419&Itemid=377&lang=en (Accessed: 12 Jul 2015).

Inter-governmental Oceanographic Commission. 2012. History of GEBCO. Available at www.gebco.net/about_us/project_history/#grids (Accessed: 28 October 2012).

NP100. 1999. *The Mariner's Handbook*, Taunton: UKHO.

Southern, R. M. 1938. *The Admiralty Manual of Hydrographic Surveying*, London: HMSO.

United Nations Division for Ocean Affairs and the Law of the Sea. 2012. Available at www.un.org/Depts/los/convention_agreements/convention_historical_perspective.htm (Accessed 29 October 2012).

31

MARINE PROTECTED AREAS AND OCEAN PLANNING

Tundi S. Agardy

Special considerations in managing ocean space

The global ocean is stressed from all directions. Each passing day brings news of marine degradation, continued over-exploitation, heightened conflict, and the ravages of climate change. And while there are small-scale success stories, reports from one commission after another suggest that while we may be winning a battle now and again, we are losing the war on keeping the seas (and by extension, the planet) safe.

The continuing decline in the health of the global ocean and its coasts comes with risk to human well-being everywhere. While coastal and marine ecosystems are naturally dynamic, they are now undergoing more rapid change than at any time in their history (Millennium Ecosystems Assessment 2005, cited in Agardy 2010). Transformations have been physical, as in the dredging of waterways, infilling of wetlands, and construction of ports, resorts, and housing developments, and they have been ecological, as has occurred with declines in abundances of marine organisms such as sea turtles, marine mammals, seabirds, fish, and marine invertebrates. The dynamics of sediment transport and erosion-deposition have been altered by land and freshwater use in watersheds; the resulting changes in hydrology have impacted coastal dynamics. These pressures, together with chronic degradation resulting from land-based and marine pollution, have caused significant ecological changes and an overall decline in many ecosystem services.

In order to safeguard the global ocean, keep the flow of natural goods and ecosystem services flowing, and reduce risks borne from natural catastrophe, there must first be recognition that there are special challenges inherent in managing the marine environment, which include:

- the reality that the oceans are opaque, contain ecosystems with nebulous boundaries, and often have overlapping jurisdictions that make agency responsibilities and opportunities difficult to determine;
- the undeniable fact that much of ocean health and productivity is linked to terrestrial and wetland ecosystem condition, therefore planning and management must be done systematically in concert with land use management;

- the difficulties inherent in defining, mapping, quantifying, and valuing marine ecosystem services, despite the fact that marine and coastal ecosystems may provide even more natural capital and services of value than terrestrial ecosystems;
- the limited options available for effective management of priority areas, given the need for inter-agency cooperation, surveillance and enforcement, and geographically large-scale coordinated and collaborative action.

These special challenges have made management of ocean space difficult, constraining our ability to protect marine areas for the goods and services they provide, even as those goods and services are increasingly recognized as critical to human well-being the world over (Agardy *et al.* 2011a).

Many have argued that marine protected areas (MPAs) are the answer. But discrete protected areas, while meeting many varied objectives when successfully designed and implemented, can only be islands of protection in a sea of trouble (Agardy *et al.* 2011b). In fact, as Giuseppe Notarbartolo-di-Sciara has pointed out in this volume (Chapter 9 on Marine Conservation), MPAs can really be thought of as an admission of our failure to adequately manage oceans comprehensively and effectively. Lacking good comprehensive management, we are left with small areas of special value that we try to maintain, while the surrounding context deteriorates.

Still, effectively protecting small bits of the coasts and oceans does bring benefits. MPA establishment can raise awareness about special places in the sea, and that attention can generate the political will to do more (Agardy 1997; Gubbay 1995). MPAs provide necessary control sites for monitoring the condition of marine and coastal ecosystems, for understanding the functioning and the value of marine ecosystems, and for determining whether marine management is meeting its objectives. MPAs can generate funds for broader marine conservation and education, and—perhaps most importantly—they can provide the foundation for broader scale marine spatial planning and subsequent ocean zoning (Agardy *et al.* 2012). It is in the latter that the greatest hope may lie for protecting ocean space and the ecosystems supported by the ocean environment.

Marine policies to control access to ocean areas and manage the use of marine resources have grown exponentially as coastal jurisdictions have expanded, and as the global community has taken increasing interest in those high seas areas that are beyond national jurisdictions (Zacharias 2014). Grounding those policies to particular places is the domain of spatial planning, the end result of which includes marine protected areas and reserves, areas designed for exclusive use by maritime sectors such as shipping, mining, energy development, fisheries, and recreation, and comprehensive ocean zoning that accommodates multiple uses in a single strategic framework.

Marine protected areas

Marine protected areas are designations of a "clearly defined geographical space, recognized, dedicated and managed, through legal or other effective means, to achieve the long-term conservation of nature with associated ecosystem services and cultural values" (IUCN 2012). MPAs, and to a lesser degree networks of MPAs, have been used for decades to protect discrete patches of coastal and/or marine habitat, and particular populations of marine species. MPA objectives run the gamut from conservation/preservation to expansion of marine tourism to maintenance or even enhancement of fisheries production. Increasingly, MPAs are formed to allow local communities and user groups such as fishing cooperatives more say in the management of marine areas upon which they depend (Agardy, 1997).

Coastal and marine protected areas have been around for a long time, but until recently were created in an ad hoc manner with minimal and usually insufficient enforcement. However, in the last fifteen years the science behind site selection, MPA design, and evaluation of MPA effectiveness has grown, coalescing into a field of inquiry increasingly accessible to planners (Halpern and Agardy 2013). Systematic conservation planning has matured into an applied multidisciplinary field, providing the science backbone to support site selection of MPAs, MPA and MPA network design, and development of both zoning within MPAs and management plans to support the zonation (Devillers *et al.* 2014; Fox *et al.* 2012; Game *et al.* 2008).

MPAs come in a wide variety of shapes, sizes, regulatory regimes and governance structures. In 1992, when the seminal IUCN booklet on MPAs was published (Kelleher and Kenchington 1992), there appeared to be significant consensus on what constituted a legitimate and effective MPA. Since then, the issue of what defines an MPA has gotten cloudier as more types of MPAs have been tried. Complex semantics have created confusion as environmental groups advocate for more protected areas and as countries commit to establishing them (Halpern and Agardy 2013).

Assessing the number of MPAs, the areal extent of their coverage, and the degree of protection afforded by MPAs the world over is made complex by the wide variety of attitudes around what qualifies as an MPA, and even more so, what constitutes an effective protection regime within an MPA. Nonetheless, consensus has it that as of 2012 there were over 5000 marine protected areas across the globe. In general the more recently established MPAs are more effective and efficient (Roberts *et al.* 2003; Wood *et al.* 2008); however it must also be noted that some of the most iconic examples of MPA success have been around for many decades, such as the Great Barrier Reef Marine Park in Australia (Agardy 1997).

Despite improvements in MPA design, effectiveness, and ease of establishment, in 2010 less than 1.2 percent of the ocean was within an MPA and a tiny fraction of that within fully-protected, no-take marine reserves (Toropova *et al.* 2010). In this decade the numbers increased dramatically with the creation of several very large MPAs: in Chagos, Phoenix Islands of Kiribati, NW Hawaiian Islands, and New Caledonia. This drive for new MPA creation is occurring in the face of increasing degradation of the ocean and coasts from direct and indirect effects of human activity, and increasing demand for areas allocated to new uses such as renewable energy and offshore aquaculture. However, even with these monumental (and perhaps unenforceable) protected areas, the total area within MPAs globally hovers around 2–3 percent (Halpern and Agardy 2013).

The Guidelines for Applying the IUCN Protected Area Management Categories to Marine Protected Areas (IUCN 2012) is thus a useful standard and a good basis for getting international agreement on terms and approaches for MPA planning. Yet irrespective of which type of MPA is utilized, the path of least resistance is still too often establishing MPAs opportunistically, as opposed to strategically locating MPAs where they are most needed to maintain ecosystems and support human well-being (Agardy *et al.* 2011a; Devillers *et al.* 2014). However, where successful, MPAs demonstrate that natural and social science can be used as a solid foundation for: 1) siting MPAs—in other words determining the best location for a protected area or a network of MPAs; 2) zoning within MPAs to allow for a variety of uses and a regulatory structure that can prevent overuse or conflicting uses; and 3) creating governance structures that are appropriate to the context, allow as much participatory planning and management as possible, and ensure that ecological processes and thus the delivery of ecosystem services are not undermined.

MPAs are often expected to also serve regional (i.e., outside their boundaries) conservation goals and fisheries objectives, both of which require MPAs to be connected to each other via larval dispersal and fish movement and to the surrounding areas where fishing occurs (Halpern and Agardy 2013). In both cases larval dispersal from MPAs is essential, and for fisheries objectives adult spillover across reserve boundaries is also important and can be enhanced through the synergistic effects of MPA networks (Brock *et al.* 2012). This potential is supported by scientific theory; however it is also clear that the potential can never be realized unless regulations are upheld and enforcement encourages compliance (Guidetti *et al.* 2008).

MPAs provide many benefits, not the least of which is raising public awareness about the global ocean, the special marine places that still exist, and the opportunities we have to safeguard the seas. Protected areas enhance fisheries production, protect habitat from degradation, and help reduce risks to property and livelihoods. MPAs also serve an important function as sentinel places, providing control sites for scientific study that enhance our understanding of marine ecology and marine management. Improving MPA performance requires both adequate design and effective management implementation—both are discussed in the following sections.

MPA siting and design

Scientific advances have allowed for a move in MPA planning away from ad hoc designations to more strategic processes. Three communities have contributed to this shift from opportunistic to systematic: 1) academia, which has collectively contributed models to guide planning, empirical data to support the models, and monitoring of programs to assess progress towards implementing truly effective MPAs; 2) the environmental community, including inter-governmental organizations, non-governmental organizations (NGOs), and community groups, which has led multi-stakeholder planning processes and catalyzed action at the regional and local levels; and 3) government agencies, which have tackled MPA and MPA network design within their jurisdictions, working closely with both of the aforementioned communities (Halpern and Agardy 2013).

In order that MPAs meet their potential to be fully effective, social and economic as well as natural variables must be considered in their design and implementation (Halpern and Agardy 2013). The last decade has reflected this awareness with increased focus on the economic dimensions of establishing MPAs, in particular on the prospective impacts on local fishermen or on general social science to support MPA and marine managed area design (Fox *et al.* 2012).

In determining where MPAs should be sited, planners generally focus on biological attributes and values, on threats to biota, or some combination of both. Systematic conservation biology

---

Box 31.1 Marine Conservation Zones in the UK

In November 2013, the UK government finalized assessment and consultations on marine areas of interest. Twenty-seven sites in UK waters were listed as Marine Conservation Zone sites in England and Northern Ireland. According to the UK Department of Environment, Food, and Rural Affairs (Defra), each MCZ is established by a legal order made by Defra under section 116(1) of the Marine and Coastal Access Act 2009 (MCAA). Each order designates an area as an MCZ and defines that area, lists the features being protected within that area, and specifies the conservation objective or objectives of the MCZ.

Generally for a habitat, favorable condition is defined as: its extent is stable or increasing; and its structures and functions, its quality, and the composition of its characteristic biological communities are such as to ensure that it remains in a condition that is healthy and not deteriorating (DEFRA 2013). England and Northern Ireland combined their planning; Scotland and Wales undertook complementary processes.

would dictate that connectivity and ecological drivers are quantified, but in fact most planning processes claim to rely on these concepts but then plan without using these data or models (Devillers *et al.* 2014). Biological attributes typically include species richness, total or trophic level abundance or biomass, average organism size, and primary productivity (Agardy 1997). For MPA networks, representativeness is also important to ensure that the full suite of regional biodiversity is sufficiently included (for instance, Kelleher *et al.* 1995). One recent example is provided by the UK's Marine Conservation Zone identification process (see Box 31.1—taken from Agardy *et al.* 2012).

Increasingly, MPA locations are being selected through elaborate stakeholder engagement processes that focus on balancing human needs with conservation needs. In the developed world, one of the best examples of this is the Marine Life Protection Act (MLPA) in California that recently completed a nearly ten-year process to establish a network of MPAs along the entire coast of the state (Halpern and Agardy 2013). The process required meeting minimum ecological criteria, explicitly balanced with efforts to minimize costs to fishermen and other resource users, while optimizing conservation (Roberts *et al.* 2003). While process did result in MPA designs that are expected to achieve agreed objectives, it should be noted that it took California three tries to get this balance right, and it wasn't until stakeholder engagement was seriously attempted that the public accepted the science-based designs put forward by planners (Agardy *et al.* 2003).

Incorporating ecosystem services into MPA design

Ecosystem services are the benefits provided by healthy, well-functioning environments. Such services include provisioning of food and water resources, as well as regulating and supporting functions such as flood control, waste management, water balance, climate regulation, and other processes. Human reliance on these ecosystem services is significant—although we rarely recognize the value of ecosystem services until they are lost. The oceans and coasts provide a great many of these critical but undervalued services—supporting not only coastal inhabitants but all life on the planet.

Nature's greatest benefits came largely from coastal systems. Coastal wetlands maintain hydrological balances, recharge freshwater aquifers, prevent erosion, regulate flooding, and buffer land from storms. Marine habitats supply us with food, recreational opportunities, pathways for transport, places to do research and learn, and spiritual values. Regulating services arising from these and other coastal habitats also include disease regulation (prevention of spread of water-borne pathogens, for instance) and contribute to overall ecological resilience in the face of climate change. Some of these many ecosystem services are displayed in Figure 31.1.

Such valuable ecosystem services are provided by a diverse array of coastal and marine habitats, including mangrove forests and fringe, coral reefs, seagrasses, saltmarsh, oyster reefs, rock reefs, vast productive bank areas, and kelp forests, as well as pelagic habitats, sea mounts, and upwelling areas. These coastal and marine habitats are functionally or physiologically linked to each other and to other aquatic and terrestrial habitats. Too they are immutably linked to economic activity and human well-being. Figure 31.2 illustrates these feedback loops and benefits flows. In order that mangrove forests continue to provide nursery grounds for commercially and recreationally important fish populations, the two-way linkages between mangrove and offshore habitats such as seagrass beds, coral reefs, and sea mounts must be maintained. Similarly, offshore systems such as coral reefs create the conditions necessary for inshore systems such as seagrasses to thrive; while mangroves and saltmarsh act to trap sediments and nutrients that might smother or degrade seagrasses. The delivery of goods and services from natural systems is dependent not only on the condition of the habitat but also its functional linkages to associated

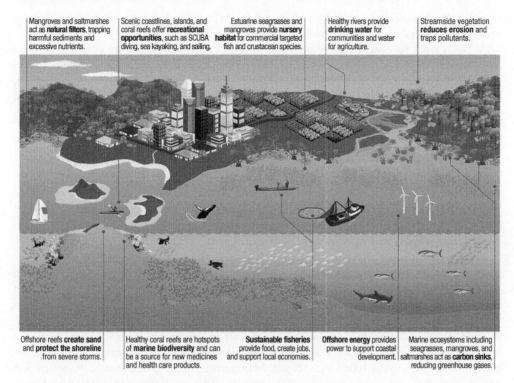

Mangroves and saltmarshes act as **natural filters**, trapping harmful sediments and excessive nutrients.

Scenic coastlines, islands, and coral reefs offer **recreational opportunities**, such as SCUBA diving, sea kayaking, and sailing.

Estuarine seagrasses and mangroves provide **nursery habitat** for commercial targeted fish and crustacean species.

Healthy rivers provide **drinking water** for communities and water for agriculture.

Streamside vegetation **reduces erosion** and traps pollutants.

Offshore reefs **create sand** and **protect the shoreline** from severe storms.

Healthy coral reefs are hotspots of **marine biodiversity** and can be a source for new medicines and health care products.

Sustainable fisheries provide food, create jobs, and support local economies.

Offshore energy provides power to support coastal development.

Marine ecosystems including seagrasses, mangroves, and saltmarshes act as **carbon sinks**, reducing greenhouse gases.

Figure 31.1 Schematic showing coastal and marine ecosystem services

Source: Agardy *et al.* 2011a.

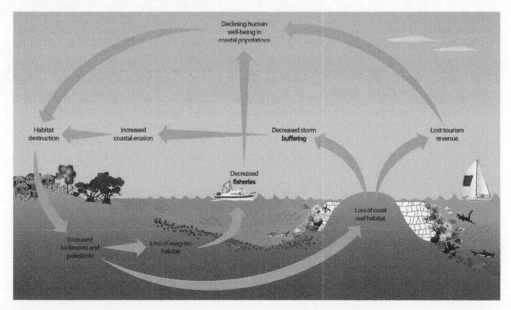

Figure 31.2 Feedback loops demonstrate connectivity among terrestrial and marine habitats, as well as between habitats and human well-being

Source: Agardy *et al.* 2011a.

habitats. This connectivity must be recognized for MPA planning to successfully protect the ecosystem services so valued by coastal communities and nations.

While it is not necessary to undertake a full assessment of all ecosystem services, nor to perform a valuation exercise to determine the ecosystem services provided, ecosystem services information is now widely used in siting and designing MPAs. Ecosystem services' identification, assessment, and valuation has thus been used as a priority-setting methodology for MPA designation and land use/sea use planning, in places ranging from Abu Dhabi to Zanzibar, from temperate to tropical regions across the global ocean.

Managing MPAs

As mentioned previously, MPAs run the gamut from small no-take reserves to very large, zoned multiple use areas. Simple no-entry reserves and no-take reserves are generally easy to monitor, enforce, and manage. However, most MPAs are designed to meet multiple objectives and address a wide array of uses, an embodiment of ecosystem-based management principles (Agardy *et al.* 2011a), and multiple use areas are inherently more complex and challenging to manage.

One method for addressing the challenge of multi-objective MPAs is to develop a zoning plan that determines spatial allocation of uses across the landscape/seascape (see for instance Agostini *et al.* 2012), and then develop separate plans for different sectors. In the Florida Keys National Marine Sanctuary in the south-eastern United States, which covers nearly 10,000 square kilometers and is heavily used by fishers, divers, boaters, beachgoers, tourists, researchers, and local businesses and individuals, the Marine Sanctuary has developed separate sectoral action plans. These include the following: Science Management and Administration, Research and Monitoring, Education and Outreach, Volunteer, Regulatory, Enforcement, Damage Assessment and Restoration, Maritime Heritage Resources, Marine Zoning, Mooring Buoy, Waterway Management, Water Quality, Operations, and Evaluation. Each plan has specific objectives, with a means for monitoring whether the management is achieving those objectives (Agardy 2010).

Equally important as setting explicit and measurable objectives in MPA success is having the resources to monitor, both for enforcement and to assess MPA effects. Far too many 'paper parks' exist where space has been gazetted into a protected area but insufficient resources are provided to enforce regulations and monitor activities within boundaries (Christie and White 2007; Guidetti *et al.* 2008). In these cases, stakeholder expectations are often set too high, creating disappointment when MPAs do not perform as promised (Agardy *et al.* 2011b; Charles and Wilson 2009). Failing MPAs not only waste opportunities for effective management, but they can embitter and embolden MPA opponents to fight future designations (Agardy *et al.* 2003).

Other challenges and needs exist for assessing how species and ecosystems respond to MPA regulations and overall protection. The challenges arise when trying to determine what effect can be attributed to the MPA (this often requires adequate monitoring prior to MPA creation), and the needs arise in being able to report to stakeholders and managers that conservation actions are meeting goals and expectations (Halpern and Agardy 2013).

Not surprisingly, larger MPAs that employ zoning can provide more effective and efficient protection while accommodating appropriate uses and addressing thorny issues such as displacement effects (Agardy 2010; Sale *et al.* 2014). Australia's Great Barrier Reef Marine Park (GBRMP) is still considered the global leader in using science for zoning in order to achieve specific management objectives (Fernandes *et al.* 2005). The spatial plan that initially established the GBR marine protected area, and the subsequent re-zonings and extensions of influence as the park gained World Heritage status, have all been done under a single planning and management authority, the Great Barrier Reef Marine Park Authority (GBRMPA).

Global attention to some of the iconic or better known MPAs, along with the increasing number of marine World Heritage sites, suggests that the global public values many marine areas as special and worthy of protection. Yet as already stated, some conservationists caution that the fact that we need to establish discrete marine protected areas at all is an admission of our inability to protect the seas. Whether or not this is true, many MPAs fail to achieve their stated objectives and might be more successful if conceived in tandem across a wider regional network, or placed in the context of broader spatial planning and comprehensive ocean zoning (Agardy 2010).

Particularly controversial are the very large MPAs that have characterized recent designations. Fueled in part by a race to meet commitments, such as the commitment by nations party to the CBD to protect 10 percent of their marine jurisdictions by 2010 (now extended to 2020 since targets were not being met), these large MPAs face huge challenges in implementation—not the least of which are unenforceability and excessive potential costs of management (Devillers *et al.* 2014; Leenhardt *et al.* 2013). Political considerations often trump ecological ones in these notably large protected areas, suggesting that sites for some large MPAs are chosen not for their ecological or social value, but rather for their expediency in meeting targets (Agardy *et al.* 2011b; Agardy *et al.* 2003). Nonetheless, champions of very large reserves argue that even imperfect MPA designations are better than nothing, and that by representing more than three-quarters of the world's managed marine areas, very large MPAs are a powerful tool for arresting the decline in marine biodiversity (Toonen *et al.* 2013).

Regardless of drivers and motivations for MPA establishments, the main underlying factor in whether MPAs are successful in meeting the objectives for which they were designed is whether governance is suitable, appropriate, and efficient (Jones 2014). This is of course true in any marine or coastal management endeavor, since control of impacts—both direct and indirect—requires a different approach in the marine commons than similar management on land. Governance assessments early in the MPA planning process can indicate what kind of pitfalls lie in store for MPA management agencies, and can direct attention to new governance arrangements that can ensure fair and equitable access to marine areas while allowing management to protect ecological processes and productivity (NRC 2007).

Governance considerations should be used early in planning to help set achievable objectives for management, so that MPAs will be maximally effective. MPA governance must be tailored to the specific socio-economic and political characteristics of the place—there is no one-size-fits-all approach to governance (Jones 2014). Institutional arrangements that include a wide group of government agencies and institutions of civil society are in many cases the best frameworks for governing MPAs, allowing effective and non-partisan evaluation of their performance, and setting the stage for management that is as adaptive as possible as social and environmental conditions change.

Financing MPAs

For marine protected areas to provide a lasting solution to the challenges facing particular regions of oceans and coasts, sustainable financing is needed to support day-to-day management, coordination, and information exchange between agencies, as well as the continual adaptation that good management requires. This does not only mean securing budgets for the government agencies typically involved in marine management—the private sector can be tapped as well.

Private sector investments in coastal conservation can finance MPAs, as is the case with the privately owned and operated Chumbe Island Marine Park in Zanzibar, United Republic of Tanzania. Revenue from user fees collected by the park's private owners covers 100 per cent of the costs of management (monitoring, enforcement, outreach, maintenance, etc.). Other examples include less direct private sector protection of services, including a range of funding flows that originate in the private sector. These include developer-financed conservation or restoration/rehabilitation projects, such as those undertaken as part of no-net-loss of wetlands regulations. It also includes public/private partnerships such as municipal governments teaming up with chambers of commerce, or private financing of public sector resource management—such as the generation of conservation funds through licensing fees (fishing and hunting, for example).

There are many other mechanisms that provide opportunities for sustainable financing. Funding can come from a share of lottery revenues, dedicated revenues from wildlife stamps, tourist related fees, fees for eco-labeling and certification, and fishing licenses or fishing access agreement revenues. There can also be fees for non-renewable resource extraction, fines for illegal activities, campaigns to establish trust funds, fees for bio-prospecting, and income derived from local enterprises (such as the sale of handicrafts). An example of indirect payments for coastal and marine ecosystem protection is the growing movement of communities hiring watchdogs to monitor compliance with existing pollution and/or fishing regulations, and to publicly blow the whistle when infractions occur. Thus governments need not shoulder the burden of coastal and ocean management alone.

At the same time, there is growing attention to more direct involvement of markets in protecting ecosystem services. Recognition of the immense value of ecosystem services has opened the door to innovative approaches to conservation and greater engagement of the private sector. Market-based approaches to marine conservation include coastal payment for ecosystem services (PES) systems and associated market offsets. These have the potential to achieve better and more cost-effective conservation outcomes than currently result from non-market-based projects seeking to isolate coastal areas from human encroachment. By clarifying the linkages between ecological function, ecosystem service delivery, and market incentives, PES systems can be a useful tool for people operating to lower risk and manage projects.

MPA networks

Recognizing that one-off designations cannot possibly protect marine species that are widely dispersed or highly migratory, MPA planners have increasingly turned to systematically designed MPA networks (Olsen *et al.* 2013). MPA networks provide a way to link broader global and regional marine policies (such as agreements to protect threatened marine species such as whales, sea turtles, seabirds, etc.; or regional agreements to conserve biodiversity) with the site-based and more localized conservation measures—in effect linking top-down and bottom-up initiatives (Agardy, 2005).

To guide MPA network design, the Center for Environmental Cooperation (CEC)—the environmental side arm of the North American Free Trade Agreement between Canada, the U.S., and Mexico—published a report entitled "Scientific Guidelines for Designing Resilient MPA Networks" (Brock *et al.* 2012) which further elaborates the science behind MPA site selection. The four main principles or guidelines discussed include 1) protect species and habitats with crucial ecosystem roles, 2) protect potential carbon sinks, 3) protect ecological linkages and connectivity pathways for a wide range of species, and 4) protect the full range of bio-diversity present in the target area.

An example of network design that addresses all four principles is the California's Marine Life Protection Act (MLPA) planning process, which established a network of MPAs throughout State waters. In some senses the MLPA has become a global benchmark for both how not to do a large MPA network design process (it failed the first two implementation attempts) and how to do it with great success (demonstrated with the third attempt) (Halpern and Agardy 2013). Other MPA networks developed within a strategic framework, such as in Belize, result in effective protection of high value, high biodiversity areas. Increasingly, marine planners are incorporating ecosystem resilience to climate change in the way they design MPA networks (McLeod *et al.* 2009; Magris *et al.* 2014).

In the Mediterranean Sea region, MPA network development proceeds differently, focusing on representativeness of habitats to secure a comprehensive system of protected areas. There, MPA practitioners, government planning agencies, the Parties to the Barcelona Convention, and conservation organizations concerned about underrepresented areas, habitats, and species within the Mediterranean MPA network (Abdulla *et al.* 2009) have jointly undertaken MPA network planning. In 2012, researchers performed analyses to show that systematic conservation planning to identify new sites with greater representation was possible (Giakoumi *et al.* 2011), and, by designating these areas in an MSP framework, could achieve the goals of MPA network representativity and biodiversity protection (Portman et al, 2013).

Networks of MPAs thus utilize priority areas to either enhance the protection of a single target site (involving source/sink modeling, or looking at habitat connectivity as shown in Figure 31.2, for instance), or are used to create fully representative systems of protection. The WCPA/IUCN (2007) guidelines on MPA networks list eight design considerations: 1) representativeness, 2) replication, 3) viability, 4) precautionary design, 5) permanence, 6) resilience, 7) connectivity, and 8) sufficient size and shape to meet objectives (WCPA/IUCN 2007). At this point in time, however, it should be noted that the creation of MPA networks for biodiversity protection, fisheries management, and habitat representation is largely theoretical, with few such networks actually having been implemented. With the growing popularity of large-scale marine spatial planning (MSP), it seems that MPA network design has gotten swallowed up in broader spatial planning to allocate ocean space for ocean use and conservation simultaneously.

Marine spatial planning and ocean zoning initiatives

Marine spatial planning is an approach that allows for improved decision making about what kind of human activities should occur where, in order to meet ecological, economic, and social goals. MSP is forward-looking and informed by predefined goals, objectives and policies, and is increasingly used to improve management and reduce conflicts, either between direct users of marine and coastal resources and space, or between institutions playing a role in managing activities impacting those resources and areas (Agardy *et al.* 2012; White *et al.* 2012).

The strategic MSP process creates blueprints for spatially explicit use — typically producing many different marine spatial plans that can then be compared for costs and benefits, or potential trade-offs. Importantly, the outputs of MSP are not maps cast in concrete or zones fixed in perpetuity, but participatory processes that allow for dynamic allocation of space and resources that can be changed as conditions (or human needs) change.

A plethora of spatial planning is underway around the world, at various scales and in order to achieve various goals and objectives. Such spatial planning processes include not only marine or maritime (the term preferred by EU countries) spatial planning and zoning, undertaken at the local, national, regional, and international level, but also integrated coastal zone planning at the national or subnational level, the design of marine protected area networks, and systematic conservation planning at all geographic scales. The focus of this chapter is on national or subnational MSP that results in ocean zoning and corresponding management plans.

Like MPA planning, MSP requires the identification of priority areas, which can then be managed or protected in a broad-based spatial plan or specifically within an MPA designation. MSP can build from existing MPAs or MPA networks, though it is likely that additional protected areas or zones may be needed to achieve MSP goals. Intrinsic to MSP is the concept of ecosystem-based management (EBM), which reiterates the need to think about priority areas in the broader context and necessitates the linking of marine planning with land use planning and watershed management (Álvarez-Romero *et al.* 2011).

MSP planning processes

The process of marine spatial planning that leads to spatial allocations or ocean zoning is itself valuable, not only because of its valuable end-product. Like creating a mathematical model, undertaking the steps of the planning process for zoning provides insights. The planning process forces identification of stakeholders, recognition of connections between use and condition of the system, determinations of sustainable limits of resource use and development of management that is as strategic as possible (Agardy *et al.* 2011b). At the core of MSP are the tacit assumptions that some areas are more important than others for achieving certain goals, and that this relative importance drives the establishment of spatially explicit rules and regulations.

Planning and then implementing ocean zoning through MSP is about more than creating maps—it is also about meeting the challenges of bounding ecosystems to determine scale and scope of management; assessing ecosystem conditions, threats and drivers; appraising management needs; evaluating trade-offs and choices in order to determine optimal management through a zoning scheme; and planning a monitoring regime to determine efficacy of management and generate, over time, the information necessary to make adjustments to zoning and its corollary regulations (Agardy 2010).

MSP that leads to ocean zoning process can be summarized as having the following ten steps, whether undertaken in a formal or an informal manner (Agardy 2010 as adapted from Agardy 1997, pp. 45–46):

1. Identify and involve all stakeholder groups, to the extent practically feasible, in a visioning exercise.
2. Set realistic objectives through a participatory process involving relevant stakeholders.
3. Study the area (using all applicable science, as well as local knowledge) to determine threats, as well as impediments to realizing objectives.
4. Develop outer bounds of the zoning area to reflect objectives and management feasibility.
5. Develop a preliminary zoning plan to accommodate different uses – if multiple-use is a goal.
6. Amend zoning to reflect user group expectations and needs.
7. Formulate a management plan that stipulates permitted uses and levels of use in each zone, in order to address threats (mitigation) and accomplish objectives.
8. Develop necessary regulations for each zone and develop incentives to foster voluntary compliance, in order to carry out management.

9. Monitor to see if objectives are being met over time.
10. Amend management as necessary to practice adaptive management.

In order to support assessment and then planning of spatially explicit regulations, whether through comprehensive zoning or not, many analytical and decision support tools are available. Most commonly Geographic Information Systems (GIS) are used to understand the condition of ecosystems to be managed, spatial distribution of human uses and impacts, and evaluation of potential marine zoning and management plans. Sea Sketch is a popular analytical software developed by Will McClintock (now at UC Santa Barbara) that allows rapid quantitative comparison of different zoning patterns, in order to help decision makers and the public understand possible trade-offs that exist with any particular zoning plan. Decision-support tools such as MARXAN, on the other hand, optimize zoning patterns according to predetermined operating principles (such as, protect 20 per cent of the area as no-take; protect high biodiversity and high endemism areas; allow traditional fishing in established marine tenure areas; retain accessible reefs for dive tourism; etc.) (Watts *et al.* 2009). The availability of these and other tools creates immense opportunities for planners to practice rigorous, systematic, and comprehensive MSP and zoning.

MSP opportunities and constraints

A report commissioned by the Global Environment Facility's Scientific and Technical Advisory Panel, at the request of the Convention on Biological Diversity (Agardy *et al.* 2012), presents lessons learned about MSP at all scales. The findings confirm that the theoretical basis for MSP is well-established, but the practical execution of MSP is still in its infancy. The report reviews conventional planning processes, identifies innovative new tools, and discusses the potential MSP has—as yet not fully realized—in aligning conservation and development interests while protecting vital ecosystems, the services they deliver, and the biodiversity they support.

The CBD report suggests that one of the important keys to success includes truly participatory planning, and at the same time using planning approaches appropriate to the particular circumstances of the place (this includes using scientific information as well as traditional ecological knowledge to support management plans and regulations). Another essential element is having a supportive legislative framework in place, with a means to determine priorities (based on the best available science), and a hierarchical system to clearly establish goals, objectives, and strategies for MSP. Governance of management that flows from marine spatial plans is at least as important as effective planning of MSP—so the current undue emphasis on planning as opposed to implementation needs to be overcome.

Since MSP inexorably deals with multiple, and sometimes, conflicting or competing uses of ocean space, the process requires bona fide participation from affected and affecting parties. Stakeholders should be engaged not so as to react to plans drawn up by a particular constituency (usually scientists), but rather to meaningfully participate in articulating the vision (why MSP is being used—to achieve what?), in stipulating clear objectives for management, and in weighing options for spatial allocation and management of uses (Olsen *et al.* 2014).

Despite these advances and the wealth of literature and websites that have been put forward on MSP (see especially guides by Beck *et al.* 2009; Ehler and Douvere 2007; Foley *et al.* 2010; and MarViva 2012), practical guidance on how to best implement MSP has remained somewhat elusive. In large part this is because much of that information is theoretical and is either so generic as to provide little specific direction in how planning should proceed in any given place, or is so specific to the circumstances of a place that the lessons learned are not easily transferable.

There are many examples of MSP that begin with the identification of priority areas in the development of marine spatial plans, at all scales. It is in this aspect that MSP aligns most closely with MPA planning. Two examples that are particularly illustrative are national level MSP in Norway and in Belgium. Norway's MSP is notable because fisheries are included in the mix of uses being allocated to specific spaces, and fisheries regulations are enacted in special zones to protect priority areas. Belgium is noteworthy because MSP rests on the identification of "biological valuation"—which uses detailed and quantifiable criteria for identifying areas of interest (Agardy 2010). These marine planning efforts are followed up with effective, ecosystem-based management, post-implementation, that involves use of zonation, permitting, and performance-based management.

Given the obvious benefits and efficiencies that can flow from marine spatial planning and comprehensive ocean zoning, it may be surprising that not all coastal nations have committed to developing and implementing spatial management plans. Part of the hesitancy may have to do with the culture of marine use, and the general aversion maritime users (including, but not restricted to, the fishing industry) have about being regulated in any way. But domestic legal regimes may play a role as well.

In the U.S., public lands law and ocean law both "represent legislatively-created mechanisms for resolving resource allocation disputes among people who seek to use public property, or the resources located on public property, in incompatible ways" (Eagle 2013), but whereas the public accepts the role of government in being the steward for public lands, it seems less willing to accept the role when it comes to oceans commons. And although almost half of public lands in the U.S. fall under "dominant use" laws allowing a single use, or narrow set of compatible uses, the entire federal waters, covering almost four times the collective area of public lands, falls under multiple use laws. According to Eagle and co-authors (2008), decisions made about the allocations of space or resources are made effectively by a single management authority that balances competing uses. For the oceans, the lack of a single federal management authority means that such a delicate balancing act is not possible, and invoking dominant use laws in federal waters would likely necessitate the use of ocean zoning, as the logical outcome of marine spatial planning. Ocean zoning allows responsible authorities to display what can be done where, what critical areas need to be protected, and where multiple uses can be accommodated (Agardy 2010; Ogden 2010). Ocean zoning also facilitates clearer understanding of jurisdictions.

However, some nations seem to have stopped marine spatial planning dead in its tracks by not carrying visioning and trade-off discussions through to the development of a zoning plan. In part this may be due to misconceptions about ocean zoning that include myths such as:

- zoning leads to segregation of all users, and privatization of the commons;
- zoning plans are cast in stone, and therefore quickly become obsolete;
- zoning cannot be done in the absence of property rights;
- zoning is pro-business, anti-conservation—or conversely, zoning is just a way to deny access to users in order to protect nature;
- zoning allows governments to exert unnecessary control on maritime users.

These myths have been largely dispelled by putting MSP into practice, with the resultant spatial management and zoning that achieve common objectives in a fair and equitable manner (Kenchington and Day 2011).

MSP and the resultant allocation of ocean space for a carefully coordinated variety of marine uses is now spreading rapidly around the globe. Charles (Bud) Ehler, formerly of the U.S. National Oceanic and Atmospheric Administration (NOAA) and now working in the international arena, has tracked MSP and has determined that as of 2013 almost 10 per cent of the world's exclusive economic zones were covered by marine spatial plans. And many more countries are taking on comprehensive MSP and ocean zoning.

In fact, the European Union recently issued a draft Directive for MSP that requires member states to develop plans for maritime activity, in order to promote "smart blue growth." And in 2012, the Conference of Parties to the Convention on Biological Diversity adopted a set of recommendations urging all signatory countries to engage in MSP processes. Marine spatial planning has now seemingly eclipsed coastal planning and integrated coastal management, as well as MPA planning, as the key to a sustainable ocean future.

Conclusions

The use of ocean space can and must be addressed in a systematic and collaborative manner. MPA and MPA network planning are small steps in that direction; MSP takes us a bit farther, and comprehensive MSP and ocean zoning farther still.

Marine spatial planning and consequent ocean zoning can grow out of existing use patterns, with the end result essentially codifying the already existing segregation of uses. This was largely the case in the original Great Barrier Reef Marine Park zoning plan, especially that of the first zoning done in the Northern Section (Agardy 2010). Alternatively, zonation may be based on the relative ecological importance of areas within the region, and inherent vulnerabilities of different habitats or species. This is the process that drove Belgium's biological valuation exercise, and much of the systematic conservation planning put forward by academics and research institutions in different parts of the world. In stark contrast, zonation may also be based on a kind of conservation-in-reverse process, whereby areas NOT needing as much protection or management as others would be highlighted. Such high-use zones could be "sacrificial" areas, already so degraded or heavily used that massive amounts of conservation effort would not be cost-effective, or they might be areas determined to be relatively unimportant in an ecological sense (Agardy 2010).

It is likely that future marine and maritime spatial planning efforts will use all three of these processes in concert. Existing MPAs and MPA networks will form an important foundation for all zoning plans, regardless of the process or criteria used, because they are de facto precursors of a certain kind of zone. Zoned ocean areas within a region could be administered by a variety of means—by a single overseeing state or federal agency that designs the zoning plans, by a coordinating body that ties together areas variously implemented by different government agencies, or by an umbrella framework such as Biosphere Reserves of UNESCO (the United Nations Educational, Scientific, Cultural Organization and its Man and Biosphere Programme). The latter has benefits in that local communities become a part of the network, ecologically critical areas are afforded strict protection while less important or less sensitive areas are managed for sustainable use, and the biosphere reserve designation itself carries international prestige (and can be used to leverage funds). For shared coastal and marine resources, regional agreements may prove most effective, especially when such agreements capitalize on better understandings of costs and benefits accruing from shared responsibilities in conserving the marine environment (see discussion by Lee Kimball, Chapter 5 this volume).

Through such large-scale spatial planning and management, goals such as biodiversity conservation, conservation of rare and threatened species, maintenance of natural ecosystem functioning at a regional scale, and management of fisheries, recreation, education, and research can be addressed in a coordinated and complementary fashion. Such an integrated approach is an efficient way to allocate scarce time and resources to combating the issues that are most critical. Nations and/or agencies that invest in spatial planning and ocean zoning stand to reap the benefits of effective conservation and its role in providing the foundation for sustainable use and growth of the so-called Blue Economy. The alternative is to stand by as the oceans suffer the death of a thousand cuts, and the oceans act as a lifeline for humanity no longer.

References

Abdulla, A., Gomei, M., Hyrenbach, D., Notarbartolo-di-Sciara, G., and Agardy, T. 2009. Challenges facing a network of representative marine protected areas in the Mediterranean: Prioritizing the protection of underrepresented habitats. *ICES Journal of Marine Science* 66: 22–28.

Agardy, T. 1997. *Marine Protected Areas and Ocean Conservation*. RE Landes Press, Austin TX.

Agardy, T. 2005. Global marine policy versus site-level conservation: the mismatch of scales and its implications. Invited paper, *Marine Ecology Progress Series* 300: 242–248.

Agardy, T. 2010. *Ocean Zoning: Making Marine Management More Effective*. Earthscan, London.

Agardy, T., Notarbartolo-di-Sciara, G., and Christie, P. 2011b. Mind the gap: Overcoming inadequacies of marine protected areas. *Marine Policy* 35 (2): 226–232.

Agardy, T. Christie, P., and Nixon, E. 2012. Marine Spatial Planning in the Context of the Convention on Biological Diversity: A study carried out in response to CBD COP 10 Decision X/29. CBD Tech. Series 68, Montreal.

Agardy, T., Davis, J., Sherwood, K., and Vestergaard, O. 2011a. *Taking Steps Towards Marine and Coastal Ecosystem-Based Management: An Introductory Guide*. UNEP, Nairobi, 67 pp.

Agardy, T., Bridgewater, P., Crosby, M. P., Day, J., Dayton, P. K., Kenchington, R., Laffoley, D., McConney, P., Murray, P. A., Parks, J. E., and Peau, L. 2003. Dangerous targets: Differing perspectives, unresolved issues, and ideological clashes regarding marine protected areas. *Aquatic Conservation: Marine and Freshwater Ecosystems* 13: 1–15.

Agostini, V. N., Grantham, H. S., Wilson, J., Mangubhai, S., Rotinsulu, C., Hidayat, N., Muljadi, A., Muhajir, Mongdong, M., Darmawan, A., Rumetna, L., Erdmann, M. V., Possingham, H. P. 2012. *Achieving Fisheries and Conservation Objectives within Marine Protected Areas: Zoning the Raja Ampat Network*. The Nature Conservancy, Denpasar.

Álvarez-Romero, J. G., Pressey, R. L., Ban, N. C., Vance-Borland, K., Willer, C., Klein, C. J., and Gaines, S. D. 2011. Integrated land-sea conservation planning: The missing links. *Annual Review of Ecology and Evolution Systematics* 42, 381–409.

Beck, M., Ferdana, Z. Kachmar, J. and Killerlain, K.. 2009. Best Practices for Marine Spatial Planning. TNC, Arlington, VA.

Brock, R. J., Kenchington, E., and Martínez-Arroyo, A. (Eds.). 2012. *Scientific Guidelines for Designing Resilient Marine Protected Area Networks in a Changing Climate*. Commission for Environmental Cooperation, Montreal, Canada.

Charles, A., and Wilson, L. 2009. Human dimensions of Marine Protected Areas. *ICES Journal of Marine Science: Journal du Conseil* 66: 6–15.

Christie, P., and White, A. T. 2007. Best practices for improved governance of coral reef marine protected areas. *Coral Reefs* 26: 1047–1056.

Department for Environment, Food, and Rural Affairs (DEFRA). 2013. Marine Conservation Zones Designation Explanatory Note. November 2013. available at www.gov.uk/government/uploads/system/uploads/attachment_data/file/259972/pb14078-mcz-explanatory-note.pdf (last accessed July 12, 2015).

Devillers, R., Pressey, R. L., Grech, A., Kittinger, J. N., Edgar, G. J., Ward, T., and Watson, R. 2014. Reinventing residual reserves in the sea: Are we favouring ease of establishment over need for protection? *Aquatic Conservation: Marine and Freshwater Ecosystems*. Published online in Wiley Online Library (wileyonlinelibrary.com). http://onlinelibrary.wiley.com/doi/10.1002/aqc.2445/abstract (last accessed July 12, 2015).

Eagle, J. 2013. Complex and murky spatial planning. *Journal of Land Use and Environmental Law* 28: 35–59.

Eagle, J., Sanchirico, J. N., and Thompson, B. H. Jr. 2008. Ocean zoning and spatial access privileges: Rewriting the tragedy of the regulated ocean. *New York University Environmental Law Journal*. November. apers.ssrn.com/sol3/papers.cfm?abstract_id=2165984 (last accessed July 12, 2015).

Ehler, C., and Douvere, F. 2007. *Visions for a Sea Change. Report of the First International Workshop on Marine Spatial Planning*. UNESCO Intergovernmental Oceanographic Commission, Paris.

Fernandes, L., Day, J. O. N., Lewis, A., Slegers, S., Kerrigan, B., Breen, D. A. N., Cameron, D., Jago, B., Hall, J., Lowe, D., Innes, J., Tanzer, J., Chadwick, V., Thompson, L., Gorman, K., Simmons, M., Barnett, B., Sampson, K., De'Ath, G., Mapstone, B., Marsh, H., Possingham, H., Ball, I. A. N., Ward, T., Dobbs, K., Aumend, J., Slater, D. E. B., and Stapleton, K. 2005. Establishing representative no-take areas in the Great Barrier Reef: large-scale implementation of theory on marine protected areas. *Conservation Biology* 19: 1733–1744.

Foley, M. M., Halpern, B. S., Micheli, F., Armsby, M., Caldwell, M. R., Crain, C. M., Prahler, E., Rohr, N., Sivas, D., Beck, M. W., Carr, M. H., Crowderg, L. B., Duffy, J. E., Hacker, S. D., McLeod, K. L., Palumbi, S. R., Peterson, C. H., Regan, H. M., Ruckelshausm, M. H., Sandifer, P. A., and Steneck, R. S. 2010. Guiding ecological principles for marine spatial planning. Marine Policy 34: 955–966.

Fox, H. E., Mascia, M. B., Basurto, X., Costa, A., Glew, L., Heinemann, D., Karrer, L. B., Lester, S. E., Lombana, A., and Pomeroy, R. 2012. Reexamining the science of marine protected areas: linking knowledge to action. *Conservation Letters* 5: 1–10.

Game, E. T., McDonald-Madden, E., Puotinen, M. L., Possingham, H. P. 2008. Should we protect the strong or the weak? Risk, resilience, and the selection of marine protected areas. *Conservation Biology* 22: 1619–1629.

Giakoumi, S., Grantham, H. S., Kokkoris, G. D., Possingham, H. P. 2011. Designing a network of marine reserves in the Mediterranean Sea with limited socio-economic data. *Biological Conservation* 144: 753–763.

Gubbay, S. (Ed.) 1995. *Marine Protected Areas: Principles and Techniques for Management*. Chapman & Hall, London.

Guidetti, P., Milazzo, M., Bussottia, S., Molinari, A., and Agardy, T. 2008. Italian marine reserve effectiveness: Does enforcement matter? *Biological Conservation* 141: 699–709.

Halpern, B., and Agardy, T. 2013. Ecosystem based approaches to marine conservation and management. Ch. 21, pp. 477–493. In M. D. Bertness, J. F. Bruno, B. R. Silliman and J. J. Stachowicz (Eds.). *Marine Community Ecology and Conservation*. Sinauer Associates, New York.

IUCN. 2012. *Guidelines for Applying the IUCN Protected Area Management Categories to Marine Protected Areas*. IUCN, Gland Switzerland

Jones, P. J. S. 2014. *Governing Marine Protected Areas: Resilience through Diversity*. Earthscan, Abingdon.

Kelleher, G. G., and Kenchington, R. 1992. *Guidelines for Establishing Marine Protected Areas*. IUCN, Gland, Switzerland.

Kelleher, G. G., Bleakley, C. J., and Wells, S. 1995. *A Global Representative System of Marine Protected Areas*. Four Volumes. World Bank Group, Washington DC.

Kenchington, R. A., and Day, J. C., 2011. Zoning, a fundamental cornerstone of effective Marine Spatial Planning: Lessons learnt from the Great Barrier Reef, Australia. *Journal of Coast Conservation* 15: 271–278.

Leenhardt, P., Cazalet, B., Salvat, B., Claudet, J., and Feral, F. 2013. The rise of large-scale marine protected areas: Conservation or geopolitics? *Ocean and Coastal Management* 85: 112–115.

McLeod, E., Salm, R., Green, A., and Almany, J. 2009. Designing marine protected area networks to address the impacts of climate change. *Frontiers in Ecology and Environment* 7: 362– 370.

Magris, R. A., Pressey, R. L., Weeks, R., and Ban, N. C. 2014. Integrating connectivity and climate change into marine conservation planning. *Biological Conservation* 170: 207–221.

MarViva. 2012. *MSP: A Guide to Process and Methodological Steps*. MarViva, San José, Costa Rica.

NRC (National Research Council/US). 2007. *Increasing Capacity for Stewardship of Oceans and Coasts: A Priority for the 21st Century*. National Academies Press, Washington DC.

Ogden, J. C. 2010. Marine spatial planning (MSP): A first step to ecosystem-based management (EBM) in the Wider Caribbean. *Revista de biología tropical* 58 (3): 71–79.

Olsen, E., Johnson, D., Weaver, P., Goni, R., Ribeiro, M. C., Rabaut, M., Macpherson, E., Pelletier, D., Fonseca, L., Katsanevakis, S., and Zaharia, T. 2013. *Achieving ecologically coherent MPA Networks in Europe: Science Needs and Priorities*. European Marine Board, Ostend, Belgium.

Olsen, E., Fluharty, D., Hoel, A. H., Hostens, K., Maes, F., and Pecceu, E. 2014. Integration at the Round Table: Marine spatial planning in multi-stakeholder settings. *Plos One* 9(10): e109964.

Portman, M. E., Notarbartolo-di-Sciara, G., Agardy, T., Katsanevakis, S., Possingham, H. P., and di Carlo, G. 2013. He who hesitates is lost: Why conservation in the Mediterranean is necessary and possible now. *Marine Policy* 42: 270–279.

Roberts, C. M., Andelman, S., Branch, G., Bustamante, R. H., Carlos Castilla, J., Dugan, J., Halpern, B. S., Lafferty, K. D., Leslie, H., and Lubchenco, J., 2003. Ecological criteria for evaluating candidate sites for marine reserves. *Ecological Applications* 13: 199–214.

Sale, P., Agardy, T., Ainsworth, C. H., Feist, B. E., Bell, J. D., Christie, P., Hoegh-Guldberg, O., Mumby, P. J., Feary, D. A., Saunders, M. I., Daw, T. M., Foale, S. J., Levin, P.S., Lindeman, K. C, Lorenzen, K., Pomeroy, R.S., Allison, E. H., Bradbury, R. H., Corrin, J., Edwards, A. J., Obura, D. O., Sadovy de Mitcheson, Y. J., Samoilys, M. A., and Sheppard C. R. C. 2014. Transforming management of tropical coastal seas to cope with challenges of the 21st century. *Marine Pollution Bulletin* 85: 8–23.

Toonen, R. J., Wilhelm, T. A., Maxwell, S. M., Wagner, D., Bowen, B. W., Sheppard, C. R. C., Taei, S. M., Teroroko, T., Moffitt, R., Gaymer, C. F., Morgan, L., Lewis, N., Sheppard, A. L. S., Parks, J., Friedlander, A. M., and The Big Ocean Think Tank. 2013. One size does not fit all: The emerging frontier in large-scale marine conservation. *Marine Pollution Bulletin* 77: 7–10.

Toropova, C., Meliane, I., Laffoley, D., Matthews, E., and Spalding, M. (Eds.). 2010. *Global Ocean Protection: Present Status and Future Possibilities*. IUCN, Gland, Switzerland.

Watts, M. E., Ball, I. R., Stewart, R. S., Klein, C. J., Wilson, K., Steinback, C., Lourival, R., Kircher, L., and Possingham, H. P. 2009. Marxan with zones: Software for optimal conservation based land- and sea-use zoning. Environmental Modelling Software 24: 1513–1521.

White, C., Halpern, B., and Kappel, C. V. 2012. Ecosystem service tradeoff analysis reveals the value of marine spatial planning for multiple ocean uses. PNAS. www.academia.edu/1757142/Recasting_shortfalls_of_marine_protected_areas_as_opportunities_through_adaptive_management (last accessed July 12 2015).

Wood, L. J., Fish, L., Laughren, J., and Paul, D. 2008. Assessing progress towards global marine protection targets: Shortfalls in information and action. *Oryx* 42: 340–351.

WCPA/IUCN. 2007. Establishing networks of marine protected areas: A guide for developing national and regional capacity for building MPA networks. Non-technical summary report. IUCN, Gland, Switzerland.

Zacharias, M. 2014. *Marine Policy: An Introduction to Governance and International Law of the Oceans*. Earthscan, Abingdon.

Regional developments:
key core maritime regions

32

MARITIME BOUNDARIES

The end of the Mediterranean exception

Juan Luis Suárez de Vivero and Juan Carlos
Rodríguez Mateos

Introduction: a pioneering region and an exception

At the same time that it was a basin pioneering the adoption of regional marine management instruments (Mediterranean Action Plan, 1975; Barcelona Convention, 1976), the Mediterranean Sea was characterised by coastal States showing restraint in declaring jurisdictional rights beyond the territorial sea for a (relatively) long period of time when the creation of extended jurisdictions proliferated in the world's oceans (see Chapter 2, Changing geopolitical scenarios).

However, the turn of the century produced a historic sea change. Jurisdictions and changes in maritime areas' legal statuses were progressively declared in an overt transition from what might be referred to as a regime of *mare liberum* (75 per cent High Seas in 1970) to one of *mare clausum* (29 per cent High Seas in 2012). If, when the United Nations Convention on the Law of the Sea was passed (UNCLOS, 1982) only two States – out of a total of 21 – had declared some kind of extended jurisdiction, three decades on there are seventeen States with exclusive economic zones, fishing zones or protected ecological zones and a wide range of jurisdictional regimes existing alongside each other in a relatively small basin (2.5 million km^2).

In the three decades that have gone by since UNCLOS was passed some quite profound changes have taken place in the Mediterranean basin in the domains of geopolitics and the environment. In some cases these have been the result of events on the global scale, such as the end of the Cold War and its bloc policy, in which the Mediterranean Sea was deemed to be the scenario of a possible clash between the powers that dominated. Freedom of navigation and passage through straits were therefore key factors in the maritime environment of this basin. If at that time the jurisdictional geography was set against the backdrop of marked global geostrategic interests, at the current time it is the regional and local interests that induce us to review the jurisdictional issue from new perspectives and re-examine the way that marine governance might be affected. The Mediterranean's gradual maritime nationalisation – in the sense that Lucchini and Voelckel (1978) coined the expression – can be said to have taken place in parallel with the progressive presence of the European Union: two Mediterranean States in 1952, three in 1973, four in 1986, seven in 2004 and eight in 2013, with growth burgeoning in the new century. At the current time,[1] the jurisdictional importance of EU member States accounts for 36 per cent of the basin's maritime surface area, giving weight to the idea of a possible 'Europeanisation' process of this sea (Radaelli, 2004). In this respect, advances in the construction of a Mediterranean political and institutional framework have much greater links to the north bank and the major stronghold that is the European Union (EU), which unites and bonds an as yet not too large a number of coastal States (nine with the UK in 2013) but one that could expand to as many as twelve (Montenegro and Turkey, candidates; Albania and Bosnia-Herzegovina, potential candidates). A second strategic bulwark in the region that similarly contributes to uniting the northern shore is the North Atlantic Treaty Organisation (NATO) with eight Mediterranean States among its twenty-eight members. There are a number of organisations (Arab Maghreb Union (AMU)), the African Union, the Arab League, the Organisation of the Islamic Conference) on the southern banks and in the south east but they do not have the same degree of internal integration. This (greater integration on the northern bank and little cohesion on the southern shore) means that it is the EU that is spearheading the only inter-regional alliances around the basin with its European Neighbourhood Policy and, indirectly, through the Union for the Mediterranean, while in NATO the so-called Mediterranean Dialogue has been established which includes five southern bank countries (Algeria, Egypt, Israel, Morocco and Tunisia). This political composition of the region has been reflected and set out in the construction of marine governance, which can only be conducted in the framework of cooperation between the States that project their sovereignty and/or jurisdictional rights over the waters and seabeds, and with the involvement of third States while waters under the regime of the High Seas still remain. Recent EU initiatives in the domain of marine governance (Integrated Maritime Policy, 2007; Marine Strategy Framework Directive, 2008) including an action designed to promote an integrated maritime policy in the Mediterranean (COM(2009) 466 final) make the EU the leading institution in promoting marine governance in the region, and the only one with the economic, technical and political ability to coordinate and integrate the actions of the twenty-one coastal States and cooperation with third States given the 'institutional ambiguity' (van Leeuwen *et al.*, 2012) that afflicts the southern shore.

The progressive shaping of the Mediterranean Sea's maritime jurisdictional map can be interpreted as the setting down of a series of geopolitical events, of which the following can be highlighted:

- The finalisation of the region's political map conclusively defining coastal States' respective maritime territories. In some cases (Greece, Turkey, Cyprus) this process will still require long negotiations and current unyielding hard-line positions to be addressed, but good progress is being made towards defining a new territorial reality.
- Without underestimating the grave crisis in the Middle East and its associated risks, the Mediterranean is evolving towards a scenario dominated by inter-regional issues due to the shift of major strategic interests towards Asia and the consequent loss of the relative import-ance of trade flows in the basin and, therefore, of the primacy of freedom of the seas over local interests. The profusion of jurisdictional claims made since the turn of the century can be taken as an indicator of the weakening of the global position of the Mediterranean since the end of the Cold War.
- Marine governance in the basin is entering a transitional phase leading towards the Euro-pean Union hegemony, which could result in United Nations institutions, such as the Mediterranean Action Plan and the Barcelona Convention, losing some of their importance and, consequently, the multilateral relations model becoming weaker to the advantage of bilateral relationships dominated by European institutions. The nationalisation of Mediterranean maritime space can be interpreted as a favourable factor for marine gov-ernance as all the waters (seabed and subsoil included) will come under the responsibility

of coastal States and be subject to the corresponding norms and planning instruments. However, this consideration is only feasible with solid cooperation structures that are appropriate for a fluid environment. Progress may be made in EU territory as a result of its maritime policy being developed further, but this progress will be held back if no moves are made to shorten the North–South divide.

Proliferating maritime boundaries are an indicator of the institutional change that is taking place in the Mediterranean basin. While the EU continues to expand and forges ahead in constructing policies for maritime governance, institutional development has stagnated in third States that have no political structures to balance the weight and size of the EU, perpetuating their dependence. How this scenario evolves in the future is linked to the way that the crisis in Europe develops; if it continues as it is, it will open the door to a new era in Mediterranean relations.

The post-UNCLOS Mediterranean: some geopolitical keys

During the Cold War the Mediterranean was both a bone of contention for the two superpowers of the time and a sea dominated by the legal regime of the High Seas, but today it would seem that 'domestic' and intra-regional problems are gaining in importance and becoming both the causes and consequences of a number of jurisdictional claims, (still unresolved) boundary disputes and growing maritime nationalism. For this reason, no analysis of the Mediterranean in geopolitical terms can be strictly limited to the maritime area. As such, and even though the major challenges that managing this sea present are already per se significant, this section seeks to outline aspects that pertain to what has been referred to as the 'wider Mediterranean', especially as the area interacts with neighbouring regions (the Caucasus, Central Asia, the Middle East and the Maghreb) and because it continues to be a strategically important area for outside powers, such as the USA and Russia (Echeverría, 2011). The following geopolitical analysis highlights both the most positive (cooperative) and the less reassuring (spatial de-structuring, inequalities, and disputes) elements as both explain the complexity of the region (Serrano Martínez, 2007).

From the 1970s onwards and especially during the whole of the 1980s a number of European countries embarked upon a period of political dialogue with Mediterranean countries, regaining an old tradition (that had been broken during the Cold War, when the North American position had been sided with) of European Mediterranean policies that had existed prior to the Second World War. Progress made in this Euro-Mediterranean rapprochement process was embodied in the so-called Euro-Arab Dialogue and in the EEC's Global Mediterranean Policy, both of which benefited from the beginning of the Peace Process in the Middle East. Against this backdrop and along with dialogue in the Mediterranean driven by European institutions and countries, there were other developments, such as Greece, Spain and Portugal joining the EEC, changes in the concept of regional security that resulted in talks between the WEO and NATO and countries on the Mediterranean's eastern and southern banks, and tentative attempts to set up regional organisations on the southern shore (the Arab Maghreb Union).

Other events, such as the wars in the Balkans and, more especially, the Gulf War, and their aftermaths meant that some of the achievements in regional cooperation unfortunately took a backward step in the 1990s (Amin, 1994). However, from that decade on there has been an ongoing parallel process of Euro-Mediterranean cooperation in the fields of science and the environment: the MEDSPA Programme, various scientific projects, the Mediterranean Environmental Programme, Euro-Mediterranean environmental and fisheries cooperation (1990 Nicosia Charter, 1992 Cairo Declaration, 1994 Heraklion Declaration) and, indeed, the

Mediterranean Action Plan itself and the Barcelona Convention (both renewed in 1995). This foundation stone, which could be referred to as 'maritime-environmental', and the economic and political foundations represented by the Euro-Mediterranean Association and other formal (in NATO, the WEO and OSCE) (Sánchez Mateos, 2002) and informal (the Mediterranean Forum, 5+5 Dialogue) political dialogue, were rays of hope in an area historically beset by upheaval.

Be that as it may, the Mediterranean is no longer a stage for the typical polarised tensions of the Cold War but has become an area of more local disputes, either linked to demography (south–north migratory flows), or the environment (pollution, eutrophication, loss of biodiversity, inter alia). Naturally, there are also latent tensions or even open conflict on the international political plane (boundary disputes, the remnants of colonialism, internal tensions with knock-on effects on neighbouring countries, etc.) (Sanguin, 2000, 15–24). This third type, strategic–political conflict, can have causes rooted in history (traditional disputes) or sparked by more recent events (current disputes) and is the cause of major instability in the region. In many of these cases of latent or open conflict the ultimate reason is control over land and its resources (water and other natural resources) (Bethemont, 2002). The sea is not devoid of these issues either, and disputes there can be linked to the control of the waters and the seabed and, consequently, of living and non-living marine resources.

Instability in the Mediterranean that used to be seen from the European shore as rooted in the southern shore's lesser economic development, intense demographic development and the rise of fundamentalist movements can today transmute into instability caused by new movements making demands at the heart of the Maghreb countries. The key issue is whether this internal instability will eventually give way to positions that show greater respect to international law and a greater willingness to take part in cooperative processes in the Mediterranean, or whether they will lead to further territorial/border claims (something that is not unknown in northern bank countries, either) as a pretext for nationalism and a means of diverting public attention away from internal social problems. If the southern Mediterranean countries were to go down this second route not only would they stifle the aspirations of civil society, they could further complicate the already complex geopolitical context in the Mediterranean, turning the sea into an area of claims, unilateral standpoints and contention. At the same time, neither would it be positive if the scenario was one of the countries on the European bank using jurisdictional claims as yet more grounds for dispute, and maritime boundaries and borders as protective barriers against 'the threat from the South' (migration, security, etc.).

The grand and ambitious Euro-Mediterranean political cooperation/association processes seem to have floundered in the face of such a complex and potentially conflictive situation, but the opportunity remains of making good use of instruments that, although they seem on the surface to be more modest and restricted to specific subjects (science, the environment, fishing, etc.), are none the less useful for that. Perhaps the key to cooperation in the Mediterranean lies in finding common ground around the management of areas and resources with a natural and economic value rather than getting lost in interminable arguments about civilisations, culture and beliefs. The fact that the Mediterranean has seen its leading role in the world geopolitical arena somewhat diminished takes no interest away from the area, as it can still be the subject of disputes and desires both from the countries around its shores and powers further afield. It should be remembered that the Mediterranean is on the periphery of Eurasia which has always made it something of an unstable and dangerous area, and also one of major geopolitical and geostrategic importance.

The jurisdictional puzzle

The implementation of UNCLOS III, which came into force in 1994, meant that the Mediterranean Sea could be regarded as an exception to the widespread process to extend jurisdictions that was initiated in the 1970s. However, recent years have seen a turnaround with extended jurisdictions being declared, resulting in a reduction in the area of water that comes under the High Seas regime.

Up to the 1970s, States in the Mediterranean had basically been interested in either extending their territorial seas up to a maximum 12 mile limit,[2] or declaring fishing zones with a similar outer limit. During the same decade agreement was reached on most of the bilateral agreements on continental shelf delimitation (i.e. after the 1958 Geneva Convention) and on the establishment of straight baselines (Suárez de Vivero and Rodríguez Mateos, 2006).

The few exclusive economic zone (EEZ) declarations that have been made in the Mediterranean were also late in coming. Up until the 1990s only two countries had decided to create an EEZ. This lack of an exclusive economic zone is what enabled the high seas (a jurisdiction determined by exclusion), which constituted 45.9 per cent of the whole basin (up to 2011), to play such a major role, an anomaly that was almost unique to the Mediterranean. However, from the 1990s onwards Mediterranean coastal States began declaring fisheries protection zones and/or ecological protection zones, instead of EEZs, with the aim of protecting resources (primarily fisheries resources). These two legal concepts only benefit from part of the legal regime that UNCLOS envisages for the EEZ. Contiguous archaeological zones have also been created (Algeria, Cyprus, France, Italy and Tunisia) to protect submerged cultural heritage. The delimitations between the various jurisdictions in the Mediterranean have given rise to what could be called 'jurisdictional asymmetry' as a result of the heterogeneous legal regimes of the adjacent maritime areas: exclusive economic zones, fishing zones, fisheries protection zones and ecological protection zones. Jurisdictions analogous to the EEZ imply exercising rights of sovereignty in specific issues that may not be the same for neighbouring States, and therefore contribute to disputes.

The obvious result for the Mediterranean in broad terms is the shrinking of the high seas, which at the present time (2013) have been reduced to 29.2 per cent in the wake of France's recent 2012 declaration of an exclusive economic zone (previously an ecological protected zone) and the declaration of an ecological protected zone in Italian waters (2011). This has ushered in a new period characterised by jurisdictional expansion – leading to an 11 per cent increase in the surface area devoted to EEZs, driven by Mediterranean coastal States' increasing interest in controlling and exploiting marine (biogenetic and energy) resources that are now accessible as a result of new technological advances (Instituto Español de Oceanografía, 2006).

To date, ten States in the Mediterranean (Spain, France, Morocco, Tunisia, Libya, Egypt, Israel, Lebanon, Syria and Cyprus) have declared EEZs in addition to their territorial seas and contiguous zones. Eight of the remaining Mediterranean coastal States have declared some type of jurisdiction. Another five States (Bosnia-Herzegovina, Montenegro, Albania, Greece and Turkey) have declared no other jurisdictional rights beyond their territorial seas (Figure 32.1).

Two historic bays have also been declared (the Gulf of Taranto in Italy and the Gulf of Sidra in Libya) that, in turn, are both internal waters and protested by third countries. The majority of States (twelve) have laid down straight baselines as opposed to nine that have not.

The importance of the European Union in the waters of the Mediterranean is attested to by the area of jurisdictional waters that belong to the nine (with UK) member countries. These make up approximately 51.02 per cent of the whole surface area under national jurisdiction in the sea. Only three EU countries have declared EEZs (Spain, Cyprus and France). However,

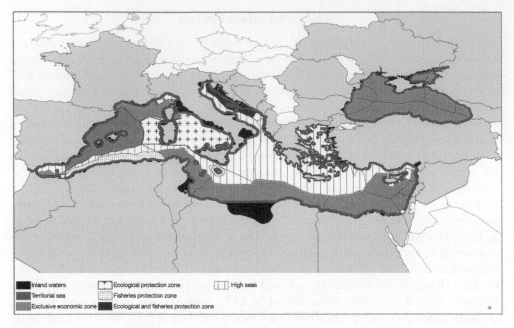

Inland waters

Territorial sea

Exclusive economic zone

Ecological protection zone

Fisheries protection zone

Ecological and fisheries protection zone

High seas

or

Figure 32.1 Maritime jurisdictions of the Mediterranean and Black Seas
Source: Author.

four States belonging to the EU, Malta, Slovenia and more recently Italy and Croatia, have declared some type of jurisdiction beyond the territorial sea (fishing zone, ecological protection zone and fisheries and ecological protection zone), while Greece has declared no jurisdictional rights beyond its territorial sea (Figure 32.2). In quantitative terms, member countries' jurisdictional waters could extend to 67.04 per cent of the basin's surface area if the jurisdictional waters of States in the process of joining and EU candidate States – Turkey, Montenegro and Albania – along with the whole of the waters that Italy and Greece are able to declare were to be included (in which case the current size of the high seas could be reduced by 35 per cent (median lines no high seas)).

The number of States located around the sea and its geographical characteristics give rise to a considerable number of boundaries between the various maritime jurisdictions. Interactions between adjacent and opposite States in the Mediterranean produce twenty-nine boundary contacts with different types of delimitations between the maritime jurisdictions, although there are only fifteen agreed boundaries, reflecting the conflict inherent in the delimitation of maritime areas.

The oldest boundary treaty dates back to 1968 (Italy–Yugoslavia) and the most recent, which was signed by Cyprus and Israel in 2010, came into force in 2011 and was the first agreement on EEZ delimitation. Sixty per cent of the agreements made during the forty years separating the first and the last were signed during the 1970s and 1980s. Eight of the fifteen agreements on limits concern the continental shelf (Figure 32.1). This is due to the fact that States do not need to make any express claim to rights of jurisdiction over the continental shelf and, given the basin's geographical characteristics, all the seabed and subsoil is under national jurisdiction (Scovazzi, 2012). However, the overlying waters may be subject to the high seas regime if the corresponding coastal State has not declared jurisdictional rights over the waters lying outside its territorial sea.

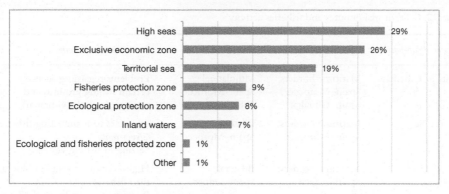

Figure 32.2 Distribution of maritime jurisdictions in the Mediterranean Sea (%)
Source: Author.

Italy is the State that has subscribed to most agreements (eight). This can be explained by the relative position that it occupies as a peninsula in the middle of the basin and the creation of new States out of the old Yugoslavia, with which Italy signed the first agreement on maritime boundaries. There are six States (Bosnia-Herzegovina, Israel, Lebanon, Morocco, Syria and Turkey) that have not subscribed to any agreements on limits.

There are secondary factors that can have indirect consequences for the delimitation of maritime jurisdictions, such as historical territorial disputes, and these are exacerbated by conflict spreading to the marine domain. In other cases, the process of expanding sovereignty over the seas has given rise to new disputes due to overlaps between jurisdictions (which refers us back to the jurisdictional asymmetry described above) and the creation of new boundaries (Table 32.1).

Marine governance and Europeanisation of the Mediterranean region

Although some of the accepted uses of the term 'Europeanisation' basically tie in with internal European Union aspects linked to the process of constructing Europe,[3] here emphasis is placed on the uses with a more international dimension, in other words, the consideration of Europeanisation as changes in external territorial boundaries (one of the results of progressive EU expansion) and also Europeanisation as exporting forms of political organisation and governance that are typical and distinct for Europe beyond the European territory (Olsen, 2002), which is a dissemination process that would seem to be growing more and more around the Mediterranean (Kocamaz, 2012).

The importance and power of the EU in Mediterranean waters is reflected in the size of member States' jurisdictional waters. The surface area of EU member States' jurisdictional waters in the Mediterranean amounts to 905.704 km^2, which represents approximately 51 per cent of all national jurisdictions in the waters of the Mediterranean (Suárez de Vivero, 2012). This, and the dominant financial and technological capabilities of these countries, will determine the future management of this regional sea. In all probability, the EU will have a major and growing presence in forums and institutions devoted to developing regional policies.

This situation is not entirely new, however. For decades, the EU has been taking action in two large parallel and complementary fields: i) an area that could be called general politics and ii) the more strictly maritime area (Table 32.2).

Table 32.1 Territorial disputes and fisheries activity

Location (Sub-basin)	Type of dispute	Involved states	Importance for fisheries
Straits of Gibraltar	Maritime borders: Spain–Morocco Spain–Gibraltar	Multilateral Spain–Morocco– Gibraltar	Problems regarding access to certain fishing grounds and seizing of vessels (Gibraltar)
Alboran Sea	Maritime borders: Spain–Morocco	Bilateral Spain–Morocco	Obstacles to formulating fisheries agreements Seizing of fishing vessels
Gulf of Lion	Maritime borders: Spain–France	Bilateral Spain–France	High-economic value of blue fin tuna fisheries Restricted access beyond 12 nm for third country fleets
Mammellone	Italy Historical fishing rights	Bilateral	Resource protection
Straits of Sicily	Maritime borders: Malta–Libya	—	Irrelevant for fishing
Sicily Channel	Maritime borders: Malta–Italy	—	Irrelevant for fishing
Gulf of Gabes	Maritime borders: Tunisia–Libya	—	Irrelevant for fishing
Gulf of Sidra	Maritime borders: Libya–Italy	Bilateral	The waters claimed by Libya have high fisheries value, especially for tuna fishing, which has already caused disputes between the two States
Italy–Yugoslavia (Adriatic Sea)	Yugoslavia Exclusive fishing zone	Bilateral	Relevant for fishing
Bay of Piran (Adriatic Sea)	Outlet to the high seas for Slovenia	—	Irrelevant for fishing
Klek-Neum Bay (Adriatic Sea)	Historical rights in Klek-Neum bay and maritime outlet to the Adriatic for Bosnia-Herzegovina	—	Irrelevant for fishing
Island of Kastelorizo Megisti (Aegean Sea)	Maritime borders: Turkey–Greece	Bilateral Turkey–Greece	The waters between Greece and Turkey have high fisheries value
Israel-Gaza	Maritime borders: Israel–Gaza	Bilateral Israel–Gaza	Gaza's fisheries activities are restricted in adjoining waters
MERIDIAN 32°16′18″	Maritime borders: Cyprus–Egypt– Turkey	—	Irrelevant for fishing
Akrotiri and Dhekelia	Maritime borders: Cyprus–UK	—	Irrelevant for fishing

Source: Author.

The first sphere of EU action is markedly geopolitical in nature and has been embodied in such ventures as EU–Mediterranean third country bilateral agreements, the European Neighbourhood Policy, the Euro-Mediterranean Partnership and the Union for the Mediterranean. Although more directed towards shaping international cooperation initiatives for the whole of the Mediterranean region, it lays down a political and institutional framework that could drive and benefit actions in the strictly maritime sense. This is what is occurring, for example, with the support that some European institutions (European Commission, European Investment Bank) are giving – through the Facility for Euro-Mediterranean Investment and Partnership (FEMIP) – to Mediterranean maritime cooperation in a number of areas: social aspects (employment), maritime surveillance (regarding environmental problems, illegal immigration and illicit traffic) and maritime infrastructure (basically linked to transport).[4]

The Union for the Mediterranean is especially interesting. This has meant the relaunch of the EU's Mediterranean vocation, and a new and more ambitious institutional framework, and being able to include a wide range of very different Mediterranean countries. It could also be the first step in regional cooperation in a range of areas, with maritime space involved in two of these in new blue economy-related uses: de-pollution of the Mediterranean (including coastal and protected marine areas) and Maritime Highways (with particular attention to cooperation in the field of maritime security and safety).[5]

In other respects, various Euro-Mediterranean and European Neighbourhood Policy multilateral programmes spotlight the problem of the EU's external borders. This is the case of the Euromed Migration I programme, for example, a major part of which is given over to border management, or the important European Neighbourhood & Partnership Instrument-endorsed group of programmes on cross-border cooperation issues in the Mediterranean region (especially with reference to the Mediterranean maritime basin – Mediterranean Programme – and the Spain–Morocco North and Italy–Tunisia sea routes).[6]

With respect to policies and actions that focus more specifically on the marine environment, the EU is involved in a range of multilateral regional initiatives in the Mediterranean[7] – MAP system, General Fisheries Commission for the Mediterranean, Mediterranean Strategy for Sustainable Development – but it has also been driving and taking a leading role in fisheries-related initiatives,[8] in environmental affairs (Environment Strategy for the Mediterranean (COM(2006) 475 final)) and in maritime management issues (2009 proposal for the application of the Integrated Maritime Policy to the Mediterranean (COM(2009) 466 final)). This incursion by the EU into Mediterranean maritime policies also manifests itself in the role that member States play when they transpose and implement EU legislation (fisheries regulations in the area, Marine Strategy Framework Directive) and put in place COPEMED (FAO) initiative sub-regional projects.

In both these fields of action (the field of international macro-politics and that of maritime policies *sensu stricto*) the European Union is evidently fashioning a Mediterranean regional system Europeanisation process, not only through the leading role that the European community takes, but also due to the fact that most of the Mediterranean coastal States converge with the EU's view on legislation, and assume its regulations and directives as their own. There is, as it were, a kind of unilateral projection of community legislation towards Mediterranean third countries and this creates the idea of solely European governance in the Mediterranean and boosts Europe's image as the normative power.[9] However, it is also true that this process is occurring as part of a political system of complex bilateral and multilateral relations and ground rules that come from several different political-administrative tiers (global, European, regional). While taking it for granted that there is European leadership in the Mediterranean, this leadership is spread thinly and limited by the very complexity and degree of interaction produced in the region

(Barbé, 2010). It should also be highlighted that other countries sometimes refuse to adopt these European ground rules (Youngs, 2010, 21; Barbé, 2012); for them, to use Radaelli's (2004) expression, Europeanisation is not the solution, but the problem. They carry on bureaucratically in their own way and with their own organisation, and this does not make the role of leader that Europe seeks to hold in the region any the easier.

Table 32.2 Ways that the EU participates in Mediterranean governance

Marine management policies	Management policies concerning European marine space	– EU environmental strategies: Strategy for the protection and conservation of the marine environment; Proposal of European environmental strategy for the Mediterranean; funding instruments for the environmental protection (MEDSPA Programme/LIFE Programme) – European Territorial Strategy – Common coastal strategy – Integrated Maritime Policy; Integrated maritime policy for the Mediterranean – Common Fisheries Policy; Fishery regulations in Mediterranean waters; EU Action Plan for the conservation and sustainable exploitation of fishery resources in the Mediterranean Sea
	Euro-Mediterranean management policies	– Participation in MAP and Barcelona Convention – MEDSPA Programme – Scientific projects in cooperation with third countries – Support for environmental programmes in the region – Cooperation in the field of underwater cultural heritage – Euro-Mediterranean environmental and fishery cooperation (Nicosia Charter, Cairo Declaration, Heraklion Declaration)
General Euro-Mediterranean cooperation policies	Promoted by EU	– Euro-Arab dialogue – EU Mediterranean policy – EU–Mediterranean third country accords – European Neighbourhood Policy – '5+5' Group – Euro-Mediterranean Partnership – Union for the Mediterranean
	Private initiatives	– Mediterranean Forum – Inter-parliamentary Conference on Security and Cooperation
	Other political dialogue	– Security dialogue between Western organisations (NATO, WEU) and Mediterranean third countries – Conference on Security and Cooperation in the Mediterranean (CSCM)

Source: Prepared by author.

Notes

1   All calculations included in this chapter are based on data available at the United Nations (Division for Ocean Affairs and Law of the Sea, DOALOS) until 31 December 2013.
2   Some countries still adopt lesser limits due to complex geographical and political situations. These include Greece and Turkey in the Aegean Sea, which adopt a 6 nm limit, and Gibraltar.
3   Europeanisation as the development of institutions of governance at the European level, as central penetration of national and sub-national systems of governance, as a political project aiming at a unified and politically stronger Europe (Olsen, 2002).
4   See website: www.enpi-info.eu/mainmed.php?id=24624&id_type=1&lang_id=450 (accessed 1 December 2014).
5   Joint Declaration of the Paris Summit for the Mediterranean (Paris, 13 July 2008). See website: http://ec.europa.eu/research/iscp/pdf/paris_declaration.pdf (accessed 1 December 2014).
6   See website: http://ec.europa.eu/world/enp/pdf/country/enpi_cross-border_cooperation_strategy_paper_en.pdf (accessed 1 December 2014).
7   In this context it is interesting to refer to the 1990 Nicosia Charter and the 1992 Cairo Declaration on Euro-Mediterranean cooperation on environmental issues in the Mediterranean basin.
8   1994 Heraklion Euro-Mediterranean Conference/Declaration on the conservation and management of Mediterranean fishery resources; 2002 EU Action Plan for the conservation and sustainable exploitation of fishery resources in the Mediterranean Sea in the framework of the Common Fisheries Policy (COM(2002) 535 final); Council Regulation (EC) No. 1967/2006 of 21 December 2006, concerning management measures for the sustainable exploitation of fishery resources in the Mediterranean Sea, *European Union Official Journal*, L 409, 30.12.2006.
9   This normative power could have three meanings: EU intervention in global affairs to support norms and not interests; 'ideational power' or 'soft power'; the ability to define notions of what constitutes 'normal' in international relations (Manners, 2002).

References

Amin, S. (1994): 'Algunos problemas de política internacional relativos a la Región Mediterránea y el Golfo', in García Cantús, D. (ed.): *El Mediterráneo y el mundo árabe ante el nuevo orden mundial*, Valencia, Universitat de València, pp. 15–32.

Barbé, E. (Dir.) (2010): *La Unión Europea más allá de sus fronteras. ¿Hacia la transformación del Mediterráneo y Europa Oriental?*, Madrid, Tecnos.

Barbé, E. (2012): 'The EU and the emergence of a post-Western world: in search of lost prestige', *Revista CIDOB d'Afers Internacionals*, 100, pp. 91–112.

Bethemont, J. (2002): 'La Méditerranée, espace, enjeux et conflits', *L'information géographique*, 1, pp. 18–33.

Commission of the European Communities (2006): *Establishing an Environment Strategy for the Mediterranean*, COM(2006) 475 final, Brussels, 5.9.2006.

Commission of the European Communities (2009): *Towards an Integrated Maritime Policy for Better Governance in the Mediterranean*, COM(2009) 466 final, Brussels, 11.9.2009.

Echeverría, C. (2011): 'Los entornos sociopolítico y diplomático', in *El Mediterráneo: Cruce de intereses estratégicos*, Monografías del CESEDEN, No. 118, Madrid, Ministerio de Defensa, pp. 17–33.

Instituto Español de Oceanografía (2006): La bioprospección de los recursos genéticos y su exploración, in *Revista electrónica del Instituto Español de Oceanografía*, no. 3 March–April, pp. 15–18.

Kocamaz, S. (2012): 'Europeanization through the Union for Mediterranean Policy: The dynamics of Europeanization in the Mediterranean Partner States', International Conference: Comparing and contrasting 'Europeanization': concepts and experiences, Athens (Greece), 14–16 May 2012. Available from: www.idec.gr/iier/new/Europeanization%20Papers%20PDF/KOCAMAZ%20EUROPEANIZATION%20THROUGH%20THE%20UNION%20FOR%20MEDITERANEAN%20POLICY.pdf (accessed 1 December 2014).

Lucchini, L., Voelckel, M. (1978): *Les états et la mer: le nationalisme maritime*, Paris, La Documentation Française.

Manners, I. (2002): 'Normative Power Europe: A Contradiction in terms?', *Journal of Common Market Studies*, 40 (2), pp. 235–258.

Marine Strategy Framework (2008): Directive 2008/56/EC of the European Parliament and of the Council of 17 June 2008 establishing a framework for community action in the field of marine environmental policy (Marine Strategy Framework Directive) L 164/19 (25.6.2008).

Olsen, J. P. (2002): 'The Many Faces of Europeanization', ARENA Working Papers, WP 02/2002. Available from: www.sv.uio.no/arena/english/research/publications/arena-publications/workingpapers/working-papers2002/ (accessed 1 December 2014).

Radaelli, C. M. (2004): 'Europeanization: Solution or Problem?' *European Integration online Papers (EIOP)*, 8 (16), pp. 1–23. Available from: http://eiop.or.at/eiop/texte/2004–016.htm (accessed 1 December 2014).

Sánchez Mateos, E. (2002): 'La Seguridad Global en el Mediterráneo', *Revista CIDOB d'Afers Internacionals*, 57–58, pp. 7–28.

Sanguin, A.-L. (ed.) (2000*): Mare Nostrum: Dynamiques et mutations géopolitiques de la Méditerranée*, Paris, L'Harmattan.

Scovazzi, T. (2012): *Policy Brief. Maritime Boundaries in the Eastern Mediterranean Sea*, in Eastern Mediterranean Energy Project. German Marshall Fund of the United States (GMF).

Serrano Martínez, J. M. (2007): 'Geopolítica del Mediterráneo occidental. Consideraciones generales', *Anales de Historia Contemporánea*, 23, pp. 51–98.

Suárez de Vivero, J. L. (2012): *Fisheries cooperation in the Mediterranean and Black Seas*. Brussels: European Parliament, Directorate General for Internal Policies, Policy Department, Fisheries. Available from: www.europarl.europa.eu/committees/en/studiesdownload.html?languageDocument=EN&file=78711 (accessed 1 December 2014).

Suárez de Vivero, J. L. and Rodríguez Mateos, J. C. (2006): 'Maritime Europe and EU enlargement. A geopolitical perspective', *Marine Policy*, 30, 167–172.

van Leeuwen, J., van Hoof, L., van Tatenhove, J. (2012): 'Institutional ambiguity in implementing the European Union Marine Strategy Framework Directive', *Marine Policy*, 36 (3), pp. 636–643.

Youngs, R. (2010): *Europe's decline and fall: The struggle against global irrelevance*, London, Profile Books.

33

MARINE SPATIAL PLANNING IN THE UNITED STATES

Triangulating between state and federal roles and responsibilities

Stephen B. Olsen, Jennifer McCann and
Monique LaFrance Bartley

Introduction

Marine Spatial Planning (MSP) is an expression of comprehensive spatial planning that identifies use-priority areas and allocates user rights within geographically defined zones (Sanchirico *et al.*, 2010). MSP may therefore be viewed as an extension of land use zoning into marine areas and works to separate incompatible uses and thereby reduce conflicts among users (Sanchirico, 2004; Chowder *et al.*, 2006; Agardy, 2010). MSP in the United States is seen as an expression of ecosystem-based management that calls for:

> analysis, planning and decision making that considers the entire ecosystem, including humans, and evaluates the cumulative impacts of diverse human activities in order to regulate human activities in a manner that maintains or restores an ecosystem to a healthy, productive and resilient condition that provides the services that humans want and need.
>
> *(McLeod et al., 2005)*

In the United States, MSP requires a high degree of collaboration between governmental agencies at the state and federal levels. As a federation of semi-autonomous states, much policy making, planning, and decision making is the prerogative of each state. States have authority over their territorial sea, an area of marine waters usually extending out three nautical miles from their coastline. However, federal authorities take precedence over such topics as defense, navigation, and interstate commerce within territorial seas, in addition to having lead authority for planning and decision making in waters overlying the Outer Continental Shelf and exclusive economic zone (EEZ) extending out 200 nautical miles. The political, cultural, and governance traditions of the United States provide a context for MSP with interesting differences to its practices in Europe and other regions (Collie *et al.*, 2013).

The importance of citizen participation in government plays a central role in shaping the processes of spatial planning in the United States. Federal procedures, and in many cases state legislation, call for ample opportunity for public review and comment on proposed governmental policies, plans, and actions. Pending coastal and marine development or conservation decisions must be made known and the record of the decision-making process is typically available to the public. The identification of stakeholders and the solicitation of their views can be a lengthy, at times cumbersome process, but it is an essential feature of coastal governance in the U.S. When the stakeholder process works well, the outcome is trust among the responsible agencies and the stakeholders and a high level of voluntary compliance with the policies and regulations that are developed. When an effective stakeholder process is not established and decision making is not transparent, the process of evaluating a development proposal and reaching a decision on whether, and under what conditions, to permit a proposed activity may consume many years of expensive, often acrimonious debate and complex legal maneuvers (Chowder *et al.*, 2006; Douvere, 2008).

The evolution of coastal and marine management in the United States

The challenges posed by the planning and management of coastlines and marine waters have been shaped by a sequence of national commissions among which the 1969 report of the Stratton Commission "Our Nation and the Sea" remains a benchmark. The Stratton Commission's recommendations led to: (1) the establishment of the National Oceanic and Atmospheric Administration (NOAA) within the Department of Commerce; (2) a revised approach to the management of fisheries; and (3) setting the stage for major investments in the restoration and protection of coastal and marine habitats. The report recommended the enactment of a national coastal zone management program to encourage states to develop comprehensive planning and decision-making programs that would integrate across sectors and address the unique development pressures and environmental features of coastal zones. The Stratton Commission's recommendation led to passage of the Coastal Zone Management Act (CZMA) in 1972 and a unique set of incentives to encourage coastal states to prepare management plans and policies that would meet standards set by the newly created federal Office of Coastal Zone Management (subsequently the Office of Ocean and Coastal Resources Management) within NOAA.

The limitations on federal authority over individual states have led the federal government to rely upon incentives to encourage cooperation between the federal government and state-level counterpart agencies and programs. The Federal CZMA is a prime example. Participation in the CZMA program by individual states is voluntary and the Act provides two major incentives to encourage states to engage. The first is the provision of federal funds, initially for planning a state coastal zone management program, and then as sustained annual grants for the implementation of programs that have been approved by NOAA Office of Ocean and Coastal Resource Management as meeting the federal standards. The second incentive is the "federal consistency clause" that requires that "each federal agency conducting or supporting activities directly affecting the coastal zone shall conduct or support those activities in a manner which is, to the maximum extent practicable, consistent with approved state management programs" (CZMA, 1972). The consistency clause may be applied to federal actions within the state's coastal zone, as well as to federally licensed or permitted activities in the adjoining Outer Continental Shelf and EEZ when these actions can be demonstrated to directly affect resources and activities within a state's coastal zone. The consistency provision is designed to encourage a state CZM program and the lead federal agency to work together when assessing the impacts and making permitting decisions on proposals for activities subject to the environmental impact assessment process in marine waters. It has proved to be a powerful tool for states wishing to influence actions in federal waters off of their coastline.

Three decades after the Stratton Commission and the passing of the CZMA, the Oceans Act of 2000 called for another national ocean commission, the United States Commission on Ocean Policy, to make recommendations for a comprehensive and coordinated national ocean policy. Released in 2004, this Commission's report, An Ocean Blueprint for the Twenty-first Century, documented the declining conditions of the nation's oceans and coasts and called for a new governance framework based upon the principles of ecosystem-based management. Complementary recommendations had been made by the Pew Oceans Commission a year earlier, a private effort conducted in parallel with the Commission on Ocean Policy (Pew Oceans Commission, 2003). Like the Stratton Commission, both the Commission on Ocean Policy and Pew Oceans Commissions found that the continuing degradation of coastal resources and the fragmentation of responsibility and of programs designed to address coastal and marine issues remain major national challenges. Specifically, the National Oceans Commission's assessment as it relates to the marine and coastal governance system includes the following:

- While management of the nation's coastal zone has made great strides, further improvements are urgently needed, with an emphasis on ecosystem-based approaches that consider environmental, economic, and social concerns.
- The many entities that administer conservation and restoration activities largely operate independently of one another, with no framework for assessing overall benefits from an ecosystem perspective. The multitude of disjointed programs prohibits a comprehensive assessment of the progress of conservation and restoration efforts and makes it difficult to ensure the most effective use of limited resources.
- Management approaches have not been updated to reflect the complexity of natural systems, with responsibilities dispersed among a confusing array of agencies at the federal, state, and local levels.
- A pervasive problem for state and local managers is a lack of sufficient, reliable information on which to base decisions.

Both Commissions recommended integrated multi-use ocean planning as the best means for addressing these challenges. The Commission on Ocean Policy proposed a National Ocean Council, linked to the President's Office, be created to "provide high-level attention to ocean, coastal, and Great Lakes issues, develop and guide the implementation of appropriate national policies, and coordinate the many federal departments" (U.S. Commission on Ocean Policy, 2004).

In response to the proposal put forth by the Commission on Ocean Policy, in 2009, the first year of his first term, President Obama established an Interagency Ocean Policy Task Force that identified nine national priority objectives. One of these objectives called for the implementation of ecosystem-based coastal and marine spatial planning and management "designed to decrease user conflict, improve planning and regulatory efficiencies, decrease associated costs and delays, engage affected communities and stakeholders, and preserve critical ecosystem functions and services" (Council on Environmental Quality, 2010). The next year President Obama promulgated, as an Executive Order, a National Ocean Policy that established a framework for coastal and marine spatial planning (CMSP) as an expression of integrating ecosystem-based management. The CMSP therefore calls for analyzing current and anticipated ocean uses, assessing their impacts and identifying areas most suitable for specific types of activity.

However, before President Obama's efforts, in 2007–2008 respectively, three states, Massachusetts, Rhode Island, and Oregon, recognized the need to engage in MSP in response to the pressures for offshore development and its associated opportunities and impacts. These states recognized that significant new forms of human activity, including technologies designed to harness renewable sources of energy, required detailed planning in marine waters where a diversity of activities already compete for space.

In 2010, the Minerals Management Service (MSS), the Federal offshore energy management agency responsible for: (1) promoting resource development; (2) enforcing safety regulations; (3) maximizing revenues from offshore operations, was divided into three entities by the Secretary of the Interior, with each entity now responsible for one of these roles. One of these, the Bureau of Ocean Energy Management (BOEM), assumed responsibility for ensuring that offshore resources on the Outer Continental Shelf under United States jurisdiction are developed and managed in an economically and ecologically appropriate manner. BOEM's goal is to achieve efficient, safe and environmentally sound ocean energy development within a context of multiple uses of the Outer Continental Shelf. Renewable energy activities under BOEM's jurisdiction include commercial energy production from wind. BOEM's Smart from the Start Program has defined a new leasing and permitting process for renewable energy projects. This program calls for BOEM to identify potential Wind Energy Areas and gather information on the areas identified.

The second step is for BOEM to prepare an Environmental Assessment for selected lease tracts. Leasing notices are published and, if developers express interest, a lease sale is scheduled. A developer who wins a lease through an auction process will then conduct surveys in their leasehold and submit a Site Assessment Plan to BOEM for review and approval. If, within five years, the leasee decides to proceed to commercial development, it must submit a Construction and Operational Plan. BOEM then prepares an Environmental Impact Assessment and, if it finds no significant impact, will approve the operational plan for a 25-year term in which to construct the facilities and generate electricity. After 25 years, either decommissioning is required or the lease may be renewed. Each step provides for consultation with stakeholders and public input through both formal and informal processes. For instance, BOEM has assembled regional (Atlantic, Pacific and Alaska) intergovernmental task forces that bring together affected states, local and tribal governments, and federal agencies to discuss renewable energy options, exchange data, and provide for a forum to discuss leasing and permitting options.

Although many parties, including the majority of the coastal states, see value in MSP there are those in the U.S. who believe strongly that MSP is a needless constraint on free enterprise that will complicate and constrain the advance of human activities in marine areas. For example, The Hill's Congress Blog includes an entry made March 30, 2012 by the Chairman of the House subcommittee on Fisheries, Wildlife, Oceans and Insular Affairs (Fleming, 2012) that states:

> President Obama's National Ocean Policy takes zoning to a massive scale, giving Washington pencil pushers more power to decide what activities are acceptable in the ocean zones they create . . . when federal agencies are authorizing activities, the converse can be assumed: they will close off other activities, and limit authorized activities only to approved zones. The uncertainty that results will further limit economic growth . . . The litigation and court challenges will be endless, and the permits that fishermen and coastal businesses need [if they are] to continue making a living will be hard to come by.

Such differences in opinion on the merits of MSP are reflected in the degree to which it is being advanced in different regions and states.

The status of MSP in United States waters

A simplifying five phase framework (Figure 33.1) for tracing the formulation of a coastal or marine management program put forward by GESAMP (Joint Group of Experts on the Scientific Aspects of Marine Environmental Protection) (GESAMP, 1996) can be applied to trace the evolution of MSP initiatives. Phase 1, "Issue Identification and Assessment," is concerned with defining the objectives of an MSP initiative, establishing the spatial boundaries of the area to be addressed, and securing the necessary resources (funding, personnel) for assembling a multi-use spatial plan. Initiatives in Phase 2, "Program Preparation," are engaged in the spatial analysis of resources and activities in the planning area in order to guide management policies, delineate use zones and associated regulations. Phase 2 requires the integration of existing data and may require new research (environmental, economic, social) to fill gaps. In the United States, phase 2 is the period when consultation with stakeholders and public involvement is most intense. This phase culminates with the completion of a multi-use zoning plan and the articulation of the policies and regulations by which it will be implemented. Phase 3, "Formal Adoption and Funding," is when the MSP is the subject of the formal approval process that, if successful, provides the plan with the legal status and the resources required to make it operational. Phase 4, is the implementation of a MSP that has received the necessary governmental endorsements. Phase 5, "Evaluation," is the transition from one generation of MSP to another. This is when the degree of success in achieving MSP goals, the lessons learned from a period of implementation, and changing conditions in the MSP are the basis for making adjustment to the program.

As of mid 2015, at least 23 US states and/or US territories were engaged in either state or regional MSP efforts. The degree of involvement varies widely with some playing a minor role in regional meetings while others see advantages to taking a lead role in discussions and information gathering in their region or state. Table 33.1 shows the current status of initiatives that are underway in the six states that have expressed the greatest interest in MSP. Three of these states (Rhode Island, Washington, and New York) are addressing areas that include both their territorial sea and adjacent Federal waters. Five states (Massachusetts, Oregon, and Hawaii) are limiting MSP initiatives to state waters. States in Phase 1 (Hawaii) are in the process of defining the scope and approach to MSP. Both states have: (1) chosen to focus on state waters, but have yet to define a precise planning area; (2) secured some financing for their respective efforts; and (3) are in the initial information gathering stage. The three states in Phase 2, Washington and New York, are assembling MSPs for their territorial waters and are committed to delineating use zones. These states have compiled the relevant existing data, are conducting new research in support of their MSP initiative, and are engaged in a variety of strategies designed to encourage stakeholder participation. Rhode Island, Massachusetts, and Oregon have proceeded to Step 4 and have formally approved MSP in places that feature a multi-use zoning scheme.

Marine spatial planning is also underway at the regional scale. The National Ocean Policy has led to the establishment of nine regions (Northeast, Mid-Atlantic, South Atlantic, Gulf of Mexico, Caribbean, Great Lakes, West Coast, Alaska, and the Pacific Islands). The boundary of these regions extends from the mean-high water line to the outer limit of the EEZ. Within each region, a Regional Planning Body (RPB) has been assembled composed of federal, state,

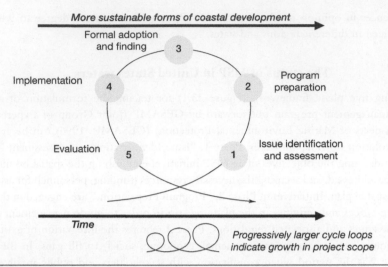

More sustainable forms of coastal development →

Time →

Progressively larger cycle loops indicate growth in project scope

Step 1 Stage Setting
• Define issues to be addressed, goals and spatial boundaries of the Ocean SAMP area
• Define scope of research to be undertaken in support of the Ocean SAMP
• Negotiate the agreements that provide a formal mandate and funding for a Ocean SAMP process
• Design and widely advertise the stakeholder and public involvement process
• First round of stakeholder consultations and public events

Step 2: Preparation of the SAMP
• Compile existing geographic information in a GIS system
• Conduct field research on unknowns and uncertainties important to Ocean SAMP preparation
• Assemble local and traditional knowledge from fishers, sea-pilots, recreational boaters and Indian Tribes
• Apply a Technology Development Index (TDI) to identify potentially suitable sites for wind turbines based on engineering and power production characteristics
• Develop findings and policies for each Ocean SAMP draft chapters
• Involve stakeholders in the assembly of each of the 11 draft Ocean SAMP chapters

Step 3: Formal Adoption
• Adoption of each refined Ocean SAMP chapter following public hearing(s)
• Adopt the full Ocean SAMP by the RI Coastal Council following additional public hearing(s)
• Approval of the portion of the Ocean SAMP within state waters as an element of the state's Coastal Zone Management Program by NOAA
• Endorsement of the Ocean SAMP as a basis for the review of permit applications in the portions of the SAMP that apply to federal waters
• Secure funding for Ocean SAMP implementation

Step 4: Implementation
• Process permits for new activities within the Ocean SAMP boundaries
• Implement mechanisms for coordination of state and federal agencies and adaptation to new knowledge
• Implement consultation procedures when permit applications are evaluated
• Monitor and enforce performance standards for permitted activities
• Monitor selected ecosystem variables and the impacts of selected human activities

Step 5: Evaluation
• Assess the degree to which the Ocean SAMP has achieved its goals and the success of its strategies
• Adapt the Ocean SAMP in response to experience gained from its implementation and to changes in social and environmental conditions
• Revise Ocean SAMP policies and procedures as necessary following the Coastal Council program amendment process

Figure 33.1 The five-phase policy cycle and the key actions associated with each phase as adapted to Rhode Island's Ocean SAMP

Source: Adapted from GESAMP, 1996.

Table 33.1 Current Status (mid 2013) of multi-use marine spatial planning by individual coastal states that are actively engaged in planning or have advanced to MSP implementation (GESAMP Phases 2 to 4)

	Massachusetts	Rhode Island	New York	Washington	Oregon
What is the lead agency for the MSP effort?	Executive Office of Energy and Environmental Affairs—Office of Coastal Zone Management	Rhode Island Coastal Resources Management Council	New York State Department of State	Washington Department of Ecology	Oregon Department of Land Conservation and Development
Does the boundary include state or federal waters, or both?	State	Both	Both	Both	State
What is the size of boundary area (sq. km)?	5,549	3,800	42,787	coast extending to 400 or 700 fathoms depth (26,000)	3,269
Is siting for renewable energy one of the major MSP drivers?	Yes	Yes	Yes	Yes	Yes
Is there stakeholder involvement in MSP planning?	Yes	Yes	Yes	Yes	Yes
Is financing for MSP secured?	Yes	Yes	Yes	Yes	Yes
Is new research for MSP being undertaken?	Yes	Yes	Yes	Yes	Yes
Is information being assembled in GIS formats?	Yes	Yes	Yes	Yes	Yes
Have use zones been identified?	Yes	Yes	Yes	No	Yes
Has the MSP approval process been completed?	Yes	Yes	No	No	Yes
Is the MSP being implemented?	Yes	Yes	No	No	Yes
Has the MSP been evaluated and revised?	Yes	Yes	No	No	No

Source: Authors.

and tribal representatives. The RPBs have a coordinating function and no regulatory authority. Their objective is to develop an integrating regional approach to ocean management based upon a shared data base on marine resources and activities. The RPBs have an uncertain funding base (Burger, 2011). Some have received federal funds and benefit from in-kind support from state agencies, academic institutions, tribes, and non-profit organizations.

While some RBPs have been formed and are actively engaged in developing regional ocean plans, other regions have expressed less interest. In the West Coast region, for instance, the Pacific RPB is in Phase I, but has been faced with a complex hurdle in determining how to best involve the more than 150 tribes in the region. Alaska has decided not to engage in MSP. The Northeast and Mid-Atlantic RPBs appear to be the most advanced. Both of these RPBs have developed and presented goals and objectives for public review and comment, and have completed extensive data collection to address issues that will help develop and improve plans.

Case study: Rhode Island's approach to marine spatial planning: the Ocean Special Area Management Plan (Ocean SAMP)

The state of Rhode Island adopted powerful coastal zone management legislation in 1971, a year before the federal Coastal Zone Management Act was adopted. The Rhode Island Coastal Resources Management Act created a 17 member Coastal Resources Management Council (hereafter Coastal Council) and granted it broad powers to both plan for, and to regulate through a permitting program, activities that alter the state's 413-mile coastline and its three-mile territorial sea. In addition, the Coastal Council has jurisdiction over specified activities anywhere in the state if they are shown to have an impact on the "the environment of the coastal region." The legislation declares that it is the policy of the State of Rhode Island:

> to preserve, protect, develop and where possible restore coastal resources for this and succeeding generations; through comprehensive long-range planning and management designed to produce the maximum benefit for society; and that preservation and restoration of ecological systems shall be the primary guiding principle by which alteration of coastal resources will be measured, judged and regulated.

This statement may be considered as an early expression of the ecosystem based approach called for by the Obama marine policy.

Rhode Island's Coastal Zone Management (CZM) program has evolved through two distinct generations of planning that have progressed through the five steps of the GESAMP policy framework. Both the first and second generation programs received federal approval and brought the benefits of federal financial assistance for implementing the program as well as activation of the federal consistency clause. The first generation won federal approval in 1978 as a permitting program that relied upon a case-by-case impact assessment of applications for permits for the many activities under the Coastal Council's jurisdiction. This approach proved to be inefficient and produced decisions viewed as inconsistent in instances where the Council concluded that an alteration that was found to be suitable in one part of the state was judged to be unsuitable in another setting. This prompted a redesign of the state-wide regulatory process and the approval of a second generation program in 1983 (Olsen and Seavey, 1983). This second generation program features the application of a zoning system that assigns one of six Water Use Categories (Table 33.2) to all marine waters within the State of Rhode Island's three-mile territorial sea. Each Water Use Category defines permitted and prohibited activities and

Table 33.2 Rhode Island's 1983 water use zones

Use Zones	Description	Restrictions
Type 1: Conservation Areas	Waters (1) within designated wildlife refuges, (2) undisturbed areas of high scenic or natural habitat value (3) water areas unsuitable for structures due to exposure to flooding/erosion	All forms of in-water construction prohibited; grading, construction on adjacent shoreline prohibited
Type 2: Low-Intensity Use	Waters with high scenic value that support low intensity boating and residential uses	Improvement dredging and filling, industrial/commercial developments prohibited, residential docks and shorefront protection may be permitted
Type 3: High-Intensity Boating	Intensely used water areas for recreational boating and associated shorefront marinas, boatyards	Recreational boating structures and associated services take precedence
Type 4: Multipurpose Waters	Open water areas supporting a variety of commercial and recreational uses and of good habitat value	Important fishing grounds protected, aquaculture, pipelines, cables etc. may be permitted if the balance of mixed use is maintained
Type 5: Commercial and Recreational Harbors	Waters adjacent to designated harbors supporting a mix of commercial, recreational uses	Precedence given to water dependent and water enhanced commercial and recreational uses
Type 6: Industrial Waterfront and Commercial Navigation Channels	Waters extensively altered by adjacent commercial and industrial activities including commercial fishing ports, naval bases, commercial shipping ports	Precedence given to high intensity commercial and industrial uses and associated water dependent and water enhanced services

Source: Authors, from www.crmc.ri.gov/regulations/RICRMP.pdf.

sets detailed standards that specify the types of alterations that can be made on the shoreline adjoining each water use category. Seventy percent of the state's waters are classified as Type 1, where no alterations are permitted, and Type 2, where only very limited construction and shoreline alterations may be permitted. Detailed performance standards are specified for all types of coastal alterations and activities under the Coastal Council's jurisdiction. This marine zoning feature of the Rhode Island program has proved effective in guiding both coastal and marine development for three decades and has generated a body of experience directly applicable to designating geographically defined areas for different types and intensities of use in the offshore Ocean SAMP area.

As a complement to the 1983 Water Use Category zoning scheme and permitting standards, the Rhode Island CZM program began developing Special Area Management Plans (SAMPs) as the program's response to those coastal areas where environmental issues and competing human activities create complex situations that demand a comprehensive ecosystem analysis and a planning and policy making process for a spatial area that typically extends beyond areas and activities subject to the Coastal Council's direct regulatory authority. The first SAMP, a revitalization plan for Providence Harbor, was adopted by the Coastal Council in 1983. In the subsequent 30 years, six additional SAMPs have been developed, formally adopted by the Coastal Council, and incorporated into the federally approved state CZM program. All SAMPs call for investments in targeted research on the key management issues and sustained engagement with all interested stakeholders, and follow an open and transparent process. Each SAMP results in a detailed spatial plan, associated regulations and an agenda of actions and investments to be taken by the relevant municipalities, state and federal government agencies with responsibilities and interests in the specified area. A detailed description of the Ocean SAMP process and structure may be found in Olsen *et al.* (2014).

The genesis of the Ocean SAMP

Rhode Island's most recent venture into marine spatial planning, the Ocean SAMP, was prompted in mid 2008 by Rhode Island's then-Governor's interest in promoting wind energy as a means for honoring the state's commitments to reduce its dependence on fossil fuels while attracting a new industry to the state at a time of slow economic growth and high unemployment. Early studies of wind as a resource showed that the best sites for wind turbines were offshore. The challenge was to identify the most suitable wind development sites following a planning and analytical strategy that would minimize impacts on existing activities and reduce the typically expensive and time-consuming process generated by the federal Environmental Impact Assessment (EIA) process. Former experience with the EIA process for selecting disposal sites for dredged materials in nearby Rhode Island Sound had consumed years of conflict and sequences of expensive studies. In nearby Massachusetts, litigation over a proposed Cape Wind wind turbine facility to be built in federal waters on Nantucket Shoals had by 2009 dragged on for more than a decade at a cost to the developer in excess of $40 million (Williams and Whitcomb, 2007). In this case the developer selected an area in federal waters and the public and stakeholder consultation process has centered upon the federal government's impact assessment process. In contrast, the SAMPs prepared by the Rhode Island coastal program place the Coastal Resources Center at the University of Rhode Island as the unit responsible for the stakeholder process during Phases 1 and 2. The suggestion was made that the CZM program's SAMP approach offered Rhode Island an alternative that would streamline the regulatory process, put the state of Rhode Island in a lead role for directing future development, and employ a transparent process that would result in the development of a comprehensive ecosystem-based MSP that drew upon the best available science. An agreement was negotiated among the Coastal Council, the Rhode Island Economic Development Corporation, and the University of Rhode Island to produce the Ocean SAMP within a 24-month timeframe.

While the Ocean SAMP came about in response to interests in offshore energy development, this MSP was designed to put in place a process and plan that could address any new proposed activity in the marine waters adjacent to the state's coast.

We now trace the development of the Ocean SAMP as it proceeded through the phases of the GESAMP framework.

Phase 1: State setting

The goals of the Ocean SAMP were to:

- streamline the permitting process;
- promote and enhance existing uses;
- encourage marine-based appropriate economic development; and
- maintain and restore the ecological integrity and resilience of the biophysical and socio-economic systems in the Ocean SAMP area.

The Ocean SAMP planning area included all areas off the state's coast that were potentially suitable for a wind farm. Since the anticipated developer proposed placing wind turbines on lattice platforms similar to those used by the offshore oil and gas industry, a planning boundary extending out to the 60-meter depth contour was selected. For pragmatic reasons, the eastern and western boundaries of the area were defined as those established through existing agreements with the neighboring states of Massachusetts and Connecticut. Waters within 152 m (500 feet) of the ocean shore zoned in 1983 are excluded from the Ocean SAMP. So defined, the Ocean SAMP area encompasses a marine area of 3,800 square kilometers (1,467 square miles) and exists primarily in federal waters beyond the state's territorial sea (Figure 33.2).

This area contains important commercial and recreational fishing grounds and for much of the year is used by thousands of recreational crafts with homeports in Rhode Island and neighboring states. Ferries, tankers, and other vessels make thousands of trips across the Ocean SAMP area each year. Migratory fish, marine mammals, birds, and sea turtles are seasonally present but information on their numbers, seasonal distribution, and migratory routes were absent or rudimentary at the onset of the Ocean SAMP. The seafloor and sub-surface of the Ocean SAMP area is heterogeneous, the product of submergence of a glaciated landscape containing recessional moraines, outwash plains and valleys, and occasional deep "holes." Since wind turbines have to be secured to the seafloor and the costs and ease of construction vary greatly with bottom type, a focus of the Ocean SAMP research was to gather information on sub-surface geology in the areas that appeared to be potential candidate sites for wind farm construction. It was considered important, for example, to identify submerged glacial moraines and areas of glacial till that are likely to contain large boulders that would complicate the process of stabilizing the pilings of a platform and could significantly increase construction costs.

To achieve the goals of the Ocean SAMP and increase knowledge of the area, the plan was designed to identify and map the spatial distribution and patterns of environmental parameters (biological, geological, and physical), cultural features, and current human activities (e.g. fishing grounds, transportation routes, recreational activities), as well as to place current conditions in the context of long-term trends and projections—including those related to climate change.

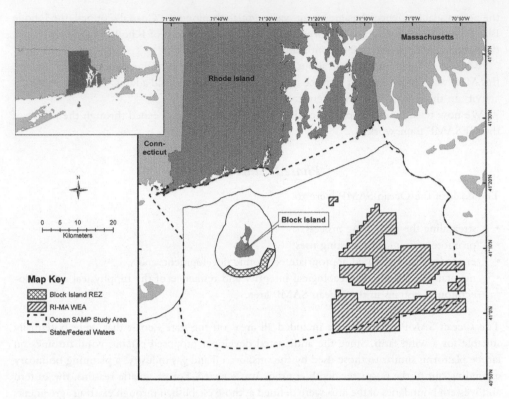

Figure 33.2 Sites identified as potentially suitable for placement of wind turbines in state and federal waters. The Renewable Energy Zone (REZ) was found by the Ocean SAMP as the most suitable site for a pilot-scaled wind farm in state waters. The WEA, or Wind Energy Area, was identified by BOEM as a suitable site for energy production from wind and will be the subject of a lease sale in 2013. The WEA coincides with the Area of Mutual Interest identified by the Ocean SAMP as a potentially suitable area for wind turbines in federal waters.

Sources: Figure credit: Monique LaFrance. Data sources: RI Ocean SAMP, BOEM.

To support this endeavor, the majority of what eventually became a $7.6 million budget assembled from state and federal sources was allocated for new data (environmental, economic, social) acquisition, the analysis of which featured a sophisticated geographical information system (GIS). As a result, the primary data products from the Ocean SAMP were maps describing and characterizing the various natural resources and human uses. This work was undertaken by an interdisciplinary team largely drawn from various departments within the University.

A key feature of the Ocean SAMP was a robust stakeholder process that engaged the many stakeholders in a sustained and informed dialogue. The State recognized the success of the plan also lay in building trust in the process itself and support for the final plan and conclusions reached on potentially suitable sites for wind turbines. This consultative and outreach effort absorbed approximately one-fifth of the total budget.

At the onset of the Ocean SAMP in mid 2008, a public education and involvement process was established, featuring seminars and public workshops and a user-friendly website designed to respond to the issues identified by the public as well as specific stakeholder groups. Issues identified by stakeholders also served as the foundation for both the outreach/communication strategy as well as the Ocean SAMP document table of contents. The consultation process was designed to create an informed and supportive constituency for both the process by which the Ocean SAMP would be assembled and the content of the resulting policies and plan of action. Several consultative groups were established, including a Science Advisory Committee, a state and federal agency Technical Advisory Committee, and a Stakeholder Consultative Group that held monthly meetings to present and discuss new research findings, and discuss the scope and contents of the Ocean SAMP chapters reviewed. Technical Advisory Committees (TACs) were also established for each of the 11 chapters of the Ocean SAMP document. The TACs were composed of scientists, government agency representatives, and resource users selected for their knowledge of the topics addressed within a given chapter.

Phase 2: Preparation of the Ocean SAMP

With the fundamental building blocks of Phase 1 in place, Phase 2 was devoted to developing the substance of the Ocean SAMP. The GIS system permitted overlaying and integrating scientific and traditional knowledge in an engaging, visually accessible manner. The patterns of human activity required consulting with those with detailed traditional knowledge—such as commercial and sport fishers, sea pilots, and organizers of yachting regattas—to prepare comprehensive maps of the spatial and temporal distribution of these activities. Field research was framed around sets of issues. Environmental and engineering issues concern the distribution and abundance of offshore bird populations, marine mammals, and the likely impacts of climate change on the ecology of the Ocean SAMP area. Social issues include the identification of the most important fishing grounds and the anticipated economic impacts of the construction of a wind farm.

As the layers of spatial information accumulated a second tier of analysis was undertaken that considered the engineering and economic attributes of the wind turbines using a Technology Development Index (TDI) to identify areas that present major engineering and economic constraints (Spaulding *et al.*, 2010). For wind turbine construction, the TDI is defined as the ratio between the technical challenge of construction and the power production potential. The potentially suitable areas were gridded and an analysis performed for each grid. The grids with the lowest TDI were classified as the most suitable sites for a proposed facility. Since this form of analysis requires calculating numerical estimates for each variable, and the uncertainties for some variables are relatively high, simulations could be performed deterministically using a Monte Carlo method to understand the impacts of such uncertainties on the conclusions reached. In the case of wind turbines, this required identifying the areas where the winds are insufficient and where there are major constraints to their construction. The TDI analysis identified the potential sites for a wind farm that are not constrained by significant human activities or natural resources, contain sufficient wind, are relatively close to a connection to the electrical grid, and do not appear to contain significant construction constraints—such as boulder fields and shallow depth to bedrock. Sites determined to be potentially suitable still require additional research to confirm these initial conclusions. Though the TDI analytical method was used to identify potentially suitable sites for a wind farm, it could in the future be applied to analyze siting issues raised by any other proposed activity or structures including, for example, an offshore aquaculture operation or the placement of a submarine cable.

The area that was found most suitable through the TDI analysis within state waters was classified as the Renewable Energy Zone (REZ) and is located to the southeast of Block Island fully within state waters, though bordering the boundary between the three-nautical mile territorial sea under state control and waters under federal jurisdiction. Once this area was identified, it was vetted through the public stakeholder process. Due to comments by the charter fishing industry, the REZ was moved slightly to the southeast so as to minimize the impact from this industry. The TDI analysis also identified an area within federal waters referred to as the "Area of Mutual Interest" (AMI) between Rhode Island and Massachusetts that is considered by both states to be the most suitable site for a wind farm (refer to Figure 33.2).

To achieve the goal of "protecting and enhancing existing activities," the Ocean SAMP identifies Areas Designated for Protection (ADPs) and Areas of Particular Concern (APCs). The Ocean SAMP identified one ADP, which includes all waters within the 20m depth contour, since this belt of shallow waters has been found to be of exceptional value to sea ducks. All forms of large-scale construction are prohibited within an ADP. Areas classified as an APC encompass 53 percent of the Ocean SAMP area and include:

- areas with unique or fragile physical features or important natural habitats;
- areas of high natural productivity;
- areas of significant historical or cultural value;
- areas of substantial recreational value;
- areas important for navigation, transportation, military, or other human uses;
- areas of high fishing activity.

The Ocean SAMP policies state that any application for a permit in areas designated as an APC must demonstrate (1) that there are no practicable alternatives that are less damaging in areas outside the APC and (2) that the proposed project will not result in a significant alteration to the values and resources of that APC.

Phase 3: Formal adoption

Phase 3 of the GESAMP management framework is concerned with the formal adoption of the Ocean SAMP and is the bridge between planning and implementing the plan and its associated policies and regulations. It was decided that the Coastal Council would adopt the 11 chapters of the Ocean SAMP document chapter-by-chapter over a six-month period. The reasoning behind this decision was twofold: (1) each chapter of the Ocean SAMP engaged a somewhat different group of stakeholders; and (2) it was important to gauge reactions to the information and the policies developed for each topic. From a practical standpoint, the intense MSP effort resulted in a document of over 1,000 pages, which would have been unrealistic to present in its entirety to stakeholders for comment and review. The adoption process required proceeding through the sequence of informal public workshops, formal public hearings, refinements, and formal adoption by the Coastal Council for each chapter as set forth by the widely distributed Ocean SAMP calendar. When all the chapters had proceeded through this process some adjustments were made to the document as a whole. After three hearings on the refined version, the Ocean SAMP was approved by the Coastal Council on October 19, 2010, within the timeframe called for at the beginning of the process two years earlier.

Upon approval by the Coastal Council, the Ocean SAMP sought federal approval by NOAA for the portion of the Ocean SAMP within state waters to become an element of the state's CZM program. During the planning phase, the Ocean SAMP team worked closely with NOAA to assure that the policies and regulations were consistent with the mandate and authority of both the Coastal Council and the federal coastal and ocean management program. As a result, federal approval of the Ocean SAMP in May 2011 was processed as a routine program change. Three months later, through a process termed the Geographic Location Designation (GLD), NOAA identified the federal activities that will be subject to the CZMA's consistency clause.

The federal activities identified include those addressed by the Ocean SAMP. The result is that the information, policies, and use zones included in the Ocean SAMP will be a basis for all decision making by federal agencies of government. This sets the stage for collaborative planning and decision making in the SAMP area with the Rhode Island coastal management program.

In 2012, BOEM identified the AMI, the area in federal waters that the Ocean SAMP TDI spatial analysis had concluded was the most suitable area for a wind farm, as the area where leases for wind energy development would be offered. From then on the AMI has been referred to as the "Wind Energy Area" (WEA). The close working relationship between the Ocean SAMP team and various fishing operations in this area led to a detailed understanding of how the WEA is utilized and concerns arose over the leasing of portions of the WEA of high value to several fisheries. In response, the Ocean SAMP Fisherman's Advisory Board developed a map showing areas of high fishing intensity within the WEA. This resulted in BOEM removing from consideration lease blocks within the WEA identified as important fishing grounds. An auction for the remaining lease blocks within the WEA will be held by BOEM in July 2013. This auction will be BOEM's first lease auction for wind energy development in the country.

Phase 4: Implementation

The critical issue for Phase 4, which is yet to be determined, is whether the process of assembling the Ocean SAMP and winning its approval at both the state and federal level will indeed simplify the process of siting a new activity, such as a wind farm, in coastal waters already crowded with human activities. Will the existence of the Ocean SAMP prevent the years of delays and legal actions that have characterized the offshore permitting process as illustrated by Cape Wind? Assessing the degree to which the Ocean SAMP and concurrent revisions to the federal leasing and permitting process have created a better context for decision making on proposals for renewable energy offshore facilities will be known only after the responsible state and federal agencies have acted upon proposals for wind farms within the Ocean SAMP area. It will then be apparent whether the application of the SAMP approach to a marine area spanning both state and federal jurisdictions has indeed produced a more efficient and less contentious decision-making process. As of mid 2015, the evidence is that the permitting process for the pilot scale wind farm (five turbines) within state waters off Block Island in the REZ is proceeding as expected. The greatest resistance to this proposal has been focused on the high cost of the energy that would be produced. The fact that BOEM held its first ever auction for leases for energy generated from wind in the Ocean SAMP-defined WEA, and that construction of the Block Island wind farm has begun, suggests that the Ocean SAMP process may indeed generate the desired outcomes.

Phase 5: Evaluation and adaptation

All Rhode Island SAMPs are designed for adaptive management that provides for responses to new information, new issues, and the experience gained from implementing a course of action. The Ocean SAMP has already been amended on five occasions since it was adopted in 2010. Such updates and revisions are incorporated through a formal amendment process. In 2013, the Ocean SAMP will be evaluated to assess the degree to which it has succeeded in achieving its goals and objectives. According to this review, the Ocean SAMP is achieving its goals and has strengthened both conservation and development, while streamlining the regulatory process.

Looking to the future: barriers and bridges to MSP in the United States

The Rhode Island experience suggests that the decision of an individual coastal state to extend its coastal and marine planning and management functions into adjoining federal waters can be an effective way to negotiate where new offshore activities may be accommodated. While the Regional Planning Bodies provide for important coordination and information sharing that draws together a number of contiguous states, they have no regulatory authority and an uncertain funding base. State-based marine spatial planning that extends into federal waters, as illustrated by the Ocean SAMP, appears to offer a stronger model for proactive ocean planning (Burger, 2011). This is because the Coastal Zone Management Act bridges the regulatory divide between state and federal waters by granting states the authority to review projects in federal waters for consistency with state coastal management plans that have been previously approved by the federal government. While the federal government retains the final word on federal permits, the federal consistency provisions can place such decisions within an explicit context of detailed spatial planning and policy making. States that take the initiative to engage in an MSP planning process that involves all stakeholders and is constructed upon consideration of local conditions, local preferences, and locally generated scientific knowledge provide an attractive alternative to a more centralized approach. The federal Coastal Zone Management Act provides states and federal agencies with a place to negotiate inter-jurisdictional issues and together develop marine spatial planning arrangements that have immediate legal force.

The experience of Rhode Island suggests that the formulation of an MSP by individual states as an expression of the ecosystem approach to planning and decision making need not be so all encompassing, complex, and time consuming as to be practically untenable. To the contrary, MSP may prove to make offshore permitting a transparent and efficient decision-making process that benefits both existing activities and new developments while promoting an ecosystem stewardship ethic.

References

Agardy, T. 2010. *Ocean Zoning: Making Marine Management More Effective.* Earthscan, London.

Burger, M. 2011. Consistency Conflicts and Federalism Choice: Marine Spatial Planning beyond the States' Territorial Seas. *Environmental Law Reporter* 4, 10602–10614.

Chowder, L., G. Oshrenko, O. Young, S. Airames, S. Norse, E. Baron, N. Day, J. Douvere, F. Ehler, C. Halperin, B. Langdon, S. McLeod, K. Ogden, J. Peasch, R. Rosenberg, J. Wilson. 2006. Resolving Mismatches in U.S. Ocean Governance. *Science* 313, 617–618.

Coastal Zone Management Act (1972) Section 307 (16 U.S.C. section 1456). http://coast.noaa.gov/czm/act/sections/?redirect=301ocm (last accessed July 19, 2015).

Collie, J. S., W. L. Adamowicz, M. W. Beck, B. Craig, T. E. Ellington, D. Fluharty, J. Rice, J. N. Sanchirico. 2013. Marine Spatial Planning in Practice: Estuarine, Coastal and Shelf. *Science* 117, 1–11.

Council on Environmental Quality. 2010. Recommendations of the Ocean Policy Task Force.

Douvere, F., 2008. The Importance of Marine Spatial Planning in Advancing Marine Ecosystem-based Management. *Marine Policy* 32, 762–771.

Fleming, J. 2012. National Ocean Policy is an executive power grab—*The Hill*'s Congress Blog. *The Hill*. N.p., 30 Mar. 2012. http://thehill.com/blogs/congress-blog/energy-a-environment/219309-national-ocean-policy-is-an-executive-power-grab (last accessed: June 18, 2013).

GESAMP, 1996. *The Contributions of Science to Integrated Coastal Management.* GESAMP Reports and Studies No. 61. Rome.

McLeod, K. L., J. Lubchenco, S. R. Palumbi, A. Rosenberg. 2005. Scientific Consensus Statement on Marine Ecosystem-Based Management (signed by academic scientists and policy experts with relevant expertise). *Communication Partnership for Science and the Sea.* www.compassonline.org/sites/all/files/document_files/EBM_Consensus_Statement_v12.pdf (last accessed: April 2, 2013).

Olsen, S., G. Seavey. 1983. *The State of Rhode Island Coastal Resources Management Program, As Amended.* R.I. Coastal Resources Management Council, Providence, Rhode Island.

Pew Oceans Commission. 2003. *Charting a Course for Sea Change: A Report to the Nation.* Pew Trusts, Washington D.C.

Sanchirico J. N. 2004. Zoning the Oceans. In: R. Morgenstern, R. Portney (Eds.). *New Approaches on Energy and the Environment: Policy Advice for the President.* Resources for the Future Press, Washington D.C., pp. 114–119.

Sanchirico, J. N., J. Eagle, S. Palumbi, B. H. Thompson. 2010. Comprehensive planning, Dominant Use Zones, and User Rights: A New Era in Ocean Governance. *Bulletin of Marine Science* 86, 273–285.

Spaulding, M. L., A. Grilli, C. Damon, G. Fugate. 2010. Application of Technology Development Index and Principal Component Analysis and Cluster Methods to Ocean Renewable Energy Facility Siting. *Journal of Marine Technology,* Special Edition on Offshore Wind, 4(1), 8–23.

Stratton Commission on Marine Science, Engineering and Resources. 1969. *Our Nation and the Sea: A Plan for National Action.* United States Government Printing Office. Washington D.C.

U.S. Commission on Ocean Policy. 2004. *An Ocean Blueprint for the Twentyfirst Century. Final Report.* USCOP. Washington D.C.

Williams, W., R. Whitcomb. 2007. *Cape Wind: Money, Celebrity, Class, Politics, and the Battle for Our Energy Future on Nantucket Sound.* Public Affairs Books, New York.

34

THE EAST ASIAN SEAS

Competing national spheres of influence

Sam Bateman

Introduction

The maritime geography of East Asia is extremely complex. The East Asian seas are the greatest concentration of seas in the world covered by the regime of enclosed and semi-enclosed seas in Part IX of the 1982 UN Convention on the Law of the Sea (UNCLOS). However, economic, political and military competition is in many respects more intense in these seas than elsewhere in the world's maritime domain, and this has prevented development of cooperative regimes for managing these seas and their resources.

The complex maritime geography of the region drives both competition and the need for cooperative management. Achieving straight-line maritime boundaries and clear sovereign jurisdiction at sea and over marine resources is a difficult task. Examples of all the contentious issues with the contemporary law of the sea can be found in the region, including 'creeping jurisdiction' with for example, use of excessive territorial sea straight baselines and claims to an outer continental shelf, and 'thickening jurisdiction' with an over-reading of coastal state rights in the exclusive economic zone (EEZ). These factors markedly complicate the management of regional seas.

Many of the trends with achieving effective cooperation to manage the seas of East Asia are in the wrong direction. Sovereignty disputes seem as far from solution as ever. Attempts to implement the UNCLOS Part IX obligations to cooperate in the management of enclosed and semi-enclosed seas are not achieving the desired outcomes.

Regional countries are becoming even more fixated on their claims to sovereignty over offshore features and maritime space. Competition over national sovereignty has become the main obstacle to effective governance of the East Asian seas.

Maritime geography

The map in Figure 34.1 shows the maritime geography of East Asia. This is perhaps the most complicated area of maritime geography in the world. The area covered by Figure 34.1 is roughly 6,500 kilometres east-west and nearly 9,000 kilometres north-south. It measures about 57 million square kilometres and includes countless islands. As a consequence of the extended maritime jurisdiction allowed by UNCLOS, a large part of the maritime domain in this area is enclosed as the EEZs or archipelagic waters of coastal and archipelagic States with relatively few areas of high seas.

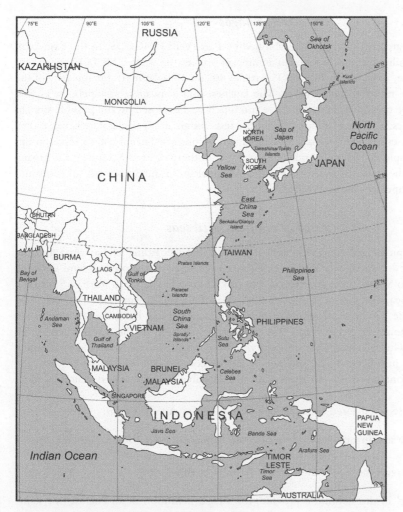

Figure 34.1 The maritime geography of East Asia

Source: Map drawn by I. Made Andi Arsana, Gadjah Mada University, Yogyakarta, Indonesia.

The coastline of continental East Asia stretches roughly 15,000 kilometres from the Sea of Okhotsk in the North to the Malay Peninsula in the South. Significant features of this coastline include the peninsulas (primarily the Kamchatka, Korean, Shandong and Malay Peninsulas), and the concave gulfs and bays (mainly the Yellow Sea, Bohai Sea, Gulf of Tonkin and the Gulf of Thailand).

Another significant feature of the maritime geography of East Asia is the chain of off-lying archipelagos and islands, stretching from Sakhalin and the Kamchatka Peninsula through the Japanese archipelago and the Philippines archipelago to the Indonesian archipelago and northern Australia. The number of islands and groups of islands lying between the continental coastline and the outer archipelagic chain, such as Taiwan and the Senkaku, Paracel and Spratly Islands, further complicates the situation. Achieving straight-line maritime boundaries and clear sovereign jurisdiction over maritime areas and marine resources in such a region is an extraordinarily difficult task. This situation establishes competing national spheres of influence that markedly complicate marine environmental management, including the exploitation of the marine resources.

Archipelagos

The regime of the archipelagic State is established by Part IV of UNCLOS. In the East Asian region, only the Philippine and Indonesian archipelagos meet the strict UNCLOS criteria for an archipelagic State, as set out in UNCLOS Articles 46 and 47, relating to the integrity of the archipelago and the ratio of land to water. The Japanese and Taiwanese archipelagos, as well as Singapore, fail the criteria because they have too much land enclosed by any possible straight archipelagic baselines. Smaller island groups fail the tests for an archipelagic State because they are off-lying territories of countries that are not themselves comprised mainly of islands or groups of islands. For example, the Mergui Archipelago off the southern coast of Myanmar, the Hoang Sa Archipelago in the Paracels, and the Trong Sa Archipelago in the Spratly Islands could not be classified as archipelagos under UNCLOS.

Enclosed or semi-enclosed seas

This geographical picture of concavity along the continental coastline of East Asia and the numerous off-lying archipelagos and islands creates a large array of enclosed or semi-enclosed seas. From North to South, these seas are: the Sea of Okhotsk, Sea of Japan (or East Sea to the Koreans), Yellow Sea, East China Sea, South China Sea, Gulf of Thailand, Java Sea, Sulu Sea, Celebes Sea, and the Timor and Arafura Seas, as well as several seas lying entirely within the Philippine and Indonesian archipelagos. The geography of some of these seas, such as the extremely concave Gulf of Thailand, is complex with two or more littoral States and overlapping claims to maritime jurisdiction. The situation is further complicated when a sea, such as the South China Sea, contains islands of varying sizes that are subject to sovereignty claims by more than one State.

An enclosed or semi-enclosed sea is defined by UNCLOS Article 122 as:

> a gulf, basin or sea surrounded by two or more States and connected to another sea or the ocean by a narrow outlet or consisting entirely or primarily of the territorial seas and exclusive economic zones of two or more coastal States.

The second last 'or' in this definition provides for a legal definition of an enclosed or semi-enclosed sea as one consisting entirely or primarily of the territorial seas and EEZs of two or more countries. This means that to be classified as enclosed or semi-enclosed, a sea does not have to be wholly surrounded or locked in by land. A gulf or a bay can also meet the criteria.

International straits

The maritime geography of East Asia with the off-lying island chain creates a large number of straits, as well as other 'choke points' for shipping. These straits occur both along the coast of mainland Asia where the island chain presses close into the coast (e.g. the Tsushima Strait, the Taiwan Strait and the Straits of Malacca and Singapore), and through the off-lying islands (e.g. the Tsugaru Strait through the Japanese archipelago, the San Bernadino Strait in the Philippines, the Balabac Strait North of Borneo, and the Sunda and Lombok Straits through the Indonesian archipelago). Many of these straits are important for international shipping both for the ships making passage along the East coast of Asia between Southeast Asia and Northeast Asia, and for those heading into mainland Asian ports from the Americas and Oceania.

Part III of UNCLOS establishes a special regime applying when a strait used for international navigation is wholly or partly contained within the territorial sea of one or more States.

This regime potentially applies to several straits in the East Asian seas, most notably to the Malacca and Singapore Straits.

UNCLOS Article 43 is particularly relevant to our consideration of maritime cooperation in East Asia. This provides that user States and States bordering a strait should by agreement cooperate in the establishment and maintenance in a strait of necessary navigational and safety aids or other improvements in aid of international navigation; and for the prevention, reduction and control of pollution from ships.

This cooperation has been implemented in the Malacca and Singapore straits through the Cooperative Mechanism for Safety of Navigation and Environment Protection in the Straits of Malacca and Singapore agreed at a meeting in Singapore in 2007 sponsored by the International Maritime Organization (IMO) (Beckman, 2010). It consists of three elements – a Cooperation Forum, an Aids to Navigation Fund, and specific Projects. The Cooperation Forum is intended to promote open dialogue and discussions between the littoral States, user States and other stakeholders interested in the safety of navigation in the straits. The Aids to Navigation Fund accepts voluntary contributions from user States and other stakeholders to finance navigational safety and environmental protection by maintaining and replacing aids to navigation such as lighthouses and buoys.

Maritime zones

Table 34.1 shows the size of the maritime zones of East Asian countries (with maritime zones comprising EEZs, continental shelves, archipelagic waters and territorial sea), along with the ratio of maritime zones to land area. The latter ratio provides a crude indicator of a *maritime state* if this ratio is comfortably above 1:1. In absolute terms, Indonesia, Philippines, Japan and China have gained the largest maritime zones under UNCLOS although the ratio of maritime jurisdiction to land area for China is quite small, being distorted by China's continental nature and large land area.

The data in Table 34.1 also suggests the major 'winners' in East Asia under UNCLOS in terms of additional maritime jurisdiction. The big 'winners' have been the archipelagic and island States: Japan, Philippines, Indonesia and, to a lesser extent, Taiwan. Singapore is the exception

Table 34.1 Maritime zones of East Asia

Country	Land area (sq. km)	Maritime zones (sq. km)	Maritime/Land area ratio
Japan	370,370	3,861,000	10.4
North Korea	121,730	129,650	1.1
South Korea	98,400	348,478	3.5
China	9,600,000	1,355,800	0.1
Taiwan	32,360	392,381	12.1
Philippines	300,000	1,891,247	6.3
Vietnam	332,556	722,337	2.2
Thailand	414,001	324,812	0.6
Malaysia	332,649	475,727	1.4
Singapore	588	343	0.6
Indonesia	1,904,342	5,409,981	2.8
Brunei	5,765	24,352	4.2
Cambodia	181,041	55,564	0.3

Source: Based on Buchholz (1987, Table 6).

here as, despite being an island State, it is at a significant geographical disadvantage as a result of being locked in by its two neighbours, Malaysia and Indonesia.

The 'losers' under UNCLOS in East Asia are Singapore and the continental states with much of their coastline along gulfs or semi-enclosed seas. North Korea, Thailand and Cambodia are in this category. These are sometimes referred to as 'zone-locked' because they are locked in by the maritime zones of other countries. However, before assuming that these countries have relatively fewer maritime interests and concerns with the marine management, a wider view should be taken of their maritime interests. For example, Thailand, despite having a relatively small area of maritime jurisdiction, is a major distant water fishing nation (DWFN) with its fishing industry heavily dependent on access to foreign EEZs. This might help explain why Thailand did not ratify UNCLOS until 2011.

Marine resources

The marine environment of East Asia is extraordinarily complex and rich in marine resources, both living and non-living. Most East Asian countries depend on the sea for foodstuffs, trade and longer term economic prosperity and security, but there is growing concern over the depletion of fish stocks, the degradation of the marine environment and the destruction of habitats, such as seagrass beds, mangroves and coral reefs. Most coastal states in the region are looking towards possible oil and gas reserves beneath the sea for future economic prosperity and are investing heavily in offshore oil and gas exploration and exploitation.

Economic and population growth combine to increase demand for the resources of the sea. Population growth is insidious in Asia. Higher per capita incomes mean higher demand for seafood, while industrial development and rapidly growing manufacturing sectors lead to higher demand for energy and a greater reliance on foreign coal, oil and gas. These energy trends provide strong incentives for higher levels of offshore exploration for new sources of production, including for natural gas, the demand for which in the region has increased dramatically, particularly in China, Japan and South Korea.

Unfortunately, the fact that offshore resources are either believed to be so rich, particularly in the case of hydrocarbons, or have already been subjected to gross over-exploitation (in the case of many fish stocks) has led to sovereign rights over marine resources becoming problematic. It is a fiercely competitive process with emotive, nationalistic overtones, sometimes quite disproportionate to the proven value of the resources involved. Despite the rich potential of marine resources in East Asia, the development of these resources is troubled by major jurisdictional problems, and 'beggar thy neighbour' attitudes that have led to over-fishing, and marked degradation of natural habitats of coral reefs, mangroves and seagrass beds. Marine pollution originating from the land is a serious problem in the region. The preservation and protection of the East Asian marine environment, the conservation of species, and the exploitation of marine resources are seriously complicated by conflicting and overlapping claims to maritime jurisdiction and the lack of agreed maritime boundaries.

Living resources

The waters of Southeast Asia have an abundance of coral reefs, mangroves and seagrass beds that support a rich array of marine animals and plants – among the most diverse marine flora and fauna in the world. Northeast Asian waters have an enormous profusion of fish species caused by their particular climatic conditions and ocean currents, making them among the most productive in the world for fishing.

Most East Asian countries have a high dependence on fish and related marine products as a source of protein. Apart from the importance of subsistence fishing and domestic market supply, many East Asian countries also look to the living resources of their EEZs as an important source of export income, and some are leading DWFNs, fishing both for the domestic market and for export. Fish and fishing are important political issues in the region and fisheries management and access arrangements are major factors in regional relations. The long-standing DWFNs of Japan, South Korea, Taiwan and Thailand have been affected by tighter access controls by coastal states and international arrangements to control straddling stocks and highly migratory species, including on the high seas.

China is an emerging DWFN with the level of Chinese longline fishing activity in the Western and Central Pacific increasing significantly (Zhang, 2012). Many of these are small vessels, about 20 metres long, catching sashimi grade tuna for the Japanese market with transhipment to Tokyo through Guam and other Pacific island ports. China has increased its distant water effort through the acquisition of used purse seiners and long liners from Japan, Korea and Taiwan.

While shared stocks and fishing grounds create some commonality of interest, they also present a potential source of tension, particularly when jurisdictional problems and resource access issues are present. Numerous incidents have occurred in the region when fishing boats have been fired upon or even sunk. It is not unusual for fishing vessels to be escorted by naval vessels when undertaking fishing operations in disputed waters and this situation poses the risk of direct confrontation between the naval forces of regional countries.

Fisheries are in crisis in most areas of the world, and East Asia is no exception. While over-fishing is the major problem, this problem has been exacerbated in East Asia by the rapid and largely unquantified loss and degradation of coastal habitats with serious implications for sustainable development, fisheries and their management. Fish stocks are affected by ecologically harmful practices such as the clearing of mangroves to provide areas for aquaculture, the exploitation of coral reefs, the destruction of seagrass beds, and coastal development generally. Other problems include the sedimentation of estuaries, land-based pollution, dynamiting, the use of very fine mesh nets, and other forms of illegal fishing. Aquaculture was initially seen as a solution to the problem of depleted fish stocks, but many development programmes for aquaculture have been disappointing and have also caused problems of pollution and natural habitat destruction.

Non-living resources

The offshore areas of most East Asian countries have some hydrocarbon potential (Schofield, 2011). Southeast Asian waters are underlain by sedimentary basins that are either producing oil and gas already, or possibly contain further major hydrocarbon deposits. In Northeast Asia, the Yellow Sea and East China Sea basins are areas of hydrocarbon potential and good prospectivity (Valencia, 1992, p. 84). In addition to hydrocarbons, the Sea of Japan and the East China Sea are also favourable locations for submarine metallic sulphides including copper, zinc, lead, nickel, cobalt, manganese, iron, gold and silver associated with geological faults and spreading zones.

With concern growing for the increased demand for and security of energy supplies, most coastal States in the region are preoccupied to some extent with staking offshore claims and exploring for oil and gas. As the numbers of offshore oil and gas installations increase in the region, these potentially create multiple use conflicts particularly with navigation and fishing, as well as posing problems associated with the disposal of offshore installations and for installation safety and security generally.

The exploitation of non-living marine resources in East Asia is severely hampered by unresolved offshore boundaries and overlapping claims to sovereignty that inhibit development (Schofield *et al.*, 2011). Overlapping concessions are relatively common throughout the marginal seas of East Asia, including in the South China Sea between China and Vietnam, in the East China Sea between China, South Korea and Japan, and in the Sea of Japan (or East Sea) between Japan and South Korea.

The legal status of regional seas

Most regional seas are 'semi-enclosed seas' covered by Part IX of UNCLOS. Use of the words 'should co-operate' and 'shall endeavour' in Article 123 of UNCLOS places a strong obligation on the littoral States to coordinate their activities as defined in the sub-paragraphs of that article. While resource management, the protection of the marine environment and marine scientific research are mentioned specifically as areas for cooperation, the opening sentence of Article 123 creates a more general obligation to cooperate.

It is also relevant to note that these seas are not 'international waters'. Rather they comprise the EEZs of the several littoral countries that have significant rights and duties in the seas, as set out in UNCLOS Part V defining the EEZ regime. The EEZ is a separate type of maritime zone (*sui generis*) subject in accordance with UNCLOS Article 55 to its own specific legal regime.

The use of the term 'international waters' by the US goes dangerously close to taking the world back to a pre-UNCLOS era when the US and other maritime powers argued that the extended offshore resources zone (which became the EEZ) should be an extension of the high seas while coastal states tended to see it as an extended territorial sea. The solution was an EEZ that is *sui generis*, i.e. a zone all of its own neither high seas nor territorial sea. Use of the term 'international waters' derogates from the agreed nature of the EEZ.

The US includes EEZs within its operational definition of 'international waters' because, in accordance with UNCLOS Article 58(1), other States have the freedoms of navigation and overflight in the EEZ of a coastal State, as well as the freedom to lay submarine cables and pipelines, and other internationally lawful uses of the sea related to those freedoms. However, UNCLOS Article 58(3) requires that, in exercising these freedoms, other states should have due regard to the rights and duties of the coastal state, but in practice, it is proving very difficult to define an operational test to distinguish between an action that has due regard to the rights and duties of the other party, and one that does not (Bateman, 2011, p. 7).

Several clashes have occurred in the East Asian seas between US and Chinese vessels and aircraft. The US argues that its military activities in China's EEZ are an exercise of the freedoms of navigation and overflight while China argues that some of these activities do not have due regard to its rights in its EEZ. Other states, including Malaysia and Thailand, appear to share China's view. As virtually all East Asian maritime space is the claimed EEZ of one country or another, these fundamental differences over rights and duties in an EEZ have significant implications for the maintenance of good order at sea in the East Asian seas.

Obligations to cooperate

There are still no effective regimes for cooperative marine management and good order in the East Asian seas. This is despite the obligation of all countries bordering a body of water, such as the South China Sea, to cooperate in accordance with Part IX of UNCLOS, to which all the littoral countries are parties.

The Council for Security Cooperation in the Asia Pacific (CSCAP) has produced Guidelines for Maritime Cooperation in Enclosed and Semi-Enclosed Seas and Similar Sea Areas of the Asia Pacific (CSCAP, 2008). The guidelines are a set of fundamental, non-binding principles to guide maritime cooperation in the enclosed and semi-enclosed seas of the region, and to help develop a common understanding and approach to maritime issues in the region.

In the case of the South China Sea, the 2002 Declaration on the Conduct of Parties in the South China Sea (DOC) identifies the following activities as requiring cooperation pending a comprehensive and durable settlement of the sovereignty disputes in this sea:

a) marine environmental protection;
b) marine scientific research;
c) safety of navigation and communication at sea;
d) search and rescue operations; and
e) combating national crime, including, but not limited to, trafficking in illegal drugs, piracy and armed robbery at sea, and illegal traffic in arms.

(ASEAN, 2002, Article 6)

The demands for effective management regimes in regional seas will become more pressing in the future. Volumes of shipping traffic will continue to increase with greater risks of ship-sourced marine pollution and shipping accidents. There will be increased pressure on marine resources, both living and non-living, as well as growing concern for the protection and preservation of the sea's sensitive ecosystems and marine biodiversity.

Regional seas programmes set up under the auspices of the UNEP Regional Seas Programme have had few outcomes. The Northwest Pacific Action Plan (NOWPAP) and the Coordinating Body on the Seas of East Asia (COBSEA) are not well supported. A separate programme, the Partnerships in Environmental Management for the Seas of East Asia (PEMSEA) arrangement has had more success but has generally kept clear of projects in disputed areas.

Elsewhere in the world, attempts to implement the UNCLOS Part IX obligations to cooperate in the management of enclosed and semi-enclosed seas have not achieved the desired outcomes. Experience in the Mediterranean and Caribbean seas, and so far with the East Asian Seas Action Plan steered by the COBSEA give few grounds for optimism that successful outcomes will be achieved (Hu, 2010, pp. 281–314).

President Aquino of the Philippines has called for the South China Sea to become a Zone of Peace, Freedom, Friendship and Cooperation (ZOPFF/C), which would provide a framework for collaborative activities (Thayer, 2014). China has also proposed the establishment of technical committees on three areas: marine scientific research and environmental protection; navigational safety and search and rescue; and combating transnational crime at sea (Wain, 2011). All three are areas where cooperation is urgently required but ASEAN has not accepted the Chinese proposal and wants an agreed Code of Conduct first. Unfortunately, the drive for a Code of Conduct has diverted attention from the very important and necessary requirement for cooperation in the sea. Thus one of the 'means' of achieving cooperation (i.e. the Code of Conduct) has become an 'end' in itself. More focus is required on the 'ends'.

Declaration on conduct of parties

The DOC is agreed 'soft law' that invites littoral countries to cooperate on certain marine activities. It was a pragmatic move to put the disputes in the background and bring ASEAN China economic ties to the fore (Chung, 2009, p. 95). However, it is non-binding and falls short of constituting a successful regime for providing security and cooperative marine management in the South China Sea. It was the outcome of a long process of negotiation that began in earnest with the 1992 ASEAN Declaration on the South China Sea that followed in the aftermath of potentially destabilizing naval clashes in the Paracels and Spratlys.

The DOC commits parties to peaceful modes of dispute settlement, the application of international law, the need for building up confidence and trust, and recognition of the freedoms of navigation and overflight in the South China Sea. While it has been successful until recently in containing disputes and tensions in the South China Sea, it has not contributed to cooperative activities in the way that was hoped, or led to appropriate confidence-building measures (CBMs).

After the verbal confrontations at ARF meetings and elsewhere during 2010, there has been some improvement in negotiations between ASEAN and China. At the ARF meeting in Bali in July 2011, China and ASEAN agreed to guidelines for developing a code of conduct between the claimant countries in the South China Sea (ASEAN, 2011). The Guidelines met with a mixed reception. Some have seen them as disappointing and not going far enough, particularly with easing tensions and resolving the disputes, while others welcomed them as sound progress towards greater stability in the South China Sea. It is also noteworthy that the Guidelines do not refer specifically to cooperation for management of the sea but rather identify confidence building measures as the initial activities to be undertaken under the ambit of the DOC.

This is yet another indication of how the focus with the South China Sea has shifted from cooperation to one of sovereignty and dispute resolution. Of course it could well be argued that cooperation is a 'building block' for dispute resolution rather than the other way around, i.e. confidence building is a prerequisite of cooperation.

Sovereignty disputes

The East Asian seas contain numerous islands, some of which are the subject of disputed sovereignty claims. Figure 34.1 shows the main disputed islands in East Asia. This section briefly discusses the problems of maritime boundaries in the main seas of East Asia (including the issue of disputed islands that lie in these seas) but does not embark on a detailed examination of the rival claims. That requires a detailed technical examination of legal and geographical factors that is beyond the scope of this chapter.

This chapter's intention is to demonstrate the overall complexity of maritime boundary making in the East Asian seas. This will support a general conclusion that in many parts of these seas, straight-line maritime boundaries may in fact never be agreed in view of the impossibility of political agreement being reached between the several parties involved. Bilateral agreements are feasible even though they sometimes might disadvantage third parties. Trilateral or even quadrilateral agreements become much more problematic.

Sea of Okhotsk

The Northern part of the Sea of Ethos is entirely Russian, being bounded by the Russian mainland and the Kamchatka Peninsula, but the Southern part of the sea comprises an enclosed or semi-enclosed sea where Russia and Japan share sovereignty and need to delimit maritime boundaries. The situation is confounded, however, by disputed sovereignty over a group of four islands and some islets at the southern end of the Kuril Island chain to the North of Hokkaido (known in Japan as the Northern Territories). The Soviet Union occupied these islands towards the end of World War II although Japan had always previously occupied them. Sovereignty over the islands is important as at stake is a large EEZ with rich fisheries and potential minerals and petroleum and they also have a high profile in terms of the bilateral political relationship between Japan and the Russian Federation. Domestic politics play a large role in formulating the policies of the two countries towards the sovereignty dispute (Miller, 2012).

Sea of Japan (East Sea)

The Sea of Japan (or East Sea as it is known to the Koreans) is a difficult area for maritime boundary delimitation. This is due to the number of countries involved (i.e. Russia, the two Koreas and Japan), North Korea's extreme claims, and the sovereignty dispute between Korea and Japan over the Takeshima islands (or Tok-do islands in Korean). These latter islands consist of two small rocky islets eight miles east of South Korea's Ulleung-do and 31 miles northwest of Japan's Dogo islands.

Sovereignty over the islands will have a significant impact on the maritime boundary between South Korea and Japan, and also with the boundaries between these countries and North Korea, while the latter remains an independent country. The adjacent waters are believed to have some oil and gas potential (Valencia, 1992, p. 82).

East China Sea

The major complication with effective management of the East China Sea is the issue of sovereignty over the Senkaku islands (or Diaoyu islands in Chinese). This dispute has become an extremely vexed issue in relations between China and Japan. The islands are situated about 90 nm northeast of Taiwan. They are a group of five islands under the control of Japan but are also claimed by China and Taiwan. China does not have a strong historical claim. Austin believes that China was pushed into the claim to the Senkakus for the first time in 1970 in response to actions by Taiwan and that its claim is 'probably inferior to that of Japan and may even be insufficient to demonstrate any effective exercise of sovereignty by China over the Senkaku Islands at any time' (Austin, 1998, p. 4).

Sovereignty over the Senkaku islands is important because it could support a claim to the adjacent continental shelf that is believed to contain rich petroleum deposits. Even if sovereignty over the islands could be agreed, delimiting maritime boundaries in the area would still prove difficult as there is a dispute between China and Japan as to whether there is a continuous continental shelf in the region or whether there is a trough in the shelf close to the Ryukyu chain. The former position would favour Japan with a median line as possibly the appropriate boundary while the latter situation could support China's claim to a broad continental margin as its continental shelf.

South China Sea

The situation in the South China Sea is the most notorious and problematic of all the jurisdictional problems in the East Asian seas (International Crisis Group, 2012). Virtually all the geographical features within the South China Sea are subject to disputed sovereignty claims by various countries. The South China Sea is the largest of the marginal seas of East Asia and the region is geologically and oceanographically complex. Most of the area around the Spratly Islands is very deep water beyond the physical continental shelf and will therefore be much more costly to exploit.

China, Taiwan and Vietnam claim sovereignty over the Spratly and Paracel islands in the South China while individual islands and reefs are claimed by Brunei, Malaysia and the Philippines. Until the 1980s the disputes over islands in the South China Sea was largely a China–Vietnam affair but then Malaysia and the Philippines also started pressing their claims. Generally, the level of attention to the conflicting claims in the South China Sea has increased in line with estimates of the area's resource potential with very little attention being given to sovereignty until the 1960s and 1970s.

The Chinese position is that these disputes can be settled only by the parties concerned negotiating on a bilateral basis, not regionally or multilaterally or internationally. Indonesia has a close interest in the situation as it has sovereignty over the Natuna Islands and will eventually have to be involved in maritime delimitations in the area.

The Chinese claim to the Spratly Islands is based mainly on historical occupation and usage although it is possibly weakened by some lack of continuity (Kaye, 1998, p. 19). China's claim to waters in the South China Sea has been subject to considerable speculation and mis-reporting. This is largely caused by differing interpretations of the dotted line that appears on some maps enclosing most of the South China Sea (see Figure 34.2). This line is referred to in Chinese literature as the 'traditional maritime boundary line', 'the southernmost frontier', 'territorial limit', and so forth (Gao, 1994, p. 346), but the legal nature of the line has not been clarified.

Some Western writers have suggested that this dotted line claims the whole area as Chinese territorial waters or 'historic waters' (Thomas and Dzurek, 1996, p. 308), but this is not believed to be the case. As suggested by Zhiguo Gao, 'A careful study of Chinese documents reveals that China never has claimed the entire water column of the South China Sea, but only the islands and their surrounding waters within the line' and that 'the boundary line on the Chinese map is merely a line that delineates ownership of islands rather than a maritime boundary in the conventional sense'(Gao, 1994, p. 346). In effect China is saying that all the islands and reefs within the dotted line are Chinese and that these generate maritime zones as allowed by international law.

Both China and Vietnam claim the Paracels but the Chinese claim is usually regarded as superior. Unlike the Spratlys with the claims of different countries to various islands, the Paracel island group is usually considered as a whole as a sovereignty issue. After a thorough analysis of the two claims, Austin reached the conclusion that North Vietnam's recognition of Chinese sovereignty over the Paracels in 1958 and its lack of protest between 1958 and 1975 create an estoppel in respect of South Vietnam's claim to the islands and this probably still stands against the current Vietnamese claim (Austin, 1998, p. 130). Except for its impact on the South China Sea as a whole, the Paracels are generally regarded as a bilateral matter between China and Vietnam.

On the basis of historical usage, Vietnam claims the whole of the Spratly and Paracel groups and all its continental shelf, as well as an extensive area of the South China Sea, although the limits of the claim have not been clearly identified by coordinates. Vietnam's claim is weakened by the consistent recognition of Chinese sovereignty over the Spratlys by North Vietnam during the period between 1951 and 1975 (Kaye, 1998, pp. 21–22).

The Philippines claim is based on the so-called proximity principle and on the discovery of the islands concerned by a Philippine explorer, Thomas Cloma, in the 1950s. It is 'in some ways the strangest of all the arguments advanced by the protagonists' largely because of the bizarre nature of Cloma's original claim and the subsequent reaction of the Philippine Government (Kaye, 1998, p. 21). The Malaysian claim is clearly defined by coordinates showing the extent of its claim to continental shelf in accordance with international law. Malaysia claims those islands and reefs that it considers to be situated on its continental shelf. Brunei's claim is

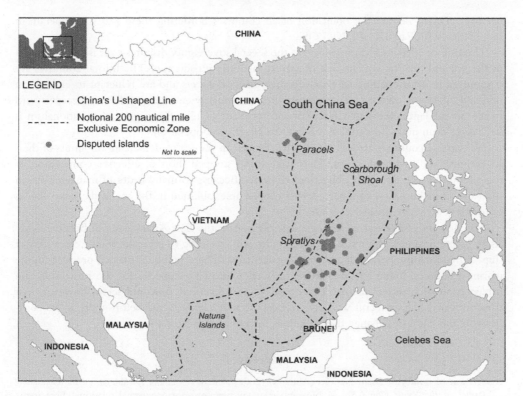

Figure 34.2 The South China Sea island groups and China's U-shaped line
Source: Map drawn by I. Made Andi Arsana, Gadjah Mada University, Yogyakarta, Indonesia.

also to an area of continental shelf, although its boundary lines are simply drawn perpendicular from the end of its land boundaries on the coast. The claims by the Philippines and Malaysia lack a historical basis and probably are 'too weak to have displaced Chinese title' (Austin, 1998, p. 161). They are also postulated on a basis of geographical contiguity but at present this is not recognized as a valid basis for a claim of sovereignty (Kaye, 1998, p. 21).

This drive for nationalization of the oceans was amply illustrated by the joint submission by Malaysia and Vietnam in May 2009 to the Commission on the Limits of the Continental Shelf (CLCS). This was a strongly nationalistic approach to resource allocation in the South China Sea. It amounted to a 'zero-sum' claim by these two countries to the seabed resources of the entire southern part of the China Sea to the exclusion of other interested parties, Brunei, China and the Philippines. Nationalistic fervour remains a powerful driving force for the sovereignty claims in the South China Sea. Access to hydrocarbons is another consideration but it is unfortunate in this regard that high expectations of the oil and gas potential of large parts of the sea are probably unjustified (Owen and Schofield, 2012).

Conclusions

It has proven difficult to develop effective regimes in the East Asian seas for cooperative marine management and good order at sea for the safety and security of shipping; the preservation, protection and conservation of the marine environment; the exploration and exploitation of marine resources; the prevention of illegal activity at sea; and the conduct of marine scientific research. This is despite the obligation of all countries bordering these seas to cooperate in accordance with Part IX of UNCLOS.

The lack of effective regimes is due to the several sovereignty disputes over islands and reefs in the sea and the consequent lack of agreed maritime jurisdiction. The littoral countries are committed to a nationalistic approach to their claimed waters and are reluctant to embark on initiatives that may appear to compromise their sovereignty and independence. Recent events in the Sea of Japan, the East China Sea and the South China Sea suggest that nationalism is becoming even more powerful as a barrier to effective cooperation.

The littoral countries are still looking for 'fences' in the sea to demarcate the limits of their sovereignty and sovereign rights. They have so far stopped short of effective cooperation or regime building despite their obligations to do so. Because so many issues of managing ocean space are trans-boundary in nature, fences cannot be established in the sea in the same way as border fences are established on land.

The demands for effective management regimes in the East Asian seas will become more acute in the future. Volumes of shipping traffic will continue to increase with greater risks of ship-sourced marine pollution and shipping accidents. There will be increased pressure on the marine resources, living and non-living, as well as growing concern for the protection and preservation of sensitive marine eco-systems and biodiversity. It would seem, however, that these pressures and concerns will have to become much more acute before there is less competition and countries change their mindsets in favour of greater cooperation.

References

Association of Southeast Asian Nations (ASEAN) (2002) *ASEAN – China Declaration on the Conduct of Parties in the South China Sea*, 4 November, www.asean.org/asean/external-relations/china/item/declaration-on-the-conduct-of-parties-in-the-south-china-sea (accessed 22 July 2015).

ASEAN (2011) *Guidelines for the Implementation of the DOC*, 5 July, www.southchinasea.com/documents/law/306-guidelines-for-the-implementation-of-the-doc.html (accessed 22 July 2015).

Austin, G. (1998) *China's Ocean Frontier: International Law, Military Force and National Development*, Allen & Unwin, St Leonards.

Bateman, S. (2011) 'Solving the "Wicked Problems" of Maritime Security – Are Regional Forums up to the Task?', *Contemporary Southeast Asia*, vol. 33, no. 1, pp. 1–28.

Beckman, R. (2010) 'Maritime Security and the Cooperative Mechanism for the Straits of Malacca and Singapore' in S. Bateman and J. Ho (eds) *Southeast Asia and the Rise of Chinese and Indian Naval Power: Between Rising Naval Powers*, Routledge, Abingdon, pp. 114–128.

Buchholz, H. J. (1987) *Law of the Sea Zones in the Pacific Ocean*, Institute of Southeast Asian Studies, Singapore.

Chung, C. (2009) 'Southeast Asia and the South China Sea Dispute' in S. Bateman and R. Emmers (eds) *Security and International Politics in the South China Sea: Towards a Cooperative Management Regime*, Routledge, Abingdon, pp. 95–109.

Council for Security Cooperation in the Asia Pacific (CSCAP) (2008) Guidelines for Maritime Cooperation in Enclosed and Semi-Enclosed Seas and Similar Sea Areas of the Asia Pacific, CSCAP Memorandum No. 13, www.cscap.org/uploads/docs/Memorandums/CSCAP%20Memorandum%20No%2013%20—%20Guidelines%20for%20Marit%20Coop%20in%20Enclosed%20and%20Semi%20Enclosed%20Seas.pdf (accessed 22 July 2015).

Gao, Z. (1994) 'The South China Sea: From Conflict to Cooperation', *Ocean Development and International Law*, vol. 25, pp. 345–359.

Hu, A. N-T. (2010) 'Semi-enclosed Troubled Waters: A New Thinking on the Application of the 1982 UNCLOS Article 123 to the South China Sea', *Ocean Development and International Law*, vol. 41, no. 3, pp. 281–314.

International Crisis Group (2012) 'Stirring up the South China Sea (I)', *Asia Report No. 223*, Brussels: International Crisis Group, 23 April.

Kaye, S. B. (1998) 'The Spratly Islands Dispute: A Legal Background', *Maritime Studies*, vol. 102, September/October, pp. 14–25.

Miller, J. B. (2012) 'It's Time for a "Grand Bargain" Between Japan and Russia', *Global Asia*, Vol. 7, No. 2, Summer, pp. 58–63.

Owen, N. A. and Schofield, C. H. (2012) 'Disputed South China Sea Hydrocarbons in Perspective', *Marine Policy*, vol. 36, no. 3, May, pp. 809–822.

Schofield, C. (ed.) (2011) 'Maritime Energy Resources in Asia: Energy and Geopolitics', *NBR Special Report #35*, Seattle: The National Bureau of Asian Research (NBR), December.

Schofield, C. Townsend-Gault, I., Djalal, H., Storey, I., Miller, M., and Cook, T. (2011) 'From Disputed Waters to Seas of Opportunity: Overcoming Barriers to Maritime Cooperation in East and Southeast Asia', *NBR Special Report #30*, Seattle: The National Bureau of Asian Research (NBR), July.

Thayer, C. (2014) 'China-ASEAN and the South China Sea: Chinese Assertiveness and Southeast Asian Responses' in Y-H. Sing and K. Zou (eds) *Major Law and Policy Issues in the South China Sea: European and American Perspectives*, Farnham, Ashgate, pp. 25–53.

Thomas, B.L. and Dzurek, D.J. (1996) 'The Spratly Islands Dispute', *Geopolitics and International Boundaries*, vol. 1, No. 3, pp. 300–326.

Valencia, M. J. (1992) 'Oil and Gas' in Morgan, J. and Valencia, M. J. (eds) *Atlas for Marine Policy in East Asian Seas*, University of California Press, Berkeley, CA, pp. 81–99.

Wain, B. (2011) 'China faces New Wave of Dispute', *The Straits Times*, 17 October, www.viet-studies.info/kinhte/NewWaveOfDispute.htm (accessed 22 July 2015).

Zhang H. (2012) 'China's Evolving Fishing Industry: Implications for Regional and Global Maritime Security', *RSIS Working Paper No. 246*, S. Rajaratnam School of International Studies (RSIS), Singapore, 16 August.

*Regional developments:
the developing periphery*

35

AFRICA

Coastal policies, maritime strategies and development

Francois Odendaal, Zvikomborero Tangawamira,
Bernice McLean and Joani November

Introduction

This chapter draws attention to the pivotal role that local communities have in the management of marine and coastal areas in Africa, and the increasingly important role that they can play not only at the local, but also the national and regional levels such as in the management of Large Marine Ecosystems (LMEs) – many of which involve several countries (Hennessey and Sutinen, 2005).

From the outset it should be recognized that local communities *do* play a role in coastal area management, whether or not this role is recognized by coastal management practitioners and donor-funded programs. Communities can likely play a role at a larger scale than any other single entity, except perhaps in small areas of the coast where industrial activities such as harbours or tourism development override community baseline activities such as fishing and other livelihood activities.

It is important to note that in the context of Africa, the term 'local communities' refers to assemblages of coastal dwellers that have had a long association with a particular stretch of coast, are fairly traditional and have a strong reliance on marine resources in order to meet their daily needs. Most such coastal communities are located in rural areas,[1] which to this day characterizes the vast majority of the African coastline.

In earlier times, artisanal fishing communities relied on traditional management systems, such as customary fishing areas, closed seasons and areas, and highly selective fishing gear. For instance woven traps were more selective in their target catch than the small-gauge nets (in some cases even mosquito netting) that are often used today. As may be expected, many of the traditional fishing methods and approaches have been replaced with more efficient technologies that are more readily available yet also less selective in their catch and more environmentally damaging. Migration from the inland to the coastal areas, and population growth have led to expanding coastal populations that have dramatically increased pressure on coastal areas. These trends demand the involvement of local communities in coastal areas management more than ever before. For instance, survey results in Zanzibar have shown that fishers have increased substantially from

18,618 fishers in 2002 to 26,666 in 2008, a third increase in a mere six years (Revolutionary Government of Zanzibar, 2009) where the coastal ecosystem is already under heavy stress. It is logical that any solution that does not include these fishers in a central way will be doomed to fail.

Two seminal insights on local communities and natural resource management have particular relevance to this chapter. The works of Garret Hardin (1968) and Robert Chambers (1983) are both highly relevant to coastal communities. Hardin's paradigmatic economic theory highlights the social dilemma, known as 'The Tragedy of the Commons' whereby the nexus between self-interest and open access to resources causes, or exacerbates resource depletion. The 'Commons' concept is often cited in discussions on sustainable development and has also been applied to discussions on other shared resources including the atmosphere, oceans, rivers, fish stocks, national parks, advertising markets, and so on. On land, the 'Tragedy of the Commons' concept has been applied to land degradation in communal grazing areas across the African continent, and beyond. In the marine realm, inshore fishing grounds, including estuaries, are often treated as open access areas, and the depletion of marine resources and degradation of these sensitive ecosystems is commonplace. Efforts are made to control overexploitation and destruction of marine commons through marine managed areas (MMAs), to varied success. Some MMAs are protected areas only in name, while many others experience overt or simmering conflicts.

Another highly relevant line of thinking comes from Robert Chambers. Since the 1980s, Chambers has advocated for putting the interests of the poor, destitute and marginalized at the centre of development processes, a recommendation that could be applied to many African coastal villages. Chambers argues that such sectors of society should be taken into account when the development problem is identified, policies are formulated and projects are implemented. Within development circles, Chambers populorized such phrases as 'putting the last first' and stressed the need for development professionals to be critically self-aware. Today's widespread acceptance of a participatory approach to development, and the generally accepted need for stakeholder involvement in initiatives that affect them, is in part due to his 1983 book *Rural Development: Putting the Last First*.

In the powerful sequel *Whose Reality Counts? Putting the First Last*, published in 1997, Chambers argues for a revision in thinking about and doing 'development', calling for a complete shift towards approaches that are more community-driven and process-orientated. He contends that we shouldn't settle for neatly designed projects executed by government agencies or big NGOs. Chambers reconceptualizes 'development' as non-linear processes that can be sensitively facilitated and supported by privileged outsiders but ultimately must be controlled and led by communities (who he terms 'lowers') themselves. Thus, 'development' should be done *by* or *with* the people concerned, rather than *for* or *to* them, and can be called local economic development or LED planning.

The status quo

Evidence of projects that have involved communities in a meaningful way and have resulted in sustainable coastal areas, especially at a large scale, is uncommon in Africa and the Western Indian Ocean (WIO) Island States. This despite the two regional agreements related to coastal management in Africa that have been adopted under the United Nations Environment Programme (UNEP) Regional Seas Program: the Abidjan Convention for Cooperation in the Protection, Management and Development of the Marine and Coastal Environment of the Atlantic Coast of the West, Central and Southern Africa Region and Protocol concerning Cooperation in Combating Pollution in Cases of Emergency (1984), and the Nairobi

Convention for the Protection, Management and Development of the Marine and Coastal Environment of the Eastern African Region and Related Protocols (1996). The objectives of the Abidjan Convention are to protect the marine environment, coastal zones and related internal waters falling within the jurisdiction of the states of the West and Central African region. The convention also contains a protocol concerning 'Cooperation in Combating Pollution in Cases of Emergency' (UNEP, 2008). The objectives of the Nairobi Convention are very similar to the Abidjan Convention and propose to protect and manage the marine environment and coastal areas of the Eastern African region (UNEP, 2010).

Both regional agreements make ample reference to the importance of the inclusion and well-being of local communities. The plethora of initiatives and huge sums of money that have flowed through implementing agencies over the previous two decades have failed to reverse the downward trend of resource degradation and increasing poverty levels along the coast. While individual projects often claim successes in project Terminal Evaluation reports, in our experience the realities on the ground often present a different picture. The reason could be that although project actions may have been met on paper, lasting results on the ground are far more difficult to achieve. It is proposed that more studies be done by the Global Environment Facility (GEF) on the true results of projects funded by them (GEF, 2006; Griffiths, 2006), and that critical examinations be expanded to the oceans and coasts programmes.

Where does the problem lie then? Much has been written about communities in terms of locally managed marine areas (LMMAs) (Agardy, 1994; Kelleher and Phillips, 1999; McClanahan *et al.*, 2006; Agardy *et al.*, 2011), usually in relation to marine protected areas (MPAs) where it soon becomes clear that conservation cannot be effective without considerations of resource users. The concept of LMMAs has taken root in practical ways, with the recent broadening of LMMA networks such as the one in Madagascar (Mayol, 2013).

While the positioning of local communities may appear obvious to some, and lip service is often paid to the plight of local communities, the reality *in practice*, is that the fundamental linkages between poverty, effective coastal zone management, and local communities are often overlooked in the design and implementation of large marine and coastal projects. Many of the parties we had discussions with agree that local community involvement simply does not feature in many LME projects, other than perhaps in a token sense. Also, while it may feature in others it is not clear what the lasting outcomes are. Again, a systematic examination of the extent of community involvement in project planning as per GEF policy (see below), accompanied by ground level investigation will undoubtedly assist in pointing the way forward.

In our perspective on coastal management in Africa, we start by briefly examining the fundamental linkages between poverty, effective coastal zone management and the imperative for local community involvement in coastal management projects. We then consider the policy of the GEF in terms of local community involvement, and specifically in terms of the International Waters (IW) focal area. The GEF (see: www.gef.org) is arguably the largest source of funding for ecosystem-based coastal projects globally and the funding provided is coordinated by a growing set of implementing agents.

The broad questions addressed in this chapter relate to the extent to which GEF-funded LMEs and other large-scale projects optimize the role of local communities in coastal areas management in Africa and how this role may be strengthened. We consider two African LME projects in which our team has had a long-standing involvement, to examine how the agenda for local community involvement has unfolded itself. Further reference is made to other large marine and coastal initiatives that we have had involvement in. Finally, practical options for the future are proposed that could lead to the greater involvement of communities in coastal management projects.

The imperative for community and stakeholder involvement

Analysis of coastal management, Integrated Coastal Areas Management (ICAM), stakeholder involvement, marine and coastal reserve planning and implementation for the past 30 years has highlighted several fundamental justifications for the meaningful involvement of local communities in coastal areas management. The fact that local communities form the bulk of the populace over vast sections of the coast, justifies their central role in ICAM – a practice that necessarily involves managing people rather than resources or ecosystems.

Local communities, at least in democratic societies, have an increasingly influential political say. This is particularly true in countries where decentralization is underway. In many of these countries national government is sensitive to the opinions of local communities. Local communities can therefore exert a positive influence on donor projects, whether in their design or their implementation approach. In reality, not all communities are aware that they have this power, and sometimes are not informed of projects.

A third reason for placing communities centrally in the agenda of ICAM projects relates to the often very severe impacts that the activities of local communities have on the resource base and the ecosystem as a whole. Poverty is characteristic of many coastal communities throughout the developing world, particularly in Africa. Many coastal communities suffer from debilitating poverty that tends to exact a heavy toll on the natural resource base on which the communities depend for their livelihood. A downward spiral is thus set in motion. This leads to the basic premise that unless coastal poverty is addressed, effective coastal management is difficult to achieve, particularly in areas of high population density where there is a direct reliance on coastal resources to sustain families on a daily basis.

The influence of poverty on resource-use patterns complicates any form of environmental management, including the establishment and functioning of MMAs and other initiatives designed to combat resource degradation. The old adage that poverty is the worst enemy of the environment rings as true as ever. Poverty, together with inappropriate development, industrial sprawl and poorly planned and managed coastal activities that are designed primarily to benefit developers and investors are leaving indelible scars on Africa's coastal and inshore areas. Poverty is a major concern of all coastal countries in the developing world and is frequently echoed in national agendas and the need to contain it is expressed in policies, national development strategies, action plans and programmes.

The role of the Global Environment Facility in promoting marine management

As far back as February 1995, the Council of the Global Environment Facility requested the GEF Secretariat to prepare a 'document proposing GEF policies for information disclosure and public participation' (Joint Summary of Chairs, p. 3). At its meeting in April 1996, the GEF Council approved the principles presented in the document *Public Involvement in GEF Projects*, as a basis for public involvement in the design, implementation and evaluation of GEF-financed projects. The document, together with subsequent policy papers, makes a convincing case that the GEF as an institution has a solid understanding of the importance of stakeholder involvement and why benefits should be accrued to local communities. The theory is sound, the intentions pure and the desire to pursue the notion of sustainable development, real.

Since 1996, the GEF has mainstreamed principles for public consumption in multiple publications and other forms of media. Stakeholder involvement and community benefits have been rallying cries for the international organization. These imperatives have been taken up by all GEF focal areas, including the International Waters (IW) portfolio, where the need is often expressed for ensuring local benefits in its LME initiatives. In the Sixth International Water Conference (IWC6), one of the key themes of the gathering: 'From Communities to Cabinet', highlighted the need to ensure that project activities are relevant to people at the local level. Today there is scarcely a coastal or marine-related donor-funded initiative that does not make ample mention of poverty in its core justifications.

The question is whether these sentiments, crystallized into a plethora of projects worth billions each year, each one with its promised interventions, have translated into reality and have made a meaningful difference on the ground. Is sustainable development in coastal areas an achievable notion or a dream that is perhaps out of reach? Has the attention been diverted from the pure intentions of the Rio Earth Summit in 1992 to what amounts to a money-spinning machinery that is diverted from the intention of 'Whose Reality Counts?'. Local communities often complain that funding destined for bettering their lives is diverted along the way. They see evidence of project funding being spent on cars, boats, teams of researchers and so forth, but they do not feel the benefit.

From dream to reality

With the highest tiers of the GEF placing a premium on stakeholder involvement and community development, the scene is ostensibly set for marine and coastal projects in Africa to reach the target. The following section examines how two large donor-funded coastal and marine projects have fared in terms of the imperative for stakeholder involvement and community benefits. The two projects are chosen based on the close and extended involvement of the authors in both projects over a number of years as participants in pre-selected partner organizations that were part of the project design and implementation. This involvement has provided a perspective to assess more clearly key obstacles that were encountered.

The BCLME Programme

The first project is the Benguela Current Large Marine Ecosystem (BCLME) Programme.[2] This initiative was funded through the GEF and implemented from the early 1990s to 2008. The initiative involved extensive research that led to the identification of key areas for intervention, outlined in a Strategic Action Plan (SAP). The focus of the second phase of the BCLME initiative involved the implementation of the Strategic Action Plan (BCLME SAP IMP).

The Benguela Current flanks three countries namely, Angola, Namibia and South Africa. It bestows a multitude of benefits at local, national and regional levels but perhaps most essentially, to local coastal communities who depend directly on natural resources for subsistence and livelihood. The growing pressure on the living marine resources of the BCLME has become increasingly apparent over the last two decades. The range of negative impacts includes decreased revenue to national governments, the closure of coastal industries and other deleterious effects on local livelihoods. As a response to the lack of comprehensive legal frameworks, poor planning, the depletion of non-renewable resources, land-based sources of pollution and downscaling of associated industries, the BCLME Programme was established.

The BCLME Programme was co-sponsored by the governments of Angola, Namibia and South Africa to develop and institutionalize an intergovernmental transboundary management process for the marine and coastal resources of the three countries. The BCLME Programme lists among its achievements, the endorsement of an Interim Agreement for the establishment of the Benguela Current Commission (BCC), and an agreement to pursue a multilateral Treaty to formalize the BCC. To achieve these goals, the three countries secured funding from the GEF for a four-year project to support the implementation of requisite governance reforms and priority outcomes identified in the BCLME Strategic Action Programme Implementation.

While the well-funded BCLME Programme pursued a range of scientific studies, linkages with other stakeholders, in particular local communities, lagged far behind. Furthermore, key players at regional and local levels, and concurrently running GEF-funded projects had little or no involvement in the BCLME Programme, in spite of the GEF being the major source of 'catalytic funding' in all of them.

By the time of the World Summit for Sustainable Development (WSSD) in 2002, the IW of the GEF had become cognizant of the void left by the lack of stakeholder involvement in the BCLME Programme, and responded in several ways. The NGO International Knowledge Management (IKM) was tasked to develop the Distance Learning and Information Sharing Tool (DLIST) for the BCLME region (visit: www.dlist-benguela.org), as part of a medium-sized project (MSP) that operated independently of the BCLME programme. The main aim was to provide training, share information and stimulate debate on coastal issues. DLIST was considered highly successful by stakeholders and through evaluation by the GEF itself. In 2004, towards the end of the first phase of the BCLME, an assessment was commissioned to identify opportunities for involvement of coastal communities in the BCLME Programme. Site visits to potential villages in Angola, Namibia and South Africa identified potential demonstration sites for promoting sustainable development.

Towards the end of the BCLME Programme Phase I, a study was also commissioned on stakeholder and community involvement. The findings of the study indicated clear stakeholder needs and aspirations as well as readiness of local communities for involvement during site visits in South Africa, Namibia and Angola. A Stakeholder Involvement Plan (SIP) was commissioned that identified stakeholders and several modes through which they could participate and contribute to the implementation of the project. Access to information and opportunities for stakeholder input and direct participation in project implementation were flagged as major issues. Taking this into consideration, the SAP IMP included ample reference to information sharing and the involvement of local communities and lower tier structures in the project plan.

Following the BCLME Programme, the second phase – the BCLME SAP IMP initiative – presented an opportunity to 'reach out and involve "new" stakeholders who were not previously involved in the BCLME Programme' (UNDP, 2008: p. 32). These new stakeholders were identified as both resource users (private sector stakeholders from sectors such as fishing, mining and offshore oil and gas) and ground level/grassroots stakeholders such as coastal community members. IKM was engaged as a project partner to facilitate the stakeholder engagement activities. The scene was set for meaningful stakeholder involvement and community development and accruement of benefits during the BCLME SAP IMP phase.

The ASCLME Project

The second project that our team was intimately involved in is the Agulhas and Somali Current Large Marine Ecosystem (ASCLME) Project, this time through an organization known as EcoAfrica Environmental Consultants (www.ecoafrica.co.za). The goal of the ASCLME project is to ensure the long-term sustainability of the living resources of the ASCLME region by introducing an ecosystem-based approach to management. The overall project objectives are to acquire sufficient baseline data to support an ecosystem-based approach to the management of the Agulhas and Somali Current LMEs, and to produce a Strategic Action Plan (SAP) for both the Agulhas Current and Somali Current LMEs. The first phase involved a sizeable community component, aimed primarily at generating Local Economic Development (LED) plans that could inform the way forward in terms of implementation of the SAP. A strong information-sharing component also formed part of an overall stakeholder participation programme, for which the DLIST web-based platform was used (see: www.dlist-asclme.org).

Nine communities in eight countries were targeted as demonstration sites (see Figure 35.1). The purpose of the demonstration sites was to pilot interventions in which local communities play a central role in developing LED plans for their villages and the surrounding areas, typically including an estuarine system or a combination with coastal forests. One demonstration site was selected in each country in South Africa, Mozambique, Kenya, Seychelles, Comoros, Madagascar and Mauritius. Two sites were selected in Tanzania (one for the mainland Tanzania and one for Zanzibar).

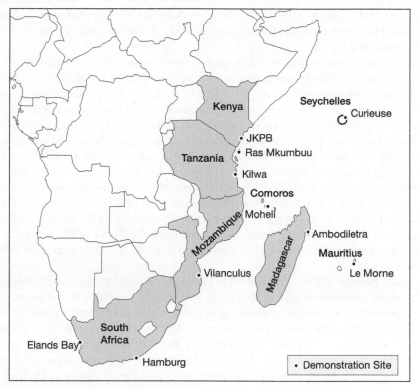

Figure 35.1 Demonstration sites for the ASCLME project
Source: EcoAfrica (2011).

The nine communities were not only spread across the east side of the coastal Africa region, but also represented a range from the semi-urban to deep rural, hence collectively generating a wealth of lessons learned that can be accessed on the DLIST ASCLME website (www.dlist-asclme.org). Although the LED planning interventions have resulted in varying degrees of success beyond planning, their results are all informative in terms of how the plans may be implemented. The approach that was taken can be readily applied by donors, NGOs, governments and any other agent that has an interest involving local communities or alleviating poverty along the coast. In this chapter we discuss the methodology, briefly outline the nine communities and focus on one demonstration site in particular as an example.

Putting the last first, in the commons

The landscape and 'Whose reality counts?'

The African coast is some 26,000 km long, stretching along 38 countries. Madagascar has the longest coastline (4838 km), followed by Somalia (3025 km), South Africa (2798 km) and Mozambique (2470 km) at least. Up to 40 per cent of the African population depend on marine resources for livelihood and live along the coasts of the continent. In the 33 sub-Saharan African coastal countries, over 50 per cent of the population live within 100 km of the coast (UNESCO, 2013). Poverty is a common thread that runs through most of these coastal communities. Many do not have much of a say over what affects their current and future livelihood. Development initiatives that do not adequately involve local stakeholders but that impact the realities surrounding them can limit their options even further. Of course not all local communities can be fully supported to achieve their development needs at the same time, but a winning approach that can be applied successfully in limited contexts that can be replicated in other communities is worth examining.

We have test driven an approach that 'gives a voice to the voiceless' yet does not add to their marginalization, namely local economic development planning. This is due to the fact that the inclusion of other tiers of society is implicit in the methodology, including planners and decision makers that have the power to drive changes that local communities do not have. The approach synergizes ground-level realities with upstream policy and planning and the aspirations of all stakeholders and places the local community centre stage. The general principles in the approach originate from the DLIST MSP Project that was deployed during Phase 1 of the BCLME Programme as an independent project to develop a knowledge management tool. In this initiative, a 'common pool of knowledge was created' through mechanisms that encouraged all sectors of society to share their views (Mabudafhasi, 2001, 2002). The inclusion of the LED projects in the ASCLME Project was made possible by embedding the DLIST toolkit in the initiative and was grounded in the perspective taken during the DLIST MSP project that located people at the centre of the process rather than simply as a side-line issue to be dealt with while the science is underway.

The main purpose of the use of the DLIST platform in these projects was to collect input from the coastal communities on how they think the management of marine and coastal resources can be improved and what their priorities are in terms of planning for future interventions. The DLIST team used the opportunity provided by the ASCLME and more recently, the BCLME SAP IMP initiatives to facilitate the creation of development plans for each of the ten demonstration site communities (see Figure 35.1). These development plans are designed as free-standing community development documents and are also used to give input to the planning documents for the ASCLME Project.

Why then the LED approach? LED planning is an approach that places the community at the centre and highlights their daily realities, needs and aspirations. Different from the usual participative rural planning initiatives, the approach involves all stakeholders, from national government all the way to the community itself with all of its own multiple ground level dimensions and stakeholder groups. Multiple inputs are generated into a collective plan that is also informed by the biophysical and other realities of the immediate area. It therefore places the community centrally in planning, but also involves all other stakeholders while considering the nature and dynamics of the resource base. Hence a rich interface between resources, the local population, policy, governance and even science is constructed. Most importantly, the plan is of direct relevance to the community as developed with their direct input rather than being imposed from the outside.

The purpose of LED planning

According to the World Bank, the purpose of local economic development is to build up the economic capacity of a local area in such a way that its economic future can be improved, with a positive impact on the quality of life for all inhabitants. It is a process by which public, business and non-governmental sector partners work collectively to create better conditions for economic growth and employment generation (World Bank, 2006).

A good LED Plan typically calls for improved infrastructure and services; the creation of sustainable business opportunities based on the strengths and potential of the area; better resource use in the commons; and employment or livelihood that is related to opportunities that may exist or could be developed in the area.

A well-defined LED Plan that addresses short-term as well as long-term development opportunities will help to secure and steer sustainable development at the local level in a way that will also benefit the region and the country at large (such as promoting diversification of the tourism industry). Priorities identified in the LED Plan can also provide guidance in terms of Corporate Social Responsibility (CSR) contributions, government infrastructure and social spending. The LED Plan may also inform certain private sector initiatives. A good LED Plan can also assist in unlocking donor-funded interventions. Without a LED Plan, development at the local level runs the risk of ending up consisting mostly of ad hoc interventions. Worldwide, LED guidelines are scattered through many documents and development initiatives.[3]

Application of the LED approach within LME projects

Criteria for site selection

Due to the need to select a specific number of demonstration sites within the ASCLME project, criteria were devised to identify sites that would have the highest likelihood of success. The selection of demonstration sites at which to implement the LED planning initiatives in the ASCLME project, were based on seven key criteria (see Table 35.1).

The criteria can obviously be adapted to the overall priorities of a particular project. For instance, a project that has a strong relevance for coastal tourism will have tourism-related criteria. The COAST Project, funded through UNIDO, focuses primarily on achieving sustainable development through tourism in nine African countries. It works through a multi-stakeholder Demonstration Site Management Committee in each country, which in theory is a good mechanism to improve stakeholder involvement, and in practice this is achieved to a certain degree in at least three of the countries where our team work. However, for tourism to become truly sustainable and local people to benefit more, the initiative has to reach grassroots level (COAST, 2013). Applying the LED approach would be one way to do so, especially since tourism involves multiple-sectors and has a diversity of trickle-down effects.

Building a solid foundation for the process

The LED planning process follows a method that is similar to a SWOT analysis, whereby strengths, weaknesses, opportunities and threats of a particular site are evaluated. The planning process also creates a platform for the community members to formulate a common vision for future development. The vision serves as the guide for the future, while the LED planning process highlights particular steps towards achieving the vision for each demonstration site. Each LED planning process followed the same overall method, with a slight variation in special interest

Table 35.1 Criteria for demo site selection

Criteria for demo sites selection in each ASCLME country
1   Physical location of the community in or near biodiversity and ecosystems 'hotspot'[a]
2   Dependency of the communities directly on natural resources for food security, livelihood and to sustain cultural and traditional values
3   Communities express an interest in working with the project team to identify alternative livelihood and resource management
4   Strong and growing network of partners with a direct interest in helping the community along and inter-organizational linkages
5   All government levels expressing support for the initiative
6   High probability of replication exists (for social, economic, bio-geographic or other reasons)
7   Other important factors that enhance sustainability (e.g. the presence of a national park, World heritage site, etc.)

[a] A protected area (PA) near the selected demo site community can benefit the site in many ways as there are often conflicts between local communities and resource managers that can be facilitated through improved communication and planning. Initiatives related to PAs may also increase the chances of sourcing additional funding for project implementation.

Note: These criteria were approved by the ASCLME Steering Committee during the selection of the demonstration sites.

groups and numbers of meetings. Figure 35.2 outlines the process and methodology followed for the Le Morne demonstration site in Mauritius (LMHTF, 2014), which is the site where tremendous and growing success has been achieved.

As seen in the flow-chart describing the LED planning process in Figure 35.2, the process begins with a community meeting in which all interested community members participate. The purpose of the planning process is explained and agreed upon. The different stakeholder groups that exist in the community are identified. During the first or the second community meeting, the community also selects a small group of representatives (women and men from a range of the different stakeholder groups) to be part of the 'planning committee'. This committee can be assisted throughout the process by the donor-funded project, and provides an opportunity to integrate science, policy and the needs and aspirations of the community.

Developing a vision

The identified stakeholder groups met to list weaknesses and opportunities for good development in the community as well as to develop a draft vision to outline how they would like to see their community developing in the future. Once all groups have created their vision and have listed their weaknesses and opportunities they can be merged into an overall vision for the community.

An example of one such vision is from the Le Morne Demonstration site:

> We envision Le Morne to be a safe place to live, where there are adequate facilities and infrastructure. Our natural resources, both on land and in the sea, and our cultural environment should be respected, kept clean and developed in a wise manner so the local population can benefit, and tourists will be attracted to our village. Our human resource should be developed through capacity building opportunities and access to

equal education, leading to flourishing business opportunities. Development in Le Morne should always be mindful of our heritage values and way of life that makes our village so special.

Interestingly, the visions for the ten ASCLME Demonstration Sites were all rather similar in terms of showing a strong emphasis on marine resources.

Community members can also give input to the planning process by taking part in individual interviews that are conducted in conjunction to the stakeholder group consultation processes. These interviews can provide useful background information about the socio-economic situation of the community and also give further input to the development plan by listing opportunities and weaknesses.

Verifying the findings and recommendations

Once all the information has been collected, the draft LED plan incorporating the draft vision will be developed. To craft a good LED plan, information should also be included from existing research done or underway in the area. It is also essential that the LED plan is compliant with and fits into existing planning and development frameworks for the area.

Involvement of local, regional and even national government representatives increases the chances for successful mainstreaming and implementation of the final LED plan. The draft vision and initial findings of key weaknesses and opportunities, as well as possible interventions to

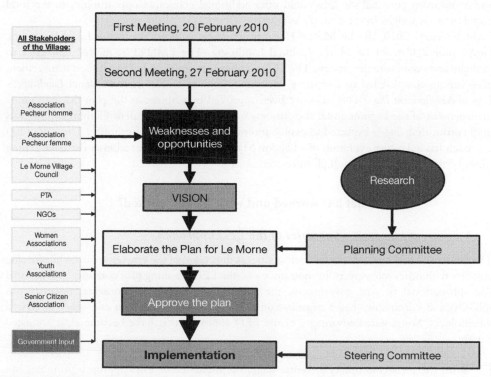

Figure 35.2 Flowchart LED planning process/methodology for Le Morne site
Source: LMHTF (2014).

address the identified problems, must then be presented to the entire community in a meeting to ensure that the information from the community has been correctly interpreted. Here, community members will have the opportunity to finalize the vision and agree on rankings of weaknesses and opportunities and to give more input on the proposed interventions. It is only after this meeting that the LED plan can be finalized and implementation can begin.

In Figure 35.2, each box illustrates respective points at which public meetings were held and input from stakeholder groups was solicited and fed into the process. The figure also shows the stage at which the vision was formulated, when the plan was elaborated and approved and when the planning and steering committees for the plan development and implementation were established.

Le Morne, Mauritius

The Le Morne Cultural Landscape (LMCL), inscribed as a World Heritage Site in 2008, is located on the southwest tip of Mauritius. It is one of the least developed coastal areas in Mauritius and its foothills represent some of the last remaining pristine coastal landscapes on the island. This landscape also contains heritage that is universally important, and is of particular significance to the people of Mauritius.

The economy of the area ranges from subsistence and small-scale fisheries, to an increasing number of local guesthouses and five higher end hotels. The small villages in the area are economically depressed. Inhabitants are primarily dependent on low-income jobs elsewhere due to a poor standard of formal education. Nonetheless, considering its exceptional heritage status and ecotourism potential the area could offer additional economic opportunities to the local population, as well as to the country at large.

In February 2010, the Le Morne Heritage Trust Fund commissioned a Local Economic Development Plan for Le Morne Cultural Landscape. The LMHTF requested EcoAfrica to facilitate and coordinate the process. The LED planning process soon led to a lagoon management planning process that led to a separate document, namely the *Le Morne Cultural Landscape – Lagoon Management Plan*. This has since been approved by Cabinet as the plan that will guide management of the Lagoon under the purview of a Lagoon Management Committee that was duly constituted, and is expected to evolve soon into a management authority. Thus, the LED approach has led to the evolution of a Lagoon Management Plan that addresses the far-reaching issues of this 'open access' multiple-user commons.

What has worked and what hasn't worked?

Successes of the LED approach

The LED planning approach places communities at the centre of the emerging plan while forging linkages with other role players. In most instances, the LED planning processes included national and sub-national tiers of government, the private sector, non-governmental organizations (NGOs) and community-based organizations (CBOs), donors, scientists and other interested stakeholders. Some stated advantages of the LED approach that have become clear are listed below:

1   With the local community at centre stage, the multi-sector LED approach promotes not only LED but also valuable 'bottom-up, top-down' government linkages. This causes an open flow of information that is a prerequisite to participatory democracy and policy

development, which are cornerstones of good governance. Le Morne now enjoys open linkages to ten government Ministries by way of the LED Steering Committee. In addition, the Lagoon Management Committee established links among stakeholders within the multiple-user, multiple-owner area that is now under an emerging management regime in which all stakeholders are involved.

2   The communities respond very well to this approach, as it is highly relevant to their circumstances. In many instances it gives a 'voice to the voiceless', and assists the community in expressing their needs in a coherent manner. Their needs can now be taken up readily in government-funded programmes. The LED plans have also strongly informed the ASCLME SAP that forms the foundation for the upcoming Phase 2 of the ASCLME Project.

3   The LED approach makes it easy to find and strengthen linkages to national priorities. The Mauritius demonstration site LED Plan has strong resonance with national agendas, priorities and action plans, including the well-resourced Maurice Ile Durable Implementation Plan. The LED plan may also become a statutory instrument in terms of the Planning and Development Act of 2004.

4   The results of LED planning generally include a wide range of imperatives that pertain directly to management of the terrestrial and aquatic coastal ecosystems and marine resources. It also facilitates the identification of possible solutions and enlists partners in the process. Rather than being a donor-imposed plan, its origins are in the community, yet it connects easily with donor (and national government) agendas and in fact helps to anchor and concretize these agendas at a practical level.

5   The LED approach further helps to contextualize and thereby promote community concerns in broader frameworks. For instance, three projects make up the overall ASCLME Programme.[4] A single LED plan generates actions that pertain to all three projects. This is primarily because the LED plan has the local community at its core. Thus, even though the three projects have widely divergent origins, they can now be tied together easily in a manner that bridges inter-sectoral gaps and also reaches the entire distance along the vertical axis of governance.

In short, the LED approach has worked well thus far. This is not to say that the approach is a perfect one, or was applied faultlessly in all the demo sites, far from it. But building on the 'lessons learned' from the nine demonstration sites, the approach can be applied more effectively to future sites. In our experience the approach far supersedes ad hoc projects to alleviate poverty, and yields better results than expensive consultant studies and reports that often struggle to find a home in the real world. Typically a LED plan can be completed for $50,000 or less, even in remote areas where travel is expensive.

This is not to say that the LED process is the 'end all' either. A recent article by the World Bank (2013) discusses successes in involving community groups in fisheries management initiatives through the West African Regional Fisheries Program. The article discusses how along Senegal's coast, fishing communities have made significant progress in developing fisheries management, where the communities themselves take responsibility for managing the health of their natural resources and marine environments. It states that 'co-management areas' are being established that are run entirely by the local fishing community. In Monrovia, the Community Sciences for Coastal and Inshore Marine Resources Program is working with the Ministry of Agriculture's Liberian Bureau of National Fisheries (BNF) to engage volunteers to use basic science methods to periodically monitor key indicators of ocean and fishery health. The information is then collected by the national government to help inform policy decisions. In Sierra Leone, the WARFP is working with the Ministry of Fisheries to improve surveillance and prosecution of illegal fishing vessels. Here, fishers in the marine villages are learning sustainable fishing practices, methods to improve fish processing and options for reducing damaging practices (World Bank, 2013).

Pitfalls in stakeholder engagement

The experience from the stakeholder engagement activities of the BCLME SAP IMP programme provide a useful example of some of the challenges encountered as well as opportunities for improvement.

Despite well-meaning goals of stakeholder engagement during the BCLME SAP IMP Programme planning and early efforts to expand stakeholder involvement as per the project plan, serious challenges were encountered by IKM in implementation of project activities. Erratic project management and delays in decision making and release of funds hampered field visits to potential demonstration site communities. Despite these challenges, initial achievements included a highly successful multi-sectoral stakeholder workshop held in Cape Town, South Africa. Participants at the workshop expressed great interest in and support for the BCLME SAP IMP initiative, the draft Benguela Current Convention and the BCC activities and highlighted some key concerns including a lack of meaningful involvement of local stakeholders in the project and in issues of governance of the shared resources of the BCLME.

A modification of the project priorities after only one year of implementation of the stakeholder engagement activities signalled the death knell for any further stakeholder involvement and participation, particularly at community level. The planned outreach to a broader stakeholder base was never achieved due to the unilateral decision by the project management to sacrifice stakeholder engagement and in particular to abandon the planned community involvement activities. An issue of unforeseen budgetary expenditure in other components of the programme resulted in the re-orientation of programme budget. While this was no doubt necessary for recouping priority aspects of the programme following over-expenditure, the sacrifice of stakeholder engagement in the project represented a serious disregard for the stakeholder engagement priority that was identified during the development of the second phase of the BCLME programme. IKM cautioned that in order for the efforts of the BCC to be sustainable or to have any meaningful impact, there is a critical need to foster a far greater synergy with stakeholders at the local level.

Strong opposition by IKM to the decision to forego stakeholder engagement in the SAP IMP initiative yielded some minor results. Yet it was only in the last months of the project that the project activities were approved to salvage some of the momentum gained through impactful participatory processes that could be built on in future phases of the BCLME initiative. A much-reduced budget was allocated for curtailed stakeholder interventions in each of the three countries. In South Africa, this involved a shortened process for identifying local economic development opportunities in the demonstration community that was selected by the South African government prior to the reduction of stakeholder engagement budget. IKM worked with the local fishing community in Elands Bay on the Cape West coast to identify needs and prioritize local economic development aspirations.

Another effort for promoting stakeholder engagement before the end of the programme included a fruitful BCC Youth Summit held in June 2013. This gathering resulted in the formation of the Benguela Youth Ocean Network (BYON), which, with the appropriate support and facilitation, has the potential to anchor the BCC to future constituencies and provide an essential link for building capacity in the region. An opportunity has therefore been established through the BYON for the BCC to engage in a meaningful way with stakeholders in the region and to foster expertise in the field. Even so, the momentum with stakeholder involvement had been lost. Thus far, the BCLME SAP IMP can only be described as a great opportunity missed for stakeholder participation, the realities on the ground being a far cry from the GEF policy documents.

Despite disputes by IKM at the project reorientation and the much-reduced budget allocation to overall stakeholder engagement, calls for an investigation into the failure of the programme to support local stakeholder engagement went unheeded. Interestingly, Chambers highlights that greater recognition of the poor requires personal, professional and institutional change. Thus, if the design of a large programme incorporates a strong emphasis on community benefits, there is a need for support from both the implementing agent and the project management team to ensure that the needs of communities are adequately served as per the intended project goals and objectives. In the case of the BCLME SAP IMP Programme this synergy of supporting factors was lacking and local stakeholder involvement failed to materialize despite being highlighted as a priority in the Project Document.

The lack of consideration of the broader stakeholder base and particularly of local stakeholders, despite the significant emphasis on this during the project preparation, will most certainly impact the social sustainability of efforts of the BCC in the long term. There is a need to identify a proper framework for the integrated involvement of all stakeholders into the governance of the BCLME. The continued lack of any discernible results or impacts from the BCLME initiative on local communities raises serious concerns about the long-term relevance of the initiative and future investment into the BCC by the donor community should be contingent on the prioritization of broader and more meaningful stakeholder engagement, as well as tangible benefits to local communities.

Lessons from the LED planning processes

The following points list some of the main lessons emerging from the LED planning processes from the DLIST ASCLME project. First, the LED planning process provides ample opportunity for the recasting of ground-level agendas into formats that can be driven upstream. This process yields strong input into the LME project that is specific, and can be double-checked and augmented with further input from special interest groups (such as fishers) and government departments. LED plans easily fit within and influence larger-scale planning frameworks to empower communities. A good example is the synergy between the ASCLME demonstration site in Mauritius and the Maurice Ile Durable policy implementation programme, which not only funds parts of the LED Plan but has started to support replication in further communities.

LED planning involves a range of stakeholders, including the different tiers of government, and places community interests at the centre. A good LED Plan enables parties along the vertical governance axis to participate in 'bottom up, top down' planning in a proactive and dynamic way. As such, the LED planning process relies on studies, government positions and expert opinion, but more importantly, it relies specifically on community-identified needs and aspirations, strong and weak points, and issues and obstacles.

LED Plans establish an action plan and a template for funding that facilitates the involvement of diverse parties and shares the objective of improving the situation on the ground through improved use of natural and other resources. Communities are easily able to support the LED planning process and understand that the LED Plan is a result in itself. The LED planning process provides the community with a development plan that they can present to any other entities, but can also start to implement themselves.

The LED planning process generates baselines that can be used to measure the efficacy of interventions over time. In several sites the communities called for environmental assessments or situational analyses to inform planning, and generously contributed their indigenous knowledge. Furthermore, a good LED framework is not a once-off occurrence, but provides opportunities for adaptive management.

Finally, the LED planning approach provides an effective tool for incorporating diverse initiatives into a common local agenda that places the welfare of the community and sound environmental management front and centre. At all sites it was possible to create links between ground level entities and higher-level structures.

Trends and conclusions

The overall trend is that local communities are receiving neither adequate nor appropriate attention when it comes to marine and coastal management projects, at least in the Africa region. The reasons for this are not often clear-cut. Various possible explanations or combinations of explanations have emerged from our experiences and from discussions with many different parties, including project managers, members of steering committees and representatives of the various UN agencies. The perspectives below are therefore seldom explicitly stated but are often informally acknowledged. By creating a discourse around them, it is hoped that development initiatives can be improved and made more relevant and sustainable over the longer term.

Many large projects, not only those funded by the GEF appear to pay lip service to local stakeholder involvement and community benefits. Even when listed as priorities, these imperatives are often poorly elaborated in project designs and as a result, receive lower priority during project implementation than other activities.

One reason for this may be that community development is often regarded as a simple matter, where ad hoc interventions are designed without consultation or a proper understanding of the project context. On the other hand, science is considered complex and as belonging purely to the domain of scientists. Both views persist but should be challenged. The importance of science is understood widely by communities. This is evident in the request for scientific research by communities during several LED planning processes to inform decision making. Local stakeholders are also able to contribute valuable indigenous knowledge, understanding or simply ground-level observations that are important to planning process. As a stakeholder once stated during discussions about climate change detection: 'the small people see it first'.

It is important to note that community development is not a simple matter. Chambers argues for a revolution in thinking about and doing 'development', calling for a complete shift towards approaches that are more community-driven and process-orientated. It is argued that we shouldn't settle for neatly designed projects that are executed by government agencies or big NGOs. Chambers re-conceptualizes 'development' as messy, dynamic and non-linear processes that can be sensitively facilitated and supported by privileged outsiders (who he terms 'uppers') but that must ultimately be controlled and led by communities (who he terms 'lowers'). So 'development' should be done *by* or *with* the people concerned, rather than *for* or *to* them.

Another possible explanation of the lack of attention to local community development is that some projects, if not the overall IW agenda, are primarily driven by scientists and technocrats. This is not to suggest that science should not play a role. In large projects such as the GEF initiatives, a healthy balance is needed that promotes integrated, multi-sectoral collaboration and stakeholder involvement. This would prevent a narrow focus on research agendas rather than broad-based, integrated approaches to coastal and marine governance. Writing communities into a project document design is one thing, successful practical implementation requires significantly more attention and effort. If the project management does not prioritize or facilitate meaningful stakeholder involvement, it will simply not be adequately achieved. Experience in the BCLME Programme has indicated that science and higher-level political processes have dominated the project agenda to the almost complete exclusion of local stakeholder involvement and community development. This state of affairs came into being despite the stated prioritization of stakeholder involvement in project development processes and documents.

For the ASCLME Project, which is currently nearing the finalization of its planning phase, there exists a real possibility to bridge the gap between the aspirations of sustainable community development and reality. This is not, however, a foregone conclusion and the end result will depend to a large extent on the degree to which the project management team supports this priority. While initially community development was omitted from project planning, it has since been corrected and the project has the potential to support significant positive changes among community development initiatives in the Western Indian Ocean region.

An apparent paradox remains. Development projects should necessarily align strongly with country priorities. If large projects claim to be country-owned, and if communities hold significant political power in decentralizing countries, why then are local priorities not considered in deliberations of the project steering committees that consist largely of government representatives? The question remains as to whether these steering committees are correctly composed to include relevant representatives that are interested in and able to support a community agenda? If not, how then should appropriate and representative bodies be established to ensure efficient and effective project implementation that achieves the stated objectives?

Even in cases where representation is achieved in steering committees, logistical challenges prevent meaningful deliberations and informed decision making. Projects can thus end up being driven by other agendas. Empowerment of members is also essential to ensure adequate participation and avoid instances where project implementation is driven largely to meet the project management requirements rather than to achieve meaningful change.

Yet, steering committees are the most appropriate model to ensure the collaborative interaction between development partners, implementing agencies and the intended beneficiary country. Their usefulness in keeping projects on track will be vastly increased if members receive good inductions, if strong stakeholder involvement is cultivated and open lines are established between the intended beneficiaries and the steering committee members who are supposed to 'represent' them. Without that, projects and initiatives may continue to flounder. Representation of active and informed NGOs and CBOs on the steering committee would also be beneficial.

Lastly, but most importantly, in spite of the principles and policies of stakeholder involvement having long been adopted by the GEF, their application still requires stronger enforcement and transparent and objective evaluation. A mechanism is needed to strengthen the accountability of the project implementing and executing agencies to ensure proper process and effective problem solving and conflict management.

Development partners will inevitably take note of the generally weak stakeholder involvement efforts in many projects, and recognize the challenges to realizing meaningful community benefits. However, the GEF should be proactive and carefully examine current coastal and marine projects to identify weaknesses and areas for improvement. The spirit of the discussion herein is intended to be constructive and a call for a greater effort in addressing the needs and aspirations of local coastal communities in a way that recognizes and supports their central role in coastal and marine governance.

Acknowledgements

Our sincere thanks are extended to the ASCLME Project and its participants in eight East African and WIO Island States, the BCLME SAP IMP in South Africa, Namibia and Angola, the DLIST users who now are from over a dozen countries, the COAST Project in Kenya, Tanzania and Mozambique, IW:Learn, the NACOMA Project in Namibia, and the United Nations Development Programme (UNDP), the United Nations Environmental Programme (UNEP), the United Nations Industrial Development Organization (UNIDO), the World Bank for entrusting projects or components of projects to our teams. We also specifically would like to extend our appreciation and gratitude to a host of individuals involved in other projects and agencies who have freely shared perspectives with us, and who are part of a growing group of people who want to make things better.

Notes

1   This is not to say that urban and peri-urban areas do not have local communities. In fact, all of the developing world coastal cities do have artisanal or marginalized local communities that depend on the marine resources to feed themselves; we have seen them in many large coastal cities in Africa and South America, often living 'off the grid' between the high water mark and the coastal road, this being true of Asia as well.

2   The authors have been involved in the BCLME Programme since the first phase of the initiative. Involvement of the authors in the second phase of the programme, the SAP IMP, was through the NGO International Knowledge Management (IKM) (www.ikm.org.za) which was a pre-selected partner.

3   Examples include: i) Tourism and Local Economic Development (see: www.pptpartnership.org); ii) National Responsible Tourism Guidelines for South Africa (see: www.icrt.org); iii) Local Economic Development Guidelines (see: www.owda.org).

4   The three projects in the overall ASCLME Programme are WIO-LaB, SWIOFP and the ASCLME Project. See: www.asclme.org.

References

Agardy, T. M. (1994) Advances in marine conservation: the role of marine protected areas. *Trends in Ecology and Evolution* 9(7): 267–270.

Agardy, T. M., Davis, J., Sherwood, K., Vestergaard, O. (2011) Taking Steps toward Marine and Coastal Ecosystem-Based Management – An Introductory Guide. UNEP. Internet material: www.unep.org/pdf/EBM_Manual_r15_Final.pdf. Accessed 28 February 2014.

Chambers, R. (1983). *Rural Development: Putting the Last First*. Longman, Harlow, pp. 108–139.

Chambers, R. (1997). *Whose Reality Counts: Putting the First Last*. Intermediate Technology Publications. London.

COAST. (2013) Collaborative Actions for Sustainable Tourism (COAST) Project: 4th COAST Project Newsletter. Internet material: http://coast.iwlearn.org/en/en/4th%20COAST%20Project%20Newsletter%20-%20French%20Edition.pdf. Accessed 3 March 2014.

EcoAfrica. (2011) DLIST Progress Report for the period 31/07/2010–30/11/2011. Unpublished report to the ASCLME.

GEF. (2006) *The Role of Local Benefits in Global Environmental Programs*. Evaluation Report No. 30. GEF Evaluation Office. Washington, DC.

Griffiths, T. (2006) The Global Environment Facility and its Local Benefits Study: A Critique. A report prepared for the Third GEF Assembly, Cape Town, August 29–30. Forest Peoples Programme. Cape Town.

Hardin, G. (1968) The tragedy of the commons. *Science* 162: 1243–1248.

Hennessey, T. M. and Sutinen, J. G. (eds). (2005) *Sustaining Large Marine Ecosystems: The Human Dimension, Volume 13*. Elsevier Science. Oxford.

Kelleher, G. and Phillips, A. (eds). (1999) *Guidelines for Marine Protected Areas*. Best Practice Protected Area Guidelines Series No. 3. World Commission on Protected Areas (WCPA) of IUCN – The World Conservation Union. International Union for the Conservation of Nature and Natural Resources (IUCN). Gland, Switzerland and Cambridge, UK.

LMHTF. (2014) Local Economic Development Plan for the Le Morne Cultural Landscape, 2013. Internet material: www.dlist-asclme.org. Accessed 21 July 2015.

Mabudafhasi, R. (2001) Global Conference on Oceans and Coasts at Rio + 10: Concluding Remarks. Paris. 7 December 2001.

Mabudafhasi, R. (2002) *The Role of Knowledge Management and Information Sharing Capacity Building for Sustainable Development: An Example from South Africa*. The International Bank for Reconstruction and Development/The World Bank. Washington, DC.

McClanahan, T. R., Marnane, M. J., Cinner, J. E. and Kiene, W. E. (2006) A comparison of marine protected areas and alternative approaches to coral-reef management. *Current Biology* 14: 1408–1413.

Mayol, T. R. (2013) Madagascar's nascent locally managed marine area network. *Journal of Madagascar Conservation and Management* 8(2): 91–95.

Revolutionary Government of Zanzibar. (2009) *The Status of Zanzibar Coastal Resources: Towards the Development of Integrated Coastal Management Strategies and Action Plan*. Prepared by the Department of Environment through support from Marine and Coastal Environmental Project (MACEMP).

UNDP. (2008) Implementation of the Benguela Current LME Strategic Action Programme for Restoring Depleted Fisheries and Reducing Coastal Resources Degradation. Project Document. Internet material: http://iwlearn.net/iw-projects/3305/project_doc/view. Accessed 3 March 2014.

UNEP. (2008) Abidjan Convention for Cooperation in the Protection, Management and Development of the Marine and Coastal Environment of the Atlantic Coast of the West, Central and Southern Africa Region and Protocol concerning Cooperation in Combating Pollution in Cases of Emergency. Internet material: http://hq.unep.org/easternafrica/AbidjanConvention.cfm. Accessed 3 March 2014.

UNEP. (2010) Amended Nairobi Convention for the Protection, Management and Development of the Marine and Coastal Environment of the Western Indian Ocean. Internet material: www.unep.org/nairobiconvention Accessed 3 March 2014.

UNESCO. (2013) UNESCO/Flanders Funds-in-trust for the support of UNESCO's activities in the field of science. ODINAFRICA-IV: Background. Internet material: http://fust.iode.org/oceans/odinafrica/odinafrica-iv-bacground. Accessed 3 March 2014.

World Bank. (2006) *Local Economic Development: Developing and Implementing Local Economic Development Strategies and Action Plans*. The World Bank. Washington, DC.

World Bank. (2013) West Africa: Fishing Communities Restore Health to Ocean Habitats. Internet material: www.worldbank.org/en/news/feature/2013/06/05/west-africa-fishing-communities-restore-health-to-ocean-habitats. Accessed 4 July 2013.

36

SOUTH PACIFIC AND SMALL ISLAND DEVELOPING STATES

Oceania is vast, canoe is centre, village is anchor, continent is margin

Peter Nuttall and Joeli Veitayaki

In this chapter, the discussion focuses on critical aspects of life in Small Island Developing States (SIDS) in the Pacific Ocean, a unique water-based region, ancient home to voyagers, Islanders and villagers, a place where small is still beautiful but where unprecedented levels of change threaten the existence of countries and communities. Pacific peoples are known for their patience, generosity and resilience, traits honed by millennia of close association and intimacy with their ocean and island homes. These traits have allowed them to live with minute resource bases and a changing environment for thousands of years. However, contemporary changes such as global warming, acidification, environmental degradation, globalisation and rampant consumerism promise a gathering tropical cyclone or tsunami of magnitude greater than anything the Pacific Islands have ever faced.

Pacific SIDS have resource management responsibilities over a significant portion of the world's ocean space. Although they have jurisdictional rights over large ocean areas that are rich in resources including fisheries, gas, seabed minerals and renewable energy, many are vulnerable to the conquest of the sea, which is predicted to worsen with climate change. Many of these states are not benefiting fully from the marine resources within their EEZs due to inadequate technical and management capacity and limited financial and physical resources.

The Barbados Programme of Action (BPOA) from the 1994 United Nations Conference on the Sustainable Development of SIDS outlined priorities for action. It was followed by the Millennium Development Goals (MDGs) in 2000, the World Summit on Sustainable Development (WSSD) in 2002, and the 2005 Mauritius Strategy (MSI) for the further Implementation of the BPOA and the five-year review of the Strategy undertaken in 2010 (www.sidsnet.org). In spite of all plans and commitments, few tangible accomplishments have been witnessed in SIDS. It is evident that the huge marine areas surrounding SIDS present development challenges and vulnerability to natural and environmental disasters (Ashe, 1999).

The Pacific Ocean covers nearly one-third of the earth's surface, and is one of nature's greatest carbon sinks (UNESCAP, 2010, p. 9). It affects the climate, ocean currents and the complex ecosystems it hosts. This engine-room of the earth's climate and mainstay of Pacific Island economies must be cared for and allowed to continually provide ecological and economic services for Pacific Islanders and humanity in the future.

Biodiversity loss in Pacific Island Countries (PICs) is ranked among the highest in the world but little is known of the impacts on marine life despite it being the subject of a number of research initiatives. In many parts of the region, changing conditions are expected to reduce ocean productivity in the future. It is therefore critical that Pacific Islanders better understand the state of their ocean and how it supports life now and in the future to effectively manage and develop it, protect its health and benefit from its potential.

Under the 1982 United Nations Convention on the Law of the Sea (UNCLOS), PICs jointly hold access rights and management responsibilities over 30 million square kilometres of the Pacific Ocean. UNCLOS has enormously increased the maritime areas of the PICs. While the new wealth and resources associated with these extended areas are untapped, the burden that they place on the custodians is overwhelming. Pacific SIDS have established regional organisations to assist them with advice, development activities, education and training on pertinent issues determined by member countries.

Growing population pressures, pollution and alteration of sensitive coastal environments, greater pressures on reef and lagoon fisheries, increasingly lucrative opportunities to give foreign fleets access to the fisheries, dwindling resources, emphasis on economic development, and the development of new technology in aquaculture, postharvest fisheries, ornamental and aquarium trade and energy all increase the demand on trained human capacity.

Pacific SIDS

The Pacific Islands region contains 15 SIDS: Cook Islands, Federated States of Micronesia (FSM), Fiji, Kiribati, Nauru, Niue, Palau, Papua New Guinea (PNG), Republic of the Marshall Islands (RMI), Samoa, Solomon Islands, Timor Leste, Tonga, Tuvalu and Vanuatu.

Land is scarce, constituting less than 2 per cent of the total area, less than 0.4 per cent if PNG is omitted. Four states have less than 30 square kilometres each. Several are either made up wholly of atolls or largely of atolls and coral islands. There are at least 11 square kilometres of ocean for every coastal Pacific Islander (Anderson *et al.*, 2003, p. 2) making the Pacific one of the most remote and far flung regions in the world (AusAID, 2008, p. 1).

The region is immense and diverse, varying from the larger, mineral and natural resource rich, high volcanic islands of Melanesia, home to fast-growing millions, some of whom will go a lifetime without ever seeing the ocean, to tiny atolls less than five metres above sea level with less than 2000 inhabitants and only a handful of terrestrial flora species. The region is home to a quarter of the world's languages and living cultures. It contains newly independent democratic nation states, ancient monarchies, states of superpowers, territories of Europe, dependencies, military dictatorships and states in 'close association' with superpowers. The region has been fought over repeatedly and, since World War II, has been rained with more nuclear bombs and radiation by the US, France and Britain than anywhere else on earth. It has seen bitter wars of conquest and civil insurrection, minute by global standards but with the heaviest per capita casualties as witnessed in Timor Leste and Bougainville. Pacific SIDS now face the ignominy of being the first region on earth where the carelessness of human beings as a species is proven as whole countries sink because of anthropocentrically generated climate change, little of which has been of their making.

Table 36.1 Physical capital, population, density, growth rate per capita

Country/Territory	Land Area (km²)	Mid Year 2013 Population (estimate)	Density (Persons/km²)	2013 Growth Rate (%)	2013 GDP (US$,000)	GDP per capita (US$)	Year
American Samoa	199	56,500	284	-0.3	615,000	9,333	2010
Cook Islands	237	15,200	64	-0.5	272,769	17,565	2011
FSM	701	103,000	147	0.3	310,213	3,031	2011
Fiji	18,333	859,200	47	0.8	3,099,191	3,639	2011
Guam	541	174,900	323	0.3	4,577,000	25,420	2010
Kiribati	811	108,800	134	2.2	170,542	1,651	2011
RMI	181	54,200	299	0.4	173,700	3,158	2011
Nauru	21	10,500	499	1.8	85,337	8,379	2010–11
Niue	259	1,500	6	-0.2	22,857	15,807	2011
New Caledonia	18,576	259,000	14	1.9	9,093,963	36,405	2010
N. Marianas	457	55,600	122	-2.5	733,000	11,622	2010
Palau	444	17,800	40	-1.9	212,903	10,314	2011
PNG	462,840	7,398,500	16	2.3	127,000,000	18,437	2011
Pitcairn	47	60	n.a.	n.a	n.a.	3,385	2005
French Polynesia	3,521	261,400	74	1.8	7,200,000	26,667	2011
Samoa	2,934	187,400	64	0.8	675,729	3,680	2012
Solomon Islands	28,000	610,800	22	2.8	927,390	1,676	2012
Tokelau	12	1,200	98	0.9	–	–	
Tonga	749	103,300	138	0.2	470,669	4,557	2011–12
Tuvalu	26	10,900	420	1.1	38,178	3,407	2011
Vanuatu	12,281	264,700	22	2.5	760,097	3,099	2011
Wallis and Futuna	142	12,100	85	-2.1	183,181	12,324	2005

Source: SPC Pocket Statistical Summary 2013b.

All PICs are members of the Alliance of Small Island States (AOSIS) and, except for Timor Leste, the Pacific Islands Forum (PIF). Some of these countries are among the poorest in the world, with Kiribati, Samoa, Solomon Islands, Timor Leste, Tuvalu and Vanuatu currently classified as Least Developed Countries.

The estimated population of just over 10 million people in 2013 makes Oceania numerically minute on the global scale. PNG has the largest population with 7.39 million. Half have populations of less than 100,000 and Pitcairn is the smallest with 60 inhabitants (SPC, 2013b). While there was no urban tradition before European colonisation, the region today averages 53 per cent in urban settlements. Nauru, Northern Marianas and Guam are now without any significant rural populations (SPC, 2013a).

A significant proportion of national incomes come from aid and remittances. SIDS suffer from diseconomies of scale in production and exchange of goods and services, remoteness from export markets and high vulnerability to natural disasters and climate change. In addition, there is a high degree of economic and cultural dependence on the natural environment and primary commodities.

The combined value of Pacific SIDS' GDP is around US$156 billion. PNG has the largest economy (US$127 billion in 2011), the second largest is New Caledonia (US$9 billion in 2010) then French Polynesia (US$7.2 billion, 2011), Guam (US$4.5 billion, 2010) and Fiji (US$3 billion, 2011) (Table 36.1, Figure 36.1). PNG accounts for over 80 per cent of the region's GDP. The remaining Pacific SIDS have very small economies ranging from Solomon Islands' US$927 million to Niue's US$22 million (SPC, 2013b).

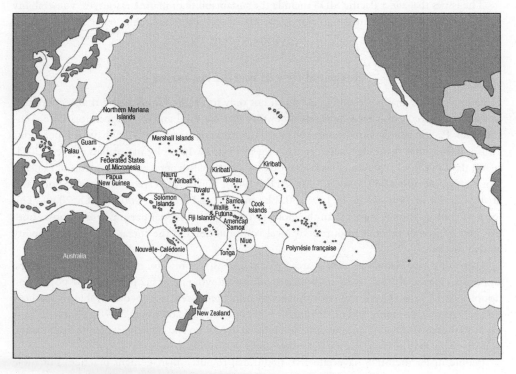

Figure 36.1 Pacific EEZs
Source: Map reproduced with the permission of SPC.

Pacific SIDS will not achieve the MDGs by 2015, particularly the targets for 'universal primary education, reducing child mortality by two-thirds and halving the proportion of people without access to water and sanitation' (AusAID, 2008, p. 3). The MDG framework has been criticised for not recognising 'key underlying determinants of well-being such as family and societal cohesion' and 'neglecting the importance of economic development and the private sector' (Coates, 2009, p. 29). In addition, the exclusion of women's contribution and the failure to account for subsistence productions greatly disadvantage PICs.

Poverty is worsening. More than 80 per cent of the region's population lives in the four poorest countries: Kiribati, PNG, Solomon Islands and Timor Leste. Data on poverty is limited, but alarming. Three national surveys in Fiji show that poverty rose from 11.4 per cent in 1977 to 34.4 per cent in 2002. Underemployment was at 22 per cent for males and 35 per cent for females. This increased to 26 per cent and 45 per cent respectively in rural areas and to 45 per cent and 67 per cent for young men and women under the age of 20 (AusAID, 2008, p. 3).

It is difficult to look past the enormous obstacles posed by long distances, tiny populations, minute economies and minimal resources. Remoteness has significant economic, environmental and social impacts: high fuel costs and low economies of scale make the cost of developing and maintaining essential infrastructure high. Small populations offer a narrow range of resources and skills that limit capacity. Small populations and land areas create limited markets, reducing people's income earning potential (Redding and Venables, 2004). SIDS are less open to international trade, due either to protectionist policy choices or to remoteness causing transport costs to be a larger source of 'natural protection' and isolation (Gibson and Nero, 2007).

The issues that affect Pacific SIDS include the sustainability of marine resources, vulnerability to natural disasters and climate change, dependence on fossil fuel, and social and economic transition.

Sustainability of marine resources

Sustainable use of marine resources is critical not only to Pacific Islanders and their economies; but also to the coastal habitats and resources that provide food and livelihoods for these communities. Pacific Islanders are heavily reliant on their marine resources. They have fish per capita consumption rates that range between 16.9 kg in PNG to 181.6 kg in Kiribati (Gillett, 2011, p. 83). In Tuvalu, each person consumes about 500 grams of fish per day; the residents of Funafuti alone consume about 730 tons per year (ADB, 1994). Fijians' per capita fish consumption was around 26 kg in 1986, 47 kg in 1990, 50.7 kg in 1997 and 33 kg in 2002 (FAO, 2009). This level of consumption is higher than the global average of 16.5 kg per person, which shows both the importance of fisheries resources to local communities and the threat to the sustainability of fish stocks.

Increasing population in urban areas places intensive pressure on all marine resources. In South Tarawa, Kiribati, with an estimated annual growth rate of 5.2 per cent, the population doubles in 13 years. Given the enormous population pressures, it is inconceivable to see how South Tarawa's economy and environment will cope with an additional 36,700 people in nine years (Haberkorn, 2004). The same situation is faced in Majuro and Funafuti where the population density rivals that of Hong Kong and other cities in Southeast Asia.

People in Pacific SIDS have effective traditional resource management practices but these alone are insufficient to the people who are now staring at depleting resources, altered environments and increasing demands that threaten their food security.

Vulnerability to natural disasters and climate change

PICs are among the most vulnerable regions in the world to natural disasters such as cyclones, earthquakes, floods, drought and tsunami. The sediment load through Rewa River floods was estimated at an average of 107 tonnes/yr (Hasan, 1986). The estimated soil loss in the Rewa River catchment was about 34–36 tonnes/hectare/year (Morrison, 1981; Hasan, 1986; Nunn, 1990). According to Nunn (1990), the losses in the four main tributaries of the Rewa River were: 30 t/ha/yr in the Wainimala, 69 t/ha/yr in Waidina, 24 t/ha/yr in Wainibuka and 79 t/ha/yr in Waimanu. Consequently, the Fiji Government since 1983 was spending about US$6 million annually on dredging to alleviate the problem of flooding in the Rewa and other rivers (Togamana, 1995).

Between 1950 and 2004, extreme natural disasters accounted for 65 per cent of the total economic impact of disasters on the region's economies. Ten of the 15 most extreme events reported over the past 50 years occurred in the last 15 years (UNESCAP, 2010, p. 10). Climate variations and extremes disrupt food production, water supply and economic development. 'Events during the last decade have demonstrated that vulnerabilities remain high and efforts to build resilience have been insufficient' (UNESCAP, 2010, p. 10) so PICs are continuously rebuilding and recovering from disasters, spending millions of dollars otherwise earmarked for development activities.

Projections are bleak as primary sectors (agriculture, fisheries, forest) and water are all being negatively impacted by human activities (UNESCAP, 2010, p. 10). Climate change will worsen these devastating and economically crippling impacts. While Pacific Islanders have historically shown strong resilience and adaptability in the face of environmental threat and change, the immensity and immediacy of the effects of climate change will make adaptation insufficient (Barnett, 2002).

It is the scale and irreversibility of the effects of Greenhouse Gas (GHG) emissions combined with the total inability of any local measures to mitigate the problem that makes climate change a threat over and above all others. The minute contribution of GHG by PICs, estimated by the South Pacific Regional Environment Programme (SPREP) to be 0.03 per cent of global totals (Hay, 2002), makes any mitigation taken by PICs symbolic, no matter how successful.

Unfortunately, Oceania's concerns are almost unheard on the global stage, drowned out by larger states, superpowers and alliances whose consumption-based development and security interests easily outweigh any PIC's voice. Failure to gain international consensus on emissions reduction at Copenhagen, Mexico, Rio+20, Warsaw, Bonn, reinforces the futility of SIDS expecting sufficient response from developed countries. Ironically, Pacific Islanders, along with indigenous communities at the poles, will be the first and the worst affected victims (Barcham *et al.*, 2009; Merson, 2010).

The uniqueness of PICs, which are ocean-centred, non-continental, scattered islands, village-dominated, culturally rich but economically poor, requires adaptation measures tailored to their needs and conditions if they are to be effective and durable. Barnett (2001) warned more than a decade ago that continental-centred or focused solutions may prove inadequate or inappropriate.

Adaptation strategies in Oceania are diverse in range and scope, reflecting the variation in local conditions. At one extreme is potential migration of entire PICs to new (and possibly radically different) homelands (Burson, 2010). Long-term integrated programmes of change for many critical sectors have yet to materialise. According to Nunn (2010, pp. 233–234):

> There has been a failure on the part of most regional agencies serving the PICs to develop proactive plans independent of either international or national agendas that

take into account either the special needs of PICs or the importance of developing adaptive solutions that acknowledge their singular cultural and environmental contexts. Instead such agencies have been largely reactive, uncritically imposing the priorities of international organizations on Pacific Island nations and focusing on short-term pilot studies rather than mainstreaming the lessons learned from these.

Much of the global concern about climate change impacts on Oceania is focused on the plight of atoll dwellers, a view that is often expanded as representative of the whole region. While such concerns are real and pressing, this narrow definition ignores the wide diversity of Pacific Islands and the effects, although it has proved useful in leveraging global attention. Barnett and Campbell (2010, p. 155) argue that 'representation of the Pacific Islands as extremely vulnerable may have created the illusion that adaptation is pointless, and denies the resilience, agency, capacity, and potential that Pacific Island communities have and can be made an adaptation response'.

The Secretariat of the Pacific Community (SPC), SPREP, USAID and ADB have all produced plans for addressing climate change that have been criticised for being top down and insufficiently community/recipient focused. PICs are frustrated by processes that are agency and donor driven, the high degree of overlap and insufficient collaboration. Oxfam highlighted the institutional rigidity of donor organisations that hinders cooperation and interagency collaboration (Maclellan *et al.*, 2012). AusAID's 2012 assessment of multilateral agencies agrees that there is a need to reduce the duplication of programmes. Donor organisations point to the lack of capacity of recipients to absorb available funding, to extend into new areas and expand programmes, and a lack of transparency and accountability (Barnett and Campbell, 2010).

Programmes funded by international donors at times underline the disjuncture between real need and donor priorities. For example, all of the $US535 millions committed by donors at the 2013 Pacific Energy Summit in Auckland to reduce the region's dependence on imported diesel was allocated for electricity generation (which uses some 17 per cent of the region's imported fuel) while transport (which accounts for more than 70 per cent of fuel use) was simply ignored (Holland *et al.*, 2014). This type of recipe for disaster needs to be rectified.

Fossil fuel dependency

The Pacific is the world's most imported-fuel-dependent region with 95 per cent dependency (99 per cent if PNG and Fiji are excluded). Imported fossil fuels account for 8–37 per cent of total imports raising critical issues of fuel price and security of supply (Woodruff, 2007). In 2011, fuel imports cost PICs more than $US1.3 billion, which represents a major drain on economies and has a crippling effect on national budgets and revenues and impacts all key productive sectors in the region (UNESCAP, 2010). Sea transport is entirely dependent on, and is one of the largest users of, imported fossil fuels in the region.

In the past decade there has been an increasing and concentrated effort by PICs supported by international donors to introduce a range of renewable energy technologies to substitute fossil fuel use for electricity generation. Despite the electricity sector being a minority user of fossil fuel for PICs, several donors have seen this as the 'low hanging fruit'. There is insufficient analysis to demonstrate that these measures are ultimately cost effective and sustainable.

Social and economic transition

Pacific Islanders have developed enormous resilience that derives from their access to communal land, strong cultural identity, and systems of community governance. Such resilience is supported today through strength of kinship ties, sharing of communal resources, and cultural obligations of reciprocity (Bayliss-Smith *et al.*, 1988; Coates, 2009, p. 30; Veitayaki *et al.*, 2011). Unfortunately, this coping strategy and survival mechanism is quickly eroding as a result of the social and economic transformation taking place.

Customary roles and duties are less clear and effective today. Traditional tenure systems and resource management strategies have gradually eroded due to colonisation (Govan, 2009, p. 25) and are now undergoing violent changes of unknown rapidity. While traditional roles and resource-use systems within communities are often still well defined, leadership structures, protocol, respect, practices and beliefs are changing and are increasingly questioned (Vunisea, 2002).

More people are moving into urban centres while those in urban areas are moving to the Pacific margins in Auckland, Brisbane, Los Angeles and further in search of a better life. Exposure to consumerism and international information technology is leading to the replacement of traditional diets with canned and processed foods, traditional materials with throwaway goods and traditional values with populist global cultures.

The erosion of traditional knowledge and culture has led to increased vulnerability and dependency. Historically, sailing technology was an integral part of daily life in PICs, essential for social interaction, transport, warfare, trade and fishing. Sea passages were not feared barriers but exploited highways, the basis of connectivity and maintenance of kinship and exchange. Using learned knowledge of seafaring, navigation and ship design and construction, Pacific people made this ocean their home. These early Pacific Islanders did not see themselves as small, weak or vulnerable. As Hau'ofa (1993, p. 7) chides, Pacific Islanders 'did not conceive of their world in such microscopic proportions. . . . their world was anything but tiny'.

It is ironic that contemporary transport development that now connects the world and supports globalisation has left many Pacific Islanders isolated. Dependence on fossil fuel driven ships, high cost of fuel and loss of traditional sailing culture have made outlying island communities vulnerable and a burden on their resource-strapped governments.

The way forward

In 1993, Hau'ofa's regularly quoted 'Our Sea of Islands' challenged Pacific Islanders to recast themselves as big people at home on a big ocean with a big history. He reasoned that Pacific Islanders were connected rather than separated by the sea and that far from being sea-locked peoples marooned on coral or volcanic tips of land, Pacific Islanders formed an oceanic community based on vessels and voyaging. The people were connected across the ocean.

Pacific SIDS need to determine their own development pathway to allow them to live in their countries with the challenges they face. Some areas suggested for the future development of Pacific SIDS include regional policy development, better use and non use of resources, disaster risk reduction and climate change adaptation, use of renewable energy, partnerships and capacity building.

Regional policy developments

Pacific Island leaders' endorsement of the Pacific Islands Regional Ocean Policy (PIROP) and its presentation at the WSSD illustrate the regional effort to safeguard a 'healthy ocean that sustains the livelihood and aspirations of Pacific Island communities' and provide a principled approach to responsible ocean governance in the region. Sadly, no national ocean policy has been formulated in the region.

PICs are now focusing on 'Pacific Oceanscape', which emphasises resource conservation triggered by the collaboration between the Government of Kiribati, the New England Aquarium and Conservation International that resulted in the declaration of the Phoenix Islands Protected Area (PIPA) as the world's largest MPA at that time. The competition to declare the world's largest MPA in the Cook Islands, Niue and New Caledonia, the appointment of an Ocean Commissioner at the PIF and the announcement of financial support from Australia and NZ have enhanced the initiatives to better manage the ocean that are important to Pacific Islanders.

There was commitment at both Barbados and Mauritius to implement the Global Programme of Action for the Protection of the Marine Environment from Land-based Activities to reduce, prevent and control waste and pollution and their health-related impacts. The BPOA for the Sustainable Development of SIDS remains the blueprint for national and regional development that takes into account the economic, social and ecological aspects that are the pillars of sustainability.

At the 2012 Pacific Forum, Cook Island's Prime Minster Henry Puna challenged other PIC leaders to consider a rethink of shared identity within the Pacific saying: 'it is time that we break the mould that defines us too narrowly and limits us in any way'. Puna called for a recasting of regional identity to one of 'Large Ocean Island States'.

> Our large ocean island states should demonstrate – now more than ever – renewed commitment to define our future in our own terms. Our intimate and connected relationship is built from a deep spiritual bond and translated across an expanse of ocean in unique and traditional ways.
>
> *(Cook Islands Herald, 21 March 2012)*

The establishment by Fiji of its Green Growth Framework is fascinating because it is aimed at restoring the balance between the three pillars of sustainable development that the Fiji Government fears is under threat with the current approaches that emphasise economic development over social and environmental well-being (Ministry of Strategic Planning, National Development and Statistics, 2014, pp. 4–5). The Framework is to provide the process to ensure that development in years to come is sustainable and maintains Fiji's environment. It offers the opportunity in which government, nongovernment, private sector, faith-based organisations, the media, urban and rural communities and individuals alike can be engaged in the pursuit of sustainable development (Ministry of Strategic Planning, National Development and Statistics, 2014, pp. 4–5). 'The Green Growth Framework is the first of its kind for Fiji' and is the impetus to take the country into the uncertain future.

Better use and non-use of resources

PICs' unique resilience is related to people's close relationship with their environment and with each other. From minerals and forests in PNG and Solomon Islands to fisheries in Kiribati and Tuvalu, the challenge is to determine the optimum level of use that produces the best return in ecological, social and economic terms. Pacific Islanders are taking measures to protect their resources by combining their customary and contemporary resource management arrangements to pursue the type of developments that improve their livelihoods and well-being while ensuring a vibrant environment capable of continually supporting life (Pearce and Warford, 1993, p. 8).

Governments, regional organisations and civil society must work collectively to ensure that the exploitation of natural resources, focus of investment, orientation of technological development and institutional changes are consistent with present and future needs of PICs and their people (Cicin-Sain, 1993, pp. 15–6). Issues such as corruption, illegal, unreported and unregulated fishing, pollution and poorly planned developments must be appropriately addressed. PICs must be:

> guided by a basic philosophy which emphasises development to improve the quality of life of the people (assuring equity in the distribution of benefits flowing from development) and development that is environmentally appropriate, making proper use (and sometimes non use) of natural resources and protecting essential ecological processes, life support systems and biological diversity.
>
> *(Cicin-Sain, 1993, p. 17)*

Pacific Islanders must pay attention to the assimilative capacity of their environment, which is limited and must not be exceeded. Waste products and toxic substances are affecting the resilience and adaptability of many islands' biotic systems. With the rapid changes taking place across the region, everyone must remember the uncertainties we are dealing with, where ecological change represents the greatest threat to the system, and the extent to which the poor are paying the cost of environmental degradation (Hamilton, 1997, p. 30). Sound ecological, social and economic policies, democratic institutions responsive to the needs of the people, the rule of law, anti-corruption measures, gender equality and an enabling environment for investment are the basis for sustainable development.

Palau is providing international leadership in having vibrant and healthy coral reefs as the centrepiece of sustainable development that supports strong and robust economies. Working under the Micronesia Challenge, a Northern Mariana Islands, FSM, Guam, RMI and Palau initiative to protect 30 per cent of their coral reefs and 20 per cent of their forest resources by 2020 (themicronesiachallenge.blogspot.com) and contribute to global coral reef conservation targets, these countries have heightened marine resource management, solicited much-needed funds to support local initiatives and advocated the importance of taking appropriate action at all levels of governance. In addition, Palau has declared a shark sanctuary because live sharks are worth a lot more to its marine-based tourism industry than the price of the fins to its fishers. In a keynote address to a United Nations meeting on 'Healthy Oceans and Seas' in February 2014, Palau President, Tommy Remengesau Jr. announced his country's plan to outlaw commercial fishing in its waters once current fishing contracts in the country expire (Molland, 2014).

Disaster management and climate change adaptation

Managing natural disasters and climate change is important to reduce the economic, cultural and ecological costs of such events. Although Nunn (2007) highlighted the high resilience that allowed Pacific Islanders to quickly recover from natural disasters, this independence needs to be emulated today if the costs of damages from disasters are to be minimised and the recovery less demanding. As Maclellan *et al.* (2012, p. 6) explain, 'Pacific communities face no option but to adapt if they are to build a resilient future'.

Disaster and climate change adaptation needs to be at levels where important decisions about social organisation are made to be successful (Barnett and Campbell, 2010, p. 178). This is mostly in the villages where 'community-based approaches are likely to offer the most effective approach to adaptation, as these can avoid the pitfalls of externally imposed and top-down projects which underestimate local capacities and ignore local particularities' (Barnett and Campbell,

2010, p178). Adaptation programmes 'must be consistent with the values, needs, and rights of affected communities' while local people should be asked about the 'support they need, rather than being told what they should receive' (Elliot and Fagan, 2010, p. 82).

At the Sixth Pacific Platform for Disaster Risk Management workshop in Suva, in 2014, with the theme 'The Way Forward: Climate Change and Disaster Resilient Development for the Pacific', the stakeholders used the holistic approach to formulate communication protocol to use during disasters (Naleba, 2014). Sharing national information and experiences will enhance disaster and climate change resilience and sustainable development among PICs.

People must act at all levels to climate-proof their surroundings and activities. President Anote Tong of Kiribati leads the way in calling for all to take a moral responsibility in the fight against climate change to ensure the well-being of all (*USP Beat*, 2012, p. 7). At their level, villagers of Vunidogoloa in Cakaudove, Fiji, have worked with Government to relocate their village away from the encroaching shoreline (Silaitoga, 2014) while others are rehabilitating their coastal habitats and using adaptive arrangements to live with climate change. Pacific Islanders are not giving up and are trying to adapt to prevailing conditions.

Renewable energy and maritime transport

Reducing the region's dependency on imported fuel is more practical and a higher priority than reducing emissions caused by burning that fuel. The available options are to increase the efficiency of current users, reduce fuel consumption (which would come at a high social and development cost), or introduce or increase the use of alternatives. In many cases, there is a serendipitous correlation between the use of alternatives and emissions reduction.

The lack of adequate policy and financing are major constraints to developing more appropriate sea transport for PICs (Prasad *et al.*, 2013; Nuttall *et al.*, 2014). Despite the logic of renewable energy shipping as a priority for adaptation, shipping projects are generally considered only as mitigation measures. Renewable energy shipping does not meet the criteria for many mitigation funds because it would not be displacing fuel used for electricity generation (Nuttall *et al.*, 2014), the current priority set by donors. This needs to be addressed. Investment in research and development to prove commercial viability of renewable energy vessels is a priority.

ADB's 2007 shipping overview is typical of expert opinion that shipping is best left to private investment and the market. The marginal nature of shipping has always meant that financing, either for governments or private operators, is difficult and the current global economic environment has only exacerbated this. The assumption that the private sector acting alone or unaided is best situated to provide services needs to be re-examined. There is room for governments and agencies to provide access to vessel and industry financing (e.g. providing loan security and preferred operator status for renewable energy powered or retrofitted vessels) (Holland *et al.*, 2014).

Although no analysis has been undertaken, we surmise that the largest proportion of emissions from shipping in the Pacific does not come from ships owned by, trading with or otherwise benefiting PICs but from large vessels transiting Pacific waters. It would be logical to argue for a compensation mechanism levied on such shipping to support PICs' move to a lower carbon footprint.

Given the importance of marine transportation in SIDS, having appropriate training is critical to meet human capacity needs in this area. Pacific Islanders are naturally talented and skilled maritime workers who must meet international standards to continue to be a part of the global industry. Training in safety practices can improve performance and efficiency.

Partnerships

South–South cooperation such as practised in PICs is critical at bilateral, sub-regional and regional levels. Areas that benefit from such arrangements include: marine resource management; information and communication technology; trade and investment; capacity building; disaster management; biodiversity; food, water and energy security; health and education; culture; youth and gender equality.

Regional institutions such as PIF, Pacific Islands Forum Fisheries Agency (FFA), SPREP, SPC and The University of the South Pacific (USP) are assisting PICs to meet their needs and obligations in accordance with the agreements, treaties and conventions they have signed and ratified. PIF looks after political and policy matters while FFA guides management and development of the region's valuable fisheries resource and coordinates the array of regional instruments that PICs have formulated to control and manage their maritime zones and resources (Veitayaki, 1994, 2005; South and Veitayaki, 1998). SPREP assists with formulation of programmes to manage, develop and foster appropriate responses to environment issues while SPC advises on social, economic and cultural issues. SPC's marine resources section provides scientific advice on the status of inshore fisheries (Dalzell and Adams, 1995a, 1995b; Dalzell *et al.*, 1996), the state of the region's tuna resources and the status of coral reefs (Grigg and Birkeland, 1999; Wilkinson, 1999). SPC's SOPAC Division assists with EEZ boundary delineation, coastal protection and sustainable living in small island environments while USP offers tertiary education, research and consultancy services (Crocombe and Meleisea, 1998; South and Veitayaki, 1999).

These organisations prioritise the pooling and sharing of resources and adaptive management. PICs also need innovative approaches to optimise their benefits from the use of their tuna resources (Ram-Bidesi, 2011). The successes of the Parties to the Nauru Agreement in markedly increasing their licensing income from the adoption of the Vessels Day Scheme and their closure of the High Seas pockets demonstrate this point.

Capacity building

Capacity building must emphasise the prioritisation of need, identification of appropriate training and education, design and delivery of programmes in a timely fashion, and the rationalisation of training responsibilities and effort. While institutions such as USP must decide on the capacity building programmes they provide, these must be based on countries' needs, which are continually evolving. At the moment, all the regional organisations provide training that is based on organisational priorities. In many cases, donor-funded projects include only superficial capacity building to meet requirements of individual projects. Rather than addressing the problems, this arrangement only perpetuates the lack of capacity.

Capacity building is a cross-cutting issue that is a critical prerequisite to the success of any project. Training institutions must address it with vigour and rigour. They must examine the wide range of issues that confront PICs and provide the conduit between the needs of local communities and governments and scientific research.

Highly qualified tertiary teachers, trainers and researchers are needed to share and gather the knowledge and data needed to support sustainable development in PICs. Marine resources need to be identified, assessed and used sustainably. This demands good understanding of the marine environment and its interrelationships. Maritime zone boundaries have to be delineated and negotiated, and resources must be equitably distributed. International resource management instruments have to be ratified and implemented while governance issues and economic arrangements must be put in place.

Training is needed in areas such as statistics, economics, sustainable technologies, and ocean law and policy. Critical skills demanded include analytical thinking and problem solving; ability to present clearly and concisely in English; to gather, interpret, manage and present data; to analyse and critically evaluate options and understand concepts such as integrated approaches, good governance, accountability, economic viability, sustainability and conflict resolution.

Training and education are long-term processes that must be nurtured all the time. Lessons from past training and education activities must be incorporated into contemporary training to meet future demands. Partnerships forged with organisations and institutions within the region and globally can provide welcome opportunities in all areas of capacity building. Such partnerships should continue to expand and publicise useful and innovative methods that should be mainstreamed within the Pacific Islands and beyond.

Conclusion

The future of Pacific SIDS depends heavily on how well the issues examined above are addressed. Living in the world's largest ocean offers inherent challenges as well as opportunities that can only be realised if smart, innovative and painful decisions are taken. This will require that the PICs commit to work together to implement the plans of action they have agreed to and to continue to look for local solutions to their issues. The regional governments must take the lead, while securing the contributions of development agencies, NGOs and the private sector.

The collaborative work now adopted in the Pacific is logical for the national governments that do not have the capacity to have their own people attend to required jobs. While the PICs are helping each other, national governments must commit their own resources to address their issues themselves. Environment departments within some of these countries need the resources to conduct their activities diligently. The countries of the region need to guard their national interests and keep them aligned with international conventions they are party to. As more demands and higher expectations are required of environmental resources, relevant government agencies need to be strengthened with adequate resources and clearer mandates.

The challenge for Pacific SIDS is to appropriately address the needs of their people using the advice and support from its international partners. The countries must develop and strengthen their national capacity to ensure that the regional effort supported by the international community is taken through to local communities who are the owners and guardians of environmental resources. Development projects should be stringently assessed and evaluated, while funding should be provided only for those who have helped themselves. This would require good, transparent, accountable and just governance. Development projects must be related to national priorities and not those for which funds and assistance are available.

Fiji is now focusing on providing A Better Fiji For All and is using sustainable development to achieve its goal. The country aims to use effective resource management practices to unleash the development opportunities that will then benefit the people, the environment and the economy. This vision is now adopted after four decades of pursuing economic development that has delivered worsening poverty, degraded environmental resources and stunted economic growth. It is time to change and time for PICs to articulate the plans of action for sustainable development that have been there since 1994. Customary and community-centred conservation and contemporary, science-based and government-led resource management arrangements should be used to implement the plans. Sustainable development is the best and only available option for PICs.

References

ADB (Asian Development Bank) (1994) *A Study of the Fisheries Sector of Tuvalu*. Asian Development Bank, Manila.

ADB (Asian Development Bank) (2007) *Oceanic Voyagers: Shipping in the Pacific*. Asian Development Bank, Manila.

Anderson, E., B. Judson, S. Fotu, and B. Thaman (2003) *Marine Pollution Risk Assessment for the Pacific Islands Region (PACPOL Project RA1), Volume 1: Main Report for Marine Pollution Programme*. South Pacific Regional Environment Program (SPREP), Apia.

Ashe, J. W. (1999) Small Island Developing States and global climate change: overcoming the constraints. *Natural Resources Forum* 23 (3): 187–194.

AusAID (2008) *08 Pacific Economic Survey: Connecting the Region*. Pacific Affairs Group, Canberra.

Barcham, M., R. Scheyvens and J. Overton (2009) New Polynesian triangle: rethinking Polynesian migration and development in the Pacific. *Asia Pacific Viewpoint* 50 (3): 322–337.

Barnett, J. (2001) Adapting to climate change in Pacific island countries: the problem of uncertainty. *World Development* 29(6): 977–993.

Barnett, J. (2002) Environmental change and human security in Pacific island countries. *Development Bulletin* 58: 28–32, ANU, Canberra.

Barnett, J. and J. Campbell (2010) *Climate Change and Small Island States: Power, Knowledge and the South Pacific*. Earthscan, London.

Bayliss-Smith, T. P., H.C. Brookfield, R.D. Bedford and M. Latham (1988) *Islands, Islanders and the World: The Colonial and Post-Colonial Experience of Eastern Fiji*. Cambridge University Press, Cambridge.

Burson, B. (ed.) (2010) *Climate Change and Migration: South Pacific Perspectives*. Institute of Policy Studies, Victoria University, Wellington.

Cicin-Sain, B. (1993) Sustainable development and integrated coastal management. *Ocean and Coastal Management* 21 (1–3): 11–43.

Coates, B. (2009) Getting serious about achieving the Millennium Development Goals in the Pacific: strengthening economic development. *Policy Quarterly* 5 (3): 28–37.

Crocombe, R. and M. Meleisea (eds) (1988) *Pacific Universities, Achievements, Problems and Prospects*. Institute of Pacific Studies, University of the South Pacific, Suva.

Dalzell, P. and T. J. H. Adams (1995a) South Pacific Commission and Forum Fisheries Agency Workshop on the Management of South Pacific Inshore Fisheries, Manuscript collection of country statements and background papers Vol. I. Integrated Coastal Fisheries Management Project Technical Document No. 11. South Pacific Commission.

Dalzell, P. and T. J. H. Adams (1995b) South Pacific Commission and Forum Fisheries Agency Workshop on the Management of South Pacific Inshore Fisheries, Manuscript collection of country statements and background papers Vol. II. Integrated Coastal Fisheries Management Project Technical Document No. 11. South Pacific Commission.

Dalzell, P., T. J. H. Adams, and N. V. C. Polunin (1996) Coastal fisheries in the Pacific Islands. *Oceanography and Marine Biology: an Annual Review* 34: 395—531.

Elliott, M. and D. Fagan (2010) From community to Copenhagen: civil society action on climate change in the Pacific. In B. Burson (ed.) *Climate Change and Migration*. South Pacific Perspectives Institute of Policy Studies, Victoria University, Wellington, pp. 61–88.

FAO (2009) National Fishery Sector Overview – Fiji. Fishery and Aquaculture Country Profile. FAO. FID/CP/FJI: 1–13.

Fiji Times (26 October 2012) www.fijitimes.com/ (accessed 2013).

Gibson, J., and K. Nero (2007) Are the Pacific Island Economies Growth Failures? Geo-political assessments and perspectives. Report for the Pasifika Project, Institute of Policy Studies, Victoria University of Wellington, Wellington.

Gillett, R. (2011) *Fisheries of the Pacific Islands Regional and National Information*. RAP Publication 2011/03. FAO, Regional Office for Asia-Pacific, Bangkok.

Grigg, R. W. and C. Birkeland (1999) *Status of Coral Reefs in the Pacific*. Sea Grant, Hawaii.

Govan, H. (2009) Status and potential of locally-managed marine areas in the South Pacific: Meeting nature conservation and sustainable livelihood targets through wide-spread implementation of LMMAs. SPREP/WWF/WorldFish-Reefbase/CRISP. Online at www.crisponline.net/ Portals/1/PDF/0904 C3AGovanMMAs.pdf (accessed 24 April 2011).

Haberkorn, G. (2004) *Current Pacific Population Dynamics and Recent Trends*. SPC, Noumea.

Hamilton, C. (1997) Foundations of ecological economics. In M. Diesendorf and C. Hamilton (eds) *Human Ecology, Human Economy: Ideas for an Ecologically Sustainable Future*. Allen & Unwin, NSW, pp. 35–63.

Hasan, M. R. (1986) Hydrology of Rewa and Ba watersheds. Unpublished report, UNDP/FAO for Fiji Ministry of Primary Industries. 159 pp.

Hau'ofa, E. (1993) Our sea of islands. In E. Waddell, V. Naidu and E. Hau'ofa (eds) *A New Oceania: Rediscovering our Sea of Islands*. Beake House, Suva, pp. 2–16.

Hay J. E. (2002) *Climate Variability and Change and Sea Level Rise in the Pacific Islands Region: A Resource Book for Policy and Decision Makers, Educators and Other Stakeholders*. Tokyo, South Pacific Regional Environment Programme and Japan Ministry of the Environment.

Holland E, P. Nuttall, A. Newell, B. Prasad, J. Veitayaki, A. Bola, and J. Kaitu'u (2014) Connecting the dots: policy connections between Pacific Island shipping and global carbon dioxide and pollutant emission reduction. *Carbon Management* 5 (1): 93–105.

Maclellan, N., S. Meads, and B. Coates (2012) *Owning Adaptation in the Pacific: Strengthening Governance of Climate Adaptation Finance*. Oxfam, Auckland.

Merson, J. (2010) The Environment and the Economy. Radio Interview, RNZ Smart Talk series. Online at www.radionz.co.nz/podcasts/smarttalk.rssdoco-20100103–1600-Smart_Talk_for_3_January_2010_Going_Global-048.mp3 (accessed 20 April 2014).

Ministry of Strategic Planning, National Development and Statistics (2014) A Green Growth Framework for Fiji: Restoring the Balance in Development that is Sustainable for Our Future. Draft for consideration by the Prime Minister's Summit, 2 June 2014.

Molland, J. (2014) (March 1). Meet the Island Country that Turned itself into a Giant Marine Sanctuary. Online at www.care2.com/causes/meet-the-island-country-that-turned-itself-into-a-giant-marine-sanctuary.html#ixzz2v0oWrSMu (accessed 24 April 2014).

Morrison, R. J. (1981). *Factors Determining the Extent of Soil Erosion in Fiji*. Environmental Studies Report. No.7. INR. The University of the South Pacific, Suva, Fiji, 16pp.

Naleba, M. (2014) Disaster awareness need to close the divide, says Li-Col Seruiratu. *The Fiji Times*, 3 June, p. 2.

Nunn, P. D. (1990) Recent coastline changes and their implications for future changes in the Cook Islands, Fiji, Kiribati, Solomon Islands, Tonga, Tuvalu, Vanuatu and Western Samoa. In J. C. Pernetta and P. J. Hughes (eds) *Potential Impacts of Climate Change in the Pacific*. UNEP Regional Seas Reports and Studies, 128, pp. 149–160.

Nunn, P. D. (2007) Responding to the Challenges of Climate Change in the Pacific Islands: management and technological imperatives. Intergovernmental Panel on Climate Change (IPCC) Meeting, Nadi, Fiji.

Nunn, P. D. (2010) Bridging the Gulf between Science and Society. Sumi, A. Fukushi, K. and Hiramatsu, A. (eds) *Adaptation and Mitigation Strategies for Climate Change*. Springer, Tokyo, pp. 233–248.

Nuttall, P., A. Newell, B. Prasad, J. Veitayaki, and E. Holland (2014) A review of sustainable sea transport for Oceania: providing context for renewable energy shipping for the Pacific. *Marine Policy* 43: 283–287.

Prasad, B., J. Veitayaki, E. Holland, P. Nuttall, A. Newell, A. Bola, and J. Kaitu'u (2013) Sustainable sea transport research programme: toward a research-based programme of investigation for Oceania. *The Journal of Pacific Studies* 33: 78–95.

Pearce, D. W. and J. J. Warford, (1993) *World Without End: Economics, Environment and Sustainable Development*. World Bank, Oxford University Press, Oxford.

Puna, Hon. Henry (2012) *Large ocean Island states the pacific challenge*. Opening speech of the 43rd Pacific Island Forum by the Prime Minister of the Cook Islands, Hon. Henry Puna. Cook Islands Herald Weekly Issue 608: 21 March 2012.

Ram-Bidesi, V. (2011) Economics of Pacific Tuna Management. In P.N. Lal and P. Holland (eds) *Integrating Economics into Resource and Environmental Management: Some Recent Experiences in the Pacific*. IUCN, Gland, Switzerland in collaboration with the IUCN Regional Office for Oceania, Suva, Fiji and SOPAC, Suva, Fiji., Gland, Switzerland and Suva, Fiji, pp. 87–93.

Redding, S. and A. Venables (2004) Economic geography and international inequality. *Journal of International Economics* 62(1): 53–82.

Secretariat of the Pacific Community (2013a) SPC SDD Population Data Sheet. Online at www.spc.int/sdd/index.php/en/downloads/cat_view/28-pacific-data (accessed 22 April 2014).

Secretariat of the Pacific Community (2013b) SPC Pocket Statistical Summary. Online at www.spc.int/sdd/index.php/en/downloads/doc_download/737–2013-pocket-statistical-summary (accessed 22 April 2014).

Silaitoga, S. (2014) Villagers to move into new homes. *FijiTimesOnline*, 15 January. Online at http://www. fijitimes.com/story.aspx?id=256963 (accessed 22 April 2014).

South, G. R. and J. Veitayaki (1998) The constitution and indigenous fisheries management in Fiji. *Ocean Yearbook* 13: 452–466.

South, G. R. and J. Veitayaki (1999) Global initiatives in the South Pacific: regional approaches to workable arrangements. Asia Pacific School of Economic and Management Studies Online, http://ncdsnet.anu. edu.au (accessed 22 April 2014).

Togamana, C. (1995) Nutrient Transport via Sedimentary Processes in the Rewa Catchment. Thesis. The University of the South Pacific, Suva. 123 pp.

UNESCAP (United Nations Economic and Social Commission for Asia and the Pacific) (2010) *Sustainable Development in the Pacific: Progress and Challenges.* Pacific Regional Report for the 5-Year Review of the Mauritius Strategy for Further Implementation of the Barbados Programme of Action for Sustainable Development of SIDS (MSI+5), ESCAP Sub-regional Office for the Pacific, Suva.

USP Beat (2012) Kiribati Head of State Visits USP. 11(7): 7.

Veitayaki, J. (1994) The peaceful management of transboundary resources in the South Pacific. In G. H. Blake, J. Hildesley, M. A. Pratt, R. J. Ridley, C. H. Schofield (eds) *The Peaceful Management of Transboundary Resources.* Graham & Trotman/Martinus Nijhoff, London, pp. 491–506.

Veitayaki, J. (2005) Staking their claims: the management of marine resources in the exclusive economic zones of Pacific Islands. In S. A., Ebbin, A. H. Hoel and A. K. Sydes (eds) *A Sea Change: The Exclusive Economic Zone and Governance Institutions for Living Marine Resources.* Springer, Dordrecht, the Netherlands, pp. 150–168.

Veitayaki, J., D. A. R. Nakoro, T. Sigarua and N. Bulai (2011) On cultural factors and marine managed areas in Fiji in Pacific Island heritage: archaeology, identity and community. *Terra Australis* 35: 37–50.

Vunisea, A. (2002) Community-based marine resource management in Fiji: the challenges. *SPC Women in Fisheries Information Bulletin* 11: 6–9.

Wilkinson, C. (ed.) (1999) *Status of Coral Reefs of the World: 1998.* Australian Institute of Marine Science, Townsville.

Woodruff, A. (2007) *The Potential for Renewable Energy to Promote Sustainable Development in Pacific Island Countries,* SOPAC Miscellaneous Report 692, SOPAC, Suva.

37

POLAR OCEANS

Sovereignty and the contestation of territorial and resource rights

Klaus Dodds and Alan D. Hemmings

Introduction

The title of our chapter is deliberately provocative, and to the point. Whatever claims might be made for the relative remoteness or geographical inaccessibility of the polar oceans, the Arctic Ocean and Southern Ocean are today enmeshed in a series of tensions and contradictions that bring to the fore the uneasy coexistence of political–territorial colonization, resource exploitation, scientific research and the maintenance of managerial regimes, both global and regional in scope. At worst, we may well be witnessing the unfolding of a future for the polar oceans that is characterized by ever greater resource extraction (including living and non-living resources) and territorial competition for access to areas within contested coastal waters and open seas/deep waters. Such a scenario has attracted, repeatedly, media framings such as 'last resource frontier', 'land grab' and 'scramble for territory' (for example, Bert 2012 and more critically, Craciun 2009, Nuttall 2012). At best, the polar oceans (along with other oceanic spaces) might be managed proactively with due regard to existing managerial regimes such as the Convention on the Conservation of Living Marine Resources (CCAMLR) and the United Nations Law of the Sea Convention (UNCLOS), coupled with an appreciation for the so-called 'Lisbon Principles' regarding the sustainable governance of the oceans (Costanza *et al.* 1998).

Whatever the objective differences in the circumstances of the two polar regions, and despite the obvious differences in the detail of their legal–political arrangements, in one central detail the regions face a common geopolitical challenge. In both the Arctic and Antarctic regions, we see a subset of states asserting their rights on the basis of territorial sovereignty. In the Arctic, this stems from established sovereignty over the metropolitan territory of surrounding states (the so-called Arctic five), in the Antarctic from the assertion of rights by seven territorial claimants that are not generally recognized by the international community. In varying forms, the central polar contest is the claim that particular rights adhere to these territorial states versus claims to wider global rights there. What we detect, therefore, is a select group of coastal and claimant states eager to assert their territorial-sovereign rights at the expense of other states who wish to preserve open access (or arrangements that preserve *their* access), to shape resource regimes, to protect mobility/transit rights and, where possible, to contain the sovereign rights of coastal/claimant states by ensuring that the freedom of the high seas is not compromised.

The challenges facing the world's oceans, including the polar oceans, should not be under-estimated. As Costanza and colleagues noted some 17 years ago (at the time of writing in 2015), there are at least five major challenges confronting the global community, and these include resource extraction (with over-fishing being the biggest challenge), coastal ecosystem destruction, disposal and spillage, contamination and climate change (Costanza *et al.* 1998: 198). In the intervening period, there is little to suggest that their assessment was unduly pessimistic.

The Southern Ocean faces unprecedented challenges as changing technologies, emerging global markets, ongoing commercial development in areas such as tourism and biological prospecting escalate resource activity. Such an escalation will unfold in a context in which:

> The most immediate conservation threats to species, ecosystems, and resources around the Antarctic margin are consequences of regional warming, ocean acidification, and changes in sea-ice distribution. Marine resource extraction may exacerbate these threats. Current information suggests that toothfish and krill are particularly at risk into the future, but the full extent thereof is unclear due to the lack of comprehensive understanding of their life histories and spatial dynamics and difficulties in obtaining such information.
>
> *(Chown et al. 2012, p. 158)*

Steven Chown and colleagues pose the question as to whether the Antarctic Treaty System and its constituent legal instruments such as the Protocol on Environmental Protection are up to coping with such change.

Acknowledging the physical and environmental complexities of the polar oceans is vital because we want to insist upon the Arctic and Southern Oceans being resolutely experiential, material and representational. These oceans are not simply blank spaces on a world map that merely divide and connect the continental spaces of the planet. Our understanding of the polar oceans is derived in part from the stories that we tell about these spaces but also through engagement with it: the physical encounters with sea ice and sea currents (e.g. polar front zone), the circulating presence of flora and fauna (e.g. fish, whales and birds), the deployment of metaphors and analogies (e.g. the Arctic Ocean as a polar Mediterranean, the Southern Ocean as a last frontier), the historical legacies of exploration and exploitation (e.g. the search for oceanic passages) and the affective allure of vast spaces (e.g. the high Southern Ocean as a global commons).

In this chapter, we initially explore the Arctic Ocean and Southern Ocean separately in order to tease out some of the distinct geographical, legal and political dimensions to be found within each region. In these sections, we clarify our geographical terms of reference, while mindful that boundaries are best thought of as either blurred and/or fuzzy. After a concise analysis of both oceanic regions, our conclusion brings together those strands and returns to common problems and challenges facing the polar oceans. The subtitle is deliberate therefore because we end on a pessimistic note. A note that pays due emphasis to the intensification of what we would describe as profoundly territorial. An eagerness on the behalf of coastal/claimant states to maximize their rights over territory, ocean space and the resources therein, while other states and other actors either seek access to these regions of the world or utilize regimes such as the Arctic Council (AC) and Antarctic Treaty System (ATS) to internationalize governance and regulatory structures to create space within which their own interests may be argued and (as they hope) realized. While there is plenty of evidence of non-state actors operating in and with the polar oceans (including fishing companies, tour operators, environmental organizations, scientific organizations, and oil and gas corporations), we believe that understanding the contemporary behaviour of states and their relationship to territory and resources is essential.

The Arctic Ocean

The Arctic Ocean is usually considered to be the smallest of the world's oceans. The area under consideration is estimated to be around 14 million square kilometres and is defined by the International Hydrographic Organization (IHO) as protruding no further south than 60 degrees North (just beneath the southern tip of Greenland). It is an ocean bordered by the five coastal states of Canada, Greenland/Denmark, Norway, Russia and the United States (see Figure 37.1). In recent years, the Icelandic government has made considerable diplomatic efforts to be recognized as the sixth Arctic Ocean coastal state, largely in response to fears that the Arctic 5 might have been seeking to establish an informal inter-governmental forum to develop points of common interest (Dodds and Ingimundarson 2012). Geographically, however, there are a number of access points for shipping (and the movement of bio-physical currents and materials such as ice) including the Bering Sea/Strait, the Barents Sea, the Greenland Sea, the Norwegian Sea and the Labrador Sea (see Figure 37.1). As the distribution and thickness of sea ice alters, with particular regard to the possibility of regular ice-free summer seasons, so speculation has mounted that the Arctic Ocean will become ever more accessible to transit shipping, resource exploitation, and possible competition over control. The analogy of the 'Mediterranean' has been widely used to convey a sense of the Arctic Ocean as transit space, and a place where countries/actors/interests increasingly rub up against one another (Holmes 2012). For some commentators, the prospect of an ice-free Arctic Ocean is regarded as a strategic game-changer, comparable to the opening of the Panama Canal in 1914. While for others, greater emphasis is given to the Arctic Ocean as a resource frontier, as underwater territory and as a space to be 'saved' from exploitation.

In order to make sense of some of the changes affecting the Arctic Ocean, we briefly explore the role/interests of the coastal state and for the sake of convenience the role/interests of others including the international community. Even if this division is a touch contrived, the point is to highlight a tension between what one might think of as three pressure points – the territorial (coastal state), the international (third parties/non coastal states including environmental organizations) and the indigenous (northern communities with due recognition of internal/circumpolar diversities). Towards the end of this section, however, we will briefly conclude with an assessment of how the premier inter-governmental forum, the Arctic Council, is seeking to manage these pressure points. As with the Southern Ocean, it is probably best seen as very much a work in progress and one in which there is no guarantee of more pacific/collaborative/cooperative behaviour by interested parties.

The coastal state

The prospect of impending competition between coastal and third parties/non-coastal states manifests itself in a variety of ways. One starting point is to focus on the recently published Arctic strategies of the Arctic 5/6 and other 'Arctic states' such as Sweden and Finland (for one overview, Heininen 2012). Strikingly, in the 1990s, there were only two countries authoring Arctic strategies namely, Canada and Norway. The latest Arctic strategy was released by Sweden in 2011, and followed other Arctic countries in scoping their interests in the Arctic region and rehearsing their regional/geographical credentials including emphasis on the long-term presence of indigenous/Northern communities. While the underlying reasons and particular form of each strategy/strategies vary according to national/domestic considerations, they do share one common characteristic and that is to articulate a vision in which the Arctic coastal states advance their sovereign and security-related interests.

Figure 37.1 The Arctic and the Arctic Ocean

Source: CIA from: www.zonu.com/fullsize-en/2009–09–18–7442/Arctic-political-map-2001.html.

The Arctic Ocean, however, is a complex legal–geopolitical space, and those Arctic strategies are underpinned by a commitment of the Arctic coastal states to the principles of UNCLOS. This was emphatically affirmed in what was termed the 2008 Ilulissat Declaration. Signed in Greenland, the representatives of the five coastal states noted that:

> By virtue of their sovereignty, sovereign rights and jurisdiction in large areas of the Arctic Ocean the five coastal states are in a unique position to address these possibilities and challenges . . . Notably, the law of the sea [note the reference to law of the sea and not UNCLOS because of US non-ratification] provides for important rights and obligations concerning the delineation of the outer limits of the continental shelf, the protection of the marine environment, including ice-covered areas, freedom of navigation, marine scientific research, and other uses of the sea. We remain committed to this legal framework and to the orderly settlement of any possible overlapping claims.
>
> (Denmark 2008)

Under the terms of UNCLOS, the five Arctic Ocean coastal states have undisputed territorial seas (no more than 12 nautical miles from the baseline), an exclusive economic zone (EEZ, no more than 200 nautical miles from the baseline including water column, seabed and subsoil) and a continental shelf, comprising the seabed and subsoil of the submarine areas that extend beyond the territorial sea. The rights of the coastal state in the territorial sea include sovereignty over airspace and seabed and subsoil beneath the territorial sea. The coastal state also enjoys, in the EEZ, sovereign rights to explore, exploit, conserve and manage natural resources including fish, minerals and other commodities including oil and gas. The coastal state, in the context of the EEZ, is under the terms of UNCLOS permitted to undertake marine scientific research and establish protocols and regimes relating to the protection and preservation of the marine environment.

Where there has been a great deal of interest is in the area of the outer continental shelf. Those remote regions of the seabed where coastal states, under Articles 76 and 77 of UNCLOS, may submit materials to the UN-based Commission on the Limits of the Continental Shelf (CLCS) regarding a possible extension of sovereign rights. In order to do so, coastal states such as Denmark and Norway have invested heavily in the mapping, surveying and charting of various regions of the Arctic Ocean in order to substantiate their contention that there are identifiable submarine geological prolongations that extend beyond 200 nautical miles. Every coastal state enjoys, under the terms of UNCLOS, a potentially realizable continental shelf stretching 200 nautical miles from the coast. Under Article 76, there is the possibility of extending sovereign rights even further over the continental shelf regions. Depending on the technical criteria selected by coastal states, the outer limits of continental shelves might extend beyond 350 nautical miles if there is sufficient evidence relating to the topography and composition of the seabed.

Any relevant information is submitted to the CLCS and a panel of experts then evaluates and judges the submission of a coastal state. It is a lengthy and time-consuming process, and it is expensive. Millions of dollars, krone and roubles have been spent on collecting the required geological, bathymetric and geo-physical information in order to collect a so-called recommendation from the CLCS. The CLCS has no legal competence so the recommendation, in the case of the Arctic Ocean, requires coastal states to work together to secure final delimitation of the extended continental shelves.

At present the exact limits of the extended continental shelves of the coastal states are a work in progress. Norway has established its outer continental shelves, and did so following a recommendation from the CLCS in 2009, followed up by an agreed delimitation of the previously contested Barents Sea involving itself and Russia. Canada, Denmark and Russia are due to submit materials in 2013 or 2014 to the CLCS and the United States because it is not a party to UNCLOS cannot formally participate even though it has been collecting relevant materials via its Extended Continental Shelf project. Notwithstanding the eventual submissions of the other Arctic coastal states, the CLCS will have to adjudicate on the materials, and determine whether the continental shelf does extend all the way to the central Arctic Ocean.

There are two reasons why this time-consuming and expensive process matters. First, for the coastal states the delimitation of extended continental shelves is widely regarded as noteworthy both in sovereign-resource terms but also in more nationalistic/historic terms. Norway's so-called High North strategy is predicated on an assumption that the country will be able to use this geological–legal–cartographic process to plan and administer further exploitation of resources in the future. As the then Norwegian foreign minister, Jonas Gahr Store noted:

This establishes a clear division of responsibility and creates predictable conditions for activities in the High North. It confirms that Norway has substantial rights and responsibilities in maritime areas of some 235,000 square kilometres. The recommendation is therefore of historic significance for Norway . . . The recommendations provide a basis on which Norway can establish the limits of its continental shelf in the High North. This is a precondition for future resource management, creates a firmer basis for investments and is an effective implementation in the High North of the legal order for the oceans set out in the Law of the Sea Convention.

(Norway 2009)

Second, the delimitation of continental shelves in the Arctic Ocean will have implications not only for the international community but also for regional governance involving Arctic and non-Arctic actors (see below). One aspect worth noting is that coastal states have carried out joint mapping of the Arctic Ocean, recognizing explicitly that it is in their collective interest to work together to help establish their fullest sovereign rights.

Mapping and defining the underwater territory of the Arctic Ocean is the most important development affecting coastal states. There are some areas of controversy, however. This mapping exercise is not, and never has been, politically innocent. There are uncertainties regarding the central Arctic Ocean, and the Norwegian archipelago of Svalbard. In terms of the latter, Svalbard under the terms of the 1925 Spitzbergen Treaty is Norwegian territory but there could be arguments from other signatories over how Norway might exercise its sovereign rights over the continental shelf. The treaty governing Svalbard allows for other signatories such as Russia to exploit resources in an equal manner to any Norwegian individual or enterprise. But does the Treaty apply beyond 200 nautical miles from the Svalbard baseline? Norway believes that the continental shelf appurtenant to Svalbard is Norwegian territory and that it enjoys sovereign rights.

In the case of the central Arctic Ocean, the three main parties, Canada, Denmark and Russia will have to wait for their submissions to be examined by the CLCS and associated recommendations. Any submarine areas beyond national jurisdiction (i.e. beyond outer continental shelves of the Arctic Ocean coastal states) will in effect become 'the area' and any resources (with the exception of non-mineral resources) found on the seabed and/or sub-soil belong to the international community. The International Seabed Authority will act on behalf of the international community and any value derived from such resources is to be shared. It is likely that 'the area' in the context of the Arctic Ocean may well be physically modest, given the desire of the coastal states to extend, as far as legally possible, their sovereign rights to the seabed.

According to the US Geological Survey's assessment of undiscovered resource potential, there is every reason to believe that the vast majority of oil and gas potential lies in the undisputed exclusive economic zones of the coastal states.

Extra-territorial parties and non-coastal states

Under the terms of UNCLOS, non-coastal states have rights relating to navigation and access to the Arctic Ocean. Regardless of the size of 'the area', the global common is the high seas and the seabed and subsoil beyond national jurisdiction. Other states such as China, the UK and South Korea enjoy rights to innocent and transit passage. One area of controversy relates to the Northwestern Passage and the Northern Sea Route where Canada and Russia respectively believe that they have the right to impose rules and regulations on foreign vessels regarding transit passage. Other parties such as the United States believe that the Northwestern Passage is an international strait, and not part of Canadian 'internal waters' and do not believe, for example, that they have to seek the consent of the Canadian government before their registered vessels enter the Passage. While the introduction of a mandatory Polar Code might well help to clarify shipping and navigation standards, these maritime passages are becoming increasingly significant in terms of potential shipping routes and areas of interest regarding access and control.

The analogy of the 'polar Mediterranean' brings to the fore competing understandings of the Arctic Ocean. On the one hand, the idea of this body of water being a transit zone highlights mounting interest from extra-territorial parties in its potential to redirect commercial traffic from Europe and Asia. On the other hand, coastal states have been eager to assert their authority over territorial waters and exclusive economic zones. So the Arctic Ocean as a Mediterranean-like space highlights both potential to connect places and opportunities to divide coastal and non-coastal states. And perhaps one way to explain the exponential growth of interest in Arctic strategies and seabed mapping lies in this unease felt by the coastal states of mounting interest in the Arctic Ocean by other parties, including those located hundreds or even thousands of nautical miles away.

Another manifestation of this interest in a changing Arctic, from a proverbial frozen desert to a maritime region, lies in environmental movements such as Greenpeace launching their own campaigns to re-cast the Arctic Ocean as a sanctuary. As with the Southern Ocean, Greenpeace has argued that:

> The fragile Arctic is under threat from both climate change and oil drilling. As climate change melts the Arctic ice, oil companies are moving in to extract more of the fossil fuels that caused the melt in the first place. But above the Arctic Circle, freezing temperatures, a narrow drilling window and a remote location mean that an oil spill would be almost impossible to deal with. It's a catastrophe waiting to happen. Greenpeace is working to halt climate change and to stop this new oil rush at the top of the world.
>
> *(Greenpeace 2012)*

Here the idea of the Arctic being 'fragile' rather than 'accessible' is critical in mobilizing a different representation of this ocean. Coupled with fears about intensifying resource exploitation, Greenpeace and others are demanding that a moratorium be established so that the Arctic Ocean is not imperilled by the northern equivalent of a Deepwater Horizon disaster. One reason why the Arctic Council has addressed recently mandatory search and rescue regulations (2011) and oil spill response planning (2012) is in large part a response to this pressure to be seen to be acting on the possibility of the Arctic Ocean being imperilled by possible resource/shipping disasters. Both coastal states and other interested parties including indigenous peoples' representatives (Permanent Participants) and observers (such as China, the UK and South Korea) have recognized the need to develop common rules for these areas of common interest.

In that sense, such governance developments affecting the Arctic Council remind us that the Arctic Ocean is never reducible to simple analogy. It is a complex space composed of multiple marine environments and competing demands including indigenous fishing and hunting. The Arctic Ocean is a lived space, and reminds us that different actors have vested interests in high-lighting the connectivity and division within and beyond the region itself.

Indigenous

Finally, and briefly, it is important to acknowledge that 4 million people live north of the Arctic Circle. Over the last 30 years, indigenous peoples/northern communities have become increasingly active both within and beyond national territories. In 2009 a Circumpolar Inuit Declaration on Sovereignty in the Arctic was issued that repositioned the Arctic Ocean as integral:

> Inuit live in the Arctic. Inuit live in the vast, circumpolar region of land, sea and ice known as the Arctic. We depend on the marine and terrestrial plants and animals supported by the coastal zones of the Arctic Ocean, the tundra and the sea ice. The Arctic is our home.

Strikingly, in the Declaration there is rather more reference to the UN Declaration on the Rights of Indigenous Peoples and less on UNCLOS, as if to remind others including southern governments in Copenhagen, Moscow, Ottawa, Oslo, Washington DC that the Arctic Ocean needs to be considered to be part of a 'vast, circumpolar region of land, sea and ice' with Inuit being seen as 'partners' because 'Issues of sovereignty and sovereign rights in the Arctic have become inextricably linked to issues of self-determination in the Arctic. Inuit and Arctic states must, therefore, work together closely and constructively to chart the future of the Arctic'.

While the implications of such a Declaration are still to be understood, it stands as a challenge to those Arctic Ocean coastal states and their governments. The Arctic Ocean is changing and it is a maritime region that remains poorly understood notwithstanding decades of scientific research. The 2007–9 International Polar Year witnessed indigenous peoples working with scientists on various projects designed to better understand the varied marine and terrestrial environments. What the Declaration skilfully reflected upon was the desire of coastal states to further territorialize their interests (through ever greater sovereignty and security related discourses, performances and practices) while being less mindful of the rights and needs of indigenous peoples. As is well understood, the Arctic Ocean has long attracted its fair share of speculators, traders, financiers and sailors eager to cross icy passages, exploit the seas and oceans and establish trading centres.

The Southern Ocean

There are more variants on the area signified by 'Southern Ocean' than is the case with the Arctic Ocean. While all take the southern boundary as the coastline of the Antarctic continent, its northern limits may be taken as: the Subtropical Front at 40° South, effectively the southern coastline of Australia (e.g. AAD 2002); 60° South (a proposal of the International Hydrographic Organization – IHO 2002 – that is not in force); or the Antarctic Convergence or Polar Front (e.g. UNEP 2005). We here use the last, which has the advantage of applying to the area most commonly understood as 'Antarctic' in contemporary science, and is taken as the basis for the area of application of the Antarctic Treaty System's (ATS) primary marine instrument, the 1980 Convention on the Conservation of Antarctic Marine Living Resources (CCAMLR) (see Figure 37.2). All the other ATS instruments' areas of application fall within this area. This 'Southern Ocean' is thus sensibly bounded in scientific, geopolitical and legal terms. Extending over approximately 32 million km² (Croxall and Nicol 2004), the Southern Ocean is just under 10 per cent of the global marine area and more than twice the area of the Arctic Ocean.

At the heart of the geopolitical problem is the unresolved nature of territorial sovereignty on the Antarctic continent and surrounding islands south of 60° south latitude. In that part of the Southern Ocean north of 60°, around the sub-Antarctic islands, sovereignty and jurisdiction are not generally contentious (but note the Falklands/Malvinas and South Georgia as exceptions). But south of 60°, the familiar constructs of marine management elsewhere – the concept of the Coastal State, and its rights and duties in relation to, inter alia: territorial seas, exclusive

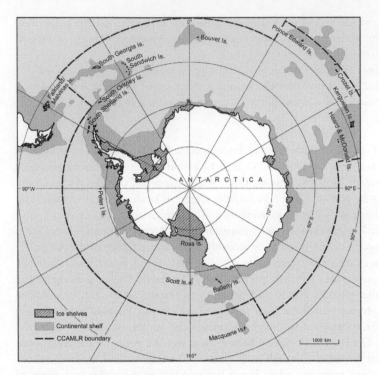

Figure 37.2 Antarctica and the Southern Ocean
Source: Alan D. Hemmings.

economic zones and the Continental Shelf *sensu* the Law of the Sea Convention (UNCLOS) – have not been readily applicable. But before turning to these complexities, we need first to note the profound transformations of the biophysical environment of the Southern Ocean due to climate change; and the current levels of human activity in the region.

Climate change research in Antarctica, now a major focus for the science project embedded from inception in the ATS, is not only a major contributor to understanding transformations at the global level, but has documented and projected significant effects across continental Antarctica and the Southern Ocean (Turner *et al.* 2009, 2014). For the latter these effects include, *inter alia*: ocean warming, marked summer salination, indications of a southward shift of the Antarctic Circumpolar Current, reductions in sea-ice extent in the Bellingshausen Sea and increases in the Ross Sea, and failure of the expected post-whaling increase of krill stocks which may be related to reduction of sea-ice in the Bellingshausen Sea (Turner *et al.* 2014: 247). Future projections show 'a minimum in sea-level rise in the Southern Ocean and a maximum in the Arctic Ocean', possibly some ocean circulation effects, and potentially significant changes in some (but not all) marine biota (*ibid*: 251–252). On present indications the near-medium term effects on navigation and available routes in the Southern Ocean, even in waters close to the Antarctic, are not yet as profound as those unfolding in the Arctic Ocean. In any case, geography and the very much lower levels of current and projected human activity in the far south, mean the circumstances around Antarctica are different from those affecting the Arctic. Nonetheless, climate change has proven a problematical political issue within the ATS, indicating a 'mismatch between "securing the Antarctic" (both continent and the Southern Ocean in terms of physicality and physical impacts) and "securing the Antarctic regime" and its core values' (Chaturvedi 2012: 283).

However, the Southern Ocean is already the scene of some substantial activities: scientific and logistics activity supporting not only marine research but research on the continent; tourism; biological prospecting; and marine harvesting. Each of these activities is not only affected by regional climate change effects, but has to be managed in the context of the jurisdictionally peculiar and complex circumstances of the Antarctic geopolitical arrangements.

Marine harvesting is primarily focused on krill and toothfish fishing under CCAMLR regulation (with some Illegal, Unreported and Unregulated (IUU) activity outside CCAMLR control) and the death throes (in every sense) of Special Permit whaling under the International Convention for the Regulation of Whaling (ICRW) and its International Whaling Commission (IWC). The total reported catch of krill in the 2010/11 season was 180,992 tonnes. While below the peaks of the 1980s and early 1990s prior to the collapse of the Soviet Union, when annual catches were often in the 300–400,000 tonne range, this catch has now been trending upwards for a decade (CCAMLR Commission 2012). With toothfish, the total reported catch in the 2010/11 season was 14,669 tonnes (CCAMLR Commission 2012), a level that seems to have been in slight annual decline over the past decade.

Putting a financial value on the overall sanctioned catch within the CCAMLR area (effectively the Southern Ocean as used here, minus the IUU catch value) is complex, but seems likely to be currently around US$ 640 million per annum.[1] Whaling presently sees Japan killing 400–900 whales a year (fewer in recent years due to protest activity), without generating any income and reportedly with an annual subsidy of around US$ 9.78 million (IFAW 2013). There is no reliable source for an economic value for biological prospecting in the Southern Ocean, which is presently only a small part of a wider (and similarly un-costed) biological prospecting activity across the Antarctic continent, generally conducted as part of national Antarctic programmes' ordinary scientific research activities. Antarctic tourism now has an estimated annual turnover of US$ 300 million per annum (MercoPress 2013). Alongside these figures, the cumulative expenditure for national Antarctic programmes appears to be around US$ 950 million per annum.[2] So, while these (ball-park) figures hardly bear comparison with economic activity anywhere else, they are not slight amounts. The Antarctic now has an economic dimension, likely above US$ 2 billion annually, of which a large (but presently imprecisely known) part attaches to Southern Ocean activities. This is a significant and growing factor in the geopolitical mix in the age of globalization in the far south (see Hemmings 2007, 2009a).

Antarctic territorial claimant states

Seven states (i.e. Australia, Argentina, Chile, France, New Zealand, Norway and the United Kingdom) claim Antarctic territory (Figure 37.3), and two 'semi-claimants' (Russia and the United States) reserve a basis of claim without presently asserting claims (Hemmings 2011: 14). Only five of the claimants cross-recognize each other's claims (on territorial claims generally see Hemmings 2012 and references therein). But, notwithstanding the apparent non-recognition of their claims by 188 UN Member States, all seven claimants see themselves as coastal states, and 'seeing' oneself as possessing such rights – whatever the actual and immediate constraints of geopolitics – may predispose one to seeking resource access (Hemmings 2009b). Consistent with their positions on territorial claims, parties to the Antarctic Treaty are able, thanks to its Article IV,[3] to hold different positions on the validity of this. But, for the 188, the usual position is that because of the special circumstances of Antarctica, the high seas extend to the coastline of the continent and there are no coastal states *sensu stricto*. On this basis, management of human activity in the part of the Southern Ocean within the Antarctic Treaty Area needs to be conducted collectively, through ATS instruments and norms, subject to those parts of UNCLOS that can

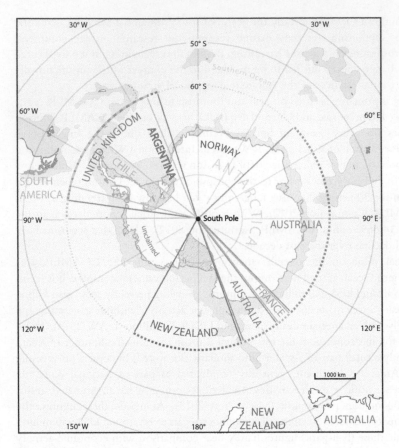

Figure 37.3 Territorial claims in Antarctica
Source: Alan D. Hemmings.

be applied there. But this creates a tension, and a sort of ongoing competition between internationalist (or at least Antarctic regime) and claimant interests. The latter seek constantly to maximize national benefit, to give effect to supposed coastal state rights, to the extent that this can be done without impairing the collective arrangements of the ATS. They cannot afford to undermine the ATS since, in various ways, it has sheltered their arguably weak territorial positions. Article IV in fact has allowed claimants to continually reflag their territorial (and consequential coastal state) pretensions since it also allowed other states to repudiate this, and thus provided a perennial safety valve for geopolitical heat in the region. The question is, whether this valve will continue to function, as resource access becomes both technically more feasible, and more attractive economically (for an overview of the challenges from diverse resource activities, see e.g. Joyner 2009).

Article IV has seemingly been no impediment to claimants asserting classical attributes of a territorial sovereign/coastal state in relation to the high Southern Ocean. These include territorial sea (New Zealand),[4] EEZ (Australia),[5] EEZ-like (Argentina and Chile),[6] or other constructs such as 'fisheries waters'. As Kraska (2011: 293) observes in relation to the Australian EEZ declaration, there is generally no attempt to apply these areas to foreign fishing vessels. The declarations are largely symbolic, but that is evidence of both the potency of the national sense of entitlement, and the enduring importance states attach to flag waving around Antarctic sovereignty – perhaps particularly in relation to their domestic constituencies (see e.g. Dodds and Hemmings 2009). The areas involved are often considerable. Australia, with one of the largest metropolitan EEZs at 8.2 million km^2 adds a further 2 million km^2 through proclaiming an EEZ off what it calls the Australian Antarctic Territory (Geosciences Australia 2010).

In that part of the Southern Ocean north of 60° South a necklace of sub-Antarctic islands deliver (aside the South Georgia and South Sandwich groups) uncontentious coastal state rights and substantial EEZs, including e.g.: 410,722 km^2 for Australia's Heard and McDonald group (Geosciences Australia 2010), 574,558 km^2 for France's Crozet group (Sea Around Us Project 2013a) and 473.380 km^2 for South Africa's Marion and Prince Edward group (Sea Around Us Project 2013b). The problematical South Georgia and South Sandwich groups under British jurisdiction but also claimed by Argentina generate an EEZ of 1,449,532 km^2 (Sea Around Us Project 2013c). Clearly the areas actually or potentially appropriable within EEZs (or other maritime zones) in the Southern Ocean are large and significant in relation to marine harvesting activity. Even where management of fisheries is internationalized, as it is under CCAMLR, sub-Antarctic coastal states exercise effective control in their EEZs and have *in extremis* a national opt-out clause for these areas under the so-called 'Chairman's Statement' appended to the CCAMLR instrument (CCAMLR Commission 1980).

Finally, in relation to territorial claimants, has been the issue of the extended continental shelf. As with EEZs, the dilemma here is that Antarctic territorial claimants see themselves as coastal states, to whom particular rights and duties in relation to the continental shelf attach, whereas the majority of states who reject the territorial claims deny that these states have that status at all. So, while it has proven impossible to assign continental shelf to states around the Antarctic coast of the Southern Ocean, the claimants have done their best to minimize the offence they suffer! Four options have presented themselves in relation to the 'need' to submit data to the Commission on the Limits of the Continental Shelf (CLCS) within ten years of UNCLOS entering into force for one's state (Hemmings and Stephens 2010: 314–316):

1   lodge the data for your Antarctic extended continental shelf without qualification – relying on the fact that while you will not have compromised your options, the rules of procedure of the CLCS (or active protests by other states) will ensure that your submission is not progressed, and thus contestation with other states avoided. Argentina has followed this route;

2   lodge the data, but yourself request CLCS not to process it for the time being. Australia and Norway have followed this route;

3   collect the data but, while telling everyone you have collected it don't submit it to the CLCS – but indicate that you might do so in the future. France, New Zealand and the United Kingdom have followed this route; and

4   submit only what is termed 'Preliminary information', thus reserving your position for the future (one of the three options above). Chile has followed this route.

While these approaches have contained the challenge of squaring the circle in relation to continental shelf issues in the Antarctic Treaty Area, it has not entirely avoided contention since a number of states (all of them Antarctic Treaty Consultative Parties) have felt it necessary to lodge notes with the UN Secretary General reiterating their repudiation of the territorial basis upon which the various CLCS submissions have been predicated (details in Hemmings and Stephens 2010). Again, analogously with the EEZs around sub-Antarctic islands north of the Antarctic Treaty Area (with, as expected, the exception again of those islands disputed between the UK and Argentina), the situation has been less contentious. This is ironic since the formally straightforward process in relation to two of these islands, Australia's Macquarie Island and the Heard and McDonald Islands, has resulted in CLCS sanction for extended continental shelf from these islands that actually extends into south of 60° South, and thus into the Antarctic Treaty. For the first time ever, we see a set of rights – those associated with activity on the continental shelf, including resource extraction – assigned to a particular state beyond the collective responsibility of the ATS (Hemmings and Stephens 2009: 288).

Non-claimant states

Confining ourselves to the 30 states (plus the EU) which have decision-making capacity at either ATCMs (28 states) or CCAMLR Commission (25 states and the EU), it is obvious that claimant states are in a significant minority, even if we include the two semi-claimants. However, the claimants/semi-claimant states remain the principal drivers of the geopolitical arrangements in the Southern Ocean. There are various factors enabling this: including domestic engagement and proximity for southern claimants, and their critical role in logistic support across all activities; relatively high expenditures on Antarctic programmes; consequential high contribution to scientific and technical activities; the enduring power of the United States; and the evident particular commitment that a sense of territorial entitlement engenders. Other non-claimant states are able to match this commitment – Germany, Japan, South Korea – but they are a small subset of the participating community. For the majority of other states, while their Antarctic research activities, or particular resource interests, may be pursued with first-rate capabilities, this is not matched by influence within the ATS, and that is the forum wherein the southern 'Great Game' is largely played out. For all the apparent anxiety (here as elsewhere) about the growing power of India and China (and what *they* are 'really' after), neither of these states is yet a really active influence in ATS fora, despite becoming Antarctic Treaty Consultative Parties in 1983 and 1985 respectively. Sanjay Chaturvedi has noted the gap that remains between their operational capacities and institutional and diplomatic projection (Chaturvedi 2013).

Overall, it presently appears that so long as the non-claimants are not disadvantaged vis-à-vis claimants in relation to such resource and other activities that are underway in the Southern Ocean, they are disinclined to expend effort on any direct challenge to the territorial and associated geopolitical verities of the regional arrangements. The difficulty may be that so central are the interests of the claimants – or at least the *containment* of the territorial problem – to the ATS, that a challenge to the modalities of the claimant position is seen as inevitably a challenge to the architecture of the ATS itself. Whatever the immediate indifference or reluctance, there must be some reasonable expectation that the present governance arrangements around the Southern Ocean will be reassessed in the foreseeable future. The diversification and intensification of resource activities, the increasing economic value of these activities, the increasing number of players *within* the ATS, and (one presumes) at some point a reinvigoration of a wider inter-national debate about rights, access and benefit sharing involving state and interests not presently represented, will surely require some rejustification or rejigging of the regional arrangements (Hemmings 2013). The Southern Ocean may be a way down some policy makers list of priorities, but some significant justice and equity issues remain to be resolved in this large part of our common planet.

Concluding observations

Our chapter has drawn attention to the fact that the two polar oceans are complex spaces. There are some similarities that are worth highlighting – territorial and coastal states are active in terms of cementing their sovereign rights and interests in both spaces. Both regions have experienced, and continue to experience, resource-related pressures and accompanying discourses stressing sovereignty, security and stewardship. For those coastal/claimant states, one key driver remains – anxieties about extra-territorial powers, especially China and India. In both the Arctic and Antarctic Oceans there is no shortage of commentary warning about nefarious planning and plotting regarding mineral resources and strategic potential. As a consequence, the governance of these spaces remains a work in progress that will no doubt have to take into account global legal, geopolitical and even socio-cultural contexts. These are demanding regions.

But it is also important to note that there are important differences. The Antarctic's activity levels are less advanced than the Arctic, which is an ocean surrounded by inhabited landmass with a great deal of the seabed and water column under some sort of jurisdictional control by coastal states. Public interest is arguably higher in the Arctic and deeply connected to climate change, energy security and indigenous peoples and their rights. But these things may well change and over the last hundred years there has been no shortage of commentators imaging radically different futures for both oceans and nearby coastlines.

Notes

1 This is based on a doubling of the calculated wholesale price per tonne for the total commercial catch in 2009/10. ADH is grateful to Professor Denzil Miller, formerly CCAMLR Executive Secretary for providing the value estimates upon which this calculation is based.
2 The total for the 22 (of 28) Antarctic Treaty Consultative Parties whose national Antarctic programmes have lodged figures on the website of the Council of Managers of National Antarctic Programs converts to approximately US$ 905,450,000 (COMNAP 2012). The six remaining states seem likely to take the overall figure beyond US$ 950 million.
3 Article IV asserts, inter alia, that parties are not renouncing any previously asserted right or claim, renouncing or diminishing any basis of claim, or prejudicing their position on recognition or non-recognition of rights, claims or basis of claim.
4 See Rothwell 2002: 117–118.
5 Via its Maritime Legislation Amendment Act 1994, protested by the US: see US Department of State 2000: 9.
6 Bush 1982 at pages 72–73 and 448–450 respectively; CIA 2013.

References

AAD (Australian Antarctic Division). 2002. What is the Southern Ocean? *Australian Antarctic Magazine*, Issue 4. Available at www.antarctica.gov.au/about-us/publications/australian-antarctic-magazine/2001–2005/issue-4-spring-2002/what-is-the-southern-ocean (accessed 25 May 2013).

Bert, M. 2012. The Arctic is now: Economic and national security in the last frontier. *American Foreign Policy Interests: The Journal of the National Committee on American Foreign Policy* 34: 5–19.

Bush, W. M. 1982. *Antarctica and International Law: A Collection of Inter-State and National Documents*, Volume II. London: Oceana Publications.

CCAMLR Commission (Commission for the Conservation of Antarctic Marine Living Resources). 1980. *Statement by the Chairman of the Conference on the Conservation of Antarctic Marine Living Resources*. Available at www.ccamlr.org/en/organisation/camlr-convention-text#Chair (accessed 28 May 2013).

CCAMLR Commission (Commission for the Conservation of Antarctic Marine Living Resources). 2012. *Report of the Thirty-First Meeting of the Commission*. Hobart: Commission for the Conservation of Antarctic Marine Living Resources.

Chaturvedi, S. 2012. The Antarctic 'climate security' dilemma and the future of Antarctic governance. In Alan D. Hemmings, Donald R. Rothwell and Karen N. Scott (eds), *Antarctic Security in the Twenty-First Century: Legal and Policy Perspectives*: 257–283. Abingdon: Routledge.

Chaturvedi, S. 2013. India and Antarctica: Towards post-colonial engagement?. In Anne-Marie Brady (ed.), *The Emerging Politics of Antarctica*: 50–74. Abingdon: Routledge.

CIA (US Central Intelligence Agency). 2013. *The World FactBook: Antarctica*. Available at www.cia.gov/library/publications/the-world-factbook/geos/countrytemplate_ay.html (accessed 27 May 2013).

Chown, S. L., Lee, J. E., Hughes, K. A., Barnes, J., Barrett, P. J., Bergstrom, D. M., Convey, P., Cowan, D. A., Crosbie, K., Dyer, G., Frenot, Y., Grant, M., Herr, D., Kennicutt II, M. C., Lamers, M., Murray, A., Possingham, H. P., Reid, K., Riddle, M. J., Ryan, P. G., Sanson, L., Shaw, J. D., Sparrow, M. D., Summerhayes, C. Terauds, A. and Wall, D. H. 2012. Challenges to the future conservation of the Antarctic. *Science* 337: 158–159.

COMNAP (Council of Managers of National Antarctic Programs). 2012. Our Members. Available at www.comnap.aq/Members/SitePages/Home.aspx (accessed 27 May 2013).

Costanza, R., Andrade, F., Antunes, P., van den Belt, M., Boersma, D., Boesch, D. F., Catarino, F., Hanna, S., Limburg, K., Low, B., Molitor, M., Pereira, J. G., Rayner, S., Santos, R., Wilson, J. and Young, M. 1998. Principles for sustainable governance of the oceans. *Science* 281: 198–199.

Craciun, A. 2009. The scramble for the Arctic. *Interventions* 11: 103–114.

Croxall, J. P. and Nicol, S.. 2004. Management of Southern Ocean fisheries: global forces and future sustainability. *Antarctic Science* 16: 569–584.

Denmark. 2008. *The Ilulissat Declaration*. Arctic Ocean conference, Ilulissat Greenland 27–29 May. Available at: www.oceanlaw.org/downloads/arctic/Ilulissat_Declaration.pdf (accessed 1 December 2014).

Dodds, K. and Hemmings, A. D. 2009. Frontier Vigilantism? Australia and contemporary representations of Australian Antarctic Territory. *Australian Journal of Politics and History* 55(4): 513–529.

Dodds, K. and Ingimundarson, V. 2012. Territorial nationalism and Arctic geopolitics: Iceland as an Arctic coastal state. *The Polar Journal* 2: 21–37.

Geosciences Australia. 2010. Oceans and Seas: Australian Exclusive Economic Zone. Available at www.ga.gov.au/education/geoscience-basics/dimensions/oceans-and-seas.html (accessed 26 May 13).

Greenpeace. 2012. Save the Arctic. Available at: www.greenpeace.org.uk/arctic (accessed 1 December 2014).

Heininen, L. 2012. State of the Arctic Strategies and Policies – A summary. In Lassi Heininen, Heather Exner-Pirot and Joel Plouffe (eds), *Arctic Yearbook 2012*. Available at www.arcticyearbook.com/images/Articles_2012/Heininen_State_of_the_Arctic_Strategies_and_Policies.pdf (accessed 1 December 2014).

Hemmings, A. D. 2007. Globalisation's cold genius and the ending of Antarctic isolation. In Lorne K. Kriwoken, Julia Jabour and Alan D. Hemmings (eds), *Looking South: Australia's Antarctic Agenda*: 176–190. Sydney: Federation Press.

Hemmings, A. D. 2009a. From the new geopolitics of resources to nanotechnology: emerging challenges of globalism in Antarctica. *The Yearbook of Polar Law* 1: 55–72.

Hemmings, A. D. 2009b. Beyond claims: towards a non-territorial Antarctic security prism for Australia and New Zealand. *The New Zealand Yearbook of International Law* 2008 6: 77–91.

Hemmings, A. D. 2011. Why did we get an International Space Station before an International Antarctic Station? *The Polar Journal* 1(1): 5–16.

Hemmings, A. D. 2012. Security beyond claims. In Alan D. Hemmings, Donald R. Rothwell and Karen N. Scott (eds), *Antarctic Security in the Twenty-First Century: Legal and Policy Perspectives*: 70–94. Abingdon: Routledge.

Hemmings, A. D. 2013. Rights, expectations and global equity: re-justifying the Antarctic Treaty System for the 21st century. In Richard Powell and Klaus Dodds (eds), *Polar Geopolitics: Knowledges, Resources and Legal Regimes*. Cheltenham and Northampton MA: Edward Elgar.

Hemmings, A. D. and Stephens, T. 2009. Reconciling regional and global dispensations: the implications of subAntarctic extended continental shelf penetration of the Antarctic Treaty area. *New Zealand Yearbook of International Law* 6: 273–291.

Hemmings, A. D. and Stephens, T. 2010. The extended continental shelves of sub-Antarctic Islands: implications for Antarctic governance. *Polar Record* 46(4): 312–327.

Holmes, J. 2012. 'Open seas' Foreign Policy blog 29 October 2012. Available at: www.foreignpolicy.com/articles/2012/10/29/open_seas (accessed 1 December 2014).

IFAW (International Fund for Animal Welfare). 2013. *The Economics of Japanese Whaling: A Collapsing Industry Burdens Taxpayers*. Yarmouthport MA: IFAW and available at www.ifaw.org/sites/default/files/economics-of-japanese-whaling-japan-ifaw.pdf (accessed 27 May 2013).

IHO (International Hydrographic Organization). 2002. Chapter 10: Southern Ocean and its Sub-Divisions. In *Names and Limits of Oceans and Seas*. Special Publication No. 23, 4th Edition, June 2002.

Available at www.iho.int/mtg_docs/com_wg/S-23WG/S-23WG_Misc/Draft_2002/S-23_Draft_2002_SOUTHERN_OCEAN.doc (accessed 21 May 2013).

Joyner, C. C. 2009. Challenges to the Antarctic Treaty: Looking back to see ahead. *New Zealand Yearbook of International Law* 6: 25–62.

Kraska, J. 2011. *Maritime Power and the Law of the Sea: Expeditionary Operations in World Politics*. New York: Oxford University Press.

MercoPress. 2013. Antarctica received an estimated 35,000 visitors this 2012/12 season says IAATO. April 30th 2013. Available at http://en.mercopress.com/2013/04/30/antarctica-received-an-estimated-35.000-visitors-this-2012–13-season-says-iaato (accessed 27 May 2013).

Norway. 2009. Extent of Norway's continental shelf in the High North clarified, 15 April 2009. Available at www.regjeringen.no/en/dep/ud/press/news/2009/shelf_clarified.html?id=554718 (accessed 1 December 2014).

Nuttall, M. 2012. Introduction: politics, science and environment in the polar regions. *The Polar Journal* 2: 1–6.

Rothwell, D. R. 2002. The Law of the Sea and the Antarctic Treaty System: rougher seas ahead for the Southern Ocean? In Julia Jabour-Green, and Marcus Haward (eds), *The Antarctic: Past, Present and Future*: 113–125. Hobart: Antarctic CRC Research Report #28.

Sea Around Us Project. 2013a. EEZ waters of Crozet Isle (France). The Pew Charitable Trusts. Available at www.seaaroundus.org/eez/896.aspx (accessed 28 May 2013).

Sea Around Us Project. 2013b. EEZ waters of Prince Edward Isle (South Africa). The Pew Charitable Trusts. Available at www.seaaroundus.org/eez/711.aspx (accessed 28 May 2013).

Sea Around Us Project. 2013c. EEZ waters of South Georgia and Sandwich Isle (UK). The Pew Charitable Trusts. Available at www.seaaroundus.org/eez/239.aspx (accessed 28 May 2013).

Turner, J., Bindschadler, R., Convey, P., di Prisco, G., Fahrbach, E., Gutt, J., Hodgson, D., Mayewski, P. and Sumerhayes, C. (eds). 2009. *Antarctic Climate Change and the Environment*. Cambridge: Scientific Committee on Antarctic Research.

Turner, J., Barrand, N. E., Bracegirdle, T. J., Convey, P., Hodgson, D. A., Jarvis, M., Jenkins, A., Marshall, G., Meredith, M. P., Roscoe, H., Shanklin, J., French, J., Goosse, H., Guglielmin, M., Gutt, J., Jacobs, S., Kennicutt II, M. C., Masson-Delmotte, V., Mayewski, P., Navarro, F., Robinson, S., Scambos, T., Sparrow, M., Summerhayes, C., Speer K. and Klepikov, A. 2014. Antarctic climate change and the environment: an update. *Polar Record* 50: 237–259.

UNEP (United Nations Environment Program). 2005. The Antarctic Convergence. UNEP/GRID-Arendal. Available at www.grida.no/graphicslib/detail/the-antarctic-convergence_26e0 (accessed 22 May 2013).

US Department of State. 2000. *National Claims to Maritime Jurisdiction: Limits in the Seas*, No. 36 (8th Revision). Washington DC: Department of State.

38

THE WORLD OCEAN AND
THE HUMAN FUTURE

Hance D. Smith, Juan Luis Suárez de Vivero and
Tundi S. Agardy

Introduction

In hypothesising future relationships between people and the sea, this chapter returns to the major themes discussed in Chapter 1, namely, economic and technological development as defined in terms of both historical and geographical contexts; and governance and social change, including the cultural worlds of the sea; both reconsidered with change in the marine environment itself.

The stages of development

The 50–60 year stages of the Kondratieff cycle illustrate the development of ocean resources and management. The late twentieth century stage gradually petered out in the 1990s, and the world is now embarked upon the early twenty-first century stage, currently undergoing profound change, among other things, in the patterns of use and management of the marine environment.

The sea use groups most closely aligned to the broad pattern of economic development are the service industries, including commercial shipping and ports; naval industries; waste disposal; the marine leisure industries; marine science and education; and marine conservation. These are subject in varying degrees to shorter timescale cycles, but on a rising trend (Ho-Chun Fang *et al.* 2013). There is no reason at present to doubt that long-run economic expansion will not run its course until around 2030, with subsequent decline and restructuring of economic activity continuing until at least 2050. As well as increasing overall levels of economic development worldwide, the world will not pass through the global demographic transition until at least the beginning of the second half of the twenty-first century, so that world population will continue to increase, albeit at a decelerating rate overall, expanding from the present level of just over seven billion to a figure in the order of 9.6 billion by the middle of the twenty-first century (United Nations 2013).

In the development of marine mineral and energy resources, long timescales of decision making are fundamental. The exploitation of offshore hydrocarbons remains pivotal (Odell 2004), and the lifetimes of offshore fields, power stations and renewable technologies are of a similar order to the long wave stages. The production lifetimes of fields currently in use have some considerable way to run, despite short-term fluctuations in the price of hydrocarbons. Investment decisions in deep water production have already been taken that will have implications beyond the middle of the present century. The balance between oil and coal on the one hand, and natural gas on the other has begun to swing in favour of the latter, not only for electricity generation, but also for sea transport. The development of marine renewables, however, remains more problematical, as these bear high costs relative at least to current offshore hydrocarbon production and utilisation. Only offshore wind is substantially developed in the core regions of north-west Europe and the north-east United States. Tidal power may be the next technological development, but it remains at an early stage, and suitable locations are very limited. Wave power is similarly at a very early stage as far as commercial development is concerned, although a much larger number of locations could be used. Meanwhile the global demand for energy is set to rise for the foreseeable future, in this case way beyond the next Kondratieff cycle, to the end of the century at least.

The exploitation of the living resources of the sea has been surrounded by prognostications of gloom and doom more or less continuously since the industrial revolution, and especially during the late twentieth century stage of development. In the first decade of the twenty-first century an influential scientific paper predicted the total collapse of global commercial fish resources by 2048 – notably around the end of the present stage – given the then levels of exploitation (Worm *et al.* 2006). Although heavy fishing pressure on the top predators in marine ecosystems continues apace (Christensen *et al.* 2014) in the manner of that of the whale and seal populations of earlier stages, there is mounting evidence in the core regions of Western Europe and North America that many commercial fish populations are now stabilising in balance with fishing (Cardinale *et al.* 2013; Oremus *et al.* 2014). However, the scourge of IUU fishing in distant water fisheries remains a potent threat, especially for the open ocean beyond national jurisdiction, and in the EEZs of the least developed world, notably off African coasts and probably parts of the Pacific and Indian Oceans.

Thus by the end of the present stage it may be that global resources of wild fish will be sustainable to a limited degree, probably only adjacent to the developed state systems of governance in the northern hemisphere. Elsewhere, the pressures of continually increasing populations, inadequate governance systems and incomplete scientific knowledge present a bleak outlook for sub-tropical and tropical oceans both within and beyond state jurisdiction. Large Marine Ecosystems will have been fundamentally altered everywhere. Meanwhile the present technologies of mariculture remain crucially dependent upon the secure supply of wild fish used as feed and, barring development feedstock from non-marine sources, will also be limited, especially as wild fish are also used in agricultural production, thus further increasing the pressure on local trophic level fisheries. However, demand for fish as food will still be running ahead of supply under the impetus of rising population and living standards. The inescapable conclusion is that the role of fish in the human diet will decline markedly, even if fully effective fisheries management were to extend throughout the world ocean.

While it is reasonable to surmise that present trends can be coupled with the timescales of decision making to predict the uses of the sea, it is particularly difficult to foresee the next large-scale restructuring of the global economy. What is possible is to foresee continually increasing demand for food, mineral and energy resources – not only until the global population stabilises, but also beyond that until that population reaches a stage in which all world regions may be said to be fully developed. Until then the pressure on the world ocean can only increase.

This in turn raises the question of the future of the stages themselves. Kondratieff cycles have been associated with the processes of economic development throughout modern European history, across a period of half a millennium (Korotayev and Tsirel 2010). However, global industrialisation, including sea use industrialisation is not yet complete. If and when it is, it may be that the intensity of these waves will decline, perhaps in the twenty-second century.

The worlds of the sea

The circumstance of the possible decline of Kondratieff long waves as industrialisation comes to an end draws attention to the even longer stages in human history. Here the focus shifts from economic development, which is only one element in these long stages, to fundamental dimensions of human society and the human mind, including especially systems such as states that can safeguard the public good; it is possible to envisage the emergence of world views capable of reconciling the inherent conflicts in the human situation, ranging from the world's great religions to philosophies based on humanism and scientific understanding, to world views based on those of pre-industrial societies that remain remarkably widespread, despite the forces of globalisation. All of these social forces have the potential to re-order human societies in a variety of ways that are very long term in nature, extending far beyond an individual human lifespan (Galtung *et al.* 1979; Devezas and Corredine 2001).

More prosaically, the transitions between Kondratieff stages can put immense pressure on governance and human societies, including much greater risks of war, and imposition of fundamentalist political and religious philosophies. In this, it is not really possible to foresee beyond the present stage: indeed it is probably even more difficult to forecast social and political change than the economic–technology interactions lying at the heart of these stages.

The present transition and those before it can be seen as associated with crises within the state and of governance, including ocean resource management. Thus within the West or Occidental culture region noted in Chapter 1, and indeed also in the East Asian culture region the state is economically overextended, needing cutbacks in its functions that may legitimately apply to marine affairs. This is happening, even despite the desire to extend state jurisdictions even further seawards. This has the effect of creating problems with the management of living resources and conflicts among sea uses. These stresses have also sharpened the political pressures between mixed economies on the one hand, and state controlled economies on the other. Meanwhile, in the economically more intermediate developing culture regions of Southeast Asia and South Asia, democratic forms of governance are similarly set against state control, although less so, with the economic resources of the state available for governance being much more constrained. What in the West would be regarded as corruption is thus inevitably widespread in the public sphere: state services are more likely to be paid for directly, rather than through taxation. Finally, the tribal societies of the world's great deserts, rain forests and in some island nations are still struggling to adopt any really meaningful state system, with even more severe limitations on the economic and human resources required for governance in general, and maritime governance in particular. In summary, the future of the Westphalian state system itself in ocean governance may be questioned, at least beyond its European heartland (Kissinger 2014).

Transcending these difficulties is the growing importance of partnerships among states, the private sector, and civil society groups in maritime governance, especially relating to the most challenging problems posed by fisheries management, marine spatial planning, coastal zone management and marine conservation. These efforts are much more fully developed in the Western culture region, and in tropical regions under varying degrees of Western influence. Such an approach ultimately questions the traditional role of the state in maritime governance, despite the prominence of states in the provision of governance as espoused in the Law of the Sea Convention. A further important development related to the foregoing is the continuing strengthening of regional approaches in fisheries management and marine environmental management, including marine conservation. These are most evident within the Occidental culture region in the northern hemisphere, but signs of this are emerging elsewhere as well.

There is every prospect that partnership approaches will continue to strengthen in the first half of the twenty-first century, not only in mixed economy states, but even also in centrally controlled and least developed states.

The world ocean

And what of the marine environment itself? The most important physical changes concern global and regional temperature changes in the coupled ocean–atmosphere system, and global and regional changes in sea level. In the history of the evolution of the world ocean, human existence has been very short indeed – only the later part of the Pleistocene ice age is concerned, and within that especially the later interglacials. As far as natural temperature changes are concerned, the most striking aspect on the timescale of the Pleistocene is that it is only 6,000 years since global sea level stabilised at its present level, and the current ice coverage on land and at sea more or less stabilised with it, all of which points to yet another interglacial.

In the context of industrialisation of the world ocean, which is a major driver in contemporary ocean resources use and management, the most notable climate fluctuation affecting both land and sea has been the pronounced cooling termed the 'Little Ice Age' which developed at the end of the European Middle Ages half a millennium ago and reached a low point in the mid-seventeenth century, only to disappear around the middle of the nineteenth century (Fagan 2000). By far the strongest signature of the Little Ice Age appears to be found in the North Atlantic region (Dawson 2009), together with the adjacent Arctic Ocean region which may well have been accentuated by the geographical configuration of land and sea relative to both oceanic and atmospheric circulation. The outcome included pronounced deterioration in weather and climate, including cooling, a prolonged and severe storminess, sufficient to cause catastrophic failures in harvests on land and in fisheries at sea, but not enough to seriously disrupt the long-stage economic development of the North Atlantic region at the time.

The warming of the global atmosphere appears to have accelerated from the industrial revolution stage onwards, but especially during the late twentieth century stage to the present (IPCC 2013–2014). Again the signature appears strongest in the North Atlantic and Arctic Ocean regions. This late warming has been laid at the door of human agency, but is worth noting that, although there is limited agriculture now taking place along parts of the coast of south-west Greenland, it is doubtful that this region is as warm as it must have been in the warmer European Early Middle Ages, when Norse settlers colonised western Greenland, a colony that lasted until the beginning of the Little Ice Age climatic deterioration (Fitzhugh and Ward 2000). Although sea ice cover in winter around Iceland has largely disappeared, it is worth remembering that as late as 1968 British trawlers were lost at Iceland due to icing of superstructures, masts and rigging (Holland-Martin 1969). Meanwhile, most recently the Arctic Ocean has rapidly lost much of its ice cover in summer. If this continues, during the early twenty-first century stage there will be profound consequences for global shipping through the maintenance of much shorter sea routes all year round connecting the core regions on both sides of the North Atlantic with those on both sides of the North and Central Pacific.

Climate change that includes accelerated warming of the ocean, and attendant sea level rise and ocean acidification has profound consequences for ocean life. Fish are very sensitive to small changes in temperature, and commercial fish stock distributions are changing on very short timescales of just a few years, well within the timescales of the economic changes (Perry *et al.* 2005). Again, there are notably strong signatures in the North Atlantic. A striking contemporary example is the North East Atlantic mackerel stock, which has moved further northwards in summer, leading to conflicts among the leading fishing states exploiting the stock, and the necessity to renegotiate national quotas. Climate change is sending many reef ecosystems towards a phase shift that will result in lowered productivity, biodiversity, and ecosystem services values – putting at risk the millions of people depending on reefs for livelihood and human well-being (Sale *et al.* 2014). Meanwhile, warming is also associated with increasing ocean acidification, which places at greatest risk ocean life dependent on carbonate.

In the case of sea level changes, regional short- and medium-term changes often overshadow what in human terms is a very slowly rising global sea level. Notable regional changes on the post-glacial timescale are geological in nature. Especially evident are tectonic movements associated with sea level rise or fall: in the Mediterranean such rising sea levels have submerged some of the ports of classical times. In regions formerly covered by ice sheets, isostatic compensation has resulted in both falling and rising sea levels due to flexing of the Earth's crust. One of the best examples is in the British Isles: on the east coast of Scotland there are impressive raised beaches and buried shorelines where isostatic compensation has been most rapid – the raised beaches were important sites for the industrial revolution stage of establishment of fishing villages (Gray 1978); in the extreme north – the Northern Isles; and far south – southern coasts of Ireland and England – drowned ria inlets are characteristic, some of which are very important harbours – Cobh, Falmouth and Plymouth, for example.

On the very short timescales of ocean–atmosphere changes, prolonged storminess destabilises vulnerable 'soft' coasts, including exposed sandy beaches and backing dunes; and low-lying estuaries subject to simultaneous flooding by rivers and the sea. The Little Ice Age was extensively associated with both, for example being largely responsible for the current geomorphological expression of extensive sand dune systems around the coasts of the British Isles. Two singular types of events are especially notable. The first are storm surges, where temporary substantial rises in sea levels under storm conditions cause widespread flooding of low-lying coasts: one of the most notable in recent times was the 1953 storm surge in the southern North Sea that caused enormous damage and loss of life in Holland and eastern England; and those associated with tropical hurricanes and typhoons, which have similar effects on an even greater scale, notably in the low-lying Sunderbans of the Ganges Delta in the Bay of Bengal. Similar flooding and inundation occurs in diverse regions from south-eastern Florida to the Maldives, causing economic impacts, crisis-response coastal engineering and much worry. The second are tsunami, which are associated with tectonically unstable ocean ridges and trenches subject to earthquakes that trigger these waves. Tsunami can move a long way inland: two of the most recent examples are the Indian Ocean tsunami in late 2004 that resulted in around one quarter of a million casualties; and the more recent one in the Sendai region of Honshu in Japan that resulted in tens of thousands of casualties and the destruction of a nuclear power station, causing substantial pollution from radiation.

Global or eustatic sea level rise is much slower, and insidious. However, even on IPCC predictions until the end of the twenty-first century, it is already within the purview of coastal defence decision making, which can extend from 100 to 200 years into the future. In the wake of the 1953 disaster of the breaching of the dykes in Holland by the storm surge, extensive coastal defence measures, including engineered defences such as the Thames Barrier, and managed retreat are characteristic of urban industrial coasts in the developed world, although these may well be insufficient to counter the higher estimates of sea level rise within their design lifetimes, extending to perhaps an early twenty-second century stage of economic development (Pilkey and Cooper 2014). Apart from these urban industrial core regions, regions most at risk include rural barrier island/sandy beach coasts such as those characteristic of the eastern and southern United States, southern North Sea and southern Baltic; mangrove coasts and large deltas in developing tropical and sub-tropical regions; and oceanic islands with coral reef coasts –

especially those consisting of coral atolls, which are especially characteristic of the tropical Pacific and Indian Oceans and the Caribbean Sea islands and adjacent coasts.

Human use has restructured food webs worldwide. At sea, this is in part due to physical disturbance, including bottom trawling, dredging and hydrocarbon and sea bed mineral extraction. On land and at the coast there is channelisation inland, sediment starvation in estuaries, which are a crucial component of the ocean ecosystem, caused by freshwater diversion, coastal construction and port development. These physical impacts in turn influence the stability, productivity and complexity of marine food webs that in turn affect the provision of goods and services important to humans. Also of great importance is marine and coastal pollution – the Millennium Ecosystem Assessment identified coastal pollution as being the second greatest issue affecting ecosystem services after habitat destruction.

In summary, physical changes can be considerable, and are only partly predictable on the developmentally significant Kondratieff timescale. Further, the natural world still points to an interglacial, whereas the implied impact of human activity will be to bring not only the Pleistocene ice age to an abrupt end, but also result in a mass extinction of species and resulting fundamental changes to ecosystems on both land and sea. However, such changes arguably lie beyond the timescales at which human development can be foreseen with even limited accuracy.

Conclusion

As the early twenty-first century development stage comes to a close, it is likely that substantial progress may be made towards a lower carbon energy economy. It is also within the bounds of possibility that management of marine ecosystems will stabilise the supply of at least some commercial fish stocks, provided practicable systems of law and management can be devised for Areas Beyond National Jurisdiction together with extended coastal state jurisdiction and improved management, especially for states in the sub-tropical and tropical regions. Also important will be the effective control of marine pollution and meaningful regulation of mineral and energy resource extraction. Yet even with these developments in place, the world ocean will still be under greater pressure than it is at present.

However, there remain two major caveats or complications.. The first concerns events with the potential to break down the economic–technological nexus of the development stages. Two of these risks, and possibly the greatest near-term risk overall are human: some kind of social revolution that might be caused by a major dislocation of the Westphalian state system; and major war conducted at such a scale as to cause this kind of disintegration of the present order. This might be a nuclear conflagration: it is worth noting that the lynchpin of nuclear war remains the nuclear powered and armed submarine, which remains practically undetectable.

The other risks are primarily within the natural events category, including those entwined with human activities. The first of these is the possibility or likelihood of one or more destabilising global pandemics, the probability of which is closely related to human activity at least in some foreseeable cases. Another might be a global collapse in fisheries (Jacques 2015). Lesser risks include a massive volcanic eruption on the scale of Tambora in 1816 (Wood 2014), or larger; an asteroid strike, or perhaps being caught in the tail of a large comet.

However, almost certainly more important than these risks is the development of human thinking in the field of ocean resources and management. There are in the order of twenty phrases in the English language alone that are used to specify the kind of management that is needed, not only for the world ocean, but for the planet as a whole. At present by far the most important of these is the notion of sustainable development, which has been largely coined in association with the present stage transition. It should be especially emphasised that the idea of 'development' is still an aspiration, come what may in terms of environment on the one hand, and society on the other. A much more sophisticated understanding of development applied to management of the world's ocean resources – and the planet as a whole – is required (Wallerstein 2004; Babones and Chase-Dunn 2012).

References

Babones Salvatore, J., Chase-Dunn, C. (2012) *Routledge Handbook of World Systems Analysis*. Abingdon, Routledge.

Cardinale, M., Dorner, A. A., Andersen, J. L., Casey, J., Doring, D., Kirkegaard, E., Stransky, C. (2013) Rebuilding EU fish stocks and fisheries: a process under way? *Marine Policy* 39, 43–52.

Christensen, W., Coll, M., Piroddi, C., Steenbeck, J., Buszowski, J., Pauly, D. (2014) A century of fish biomass decline in the ocean. *Marine Ecology Progress Series* 512, 155–166.

Dawson, A. (2009) *So Foil and Fair a Day: A History of Scotland's Weather and Climate*. Edinburgh, Birlinn.

Devezas, T., Corredine, J. (2001) The biological determinants of long wave behaviour in socioeconomic growth and development. *Technological Forecasting and Social Change* 68, 1–57.

Fagan, B. (2000) *The Little Ice Age: How Climate Made History, 1300–1850*. New York, Basic Books.

Fitzhugh, W. W., Ward, E. I. (2000) *Vikings: The North Atlantic Saga*. Washington DC, Smithsonian Institution Books with American Museum of Natural History.

Galtung, J., Rudeng, E., Heistad, T. (1979) On the last 2,500 years in Western history, and some remarks on the coming 500. In Burke P. (ed.) *The New Cambridge Modern History, 13: Companion Volume*. Cambridge, Cambridge University Press, pp. 318–361.

Gray, M. (1978) *The Fishing Industries of Scotland, 1790–1914: A Study in Regional Adaptation*. Oxford, Oxford University Press.

Ho-Chun Fang, I., Cheng, F., Incecik, A., Carnie, P. (2013) *Global Marine Trends 2030*. London, Lloyd's Register.

Holland-Martin, D. (1969) *Trawler Safety: Final Report of the Committee of Inquiry into Trawler Safety*. Cmnd 4114. London, HMSO.

Intergovernmental Panel on Climate Change (IPCC) (2013–2014) *Fifth Assessment Report*. Geneva, IPCC.

Jacques, P. (2015) Are world fisheries a global panarchy? *Marine Policy* 53, 165–170.

Korotayev, A. V., Tsirel, S. V. (2010) A spectral analysis of world GDP dynamics: Kondratieff waves, Kuznets swings, Juglar and Kitchin cycles in global economic development, and the 2008–2009 economic crisis. *Structure and Dynamics* 4, 1–57.

Kissinger, H. (2014) *World Order: Reflections on the Character of Nations and the Course of History*. New York, Penguin.

Odell, P. R. (2004) *Why Carbon Fuels Will Dominate the 21st Century's Global Energy Economy*. Brentwood, Multi-science Publishing.

Oremus, K. L., Suatoni, L., Sewell, B. (2014) The requirement to build US fish stocks: is it working? *Marine Policy* 47, 71–75.

Perry, A. L., Low, P. J., Ellis, J. R., Reynolds, J. D. (2005) Climate change and distribution shifts in marine fishes. *Science* 308(5730) 1912–1915.

Pilkey, O. H., Cooper, J. A. G. (2014) *The Last Beach*. Durham, NC, Duke University Press.

Sale, P., Agardy, T., Ainsworth, C. H., Feist, B. H., Bell, J. D., Christie, P., Hoegh-Guldberg, O., Mumby, P. J., Feary, D. A., Saunders, M. I., Daw, T. M., Foale, S. J., Levin, P. S., Lindeman, K. C., Lorenzen, K., Pomeroy, R. S., Allison, E. H., Bradbury, R. H., Corrin, J., Edwards, A. J., Obura, D. O., Sadovy de Mitcheson, Y. J., Smoilys, M. A., Sheppard, C. R. C. (2014) Transforming management of tropical coastal seas to cope with challenges of the 21st century. *Marine Pollution Bulletin* 85, 8–23.

United Nations Population Division. Department of Economic and Social Affairs (2013) *World Population Prospects: The 2012 Revision*. New York, United Nations.

Wallerstein, I. (2004) *World Systems Analysis: An Introduction*. Durham, NC, Duke University Press.

Wood, G. D'Arcy (2014) *Tambora: The Eruption that Changed the World*. Princeton, NJ, Princeton University Press.

Worm, B., Barbier, E. B., Beaumont, N., Duffy, J. E., Folke, C. Halpern, B. S., Watson, R. (2006) Impacts of biodiversity loss on ocean ecosystem services. *Science* 314(5800), 787–790.

INDEX

UK Spelling used in this index

1992 Rio Declaration 61, 64
2008 Ilulissat Declaration 579

ABNJ *see* areas beyond national jurisdiction
abyssal ocean hazards 357–358
AC *see* Arctic Council
accidents, offshore oil/gas 277
acidification 104–105, 111, 114–115
ADP *see* Areas Designated for Protection
Africa 261, 541–559
African Union 496
Agardy, Tundi S. 5–13, 283–295, 476–492,
    592–598
Agenda 21 61, 64, 162–163
aggregate mineral deposits 297–299
agriculture 7, 12, 34, 39–40, 182–183
Agulhas and Somali Current Large Marine
    Ecosystem (ASCLME) 546–558
air-sea-nutrient interfaces 89–107
air travel 9–10
Alegret, Juan-Luis 408–421
Alliance of Small Island States (AOSIS) 563
altered nutrient fluxes 102–105
AMI *see* "Area of Mutual Interest"
AMU *see* Arab Maghreb Union
anchoring ships 353–354
ancillary oil/gas technology 276
Angelidis, Michael O. 381–395
Anglo Saxon theories 21
Antarctic regions 234–238, 576–577, 583–589
Antarctic Treaty System (ATS) 37, 58, 577,
    583–588
Anthropocene 90, 95, 97, 99–106, 148
antifouling paints 386
AOSIS *see* Alliance of Small Island States

APC *see* Areas of Particular Concern
aquaculture 111, 118, 255–265
Arab League 496
Arab Maghreb Union (AMU) 496
archeological heritage conservation 409–410
Archipelagos, East Asian Seas 524–527
Arctic Council (AC) 577
Arctic Ocean 576–584, 589
Ardron, Jeff A. 55–72
"Area of Mutual Interest" (AMI) 520–521
areas beyond national jurisdiction (ABNJ) 58–69,
    73, 78–81, 83–84
Areas Designated for Protection (ADP) 520
Areas of Particular Concern (APC) 520
Aricò, Salvatore 310–328
ASCLME *see* Agulhas and Somali Current Large
    Marine Ecosystem
ASEAN *see* Association of Southeast Asian
    Nations
Asia 259, 524–537
assertiveness, conservation challenges 150
Association of Southeast Asian Nations (ASEAN)
    531–532, 536
Athens Charter 413–414
atmosphere 8, 11, 35, 89–107, 595–596
ATS *see* Antarctic Treaty System
autotrophy 93

bacteria, ocean health 117
Bahrain/Saudi Arabian boundaries 428
Baker, Elaine 199–211
Bangladesh/Myanmar case 444–446
Barbados Programme of Action (BPOA) 560,
    568
Barcelona Charter 414
Barcelona Convention 392–393, 495
Bartley, Monique LaFrance 507–523

Basel Convention 390–391
Basque ship 418
Bateman, Sam 524–537
bathymetric data 471–472
BBNJ Working Group 65–66
BCLME *see* Benguela Current Large Marine
    Ecosystem
Beaudoin, Yannick 199–211
Belgium/France boundaries 436–438
Belgium/Netherlands boundaries 438
Benguela Current Large Marine Ecosystem
    (BCLME) 545–546, 548, 554–557
benign application/tasks 367
Bernal, Patricio A. 33–54
biodegradable organic matter 381–382
biodiversity: civil society groups 68–69;
    conservation 55–69, 145–154; ecosystem
    governance 73–74, 77–78; environment
    conservation 147; fisheries 55–69; genetic
    resources 310–311, 313–316, 321; global
    governance 35, 41–42, 55–74, 78–79; green
    economies 204; integrated management
    163–164, 167; marine protected areas 480,
    483–485; marine spatial planning 487–489;
    protection/governance 35, 41–42, 55–72;
    scientific community 65–69; soft laws 64–65;
    South Pacific SIDS 561, 571; State ocean
    strategies/policies 35, 41–42; strategic
    environmental assessments 191–194
biofuels 288
biogeochemical cycling 89–107
biological productivity 89–107
biological pumps 96–97
Black Sea 499–500
Bodin, Örjan 232–240
BOEM *see* Bureau of Ocean Energy
    Management
boundary delimitation 425–446, 495, 499–501,
    533–534, 580–581
BPOA *see* Barbados Programme of Action
Brazil 17–19, 22–23, 26–28
BRIC countries 17–19, 22–23, 26–28
Brundtland Commission 188, 195
bulk shipping sectors 334–336
Bureau of Ocean Energy Management (BOEM)
    510
burial, cables 356–357
Burnett, Douglas R. 349–365

cables, telecommunications 349–361
Cameroon/Nigeria boundaries 441–443
Cannard, Toni 188–198
Capacity Building (CB) 229, 467–468, 571–572
carbon dioxide 91–104, 116
Carbonell, Eliseu 408–421
carbon sequestration 181
carbon transport 92–106

Carleton, Chris M. 425–448
Carter, Lionel 349–365
Cavaliere, Christina 199–211
CB *see* Capacity Building
CBD *see* Convention on Biological Diversity
CCAMLR *see* Commission for the Conservation
    of Antarctic Marine Living Resources
Chambers, Robert 542
Channel Islands/UK boundaries 438
charts, surveying 462–475
chemicals, waste disposal 382–387
China: East China Sea 112, 526, 529–530, 533,
    536; geopolitical scenarios 17–19, 21–23,
    26–28; South China Sea 270, 276, 377, 470,
    526, 530–536
Chown, Steven 577
Chuenpagdee, Ratana 241–254
circulation, global ocean 96–97
CITES *see* Convention on International Trade in
    Endangered Species
civil society groups 12, 45–46, 68–69
claimant states, polar oceans 585–588
classifying waste/pollution 381–388
CLCS *see* Commission on the Limits of the
    Continental Shelf
climate change 595–596: ecosystem global
    governance 73, 79–81; modulation 89–107;
    ocean health 112–116; ocean-nutrient-
    atmosphere linkages 89–107; shipping
    344–345; Southern Ocean 584; South Pacific
    SIDS 565–566, 569–570; subsea
    telecommunications 359; warming 104–105,
    111, 113–116, 595–596
CMS *see* Convention on Migratory Species
coastal ecosystem services 178–181, 480–482
coastal hazards 357–358
coastal leisure/tourism 396–407
coastal policies 541–559
coastal states, Arctic Ocean 578–581
coastal zone management: Africa 541–557;
    conservation 146; marine protected areas
    477–485; renewable energy 290; science &
    policy 157, 165–167; strategic environmental
    assessments 191–197; US MSP 508–522
Coastal Zone Management Act (CZMA) 290,
    508–509, 520, 522
Codes of Conduct 39–40, 55–56
Cold War 21, 28, 34–37, 45, 52
collaboration, IUU fishing 237–238
Commission for Sustainable Development (CSD)
    65
Commission for the Conservation of Antarctic
    Marine Living Resources (CCAMLR)
    234–238
Commission on Ocean Policy 509
Commission on the Limits of the Continental
    Shelf (CLCS) 27, 535, 580–581, 587–588

commitment, conservation 150–151
'commodification' 152
commodities, shipping 331–334
common heritage of mankind principle 449–459
communications 349–365
community assertiveness, conservation 150
community-based projects 255, 261–265, 541–559
Compliance Agreements 39
computer technology 320, 432
conceptualization, IUU fishing 237
Conference of the Parties (COP) 62, 193, 313–315
conservation: biodiversity 55–69, 145–154; challenges 149–154; cultural heritage 408–421; ecosystems 145–154; environment management 145–156; ethical values 149; floating heritage 409, 412–417, 419; human activities/impacts 145–154; marine protected areas 479–480; maritime heritage 408–421; underwater cultural heritage 409–413, 415–418; world ocean development 10–11
constabulary application/tasks 367, 369, 371, 373, 375–378
constituency aspects 160
continental margin mineral deposits 297–299
continental shelves: Arctic Ocean 580–581, 587–588; boundary delimitation 426–445; East Asian Seas 527, 533–535; geopolitical scenarios 19–20, 23–24; polar oceans 580–581, 587–588; State ocean strategies/policies 35–36, 43–44
Convention on Biological Diversity (CBD): ecosystem governance 74, 77–78; genetic resources 310–311, 313–316, 321; integrated management 163–164, 167; marine spatial planning 487–489; protection/governance 61–62, 64–65, 67–69; State ocean strategies/policies 41–42
Convention for the Conservation of Antarctic Marine Living Resources (CCAMLR): biodiversity 58–59; ecosystem global governance 76–77, 81, 83; polar oceans 576, 583, 585, 587–589; tourism 404
Convention for the Protection of World Cultural and Natural Heritage 415–417
Convention on International Trade in Endangered Species (CITES) 42, 56–59, 63, 404–405
Convention on Migratory Species (CMS) 56, 58, 63, 74, 78–80, 83
Convention on the Safety of Life at Sea 34–35, 40–41, 341, 472–474
cooperation obligations, East Asian Seas 530–532
COP *see* Conference of the Parties
corporations, genetic resources 322
del Corral, David Florido 17–32

cradle construction, shipping 337–338
cruise industry, tourism 397, 399–401
crustal–ocean–atmosphere factory 89–99
CSD *see* Commission for Sustainable Development
culture: cultural heritage conservation 408–421; South Pacific SIDS 566–567; world oceans human past and present 5–13, 594
current concerns/status: fisheries 217–224; hydrographic surveys 469–470; Marine Scientific Research 134–135; mineral deposits 306; renewable energy devices 284–286; US MSP 511–514
CZMA *see* Coastal Zone Management Act

data exchange/sharing 470
data management 322–323
decision making 119, 158–160
decision trees 119
Declaration on the Conduct of Parties 531–532
deep-ocean mineral deposits 296–297, 299–302, 305–308
deep seabed 449–461: genetic resources 451–459; legal issues 449–461
deep seas, State ocean strategies/policies 33–36, 39, 45, 47, 51
delimitation, boundaries 425–446, 495, 499–501, 533–534, 580–581
design, marine protected areas 479–482
developing country aquaculture 255, 261–265
development: offshore oil/gas 269–272; world oceans 5–13
diesel power 10
directives: MSFD 110, 118; science & policy 160–165
disaster vulnerability 565–566, 569–570
discounting 177
disease 109, 116–122
disposing of waste 381–395
Distance Learning and Information Sharing Tool (DLIST) 546–548, 555, 558
distant water fishing nations (DWFN) 528–529
diversity *see* biodiversity
Division for Ocean Affairs and the Law of the Sea (DOALOS) 136, 139, 315
DLIST *see* Distance Learning and Information Sharing Tool
DOALAS *see* Division for Ocean Affairs and the Law of the Sea
Dodds, Klaus 576–591
domestic norms/standards 49–50
Duarnenez, France floating heritage 419
Dumping Convention, London 146
DWFN *see* distant water fishing nations

'Earth Summits' 64–67, 162–163
East Asian seas 524–537, 594

East China Sea 112, 526, 529–530, 533, 536
East Sea (Sea of Japan) 526, 529–530, 533, 536
EBSA *see* Ecologically or Biologically Significant Areas
ECDIS *see* Electronic Chart Display and Information System
ecological impact assessments 207–208
ecological services 183–184
Ecologically or Biologically Significant Areas (EBSA) 42, 44, 62, 67–68, 78–81, 84, 147
economics: Africa 547–556; coastal ecosystem services 178–181; conservation challenges 151–153; decision making 159; ecosystems 176–187; fisheries 216–217, 227–228; genetic resources 312; green economies 199–211; offshore oil/gas 273, 278–279; planning 159; science & policy 159–161; shipping 334–335; South Pacific SIDS 566–567; world ocean development 7–13
ecosystems: biodiversity 73–74, 77–78; climate change 73, 79–81; conservation 145–154; economic values 176–187; environment management 176–187; fisheries 241, 244–245; global governance 10–11, 48–52, 73–85; governance 10–11, 48–52, 73–85; green economies 200–208; health 241, 244–245; marine protected areas 480–483; migratory species 74, 78–80, 83; Millennium Ecosystem Assessment 157, 164–165; nutrient-ocean linkages 100; ocean health 108–122; policy 48–52, 73, 157–173; science 157–173; services 157, 164–165, 176–187, 200–208, 480–482; social values 176–187; State ocean strategies/policies 48–52; strategic environmental assessments 188–197; world ocean development 10–11
EEZ *see* exclusive economic zones
EIA *see* Environmental Impact Assessments
Electronic Chart Display and Information System (ECDIS) 463, 466
Electronic Navigational Charts 463, 466
Elefant, Carolyn 283–295
emerging country geopolitical scenarios 17
emerging maritime heritage conservation 408–412
emissions 201
enclosed seas 526
endocrine disruptors 385
energy: development stages 592–593; East Asian Seas 528–530; offshore oil/gas 47, 53, 76, 157, 159, 179–182, 269–282, 528–530, 593–594; renewable energy 205–206, 283–295, 518–521, 570, 593; South Pacific SIDS 566, 570; US MSP 510, 518–521
enforcement, State ocean strategies/policies 46
engineering industry, offshore oil/gas 271

Environmental Impact Assessments (EIA) 62, 188–191, 207–208, 516
environmental impacts: mineral recovery 307–308; offshore oil/gas 76, 179–182, 276–278; pollution/waste disposal 381–395; renewable energy 290–292; subsea telecommunications 354–357
environment management: conservation 145–156; ecosystem services valuation 176–187; green economies 199–211; policy 157–175; science and policy 157–175; sea-power 371–373; shipping 339; strategic assessments 188–198; tourism 401–404; world ocean development 10–12
equidistance line principle 427–433, 437–446
Espai del Peix of Palamós, Spain 420
ethical values, conservation 149
EU *see* European Union
Europe, aquaculture 260
Europe/Mediterranean boundaries 495–507
European Commission 192–193
European Union (EU): geopolitical scenarios 28–30; Marine Strategy Framework Directive 110, 118; Mediterranean boundaries 495–496, 499–504; Naval Force 28; strategic environmental assessments 192–193; waste disposal 390; world ocean development 11
Europeanisation, Mediterranean region 495, 501–505
eutrophication 382
evolution: maritime heritage conservation 408–412; nautical charts 462–463
exchange systems: data exchange 470; gas exchange 89–107; genetic resources 322–323
exclusive economic zones (EEZ): biodiversity 59–60; boundary delimitation 433–434; East Asian Seas 524–533; ecosystem governance 75–77, 79–80; fisheries 232–235, 237; geopolitical scenarios 19, 23–24, 26; Mediterranean boundaries 499–501; mineral deposits 306; polar oceans 580, 586–587; science & policy integration 161, 171; Southern Ocean 586–587; South Pacific SIDS 563; State ocean strategies/policies 33, 35, 37, 44, 46, 51; strategic environmental assessments 192; tourism 404; US MSP 507–508, 511
exploration: hydrographic surveys 471; minerals 296, 298, 302, 306–307; offshore oil/gas 271–281
extra-territorial parties 581–582

Fadón, Fernando Fernández 17–32
Faeroe Islands/UK boundaries 439–440
FAO *see* Food and Agricultural Organization
fast carbon cycles 92, 94–96

Federal Energy Regulatory Commission (FERC) 289–290
federal funding 138
federal roles/responsibilities, US MSP 507–523
feedbacks 91, 104, 480–482
FERC *see* Federal Energy Regulatory Commission
Ferretti, Francesco 108–126
ferromanganese crusts 296, 299–302, 305–308
fibre-optic cables 350–351, 353–361
finance 138, 271, 292–293, 483–484
finfish 255–261
fish species 528–529
fisheries 215–231: Africa 541–542, 546, 552–555; aquaculture 255–261; biodiversity 55–69; boundary delimitation 439–441, 501–503; capacity building 229; cultural heritage conservation 412; current concerns 217–224; development stages 593; East Asian Seas 528–529; economics 216–217, 227–228; ecosystems 74–84, 241, 244–245; environment conservation 146; Europe/Mediterranean boundaries 501–503; exclusive economic zones 232–235, 237; food security 215–217, 228, 241, 247; governance 232–238, 241–254; human future 594; human well-being 215–217; illegal, unregulated, unreported fishing 228–229, 232–240, 375; industry status 223–224; international trade status 224; landings 217, 219–221; livelihood security 241, 246–247; nutrition 215–216; ocean health 111–121; policy frameworks 233–238; quotas 227–228; rights-based management 226–229, 233; sea-power 375; security 215–217, 227–228, 241, 246–247; social aspects 215–217, 227–228, 241, 245–246; Southern Ocean 584; South Pacific SIDS 560–561, 564–565, 567–569, 571; species composition 218–219; state ocean strategies/policies 35, 38–39, 43, 48–52; stock status 221–222; subsea telecommunications 353; sustainable development 224–229, 247–249; technology 233; world ocean development 8–13
fishing: ecosystem services valuation 182–183; green economies 201; rights 425–427; *see also* IUU fishing
flag state jurisdiction 375
fleet capacity 228
floating heritage conservation 409, 412–417, 419
fluxes, ocean–climate–atmosphere–nutrient linkages 89–94, 98–105
folklore 415
Food and Agricultural Organization (FAO): biodiversity 55–59, 69; ecosystem governance 76, 78; State ocean strategies/policies 34, 39–40, 43, 51–52

food security 215–217, 228, 241, 247
fossil fuel dependence 566, 570
France/Belgium boundaries 436–438
France/UK Channel boundaries 430–432, 434–436, 438–439
freedom of the high seas 452–453
freight rates 336–337
freights, shipping 331–334, 336–337
Fretheim, Atle 47
funding 138, 271, 292–293, 483–484
future prospects: deep seabed organisation 454–457; genetic resources 323–325; US MSP 521–522

gas exchange 89–107: pumps 97–98
gases: nutrient-atmosphere-climate linkages 89–107; *see also* offshore oil/gas
GDP *see* gross domestic product
GEF *see* Global Environment Facility
genetic resources 310–328, 451–459
Geneva Convention 426–429, 431
geo-economics 18–19
Geographic Location Designation (GLD) 520–521
geography of the sea: Africa 541–559; boundaries 425–448, 495–506; deep seabed 449–461; East Asian Seas 524–537; human future 592–598; marine protected areas 476–492; marine spatial planning 476–477, 485–490, 507–523; maritime boundaries 425–448, 495–506; Pacific Ocean 560–575; polar oceans 576–591; regional developments 495–537, 541–598; Small Island Developing States 560–575; spatial organisation 425–492; state maritime boundaries 425–448; surveys 462–475; US MSP 507–523; world oceans 592–598
"Geological Carbon Cycle" 92
geopolitical scenarios: BRIC countries 17–19, 23, 26–28; change factors 22–30; global governance 17–32; maritime scenarios 17–30; Mediterranean boundaries 497–498; naval doctrines 20–24, 28, 31; UNCLOS 18–20, 23–30; world ocean governance globalization 17–32
GES *see* good environmental status
GESAMP *see* Group of Experts on the Scientific Aspects of Marine Environmental Protection
GHG *see* greenhouse gases
GLD *see* Geographic Location Designation
global aquaculture production 256–258
global climate change 89–107
Global Environment Facility (GEF) 77, 543–557
global fisheries 215–231
global genetic resource regulations 320–321

global governance: biodiversity 35, 41–42, 55–74, 78–79; ecosystems 10–11, 48–52, 73–85; genetic resources 320–321; geopolitical scenarios 17–32; Open Ocean policies 33–54; state ocean strategies/policies 33–54; world oceans 10–13, 17–85
global ocean circulation 96–97
Global Ocean Observing System (GOOS) 66–67
global positioning systems (GPS) 401, 464–466
global renewable energy markets 288–289
global warming 104–105, 111, 113–116, 595–596
good environmental status (GES) 110
GOOS *see* Global Ocean Observing System
Gorina-Ysern, Montserrat 127–142
governance: biodiversity 35, 41–42, 55–74, 78–79; biodiversity protection 55–72; constituency 160; ecosystems 10–11, 48–52, 73–85; Europe/Mediterranean boundaries 501–505; fisheries 232–238, 241–254; genetic resources 320–321; geopolitical scenarios 17–32; law 160; marine protected areas 483; Marine Scientific Research 135–139; Mediterranean boundaries 501–505; offshore oil/gas 279–280; Open Ocean policies 33–54; pollution 389–391; renewable energy 289–290, 293; science & policy 160; sea-power 366–378; shipping 339–345; small-scale fisheries 241–254; state ocean strategies/policies 33–54; subsea telecommunications 351–353; tourism 404–406; waste disposal 389–394; world oceans 10–13, 17–85
government regulations, offshore oil/gas 280
GPS *see* global positioning systems
"Grasshopper Effect" 105
grave-decommissioning shipping vessels 339
Great Barrier Reef 404–405, 483
green economies 199–211
green national accounts 183–184
"Greenhouse Effect" 91–92
greenhouse gases (GHG): climate change modulation 89, 91–95, 100, 104; ocean health 116; shipping 344–345; South Pacific SIDS 565–566
Grisbadarna judgement 425–426
de Groot, Hugo (Grotius) 368–371
gross domestic product (GDP) 177, 179, 563
gross national income per capita 563
Grotian Era environment 368–371
Group of Experts on the Scientific Aspects of Marine Environmental Protection (GESAMP) 511, 514, 517, 520
growth rate per capita 562
guerre de course operations 374

Haines, Steven 366–378

Hardin, Garret 542
hazardous substances disposal 382–387, 391–394
hazards, subsea telecommunications 357–359
health, ocean health 108–126
Hedonic Price Method (HPM) 180
Hein, James R. 296–309
Hemmings, Alan D. 576–591
heritage: biodiversity 58, 63–64; deep seabed 449–459; Le Morne Cultural Landscape 552; maritime heritage conservation 408–421; World Heritage Convention 58, 63–64
heterotrophy 94
high seas: biodiversity 58–60, 65, 68–69; deep seabed 452–453; illegal, unregulated, unreported fishing 232–240; sea-power 368–378; state ocean strategies/policies 33–39, 45–46, 52
Hines, A. Margaret 108–126
history: ocean health 112–116; sea-power 367–373; shipping 331–334; world oceans human past and present 5–13
Honduras/Nicaragua boundaries 442–443
Honey, Kristen 108–126
HPM *see* Hedonic Price Method
human activities/impacts: conservation 145–154; ocean-climate-atmosphere-nutrient linkages 89–106; ocean health 111–116; world ocean development 5–13
human future 592–598
human health 109, 215–217, 241, 243
human well-being 215–217
hydrocarbons *see* offshore oil and gas
hydrographic surveys 462–475
hydrothermal fluid precipitation 303–305
hypoxia 111–112

ICP *see* Informal Consultative Process
illegal, unregulated, unreported (IUU) fishing 228–229, 232–240, 375
illicit trade 375
Ilulissat Declaration 579
Indian geopolitics 17–19, 21–23, 26–28
indicators, ocean health 118–122
indices, ocean health 110, 120–121
indigenous knowledge/peoples 323, 582–583
industrialization, world oceans 7–8
industry: aquaculture 255; fisheries 223–224; shipping 339–345; tourism 396–406
Informal Consultative Process (ICP) 65
institutions: ecosystem global governance 77–79, 81–83; genetic resources 322; state ocean strategies/policies 33–52
intangible maritime heritage 408–410, 414–416
Integrated Coastal Areas Management (ICAM) 544
Integrated Coastal Management (ICM) 165–167

integrated management 46–52, 157–173
Integrated Maritime Policy (2007) 28
Intergovernmental Oceanographic Commission
   (IOC) 11, 34, 43, 66–67, 77, 83, 316–317
Intergovernmental Oceanographic Commission's
   Advisory Body of Experts on Law of the Sea
   (IOC/ABE-LOS) 136–137, 317
Intergovernmental Platform on Biodiversity and
   Ecosystem Services (IPBES) 67
Internal Waters 35, 132–133, 135, 499, 543, 581
International Atomic Energy Agency (IAEA) 43
international contexts, maritime heritage
   conservation 408–421
International Convention for the Prevention of
   Pollution from Ships (MARPOL) 80–81, 146,
   341–342, 389
International Convention for the Regulation of
   Whaling (ICRW) 585
International Convention for the Safety of Life at
   Sea 34–35, 40–41, 341, 472–474
International Convention on Standards of
   Training, Certification and Watchkeeping for
   Seafarers (STCW) 341, 343
international cooperation, IUU fishing 234–235
International Council for the Exploration of the
   Sea (ICES) 11, 74
International Court of Justice (ICJ) 428–430
International Hydrographic Organization (IHO)
   462–463, 466–475, 578
international instruments/institutions:
   biodiversity 56–64; ecosystem global
   governance 73, 77–79, 81–83; Marine
   Scientific Research 134–135; State ocean
   strategies/policies 33–52
International Knowledge Management (IKM)
   544–546, 558
international law, subsea telecommunications
   351–353
International Law Commission (ILC) 56–57
International Management Plans (IMP) 170–171
International Maritime Organization (IMO):
   biodiversity protection/governance 55–61, 65;
   ecosystem global governance 80–81, 83; green
   economies 202; shipping 340–345; state ocean
   strategies/policies 34, 40–41, 43–44, 52
International Seabed Authority (ISA) 60–61, 83,
   450–452, 458, 581
international straits, East Asian Seas 526–527
international trade status, fisheries 224
International Whaling Commission (IWC) 40,
   57, 59, 585
intrinsic values, conservation 148–149
iron 101
IUCN Commission/guidelines 167–169, 478,
   485
IUU fishing *see* illegal, unregulated, unreported
   fishing

Japan Agency for Marine-Earth Science and
   Technology (JAMSTEC) 304, 321
Japan, Sea of 526, 529–530, 533, 536
Jentoft, Svein 241–254
Joint Group of Experts on the Scientific Aspects
   of Marine Environmental Protection 511,
   514, 517, 520
Jungwiwattanaporn, Megan 199–211
jurisdiction: biodiversity protection/governance
   59–69; deep seabed 449–451; ecosystem
   global governance 73, 75–81, 83–84; genetic
   resources 311–313; Marine Scientific
   Research 135–139; Mediterranean boundaries
   495–507; sea-power 376–377; UNCLOS
   23–26

Kenchington, Richard 188–198
Kildow, Judith 176–187
Kimball, Lee A. 73–85
kinetic renewable energy 284–286
knowledge management, Africa 544–546, 558
Kondratieff cycles 592–594, 597
Koss, Rebecca 157–175
Kroeker, Kristy 108–126

LaFrance Bartley, Monique 507–523
land areas, world ocean development 8–13
land-based activities: mineral mining 305–308;
   ocean health 115–116
landings, fisheries 217, 219–221
law, constituency/governance 160
Law of the Sea Convention (LOSC) *see* United
   Nations Convention on the LOSC
laying subsea telecommunication cables
   355–357
League of Nations 52, 408
LED *see* Local Economic Development
legal issues: constituency/governance 160; deep
   seabed 449–461; East Asian Seas 530; Marine
   Scientific Research 128, 132–133, 137;
   maritime boundary delimitation 427–446;
   polar oceans 576–577, 579–585, 589
leisure/tourism 204–205, 396–407
Le Morne Cultural Landscape (LMCL),
   Mauritius 550–552
De Leo, Giulio 108–126
Leverhulme Trust 255, 261–265
Libes, Susan M. 89–107
Libya/Tunisia boundary delimitation 434
licensing: IUU fishing 236–237; offshore oil/gas
   279–280
life cycles: shipping vessels 337–339; subsea
   telecommunication cables 357
liner shipping sectors 334–336
'Lisbon Principles' 576
litter, waste disposal 387–388
livelihood security, fisheries 241, 246–247

living resources: development stages 593; East Asian Seas 528–529; ecosystem global governance 76–77, 81, 83
LMCL *see* Le Morne Cultural Landscape
LMMA *see* locally managed marine areas
local community projects, Africa 541–559
Local Economic Development (LED) Plans 547–556
local knowledge, genetic resources 323
local level management, shipping 343
locally managed marine areas (LMMA) 543
Long Term Ecological Research (LTER) 119–120
LOSC *see* Law of the Sea Convention
LTER *see* Long Term Ecological Research
Lück, Michael 396–407

McCann, Jennifer 507–523
McCauley, Douglas J. 108–126
McLean, Bernice 541–559
macronutrients 92–93, 105
maintenance, ocean health 109
Malta/Libya boundary delimitation 434
manganese nodules 296, 302–303, 305–307
mariculture 41, 121, 159, 201, 217, 255–265, 593
marine genetic resources 310–328
marine leisure/tourism 396–407
Marine Life Protection Act (MLPA) 480, 485
marine pollution 40–41
Marine Protected Areas (MPA) 146–147, 360, 401–404, 476–492
Marine Scientific Research (MSR) 127–142, 311
marine spatial planning (MSP): constraints 487–489; geography of the sea 476–477, 485–490; green economies 206–207; integrated environment management 172–173; ocean zoning 485–490; opportunities 487–489; Rhode Island 512, 514–522; subsea telecommunications 360–361; tourism 401–404; United States 507–523; zoning initiatives 485–490
Marine Strategy Framework Directive (MSFD) 110, 118
maritime boundaries 425–448, 495–506: delimitation 425–446, 495, 499–501, 533–534, 580–581; legal issues 427–446; Mediterranean 495–506; spatial organisation 425–448; technical issues 427–446
maritime cultures 5–13
maritime environment geopolitics 17–30
maritime geopolitical scenarios 17–30
maritime heritage conservation 408–421
maritime safety information 463–464, 467–468, 473–474
market-oriented mechanisms 152

markets: ecosystem services valuation 176–187; green economies 200–201; renewable energy 288–289
MARPOL *see* International Convention for the Prevention of Pollution from Ships
Martone, Rebecca G. 108–126
Mateos, Juan Carlos Rodríguez 17–32, 495–506
material maritime heritage 408–421
Mauritius Strategy (MSI) 560
maximum economic yields 225
maximum sustainable yields 221, 225
MDG *see* Millennium Development Goals
MEA *see* Millennium Assessment
measuring ocean health 118–122
median line principle 426–446
Mediterranean, maritime boundaries 495–506
Mediterranean Action Plan (MAP) 392–393, 495–496, 498, 503–504
Melanesian maritime heritage 418–419
memoranda of understanding (MOU) 56
metagenomic analysis 311–312
metals 296–297, 384–385
methane 91, 93, 95, 103–104
Micheli, Fiorenza 108–126
micronutrients 92–105
microplastics 387–388
migratory species 56, 58, 63, 74, 78–80, 83
military applications/tasks 367, 369–378
Millennium Assessment (MEA) 176–177
Millennium Development Goals (MDG) 65, 560, 564
Millennium Ecosystem Assessment [MA] 157, 164–165, 476
mineral development: ecosystem services valuation 177, 179, 181–182; green economies 203–204; mining 203–204, 296–309; stages 592–593
Minerals Management Service (MSS) 510
mining 203–204, 296–309
Mitroussi, Kyriaki 331–348
'mixed economy' systems 11–12
Mizell, Kira 296–309
MLPA *see* Marine Life Protection Act
modern nautical charts 463–475
molecular forms, nutrients 93
molluscs 255–256, 258–261
MOU *see* memoranda of understanding
MPA *see* Marine Protected Areas
MSFD *see* Marine Strategy Framework Directive
MSI *see* Mauritius Strategy
MSP *see* marine spatial planning
MSR *see* Marine Scientific Research
MSS *see* Minerals Management Service
multiple human activities/impact drivers 111–112
Myanmar/Bangladesh boundaries 444–446

Nagoya Protocol 310–311, 318, 322
nanoparticles 387
national accounts 183–184
national directives 161–162
national hydrographic survey challenges 474–475
national hydrographic surveys 466–475
National Ocean Economics Program (NOEP) 179–180
National Oceanic and Atmospheric Administration (NOAA) 171, 291, 304, 344, 489, 508, 512, 520–521
national regulations 161–162, 320–321, 343
national spheres of influence 524–537
NATO *see* North Atlantic Treaty Organisation
natural disaster vulnerability 565–566, 569–570
natural hazards 357–359
natural resource management 542
nautical charts 462–475
naval activity 8–10, 12, 366–378
naval doctrines 20–24, 28, 31
naval forces 366–378
navigation 331–348, 401, 462–475
Navigation Laws and Corn Laws 11
NEPTUNE *see* North East Pacific Time-series Undersea Network Experiments
Netherlands/Belgium boundaries 438
networks: marine protected areas 484–485; subsea telecommunications 353–354
Neumann, Christian 199–211
NGO *see* non-governmental organizations
Nicaragua–Honduras boundaries 442–443
Nigeria/Cameroon boundaries 441–443
nitrogen 91–105
nitrous oxide 91, 94–95, 103–104
NOAA *see* National Oceanic and Atmospheric Administration
NOEP *see* National Ocean Economics Program
non-charting uses, hydrographic surveys 470–472
non-claimant states, polar oceans 588
non-coastal states, Arctic Ocean 581–582
non-governmental organizations (NGO) 12, 68–69, 236–237, 479, 546–547, 551–553
non-living resources, East Asian Seas 528–530
non-market values: ecosystem services valuation 177–181, 184–185; green economies 200–201
non-state actors, IUU fishing 236–237
norms/standards 49–50
North American aquaculture 260
North Atlantic 8, 30
North Atlantic Treaty Organisation (NATO) 11, 496
North East Pacific Time-series Undersea Network Experiments (NEPTUNE) 360
northern communities, Arctic Ocean 582–583
North Pacific 8

North Sea boundaries 428–430, 432–434, 436–437, 439–440
Notarbartolo-di-Sciara, Guiseppe 145–156
November, Joani 541–559
nutrients 89–107, 381–382
nutrition 215–216
Nuttall, Peter 560–575

OBIS *see* Ocean Biogeographic Information System
observatory programs 119–122, 360
Ocean Biogeographic Information System (OBIS) 322–323
ocean-climate-atmosphere-nutrient linkages 89–107
ocean dwellers 45–46
ocean health 108–126: climate change 112–116; definitions 109–111; disease 109, 116–122; ecosystems 108–122; fisheries 111–121; historical changes 112–116; human activities/impacts 111–116; indicators 118–122; indices 110, 120–121; measurements 118–122; multiple drivers 111–112; targets 118–122
ocean–atmosphere interfaces 89–107
Ocean Health Index (OHI) 110, 120–121
Ocean SAMP 512, 514–522
ocean spaces/resources 33–34
ocean thermal energy conversion (OTEC) 287
ocean zoning 485–490
Oceania: aquaculture 259–260; Small Island Developing States 560–575
Odendaal, Francois 541–559
offshore oil/gas 47, 53, 76, 157, 159, 179–182, 269–282, 528–530, 593–594: development 269–272, 593–594; East Asian Seas 528–530; economics 273, 278–279; ecosystems 76, 179–182; environmental impacts 76, 179–182, 276–278; exploration 271–281; governance 279–280; politics 273, 278–279; production 271–281; reserves 272–274; resources 272–274; science & policy 157, 159; social aspects 273, 278–279; technology 274–276
OHI *see* Ocean Health Index
oil: waste disposal 388; *see also* offshore oil/gas
O'Leary, Jennifer K. 108–126
Olsen, Stephen B. 507–523
O'Neill, Sean 283–295
open access data management 322–323
Open Ocean 33–54
Operation ATALANTA 28
ores 296–309
organic matter 94, 98–99, 381–382
Organisation of the Islamic Conference 496
organohalogenated compounds 386
Österblom, Henrik 232–240
OTEC *see* ocean thermal energy conversion

oxygen 91–104
ozone hole/layer 64, 91

Pacific Islands Forum (PIF) 563
Pacific Islands Regional Ocean Policy (PIROP)
  567–568
Pacific Ocean SIDS 560–575
paints, antifouling 386
parasites 116–118
partnerships 571, 594–595
Pendleton, Linwood 199–211
people trade 375
perfluorinated compounds 386
persistent organic pollutants (POP) 384
personal care products 386–387
pharmaceuticals 204, 386
phosphorites 298
phosphorous 92–105
phytoplankton 91–104
PIF *see* Pacific Islands Forum
piracy 343–344
PIROP *see* Pacific Islands Regional Ocean
  Policy
placer minerals 298
plankton 91–104
planning: economics 159; green economies 204;
  science 158–159; social aspects 159–160; *see
  also* marine spatial planning
plants 255–256
plastics 387–388
poisoning problems 384–385
polar oceans 576–591
policy: Africa 541–559; conservation challenges
  151; ecosystems 48–52, 73, 157–173;
  environment management 157–175; fisheries
  233–238; genetic resources 312–313,
  323–325; integrated management 157–173;
  Marine Scientific Research 135–139;
  pollution 389–391; South Pacific SIDS
  567–568; tourism 404–406; US MSP
  508–516, 522; waste disposal 389–391
politics: conservation challenges 151; deep
  seabed 449–461; ecosystem global governance
  73; genetic resources 312; IUU fishing
  235–236; Marine Scientific Research
  138–139; Mediterranean boundaries 497–498;
  offshore oil/gas 273, 278–279; polar oceans
  576–589; *see also* geopolitical...
pollution: biodiversity 55, 57, 62, 64, 68;
  ecosystems 74–77, 80–81, 177; governance
  389–391; policy frameworks 389–391;
  shipping 80–81, 146, 339, 341–342, 345, 389;
  state ocean strategies/policies 34–35, 40–41,
  52; tourism 404–405; waste disposal
  381–395
POP *see* persistent organic pollutants
population, South Pacific SIDS 562–564

positive feedbacks 91, 104
post-Grotian age 373–377
potential markets, renewable energy 288–289
poverty, aquaculture 255, 261–265
Press, Anthony J. 232–240
principle of common heritage of mankind
  449–459
private-based projects 255, 261–265, 322
production: aquaculture 255–261; offshore
  oil/gas 271–281
property rights 177
protected area/species governance 74
protecting shipping 375–376
protecting telecommunications 353–354
public awareness, conservation 149–150

quotas, fisheries 227–228

rare-earth elements (REE) 296, 298, 301–303,
  305–306
reactive nitrogen 100–101
Recreation Opportunity Spectrum (ROS)
  402–404
REE *see* rare-earth elements
regional development/trends: Africa 541–559;
  aquaculture 259–261; East Asian Seas
  524–537; human development 592–598;
  marine spatial planning 507–523; maritime
  boundaries 495–506; polar oceans 576–591;
  Small Island Developing States 560–575;
  South Pacific SIDS 560–575
regional ecosystem-based imperatives 73–84
Regional Fisheries Management
  Organizations and Agreements (RFMO/A):
  biodiversity 55, 58–61, 63, 67–68;
  ecosystem global governance 78–80, 83;
  high seas 234; state ocean strategies/policies
  38–39, 43
regional governance: biodiversity 67–68;
  ecosystem-based imperatives 73–84; genetic
  resources 320–321; shipping 343
regional hydrographic coordination 468–469
Regional Marine Plans (RMP) 48
Regional Planning Bodies (RPB), US MSP
  511–514
regional policies, South Pacific SIDS 567–568
Regional Seas Programme (RSP) 390
regulations: genetic resources 320–321; offshore
  oil/gas 280; renewable energy 289–290;
  shipping 341; state ocean strategies/policies
  50–51
Reis, Jeanette 331–348
renewable energy 205–206, 283–295, 518–521,
  570, 593
research, Marine Scientific Research
  127–140
reserves, offshore oil/gas 272–274

resources: Arctic Ocean 576–584, 589;
   development stages 592–593; East Asian Seas
   528–530; offshore oil/gas 272–274; polar
   oceans 576–591; Southern Oceans 576–578,
   582–589; South Pacific SIDS 564, 568–569;
   twentieth century 33–34
RFMO/A *see* Regional Fisheries Management
   Organizations and Agreements
Rhode Island 512, 514–522
rights-based management 226–229, 233
RIO+20 Conference 73, 81
Rio Declaration 61, 64
RMP *see* Regional Marine Plans
Roerich Pact 409
Romania/Ukraine boundaries 443–445
ROS *see* Recreation Opportunity Spectrum
Rosim, Daniele F. 108–126
Rotterdam Convention 390–391, 414–415
routes, shipping 331–334
RPB *see* Regional Planning Bodies
RSP *see* Regional Seas Programme
Russian geopolitics 17–19, 21–23, 26–28

safety: offshore oil/gas 280; shipping 339–341
Safety of Life at Sea (SOLAS) 34–35, 40–41,
   341, 472–474
Sail Training International (STI) 414
sales, second-hand shipping vessel 338–339
salinity gradient power 288
SAMP *see* Special Area Management Plan
San Juan 418
SAP *see* Strategic Action Plans
SBSTTA *see* Subsidiary Body on Scientific,
   Technical and Technological Advice
science: biodiversity protection/governance
   65–69; decision making 158–159; ecosystem
   global governance 73–81, 84; ecosystem
   management 157–173; environment
   management 157–175; genetic resources
   319–320; integrated management 157–173;
   Marine Scientific Research 127–140; planning
   158–159
Scorse, Jason 176–187
Scovazzi, Tullio 449–461
SCUBA diving, tourism 398
SDG *see* Sustainable Development Goals
SEA *see* strategic environmental assessments
sea cucumber aquaculture 255, 261–265
sea level changes 7, 79, 113, 185, 205, 297, 359,
   561, 584, 595–596
Sea of Japan 526, 529–530, 533, 536
Sea of Okhotsk 525–526, 532–533
sea-power 366–378
seafloor massive sulphides (SMS) 296,
   303–307
seawater mining 299
second-hand ship sales 338–339

Second World War 9–10, 21, 34, 59
sectorial regulations/policies 50–51
security, fisheries 215–217, 227–228, 241,
   246–247
Self Contained Underwater Breathing Apparatus
   (SCUBA) diving 398
semi-enclosed seas, East Asian Seas 526
service industries, offshore oil/gas 271
sewage 381–382
shared resources 74
shipping: anchorage 353–354; biodiversity
   55–61, 65; commodities 331–334; economics
   334–335; ecosystem global governance 75–76,
   78–82; freights 331–334, 336–337;
   governance 339–345; green economies 202;
   history 331–334; industry management
   339–345; navigation 331–348; ocean health
   111, 118; pollution 80–81, 146, 339,
   341–342, 345, 389; routes 331–334; safety
   339–341; sea-power 375–376; South Pacific
   SIDS 570; stakeholders 340, 343–345; state
   ocean strategies/policies 35, 39–41, 43–47,
   50–52; subsea telecommunications 353–354;
   tourism 397, 399–401, 404–405; training 341,
   343; vessel design/size 335–336; vessel life
   cycles 337–339; world ocean development
   8–13
SIDS *see* Small Island Developing States
siting aspects 479–480, 549
slime 104–105
Small Island Developing States (SIDS) 560–575
small-scale fishery governance 241–254
SMARO *see* Spectrum of Marine Recreation
   Opportunities
Smith, Hance D. 5–13, 269–282, 592–598
SMS *see* seafloor massive sulphides
social aspects: decision making 159–160;
   ecosystem values 176–187; fisheries 215–217,
   227–228, 241, 245–246; genetic resources
   312; justice 241, 245–246; offshore oil/gas
   273, 278–279; organisation 7, 11, 162, 195,
   569; planning 159–160; science & policy
   159–161; South Pacific SIDS 561–564,
   566–567
soft laws 64–65
Sokolow, Susanna 108–126
solar energy 286–287
SOLAS *see* Safety of Life at Sea
Solgaard, Anne 199–211
solubility, gas exchange 97–98
Somali Current Large Marine Ecosystem Project
   546–558
Soons, Alfred H. A. 128–129
sources, waste/pollution 381–388
South American aquaculture 260
South China Sea 270, 276, 377, 470, 526,
   530–536

Southern Oceans: IUU fishing 234–235; polar oceans 576–578, 582–589
South Pacific SIDS 560–575
sovereignty: Arctic Ocean 576–584, 589; East Asian Seas 524–526, 528, 530–536; Marine Scientific Research 135–137; polar oceans 576–591; Southern Oceans 576–578, 582–589
'space-based' conservation 146–147
spatial organisation: deep seabed 449–461; green economies 206–207; integrated environment management 172–173; marine protected areas 476–492; state maritime boundaries 425–448; surveys 462–475; *see also* marine spatial planning
Special Area Management Plan (SAMP) 512, 514–522
'species-based' conservation 146–147
species composition 218–219
Spectrum of Marine Recreation Opportunities (SMARO) 402–404, 406
stakeholders: Africa 544–557; genetic resources 323; shipping 340, 343–345; world ocean development 12
standards 49–51, 341, 343
State-centric perspectives 18–20
state jurisdiction, sea-power 375
state maritime boundaries 425–448
'State ocean strategies/policies for the Open Ocean' 33–54
state roles/responsibilities, US MSP 507–523
state setting, US MSP 517–519
Stead, Selina 255–265
steam engines 9
STI *see* Sail Training International
Stock, Andy 108–126
stock status, fisheries 221–222
Stockholm Convention 389, 391
straits, East Asian Seas 526–527
Strategic Action Plans (SAP) 392–393, 545–546, 548, 554–557
strategic environmental assessments (SEA) 188–198
Stratton Commission 508–509
structures, offshore oil/gas 274–276
subsea telecommunications 349–365: cables 349–361; environmental impacts 354–357; governance 351–353; international law 351–353; natural hazards 357–359
Subsidiary Body on Scientific, Technical and Technological Advice (SBSTTA) 313–315
sulphides 296, 303–307
Sumaila, U. Rashid 232–240
surveillance 46
surveys: national services 466–475; navigation 462–475; non-charting uses 470–472; spatial organisation 462–475; technology 463–466

sustainable development: fisheries 224–229, 247–249; genetic resources 323–324; South Pacific SIDS 564, 569–570; strategic environmental assessments 191–197; Tokyo declaration 162; US MSP 511–514
Sustainable Development Goals (SDG) 65, 323–324
SWOT analysis 549–550

Tangawamira, Zvikomborero 541–559
Tanzanian aquaculture 255, 261–265
targets, ocean health 118–122
taxation, offshore oil/gas 280
TBT *see* Tributyl Tin
TDI *see* Technology Development Index
technical issues: management 12, 77–78, 80, 307; maritime boundaries 427–446
technology: fisheries 233; genetic resources 316–320, 324; hydrographic surveying 463–466; offshore oil/gas 274–276; surveying 463–466; tourism 399–401; world ocean development 9–13
Technology Development Index (TDI) 519–520
telecommunications, subsea 349–365
temperature, ocean health 116
territorial claimant states 585–588
territorial rights 576–591
territorial seas: East Asian Seas 527, 530, 534; geopolitical scenarios 19; state ocean strategies/policies 35–40, 46, 48, 50
"The Fish Space" (Espai del Peix) of Palamós, Spain 420
thermal energy 287
threats to ocean health 109
Thrupp, Tara 269–282
tidal energy 284–286
tipping points 91
Tokyo declaration 162
tourism 204–205, 396–407
toxic elements 383–387
training 341, 343
transboundary effects 74
Transfer of Marine Technology 316–320, 324
transport: nutrients 101–102; South Pacific SIDS 570
Treaty of Westphalia 11
Tributyl Tin (TBT) 385
Truman Proclamation 36, 426–427
twentieth century ocean spaces/resources 33–34

UK *see* United Kingdom
Ukraine/Romania boundaries 443–445
UN *see* United Nations
UNCED *see* United Nations Conference on Environment and Development

UNCLOS *see* United Nations Convention on the Law of the Sea
UNCTAD *see* United Nations Conference on Trade and Development
underwater cultural heritage conservation 409–413, 415–418
une feuille de route 49
UNEP *see* United Nations Environment Programme
UNESCO *see* United Nations Educational Scientific and Cultural Organisation
UNFSA *see* United Nations Fish Stocks Agreement
UNGA *see* United Nations General Assembly
UNIDO *see* United Nations Industrial Development Organization
United Kingdom (UK): Channel Islands boundaries 438; Faeroe Islands boundaries 439–440; France Channel boundaries 430–432, 434–436, 438–439; Republic of Ireland boundaries 434–435
United Nations (UN): BBNJ Working Group 65–66; Informal Consultative Process 65; Oceans: biodiversity 65–66; Secretariat 43–44, 60–61, 129–131, 135; State ocean strategies/policies 33–40, 42–46, 51; Working Group 65–66, 313, 315–316, 318–320, 324, 452–459
United Nations Conference on Environment and Development (UNCED) 76
United Nations Conference on Trade and Development (UNCTAD) 37
United Nations Convention on the Law of the Sea (UNCLOS): biodiversity 55–65, 69; boundary delimitation 432–434, 436, 439; cultural heritage conservation 409–410, 412; deep seabed 449–459; East Asian Seas 524–536; ecosystem governance 75–77, 79–84; genetic resources 311–324; geopolitical scenarios 18–20, 23–30; hydrographic surveys 472–474; Marine Scientific Research 127–140; maritime boundaries 432–434, 436, 439, 495, 497–501; Mediterranean boundaries 497–501; polar oceans 576, 579–587; science & policy integrated management 161–162; sea-power 376; South Pacific SIDS 561; state ocean strategies/policies 33–46, 51; strategic environmental assessments 191–195; subsea telecommunications 351–353, 360–361; waste disposal 389
United Nations Educational Scientific and Cultural Organisation (UNESCO) 11; cultural heritage conservation 409–415, 418; ecosystem global governance 77–78, 83; state ocean strategies/policies 34, 38, 43, 52

United Nations Environment Programme (UNEP) 389, 391–392, 542–543, 558
United Nations Fish Stocks Agreement (UNFSA) 38–39, 60, 233–234
United Nations Food and Agriculture Organization (FAO): biodiversity 55–59, 69; ecosystem governance 76, 78; state ocean strategies/policies 34, 39–40, 43, 51–52
United Nations General Assembly (UNGA) 59, 64–68, 77, 83
United Nations Industrial Development Organization (UNIDO) 43
United States (US): Commission on Ocean Policy 509; geopolitical scenarios 21–22; marine spatial planning 507–523; renewable energy 289–290

Veitayaki, Joeli 560–575
Venice Charter 413–414
vessel design/size, shipping 335–336
vessel life cycles, shipping 337–339
de Vivero, Juan Luis Suárez 5–13, 17–32, 495–506, 592–598
vulnerable marine ecosystems (VME) 68

warming, climate change 104–105, 111, 113–116, 595–596
Warner, Robin 55–72
wars: Mediterranean boundaries 497–498; sea-power 367–378; world ocean development 8–9
waste disposal 381–395
wastewater disposal 382, 390
wave energy 284–286, 593
WCED *see* World Commission on Environment and Development
Wescott, Geoff 157–175
Westphalian state system 11
whale watching 398–399
whaling 40, 57, 59, 585
Wilson, Robert 462–475
wind energy 286–287, 521, 593
Wood, Chelsea L. 108–126
World Commission on Environment and Development (WCED) 162
World Heritage Convention (WHC) 58, 63–64
World Meteorological Organization (WMO) 43
world oceans: atmosphere 8, 11, 35, 89–107; biodiversity 35, 41–42, 55–74, 78–79; climate change 73, 79–81, 89–107; cultures 5–13, 594; development 5–13; economic development 7–13; ecosystems 10–11, 48–52, 73–85; environment 10–12; fisheries 8–13; geopolitical scenarios 17–32; global governance 10–13, 17–85; governance 11–13;

human future 592–598; human society 5–13; industrialization 7–8; land areas 8–13; naval activity 8–10, 12; nutrient cycling 89–107; Open Ocean policies 33–54; shipping 8–13; state ocean strategies/policies 33–54; technology 9–13
World Summit on Sustainable Development (WSSD) 64, 546, 560, 567–568

worldwide charting status 469–470
Worldwide Navigational Warning Service (WWNWS) 473

Ye, Yimin 215–231

zoning initiatives 485–490
Zuiderzee Museum, Netherlands 420–421